Addendum to The Auriferous Gravels of the Sierra Nevada of California

This Publication originally came with accompanying maps in the rear pocket. To keep costs of this publication down we have digitized these maps and one set is available for free download with the purchase of this book. To receive your free maps go to www.miningbooks.com and go to Downloadable Maps then load "Maps to accompany Auriferous Gravels of the Sierra Nevada of California" into your shopping cart. At checkout use Code "AURMAPS" to receive your free maps.

Republished by

Sylvanite Publishing
Florence, Colorado
www.sylvanitepublishing.com

CONTRIBUTIONS

TO

AMERICAN GEOLOGY.

VOL. I.

FROM THE

MEMOIRS OF THE MUSEUM OF COMPARATIVE ZOÖLOGY,

AT CAMBRIDGE, MASSACHUSETTS.

UNIVERSITY PRESS, JOHN WILSON AND SON.

1880.

CONTENTS.

———⋆———

THE

AURIFEROUS GRAVELS

OF THE SIERRA NEVADA OF CALIFORNIA.

By J. D. WHITNEY.

CAMBRIDGE:
UNIVERSITY PRESS, JOHN WILSON & SON.
1880.

PREFACE.

THE title of the work here presented to the public seems to explain its object with sufficient clearness; but a few words in regard to the character and source of the materials which have been used in its preparation are perhaps desirable. The stoppage of the State Geological Survey by the Legislature of California, in 1874, left me in possession of considerable information, both of my own collecting and that resulting from the labor of others; and, having obtained from the Regents of the State University authority to continue the publications of the Survey at my own expense, or on my own responsibility, I have already added several volumes to the list of those which had been issued at the time of the closing of the work by the Legislature. In the publication of the Botany, I have had assistance from a few generous and public-spirited residents of San Francisco. The present volume, although really a continuation of the work of the Geological Survey, is issued in quarto form, in consequence of an arrangement with the curator of the Museum of Comparative Zoölogy, at Cambridge, whereby that institution assumes a portion of the expense of publication, receiving a certain number of copies of the work for distribution among the societies and individuals on its exchange list. One volume, certainly, and probably others, will follow this, their publication having been provided for in a similar way. Those volumes which belong to divisions of the work which have been already commenced — as, for instance, the continuation of the Ornithology — will appear in royal octavo form, to match the already published volumes of the regular series of the Survey; the others will be of quarto size.

At the time of the discontinuance of the California Survey, the publication of a volume having as its subject the auriferous gravels had been for some time in contemplation, but its appearance had been delayed by a series of misfortunes, of which it is not necessary here to speak. Nothing, therefore, in this particular division of the geological work, was quite ready for the press at the time of the abandonment of the Survey by the Legislature. When, later, after considerable hesitation, I had concluded to resume the work of publication, the gravel volume was one of those with which I was most anxious to proceed. In view, however, of the great importance of the hypsometrical element in the survey of the gravel region, it appeared to me extremely desirable that our baromet-

Univ̵ersity Press:
John Wilson and Son, Cambridge.

rical observations made in that part of the State should be carefully revised and recomputed, with the aid of the table of corrections which we were then able to prepare from our three years' series of observations, which had been specially made in order that we might be furnished with the necessary data for that purpose.

This part of the work was, therefore, that which first received our attention. The tables in question were computed and published under the title of "Contributions to Barometric Hypsometry, with Tables for Use in California." Then came the laborious operation of recalculating and arranging the barometrical observations at the various stations in the gravel region, the corrections indicated by the tables being applied to them. The method of applying the corrections in question, and the reasons for their use, have been fully explained in the "Barometric Hypsometry," which was issued in 1874. A supplement, forming the fifth chapter of that work, was published in 1878, in which the practical value of our tables is demonstrated, by a bringing together for comparison and discussion of the results obtained by working over the many hundred observations which had been taken in the gravel region. In this division of the work it is shown that the use of the tables in question has reduced the error of the results, on the average, by fifty per cent; and I may here add, after having made a careful examination of the whole subject, that there is no method of observation by which barometric results can be freed from the class of errors which these tables are intended to remove, neither is there any kind of corrections which can be substituted for those furnished by our work. All these laborious computations and investigations of the barometrical data were made by Professor Pettee, or by assistants employed under his direction. It must be noticed, however, that a portion of the altitudes given in the body of this work have not had the tabular correction applied to them. Consequently they differ slightly, in numerous instances, from those published in tabular form in Appendix C, which presents in one body the results of all the later and more trustworthy determinations made in the gravel region proper.

During the time when the earlier part of our surveys in the hydraulic mining district were being made, we had the work of the three observers at the regular fixed stations, which we could use in calculating the results of our observations taken at various points in their respective vicinities. During the re-examination, last year, by Professor Pettee, however, of certain portions of the mining counties, as mentioned further on, he had not, to the same extent, the advantages of corresponding station barometers; but by skilful combination of observations of the aneroid with the mercurial barometer, of each of which kinds he had two, the latter being used as station instruments for short periods at important central points, he was enabled to secure very trustworthy results, probably but little, if at all, inferior in value to those obtained in former years.*

---

* In the Table of Barometric Altitudes (Appendix C), those results to which an asterisk is prefixed are from observations made by Professor Pettee in 1879, as well as some others, as specified in the headings of the groups of localities. All the elevations given in the list of places north of the Middle Yuba are from the same source.

The necessity of making the above specified computations, as well as the large size of the work and the number of illustrations which it required, has caused the publication of this volume to extend itself over a much longer period than was at first expected. A portion of the delay, however, has been caused by the time required for making additional explorations in the field; for, in view of the fact that the original material had become somewhat antiquated, it seemed very desirable that further investigations should be made in the gravel region, especially in the northern portion of it, where our work had never been carried on in a detailed manner. For this purpose I was fortunate in being able to secure the services of Professor Pettee, previously employed in similar work on the California Survey, and the results of his re-examination of certain districts previously investigated by himself and others, as well as of further explorations in more northern counties, will be found in Appendix A. These results were received and examined by me previous to the writing of Chapter IV, in which the theory of the gravels is discussed with some detail, although not as fully as I could have wished had space permitted.

As at present put together, the volume includes the following materials: *first*, the investigations of the Geological Survey in the gravel region, carried on at various times during its progress up to the time of its stoppage in 1874. The most important portions of this work, by far, are the detailed explorations of Messrs. Goodyear and Pettee in the fields noted in the body of the volume as having been assigned to them. Their explorations were supplemented by my own observations, which extended over the whole gravel region from Mariposa to Plumas, although necessarily not especially detailed in any one district. *Second*, the re-examination of certain districts and additional investigations of others made by Professor Pettee in 1879, as mentioned above. *Third*, the barometrical observations for altitude made in the gravel region at various times, by different observers, but chiefly by Messrs. Goodyear and Pettee. These have been carefully recomputed by Professor Pettee, since the publication of the "Barometric Hypsometry," with the use of the tables prepared for California, as already stated.

The fossil plants of the auriferous gravels have been investigated by Mr. Lesquereux, and his results, illustrated by ten double plates, properly form a supplement to the present work. It did not appear necessary that any separate memoir should be prepared in relation to the vertebrate remains, not human, of the gravels, as these had already been worked up by Dr. Leidy, and published in connection with his descriptions of other materials of a similar nature, from regions of the Cordilleras farther east and more prolific in vertebrate fossils. The scantiness of our collections in this department has been alluded to and explained in the body of the present volume.

There will undoubtedly be much hesitancy on the part of anthropologists and others, in accepting the results regarding the Tertiary age of man to which our investigations in the gravels of the Sierra Nevada seem so clearly to point. I feel, however, very strongly, that I should not have been justified in withholding such facts as came to my knowledge, and which seem to be perfectly well authenticated, merely because they conflict with generally accepted theories. The only course for me to pursue was, to examine with care the evidence offered, and, if I could find no flaw in it, to make it public. I may be permitted, in this connection, to allude to the circumstance that nearly all those who refuse to accept my conclusions as to the great antiquity of man in California, do so on the ground that the Calaveras skull was not taken from its bed by the hand of a scientific man. In so doing, they not only ignore the evidence presented by the skull itself, which is positively a fossil, and was chiselled out of its gravelly matrix in the presence of several eminent authorities, but they also reject the very full testimony from other quarters, some of which comes from men of education, and even of professional education. The body of this other evidence is so great that it does not appear to me that it would be materially weakened by dropping that furnished by the Calaveras skull itself.

At the time of beginning the present volume, it was my intention to include in it some account of the glacial phenomena and surface geology of the Pacific Coast, as being subjects closely connected with the occurrence of the gravels. After some further consideration, finding that a volume which should contain this additional matter would be inconveniently bulky, I thought it best to divide the work, and to issue the glacial and surface geology separately. This has been done, and the first part of the "Climatic Changes of Later Geological Times: A Discussion based on Observations made in the Cordilleras of North America," will appear at the same time with the concluding portion of the present work. The part first issued contains a pretty full account of the glacial phenomena of the region of the Cordilleras, and a brief recapitulation of those facts which prove, beyond possibility of doubt, that during the later geological periods a gradual desiccation of the earth's surface has been going on, and, as there is good reason to believe, is still continuing. The concluding portion of the "Climatic Changes" will be issued during the coming winter: it will contain a discussion of the facts set forth in the preceding two chapters and in the gravel volume, showing a gradual diminution in the amount of precipitation, as having been begun during a previous geological epoch and as still continuing, while, in certain regions, large areas have been covered by ice where none now remains. The endeavor will be made to trace the connection of these facts, and to show that they are manifestations of one great general cause, which has been in operation during an indefinite period, and is likely to continue to operate in the future.

J. D. WHITNEY.

CAMBRIDGE, MASS., Oct. 1, 1880.

TABLE OF CONTENTS.

CHAPTER I.

GENERAL DISCUSSION OF THE TOPOGRAPHICAL FEATURES AND GEOLOGICAL STRUCTURE
OF THE STATE OF CALIFORNIA.

CHAPTER II.

THE TERTIARY AND RECENT AURIFEROUS DETRITAL AND VOLCANIC FORMATIONS OF
THE WEST SLOPE OF THE SIERRA NEVADA.

CHAPTER III.

THE FOSSILS OF THE AURIFEROUS GRAVEL SERIES.

CHAPTER IV.

RÉSUMÉ AND THEORETICAL DISCUSSION OF THE PRINCIPAL FACTS CONNECTED WITH THE GEOLOGICAL OCCURRENCE OF THE AURIFEROUS GRAVELS.

APPENDIX.

A.

B.

C.

ALTITUDES OF POINTS IN THE AURIFEROUS GRAVEL REGION OF THE SIERRA NEVADA.

LIST OF ILLUSTRATIONS.

THE AURIFEROUS GRAVELS

OF THE SIERRA NEVADA OF CALIFORNIA.

CHAPTER I.

GENERAL DISCUSSION OF THE TOPOGRAPHICAL FEATURES AND GEOLOGICAL STRUCTURE OF THE STATE OF CALIFORNIA.

SECTION I. — *Topographical Sketch.*

IT seems a necessary preliminary to a detailed account of the Auriferous Gravels of the Sierra Nevada, that the general features of the geography and geology of the State of California should be sketched. Without such an introduction the reader of this volume will hardly be able to have a clear idea of the nature of the frame into which the special results here set forth are to be fitted. The sketch must necessarily be brief, and may the more properly be made so, because it is the expectation of the writer that, in a future volume, the topics rapidly passed over in this chapter will be much more fully discussed. Reference may be made by the student of Californian geology to the already published volume of the Survey, as also to the maps which have been issued during the progress of that work.[*]

The principal topographical features of California are so striking in their general aspect, that they could not fail of early recognition; they were first clearly indicated on Fremont's map, accompanying his Geographical Memoir, published in 1848. Well known as these features are in their outline to most

---

[*] "Geology of California, Vol. I. A Report of Progress and Synopsis of the Field-work, from 1860 to 1864." This volume will be quoted in the present work as "Geol. I." The principal maps issued are: *A General Map of California and Nevada*, on a scale of eighteen miles to one inch; *A Map of the Region adjacent to the Bay of San Francisco*, scale two miles to an inch; *Map of Central California*, on a scale of six miles to an inch, in four sheets, the whole covering an area of about 60,000 square miles of the most important and thickly settled region of the State, and about 18,000 square miles of Nevada. Of this latter map only two sheets have been finished and published; these embrace a strip of about 150 miles in width, extending across the State from east to west, having Visalia on its southern border, and Santa Rosa on its northern. The other two sheets of this map were nearly completed when the Survey was stopped.

students of the physical geography of North America, they must be here indicated with some detail, since an understanding of the nature of the Gravel deposits of the Sierra depends so much on an acquaintance with the physical peculiarities of the region over which they are distributed.

Let us assume as the central point of our picture of the topography of California its most striking feature, "The Great Valley," and from this let us then extend our view first to the region immediately adjacent to it, — its edge, so to speak, — and then to the more remote portions of the State, which indeed have but little geographical or geological connection with it, or with the subject of the present volume.

The Great Valley — the valley of the Sacramento and San Joaquin rivers — forms an area of almost level land, having a roughly elliptical form, trending about N. 30° W. and S. 30° E., and embracing about 18,000 square miles. It lies between latitudes 34° 50′ near Fort Tejon, and 40° 40′ near Shasta, having an extreme length of 450 miles and an average width of forty. The shape of the area of flat land, between the foot-hills of the mountains which enclose it, is somewhat irregular. The base of the Sierra Nevada on its east side has a pretty regular trend north of latitude 35° 30′; but south of this it bends around towards the west, and thus, meeting the Coast Range, closes up the southern end of the valley. The northern end is also closed by the convergence of the Sierra and Coast Ranges, the latter chain assuming a northerly trend from Clear Lake north, while that of the Sierra continues the same beyond the limits of the valley. Above Red Bluff, indeed, the ranges approach each other very nearly, the volcanic overflow from the Sierra coming quite down to the Sacramento at this point; while a little farther north this river almost touches the foot-hills on the eastern side of the valley, and is only a few miles distant from them on the west. A section across the valley, in the direction of its length, shows that the fall of the Sacramento River, from Redding to its mouth, a distance of 192 miles, is 556 feet; while that of the southern portion of the valley, between Kern Lake and the mouth of the San Joaquin, a distance of 260 miles, is 282 feet. The fall of the Sacramento River, within the limits of the Great Valley, is therefore about three times as rapid as that of the river system which drains the southern section. Indeed, the lower or most southeastern portion of the Great Valley, usually called the Tulare Valley, is very nearly a level plain, occupied in part by shallow lakes. Of these Tulare Lake is the largest, having an area of 687 square miles and a depth of only about forty feet.

The area of this lake, however, is quite variable, being considerably larger after a wet season, or especially when several seasons wetter than the average have succeeded each other. The whole region between Kern Lake and Fresno City, near the point where the San Joaquin having descended from the Sierra Nevada turns northward and receives the overflow from the south, is low and marshy, and one in which during the dry season the evaporation exceeds the inflow from the streams coming down the Sierra.

Sections across the Great Valley at right angles to its length show a more or less gradual slope from the foot-hills of the enclosing ranges towards the Sacramento and San Joaquin rivers. From Visalia to the north point of Tulare Lake the descent is four and a half feet per mile, for a distance of twenty-nine miles. Tulare River, from the crossing of the Southern Pacific railroad, falls at the rate of three feet per mile in eighteen miles. In the extreme southern end of the valley the fall of the land from the vicinity of Bakersfield to Tulare Lake is about five and a half feet per mile for thirty-eight miles. In the section between Firebaughs' and Hills' ferries, the levellings show that the ground falls from the foot-hills to within four and a half miles of the river at the rate of six feet per mile; thence it is nearly level to within half a mile of the river, which it then approaches with an ascent of one and a half feet per mile. At Banta's the valley is somewhat contracted, and the descent of the ground towards the river more rapid, being at the rate of eighteen feet per mile.*

The northern division of the Great Valley is decidedly narrower than the southern, and gradually diminishes in breadth as we go north. So the descent from the edge of the foot-hills to the Sacramento is more rapid than the corresponding slopes in the southern division. Thus the rise from Sacramento City to Folsom, at the base of the Sierra, a distance of eighteen miles, is nearly 200 feet. As is usually the case with large rivers in broad valleys, the Sacramento River runs on an elevated ridge, the banks of the river being decidedly higher than the strip of land adjacent on both sides for a distance of three or four miles. At Colusa this difference of level amounts to as much as twenty feet, and in heavy freshets caused by the rapid melting of the snows on the Sierra, the river discharges itself in part through sloughs into these adjacent lower areas, which become filled with water, so that a large region is sometimes submerged, the central part of the valley assuming

* These figures are taken from the " Report of the Board of Commissioners on the Irrigation of the San Joaquin, Tulare, and Sacramento Valleys."

the aspect of a narrow lake. Reference will be made farther on to some of these points, in connection with remarks on the profile of the former and present river grades in the gravel region.

The drainage into the Great Valley is very peculiar in character, and is the combined result of the climatic and topographical conditions of the region. This will be explained after a brief description of the chains of mountains which form the framework of the valley.

The Coast Ranges form the limit of the Great Valley on the western side, and extend to the Pacific Ocean, there being nowhere on that side more than a narrow space between the foot-hills and the ocean, while for a large part of the distance the steep slopes of the ranges come directly to the water's edge. A glance at the "Map of California and Nevada" will show in a few moments, much better than could be explained in many words, the peculiar character of the topography of the Coast Ranges. The inosculation of the coast mountains with the Sierra Nevada at both ends of the Great Valley has already been alluded to. The general fact is at once recognized that the Coast Ranges are made up of numerous broken and often rather indistinct chains, which on the whole maintain a pretty well marked parallelism with the coast. This parallelism, however, is often better made out from an examination of the courses of the rivers than from the position of the subordinate ranges. It will also be seen on the map, that while the coast mountains are often nearly broken through by cross fractures, giving a chance of escape for the secondary drainage, — that is, for the streams originating within the Coast Ranges themselves, — in only one place does the fracture or depression extend entirely across the whole series of chains. This takes place at the Bay of San Francisco, where is the only outlet for the entire drainage of the Great Valley. Here, in latitude 38°, the Sacramento and San Joaquin rivers unite in an extensive depression, partly occupied by low islands covered with a dense growth of "tule" (*Scirpus palestris*), and subject to overflow where not artificially protected, and partly by Suisun Bay and the Bays of San Francisco and San Pablo. Of these bays the two latter are in reality portions of one and the same thing, there being a narrow strait separating them, and they form a depression about fifty miles long, lying parallel with the general trend of the ranges, and enclosed within them ; while Suisun Bay, on the other hand, is rather at right angles to the others, trending across the ranges, and having no streams entering it, except the Sacramento and San Joaquin. It is, in fact, the half-submerged delta of these two rivers, and is partly occupied by large tule-covered islands. The passage from the Bay of San Francisco out into the Pacific — the so-called "Golden Gate" — cuts directly across the ranges, has high precipitous sides, and is, in its narrowest part, only about a mile wide. Through this contracted passage escapes the drainage of an area of 57,200 square miles.*

South of the Bay of San Francisco the course of the Salinas River, which is about 150 miles in length, and which empties into the Bay of Monterey, indicates a pretty well marked longitudinal division of the Coast Ranges for that distance, its course being almost exactly parallel with that of the coast itself. A subdivision of the region west of the Salinas is indicated by the branches of that river, which all flow for a considerable distance parallel with the range, and then turn at right angles and cross it. This is particularly well marked in the case of the Nacimiento River, as well as in that of the San Antonio. Opposite the head of the Nacimiento rises the Sur, and opposite the San Antonio, the Carmelo, the last-named emptying into the Bay of Monterey, and the Sur directly into the Pacific. East of the Salinas River, and parallel with it, is the San Benito, which empties into the Pajaro, the latter occupying a marked transverse break in the ranges opposite the Bay of Monterey, but which, as already hinted, does not extend quite across them. Thus we have four pretty well marked divisions of the Coast Ranges in this region, between the parallels of 35° and 37°, each of which is sufficiently distinct to have received a special appellation; and in the region south of the Bay of San Francisco the names, having been given by the Spanish-Mexican inhabitants, are in almost all cases those of Saints. Nearest the coast, and west of the Carmelo and the Nacimiento rivers, is the Santa Lucia Range; next in order, and between the Carmelo and the Salinas, are the Palo Escrito Hills; then, next east of the Salinas, and between that and the San Benito, is the Gavilan Range, while on the east of the San Benito, and forming the interior of the frame of the Great Valley, is a wide belt of irregular elevations, considered by us as belonging to the Monte Diablo Group. The total breadth of the Coast Ranges in this portion is about seventy miles, and the elevation of the different sub-ranges is pretty uniform, being in their culminating points from 4,000 to 6,000 feet above the sea-level.

As in Santa Barbara and Ventura counties, between the parallels of 34° and 35°, the coast trends almost exactly east and west, so here the divisions

* As computed by the United States Irrigation Commissioners on the basis of the State Geological Survey Maps.

of the Coast Ranges as well as the lines of drainage have a corresponding direction. The Santa Iñez Mountains rise here boldly from the coast to the height of 4,000 feet, while to the north of this range and parallel with it runs the river of the same name. North of this, again, we have the broad chain of the San Rafael Mountains, which connects at its eastern end with the Sierra Nevada. The Santa Maria, or Cuyama, River, whose course is in general near the 35th parallel, forms the boundary between the east and west and the northwest and southeast trending features of the Coast Ranges. The east and west trend of the Santa Iñez Range is continued through Ventura into Los Angeles County in the Sierra de San Fernando, and parallel with this latter, in close proximity to the coast, and on the 34th parallel is the Sierra de Santa Monica. Their eastern terminations are lost in the great mass of mountains which, with a southeast trend, extend on through the southern portion of California, and which are known by various names in their various subdivisions, as the San Gabriel, the San Bernardino, and the San Jacinto ranges.

North of the Bay of San Francisco, we find as far as Clear Lake, in the parallel of 39°, a general parallelism of the topographical features with the trend of the coast, the drainage being chiefly in a southeastern direction into the Bay of San Pablo. The ranges are much broken, however, in this portion, Napa Valley being the only depression of considerable length extending parallel with the coast. North of Clear Lake there is one pretty well marked dominating ridge, which holds a general parallelism with the coast, trending with it more to the northward; and, after passing the parallel of 40°, actually bending round a little to the east of north, as does the coast between Cape Mendocino and Crescent City. The drainage, however, from the main divide is in the direction of parallel lines having exactly the same trend as that portion of the coast between Punta Arena and Cape Mendocino. Eel River, a little over a hundred miles in length, is the principal one of the streams. A notice of the character of the drainage of the Coast Ranges as related to the Great Valley will be better introduced after a sketch of the topography of the other great system of mountains on the eastern side of that valley.

The unity of the Sierra Nevada is at once apparent in the single name which it bears, in marked contrast with the Coast Ranges. No one has ever thought of dividing the Sierra into groups with different names; while on the other hand the Spanish-Americans have many appellations for the different groups of what we call the "Coast Ranges," but no general one for the whole system.* It is true that the exact limits of the longitudinal extension of the Sierra have been a subject of much discussion; but this is a question of theoretical rather than of practical importance. To the ordinary Californian, and for the purposes of the present volume, the Sierra Nevada may unhesitatingly be taken as commencing at the lower or southern end of the Great Valley, in the neighborhood of Tejon Pass, and extending to Lassen's Peak, which is very nearly on the same parallel as Redding and Shasta (town) before mentioned as marking the northern limit of the valley. From Lassen's north, the prolongation of the axis of the Sierra is marked by a series of volcanic cones, finally culminating in Mount Shasta itself, which is on almost every side an isolated peak, rising 10,000 feet above its base, and with Lassen's Peak forming the advanced guard of the immense volcanic plateau, which stretches over many thousands of square miles in Northern California, Southern Oregon, and the region lying east and north of this. From Shasta (town) north, the Coast Ranges in Trinity, Klamath, and Del Norte counties assume characters, as to elevation, direction, and geological structure, intermediate between those of the Sierra and the Coast Ranges proper. This portion of the State, indeed, has been less explored by the Geological Survey than any other portion of California, and the relation between the two great ranges are far from being clearly understood.†

Beginning then at the southern extremity of the Sierra and proceeding north, we find, in general, less elevation of the chain, and at the same time greater complication of structure. But it must be borne in mind that the Sierra Nevada is in reality not only a range of mountains, but a range built up on the western edge of the great plateau or elevated mass of which the Rocky Mountains form the eastern edge. The Sierra differs from the Rocky Mountain ranges, however, in that it marks a descent almost to the sea-level, while the eastern edge of the Cordillera plateau at the east base of the Rocky Mountains is several thousand (five thousand in the latitude of Denver, Colorado) above the sea-level, towards which it is prolonged in a very gently descending incline which has no parallel on the other edge of

---

* The higher snow-covered portions of the Rocky Mountains and of the Himalaya are very commonly called by the dwellers at their bases the "Snowy Range"; this is exactly the same name which the Spanish-speaking people gave to the high mountains of California.

† The eastern slope of the Coast Ranges, north of Clear Lake, has never been explored either geographically or geologically.

the plateau. As one consequence of this condition, we have in the Sierra a chain with — so to speak — only one slope; at least the western slope of the range is of vastly more importance in every way than the eastern; indeed, the limits of the latter, from the head of Owen's River north, cannot be so easily defined. The geological and topographical conditions indicated in the preceding pages, and to be more fully enlarged on in those which follow, are well illustrated by the distribution of the population of the State of California. This is almost entirely concentrated in two groups: one, agricultural and commercial, about the Bay of San Francisco and in the valleys which connect with this great break in the Coast Ranges; the other, chiefly mining, on the west slope of the Sierra Nevada, and very much concentrated in the region between Mariposa and Plumas counties, or the very section over which the auriferous gravels occur which form the subject of the present volume.

In a rapid review of the most important topographical features of the Sierra Nevada, we naturally begin with the highest division, or that lying between the parallels of 36° and 37° 30'. In this portion of the range, the eastern slope is well defined and very narrow — it being hardly more than ten miles in width — and exceedingly steep. The descent, on the average, is over 1,000 feet per mile in that part of the Sierra which is opposite Owen's Lake. The western slope, on the other hand, is here about fifty miles in width, and has an average slope of about 250 feet to the mile. On the eastern border of the Southern High Sierra there is the well-marked depression occupied by Owen's Lake and the river of the same name, with the steep, lofty, and narrow chain to the east of this, known as the Inyo and White Mountain Range. All of these topographical features — the crest of the Sierra, Owen's Valley, and the Inyo Range — are nearly parallel with each other, and have about the same direction as the main axis of the Sierra Nevada, namely, N. 31° W. The elevation of the main crest of the Southern Sierra is from 12,000 to 13,000 feet, with numerous points exceeding 14,000, but no one — so far as known — quite reaching 15,000 feet.

The drainage of this elevated region is chiefly effected by Kern, King's, and the San Joaquin rivers; the first and last named of these run for long distances in secondary depressions parallel with the axis of the range, and then turn and break through at right angles to their former direction. The Kern has two parallel main branches, which run in corresponding longitudinal depressions, heading opposite the very highest part of the range, on the south flanks of

Mount Brewer, on the north slopes of which grand peak one of the main branches of the King's heads. The lower region between the Kern and King's rivers is drained by the Kaweah. The respective areas of catchment of these rivers are as follows: Kern River, 2,382; King's River, 1,853; Kaweah River, 608 square miles. The Kern drains the largest area of any stream flowing down the western slope of the Sierra, excepting the Feather. The San Joaquin has an area of catchment of 1,630 square miles. It has one main branch, — the South Fork, — which runs for fifty miles parallel with the range, or towards the northwest, heading on the flanks of Mount Goddard. Smaller branches, running in a southeasterly direction, head on the slopes of Mount Lyell and the cross range which runs transversely from that nodal point to the secondary parallel range which is called the Mount Clark or Obelisk Range. The Chowchilla and Fresno Rivers lie between the San Joaquin and the Merced; but they do not head up as far as the main crest of the Sierra, being lapped around on the north and south by the two last mentioned rivers. The areas of catchment of these streams are, for the Fresno, 258, and the Chowchilla 303 square miles.

Of the high region drained by the Kern and the South Fork of the San Joaquin and the intermediate rivers already mentioned, there has been only a hasty reconnaissance made by the Geological Survey. Before that had been done, in 1864, that portion of the Sierra Nevada was a *terra incognita.* It embraces the grandest and most picturesque portion of the range; but not being a mining region has had but little attention bestowed upon it by the people of the State.

The Middle High Sierra, as we may call the region which lies about the head of the Merced and Tuolumne Rivers, is much better known than the loftier and less accessible region to the south. It is high and grand, and much visited on account of its containing the far-famed Yosemite Valley, through which the Merced River runs, while Mount Dana and the High Sierra about the head of Tuolumne, being very easily reached from the Yosemite, the head-waters of that river are also well known, and the whole of this region has been pretty accurately mapped by the Geological Survey.* The respective areas of catchment of these rivers are: Merced, 1,072; Tuolumne, 1,513 square miles. The crest of the Sierra at the heads of these rivers is considerably broken into small groups of peaks, as will be seen from

* See " Map of that portion of the Sierra adjacent to the Yosemite Valley," in the Yosemite Guide-Book.

an inspection of the map, much better than can be described in words. The height of the dominating peaks in this part of the Sierra is still considerable, it being from 12,000 to 13,000 feet, and the passes range from 9,000 to 10,000 feet. The width of the chain is here quite large, it being fully eighty miles in a straight line from the edge of the foot-hills to the main divide at the head of the Tuolumne.

The descent on the eastern slope of the Sierra in this portion of the range is very steep,—fully a thousand feet to the mile. As opposite the more southern portion of the mountains we have Owen's River and Lake, without any outlet to the sea, forming a long and narrow closed basin, so here opposite the head of the Tuolumne, at the foot of the eastern crest of the Sierra, we have a large lake (Mono Lake) about fourteen miles long, and about 6,500 feet lower than the highest adjacent points of the Sierra crest. This also forms a closed basin, and from here north the drainage eastward from the summit of the range is into a series of depressions without outlet, forming a part of the Great Basin system. Indeed, from Mono Lake north it is not easy to separate the Sierra proper, on the east slope, from the Great Basin ranges. From the head of the Mokelumne River, at the grand volcanic peak of Silver Mountain, there is a continuous chain of elevations, crossed by narrow passes, running north and abutting on Carson River, forming the chain so prominently seen in looking towards the east, from Genoa and Carson City. A few miles northwest of Silver Mountain, at the head of the South Fork of the American River, the Sierra seems to divide, a spur almost equal in elevation to the main range going off to the north, and forming a very distinct range as far as the Truckee River. Between this spur and the more easterly main divide lies Lake Tahoe, a noble body of water, a little over twenty miles long, and from eight to twelve miles wide. This lake, which has an elevation of a little over 6,000 feet, is connected by the Truckee River with Pyramid and Winnemucca lakes, which belong to the Great Basin system. The main crest of the range to the east, at the head of the forks of the American River, are a little over 9,000 feet in elevation, and the passes from one to two thousand feet lower than this. Mount Stanford, just north of Donner Pass,—the one by which the Central Pacific railroad crosses the Sierra,—is 9,102 feet in elevation, or just about 2,100 feet higher than the pass which is four miles to the south.

The principal rivers which flow down the western slope of the Sierra, from the Tuolumne north, are: the Stanislaus, draining 971 square miles, and heading in the peaks to the south of Silver Mountain; then the Cala-veras, a small stream with only 389 miles of catchment area, rising near the grove of Big Trees of the same name; the Mokelumne, with 573 square miles of drainage, rising a little to the east of Silver Mountain; the Co-sumnes, with a catchment area just about equal to that of the Mokelumne, but not heading on the main divide. North of this we have the extensive drainage system of the American River with its numerous branches, or forks, the most southern of which heads to the southwest of Carson Pass, on the north slopes of the ridges of which the southern sides feed the Mokelumne; the most northerly branch, on the other hand, heads near Donner Pass, about fifty miles farther north. The drainage area of the American River — a stream of great importance with reference to the subject of the present volume — is 1,889 square miles. North of this again is Bear River, head-ing near Donner Pass, with 484 square miles of catchment area; then the Yuba, draining 1,329 square miles, also frequently to be mentioned in the course of this volume. Finally we have the Feather, with numerous branches, and a catchment area of 3,393 square miles, larger than that of any other of the streams flowing down the western slope of the Sierra. The head of the South Fork of the Feather is about four miles to the west of Pilot Peak, while the head of the North Fork is at Lassen's Peak; the distance between these two peaks being about sixty miles. From Mount Stanford north, the water-shed of the Sierra presents a very irregular line, and the range is equally broken and irregular. Most of the area in question is drained by the North Fork of the Feather and its various branches, which run in nearly parallel depressions having the general trend of the Sierra and unite with the main stream which occupies a position tranverse to this, and in the lower portion of its course, for a distance of forty miles, runs in a deep cañon, re-ceiving no important tributaries. The main axis of the Sierra is prolonged from Mount Stanford through the Downieville or Sierra Buttes, Pilot Peak, Clermont, Spanish Peak, Ben Lomond, and Butt Mountain to Lassen's Peak; while nearly the whole drainage area of the North Fork of the Feather lies to the east of this between the parallel ranges previously spoken of, which are some five or six in number, the total width of the western slope in this portion of the range being about eighty-five miles. The elevation of the dominant peaks in this region is not far from 9,000 feet; although Lassen's Peak a little exceeds 10,500. The passes are all lower in proportion to the height of the crest than farther south; they range from 5,000 to 6,000 feet in altitude. Farther details in regard to the topographical character of this part of the western slope of the Sierra may more conveniently be added after a brief notice of the geological features of the range, since it is hardly possible to understand the very considerable differences between the surface features of different portions of the region in question, without having previously learned something of its geological structure.

Something should, however, here be introduced with reference to the amount and distribution of the rain-fall on the western slope of the Sierra Nevada, and on the borders of the Great Valley in general, and especially in illustration of the manner in which precipitation is influenced by the topography, since these are questions of much interest in connection with the subjects to be discussed farther on in this volume. It is necessary in the first place to lament the almost total want of accurate statistics of precipitation, especially in the mountain regions. Some data may, however, be given of a comparative kind; and it must be borne in mind that the actual amount of rain falling from year to year is extremely capricious, so that many years' observations would be required in order to obtain an accurate general average result.*

The dominant fact in regard to California precipitation is, that it is almost exclusively limited to less than half the year. In the latitude of 38°, for all practical purposes, there is no rain during the six and a half months beginning with the first of April. Almost half the total rain-fall of the year occurs in the two months, November and December. As we go north from the latitude of San Francisco there is a tendency to a slight increase of total precipitation and to a summer rain-fall; but, as a general rule, there is a marked uniformity in the characteristics of the climate from one end of the Great Valley to the other; but in any cross section of the valley, and the ranges on either side, the influence of the topographical features of the country on the rain-fall is very marked. The rain-bearing clouds coming almost exclusively from the south and southwest, the west slopes of the Coast Ranges receive a considerable share of the moisture which they contain, and the eastern flanks of those mountains are very much drier. An inspection of the map will show this at once, accurate statistics being almost wanting. It will be noticed that on the western side of the Great

* At San Francisco the range in twenty-two years was between 49.3 and 7.0 inches; at Sacramento, in twenty-four years, between 36.4 and 4.7 inches; at Clear Lake, in six years, between 66.7 and 16.2 inches; at San Diego, in twenty-two years, between 14.8 and 4.5 inches.

Valley there is not a single stream which can be called permanent, flowing down the Coast Mountains from one end of the San Joaquin Valley to the other. All disappear soon after reaching the edge of the foot-hills, and in summer they hardly carry any water at all, even within the mountains. Nearly the same is true for the Sacramento Valley; Puta, Cache, and Stoney Creeks, however, are streams of some size; but even these, as the dry season advances, spread themselves out and become lost before reaching the Sacramento. On the western slope of the Coast Ranges the precipitation is far from regular, it being naturally greatly influenced by the lay of the land. It is said that the mean of nine years of observations at Pilarcitos Dam, about fifteen miles south of San Francisco, gives an average of fifty-eight inches of rain, while at the city it is only twenty-three inches and a half.* The amount of rain which falls on the east side of the Coast Ranges at the extreme southern part of the Great Valley must be exceedingly small. Statistics are wanting; but, judging from the character of the country, it is practically a rainless region. Going south from San Francisco, even on the west slope of the Coast Mountains, there is a marked decrease in the amount of precipitation, the quantity diminishing until we reach Lower California, where it is almost null. At San Diego it has diminished to an average of about ten inches, and is very irregular. From here south, the climate assumes a tropical character, and the rain-fall, which is exceedingly small, occurs during the summer season. On the Peninsula of Lower California the average for a long term of years would probably not exceed three or four inches.

In the Great Valley itself the rain-fall on the whole is small, except in the immediate vicinity of the opening into the Bay of San Francisco. There is undoubtedly a marked increase in the amount on any line of cross-section in going from the west towards the east, the influence of the Coast Mountains in cutting off the precipitation being of course less felt as we recede from the range. The west side of the San Joaquin and Tulare Valleys, when not artificially irrigated, is but little better than a desert. Crops may be raised on the eastern side of that river over considerable portions of the region in favorable years; but how often such years occur, on the average, we have no means of determining. The average amount of rain being small, it needs but a moderate addition to this, provided it falls at a favorable season, when the crops are just in the right state of forwardness, to enable

---

* Report of the United States Irrigation Commissioners.

the farmer to secure a bountiful harvest. In the winters from 1870 to 1873 the annual rain-fall at Visalia was as follows: for the year 1870 – 71, 6.8 inches; 1871 – 72, 10.3 inches; 1872 – 73, 7.2 inches. The crops of the year 1872 were abundant, those of the other years mentioned were failures. That the Great Valley on the whole is a dry region, is abundantly proved by the scantiness of its population. But there are here all possible facilities for artificial irrigation, with the aid of the copious streams coming down the slopes of the Sierra; although to utilize these in a suitable and economical way demands considerable capital and engineering skill, combined with intelligent action on the part of the Legislature, backed up by the good will of a law-abiding people.

The enumeration of the streams draining the western slope of the Sierra, as given in the preceding pages, indicates very clearly a considerable amount of precipitation in that region. Unfortunately there are no statistics giving the amount of rain and snow-fall for any considerable period at a sufficient number of stations to enable us to form any idea of the general average; neither are there any measurements of the amount of water carried down in the different streams, which amount is not only very variable from month to month during the year, but also from year to year. That the precipitation is almost entirely during the winter months has been already stated; that it increases with the elevation would also be inferred; and, the southern portion of the Sierra being higher than the northern, it is probable that this increased elevation about compensates for the difference in latitude. If the Sierra opposite the Tulare Valley were no higher than it is from Sacramento north, we should almost certainly have no such large streams flowing down in that region as the Kern and King's. If, moreover, the rain-fall were equally distributed throughout the year, instead of being, as it now is, limited to less than six months, the condition of things would be greatly changed, especially in relation to the mining interests of the Sierra. Almost the whole precipitation in the higher portions of the range is in the form of snow; and this usually remains in large part upon the surface, gradually melting away during the early spring and summer months, so that it forms a much more reliable source of supply than it would do if it had fallen chiefly as rain. The system of dams, reservoirs, and ditches by which the water of the High Sierra in the mining region is made available for the purposes of the miner, is very extensive, and a large amount of capital is invested in this form of improvements. The total quantity of the precipitated moisture in the

Sierra is very large in certain years, and probably quite irregular; but there is an entire want of statistics extending over any considerable number of years at any one point. The snow-fall in some winters near the summit of the range must amount to as much as sixty or seventy feet, and it accumulates to a depth of from fourteen to twenty feet in the more elevated valleys. After a winter of large precipitation, the snow lies in the passes until midsummer, and a considerable amount remains over until the next year on the highest peaks and ridges, especially on their northern slopes. When two or three dry seasons succeed each other, the crest of the Sierra becomes, apparently, almost entirely denuded of snow, as seen from a distance; still, even then, quite large patches would be found high up among the depressions and ravines near the summit.

Although, as before remarked, almost the whole precipitation, in the higher part of the Sierra, is in the form of snow, there are, even on the western slope, occasional heavy rain-falls, accompanying violent thunderstorms. Such a one occurred in June, 1867, at the head of the Yosemite Creek, raising that stream several feet in the course of half an hour, while its effects were also very perceptible in the South Fork of the Merced, as far down as Clark's Ranch. On the eastern slope, near the summit, such summer thunder-showers are not very infrequent; but their range seems to be quite limited. Farther down on that side, and generally through the very dry portions of the mountainous regions of the State there are occasional exceedingly heavy rain-falls over very small areas; these are popularly known as "cloud-bursts," and their effects are sometimes disastrous to persons caught in them, dry cañons being converted for a few minutes into raging torrents which sweep everything before them. Additional remarks on the snow and rain-fall of the High Sierra will be found farther on in this volume, in the chapter devoted to glacial phenomena.

Section II. — *Geological Age and Structure of the Coast Ranges.*

It is not intended in this place to give anything like an exhaustive account of the geology of the Coast Ranges; it would require more than one volume to do the subject justice. It will be convenient, however, for the purposes of the present work, that the reader should have a general idea of the geological peculiarities of the elevated border of the western side of the Great Valley, since occasional reference will have to be made to these in discussing the phenomena of the more recent formations of the Sierra Nevada. For further information in regard to the topics of this section the reader may consult Geology of California, Vol. I., a considerable portion of which is devoted to the Coast Ranges. The material exists for a much more elaborate account of the geology of these ranges than has as yet been published, and it is the wish and expectation of the writer that this work shall be performed. Meanwhile the brief synopsis which here follows will be of service to those who have occasion to use the present volume.

The most striking fact in regard to the Coast Ranges is, that this very extensive group of mountain chains is of comparatively very recent geological age. It is made up of Cretaceous and Tertiary strata, with no rocks older than these showing themselves in any portion of the complicated series of elevations which are properly included under the above designation. There are some areas within the Coast Ranges occupied by volcanic rocks, and others where granite and granitoid masses make their appearance on the surface. But by far the larger portion of these ranges consist exclusively of sedimentary beds, which have been bent, folded, and crushed, so as to form numerous subordinate ranges, as already indicated in the preceding section.

The Coast Ranges resemble the Appalachians in having no central axis or dominant range to which the others are subordinated. Neither is there in the Californian mountains any core of igneous rock, to the elevation of which the folding or disturbance of the sedimentary beds might be attributed. There are circumscribed areas where such rocks make their appearance; but it is evident that these cannot be considered as having played any prominent part in the structural development of the system of mountains in which they occur. The folds of the Pacific Coast Ranges differ also from those of the Appalachians in being much less regular and symmetrical in their form than are those of the system of mountains which runs parallel with the Atlantic side of the Continent. On the western edge of the country the work of mountain-building seems to have been much more rapidly accomplished than it was on the eastern. In the first place, the formations themselves are far more irregular in development and thickness than are those of the Appalachian system; they differ from the latter, moreover, in being more or less irregularly broken through by both granitic and volcanic outbursts. But the rocks of the Coast Mountains are especially distinguished by the fact, that the movements to which they have been subjected, and which have originated the complex of alternating elevations and depressions making up the system of chains known as the Coast Ranges, have been apparently sudden and sharp; so that the result may be called a crushing and breaking, rather than folding and uplifting. Hence the very great difficulty, on the Californian edge of the Continent, in getting sections of the formations exhibiting regularity of succession in the beds, and having therefore real stratigraphical value. Often, and especially in the central and northern portions of the State, the rocks for long distances are so broken up that a recognition of their real structural relations is entirely impossible. And, while there are certain regions where fossils are abundant, on the whole the Coast Ranges are not well supplied with these most important aids to stratigraphical research. The more recent members of the Tertiary are occasionally quite fossiliferous; but the older Tertiaries and the Cretaceous are, as a rule, almost destitute of the remains of either plants or animals.

There is another fact which helps explain the difficulty of co-ordinating the Coast Range geology. It is this; that the rocks have been very considerably metamorphosed, over wide areas, and that the products of this metamorphism are the same in groups of strata of quite different geological age. These masses of metamorphosed rock are often extremely irregular in outline, and we find extensive areas where the chemical changes have been carried so far as to completely obliterate the lines of stratification, although perhaps producing only a very slight change in the character of the material, but thus making the deciphering of the original position of the beds extremely difficult. The whole subject of the metamorphism of the Coast Range rocks is one of extreme difficulty, and it offers a fine field for investigation. Many hundred analyses and microscopic sections of the rocks would be required for this purpose; but the results, if the work were done by capable hands, would be of great interest, especially with reference to the manner in which magnesia has been introduced into the silicious and silico-argillaceous strata, leading to the formation of the serpentine masses which play so important a part in Coast Range geology.* That portion of the Coast Mountains which extends north from Clear Lake, and which, as before mentioned, is yet almost a terra incognita, both geologically and geographically, must be carefully worked over before we can clearly discriminate be-

* So striking is the preponderance of magnesian compounds in a portion of the Coast Ranges near the New Idria Mine, that it is familiarly known to the people as the "Magnesia Country."

tween Sierra and Coast Range in that region, the exploration of which would be of great interest, as well as of great difficulty, owing to the thick undergrowth on the mountain slopes, the heat and dryness of summer, and the remoteness from all lines of communication.

In a rapid review of the geological characters of the Coast Ranges, we may begin at the southern extremity of the State and proceed northwards. Along the coast, from the boundary to the Santa Anna River, there is, bordering the ocean, a narrow strip of unaltered sedimentary rocks, of Cretaceous, Eocene (?), Miocene, and Pliocene age. The region occupied by these rocks forms a " mesa," some ten or fifteen miles in width, which rises gradually with gentle slope to the base of a high, rough range of mountains, made up of granitic and highly crystalline rocks, of the geological age of which nothing is known. The stratified rocks resting on these older crystalline ones are somewhat disturbed in position, but not metamorphosed. The granitic mass is broken up into various groups, of which the one at the head of the San Diego, Sweetwater, and Tia Juana rivers is called the Cuyamaca Mountains; its continuation north, between the San Bernardo and the San Luis Rey rivers, is known as Smith's Mountain; and from this north to the Santa Anna River as the Santa Anna Range. Northeast of these ranges, and forming their continuation in that direction, is the lofty mass of the San Jacinto Range. To the north of the latter, and separated from it by the San Gorgoño Pass, is the mass of which San Bernardino Mountain is the culminating point; then, again, farther west the San Gabriel Range, which connects with both the Coast Ranges proper and the Sierra Nevada at the nodal point near the Tejon Pass. These granitic and crystalline masses, which resemble those of the Sierra in being more or less auriferous, are undoubtedly of much interest, but they have as yet been but imperfectly studied, and what is known of them has little bearing on the subject of the present volume. They, as well as the unaltered but somewhat uplifted sedimentary rocks at their bases, may be passed over without more notice.

Starting then from the beginning of the Coast Ranges, as popularly designated, in Santa Barbara and Ventura counties, we find the different ranges which have been previously mentioned — the Santa Monica, San Fernando, Santa Iñez, and San Rafael — to be chiefly made up of rocks of Miocene age, which are divided into two well-marked lithological groups; one a fine-grained slate or shale, often highly bituminous, and the other a rather coarse-grained sandstone, the latter being the inferior member. The Santa Monica

Range has for a portion of its length an anticlinal structure, and a central axis of granite, forming in this respect a marked contrast with most of the ranges of the Coast Mountains. The Santa Iñez Range, which is narrow and very steep, has also an anticlinal structure, with the axis of the flexure irregular, and not everywhere coinciding with the crest of the range, so that the latter appears sometimes to have a monoclinal dip; and, in places, the whole of the upper portion of the mountain has an inclination to the north, and in other places, in the opposite direction. These rocks are overlain along the coast, in places, with a slightly disturbed, nearly horizontal deposit of Pliocene beds; the underlying Miocene is here tilted up into a nearly vertical position, and very much contorted. The internal forces which elevated the Santa Iñez Range seem to have exhausted themselves in the vicinity of San Buenaventura, so that from the river of that name eastward to the head of the Santa Clara Valley, while the mountains are still high, reaching an elevation of 3,000 feet above the sea-level, the rocks composing them are, in general, but little disturbed; there are, however, local contortions and some marked changes of strike. The Sulphur Mountains, lying between the San Buenaventura and Sespe Rivers, are made up of considerably contorted strata of bituminous slates, and are well known from the numerous attempts which have been made here to strike, by boring, flowing wells of petroleum. The great mass of the San Rafael Mountains, occupying the area between the Santa Iñez and Cuyama rivers, is chiefly composed of Miocene shales and sandstones, considerably disturbed towards its northern edge, and perhaps folded, the want of any maps of this region making it difficult to connect the observations, so as to make out the structure. A section across the ranges, from San Buenaventura to Tar Springs in Tulare Valley, shows the great mass of the mountains to be made of bituminous shales and slates, not much disturbed, except locally, until towards the northern extremity of the section, where there are two sharp folds, between which runs the Cuyama River. At the north end of San Fernando Valley, and in various other places in this region, there are small areas of Pliocene gravels lying nearly horizontally on the upturned edges of the Miocene. Passing up the Santa Clara River, when we have reached the San Francisco Pass, we find the entire structure of the range changed. The Miocene rocks are turned up on edge, and in places so much metamorphosed as to be converted into micaslate. In the Pliocene gravels, which at the mouth of the San Francisco Cañon lie unconformably on the older rocks, gold has been obtained by washings at various times since 1841. The granite comes in to the northeast of the metamorphic Tertiary (and perhaps, in part, Cretaceous) rocks, and forms the divide between the branches of the Santa Clara and the Great Basin, or Mohave Desert, the edge of this granite, next the plain, being overlain with stratified beds of recent volcanic materials. This granitic belt forms the continuation of the San Gabriel Range, and connects, in the region between the Cañada de las Uvas and the Tejon Pass, with the great metamorphic and granitic masses of the Sierra Nevada, the crystalline rocks being apparently continuous, but the disturbance and upheaval of the Tertiary and Cretaceous formations not being discernible to the east of the Tejon Pass.

All these Miocene strata of the Santa Iñez and San Rafael chains belong to the east and west system of upheaval, already noticed as characterizing this part of the Coast Ranges. The principal disturbance of the strata must here have occurred at the close of the Miocene epoch, since the Pliocene is everywhere unconformably deposited on the underlying strata in a nearly horizontal position, as also quite unaltered.

Leaving Santa Barbara and Ventura counties, and passing northward into San Luis Obispo, Monterey, and Santa Clara to the Bay of San Francisco, we find the trend of the ranges, as before noticed, to be almost uniformly parallel with the coast, which here runs about N. 30° W. The same sandstones and bituminous shales with which we have already become acquainted form the mass of the mountains, until we reach a point opposite the north end of Tulare Lake, which may properly be considered as the southern extremity of the Monte Diablo Range. Here the Cretaceous group begins to form a part of the Coast Ranges, making up almost the whole of that member of them which lies east of the San Benito River. With the Cretaceous appear granitic and highly metamorphic granitoid rocks, which are developed in considerable quantity in the Gavilan Range, as well as in the Santa Lucia Mountains. This latter range extends along the coast from Point Pinos to San Luis Obispo, and, being almost inaccessible from the quantity of chaparral with which its ragged slopes are covered, has received but little attention from the Geological Survey. The Miocene Tertiary, and the Cretaceous rocks in the region north of the 36th parallel, as far as Monte Diablo, are also much metamorphosed over irregular areas; and, when so changed, the planes of stratification cannot usually be satisfactorily made out. The ranges are, however, chiefly monoclinal, the valleys sometimes occupying the bottom of synclinal curves. The Cretaceous mass of the

Monte Diablo Range is flanked by Miocene and Pliocene Tertiary on its eastern edge, the whole series having a dip towards the San Joaquin Valley, steep in the centre of the range, and gradually diminishing as we go eastward. Between the Gavilan and the Monte Diablo Range there is a heavy mass of gravel, of Pliocene age, which is distinctly stratified and turned up into an almost vertical position. All this portion of the Coast Ranges exhibits abundant evidence of very recent disturbances, lasting through the Pliocene epoch. Volcanic rocks are not found in large quantities anywhere in the Coast Ranges south of the Bay of San Francisco. There is, however, a line of protrusions of trachyte, extending from a point a little south of the west end of Pacheco's Pass to the southeast, diagonally across the Monte Diablo Range, as also some small patches of eruptive material at various points near San Luis Obispo. The granite of the Santa Lucia Range has not been sufficiently investigated to allow of an opinion being formed as to its stratigraphical relations.

North of Monterey Bay, in the region adjacent to the Bay of San Francisco, the formations have been studied with more care than has been possible farther south.* There are two well-marked ranges in this portion of the Coast Mountains. One — the Santa Cruz Range — extends from the Bay of Monterey up through the Peninsula of San Francisco; the other — the Monte Diablo Range — is separated from the first named by the depression of the Santa Clara Valley, which is continued north in the Bay of San Francisco. The Santa Cruz Range is largely made up of little altered Miocene strata, similar to those described as occurring farther south. But there is on its eastern edge, as it is prolonged through the Peninsula, a belt of Cretaceous rocks, forming the higher portion of the divide, which culminates in Mount Bache (3,790 feet high). These rocks are greatly changed from their original character by chemical action; and they are largely made of serpentine and an imperfect jasper, locally known as "red rock." It is this formation which carries the quicksilver ores so extensively worked in the vicinity of the Bay of San Francisco. This Cretaceous formation has also a narrow belt of limestone running through it, a rock not frequently occurring in California. These rocks are almost entirely destitute of fossils, although enough have been found to fix the geological age of the formation.

On the opposite side of the Santa Clara Valley the Cretaceous strata are

* A geological map of the region adjacent to the Bay of San Francisco, on a scale of two miles to an inch, has been prepared for publication.

largely developed, forming much the larger portion of the high and rough mountain region lying between Pacheco's and the Livermore passes, of which the culminating point is Mount Hamilton (4,448 feet). This central mass of highly altered Cretaceous is flanked on both sides with Tertiary, and in the division of the Monte Diablo group, which lies north of Livermore Pass, and extends thence to the Straits of Carquines, there is a large area occupied by Tertiary, the central and dominating mass of Monte Diablo itself, however, being made up of Cretaceous. North of Pacheco's Pass there are only a few patches of volcanic rock, and there is no granite in this range. On the opposite side of the bay, in the Santa Cruz Range, there are a few very limited areas of volcanic, but much larger ones of granitic or granitoid rock, which protrude through the Miocene strata in such a way as to show that the latter have been considerably disturbed and lifted up by the intrusion of the eruptive material.

Along the flanks of the Miocene near San Jose Mission, and in Alameda Cañon, there are occasional quite large masses of gravel, lying nearly horizontally on the upturned edge of the Miocene. Similar deposits, but of more limited extent, are also found at various points on the eastern edge of the Santa Cruz Range. These gravels near Mission San Jose contain bones of elephant, mastodon, and lama, as also strata filled with fresh-water shells. The more extensive beds of similar character south of Livermore Valley have thus far proved to be destitute of fossils. They are developed to a thickness of 200 feet or more, and are supposed to be of later Pliocene age. These gravels have never been productive of gold, so far as known to the writer. Some further remarks on these Coast Range gravels will be introduced in a future chapter.

North of the Bay of San Francisco the same formations occur which have been already noticed as making up the mass of the Coast Ranges to the south; but the Cretaceous becomes more and more prominent and the Tertiary almost disappears after we get to the north of Clear Lake. From San Pablo Bay north to that lake there is a large amount of volcanic material, and abundant evidence of the quite recent activity of igneous agencies. This volcanic band extends from the bay north through the centre of the ranges in numerous large and small isolated fields, so intricately distributed over the underlying metamorphic Cretaceous rocks, that it would require an accurate map on a large scale, and much research to be able to lay them down correctly. The high points of Mount Helena (3,343 feet), Cobb Mountain,

Mount Harbin, and Uncle Sam are all volcanic. To the east of Clear Lake there are to be observed many most interesting phenomena of active volcanism ; hot springs, small lakes holding boracic acid in solution, sulphur banks, and other occurrences of similar origin are well displayed over an extensive area of country. So far as our knowledge goes, there is an entire cessation of these volcanic phenomena in the region of the Coast Ranges to the north of Clear Lake.

Deposits of gravel of any importance are not usually to be found in the portion of the mountains which has just been noticed ; but there is one, which is quite extensive, in the neighborhood of the north fork of Cache Creek, about twelve miles east of Clear Lake. This deposit first shows itself at Chalk Mountain, and it extends off to the south, widening in that direction, for a distance of ten miles, and occupying a triangular area between Lower Lake, Chalk Mountain, and the head of Grizzly Gulch. It is everywhere stratified, in usually nearly horizontal layers, and is made up of moderate-sized boulders of the metamorphic rocks occurring in the region, mixed with water-worn fragments of unaltered shales and sandstones. This remarkable deposit has a thickness in places of as much as 400 feet. No fossils have been found in this gravel ; but it seems to resemble the Pliocene strata already noticed as occurring in the vicinity of San Fernando Pass, and in other places farther north. It has been deposited prior to the cessation of volcanic action in this region, for a stream of lava coming down from High Valley, a little east of the centre of Clear Lake, has flowed over it in one place.

North of the parallel of 39°, as far as the Klamath River, there is much monotony in the geological structure of the Coast Ranges. The rocks are almost exclusively Cretaceous, and often very much metamorphosed, jaspers, serpentines, and even mica-slates, occurring in large quantities, and in the most irregular manner ; but there are also many areas of quite unaltered strata. The topographical features of the country are on a large scale, the valleys being deep and precipitous, the ridges lofty and extensive. The abundance of chaparral, and the depth of the soil over a large portion of the northern Coast Ranges, makes their exploration exceedingly tedious and unsatisfactory, and the difficulties of making out the structure are greatly increased by the crumbly character of the unaltered strata, and the extreme irregularity of the metamorphic areas. It is a constant repetition of jaspery belts, imperfect serpentines, and coarse non-coherent sandstones, with extremely variable strike and dip. The inclination seems, however, on the whole, to be predominant in a southwest direction, or away from the crest of the range which is much nearer the eastern edge of the mountain mass than it is to the ocean. It would appear that this portion of the Coast Ranges has been lifted up bodily with a general inclination away from the axis of upheaval on its eastern edge, and that numerous minor undulations have been formed at the same time. The deep valleys have been eroded out chiefly by the agency of running water, and the level areas at the bottoms of these cañons are usually of very small size. They are sometimes occupied by horizontal or slightly disturbed strata of Pliocene age. Such deposits, however, are rare, and the exposures are narrow and of no great thickness; they contain an abundance of fossils. Generally, all through the region north of Clear Lake, the Tertiary rocks are extremely subordinate in quantity to the Cretaceous.

The lower part of Trinity and Klamath rivers seems to form the boundary between the Coast Ranges proper, and that portion of the Coast Mountains which appears to belong, lithologically, to the Sierra. The upper portion of Trinity River runs in a direction parallel to that of the streams flowing down the Sierra slope; but at Weaverville it turns and flows at right angles to its former course, or in a direction parallel with the ordinary drainage of the Coast Ranges. After running about forty miles towards the northwest it enters the Klamath, which river makes at its junction with the Trinity a similar change in its direction, after having followed a course transverse to the ranges, or towards the southwest, for a distance of a hundred miles or more. The South Fork of the Trinity heads on the north slope of the North Yallo Balley, and runs, from that point to its junction with the main river, in exactly the same direction as the lower part of the Klamath. All of Trinity, Klamath, and Del Norte counties is an extremely rough region. The main divide of the northern Coast Ranges becomes more and more elevated from Clear Lake north, and the rocks of which it is made up more and more crystalline, until finally an axial mass of granite makes its appearance, the prominent points rising to an altitude of 7,000 or 8,000 feet. To an observer stationed on any one of these elevations, and commanding a wide view over the region, there seems to be no physical break between the Coast Ranges and the Sierra. Scott's, the Salmon, and the Siskiyou groups of mountains seem to represent the continuation of the summit elevations of the Sierra proper, and the Trinity Mountains run into these from the south and from the Coast

Ranges proper, without its being possible at any one point to say; here the Coast Ranges end and the Sierra begins. The rocks in the region northeast of Klamath and Trinity rivers resemble those of the Sierra; they are auriferous, and often highly so. There are also deposits of gravel of considerable thickness in some of the valleys, but not associated with the volcanic rocks as they are in the region to be especially described in the present volume. But these points may be considered more suitably in connection with the description of the geological structure of the Sierra. It is sufficient, in the above brief sketch of the formations of the Coast Ranges, to have indicated, without attempting to go into any details, such of the prominent facts connected with the development of that chain as are needed for an understanding of that which follows. A *résumé* of these, with some few additional remarks, may here be appended.

First. The Coast Ranges are made up, in by far the larger part, of sedimentary strata of Tertiary and Cretaceous age. No rocks older than these have ever been discovered.

Second. There is no dominating or central axis of intrusive igneous rock in the Coast Ranges; but there are occasional small areas of granite, or of some granitoid material, to which the sedimentary rocks in their immediate neighborhood are evidently subordinated.

Third. There are in the Coast Ranges considerable areas covered by modern volcanic formations, and the activity of igneous agencies has, here and there, not entirely died out, but still manifests itself in the form of hot springs and solfataric emanations. The principal development of these volcanic rocks is between the Bay of San Francisco and Clear Lake, and nothing of the kind has been observed to the north of the lake.

Fourth. Of the Tertiary rocks, by far the larger portion is of Miocene age, and the Pliocene is everywhere quite subordinate to it in quantity and in position. The Miocene is exclusively a marine formation; the Pliocene, on the other hand, chiefly subaerial and made up of coarse gravels. There are, however, some areas of Pliocene, as in San Fernando Valley and a little to the south of San Francisco, which are well characterized by an abundance of marine fossils. Beyond Clear Lake, to the north, there is little or no Miocene, and the Pliocene occupies narrow strips in the bottoms of the deep cañons. The Eocene has not been clearly recognized in the Coast Ranges, although there are rocks in the position which this formation ought to occupy, but which seem to be destitute of fossils.

Fifth. The Cretaceous rocks occupy but small areas in the southern Coast Ranges. As we proceed north, we find them coming in, in force, in the Monte Diablo Range, gradually widening northwards, soon forming the bulk of that range, and to the north of the Bay of San Francisco becoming almost the exclusive formation. It is not abundantly supplied with fossils, except at certain localities of limited area. By far the larger portion of the Coast Range Cretaceous is entirely destitute of organic remains.

Sixth. While the larger portion of the Tertiary of the Coast Ranges is but little metamorphosed, there has been a great deal of chemical change in the Cretaceous since its deposition. The products of this metamorphism are chiefly imperfect jaspers, serpentine, and mica-slates. The latter grow more and more abundant as we go toward the northeastern extremity of the system, beyond Clear Lake. The large amount of metamorphism by the introduction of magnesian combinations is especially remarkable. These metamorphic bands and patches do not appear to follow any system in their distribution; at least, none such has hitherto been discovered. A large amount of study will have to be bestowed on these rocks before their nature will be well understood.

Seventh. The principal upheaval and disturbance of the Coast Range system must have taken place at the close of the Miocene epoch. The Miocene and the Cretaceous seem everywhere to be conformable with each other, although there is a large area of the last-named formation over which it does not appear that any more strata were ever deposited, and which we must suppose to have been elevated above the sea-level before the Tertiary began to be formed. The break between the Miocene and Pliocene is very marked in portions of the ranges; in other parts there is a complete conformability between these two formations. This is particularly the case on the inside — the eastern side — of the Coast Ranges, as is well seen in all the sections across the Monte Diablo Range to the San Joaquin plains.

It follows therefore that the disturbances which took place after the deposition of the Pliocene were somewhat local in character, and that they were chiefly felt on the side of the mountains nearest the Great Valley. The break at the Golden Gate, the predominance of volcanic rocks from there north, for a considerable distance, and the powerful disturbances of the Pliocene in the region just south of the Bay of San Francisco may be connected together as being probably the results of one general cause.

SECTION III. — *The Geology of the Sierra Nevada.*

The name SIERRA NEVADA, which is simply the Spanish for Snowy Range, has become fixed upon two chains of mountains, one in southern Spain, the other in California. The latter is, however, by far the most interesting and extensive of the two thus designated. Indeed, it may be asserted with truth that the Californian chain is excelled by few mountain ranges in the world in the variety and interest of its physical features. To the geologist especially it offers problems for investigation of a quite peculiar character, the nature of some of which it is, in good part, the object of the present volume to point out. Far inferior as a whole to the Alps in complexity of structure, and wanting as it is, in some of those accessories which need to be present in order that the height of the picturesque and beautiful may be reached, the Sierra Nevada still has features of its own, which will always command attention and cause travellers for pleasure as well as for scientific instruction to turn their footsteps in its direction. What the Californian Sierra especially lacks, to enable it to rival the Alps, is the glacial masses descending with their majestic flow from the higher down to the lower portions of the chain. Where these are absent, it is impossible to have that variety of scenery which is offered by the Alps. Moreover, the element of vegetation, which enters so conspicuously into the general effect produced by mountain scenery, is decidedly inferior in the Sierra Nevada when considered from the point of view of the picturesque. The grassy slopes above the forests, which have given their name to the Swiss range (die Alpen), are entirely wanting in the Sierra, where the arboreal vegetation extends nearly up to the summits, or on the higher ranges ceases at once, only to be succeeded by bare slopes and cliffs of rock, to which a few straggling and distorted clumps of trees attach themselves at points where they can find shelter and a foot-hold. There is also a certain monotony in the forests of the Sierra, in spite of the great size and beauty of the individual trees; the Coniferous predominate so greatly over the other kinds, that in a distant view they seem to have the entire monopoly of the surface; and, grand as the pines and firs may be, each tree being examined by itself, or impressive as the forest may appear to one merely looking at it as a forest, the foliage is too monotonous and sombre to give the variety which the cultivated eye demands in a panoramic mountain view. As contrasted with the Rocky Mountains, on the other hand, the Sierra Nevada greatly excels, both in the variety and grandeur of its scenery, and in the scientific interest challenged by its geological features.

The advantage of elevation is slightly in favor of the Sierra, at least so far as the height of the highest point is concerned. Yet it is quite remarkable that so large a number of the most prominent peaks of both systems should range in elevation between 14,000 and 14,500 feet,* only one, throughout the whole system of the Cordilleras, exceeding the last-mentioned number. In the Wind River group of the Rocky Mountains are some fine precipitous peaks; but usually the higher points are more or less rounded, and far inferior to the astonishingly pinnacled region of the Southern High Sierra. Neither are the cañons of the ranges making up the eastern side of the Cordilleras as grand as those of the Sierra; and, especially, is there nothing to compare with the Yosemite and the Hetch-Hetchy, with their magnificent waterfalls and almost, or even quite, perpendicular cliffs. Indeed, the type of the scenery represented by the Yosemite is unique in character; and it is no wonder that it has already become far better known and ranked higher than any other scenic feature of our country, unless it be Niagara Falls. The forests of the Sierra greatly exceed in majesty and variety those of the Rocky Mountains, where the trees are monotonous in character, and defective in size and beauty; these deficiencies are, however, in part made up by the great charm of the floral vegetation, which blooms almost up to the very summits of the highest peaks.

But it is especially with the geology of the Sierra Nevada that we have to do in this section, the object of which is to present such a general view of the structure of that range, as it will be necessary for the reader to have in order to understand the following chapters, which will be devoted to a more detailed account of a certain portion of the formations, namely,

* The only point in the Cordilleras which has an elevation greater than 14,500 feet — so far as yet known — is the culminating peak of the Sierra Nevada, Mount Whitney. The elevation of this, however, has not been determined otherwise than by the aid of the barometer; and, of course, the various measurements are liable to an error of a more or less uncertain amount. Three ascents have been made of Mount Whitney by members of the State Geological Corps, and the observations in each case give for the mountain a height exceeding 14,650 feet, even after applying to them the corrections taken from the tables published by the Survey (see "Contributions to Barometric Hypsometry, with Tables for use in California"). The data on which the tables in question are based do not, it is true, strictly apply to elevations of over 7,000 feet, that being the height of the highest station at which observations were made with a view to the fixing of the correction in question, which towards the middle of the day in summer is always subtractive. It is more likely that this correction is proportionally less for higher elevations, than that it is more. Hence we conclude, that it is highly improbable that Mount Whitney should be less than 14,650 feet high. Of twenty-two of the highest peaks measured in the Cordilleras, — so far as yet ascertained, — two range between 14,000 and 14,100; four between 14,100 and 14,200; four between 14,200 and 14,300; eight between 14,300 and 14,400; and three between 14,400 and 14,500 feet.

those included, for convenience, under the general term of "auriferous gravels."

In studying the Sierra Nevada, we have first to make ourselves familiar with the fact that, in so far as its geological structure is concerned, we have to do with two quite distinct series of formations. The principal mass of this chain is made up, as is usually the case with great ranges of mountains, of hard, crystalline rocks; a portion of these is evidently of intrusive origin,— that is, it has come up from beneath and forced its way through the other portion. The material which has played this part of an elevating nucleus or axis is granite. The rocks through which the granite has made its way are chiefly of sedimentary origin; but they have undergone a large amount of chemical change, by which they have often been rendered crystalline, and their original character has been so modified, that in some places it can no longer be recognized. Thus far the Sierra Nevada resembles many other mountain chains; but in the case of most of the great ranges we find the formations growing less crystalline as we recede in either direction from the central axis. We observe, also, that these less altered groups of strata are of more recent geological age, and it is usual to find in mountain chains of the first magnitude a considerable part of the geological series represented. The change from one formation to another is, however, usually a gradual one, and the whole mass is found to have partaken more or less in the processes of disturbance and uplifts by which the chain has been formed. Thus we have in the Alps a range remarkable for the number and variety of the flexures of the sedimentary strata, which are lapped over the crystalline nucleus; but these flexures extend up to and include rocks of Miocene Tertiary, as well as of Liassic, Jurassic, and Cretaceous. The same is the case with the Himalaya, and many other chains. There is, of course, in such cases, no easily recognized and marked line of division in the series of formations of which the range is made up. In the Sierra Nevada, on the other hand, there is a most decided separation of the materials forming the mass of the mountains into two groups. There is the granitic axis with the associated crystalline rocks, and an overlying mantle, if the term may be allowed, of rocks quite different in lithological character and in position, the two series being so distinct from each other, in every way, that they could by no possibility be confounded even by the most unobservant mind. The lower or interior portion is usually hard and crystalline, and if its planes of stratification can be made out, it is easily seen that the rock has undergone more or less disturbance.

as well as chemical change, since its deposition. In general the strata are very considerably elevated, and often nearly into a vertical position. This portion of the mass of the Sierra is universally known to the miners by the comprehensive term of "bed-rock," a term the origin of which is easily understood; it is the foundation on which repose the more recent superficial accumulations which form the mantle spoken of above. These latter are easily distinguished by their unconformability of position, their non-crystalline character, and the fact that they are usually not compacted into what the miner would recognize as a rock, or a material which cannot be excavated without blasting. It is true that in other mountain chains the older formations often have resting upon them in places a layer of detrital material, in the form of gravels, sands, clays, and soil; and in a range like the Scandinavian, which has been during a long period covered with snow and ice, or glaciated, as it is called, such accumulations often acquire a very considerable thickness. Indeed, the superficial detrital covering of the "rock in place," as geologists term it, is always of importance, if only from the fact that the agricultural character of the country is so largely dependent on its nature and distribution. In the case of the Sierra Nevada, however, there are several reasons why the formations in question — those which overlie the bed-rock — are of greater interest than they are in perhaps any other mountain chain. They are often of great thickness, as compared with the usual development of such detrital beds; they represent in the epoch of their formation a period of geological time of very considerable length, not being exclusively of recent or post-tertiary origin; they are quite largely made up of volcanic materials, so that it is easy to see that the period during which they have been accumulating has been one of great igneous activity; and, finally, they are, almost everywhere, more or less auriferous, containing a sufficient quantity of gold to make their working profitable. It is in reality this last-named quality which makes these superficial deposits of so much importance. If they contained no gold, we should also know much less about their mode of occurrence; for it is the twenty-five years' work of thousands of miners among these auriferous deposits which has revealed their real nature, of which hardly anything could have been made out from a study of the undisturbed surface. In giving a brief sketch of the geology of the Sierra Nevada, before entering on the more detailed study of the superficial detrital and volcanic formations, we begin with the bed-rock; and first, with some hints as to its lithological character.

The most easily recognized division of the bed-rock is into the granitic and the metamorphic; for there is so little of that which is not granitic which has not undergone considerable chemical change, that the whole series of non-granitic may, not improperly, be called metamorphic. So large a portion of this metamorphic rock has a slaty structure, that the whole series is often termed the "auriferous slate series"; or, more briefly still, the "auriferous slates"; not that the whole formation is gold-bearing, but that these are the rocks with which the gold is associated, this metal, however, occurring almost exclusively — so far as the bed-rock is concerned — in connection with the veins of quartz which are so abundantly distributed through certain portions of the series. When, therefore, in the course of the present volume, we speak of the "bed-rock slates" or the "auriferous slates," we mean, unless there is some special limitation to the contrary, to include all the non-granitic rocks of the gold region of the Sierra Nevada, underlying the imperfectly consolidated, detrital materials, which, with the accompanying volcanic deposits, form the superficial covering of so large a portion of the western slope of the range.

By far the greater part of the mass of the Sierra Nevada consists of the granite forming the axis of the chain. This rock constitutes, indeed, nearly the whole of the Southern Sierra, from Tahichipi Pass almost as far north as Mariposa County. Only just at its western edge, at the base of the foot-hills, is there a narrow belt of rock having either a schistose or laminated structure. As we go north from the point where the San Joaquin River leaves the foot-hills, we find the belt of non-granitic rock rapidly widening, and, at the same time, the line of junction between the granite and the metamorphic becoming more and more irregular. As far as the Mokelumne River there appears to be a well-marked separation between the two formations, and few isolated areas of granite lying outside of the main axial mass. Farther north, however, there are portions of the intrusive rock quite surrounded by the metamorphic. Just north of the American River, granite forms a wide belt in the foot-hills of the Sierra, while opposite to it on the flanks of the range a great spur projects nearly down to Placerville, reducing the width of the zone of metamorphic at this point to about twenty-five miles; while a little farther north it is more than twice as much as that. Through the counties north of the American River the auriferous slate series occupies nearly the whole width of the western slope of the Sierra, with occasional areas of granite enclosed in it, the axial mass being almost entirely limited to the very crest of the main divide, although widening out and occupying a broad area on the eastern slope, towards Pyramid Lake, and in the ranges to the northeast of it. In a cross-section of the Sierra from Honey Lake to Oroville there is hardly any well-marked granitic rock visible, excepting a small patch around Spanish Peak. A little farther north the volcanic sets in, forming a complete mantle over the subjacent rocks. After passing this broad spur of the great volcanic overflow which covers nearly all of Northeastern California, to the east of the Sierra, we have again, in the region lying to the northwest of the Upper Sacramento, or Pit River, granitic rocks occurring in large quantity and forming the summits of the very broken and elevated ranges which, as before stated, occupy most of the surface in Trinity, Klamath, and Del Norte counties.

As granite forms the axis of the Sierra Nevada, it naturally constitutes the greater portion of the highest summits of that range. The exceptions to this rule are chiefly to be found in those cases where these culminating points are exclusively of volcanic origin, as is the case with Silver Mountain and Lassen's Peak, or where the lava forms a thin capping on the granitic mass, as for instance on Mount Stanford, and many other points in the vicinity of Lake Tahoe. The intimate connection between the granitic and volcanic rocks is shown in many places, as will more properly be noticed farther on.

The form and picturesque appearance of mountain ranges and their dominating summits are largely dependent on the structure of the rock, or the form of the fragments and masses into which it separates as it weathers. In the case of the granitic rocks of the Sierra there is, almost everywhere in the range, a well-marked tendency to divide into laminæ or plates of varying thickness, and these plates have a concentric structure, so that the surface presents a succession of mammillated forms on a gigantic scale. This is well seen in the view from Sentinel Dome, above the Yosemite, looking across towards Mount Hoffmann. The whole surface of the country in this direction is a succession of domes, looking from a distance perfectly smooth, and wonderfully symmetrical. The Half-Dome — so stupendous an object as seen from this point — is one of these dome-shaped masses which has been split in two, one half having become engulfed at the time of the formation of the chasm at its base. Mount Hoffmann is another of these half-dome shaped masses, and in both of these, as well as in many others in the Sierra, a close examination will show that they are made up of an innumerable number of thin shells lapped over each other, like the coats of an onion.*

This structural arrangement of the granitic masses exhibits itself in a great variety of forms, according as more or less of the original dome has disappeared, either from actual engulfment or from the action of erosive forces. The "dome-structure" of the granite, as it has been called in Geol. I., is by no means peculiar to the Californian mountains; on the contrary, it is most conspicuously displayed in many other parts of the world, although perhaps nowhere on so grand a scale as in the Sierra Nevada. It was first noticed and described by Von Buch as occurring in the Harz, in the Scandinavian Range, and in many other regions.† The same phenomena have also been described and figured by Helmersen.‡

No one who has studied the Sierra Nevada with any care can doubt the eruptive origin of the granitic nucleus or axis of that chain. The remarkably homogeneous character of the rock over such a vast area adds weight to the evidence of an intrusive origin of the granite furnished by its position in reference to the sedimentary strata with which it is connected. The intimate connection of the volcanic deposits of the Sierra with the granite must also be considered in reference to this question, as will be noticed farther on. The epoch of the granitic outburst may best be indicated after having made some acquaintance with the geology of the stratified deposits through which it has made its way to the surface.

Of the lithological character of the auriferous slate series — the bed-rock of the miners, as already explained — we shall have frequent occasion to speak in the detailed descriptions of the overlying formations, to be given in the next chapter. A few general remarks in regard to the nature of the rocks of which this series is made up may, however, be introduced at this point, and it will be desirable first to give a brief synopsis of what is known in regard to their geological age.

The geological age of the auriferous rocks of the Sierra Nevada had remained, up to the time of the beginning of the work of the Californian

---

* This is well seen in the photograph of Mount Hoffmann, in the "Yosemite Book," also in several of Watkins's series of photographs of the Yosemite and its surroundings. See also Geol. I., for a number of sections and cuts illustrative of this structural arrangement of the granitic masses of the Southern High Sierra.

† See Abhandlungen der Akademie der Wissenschaften zu Berlin. 1842. pp. 57-77.

‡ Memoirs of the St. Petersburg Academy. Vol. XIV. No. 7, 1869. Studien über die Wanderblöcke und die Diluvialgebilde Russlands.

Survey, an unsolved problem. The highly metamorphic condition of a large portion of these rocks, the fact that the gold-bearing formations of the Appalachian chain, of the Ural and of Australia, had been proved to be of Palæozoic age, the repeatedly and strongly expressed dictum of an eminent authority, Sir Roderick Murchison, that gold could only be expected to be found in any considerable quantity in the oldest rocks, and the entire absence of any fossil evidence throwing light on the epoch of the auriferous belt of the Sierra Nevada,—all these facts combined had impressed geologists strongly with the belief that this formation could not be otherwise than of Palæozoic age.* The Silurian age of the gold-bearing rocks of the Pacific slope has indeed been attempted to be maintained by one of the geologists attached to the Pacific Railroad Survey, as late as 1866,† and Murchison himself, in the fourth edition of Siluria, published in 1867, endeavors to throw doubt on the evidence collected by the California Survey, and set forth in detail in 1864 and 1865, in the first volumes of the Geology and Palæontology, showing the existence of Mesozoic fossils all along the line of the auriferous belt from Mariposa to Plumas County, and the absence of all proof of the Silurian or Devonian age of any portion of the rocks of that region.‡

In view of these facts, it will be seen that it is quite desirable that the evidence with regard to the geological age of the auriferous belt of the Sierra Nevada should be passed in review. And it may be stated that, so far as

---

* Murchison, in the third edition of his Siluria, published in 1859,—the year before the commencement of the California Survey—says (page 475), "Whether, referring to ancient history, we cast our eyes to the countries watered by the Pactolus of Ovid, ... or to those chains in America and Australia which, previously unsearched, have proved so rich,—we invariably find the same constants in nature. In all these lands gold has been imparted abundantly to one class only of those ancient rocks [the Silurian], whose order and succession we have traced, or to the associated eruptive rocks." Also (on page 455), "We now know, therefore (and the recent explorations, in California, Oregon, etc., have confirmed the view), that sedimentary deposits of Silurian, Devonian, and Carboniferous age constitute the loftiest ranges and metalliferous plateaux of the American continent." It is probable that Murchison found the desired confirmation of the views he had so long sedulously upheld in regard to the Silurian age of gold in Mr. W. P. Blake's Report of a Geological Reconnaissance of California, published in 1858,—a reprint of part of his Pacific Railroad Report. He says (Introduction, page iv), "It is also probable that a great part of the rock formations of the Gold Region will ultimately be found to be Devonian and Silurian."

† Bulletin de la Société Géologique de France (2), XXIII. 552. Séance du 7 Mai, 1866.

‡ Murchison says in the fourth edition of Siluria, (page 471), "I can as yet see no valid reason to induce me to alter materially the generalization I adopted in my former edition, as derived from evidences in various parts of the world, viz. that the Silurian and associated Palæozoic strata, together with the igneous rocks which penetrated them, have been the main recipients of gold." The author, who thus

known to the writer, no positive statement has ever been made public bearing upon this subject, except by members of the Geological Survey Corps; and that no information has, since the close or last suspension of that work in 1874, been added to that previously obtained.

The first statement of the Secondary age of the auriferous rocks of the Sierra was made by the writer, in a synopsis of the results of the operations of the California Survey, under his direction, in the American Journal of Science, in the number for September, 1864. It is in these words: "While we are fully justified in saying that *a large portion of the auriferous rocks of California consists of metamorphic Triassic and Jurassic strata*, we have not a particle of evidence to uphold the theory that has been so often maintained, that all, or even a portion, of the auriferous slates are older than the Carboniferous; not a trace of a Devonian or Silurian fossil having been discovered in California, or indeed anywhere to the west of the 116th meridian." This statement still remains true so far as the non-existence of Silurian and Devonian fossils in the Sierra Nevada is concerned. Nothing older than Carboniferous has been found in the gold-bearing region proper of California, while quite a number of new localities and isolated specimens of Secondary fossils have been obtained from various points in the heart of the mining region along the western slope of the Sierra Nevada. The evidence afforded by fossils thus far collected by the Survey, with reference to the age of the gold-bearing rocks, will now be briefly set forth.

Calling to mind, again, the fact that, up to the commencement of our work in 1860, not a single fragment of a fossil from the auriferous rocks of California had ever been described or noticed in print, so that it seemed almost hopeless to expect to find any evidence bearing on the extremely interesting question of the geological age of this formation, we will mention in their historical order the principal occurrences which came to our notice. The first ray of light on this obscure subject was shed by a specimen obtained from Hon. John Conness, in 1861, and which was stated by him to have been found in the slates near Georgetown, El Dorado County. It was

stoutly adheres to his favorite theory of the Silurian age of all the auriferous rocks of the world, perhaps may have found the desired confirmation of his views in a communication made to the Geological Society of France, by Élie de Beaumont, in behalf of Mr. L. Simonin (Comptes Rendus, Tome 50, page 391), in which the latter claims to have recognized Trilobites in the auriferous slates of the Sierra Nevada, without giving any further information on the subject, or mentioning even the name of the locality where he made this fortunate discovery. To Mr. Simonin's value as an authority we shall have occasion to refer again in a future chapter.

a small fragment of a dark-colored argillaceous and slightly micaceous slate, bearing upon it the impression of an *Ammonite* ; or, if not an Ammonite, at least some chambered shell, which had an unmistakable Secondary look. As this was an isolated specimen, and the fact that it came from the rock in place at the specified locality could not be definitely ascertained, it was not regarded as being of much value, except that it strongly encouraged the expectation that, after much searching, fossils might yet be found in the slates which had so far seemed quite barren of all traces of organic life.

Meantime, the fossils obtained from the limestone in Shasta County, at Bass's Ranch, first noticed by Dr. J. B. Trask, in 1855, were examined by Mr. Meek, and referred by him to the Carboniferous,* in accordance with Dr. Trask's previously expressed opinion. This was in 1862. The region in which these fossils were found was, however, far from the gold region proper, and our knowledge of the stratigraphical relations of the rocks of the Sierra was not sufficient to enable us to judge how far the recognized age of this limestone mass might have a bearing on that of the auriferous belt.

The same was the case, to some extent, with the discoveries of Triassic and Jurassic fossils made by the Survey in Plumas County, in 1863.† The locality was so circumscribed in extent, the stratification so disturbed, and in part obliterated, that no satisfactory idea could be got of the structural relation of the beds in which these fossils were enclosed to the main body of the auriferous belt. It was, therefore, extremely desirable that fossils should be found in the heart of the gold region proper, in immediate proximity to the great quartz veins, and in such a position that there could be no possibility of error in connecting the results of the determination of their geological age with the occurrence of the larger and most important localities and rocks in which auriferous quartz was being mined. The Ammonite obtained from Mr. Conness came from a central point in the mining region; but, being a loose specimen, could not be accepted as anything more than an indication.

It was on the Mariposa Estate, in the immediate vicinity of the great quartz vein then extensively worked, that Mr. King found Belemnites in three different localities.‡ At one of these, Miss Errington, a lady residing

---

* See Geology of California, Vol. I. p. 326.

† l. c. 308.

‡ On the English trail, at Hell Hollow, and near the Pine Tree Vein, within twenty feet of this great mass of quartz. Geology of California, Vol. I. p. 482.

at that time on the Estate, afterwards found specimens of the *Aucella* and *Pholadomya* described in the Geology of California, Vol. I., by Mr. Meek, who had little hesitation in referring them to the Jurassic. This seemed highly satisfactory evidence, so far at least as the age of a portion of the auriferous slate series was concerned, and it was immediately after this that the article in the American Journal of Science was published, to which reference has already been made, and which contained the first distinct recognition of the Secondary age of a part of the heart of the auriferous belt.

After the publication of Geology, Vol. I., the work of tracing out the detailed geology of the gold-bearing rocks in the counties next adjacent to Mariposa in the north was continued by the writer with the assistance of Mr. A. Rémond, through the region lying between the Merced and the Stanislaus rivers. These explorations had for a result the discovery of numerous fossils, at various points along the line of outcrop of the same fossiliferous beds of dark-colored slate, in which they had first been discovered on the Mariposa Estate. These fossils were all unmistakably of Secondary age, and consisted of Ammonites and Belemnites, and especially of the *Aucella Erringtoni*, which latter was quite abundant in some places.*

Still farther north, near Spanish Flat, about six miles north of Placerville, in the fine-grained slates, several specimens of unmistakable Belemnites have been obtained by various members of the Geological Survey. At several other localities along the auriferous slate belt, specimens of Ammonites have been found. One was obtained at Wilkinson's Ranch near White Rock station, on the Pacific and Sacramento Valley railroad, about seven miles east of Folsom. The most interesting locality of Secondary fossils discovered in the auriferous slate series, north of Mariposa County, is at Colfax, on the line of the Central Pacific railroad. The exact place was in a cut on the road, at Station 2,777, Section 53, one mile west of the town of Colfax. Two specimens of Ammonites were obtained here, when the road was constructing, and were examined and described by Mr. Gabb, under the name of *Ammonites Colfaxii*; the description, with figure, will be found in the American Journal of Conchology, Vol. V. p. 7. Another specimen of an Ammonite was brought to the office of the Survey as having been obtained at Robinson's Ferry on the Stanislaus River.

---

* The localities in which these fossils were found extend along a line of about twelve miles in length, between the Mariposa Estate and Coulterville. They are all in the immediate vicinity of the Great Quartz Vein.

The evidence thus furnished — all of which comes from the Geological Survey — is satisfactory in this respect, that all the localities mentioned are in the very heart of the gold-mining region; those from which the *Aucella* was obtained in considerable numbers in Mariposa County are in the closest proximity to the Great Quartz Vein, and extend along its course for fully twenty miles. In this connection, it must be remembered that not a single fossil to which a Palæozoic age could by any possibility be assigned has ever been found in the mining counties, from Mariposa to Yuba. The only fossils older than Triassic which have been discovered to the west of the crest of the Sierra are those of the limestone belt, of which by far the most prominent and the most fossiliferous locality is that at Bass's Ranch, in Shasta County, described in Geology I. The fossils discovered here in abundance are of undoubted Carboniferous age. There is also evidence sufficient to determine the geological position of the limestone belt at Pence's Ranch, eighty miles southeast of Bass's; at that locality, however, there were but few species found, and the specimens obtained were in a much less perfect state of preservation than those occurring at the more northern locality; they were, however, referred without hesitation to the Carboniferous by Mr. Gabb. Somewhat doubtful traces of fossils of the same age were also obtained in a limestone belt in Genesee Valley, more than fifty miles northeast of Pence's Ranch. While there are many outcrops of limestone or marble along the course of the Sierra, from Plumas County to the extreme southeastern end of the range, there has never been a recognizable fossil found in any one of these, so that the identity in geological age of the different outcrops of the limestone in that portion of the Sierra with those farther north, although suspected, cannot be demonstrated.

The evidence obtained in the heart of the gold region — that is to say, in the counties where mining for that metal has been chiefly carried on — is satisfactory in so far as this, that proofs of the Secondary age of a portion at least of the auriferous belt have been obtained at numerous points all the way from Mariposa to Nevada County, a distance of perhaps 150 miles. Along this whole extent no Palæozoic fossils have been discovered. If the limestone belt in this region belongs, like that farther north, to the Carboniferous, there is no proof of it; so that it may be said, that there is no evidence of the existence of rocks older than the Carboniferous anywhere on the west slope of the Sierra Nevada; and that, in the principal gold-mining region, a portion at least of the rocks most intimately associated with the occurrence of the gold is known to be of Secondary age, and no proof has been obtained of the existence of any member of the geological series other than this.

When we seek to determine what part of the Secondary is represented in the auriferous slate series; or which of the numerous divisions or groups of the Secondary or Mesozoic series can be recognized in the Sierra gold-bearing formations, we find that the data are far from sufficient to enable us to arrive at any satisfactory conclusions. That no part of these rocks is Cretaceous, is known from stratigraphical evidence; the rocks of this age occur on the flanks of the Sierra in such a way, that we have no difficulty in assuring ourselves that they do not make up any part of the auriferous series. We must therefore admit that this latter is either of Triassic or Jurassic age, or that it includes more or less of strata belonging to both these periods.

Here it is desirable to introduce a few words in regard to the known occurrence of Triassic rocks on the west slope of the Sierra, as well as in other localities on the Pacific side of the Continent and in the Great Basin. The first discovery of this group of strata was made by the Geological Survey, in 1862, at various points in the Washoe Mining District, near what is now known as Virginia City, and also in the Humboldt Ranges and elsewhere in Nevada. The same formation was also recognized in Plumas County, California, in the course of the Survey explorations of the year 1863. Numerous discoveries of Triassic fossils have been made since that time in the State of Nevada, but no other one within the borders of California. The only locality of these fossils thus far discovered on the western slope of the Sierra is the one in Plumas County, mentioned in Geology I., p. 309, although a most careful search has been made by the members of the Geological Survey during several seasons of field-work not only in Plumas, but in all the other mining counties. The area over which these fossils were found in Genesee Valley is extremely limited; but a sufficient number of species were obtained, to make it clear that the formation was the same which has proved so prolific in a portion of Western Nevada, and especially in the Humboldt Ranges, and which has been traced to the north as far as the Aleutian Islands.* This group of rocks, so extensively developed on the Pacific coast, is the exact equivalent of the Alpine Trias, or the beds of St. Cassian and Hallstadt.† It is

* See Palæontology of California, Vol. I. p. 19. Report of the Fortieth Parallel Survey, Vol. IV. p. 99. Voyages à la Côte Nord-Ouest de l'Amérique, par A. L. Pinart, Vol. I. Pt. I, p. 34.

† Mr. Meek remarks (in the fourth volume of the Fortieth Parallel Survey Report, p. 8) "It is a remarkable fact that there should be at these distant western localities an immense series of deposits, con-

a curious fact that proof of the existence of this peculiar and most interesting member of the geological series should only have been found in a single locality within the State of California, although it is distributed over so wide an area just beyond the borders of that State.

At another locality in Plumas County, only a few miles from the one at which the Triassic fossils were found, a limited outcrop of a dark-red sandstone was discovered, in 1863, which contained several recognizable species of fossils, among which Mr. Meek identified the following genera: *Rhynchonella*, *Terebratula*, *Gryphæa*, *Pecten*, *Trigonia*, *Unicardium*, *Astarte*, *Myacites*, and *Belemnites*. This is a grouping of genera not known to occur in rocks older than Jurassic, or newer than Cretaceous, while they are, as usually limited, common to both of these epochs. The weight of positive evidence, derived from the specific affinities of these fossils, as far as it goes, was declared by Mr. Meek to be on the side of the Jurassic, while the negative evidence deducible from the entire absence amongst them of any strictly Cretaceous types pointed in the same direction, and this conclusion was still further confirmed by stratigraphical evidence, as already mentioned. Hence, Mr. Meek did not hesitate to describe the fossils in question as Jurassic.* We have, therefore, in close connection with each other two localities, in Plumas County, where rocks of undoubted Jurassic and Upper Triassic age occur. When we endeavor to connect these formations with those farther south in the gold-mining region proper, we get little or no assistance either from palæontological or stratigraphical considerations. The rocks of Plumas County cannot be correlated with those of Mariposa, and the paucity of fossils between those two extreme ends of the auriferous belt renders it impossible to come to any more definite conclusions with regard to the geological age of the formation than those which have been mentioned in the preceding

taining so exact a representation of the very peculiar Fauna of the Upper Trias of Europe, as exhibited in the St. Cassian, Aussee, and Hallstadt beds. For instance, there are, among the collections that have been found by different parties in these beds, the following peculiar genera, especially characteristic of the rocks of this age in Europe, viz. *Halobia*, *Monotis*, *Cassianella*, *Trachyceras*, *Archestes*, *Clidonites*, etc., directly associated with the more ancient genus *Orthoceras*. There have been also found in these beds the following *species*, closely allied to, or possibly in some cases identical with, *Halobia Lommelei*, *Monotis Salinarius*, *Ceratites Haidingerii*, *Archestes Aussceanus*, etc. There have also been found from this formation various other types of the *Ammonitidæ*, which, like those found at the same horizon in Europe, are not true Ammonites, nor Ceratites, nor yet Goniatites, as these genera have been restricted by late authors, but new generic types, sometimes intermediate in their characters between the typical forms of the above-mentioned genera."

* See Palæontology of California, Vol. I.

pages. Mr. Gabb, however, inclined to the opinion that the Colfax Ammonite was closely allied to *A. solaris*; he also considered that the specimen from Robinson's Ferry, mentioned on page 37, was identical with *A. Colfaxii*; and, chiefly on the strength of this rather unsatisfactory evidence, he thought it not unlikely that the rocks in which these fossils were found were Liassic.* It would seem probable, from the fact of the undoubted occurrence of the Triassic in Plumas County, that this formation is also represented in the auriferous belt farther south; but it does not appear that any palæontological evidence has yet been obtained which would make it possible to assert this as an established fact.

The case, then, stands thus in reference to the geological age of the sedimentary strata of the bed-rock of the western slope of the Sierra: The existence of the Alpine Trias in Plumas County is a well-established fact, and this member of the series, so widely developed in Nevada and along the Pacific coast, may make up a considerable part of the auriferous belt farther south, but there is no proof of this. The Jurassic also has been shown to occur in Plumas, and in Mariposa County, as well as at various points intermediate between these two localities. The specimens obtained in Nevada County, at Colfax, and from Robinson's Ferry, near the line between Calaveras and Tuolumne counties, are referred by Mr. Gabb, with some doubt, to the Liassic. The occurrence of Jurassic fossils at several localities along a belt of rock in close proximity with the Great Quartz Vein proves beyond a doubt the Mesozoic age of a portion, at least, of the most productive part of the auriferous belt. No rocks older than the Trias have been found anywhere on the western slope of the Sierra, except at certain localities of limestone, in Shasta and Butte counties, known to be of Carboniferous age. Through the whole range of the gold-mining region occur detached portions of what appears once to have been a continuous belt of this rock, and it would seem likely that this is the continuation of the Shasta and Butte County limestone; but no palæontological proof of this has been obtained. If Silurian

* Mr. Gabb remarks (Journal of Conchology, Vol. V, p. 5): "A large proportion of the stratified rocks of the western slope of the Sierra Nevada appear referable to the Jurassic formation, while at least one small tract in Nevada [in the mining district of Volcano, about thirty miles southeast of Walker's Lake] yields fossils of this age in a reasonably good state of preservation. From the paucity of species, and none being referable to known forms, we were unable, at the time of publication, to do more than designate the great group of the Mesozoic era to which they belonged. The discovery of two Ammonites, closely allied to known European species, together with other characteristic forms, lead us now to believe that all of the at present known Jurassic rocks of the Sierra and its vicinity belong to the Lias."

or Devonian fossils should be found on the western slope of the Sierra, it will probably be in the foot-hills, below the auriferous belt; but, thus far, the most careful search for proof of the Palæozoic age of those rocks has not been rewarded with success.

Having in the preceding pages given a synopsis of what is known of the geological age of the auriferous slate belt of the Sierra, from which it will have been abundantly evident that no subdivision of the series based on palæ-ontological grounds can be made, we may add some remarks on the litho-logical character of the formation. In the course of the succeeding chapter we shall have frequent occasion to allude to the different kinds of rock occur-ring in the various mining districts. It is, indeed, no part of the object of this volume to enter into any exhaustive account of the bed-rock; that would naturally be left for another occasion.* At present we are not prepared to do this, nor is it necessary for the understanding of the problems involved in an inquiry into the mode of occurrence of the gravel deposits. What we have to do with the bed-rock in connection with the subject here under dis-cussion is chiefly the character and form of its eroded surface, as will be seen farther on. It will be necessary, however, frequently to allude to the litho-logical character of the material thus eroded, and some general statement of the character and distribution of the different kinds of rock which make up the bed-rock series will be desirable. This, however, may be preceded by a few remarks, additional to those given when describing the granitic axis of the Sierra, in regard to the range and extent of the auriferous slate series. The most important fact in this connection is this: that the non-granitic bed-rock occupies but a narrow strip at the southern end of the Sierra Nevada, and that this widens as we proceed north, until it becomes lost under the overlying volcanic masses, which cover everything when we reach the north-ern boundary of Plumas. At the same time it will be noticed that the prin-cipal mining region is that in which the auriferous slate series is most exten-sively developed. This is natural enough; for, although the granitic rocks are not entirely destitute of metalliferous veins, these are comparatively

---

* Mr. M. E. Wadsworth is engaged, under the direction of the writer, in making a microscopical ex-amination of both the volcanic and the metamorphic rocks of the Sierra. The investigation, however, is not yet completed; but in most cases in the course of the present volume the names given to the different members of the bed-rock and gravel series are those furnished by Mr. Wadsworth after examination of the thin sections which have been prepared for this purpose. The full results of these investigations will, it is hoped, be published hereafter.

rare everywhere except in the slates which are evidently the natural "habitat" of the gold.

Just at the southern line of Mariposa County the auriferous belt widens out suddenly, and the formation becomes of importance. To the south of this, along the foot-hills of the Sierra, the metamorphic rocks can hardly be said to exist at all, the granite coming almost down to the plain. The outer edge of the granite through Fresno County is far from homogeneous in character; it resembles something half-way between granite and gneiss, and contains frequent segregations of the minerals occurring most commonly in the metamorphic rocks, such as epidote and hornblende. It is lithologically such a formation as might have been expected to result from intense metamorphic action along the edge of the granitic and the sedimentary masses. Passing into Mariposa County, we find the auriferous slate series becoming, all at once, quite well developed, and from this county north as far as Plumas, the western slope of the Sierra is the field of active mining operations, while but comparatively little of value has been found anywhere to the south of this. From the Mariposa Estate north, for a distance of about a hundred miles along the flanks of the Sierra, the order of succession of the metamorphic rocks — the auriferous belt, so called — is decidedly less obscure than it is beyond that point, and some subdivisions of the series can be traced for a considerable distance with the same lithological characters. In the region to the north of this, and especially in Sierra and Plumas counties, however, not only do we find the rocks so broken up that no one kind can be traced for any considerable distance; but they are also very largely covered by eruptive materials, so that, even if there were some regular order of succession existing, it would be extremely difficult to make it out, at least without much more detailed observations made with the aid of accurate maps on a large scale. The condition of things, in regard to solving the geological problems here presented, is in many respects like that existing in New England, where extreme scarcity of fossils and complete metamorphism of the strata have combined to render the task of working out the structural geology so difficult, that, thus far, in spite of many years of exploration, by far the larger part still remains to be done. And in New England the "bed-rock" is covered much less deeply with detrital materials than it is in the Sierra Nevada.

To give an idea of the distribution of the rocks belonging to the bed-rock series in that portion of the auriferous belt where they are least broken, and also least covered by detrital and volcanic materials, a sketch of the structure of the region between the Merced and the Stanislaus rivers may here be introduced. This is a district which has been of great importance on account of the quartz mines worked at numerous points along the line of the Great Quartz Vein, and it was also the scene of much rich placer-mining during the years when this class of deposits was so productive.

The width of the auriferous slate belt in this region is, on the average, about eighteen miles. The foot-hills at the base of the range are here made up of horizontally stratified Tertiary sandstones, eroded away so as to leave the rock in the form of low flat-topped elevations. Turning from the lower edge of the foot-hills up into the mountains, we meet at once, at a very moderate elevation above the sea-level, with the metamorphic belt of the Sierra, which at its lower edge is largely made up of slates of various composition, chiefly talcose and chloritic. The strata stand nearly vertical, usually but little covered with detritus, and the peculiar weathered outcrops of the slaty rocks, turned up on edge, and projecting a few feet above the surface at intervals along regular lines, are familiarly known to the miners as "grave-stone slates." Soon the character of the rock begins to change, and we strike a broad belt of compact, dark grayish-green, fine-grained rock, which frequently assumes a porphyritic structure, and hence was called by the Survey "porphyritic green slate," although it never assumes the finely laminated character of the argillaceous slate series. This rock occurs in high rocky ridges, generally much broken, and running about northwest and southeast. It forms the first noticeable range of elevations at the base of the Sierra, in the region at present under discussion. Mount Bullion, Juniper Ridge, Bear Mountain (on the Merced), and Merced Mountain are made up of this rock. This material, which was a great puzzle to us in the field, as to whether it was of volcanic or sedimentary origin, appears, from Mr. Wadsworth's (not yet completed) examination, to be a diabase tufa, a much metamorphosed volcanic deposit. It appears very probable that the lithological investigations now going on in regard to the Sierra Nevada rock will prove the existence in the bed-rock series of a large amount of material originally volcanic, but at present so much altered by those metamorphic agencies which have been so conspicuous in their action on the purely sedimentary beds, that their original character can no longer be recognized, except by the aid of microscopic investigation. This belt of altered volcanic rock occupies a width of from two and a half to five miles on the surface between the Merced and the Stanislaus rivers, but it is divided into two portions along its whole extent by the belt of argillaceous slates, already mentioned as containing *Lima Erringtoni* and other Jurassic fossils. This clay slate belt is also interesting as being intimately associated with the Great Quartz Vein, which is so prominent a feature of the geology of this region.

There is a depression near the centre of the metamorphic belt just described, in which the Peña Blanca and Moccassin and Woods' Creeks run, — the first-named a branch of the Merced, the two latter of the Tuolumne River. This depression is occupied by a continuous belt of argillaceous slates, having a width varying from half a mile to a mile or more. These slates are very finely laminated, homogeneous in character, of a dark-gray color, which is of a light rusty-brown at the surface. At a sufficient depth to be quite beyond the reach of atmospheric influences, this rock becomes almost black. The north edge of this belt passes, almost in a straight line, — although with minor flexures, — from the Pine Tree Mine, on the Mariposa Estate, just south of Coulterville, and about one fourth of a mile northeast of Jacksonville. On approaching the Stanislaus River it widens a good deal, and becomes more irregular in its outline. Along a considerable portion of its course the argillaceous slate belt is accompanied by a band of serpentine, which is more than a mile wide just beyond the Benton Mills, but gradually narrows down to a few yards, and disappears just beyond the Mary Harrison Mine. It then comes in again near Coulterville, and continues, with a width of from a quarter to a half a mile, to the north fork of Moccassin Creek. It again makes its appearance between Grizzly Gulch and Kanaka Creek, and here it is accompanied, on its northwestern edge, by a thin band of porphyry. All the outcrops of serpentine thus far mentioned are on the northeast edge of the argillite belt; but farther to the northwest, towards the Stanislaus, this rock widens out into patches of irregular form, having quite an extensive area, and occurring on both sides of the slates.

Associated with the serpentine is the very remarkable mass of quartz, known as the "Great Quartz Vein," which presents many features worthy of careful study, besides being important from the mines which have been worked at so many points, either in the mass of the vein itself, or in its immediate vicinity. This great lode may be traced from the Mariposa Estate to near the centre of Amador County, a distance of fully eighty miles. It is not visible on the surface at all points, but, like the limestone belt already mentioned, it occurs in detached portions, all of which, however, are nearly in one line, the direction of which is that of the trend of the formations making up the bed-rock series. In the region between the Merced and the Stanislaus, the Great Quartz Vein is almost continuous; it runs uninterruptedly between French Gulch and the main branch of Moccassin Creek, forming by its outcrop in the Peña Blanca a very conspicuous object, and having there a width of 261 feet, measured horizontally across it, its inclination being to the northeast at an angle of about 60°. Beyond Moccassin Creek it disappears for a time, but is seen again below Stevens's Bar, on the Tuolumne, and at numerous points between there and the Stanislaus. This powerful lode is made up of irregularly parallel plates of white compact quartz, and crystalline dolomite or magnesite,* more or less mixed with green talc; and these plates, which somewhat resemble the "combs" of ordinary lodes, are either in contact or separated from each other by intercalated layers of talcose slate. The quartz is chiefly developed in the central portion of the vein; and, from its color and resistance to decomposition, it gives rise to a very conspicuous outcrop, forming the crest of the hills, so that it can be readily seen from a distance of several miles. The dolomitic or magnesitic portion decomposes somewhat readily, and it becomes a kind of "gossan," or a cellular, ferruginous mass, of a dark-brown color, often traversed in every direction by seams of white quartz. The quartz is the auriferous portion of the lode, although it is far from being uniformly impregnated with gold. Most of the mines which have been worked, between the Merced and the Stanislaus, are on the northeast side of the Great Quartz Vein, either in contact with it, or in some parallel band of quartz subordinate to, or at a little distance from it. The talcose slate bands in the vein are often, themselves, more or less auriferous.

Beyond the belt of metamorphic volcanic rock, which lies to the northeast of the band of argillite and serpentine just described as containing the Great Quartz Vein, there is a wide belt of mica slate, which extends to the granitic nucleus of the Sierra. This is, in the region between the Merced and the Stanislaus, the second great series of metamorphic schists belonging to the bed-rock system. These mica schists are more or less quartzose on the southwest, and they pass into pure quartzites as we approach the granite, the proportion of argillaceous material in them gradually diminishing in that

---

* In the only specimen which has thus far been chemically examined, the supposed dolomitic portion proves to be an intimate mixture of quartz and magnesite.

direction. In some localities, however, the mica schists and quartzites are interlaminated with each other close up to the granite. Along the Merced River this belt of rock is very wide; but it narrows towards the northwest. East of Sonora there is a large area of granite enclosed in the mica slates, occupying most of the space between the Tuolumne and the South Fork of the Stanislaus. It is this to which the mining district about Soulsbyville belongs. There is a much smaller insulated area of granite at Big Oak Flat.

In this belt of mica schists and quartzites are found interstratified belts of metamorphic limestone, or saccharoidal marble, belonging to the limestone series mentioned before. On the Merced River there are two bands of this marble, one of which is almost in contact with the granite; the other at about three miles from it. Neither of these has been traced more than a few miles to the northwest, and the range of this rock seems to be broken until just before reaching the Tuolumne River. Here, almost in contact with the granitic mass mentioned before as occurring in the vicinity of Soulsby-ville, there is an outcrop of limestone which extends along in a curved line, and before reaching the Stanislaus widens out, so as to form a belt between one and two miles wide where it crosses that river. In the vicinity of Sonora, this rock is cut through by numerous dykes of diorite.

The region of which the lithological characteristics of the bed-rock series have just been sketched is, in general, quite bare of overlying gravels and volcanic materials. Just at its northwestern border, however, and before reaching the Stanislaus River, we meet with the lava-flow known as the Sonora Table Mountain, to be described farther on. There is also another flow which comes down from above, and branches out in various ramifications over the granitic mass mentioned as occurring in the Soulsbyville district. After crossing the Stanislaus, we find in following the auriferous belt along the flanks of the Sierra to the northwest, that volcanic materials, in the form of continuous flows, or in that of isolated patches, become more and more frequent, and that these and the associated gravels cover more and more of the bed-rock surface, the mining localities at the same time becoming more numerous and important. It is not necessary or possible at the present time, to go into any minute description of the bed-rock in this region. Some general statements may, however, be made in regard to the order of succession of the different groups of rocks, in the auriferous belt north of the Stanislaus. In Mr. Goodyear's notes on the Volcanic and

Gravel Formations of Placer and Amador counties, additional details will be found for that district; so also for the region still farther north such facts in regard to the lithological character of the bed-rock will be introduced as may seem to be required, when speaking of the overlying detrital masses.

An approximation to the order of succession of the various groups of the bed-rock series in the region between the Merced and the Stanislaus may also be traced nearly or quite through the whole extent of the auriferous belt from Calaveras to Plumas County. There is, in general, near the foot-hills, a thoroughly metamorphosed belt of rock, made up in part of sedimentary and in part of volcanic materials; then, extending through the central portion of the auriferous series, a band of a much more decidedly slaty rock, considerable portions of which are very finely laminated. A large part of this rock is argillaceous in character; but some of it is talcose, and it has, in places, a good deal of rather fine-grained sandstone interstratified with it. There are also extensive masses of serpentine found in this connection. Although the Great Quartz Vein cannot be positively identified any farther than the middle of Amador, yet it is pretty evident that this slaty formation is decidedly the most auriferous portion of the bed-rock series, and it abounds in quartz veins, many of which contain more or less gold. Above the slate belt proper the rocks are very varied in lithological character; but, on the whole, as in the region between the Merced and the Stanislaus, decidedly more quartzose than farther down the slope. The limestone, or marble, shows itself in many places, nowhere continuous for any great distance beyond Calaveras County, and nowhere to the south of Pence's Ranch, in Butte County, showing any indications of fossils, so far as yet known. From Mariposa through Tuolumne into Calaveras County, the limestone outcrops, when laid down on the map, make a tolerably continuous line, which keeps pretty near the edge of the granite, or well towards the upper portion of the auriferous belt. Through the counties north of Calaveras it is difficult to trace any regular order in the limestone bands, which are usually quite short and irregular in form. There are, however, indications of the former existence of two continuous belts of this rock; one near the granite, the other low down near the base of the Sierra.

A few words may here be introduced in regard to the structural relations of the bed-rock series, although this subject is not one which properly comes within the scope of the present volume. It may be stated, without hesitation, that the problem is an extremely difficult one. And it is not likely that it will be fully solved without much further research both in the field and in the laboratory. The fact has already been stated that there is no assistance to be obtained from fossils in deciphering the structure of the Sierra. The belt of argillaceous slates described as occurring in Mariposa and Tuolumne counties, in connection with the Great Quartz Vein, is undoubtedly of Jurassic age; and it may be assumed as highly probable that the finely laminated slates farther north are the continuation of the southern belt of similar lithological character. It may, perhaps, be considered as very probable that a part or all of the limestone outcrops in the bed-rock series are of Carboniferous age. But even then we are no farther advanced towards a solution of the problem. The area of Triassic rocks in Plumas County covers but a few acres, and is sharply bent at a right-angle. It is in close connection with the granite, and is almost surrounded with volcanic. Not the slightest clue can be obtained as to its relations with the associated members of the bed-rock series.

It is evident from a number of facts developed during the Geological Survey, that the rocks of the Sierra Nevada have undergone a large amount of longitudinal compression, — that is to say, disturbance occasioned by forces acting in the line of the strike of the formations. The manner in which the Triassic outcrop in Plumas is bent sharply at right-angles, as already mentioned, is a good illustration of this. A similar case is referred to, in Geology I. as exhibited by the slates on the Mariposa Estate.* The want of continuity in the limestone masses, and the great and sudden changes of strike, as well as in the form of the outcrops of this rock, point also to longitudinal compression. Although the trend of the different members of the bed-rock series is, on the whole, coincident with that of the axis of the Sierra, yet there are many sudden and sharp deflections in their strike, which runs sometimes north and south for a distance, and then again east and west, although the average direction would be not far from northwest and southeast.† It is true that a part of this apparent longitudinal compression may have been caused by the unequally distributed force exercised by the irregular mass of granite pressing in different directions according to its variations of form; but there are phenomena in the structure of the rocks of the west slope of the Sierra which seem only explicable on the theory of an actual thrusting force exerted in the direction of their strike.

* See Geology of California, Vol. I. p. 225.

† In one locality, at least, east of Big Oak Flat, near Big Creek, the mica slates have a strike of N. 70° – 75° E., which direction is at an angle of 100° or more with the usual trend of the formation.

The dip of the auriferous slate series is as difficult to account for as the variation in the strike. On the whole, in the central portion of the belt, — that is, central in relation to a transverse section of the range, — the inclination of the slates is usually nearly vertical; it may be said to vary from 75° to 85; and it is, in the majority of instances, to the northeast, or towards the crest of the range. The narrower the belt of slates, the more regular the inclination to the east. As the series widens in going towards the north, the dip becomes more irregular, and over extensive areas it is to the west; at the same time the strata make a more decided approach towards horizontality; and, on the whole, dip less steeply as we proceed towards the crest of the range. A carefully constructed section along the line of the Central Pacific railroad, on which the inclination of the strata at every outcrop was laid down, failed to furnish proof of any regular system of folds, there being frequent and rapid changes, both in the amount and the direction of the inclination of the strata, and the whole series presenting phenomena similar to that described as existing in the northern portion of the Coast Ranges, where the rocks seem to have been lifted up *en masse*. The principal apparent difference, in the case of the Sierra rocks, is, that here there has been more lateral compression, and that consequently the average dip is higher. The cause of this compression may be sought for in the presence of two enormous masses of granite on this line of section, one of them forming the crest of the Sierra, and undoubtedly the mass of the material underlying the stratified portion of the bed-rock series, while the other is seen in the foot-hills, constituting a belt some twenty miles in width, as visible on the surface, with an unknown extension under the Great Valley, where it is covered with recent detrital formations. A long and patient study of the specimens collected, with the aid of the microscope, and combined with a sufficient number of observations in the field, may, it is to be hoped, eventually furnish the necessary data for explaining the structure of the auriferous belt; at present, it does not seem possible to make any more decided statements in regard to this difficult question than those which have been given above. On a future occasion we may attempt to throw some farther light on the problem here presented.

In view of its importance as bearing somewhat directly on the special subject of the present volume, we have now to introduce a few remarks in regard to a class of rocks occurring on the west slope of the Sierra, and not belonging to the bed-rock nor yet to the volcanic and detrital formations to be described in the next chapter. These are the marine sedimentary strata found in the foot-hills at numerous points all along the eastern margin of the Great Valley. These marine deposits, like those higher up in the mountains, rest unconformably on the upturned edges of the auriferous slates. They are usually well supplied with fossils, so that there can be no doubt as to their geological position, or the nature of the medium in which they were deposited. They include strata both of Tertiary and Cretaceous age; and, although there are localities where they might be confounded with the true auriferous gravel deposits of fresh-water origin, there is usually no difficulty in distinguishing the two formations, as will be evident from the following brief sketch of the character and mode of occurrence of the marine strata.

The greatest development of these deposits is in the southern part of the Great Valley, and especially along the foot-hills between Kern and White rivers. Here this formation is from 200 to 600 feet in thickness, and it is either horizontal or has a slight dip to the west, away from the mountains, from the erosion of which its material appears to have been derived. The rock is usually a soft sandstone, chiefly made up of granitic *débris*. From Kern to King's River the older metamorphic rocks come down to the edge of the valley, and are there covered by recent detrital beds. Farther north, and extending to the Stanislaus, there is a belt of low flat-topped hills of sandstone, very much eroded, so as to leave many little isolated patches, of from a few feet up to 150 in elevation. All these deposits to the south of the American River appear to be of Tertiary age, and a portion of them were identified by Mr. Gabb as Miocene. The upper portion of some of these Tertiary hills are of fresh-water origin, containing bones of land animals and fragments of wood. There are also localities where the volcanic detrital material has found its way entirely down to the level of the Great Valley,— an occurrence which will be easily understood, after the description of the auriferous gravels and associated volcanic materials in the next chapter has been consulted.

On the American River, just below the town of Folsom, the Cretaceous formation makes its appearance, and there are other small patches of the same, well filled with fossils, in the vicinity. Farther north the Cretaceous occupies a considerable area, the best exposures being found to the north of Oroville. This formation is cut through in the foot-hills by the streams coming down the slope of the Sierra, and there are excellent exposures of it on Butte, Chico, and Cow creeks, as also at Tuscan Springs, where the beds are rich in fossils, which have been figured and described in the volumes of the Palæontology of California.

The most interesting locality of this formation is that at Pence's Ranch, in Butte County, about nine miles north of Oroville. This has been described in Geology I. (p. 209), and need only be referred to here, as illustrating excellently, not only the relations of the marine Tertiary and Cretaceous to the underlying auriferous slate series, but also the position of the volcanic formations of the west slope of the Sierra, which at the northern extremity of the Great Valley have flowed down so far as to cover the marine sedimentary beds, while in the counties farther south they have not usually reached the foot-hills. At Pence's we see the Cretaceous strata resting nearly horizontally on the rocks of the auriferous slate series, which latter dip at a high angle towards the northeast. On the Cretaceous lie, nearly conformably, beds of Tertiary age, which are probably of fresh-water origin, over these stratified masses of volcanic ashes and tufa, and, capping the whole, heavy masses of solid basaltic lava, which has been eroded away, so as to leave isolated patches with flat summits, forming the peculiar elevations known in California as "Table Mountains." The peculiarly interesting feature of this locality is, that here we have the volcanic and fresh-water detrital formations of the Sierra — the especial subject of this volume — brought into contact with the marine strata which are of more recent age than the auriferous slate series, and which, as has been described, form an almost continuous belt along the foot-hills of the Sierra. The relations of the different sets of formations to each other and to the bed-rock series are so clearly displayed at this locality, that there can be no mistaking their nature, and in the succeeding chapter we shall have occasion to present a great number of facts, by the aid of which we shall arrive at a still better understanding of the phenomena here presented.

CHAPTER II.

THE TERTIARY AND RECENT AURIFEROUS DETRITAL AND VOLCANIC FORMATIONS
OF THE WEST SLOPE OF THE SIERRA NEVADA.

SECTION I.— *A General Sketch of the Distribution and Mode of Occurrence of
the Volcanic and Gravel Formations of the Sierra.*

THE present chapter will be devoted to a detailed description of the
auriferous gravels and the associated volcanic rocks, which in the preceding
pages have been pointed out as forming a kind of mantle over the "bed-
rock" formations, and which are of so much interest from many points of
view, and especially from that of their value as containing gold in sufficient
quantity to be worked with profit at a great number of localities. Before
entering on this detailed description, however, it will be necessary to give
the reader a general idea of the mode of occurrence and distribution of the
formations in question; as, without this, he would not readily understand the
bearing of the facts presented on the problems involved in these remarkable
geological phenomena. It will also be desirable to give a brief sketch of the
methods by which these deposits are worked for the gold which they contain, since the exposition of the facts depends so much on the nature and
extent of the mining-operations through the aid of which these facts have
chiefly been brought to light. After the general introductory sketch, of
which the object has thus been indicated, the detailed description of the
deposits will be given to such an extent as the materials in our possession
will allow, and with a view not to occupy too much space with that which is
of purely local importance. The reader will then have been prepared for
a general discussion of the phenomena, in which the scientific aspect of the
problems will be those chiefly presented. In following the order thus indicated, there will necessarily be some repetition, although this will be as far
as possible avoided.

Deposits of gravel, or rolled and water-worn fragments of rock, are of
common occurrence the world over, and they are especially abundant and
important in our own country, in New England and the region of the Great

Lakes, as well as in many other parts of the world where the formations, being hard and crystalline, are not easily ground to a fine powder, and where other necessary conditions — partly climatological and partly geological — have prevailed. Some hints as to the nature of these conditions may be found in a subsequent chapter. The chief interest of the gravels of the Sierra Nevada depends, it must be admitted, on the fact that they are auriferous: were a large portion of the gravels of New England sufficiently rich in gold to render it possible that at any point there might be obtained a sufficient quantity of the precious metal to pay a handsome profit to the miner, it will be readily understood that everything relating to the mode of occurrence and the distribution of these detrital deposits would be of the greatest interest, while the face of the country would, over a great part of the surface, bear the traces of the most extensive mining operations, as does now the western slope of the Sierra Nevada.

The gravels of the California range, however important they may be from the economical point of view, present also facts of the greatest geological interest. The phenomena revealed by the many years of mining work upon these deposits are peculiar, almost unique. Most remarkable is it, that these peculiarities are repeated in a manner which almost amounts to identity, in an antipodal region, — Australia. But there are differences, of a curious kind, between the Californian and the Australian gravel deposits, and these will be pointed out farther on. At present we will proceed to describe what is to be seen in California.

Gravel deposits of considerable magnitude are not limited, even on the Pacific slope, to the Sierra Nevada. As already mentioned (pages 19, 22, 23), there are quite extensive gravel beds in the Coast Ranges. They have even, in one district (page 19), been worked to some extent for gold. Generally, however, they are not auriferous, at least not sufficiently so to encourage working. Neither have these Coast Range gravel deposits those peculiar features which make the Sierra gravels so interesting. The manner in which they have been accumulated would, it is true, form an interesting object of geological inquiry, and this problem will, at some future time, no doubt, be taken up; and some hints in this direction may perhaps be given in the course of the succeeding chapters. There are also gravel beds of considerable magnitude in the extreme southern and northern portions of the State of California, which are more or less auriferous, but in regard to which we have little or no definite information. The gravels in the cañons of the San

Gabriel Range, and in some of the valleys of the San Bernardino Mountain are, or have been, worked for gold at various times during the winter season. They are not the scene of extensive and permanent mining operations, like those hereafter to be described; but rather of the nature of the ordinary placer-mining gravels; that is, of local origin and comparatively unimportant. The deep cañons of the rivers of the extreme northern counties — especially the Klamath and its branches — contain large amounts of gravel, which have been, at times, quite extensively worked for gold, and in places with large profits. This region has never been systematically explored by the Geological Survey, but there is no reason to suppose that these gravels differ from ordinary river deposits of detrital origin, except in being on a scale of magnitude corresponding with the depth of the cañons. They have not been, up to the present time, so far as known to the writer, the object of systematic and permanent mining operations, such as those carried on in the gravel-mining region proper, and they will, therefore, not be included among the topics discussed in the present volume. Yet it will be readily admitted that both these more northern as well as the southern gravels demand a careful examination, especially with a view to forming some better idea than we now possess of their probable future importance.

The gravel region proper, or that portion of the western slope of the Sierra Nevada here to be especially described, may be said to be nearly conterminous with the belt of auriferous slates, the range of which has already been indicated in the preceding chapter. In the region where granitic rocks prevail, whether in the southern portion of the Sierra, or along the crest of the range, the gravels are local in character, and are not washed for gold. This statement, however, needs some explanation and limitation, and this will be found a little farther on, when describing the nature and range of ordinary placer-mining operations. The gravel-mining region proper may therefore be said to begin in Mariposa County and to extend into Plumas, as far as the line already indicated as marking the limit beyond which the slope of the Sierra is entirely covered with volcanic material. But all portions of the area thus designated are by no means equally important either in respect to the amount or the richness of the gravel which they contain. Mentioning the counties in their geographical order from south to north, we find the importance of the gravel deposits increasing as we go from Tuolumne to Calaveras, and thence to Amador. El Dorado, Placer, Nevada, and Sierra are the great mining counties, and there is a gradual falling off in

Butte and Plumas, partly because the rocks begin to be more and more covered with volcanic materials, and partly because of other conditions, to be hereafter explained.

It is true, in a general way, that the region of auriferous gravels is also that of quartz veins. For in the Sierra Nevada, as almost everywhere else in the world, there seems to be an intimate and not easily explained connection between the metal gold and the mineral quartz. It cannot be said that quartz is the exclusive "gangue" or accompanying vein-stone of gold; but this metal is found, with comparatively small exceptions, in the veins of quartz which intersect metamorphic rocks, and also, to some extent, in the slates which lie adjacent to such veins. A large portion of the gold actually obtained, however, throughout the world comes from the débris of such veins and rocks, formed by the operation of those eroding and disintegrating agencies which are and have been at work ever since rocks began to be formed. There are regions of the earth where gold has been procured in large quantities from "washings," — that is, from the working of detrital materials, especially of gravels and sands, and where no mining in the solid rock has ever been attempted, or indeed any veins of quartz discovered; but it is, perhaps, not unreasonable to say, that, in any region where this metal is found in the superficial detritus in even a moderate amount, there or somewhere in the vicinity metamorphic rocks would be discovered, after some search, and that these would be found to be intersected with veins or masses of quartz. Further investigations would prove, in the great majority of cases, that the gold in the detritus came from those quartz veins, or from the rocks immediately adjacent to them. Every great gold-producing district in the world furnishes a portion of the metal from *washings*, and another part from *mining* in the solid rock. There have been facts observed in California, however, which would seem to indicate a different origin for a portion of the gold there obtained, as will be mentioned farther on.

It is a remarkable fact that the veins which are productive of gold, the world over, are in a very great majority of cases, much simpler in their mode of occurrence than those which are productive of the ores of the other metals. Auriferous quartz veins are made up of quartz alone, with little or no admixture of other earthy minerals, such as calcite, brown-spar, heavy-spar, and fluor-spar, which, in ordinary veins, are of frequent occurrence, often making up a large part of the gangue. The writer has never seen, among the many quartz veins examined in the Sierra Nevada, a trace even of fluor-spar, or anything more than traces of the other minerals mentioned. It is also true that auriferous quartz veins rarely contain any large amount of the ores of the other metals, with the exception of iron. Pyrites, or the common bi-sulphuret of iron, is a very common associate of gold in the quartz veins; so common, indeed, that it may be said — at least for California — that there is hardly any productive quartz vein which has not some pyrites disseminated through it; yet the quantity of the latter is usually small, as compared with that of the quartz. It is also true, that the pyrites almost always contains gold; and that in the great majority of instances it is richer in gold than the quartz itself. Furthermore, it may be added that the other common sulphurets, those of lead and zinc, are also frequently present in productive quartz veins, although never, so far as observed in California, in large amount. It is considered a good "indication" — that is, a sign of probable richness in gold — when the vein has disseminated through it, in small particles, more or less of iron and copper pyrites, blende, or galena.

With these preliminary remarks in regard to the mineralogical character of quartz veins, we may proceed to add a few words about their association with the auriferous gravels in the Sierra. In general the geographical range of the gravels is much more extensive than that of the quartz veins, at least of such as are productive. It is true, however, that so large a portion of the bed-rock is covered by gravel and volcanic materials, that it is not easy to say where quartz veins do, or do not, exist. From the very nature of the case, if we suppose the quartz veins to be equally distributed through the bed-rock, there would be more of them discovered and worked when this was not covered by the detrital masses, which, except when artificially removed by mining operations, would effectually conceal the surface, and prevent finding of such veins as might exist. That the auriferous gravels should be found on the slope of the Sierra in positions lower than the region where quartz veins occur is what might naturally be expected, gravitation, aided by currents of water, continually impelling the auriferous particles downward. But there are also considerable deposits of gravel in positions higher up in the range than where quartz veins could be expected to be found; or, at least, higher than any even moderately productive ones have been. Here many beds of streams, especially in the southern Sierra, seem to have been worked over — and apparently with profit — in regions where no remunerative quartz mining has ever been carried on. Some of these points may, however, be discussed with more satisfaction after the presentation of the facts.

At present it will be desirable to give a brief explanation of the methods by which the detrital deposits are worked for the gold they contain, as preliminary to a general account of their mode of occurrence. The terms which are made use of by the miners in describing their operations are so intimately connected with and dependent on the character of the deposits they are working, that it would be impossible to give a clear idea of the phenomena, without a knowledge of the different methods of getting at and making available the variously situated aggregations of auriferous detrital material in which the gold is found. It is not, however, the object of the writer to furnish a practical guide to the mechanical operations of gold washing, in any of its departments; but only to convey a sufficiently clear idea of the methods employed in this kind of work to make the subsequent description of the geological conditions intelligible.

All the methods of separating metalliferous particles from the sands or gravels through which they are disseminated depend, of course, on the simple principle of giving gravitation a chance to act on the heavier material; and this is done, almost exclusively, with the assistance of water. In the case of ordinary ores of the metals occurring mixed with the gangue, or of particles of gold in the solid quartz, the material must first be crushed or stamped to a fine powder, and then, the pulverized material, or "slimes," being agitated or in some way allowed to move in the presence of water the heavier particles have an opportunity to sink to the bottom. But the variety and the complexity of the machinery which has been contrived to do this simple work, especially in those cases where the ore is not much heavier than the rock with which it is associated, and to do it with the greatest amount of efficiency, rapidity, and economy of labor, is very great. In the case of gold, however, such complication is unnecessary; and, besides, in working on any considerable scale the chemical affinity of quicksilver for gold is taken advantage of, as a most important assistance in helping to detain the precious metal. In the case of ordinary gold washing, the material operated on — sands and gravels — has already been pulverized by natural processes, and the miner has only to separate the valuable metal from the worthless portion. The simplest implement for effecting this is the pan, a sheet-iron or tin vessel with a flat bottom, a foot or more in diameter, which is partly filled with the material to be operated on, and then, with the repeated addition of water, shaken with a peculiar motion not easily described, the earthy particles being allowed to pass over the edge of the utensil, until finally only the heavy metallic portion remains behind. The next most simple instrument used in gold washing is the " rocker" or " cradle," a wooden trough about forty inches long and twenty wide, having at one end a hopper or box about twenty inches square, the bottom of which is a piece of sheet-iron pierced with holes half an inch in diameter. This is for the purpose of keeping back the coarser part of the material washed, which is thrown out of the hopper with the hand. The finer portion, aided by the rocking motion and the water, passes through the holes, and the gold is caught on the bottom of the cradle by projecting cleats, or " riffles," as they are universally called by the miners. The rapidity of the current through the machine depends on the angle at which it is placed and on the amount of water used. It is a very rough instrument, losing much of the gold except when this is very coarse ; but cheap, portable, and not requiring much water for its operation. The " tom " is a sort of enlarged cradle, although without the rocking motion, having a longer trough and more riffles, and requiring several men to work it. The " cradle " and " tom " were largely used during the early days of gold washing in California ; but they are now seldom seen. The so-called " sluice " is almost universally employed, whether the operations carried on be on a large or a small scale. The object of the sluice is to imitate nature as closely as possible, by allowing the auriferous sands and gravels to be carried by water over the bottom of a box provided with riffles which detain the gold as it settles to the bottom. A great quantity of material can thus be washed, because the sluice can be made very wide and the current allowed to flow rapidly. The loss of gold to which this rapidity of motion would otherwise give rise is prevented by making the sluice very long, that in the North Bloomfield Tunnel, for instance, being 1,800 feet in length. The size of the sluice and its length are regulated by various considerations, such as the character of the gravel to be washed, the grade, and the volume of water used. The riffles which serve to detain the gold are made of various materials and arranged in different ways. Square blocks of pine wood, laid side by side, so as to have spaces of from an inch to an inch and a half between them, are much in use ; in some districts longitudinal riffles, made by placing pieces of scantling lengthwise in the sluice, are preferred. In other localities round stones are used instead of blocks ; and sometimes stones and blocks alternate with each other at regular distances. A very important, and almost essential, aid to saving the gold in the sluice is the quicksilver with which the riffles are charged. The method and implements by the aid of which the gold is separated from the gravel are, in general, quite simple, as will have been gathered from the very brief notice given of them above. The sluice seems to have been introduced about the year 1851.

We may now endeavor to trace the course of the discoveries and developments which have taken place in California with reference to the way in which the auriferous gravels occur. Each one of the rivers flowing down the slope of the Sierra Nevada, within the limits of the auriferous belt, may be considered as a kind of natural sluice, the edges of the upturned slate rocks and the cavities resulting from their irregular decomposition or erosion having acted most efficiently the part of riffles. If we conceive of this region as once having been covered with auriferous débris, and notice that it is now traversed by streams running in deep gorges, with a very rapid descent, it will be easy to see that we have here all the requisites, furnished by natural causes, for concentrating a considerable portion of the gold within comparatively narrow limits, so that it may be easily got at. Hence it will not be surprising to learn that during the first two years of the Californian gold excitement, when perhaps 50,000 miners were at work on the flanks of the Sierra, at least nine tenths of them were engaged in what was called " river mining." By this term is meant washing the accumulations of sand and gravel along the channels of the present rivers, or in their immediate vicinity. This was done either by attacking the " bars," as the sand and gravel banks along the river are called, during the season of low water, or by " fluming" the streams, — that is, taking up the water in wooden flumes constructed along the banks for some distance so as to leave the bed of the river dry and accessible at all points. Such localities on the principal streams in the heart of the gold region were often found exceedingly rich, and they have all been washed over and over; in some cases a dozen times, until they have become almost entirely exhausted, so as to be no longer worth working, except as here and there a party of Chinese miners, content with very moderate gains, may find it worth while to attack them. Along the sides of the cañons, in rich districts, the crevices between the projecting edges of the slates, in the ravines or so-called " gulches" leading down to the rivers, were, during the early years of gold mining, scraped out with knives, and the detritus thus obtained panned out, often yielding large returns. From the spring of 1848 to 1851 nearly all the mining was of the character thus indicated, that in the river beds being called " wet diggings," and that in the ravines adjacent to the rivers "dry diggings." As localities of this kind could not furnish room for the many thousands who came to California, search began to be made on higher ground and away from the rivers, and quartz mines were also opened at a very early period in the history of the country, the outcrops of many of the veins being very conspicuous. The diggings having gradually become extended to the flats above the rivers, and to higher ground, where the gravel was much less rich than it had been in the river beds themselves, so that much larger quantities of material had to be handled in order to procure the same amount of gold, the necessity for improvement in the system of attacking the gravel became apparent. This led to the invention of the so-called "hydraulic method" of mining, which dates from the year 1852, and is to be credited to Edward E. Matteson, a native of Sterling, Connecticut, and which as at present operated, with machinery and methods greatly improved over those at first used, is of the highest value to the State of California.

As has been explained, the sluice is a contrivance by means of which an almost unlimited amount of material may be washed; it is only necessary to enlarge its size, and increase its length, giving it at the same time a proportionate grade. It is a piece of machinery which requires almost no looking after while in operation. After having run for a certain length of time, which may be several weeks or even months, a "cleaning up" takes place, the object of which is to obtain the amalgam which has become lodged between the riffles. After this has been done, the riffles are put back, those which are worn out being replaced with new ones, and the work goes on as before. Such being the case, it is evident that the principal expense in washing with the aid of the sluice will be in connection with getting the material into the head of the sluice. In the early days, gravel was often taken to the rocker on men's backs. Besides, the gravel where it occurs in heavy masses is usually more or less compacted together, so that it would have to be loosened with a pick, before it could be shovelled up to be carried off. This loosening, and the transportation to the head of the sluice, as well, is effected with great rapidity, and at a small expense. considering the work done, by means of the so-called "hydraulic method" of mining, the principle of which is exceedingly simple. It consists in throwing one or more jets of water, issuing from a pipe with great velocity, against the face of the gravel bank, which water in the first place loosens the gravel, and then washes it down into the sluice. The force with which the stream strikes the gravel is sufficient to " cut " it, as it is called, — that is, to disintegrate and break it up, so that it can be moved by the current, after it has spent its force against the bank, down into the sluices.* Of course the arrangements for effecting this must be suitably adapted to the circumstances. In the first place there must be an ample supply of water; then a sufficient slope of the ground, with an unchecked outlet, so that the sluices may be laid in a suitable position, and there be room for the " tailings," or material which issues from the end of the sluice, after leaving behind in the riffles the greater portion of the gold which it contained. It is owing to a happy combination of favorable circumstances that the system of hydraulic mining has been so successful on the slopes of the Sierra Nevada. That the peculiar set of conditions which makes hydraulic mining possible is not often met with is sufficiently proved, by the fact that this system, which seems so admirably adapted to the needs of the Californian gravel miners, has hardly been at all successful in any other region. It has been tried again and again in the Southern United States, with almost unvarying loss; and, even in Australia, where the mode of occurrence of the gold is in many respects so similar to what it is in California, there are few districts where the hydraulic method can be applied.

The first great need of the hydraulic miner is an abundance of water, and with a considerable " head," so that the stream may issue with sufficient velocity from the pipes. The conditions of rain-fall in the Sierra having been already explained, it will not be difficult to understand how an abundance of water may be secured for the miners' use. But this cannot be done without extensive engineering operations, and the expenditure of a large amount of money. Extensive reservoirs must be constructed, by building dams across the outlets of the mountain valleys, so as to impound the water coming from the melting of the winter's snow on the High Sierra, and the necessary canals — or ditches, as they are universally called by the miners — must be excavated to carry the water to the points where it is needed for use.† The long, rapid, and rather uniform slope of the Sierra, in the mining districts, makes it possible almost everywhere to carry the ditches with such a grade and in such a position as to allow the water to be taken from them at a sufficient elevation to give the necessary head at the point of working.

---

* Where, as is often the case, the gravel is too solidly compacted to be readily "cut" by the pipes, powder is freely used to shake up the mass, which is then much more easily acted on by the hydraulic jet.

† See farther on, for examples of the mode of construction, extent, and cost of such dams and ditches.

The great elevation of the important gravel masses, and the deep cañons into which the whole mining region is cut up, afford, in almost every locality, the necessary facilities for arranging the sluices and disposing of the tailings.

From the above sketch of the hydraulic method of mining, in regard to which various additional details will be given farther on, it will be evident that the deposits of gravel which are of sufficient magnitude to make it worth while to construct such costly works for making them available must be something very different in character and position from the accumulations in the beds of the present rivers. And, as it is especially the object of the present volume to throw all possible light on the mode of occurrence of the gravels worked by the hydraulic method, it will be necessary to show in what respect these deposits of auriferous material of such great extent and at so high an elevation differ from those which are found in the beds of the present rivers.

Before entering on the proposed description of the mode of occurrence of the high gravels of the Sierra, it will be necessary to give some additional facts in regard to the form and position of the present river cañons, and their relations to the general surface of the slope of the Sierra. To one standing on some point, not too elevated, but from which a good view of the surface of the country along the flanks of the Sierra may be had, its slope will appear to be quite uniform and unbroken, to one looking along a line parallel with the general trend of the range. It will seem, provided the point of view be favorably selected, as if the whole region was a gently descending plain, sloping down to the Great Valley at an angle of not more than two or three degrees. And the slope of the Sierra is — in the mining region, at least — quite moderate, for if we allow a rise of 7,000 feet from the lower edge of the foot-hills to the crest of the range, the distance between the two points being about seventy miles, the average rise is only 100 feet to the mile, which gives an angle of slope of less than two degrees. And if one ascends the Sierra, keeping on the divide between any two rivers in the mining districts, he will find himself, for most of the time at least, on what seems to be a plain with a very gentle rise.* Let the traveller, however, turn, and attempt to make his way across the country, in a line parallel with the crest of the range, and he will discover that this apparent plain is cut into

* The heaviest grade of the Central Pacific railroad, on the west slope of the Sierra, is 105 feet to the mile.

by the gorges or cañons in which the present rivers run, in a most extraordinary manner; he will find it several hours' work to descend into one of these and rise again to the general level on the other side, even if assisted by a well-beaten trail. All along the western slope of the Sierra, the streams have worn for themselves deep cañons, and it is these tremendous gorges which form the leading feature of the topography of the region. If the streams ran nearly on a level with the general elevation of the surface, the whole character of the mountain slope would be changed. This was formerly the condition of the drainage of the Sierra slope, as will be seen farther on. At present, however, if we start from any point in the foot-hills, we find the general level of the ridges, between the streams, rising much faster than the streams themselves; so that, when we have reached an elevation of from 3,000 to 4,000 feet on the divide between any two streams, the bottom of the cañons on either hand are from several hundred to two thousand feet and even more below us. The section (Plate G, Fig. 1) alluded to again farther on, illustrating the relative grades of the Pliocene and present channels of the American and Yuba rivers, well illustrates this, the broken lines indicating the levels of the beds of the rivers of the present day, between the foot-hills and the crest of the Sierra, while the unbroken lines show the position of the Pliocene river beds in the same region; and, as will be explained hereafter, the level of the Pliocene rivers is, in general, that of the bed-rock surface on the divides between the streams. The diagram, however, only includes the various branches of the American and Yuba rivers, which are those draining the most important portions of the hydraulic mining region. If we had the data for making a similar profile for Feather River, the great depth of the present river beds below the general level of the country would be shown in a still more impressive manner. On comparing, in the section given, the difference of elevation between the Pliocene Middle Fork of the American and the present bed of the same river, we see that at Michigan Bluff, for instance, this difference is about 1,800 feet, and that it remains about the same for several miles in either direction, gradually diminishing, however, as we descend the Sierra, and increasing as we ascend. At Iowa Hill, on the divide south of the North Fork of the American, the difference of elevation between the Pliocene and the present rivers is about 1,300 feet. To arrive at the actual difference of level between the divide and the river cañon adjacent, we have to add to the figures given the thickness of the accumulated gravels and volcanic deposits piled upon the bed-rock, as the section is intended to exhibit the position of the surface of the slates on which the more recent detrital materials have been accumulated. An examination of the table of altitudes given in the Appendix to this work will also illustrate the statements here made as to the great elevation of the surface between the streams above their present beds. As we go north from the South Yuba, the depth of the cañons increase. The difference of level between the Middle Fork of the Feather at Nelson's Point and the summit of the adjacent lava-capped mass of Pilot Peak is fully 3,650 feet. The elevation of Mount Clermont, on top of which is a mass of gravel covered by lava, is 3,570 feet above the valley at its base; that of Spanish Peak, also capped with gravel and lava, nearly 3,800 feet above American Valley. An excellent idea of the topography of the hydraulic mining region is got by the traveller passing over the line of the Central Pacific railroad, in descending the slope of the Sierra. After passing Blue Cañon, the slates begin to be met with, and all along below this, especially in the neighborhood of Dutch Flat, and beyond that for several miles, the road passes through a region of hydraulic mines, keeping on what seems to be a broad plateau, which has an elevation of a little over 3,000 feet above the sea-level. Suddenly, just before reaching Colfax, a sharp bend in the line, at a place called Cape Horn, brings the road bed just on to the edge of the cañon of the North Fork of the American, down into and along which there is an unobstructed view for eight or ten miles, the bottom of the cañon being about 1,600 feet below the level of the road. The effect of the scene presented to the eye from this point is extremely striking, because the spectator has not been prepared, by anything which he has previously seen, to expect to find the flanks of the Sierra so deeply cut into by the streams, which seem of insignificant size as compared with the immense troughs at the bottoms of which they run.

In view of what has been stated in reference to the great elevation of the divides between the streams in the mining region of the Sierra, it will be easily understood how the miners, beginning their operations in the lower portion of the range, at first almost exclusively limited themselves to the river beds and their immediate vicinity. Gradually, however, they extended the range of their " prospecting " on to the areas between the rivers, and followed them up until they found themselves at a much higher elevation, and working under very different conditions from those who kept to the " river diggings." With the gravels found in these higher localities the miners have become gradually familiar, as many years of workings have revealed all the facts necessary to enable them to understand the mode of occurrence of those deposits, in regard to whose real nature there was for a number of years the greatest uncertainty. The names given to the gravels in question have varied from time to time as more became known about them. At first, they were called simply "high gravels," as being at a higher level than those which had previously been ordinarily worked; then they were known as "deep gravels," because usually occurring in heavier deposits than the river gravels. As they are exclusively worked by the hydraulic process, they are often called "hydraulic gravels."* It is now necessary to explain, with some detail, the manner in which these high gravels occur, in order that the reader may understand the meaning of the terms used in describing the various localities where these are worked and the circumstances in which they are found as preparatory to a general discussion of the facts observed.

There are many localities in the gold region where the high gravels were sufficiently well exposed as to be easily recognized previous to any mining upon them; hence the earliest scientific observers noticed their existence, and commented upon their geological relations. As early as 1849, Mr. P. T. Tyson passed through the mining districts between the Mokelumne and the South Yuba rivers, and noticed the occurrence of gravels and volcanic rocks in many places; the former he repeatedly mentions under the name of conglomerates, and he supposed that they were the remains of a formation once continuous over the whole of the lower portion of the Sierra slope. He considers this formation to be not older than the Eocene, and accounts for the elevated position by the upheaval of the range since their deposition.†

From 1852 on, the nature of the high gravel deposits began to be a matter of great interest to the miners, and much was written and published in the

---

* Instead of saying "to work by the hydraulic method," the much less cumbersome — if inelegant — phrase "to hydraulick" is in general use.

† "We have evidence in the existence of sedimentary rocks near the Mokelemy River, that they have been elevated 2,000 feet at least since their formation, which is certainly not *anterior* to the Eocene period. . . . . It was during this long-continued diluvian era that denudations were most rapidly effected. It was then that the large valleys, before noticed, south of the American River, were mainly formed, by the removal of perhaps one half the area of the conglomerates, sandstone, etc., which once covered the entire surface of the flanks of the mountain (at least between the Cosumes and the Calaveras) to an extent of not less than twenty miles eastward from the valley. And, more than this, it scoured out innumerable ravines among the slates and other soft rocks beneath them, which were thus again exposed to the light of day." — P. T. TYSON, *Geology and Industrial Resources of California.* Baltimore, 1851. pp. 23 and 26.

newspapers in regard to this question, — as well as about the origin of the gold in the different formations. The idea that these immense gravel accumulations were in some way the result of ancient river action began to gain ground, although vehemently opposed by many. It is very easy at the present time to understand the cause of the great diversity of opinion among the miners. The idea of an ancient river formation was at first always connected with the existence of one river, and not of many. This arose from the fact that the high gravels were exposed, during the earlier years of hydraulic mining, chiefly along a certain belt of the Sierra, at various points of nearly equal height above the sea, exhibiting at these points very similar appearances, so that they seemed to be portions of one consecutive formation crossing the present system of rivers nearly at right-angles.

Thus Dr. J. B. Trask, who conducted a geological reconnaissance of the State, under authority of the Legislature, during the years 1852–55, in his second Report of work done in 1853, speaks of an ancient stream flowing with a breadth of about four miles, at an elevation of 4,000 feet above the sea, from Butte to El Dorado County.* This stream he calls the " Eastern Blue Range," and he describes its peculiarities with considerable detail, tracing it from the South Fork of Feather River as far as Georgetown. According to Dr. Trask, the boulders found in this range of gravel are almost exclusively quartz, surrounded by an earthy material of a deep blue color, giving a very marked character to the

---

* " It is now ascertained to a certainty that the placer-ranges extend to the east, within ten or fifteen miles of the 'Summit Ridge,' so called, of the Sierra Nevada ; and the condition in which it is found at these points are similar in all respects to that in the older or more western sections, with perhaps one exception, and that the relative age of both. There are evidences which clearly indicate a deposit of gold older than the diluvial drift of the lower or western diggings (which latter is often confounded with the drift deposits of the tertiary periods in this country), the character of which differs in almost every respect from any other deposit yet observed in this country, except in this particular range. Its direction has been traced for about seventy miles, and is found to extend through the counties of Butte, the eastern part of Yuba, Sierra, Nevada, Placer, and El Dorado ; it appears to have an average breadth of about four miles, with an elevation of four thousand feet above the sea for the greatest part of its length. From the examinations that were made upon this range, there are abundant evidences that an ancient stream flowed through this section of the country, and in a direction parallel with its then existing mountain ridges, and the extensive mining operations conducted in the southeast part of Sierra County on this range, has been the means of demonstrating this fact, which had heretofore been strongly suspected only. The outliers of its banks are very definitely marked throughout the entire length of the formation under consideration, and its former bed filled in many places with a volcanic sand and ashes, which probably accompanied its displacement." — J. B. TRASK, M. D., in *Report on the Geology of the Coast Mountains and Part of the Sierra Nevada*, Doc. No. 9, Session 1854.   pp. 61, 62.

mass, and distinguishing this deposit from all others in the State.* In his Report of the next year, 1854, Dr. Trask seems to have been led by more extended observations to abandon the idea of a river channel, and he divides the "placer mines" (including under this designation all the gravel washings of the Sierra, whether high or low) into three "ranges," the Upper or Eastern Range, the Middle Placers, and the Valley Mines.† The rationale of this division is not apparent from the description given; it seems, in view of the facts as at present developed, to be entirely artificial. In his last Report (1855), this author confines himself to the quartz mining interest, so far as the Sierra Nevada is concerned, except that he describes at some length the lava flow known as the Sonora, or Tuolumne County, Table Mountain, which he recognizes as having "followed the course of a stream, filling its bed and banks." He found both shells and leaves in the clays under the volcanic capping, which fossils he, however, considers as identical with those now living in the region.‡ Dr. Trask perceived that this lava stream must have crossed the Stanislaus River where the cañon is some 1,600 feet deep, and concludes that it took possession of a former bed of this river, displacing the latter and filling up the space between its banks. Nothing further is added in this, his last Report, in regard to the phenomena of the gravel deposits.

In 1854, Mr. W. P. Blake, one of the geologists attached to the Pacific Railroad Survey, made a tour through the gold mining districts of the Sierra Nevada, in the course of which he visited a considerable number of the most

---

* "The peculiarities which characterize this formation, and which distinguishes it from all others in the State, are the following: the boulders found throughout its entire extent are very uniform in their characters, and are composed of quartz exclusively (or nearly so) this has a bluish-watery color in the mass, highly translucent and vitreous when fractured, constituting ninety-seven per cent. of all the stones found in the deeper diggings, they are invested by a dull but deep blue earthy material highly charged with pyrites. . . . . The blue color of the drift in this range has been found to pervade all parts of this peculiar deposit wherever it occurs, its boulders maintain their character and percentage," etc. — J. B. TRASK, l. c. 62, 63.

† "In order to convey a better idea of the mining districts, they will be divided into three distinct ranges, denominated the Upper or Eastern Range, the Middle Placers, and the Valley Mines. This has now become necessary from the fact that the characteristics of these districts are as distinctly marked as are the northern, middle, and southern portions of the State. It separates also three evidently distinct periods of the geological history of this part of the continent, in which marked changes are apparent upon the surfaces that had emerged above the ocean during that epoch." — J. B. TRASK, *Report on the Geology of the Coast Mountains, etc.,* Doc. No. 14, Session of 1855. p. 72.

‡ "The fine clays contain an abundance of leaves of present existing genera and species, most of them may be found in the adjacent country distributed along the banks of the streams and in the deep ravines adjoining." — J. B. TRASK, *Report on the Geology of Northern and Southern California, etc.,* Doc. 14, Session of 1856. p. 24.

important localities from Tuolumne to Nevada County. At that time the hydraulic system of mining was already extensively used, and the miners were beginning to have some idea of the real nature of the high gravel deposits. Mr. Blake, however, did not arrive at any very satisfactory conclusions in regard to the formations in question, although correctly describing some of their peculiar features. He considered them to be partly of fresh-water, and partly of marine origin, and that these deposits once covered the whole region, and were brought into their present form and condition by erosion and denudation, which took place during the elevation of the Sierra.*

Dr. Hector, a geologist attached to the Government Exploring Expedition under the command of Captain J. Palliser (1857 – 60), on his return from British Columbia, visited the mining region of the Sierra Nevada, and briefly describes what he saw, in an article published in the Quarterly Journal of the Geological Society of London.† He speaks of the gravel as being of marine

---

* " These formations — the erupted and metamorphic rocks — form the floor or bed-rock upon which a very different series of formations is deposited. These formations consist of the auriferous drift in its various forms, and of a more uniform and extended series of nearly horizontal strata of clays, sand, and gravel. The last are of marine origin, and probably Miocene or Pliocene Tertiary. In many parts of the region they are entirely swept away, and scarcely a vestige remains ; but at other points they are found in extensive plateaux, or gently-sloping table-lands bordering the rivers, which have cut their way downwards through the strata and exposed them to view. The table-like hills or mountains seen from Knight's Ferry, on the Stanislaus, and between the Mammoth Grove and the Great Cave, are examples of these deposits. In many places they are overlaid by a stratum of basaltic lava, like that at Fort Miller, on the San Joaquin. It is most probable that the principal deposits of this great series of nearly horizontal strata, flanking the Sierra Nevada in the Gold Region, are of the same age as those from which fossils were obtained farther south along the Tulare Valley. . . . . Great changes have been produced in all these deposits by denudation and erosion during and since the elevation of the region to its present level. . . . . It seems most probable that the appearance of the gold was nearly coincident with that mighty convulsion which resulted in the elevation of a great part of the Coast Mountains and the drainage of the whole western base of the Sierra Nevada, until that time covered by the waves of a Post-Tertiary sea. At such a time denudation by floods would be most active ; and, until the newly risen continent had attained its permanent elevation, the streams and rivers must have been constantly changing their channels ; lakes must have been formed, and then drained, and a series of effects produced corresponding to those we now witness over the whole region." — W. P. BLAKE, in *Report of a Geological Reconnaissance in California.* 1858. pp. 276, 278.

† " Before leaving these shingle-deposits, which are so largely distributed throughout the mountain valleys of British North America, I may mention that in California I found these terraces ranging on the western slope of the Sierra Nevada, at least to the height of 3,000 feet. . . . . At Nevada City, where the coating of shingle-deposit has thus been cleared from the surface of the coarse-grained and soft granite which underlies it, gigantic masses were exposed on what had once been the rugged shore of an inlet, just as may be seen on a waterworn coast of the same material at the present day. . . . . The evidence we

origin, and evidently entertained views quite similar to those of Mr. W. P. Blake, supposing the erosion of the gravel deposits, which he calls "shingle terraces," to have taken place during the upheaval of the continent.*

In 1860, M. Laur, Ingénieur au Corps Impérial des Mines, was sent by the French government to California and Nevada, for the purpose of examining and reporting on the gold and silver mining interest of those States. In the course of his investigations, he spent two months among the hydraulic and quartz mines of the Sierra Nevada, and published his account of them in 1862 in the form of a Report to the Minister of Public Works. M. Laur describes the high gravel deposits under the designation of "auriferous alluvia anterior in age to the basalt." He considered them to have once extended over the whole western slope of the Sierra, and maintained that they were thrown into confusion (bouleversés) by the phenomena attendant on the eruption of the basalts.† The views of this mining engineer were extremely crude; he seems to have entirely misapprehended the nature of the phenomena and to have failed to notice some of the most striking facts connected with the occurrence of the gravel deposits, — facts which, even at that early period, had made a strong impression on the miners themselves.

During the years previous to 1860 there seem to have been many articles published in the Californian newspapers in regard to the real nature of the high gravels. Few of these can now be obtained, and most of them have never met the eye of the writer. Some extracts, however, are preserved in Mr. J. S. Hittell's Mining in the Pacific States of North America. In an

---

have respecting the age of the terrace-accumulations is very imperfect. There can be no doubt that those occupying the valleys of the Rocky Mountains, being farthest from the coast and at the greatest elevation, are the most ancient, and that from the time of their deposit till now the rearrangement of the same materials has been carried on during the gradual uprising of the continent." — JAMES HECTOR, M. D., in *Quarterly Journal of the Geological Society of London*, Vol. XVII. 1861. pp. 404, 405.

* "It must have been during the period when this denudation of the eastern plains accompanied the gradual emergence of the continent, but acting with very different results on the rocky sea-bottom and successive ranges of iron-bound coast presented by the western slope, that these immense deposits of shingle were formed and moulded into terraces." — HECTOR, l. c. p. 405.

† "Les alluvions aurifères de formation antérieure au basalte constituent par leur étendue et leur épaisseur les mines d'or les plus importantes de Californie. . . . . Cette formation s'étendit primitivement sur tout le versant occidental de la Névada; elle fut ensuite bouleversée par des phénomènes de l'époque des basaltes. En certains points, les dépôts furent recouverts par des coulées basaltiques ou des couches de tufs sous lesquelles ils sont aujourd'hui exploités. Ailleurs ils furent attaqués par de violentes érosions qui les firent disparaître sur tout ou partie de leur épaisseur." — P. LAUR, in *De la Production des Métaux Précieux en Californie. Rapport à s. Exc. M. le Ministre des Travaux Publics.* Paris, 1862.

article written by Mr. C. S. Capp, and published previous to 1859,* he describes the so-called "Blue Lead," to which allusion has already been made, and traces it from Sebastopol in Sierra County across the North Yuba, thence to the Middle Fork of that river, by Forest City, Chip's Flat, Centreville, and Minnesota. Here, he says, "it is obliterated by the Middle Fork of the Yuba, but is believed to be again found at Snow Point on the opposite side of the river; and again at Zion Hill, several miles beyond."† Of this "lead" he says, "it is evidently the bed of some ancient stream, because it is walled in by steep banks of hard bed-rock, precisely like the banks of rivers and ravines in which water now runs, and because it is composed of clay which is evidently a sedimentary deposit, and of pebbles and black and white quartz, which could only be rounded and polished as they are by the long-continued action of swiftly running water."

These views, however, met with much opposition on the part of the miners and others. Mr. B. P. Avery, afterwards editor of the Overland Monthly and United States Ambassador to China, thus wrote, in 1859, in reference to Mr. Capp's description of the "Blue Lead": "It is remarkable that the majority of our miners, who are commonly men of intelligence and practical knowledge in their pursuit, should have discarded entirely, if they ever entertained, when speculating upon the origin of our gold fields, the more rational theory of marine influence, for one of purely local causes. They overlook all the facts which go to prove a total submergence of this coast at some remote period, and settle down upon the narrow idea that the immense gravel beds which contain so large a portion of our mineral wealth, and which extend at least four hundred miles north and south, having an average breadth of probably not less than sixty miles, were deposited by rivers, which anciently ran here, and changed their channels from time to time, until they paved the whole country with cobble stones. These deposits have been cut through by modern streams, running a different course, and hence the present cañons and ridges. Of the ancient rivers, the one that deposited the blue lead has alone left distinctive marks of its course. Now, unfortunately for the plausibility of this theory, the blue lead is found all the way from the summit of the Sierra to the foot-hills. Instead of being confined to a certain altitude, and a certain line, it exists in every alti-

* This article probably appeared in the San Francisco Bulletin; it is quoted by Mr. Hittell in the book above mentioned, published in 1861, but without indication of date or origin.

† See Map at the end of this volume, which, however, does not extend north of the Middle Yuba.

tude, on the main ridge as well as on spurs of them, and even on isolated peaks."*

The current opinion of the Californian miners in regard to the origin of the high gravels, at the time of the beginning of the State Geological Survey under the direction of the writer, is probably very well stated by Mr. Hittell himself, and reads as follows: "The alluvial placers may be divided into ancient and modern. The ancient are those formed by streams which no longer exist, or have found new channels. Two very remarkable examples of the ancient stream placers are found in California; one called the Blue Lead of Sierra County, the other Table Mountain in Tuolumne County. It is supposed that the Blue Lead was once the bed of a large river, about fifty miles eastward of the present position of Sacramento River and parallel with its course. Table Mountain is a pile of basalt, standing on what was, in the remote past, the bed of a river nearly parallel with the Stanislaus. These ancient and deserted channels are not rare, and are found from a very small to a very large size. They are usually buried at a considerable depth beneath dirt and gravel. Sometimes they are found high above the level of the present streams running near them."† This statement of Mr. Hittell is much nearer the truth, as now made out from a long series of careful examinations by the Geological Survey, than anything which had been previously published, and contrasts in a marked degree with the crudities contained in the official report of a French mining engineer.

The essential fact that the high gravels of the western slope of the Sierra Nevada have to do with an ancient river system having been so clearly recognized by many of the miners, it is not difficult, from the stand-point of our present knowledge of the conditions in which they occur, to understand why this view did not meet with universal acceptance. It was always hampered with the idea that there was one great river running parallel with the crest of the Sierra, as we see in the quotation just given from Mr. Hittell's book. As long as this determination to change the whole present condition of the drainage system of the Sierra was persisted in, without accounting in any way for such a change; as long as a river forty miles in breadth had to be imagined, in order to embrace all the deposits in question, and regardless of the fact that one of its banks would have to be some thousands of feet higher than the other, so long of course the ancient river theory would continue to

* Quoted in Hittell's Mining in the Pacific States, pp. 74, 75.
† l. c., p. 63.

meet with opposition. It was necessary to make another step in advance and to show that the rivers were not one but many, and that the direction of the drainage was, when those high gravels were accumulated in the channels of the ancient streams, not essentially different from what it now is. It was necessary that the position of the gravels and the intercalated and overlying volcanic deposits should be laid down with accuracy on a map, and that the course of each separate stream should be followed up with a series of measurements of elevation, so that continuity or non-continuity could be asserted with confidence as an established fact, and not guessed at.

The work of the State Geological Survey was, for various reasons which it is not necessary here to explain, chiefly carried on during the first years in the Coast Ranges, and in 1862–63 there was only time for a rapid reconnaissance in the mining region of the Sierra. The general results of this preparatory work, in so far as the high gravels were concerned, were given in the following words in a *résumé* of the progress of the Survey, published in the American Journal of Science, for September, 1864 : "There is perhaps no subject connected with the geology of the Pacific coast in regard to which there are so many misapprehensions as there are in what has been published by geologists on the nature and distribution of the detrital deposits which are so extensively worked by the methods known as hydraulic and tunnel mining. It has been assumed that these deposits are of marine origin, and that they originally extended over the whole slope of the Sierra Nevada, — a condition of things which, were it true, it would be of vast importance for California to know ; but the real facts of the case are entirely different. In the first place these deposits are not of marine origin, as is proved by the fact that, although frequently found to contain impressions of leaves, masses of wood and imperfect coal, and even whole buried forests, as well as the remains of land animals, and occasionally those of fresh-water, not a trace of any marine production has ever been found in them. Again, these detrital deposits are not distributed over the flanks of the Sierra in any such way as they would have been if they were the result of the action of the sea. On the contrary, there is every reason to believe that they consist of materials which have been brought down from the mountain heights above and deposited in preëxisting valleys ; sometimes very narrow accumulations, simple beds of ancient rivers, and at other times in wide lake-like expansions of former watercourses ; and this under the action of causes similar to those now existing, but probably of considerably greater intensity. This deposition of detritus, for the most part auriferous, took place during the later Pliocene epoch, and not as late as the drift or diluvial period, as is abundantly proved by the character of the remains of plants and land animals which are imbedded in it.   The deposition of this auriferous detritus was succeeded, throughout the whole extent of the Sierra Nevada, by a tremendous outbreak of volcanic energy, during which the auriferous gravel was covered by heavy accumulations of volcanic sediments, ashes, pumice, and the like, finally winding up by a general outpouring of lava, which naturally flowed from the summits of the Sierra through the valleys, into the lake-like expansions, filling them up and covering over the auriferous gravels, which were to remain for ages, as it were, in a hidden treasure chamber, concealed under hundreds of feet in thickness of an almost indestructible material."

The above extract represents, very nearly, the general scope of the results arrived at by the Geological Survey, during the first two or three years of its existence, in reference to the origin of the high gravel deposits.   No detailed work had been done in the region where these occur, and the extreme inaccuracy of all the existing maps would have rendered it impossible to coördinate any such observations if they had been made.   There still remained much to be done before a thorough understanding of the phenomena in question could be arrived at, as will be evident to the reader of the succeeding chapters of this volume.   What has been accomplished since the Geology of California, Vol. I., was published, in the way of a detailed examination of the gravel region, will be set forth in the next section of this chapter, and, after a statement of the facts observed, their theoretical bearing will be discussed at some length.   In the mean time we have become, from what has been said in the preceding pages, sufficiently well acquainted with the general character of the high gravel deposits to be able to understand the meaning of the terms employed by the miners in their work, — terms which we shall have to use frequently in giving an account of our observations in the mining region.

It being well understood that the high gravels belong to a system of ancient rivers, in the former beds of which detrital material has been deposited, and which have since become in large part obliterated by accumulations of lava, and also very extensively worn away by erosion, the character of the miner's work will be in a great measure directed by these conditions.   That the portions of the gravel richest in gold should be found towards the bottom of the deposit, on or near the bed-rock surface, and in the lowest portions of that surface is what would be naturally expected from well-known conditions; the much heavier metal must be ever tending, as the accumulations of detritus are being swept onward in their course towards what would be their final resting-place, if they were carried far enough, namely, the surface of the bed-rock. These deeper regions where the miners expect to find the gravel richest, and which, when connected together, must evidently approximately mark the course of the former current or river, are always called the "channel." Whenever the miner is looking for some place rich enough to make it worth while to commence washing for gold as a permanent operation, he will say that he is "looking for the channel." Not that the richest places are always the lowest, or that the gold is limited in its range exclusively to the vicinity of the bed-rock surface; but, as a general rule, these are the conditions with which the miner expects to meet. The position of the channel, therefore, and its direction in any mining district, become a matter of great importance. It is with reference to these circumstances that all exploratory work must be laid out. The miner runs his drift — or tunnel, as a horizontal excavation is universally called on the Pacific coast — with a view to striking the channel, guided by what he knows, or thinks he knows, of its probable position, from a study of the results of adjacent mining operations; if in an unprospected region, he must in drawing his conclusions be guided by the form of the surface, or other circumstances which his previous experience has shown him to be of most value as a guide in prospecting. Everything, then, connected with the form, size, direction, and variations in character of the "channel," or the bed-rock frame which holds the gravel, is a matter of importance to the miner, and not less so to the geologist in search of the facts necessary as a basis for his theories. If we knew the exact position of the lowest surface of all the detrital masses in the Sierra, we could then reconstruct the ancient system of drainage with accuracy; but, of course, as but a very small portion of these ancient channels has been, or ever will be, revealed by mining operations, and as much that is laid open will never be examined and recorded while accessible, we can never expect to know exactly what were the relations of all parts of the old river system to each other. This will be evident enough when the facts presented in the following pages come to be examined.

If the channels of ancient rivers exist in the mining region of the Sierra, and at a high level above the present streams, there must be, in places at least, more or less of the sides or banks of these channels still in existence.

Such remaining portions are known to the miner as the "rims," or the "rim-rock." To one who knew nothing of the peculiarities of the gravel mining region, it would seem very strange to see — as may often be done — the miner running a tunnel in the hard bed-rock, where there are no external indications of gravel deposits whatever, and especially when it is ascertained that the work on such a preparatory undertaking, which can never begin to pay until gravel has been reached, may require years for its completion and cost many thousands of dollars. The miner, in such a case, if questioned as to the object of his work, would say that he was "tunnelling through the rim-rock, in search of the channel." His knowledge of the country would have taught him that there was a sufficient chance of finding a paying deposit of gravel, occupying a channel-like depression somewhere in the region towards which he was tunnelling, to make it worth while for him to risk his time and money in the manner suggested. Some of these channels in the Sierra are wonderfully well defined and deep, with a perfectly preserved rim on each side, as will be seen from the descriptions farther on.

In many places there is associated with the gravel, often overlying it, but sometimes in interstratified beds, a large amount of material of a volcanic origin. This may be either solid lava, in the form of a regular flow consolidated from a liquid condition, or it may be of a detrital character, and made up of more or less rounded fragments, which after having been ejected have been carried to a distance from the place of their origin by currents of water, and deposited as a stratified mass of volcanic conglomerate or breccia. Some of these deposits seem to consist of material which has fallen into the water, in the form of ashes and lapilli, and taken on a decidedly stratified form, somewhat resembling clay in appearance. Not unfrequently volcanic and ordinary sedimentary strata alternate with each other, or occur mixed together in the same layer, so that it is not easy to decide, without close examination, of what the deposit really consists. For the different kinds of volcanic material occurring with the gravels there are various names in use among the miners. In the regions where the fragmental character predominates, as is often the case, especially in the central mining counties, the volcanic deposits are called "cement." This is a term somewhat differently employed in different districts, but most frequently applied to rather closely compacted volcanic breccias and conglomerates, of which the different coarser materials seem to be pretty firmly cemented together by the finer portion of a very similar material. The whiter, fine-grained and homogeneous beds resulting from the consolidation of ashes and volcanic mud are usually called simply "lava." Other terms of local use will be noticed in the description of the various gravel mining districts farther on in this chapter.

Among the non-volcanic portions of the mass there may be fine clays, very distinctly stratified, sands of every degree of texture, and gravels proper or collections of rolled pebbles, — the latter of all sizes and of various materials. Some of the gravels are chiefly made up of quartzose pebbles; and, in places, boulders of this material form almost exclusively the mass of the gravel. The finely-laminated beds are known to the miners as pipe-clay; and the terms sand and gravel are employed much as the geologist would use them. The gravel which is rich enough to be worked is called "pay dirt," or "pay gravel," and, if there is a well-defined bed of it, it is often termed the "lead," or the "pay lead." If the gold is irregularly disseminated through the gravels, the term "spotted" is applied to them.

In many localities the auriferous gravels with the associated sedimentary deposits having been entirely covered by volcanic overflows, either of solid lava or of detrital materials, this "capping," as the miners term it, has never been removed by denudation, so that the underlying beds are entirely concealed, and there is sometimes a mass of unproductive rock several hundred feet in thickness piled over the gravel in which the gold occurs. By examining the maps accompanying this volume it will be seen, in comparing the relative position of the gravel and volcanic deposits, how much of the former must be covered by the latter. The eruptive materials having come from high up in the range, their relative quantity naturally increases in that direction, as will be shown in detail farther on. The presence or absence of a capping of greater or less thickness of volcanic rock over the gravel is, of course, a matter of much practical importance to the miner. In the use of the hydraulic method of mining everything above the level of the bottom of the workings must be removed by the current of water, and must be of such a nature that it can be "cut" by the pipes, either as it naturally occurs or after being shaken up by the use of powder. The solid lava, however, cannot be brought into the sluices in this way, so that it is evident that a capping of such material is a barrier to hydraulic mining. The less coherent kinds of lava can be washed down by the hydraulic jet; but material containing no gold cannot be handled without expense, so that it often becomes a question, in deciding upon the opening or the continued working of hydraulic mining ground, how thick a mass of unproductive detrital material, whether it be of volcanic or sedimentary origin, can be sluiced off without more than counterbalancing by the expense of the operation the profit to be got from working the productive portion of the deposit. When the cover of volcanic material or of unproductive gravel becomes too thick, so that the work no longer pays, the locality has to be abandoned, unless the channel is sufficiently well marked and rich enough to make it worth while to follow and work it by the method called "tunnel mining."

Tunnel mining was formerly extensively employed in California, although now quite overshadowed in importance by the hydraulic method. Indeed, the present knowledge of the extent and value of the hydraulic claims is largely due to the other and older method of attacking the auriferous deposits. Some localities, as, for instance, the Sonora Table Mountain, have been exclusively worked by tunnels. This method consists, simply, in driving or tunnelling in the channel and bringing out the pay dirt in cars, or otherwise, to be sluiced, exactly as any auriferous material obtained by river or bar mining would be. The channel must first be found by drifting or tunnelling to it, except in those rare cases where the ground has been so eroded as to bring the pay gravel to the surface. When a "lead" has been struck sufficiently rich to pay for working, it will be followed by a drift, so run as to keep in the best ground, or in the pay streak, as long as the operation proves remunerative. In case of necessity one or more shafts may be sunk from the surface to the workings, to furnish ventilation, and some claims have been and still are worked by means of shafts and drifts, the pay dirt being hoisted with a windlass through the shaft. This is the method chiefly followed in the Australian gravel mines, where the conditions are rarely favorable for using the hydraulic process. Shafts have also, in former times, been extensively sunk for the purpose of ascertaining the position of the channels, in order to fix the level at which the tunnel should be carried in.

The above general description of the mode of occurrence of the gravel deposits and of the methods by which they are worked seems to be sufficient to enable the reader to understand the detailed descriptions which are to follow in the next section of this chapter, and to which we now proceed.

SECTION II. — *Detailed Description of Portions of the Auriferous Gravel Region.*

The Geological Survey of the State of California, after a suspension of two years' duration, in consequence of the failure of the Legislature of 1869 – 70 to make any appropriation for its continuance, began work again in the spring of 1870. From that time on, during the ensuing three years and a half, a considerable portion of the resources at our command were devoted to detailed work on the western slope of the Sierra Nevada. Naturally, among the subjects to be investigated, that of the mode of occurrence of the auriferous gravels was one of prime importance. The principal difficulty in the way of executing satisfactorily this part of our task was the necessity of accurate maps on a large scale of the region to be geologically surveyed. It was evident that without such maps as a basis for the observations the results would be very far from meeting the expectations of those pecuniarily or scientifically interested in the gravel deposits. But the peculiar topography of the Sierra, together with the fact that its western slope is so densely wooded, render the accurate mapping of the region a work of great labor and expense. As much was done as was possible with the very limited means at our command, and we received essential aid from the Central Pacific Railroad Company, which furnished us with detailed maps of the region adjacent to their line, embracing a belt of perhaps ten or fifteen miles in width, running across the Sierra in the heart of the gravel mining region.* With this assistance, and with such additional information as could be derived from the very incomplete and erroneous surveys of the United States Land Office, combined with the topographical work executed by the Geological Survey, a map of the mining counties was constructed, on the same scale as that of the Bay Map, namely, two miles to an inch. This map covered an area of about nine thousand square miles, extending from Knight's Ferry on the Stanislaus to Quincy in Plumas County; it was never published, the survey having been discontinued before it could be made ready for the engraver.

In arranging for the detailed exploration of the mining counties, the western slope of the Sierra was divided into two portions by the North Fork of the American River. The region to the south of this river was assigned to Mr. W. A. Goodyear, with special instructions to make as thorough an investigation as possible of the auriferous gravel deposits, not neglecting, however, other facts of general geological interest. Mr. Goodyear was in the field, without assistance, from May 15 to December 17, 1871, and he went over all the important gravel mining districts between the North Fork and the Middle Fork of the Middle Fork of the American River, a region in which the gravels are very extensive, quite various in character, and where

---

* This cartographic work, as we were informed, cost the company more than the total appropriation for the Geological Survey during the two years of its continuance from 1870 to 1872.

they extend high up towards the crest of the Sierra. He also explored the region about the South Fork of the American, and examined all the important points in El Dorado County, especially the neighborhood of Placerville, as well as a considerable portion of El Dorado and Amador counties, as far south as the Mokelumne River. Nearly or quite all the hydraulic mining districts where any work was doing between that river and the North Fork of the American were examined by Mr. Goodyear during the course of the summer. The region north of the North Fork of the American, as far as the Middle Yuba, was assigned to Mr. A. Bowman, who took the field early in the spring of 1870. Professor W. H. Pettee joined him in July, and they together spent nearly all the remainder of the year in working out the geology, and mapping the detailed topography over the area indicated. From their surveys, aided by such information as could be obtained from the various gravel mining and ditch companies in the region, the map, which is appended to this volume, was compiled by Messrs. Hoffmann and Craven. This map, which is on a scale of one mile to the inch, extends over the area between the Middle Fork of the American and the Middle Yuba, embracing the most important hydraulic mining region of the State. The work of Messrs. Bowman and Pettee was supplemented on the north by surveys made for private companies by Mr. Hoffmann, so long chief topographer of the Geological Survey; the work thus introduced was especially important as including perhaps the most important of all the gravel deposits in the State, those between the South and Middle Yuba. A considerable number of ditch lines in the region surveyed are also laid down from information furnished by the various companies; also all the quartz and gravel mining claims which had, up to 1873, been patented or officially surveyed for the purpose of securing a patent under the United States laws are indicated upon this map.*

In regard to the extension of the gravel ranges north of the Middle Yuba, our data are far from complete. The region was traversed at different times by various members of the Survey, but never systematically explored. This is especially true for the interesting district between the Middle Yuba and Slate Creek. Plumas County was explored by the writer in 1866; but here the gravels have by no means the importance which they possess farther south.

---

* As the location of these claims is not a matter of scientific interest, it has not been considered necessary to have the map brought up to date in respect to the additional patents which have been issued since 1873.

For the district south of the Mokelumne River — the southern boundary of Mr. Goodyear's work — there are no complete surveys, although most of the mining region between that river and the Merced has been passed over and portions examined in some detail by the writer, and especially by Mr. Rémond, who made, in 1865, quite a detailed examination of the region between the Merced and the Stanislaus. The volcanic rocks and the quartz mines are of great interest in this region; but the gravels are much less important than they are farther north. It is, however, from the detrital deposits south of the Stanislaus that most of the interesting fossils, to be described farther on, have been obtained.

In giving the details of the various gravel deposits which have been examined, a beginning will be made with the district so carefully surveyed by Mr. Goodyear. His notes, however, which are in the form of a journal, are too voluminous for publication in full in the present volume, and they include much matter which has no reference to the gravel question. Besides, as some localities were visited more than once, at different times, there is necessarily more or less repetition, and supplementing of details imperfectly made out at a first examination. Mr. Goodyear also furnished a review, or general discussion, of his observations, which will be frequently referred to and quoted from in a future chapter. In giving a *résumé* of Mr. Goodyear's notes, the observations made by him in the course of his work will be arranged in what appears to be the most suitable order for acquiring a clear idea of the nature of the phenomena. The lithological character of the *bed-rock* and its stratigraphical position will first be indicated. Next the appearance and form of its *surface*, as exposed after the gravel has been washed away, will be described, and such other facts mentioned as naturally fall into this division of the subject. After giving all that seems necessary to be mentioned in reference to the bed-rock, the *gravel deposits* will be described with some detail; then will follow some details in regard to the *volcanic* materials so intimately associated with the gravel, as already mentioned: next, the *channels* will be taken up, and their position, size, direction, and grade at the different localities examined will be given; finally, the *distribution of the gold* in the gravel, its position in reference to the form and character of the channel and the rim-rock, will be discussed, with the addition of such facts in regard to the workings, from an economical and practical point of view, at the various localities visited, as may seem to be of general interest.

The order followed in giving these various details for each branch of the subject will be essentially that of Mr. Goodyear's route in the field, and may be followed on the accompanying sketch-map showing the distribution of the Volcanic and Gravel Formations over a portion of Placer and El Dorado counties. (Plate B.) This map is an approximation only for the region high up among the numerous branches of the Middle Fork of the American River, that district never having been topographically surveyed. A portion of El Dorado County, south of the Middle Fork of the American, — which separates El Dorado from Placer County, — extending as far south as Georgetown, is also to be found on this diagram; but the region east of this, at the head of the South Fork of the Middle Fork — also visited by Mr. Goodyear — has been omitted for want of the necessary data. For the immediate vicinity of Placerville — an interesting and important mining district — a special sketch-map, on a scale of three miles to the inch, is introduced.

The route pursued by Mr. Goodyear in his investigations was nearly as follows: From Colfax across the North Fork of the American to Iowa Hill, Wisconsin Hill, Elizabeth Hill, Sucker Flat, Monona, Wolverine, Strawberry, and the other important mining towns between the American and the branch of it called Shirt Tail Cañon; then to Damascus and the Forks House, near the head of Shirt Tail; then to Canada Hill, Millers' Defeat, Last Chance, and the other mining camps at the head of the North Fork of the Middle Fork of the American; from the head of the Middle Fork down to Deadwood and Michigan Bluff, respectively on the east and west sides of El Dorado Cañon, thence down the south side of the divide between the Middle Fork and Shirt Tail, through the extremely important mining localities between Michigan Bluff and Todd's Valley, including Bath, Forest Hill, the Dardanelles, and Todd's Valley. From Forest Hill the north side of the same divide was visited, including Yankee Jim's and the different branches of Brushy Cañon. Crossing the Middle Fork of the American, Mr. Goodyear's route next took him to the mining districts along the divide north of Otter Creek, to Volcanoville, and to the various localities north of Georgetown, on both sides of Cañon Creek. At this point the work was suspended for a short time, in order to refit and repair. On recommencing, a start was again made from Colfax, and some localities about Iowa Hill, Forest Hill, and Michigan Bluff were revisited. Another trip was also made to the region high up on the Middle Fork, about Last Chance and Startown; thence down the valley to Peckham Hill, and across the Middle Fork to Spanish Dry Diggings, and by way of Greenwood to Georgetown again. From the last-named place an

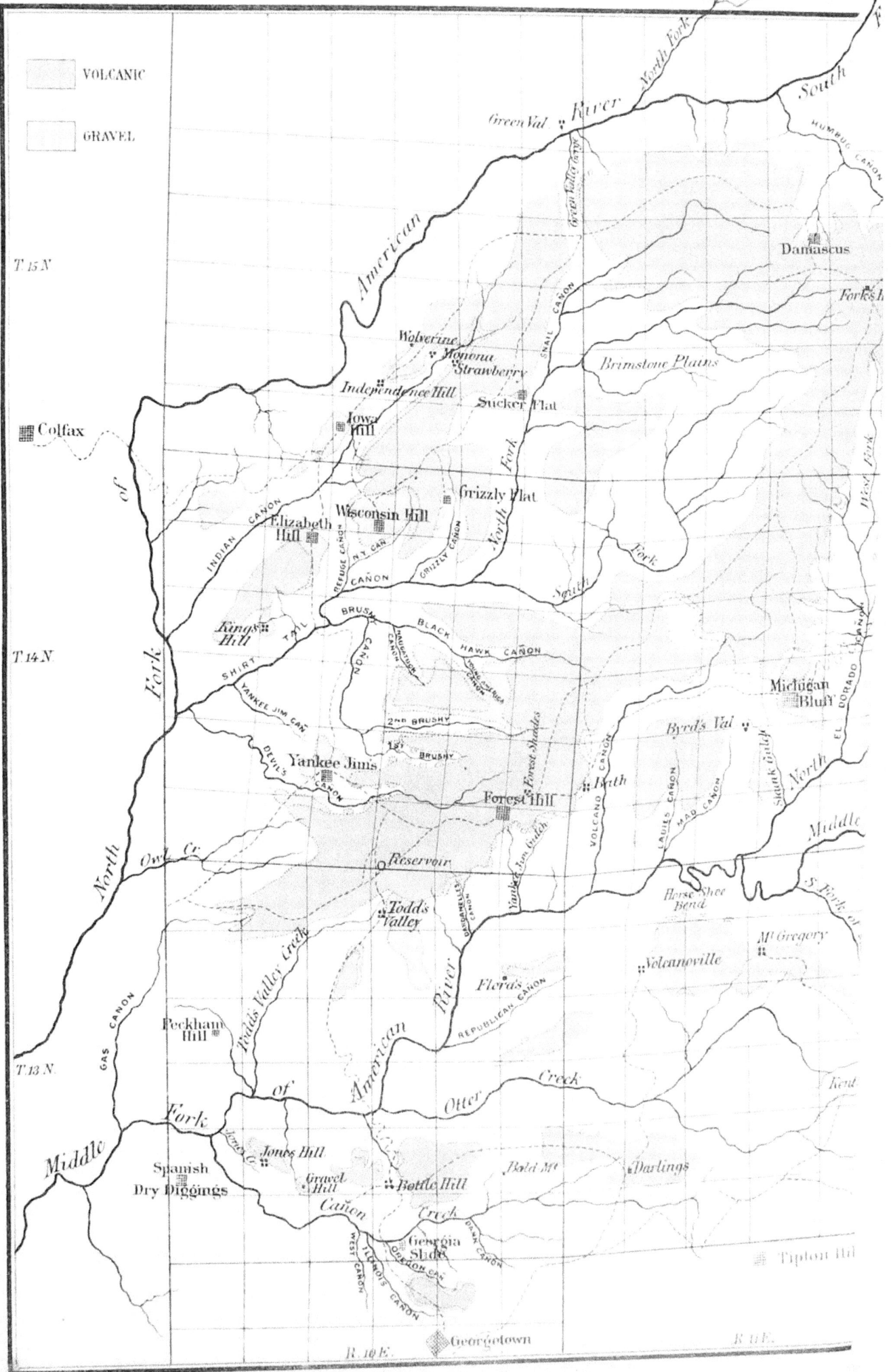

VOLCANIC

GRAVEL

T. 15 N.

American

River

North Fork

South Fork

Green Val

HUMBUG CAÑON

Damascus

Forks h

Wolverine

Monona

Strawberry

Brimstone Plains

Independence Hill

SNAIL CAÑON

Sucker Flat

Iowa Hill

Green Valley Caño

Colfax

Fork

Grizzly Flat

North Fork

Wisconsin Hill

Elizabeth Hill

GRIZZLY CAÑON

WEST FORK

INDIAN CAÑON

REFUGE CAÑON

N Y CAÑ

CAÑON

South Fork

T. 14 N.

Kings Hill

BRUSHY CAÑON

BLACK HAWK CAÑON

Fork

SHIRT TAIL

DAUGHTUCK CAÑON

YOUNG AMERICA CAÑON

Michigan Bluff

YANKEE JIM CAÑ

2ND BRUSHY

Byrd's Val

EL DORADO CAÑON

DEVILS CAÑON

1ST BRUSHY

Forest Shades

Shank Gulch

North

Yankee Jims

Bath

Forest Hill

VOLCANO CAÑON

LAULS CAÑON

MAD CAÑON

Middle

Owl Cr.

Reservoir

Yankee Jim Gulch

S. Fork of

North

Todds Valley

DARDANELLES CAÑON

Horse Shoe Bend

Mt Gregory

GAS CAÑON

Todds Valley Creek

Flords

Volcanoville

American

Peckham Hill

River

REPUBLICAN CAÑON

T. 13 N.

Fork

of

Creek

Kent

Middle

Otter

Creek

JONES C.

Jones Hill

Bald Mt

Darlings

Spanish Dry Diggings

Gravel Hill

Bottle Hill

BEAR CAÑON

Georgia Slide

WEST CAÑON

ILLINOIS CAÑON

OREGON CAÑ

Cañon

Creek

Tipton Hil

Georgetown

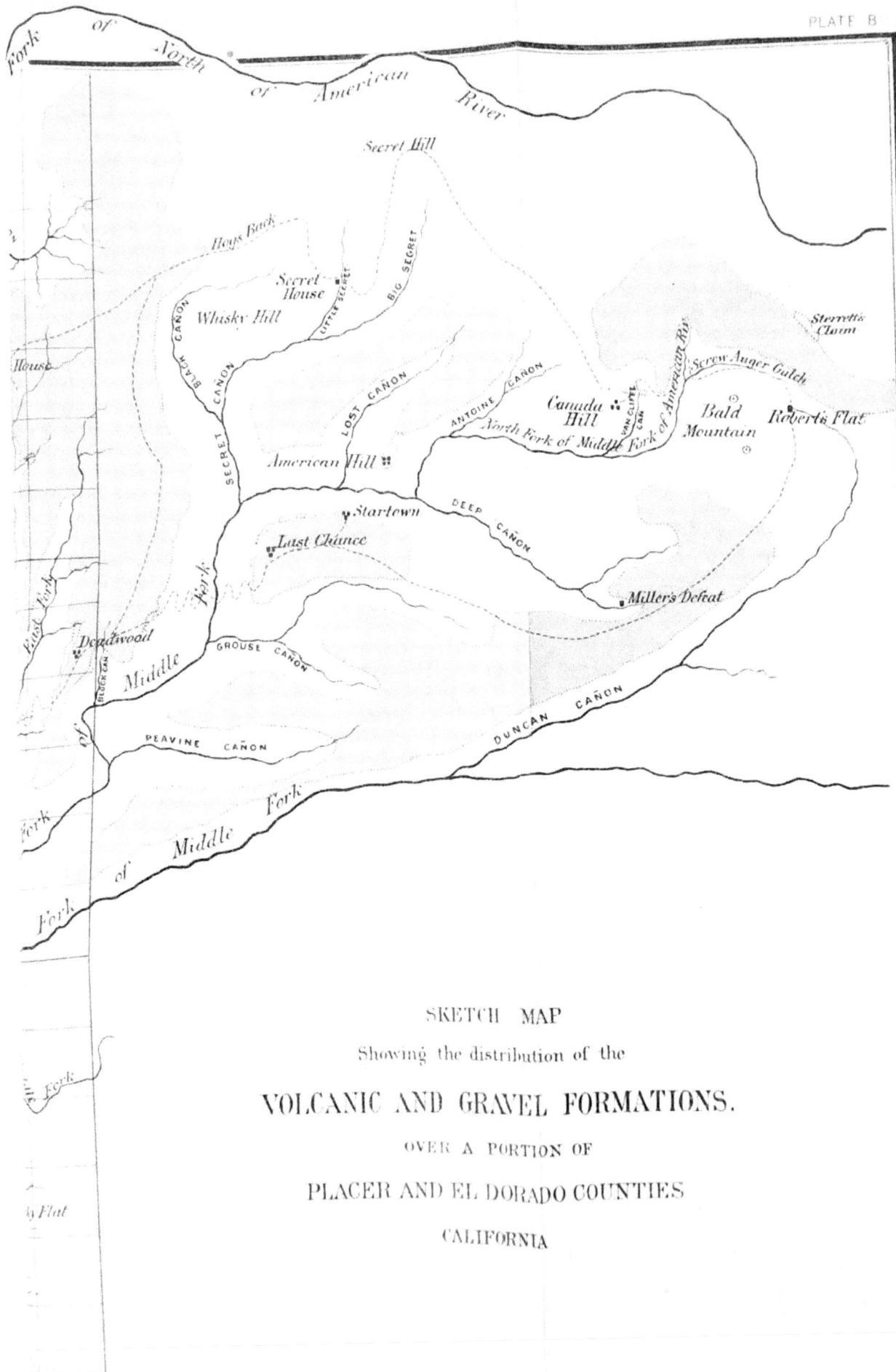

Fork of North

of American River

Secret Hill

Hogs Back

Secret House

BIG SEGRET

Whisky Hill

LITTLE SEGRET

BLACK CANON

Sterrett's Claim

House

SECRET CANON

LOST CAÑON

ANTOINE CAÑON

North Fork of Middle Fork of American Riv.

Screw Auger Gulch

Canada Hill

VAN CLIFFE CAN.

Bald Mountain

Robert's Flat

American Hill

Fork

Startown

DEEP CAÑON

Last Chance

East Fork

Deadwood

BLOCK CAN.

Middle

GROUSE CAÑON

Miller's Defeat

PEAVINE CAÑON

DUNCAN CAÑON

Fork

Fork of Middle

Fork

Fork

Fork

Flat

SKETCH MAP

Showing the distribution of the

VOLCANIC AND GRAVEL FORMATIONS.

OVER A PORTION OF

PLACER AND EL DORADO COUNTIES

CALIFORNIA

excursion was made into the region at the head of the South Fork of the Middle Fork of the American, an important mining district not included within the limits of the map. Returning to Georgetown, the explorations were continued down the slope of the Sierra to Folsom, and thence upward to Placerville, the vicinity of which important and central mining town was thoroughly examined, as well as the region at the head of the Cosumnes River, between Newtown and Grizzly Flat. The remainder of the season was given to the principal mining localities in Amador County, nearly the whole of which was gone over. Following the above geographical order, pretty nearly, the observations made by Mr. Goodyear will now be given, arranged under the various heads previously indicated.

§ 1. *The Lithological Character and Stratigraphical Position of the Bed-Rock.*

Immediately above Newcastle, on the line of the Central Pacific railroad, slates make their appearance, and the country between here and Auburn consists partly of slate and partly of granite. The slates here, so far as seen, generally strike northwesterly and stand nearly vertical. At Auburn the rocks are hard silicious slates, standing vertically or with a slight inclination to the northeast: they strike about N. 20° W. These slates are excessively hard and compact, and can with difficulty be broken. Their color varies from light green to almost black. On the stage road from Folsom to Auburn, the granite continues until within three miles of the latter place.

At Iowa Hill, at Wiessler's Claim, the bed-rock consists entirely of slate, generally very thin-bedded, argillaceous, and sometimes apparently more or less talcose. Some of these slates are hard and very tough, and are then generally of a dark bluish color; but other portions are quite soft, and then usually white. Their strike varies between N. 15° W. and N. 25° W., and they stand nearly vertical. At Wolverine, about two miles northeast of Iowa Hill, the slates strike N. 10° — 15° W., and have an almost vertical dip. At Metcalf's Claim, near Roach Hill, the strike of the slates is N. 10° — 15° W., and their dip 60° — 70° to the southwest. At the Morning Star Tunnel, on the southeast side of Indian Cañon, nearly opposite Iowa Hill, the strike of the slates is pretty uniformly N. 10° — 15° W. At Strawberry Flat, Indian Cañon, a little above Independence Hill, the bed-rock slates have a direction of N. 20° W., and dip at a high angle to the southwest. In the southwest pit at Elizabeth Hill, two miles south-southwest of Iowa Hill, the slates stand nearly vertical, with the usual predominant strike in this district of N. 15° — 20° W. At King's Hill they strike N. 35° W. and stand vertical. At Refuge Cañon, west of Wisconsin Hill, the strike of the slates is N. 50° W.; in the east pit at Elizabeth Hill, a little west of Refuge Cañon, it is N. 20° — 25° W.

At Nahor's Claim, in Green Valley Gorge, about five miles northeast of Iowa Hill, the bed-rock is entirely serpentine, and it is said that this rock can be traced for many miles north and south of this point; Brimstone Plains, about four miles south-southeast of the Green Valley Gorge, is said to be on this belt of serpentine. At the Mountain Gate Tunnel, near Damascus, four miles east of Green Valley Gorge, the bed-rock is a very fine-grained, thin-bedded clay-slate, which, if a little harder, might make good roofing-slate; its dip is nearly vertical, with a slight inclination sometimes to one side and sometimes to the other; strike N. 10° — 15° W. The color of these slates is variable, the lighter colored being the softest; the dark bluish-black varieties are often filled with small cubical crystals of pyrites. In the Lower Tunnel the slate is everywhere dark in color. At the Cement Mill Claim, near Damascus, the slate is more or less talcose, and some of it very much so.

The summit of Secret Mountain, on the road from Damascus to Canada Hill, near the head of

Secret Cañon, consists of a hard, gray quartzite, which seems to be a metamorphic form of a rather coarse-grained, quartzose sandstone. The bed-rock at Sterrett's Claim, on the left bank of Sailor's Cañon and two miles in a straight line from its mouth, near Canada Hill, is a hard slate striking N. 35° W., and dipping to the northeast at an angle of 85°. Bald Mountain, a little east of Canada Hill, is entirely made up of an enormous mass of impure quartz-rock. The boulders in the main channel at Canada Hill consist very largely of quartzose sandstone, and there is in the adjacent region a great deal of this material. The slates and sandstones about Canada Hill have a nearly east and west strike. In Miller's Defeat Cañon, about three miles south of Canada Hill, the boulders are largely of sandstone. At Yule's mine at Startown, near Last Chance, and five miles west-southwest of Canada Hill, the bed-rock varies more or less in character from a soft and light-colored slate, to a very hard and dark-colored material, which may be either slate or sandstone; much of this rock is very quartzose, as well as compact and hard. The strike of this rock is pretty uniform in direction, a little to the west of north. The rocks in the bottom of the cañon of the North Fork of the Middle Fork of the American River, below Last Chance, are metamorphic slates, dipping at a high angle to the northeast, and striking northwest.

A great deal of the bed-rock in the claims near Michigan Bluff is a very talcose slate; some of it, as at Byrd's Claim, is a kind of semi-serpentine, mixed with very soft talcose slate. In this vicinity the strike of the slates is about north and south, and they stand nearly vertical. There is a region, extending for many miles north from Byrd's Valley, among the forks of Shirt Tail Cañon, through Brimstone Plains, and Green Valley, as before noticed, where the bed-rock is almost exclusively serpentine. There is a fine locality of this rock, on the line of McKinstry's ditch, south of the North Fork of the American River, nearly east of Clipper Gap, and about a mile northwest of the "United States House." The extent covers an area of 150 feet wide by 300 or 400 feet long, over which the quality of the rock is good; but the whole mass is much larger. The stone is very beautiful, but more or less seamy. At El Dorado Hill the bed-rock is generally a talcose, and very soft and thin-bedded slate, striking a little to the east of north, and with a variable dip, as if it had been locally disturbed.

At Smith's Point, between First and Second Brushy cañons, near Yankee Jim's, the strike of the slate bed-rock is N. 10° — 15° W. The slates vary from talcose to argillaceous, and stand nearly vertical. At the Dardanelles Claim, a mile and a half southwest of Forest Hill, the bed-rock strikes N. 35° — 45° W., and stands nearly vertical; some of it is quite hard, and other portions are very soft.

In the Illinois Cañon, near Georgetown, six miles south of Forest Hill, the bed-rock is slate, which is sometimes very thinly laminated and talcose, while other portions are heavy-bedded and argillaceous, so as to pass into a fine-grained sandstone. The more talcose slates are full of cubes of pyrites, and these are said to be rich in gold. The rock is all thoroughly decomposed. These slates have a strike of N. 20° W., and dip to the northeast at an angle of about 55°; they are traversed by a system of joints, underlaying to the southwest at an angle of 45°. At Georgia Slide, near Georgetown, the bed-rock is a rather soft, decomposed slate, generally more or less talcose, and full of decomposed crystals of pyrites. Bald Mountain, between Cañon and Otter creeks, three miles north-northeast of Georgetown, is a mass of serpentine, which does not extend far to the east, being succeeded by slates in that direction. At Flora's, two miles west of Volcanoville, and a little over five miles north of Georgetown, on the extreme end of the spur between Republican Cañon and the Middle Fork of the American, the bed-rock is quite hard: it has a nearly north and south strike, and stands almost vertical. At the Buckeye-Sucker Claim, about one third of a mile southeast of Mount Gregory, near the head of Otter Creek and eight miles northeast of Georgetown, the bed-rock is a soft decomposed slate striking about N. 5° W. and standing nearly vertical. There is a very heavy body of serpentine in the vicinity of Hotchkiss's, at the southeast foot of Mount Oliver, which is about three miles south of Bald Mountain, and some of the same rock at Volcanoville. Serpentine is known to the miners in El Dorado County and the southern part of Placer as "hornblende rock."

On the road between Auburn and Colfax, at a point nearly opposite Clipper Gap, the bed-rock is porphyritic, containing large crystals of feldspar, and it is locally known as "China rock." At a point two or three miles below Colfax a little serpentine occurs, and along the road from that place to Iowa Hill, on the north side of the river (the North Fork of the American), there are outcrops of metamorphic conglomerate and imperfect serpentine.

At Spanish Dry Diggings, south of the Middle Fork of the American River, the bed-rock consists of slates and very fine-grained sandstones, generally filled with crystals of iron-pyrites, and to a greater or less extent traversed by small irregular seams of quartz. Immediately east of the Grit Claim is a heavy mass of semi-serpentine, apparently the result of incomplete metamorphic action on a fine-grained slate rock. Some of the slates here appear to be made up of an aggregation of minute acicular crystals (? tremolite or fibrolite). Near Greenwood, three miles south of Spanish Dry Diggings, there is some exceedingly fine-grained argillaceous slate, with much fine-grained sandstone, and some coarser grit-rock, which has once been, apparently, a sort of fine breccia, or conglomerate. There is also considerable porphyritic schist, locally known under the name of "China rock," as mentioned with regard to a locality near Colfax. This is a hard, fine-grained, schistose rock, filled with large feldspar crystals. In the St. Lawrence Claim, near Greenwood, the stratification of the rocks seems to have been greatly disturbed, the decomposed slates striking and dipping in various and very different directions. It is said that the "China rock" forms a continuous belt from Greenwood across the North Fork of the American.* At White & Co.'s Claim, on the tunnel, the bed-rock is thoroughly decomposed and very soft, and has occasional little seams of quartz running through it. At Pilot Hill, or Centreville, six miles south-southeast of Auburn, on the Georgetown Divide, between the Middle and South Forks of the American River, there is a large quantity of dioritic (?) rock, and hornblendic slates; portions of this bear some resemblance to gneiss. The summit of Pilot Hill has been much metamorphosed, and consists chiefly of a very hard rock through which there is distributed considerable quartz, in the form of little irregular seams of chalcedony. There is also on the summit of the hill a good deal of gossan, and the quantity of iron contained in the rock seems to be very large. At a point about half a mile, a little east of south, from the summit of Pilot Hill, there is a belt of crystalline and highly metamorphosed limestone, which at that place is not over a hundred feet in width. It is said that this limestone belt can be traced, at intervals, all the way to the South Fork of the American, in a direction of S. 34° E. "Alabaster Cave" is said to be about five miles nearly west of Centreville; while Cave Valley, on the road from Georgetown to Auburn, is above five miles north of Pilot Hill. There appears to be no continuous belt of limestone between the two localities, but only a narrow outcrop at each place. Here, as elsewhere in the region north of the Mokelumne River, the limestone outcrops are small, and cannot be easily connected with each other, so as to form a continuous line. At Powningville, three or four miles east-northeast of Pilot Hill, on the road to Georgetown, there is some granite, and some tough hornblendic rock, and this locality seems to be about on the line of demarcation between the slates and the granite, which latter then stretches on up the river as far as to the point when the road forks to go to Johntown, between Michigan Flat and Coloma. Near Alabama Flat, on Johntown Creek, the bed-rock is serpentine; and between Alabama Flat and Johntown the porphyritic material called "China rock" occurs. Just below Johntown, the bed-rock passes into thin-bedded slates which strike northwesterly and dip at a high angle, probably 80°, to the east. At Crane's Gulch, between Johntown and Georgetown, the "seam-diggings" are in slate which is considerably decomposed, and much of it talcose. Its strike is northwesterly and its dip nearly vertical; but the stratification is much contorted, and it contains irregular seams of quartz, which, however, follow pretty nearly the lines of bedding.

On the line of the road from Georgetown to Placerville, the rocks are chiefly clay slates, exceedingly thin-bedded, — the lamination being sometimes as delicate as paper. These slates usually

---

* Mr. Goodyear adds in his notes, "My experience has been, that there are no continuous belts of anything, for any great distance, in this section of country."

dip at a very high angle, sometimes inclining in one direction and sometimes in another, seeming to have been considerably disturbed. In Hangtown Hill, just south of Placerville, the bed-rock is generally soft and decomposed; in the Excelsior Claim much of it is apparently a sandstone, sometimes fine-grained, and occasionally of coarse texture and passing into a purely silicious sandstone, the quartzose grains being rounded and mixed with scales of mica. At White Rock Point, about four miles northeast of Placerville, the bed-rock is granite, much of which is soft and decomposed; it stretches southerly as far at least as Smith's Flat, distant about two miles, and appears to be an isolated patch surrounded by slates and mostly covered by the volcanic and gravel deposits. On the south side of Webber Hill, about two miles south-southeast of Placerville, the bed-rock is a hard, dark-blue argillaceous slate, standing nearly vertical, and having a strike of N. 10° W. In these slates there are well-marked veins of slaty material, in the form of a gray schist, cutting very sharply the stratification of the slates, at angles of 10° or 15°. The latter dip generally to the northeast, with an inclination of 80° to 85°, while the veins pitch at about the same angle in the opposite direction. The material which fills these so-called veins is finely laminated in a direction parallel with their sides; so that the slaty structure is nearly or quite as well marked as that of the rock in which they are enclosed. There is much pyrites disseminated through the rocks in the neighborhood of Placerville, as is often the case in other localities where the argillaceous slates occur.

In the region between Mud Springs and Shingle Springs, the bed-rock is slate, often more or less "blocky," — that is, breaking into large fragments or blocks. In some places the slate passes into a semi-serpentine. A mile or two below Shingle Springs, the rocks are gneissoid and granitoid; these continue for some distance, and gradually give place to slates and "blocky" rocks. About five miles below Shingle Springs, and a mile south of the road to Folsom, is Marble Valley, on a creek of the same name, which is a branch of Deer Creek. Here is an outcrop of saccharoidal and rather coarse-grained limestone, of a grayish-white or very light bluish color, which is distinctly stratified, with a strike of about N. 55° W., and a dip of from 80° to 85° to the northeast. Half a mile farther down the Creek is another outcrop of the same rock, with a strike of N. 10°—15° W., and, here also, standing nearly vertical. The belt is narrow, apparently but a few hundred feet wide, and cannot be traced to any great distance in the line of its strike. There are said to be several more isolated patches of limestone in a distance of two or three miles to the southeast; they are smaller, however, than those which have just been described. Along the northeast side of the limestone at Marble Valley numerous large boulders of variegated jasper were noticed. Below Shingle Springs, after passing the granitoid and gneissoid rocks, the bed-rock consists chiefly of very fine-grained metamorphic sandstone, usually breaking into "blocky" fragments. Portions of the rock are, however, argillaceous. In the bed of Willow Springs Creek, not far from where the Folsom road crosses it, the bed-rock is a fine-grained blue sandstone, which has been used for building in the vicinity; it may be seen in a wall near the north end of Willow Springs Hill. Along the road between Shingle Springs and Folsom many fragments of bed-rock are scattered, which are porphyritic, with crystals of white feldspar.

All the way between Placerville and Newtown the bed-rock is much decomposed and usually quite soft. It frequently passes into a talcose slate and occasionally becomes a well-marked soapstone. South of Placerville, on Squaw Hollow Creek, and all along the road to Fairplay, between the North and Middle Forks of the Cosumnes, the bed-rock is granite, which continues as far as Fairplay itself. From that place across the ridge to the Middle Fork of the Cosumnes, the country is all granite, generally rather soft, and containing frequent dark-colored nodules of syenite, or syenitic granite. At the Cosumnes copper mine, a heavy mass of coarsely crystalline limestone crosses the river in a northeast and southwest direction, and rests against the granite on its western side. In this rock is a cave of considerable, but unknown, extent. It is said to have been penetrated to a distance of 800 or 900 feet, without reaching its end. In the neighborhood of Grizzly Flat, situated a little to the northwest of Steely's Fork, a branch of the North Fork of the Cosumnes, the bed-rock is chiefly granite, usually soft and decomposed, but containing a good many hard bouldery masses. There are streaks and patches of slates and sandstones connected with the granite, making it a difficult problem to unravel the geology in this region. At Brownsville, there is a belt of limestone, the surface of which is worn into all sorts of fantastic forms; this rock cannot be traced for any considerable distance, and is apparently not more than a few hundred feet in width. The same limestone occurs again at Indian Diggings, on Indian Creek, where there appear to be several thin belts or patches of this rock. Many of the huge fantastic blocks into which the limestone has been eroded, or weathered, have been broken off and rolled down into the Creek below, so that it is difficult to say when the rock is in place and when not. At the last-mentioned locality the bed-rock, when not limestone, is slate and sandstone, all thoroughly decomposed and soft. This decomposed material has belts, patches, and bunches in it through which is distributed a certain quantity of rolled pebbles of quartz and other rock, the whole forming a curious *pot-pourri* of decomposed bed-rock and gravel. There are portions of the rock in which no water-worn pebbles are found; but in other places there is a great abundance of them scattered about in an irregular way. Some of the belts in which they occur are but a few feet wide, and stand nearly vertical; others are more or less inclined; others horizontal, or nearly so, while some are quite irregular in position. It is said that whenever these rolled pebbles are found distributed through the rock, the latter contains gold enough to pay for washing; and that, when they do not occur, this is not the case. It appears probable that, as the creek slowly cut its way down through the soft bed-rock, there took place, from time to time, a great deal of sliding of the ground; and that the gravel, which once overlaid the surface of the bed-rock, has thus become irregularly incorporated with its soft mass.

The most common direction of the strike of the slates in the immediate vicinity of Volcano is northeast and southwest. At the head of the Flat the slate strikes northeast; though at the foot, just below the town, it runs about north. These rocks are not unfrequently much contorted; and the outlines of the limestone masses occurring here are very irregular. In the cañon of Sutter Creek, a little way below Volcano, there is a large quantity of well-characterized talcose slate, called "soap-stone" by the miners; but there are also considerable bodies of very hard, fine grained, nearly black silicious rock, which is very tough and massive. Limestone forms the bottoms of all the workings in Indian, Jackass, and Soldier's Gulches, near Volcano; and it also makes up the lower portion of the spurs which separate these gulches. The surface of this rock, especially in the spur between Indian and Clapboard Gulches, is worn out into holes and cavernous places of the utmost irregularity of form, with fantastic pinnacles between them. These cavities are filled with gravel, and are often as much as forty or fifty feet in depth. The flat which stretches along Sutter Creek, above Volcano, is from 500 to 600 feet in width, and, for nearly its whole length, it has been extensively worked by sinking pits, raising the gravel by derricks, with whims driven by horse-power, and washing it in sluices. These pits have been worked down, sometimes, to a depth of forty or fifty feet, over a width of from 300 to 400. The surface consists of a dark-colored soil, from three to eight feet in depth, and beneath this lies the gravel, the general level of which rises but a few feet above the average level of the higher portions of the limestone, whose surface, as in other localities, is worn into the most fantastically shaped cavities. As Mr. Goodyear remarks, "No words can paint the raggedness of these excavations." [*] Similar phenomena may be observed on the right bank of the South Fork of Sutter Creek, near Aqueduct City, about six miles southwest of Volcano; at this point a considerable area of the limestone has been worked off, or uncovered by mining. The rock here, as in several other portions of the limestone belt, is intersected by numerous dykes of diorite, which traverse it in an easterly and westerly direction. These dykes vary from two to thirty feet in width; some are even wider than thirty feet. On the right bank and near the mouth of Soldier's Gulch, at Volcano, there is a locality of complex forms of chalcedony. Close by it there is a rocky point consisting entirely of a perfectly honey-combed mass of thin, flat sheets of chalcedonic and jaspery quartz. This rocky knob is from thirty to forty feet in height, and about the same in diameter.

---

[*] The claim at the upper end of Volcano Flat, where the limestone has the most ragged appearance, is called the "Upper Engine Claim."

It looks as if a body of sandstone, full of the most irregular seams, had had all these filled with chalcedony and jasper, and afterwards the rock had been dissolved away, leaving only the silicious skeleton, in the form of a honey-combed mass.

On the road from Volcano to Fiddletown, from the crossing of Dry Creek on, the rock is slate, with the regular northwesterly trend and high inclination to the east. On the road from Fiddletown to Mud Springs, on the hills to the north of Indian Creek, granitic rock occurs, and extends, for a distance of four or five miles, to the edge of the cañon of the Cosumnes River; much of it approaching mica or hornblende slate in character, but with a granitic texture. This rock continues, on the route indicated, as far as the brow of the hill on the south side of the cañon, where slates occur, some portions of which have the texture of diorite. These slates, which continue down to the Cosumnes River, have a nearly north and south (magnetic) strike; but always a high inclination, sometimes to the east, and occasionally to the west. Just below Kingsville, three miles east of Shingle Springs, on the road to Placerville, there is a body of serpentine. The rocks in this vicinity, however, are in general slates and sandstones, in a large proportion of which the stratification is nearly obliterated, although in places there are clay slates of which the bedding is perfectly preserved. There is a ridge between Latrobe and Big Cañon Creek, which looks very prominent, as seen from the hills about Placerville. Here the rocks exhibit a great variety of texture, appearing to be metamorphic forms of clayey and sandy beds, and conglomerates of small pebbles. Intercalated with these are beds of what appear to be volcanic materials, looking like diorite. Between Latrobe and Michigan Bar, on the Cosumnes, low down on the foot-hills, the bed-rock consists of hard and much altered strata, the outcrops of which are much elongated in the direction of the strike of the formation, which is about N. 15° W. It is only by the form of these outcrops that the bedding can be made out, so complete has been the metamorphism of the original material which seems to have been sandstone. In places, however, there are intercalated masses of thin-bedded argillaceous slates among these more crystalline and perhaps in part volcanic rocks. In the vicinity of Michigan Bar the bed-rock is slate, often much decomposed and dipping usually at a high angle to the northeast, although sometimes in the opposite direction.

In the Sugar Loaf, near Puckerville, which is about six miles nearly east of Forest Home, the rocks are metamorphosed and very hard slates, sandstones, conglomerates, and breccias. At a point about a mile and a half below Puckerville there is a small outcrop of limestone, and the same rock probably occurs, at a point opposite this, on the Cosumnes River. There is also a considerable quantity of imperfect serpentine in the hills, to the southwest of the Sugar Loaf, near Puckerville.

From Dry Creek to Amador Creek the rocks are almost all very hard metamorphic slates and sandstones; and in many places the stratification is only to be made out from a study of the outcrops, which are elongated in the direction of the strike of the formation. In the vicinity of Rancheria Creek there is a large quantity of porphyritic slate, containing large greenish-white crystals of feldspar, and somewhat resembling the "China rock" mentioned as occurring at Greenwood. On, or very near, the crest of the ridge next north of Sutter Creek, about half-way from the town of that name to Volcano, there is a limestone quarry, where this rock has been quarried for burning into lime. The belt, which does not appear to be more than 200 or 300 feet wide, shows at various points along a line of nearly half a mile in length, having a general northwesterly trend. The rock has been shattered and seamed in every direction, and some of the cracks thus formed have become filled with quartz. This limestone quarry is about six and a half miles from the town of Sutter Creek, and the bed-rock passed over between the two places consists of slates and schists, somewhat inclined to be porphyritic in structure. Some of the slates are thin-bedded, dark-blue, and argillaceous, and other portions are in heavy layers, or "blocky."

The bed-rock at Muletown, a little north of Ione City, is slate, passing into sandstone, generally much decomposed, and especially so under the gravel. In the adjacent ravines and gulches below the level of the floor of the gravel, it is quite hard. In some of the soft slates in this vicinity there are indications of fine lamination or cleavage, at right-angles to the planes of bedding, such as is occasionally, but rarely, seen in California, although common enough in some other districts where metamorphic slates occur. At Irish Hill, three miles northwest of Muletown, the bed-rock is also slate, but harder than at the latter place. Among the gravel hills adjacent to Irish Hill the bed-rock is completely decomposed into clay. Some of it is mottled with various colors, — brown, red, violet, purple, yellow, and pink; other portions are pure white, resembling the rock at Michigan Bar. In the bed of Dry Creek, below Dry Town, there is a large amount of pebbly conglomerate and fine breccia, both highly metamorphosed; this is seen extending over a distance of between two and three miles, along the course of the Creek, beginning a mile below the town. At the town the rock is chiefly thinly-bedded argillaceous slate, with some harder and thicker-bedded varieties intercalated. At the Potosi Mine, on the south side of Dry Creek, the rocks are argillaceous slate, which is not very hard, but nearly black and containing more or less iron pyrites disseminated through the mass. These slates stand nearly vertical, or have a very high dip to the northeast. They are very much bent and broken, especially near the surface, and are very thin-bedded. At the "Old Amador" or "Little Amador" Mine the hanging wall of the vein is a rather hard, compact, fine-grained crystalline rock (metamorphic sandstone?) filled with sulphurets distributed through it in very fine particles. It is green in color and called here "greenstone." * The foot-wall of the vein at this mine consists of the usual dark-colored, thin-bedded, fine-grained argillaceous slates. At Down's Mine, the most northwesterly one worked in 1871 at Sutter Creek, the country rock is a greenish slate, generally not very hard, and containing more or less pyrites disseminated through it. It is called "granite" by the miners. The bed-rock in the neighborhood of the great quartz mines of Amador County is mostly of the kind indicated above, — argillaceous slate, sometimes talcose, varying in hardness, and containing much pyrites disseminated through it.

§ 2. *Bed-Rock Surface under the Gravel.*

At Iowa Hill the surface of the bed-rock is often very rough and irregular. At Metcalf's Claim, Independence Hill, the bed-rock is exceedingly rough. At Elizabeth Hill the bed-rock is very uneven. At Nahor's Claim, in Green Valley Gorge, the surface of the bed-rock is smooth and water-worn.

At the Mountain Gate Tunnel, near Damascus, the surface of the bed-rock is remarkably smooth and regular; pot-holes and projecting points of rock being almost unknown.

In the Morning Star Mine, at Startown, the bed-rock is very rough; that is, uneven, with many irregular depressions; and it is generally in these depressions that the miners find the richest spots.

In the Forest Hill ridge the bed-rock has generally a pretty even surface. It is, indeed, more or less rolling; but it is rare that points project more than seven or eight feet above the general level.

At Smith's Point, between First and Second Brushy cañons, the surface of the hard bed-rock, where exposed, shows little channels cut in it by the water, in the form of furrows, which are sometimes a foot deep and eight or ten feet long. There is a general parallelism in these furrows, and most of them run in a direction about S. 75° — 85° W., across the edges of the slates.

The town of Yankee Jim's is on the southern slope of the southern rim-rock of the Big Channel. In this channel the bed-rock is exposed over a very large area, and its surface is seen to have many long furrows or channels worn into it; these vary from one to ten feet or more in length, and are generally shallow and narrow, and often somewhat crooked. The great majority of them are parallel with the direction of the main channel, that is, nearly west, crossing the stratification of the bed-rock almost at right-angles. At Indiana Hill, one quarter of a mile southwest of here, on the other hand, there are also many of these same furrows to be seen on the washed

---

* This rock, which Mr. Goodyear calls a "metamorphic sandstone," may perhaps be a metamorphic volcanic material.

surface of the bed-rock ; but there their axes are nearly parallel with the strike of the slates, or N. 10° — 15° W., and their direction points exactly to Georgia Hill, on the south or opposite side of the Devil's Cañon. It appears, therefore, that there were two channels on the bed-rock here, the first or largest one coming from the east, and the second coming from the southeast, the two forming a junction at or near the head of Yankee Jim's Cañon, which now heads just between the two channels. The true pot-holes in the rock were generally barren. In the upper or eastern portion of the Big Channel the richest pay was on the bed-rock ; but toward the lower end the surface of the bed-rock was not so rich, and the upper gravel paid much better.

At the Dardanelles and Oro claims, near Todd's Valley, the surface of the bed-rock is exceedingly rough and irregular, being filled with depressions of various shapes, some of which are from fifteen to twenty feet deep. The whole surface of the high as well as of the low bed-rock is well water-worn. Numerous small, elongated furrows are worn in the surface of the rock, which vary from a few inches to one or two feet in depth, and often several feet in length. These furrows run across the edges of the strata, and are nearly parallel with each other ; their direction is about S. 65° W., indicating that the water on the bed-rock flowed southwesterly.

In the Spring Tunnel, near Volcanoville, the bed-rock is said to have been found very uneven, so that the tunnel, which starts 150 feet below the rim-rock, and was driven for 900 feet in a direction a little west of north, passed several times alternately through bed-rock and gravel.

At a point about a quarter of a mile east of the Franklin Claim on the south side of Little Spanish Hill near Placerville the furrowings of the surface of the bed-rock are unmistakable, and run in a direction about S. 46° W. magnetic across the edges of the slates which strike N. 44° W. magnetic and stand nearly vertical.

§ 3. *Crevices in the Bed-Rock.*

In the Morning Star Claim, near Iowa Hill, a crevice was struck in the tunnel, which is said to have been 150 feet in length at the top and about eighty at the bottom, and from one to ten feet wide. Its direction was parallel with that of the strike of the slates, and it was filled with gravel mixed with fragments of the bed-rock. This crevice is said to have yielded $ 40,000.

At El Dorado Hill, near Michigan Bluff, there is in the outer portion of the hill a deep gulch in the bed-rock running nearly north magnetic and descending with a very rapid grade toward the north for some four or five hundred feet and then ending by abutting sharply against a high mass of bed-rock. This gulch is said to have been extremely rich. It is fifteen to twenty feet deep below its rims.

In the western portion of the Hook and Ladder Claim, near Placerville, there are three or four crevices in the bed-rock, running parallel with the stratification, and which are filled with gravel and little broken fragments of the slate. They have been very rich in gold, far richer than any of the gravel on the surface of the bed-rock. Not one of these crevices has yet been worked to the bottom, and their depth is therefore unknown. They extend into the adjacent location, — the Blacklock.

Near Aqueduct City there are narrow crevices in the decomposed slates, occupied by gravel, which seems to form vertical dykes as it were. These run down to depths of from twenty to a hundred feet. These gravel-filled crevices seem to have been once occupied by limestone, and this has been removed by the solvent action of water, leaving a cavity into which the gravel has later been washed.

§ 4. *Character of the Gravel Deposits.*

At Wiessler's Claim, Iowa Hill, over a considerable portion of the area worked, especially towards the southwest, the gravel resting directly on the bed-rock varies from two to ten or more feet in thickness, and is overlain by a stratum of sand, containing much mica, and called "pipe-clay" ; this ranges from ten to fifteen feet in thickness. Over this again is gravel to the top of the bank, with only thin layers of sand irregularly intercalated in it. Elsewhere there is no sand, but only gravel; this is generally yellowish-white in color, and not very strongly cemented together, the boulders and pebbles of which it is composed being entirely quartzose and metamorphic in character, without any granite or volcanic. They are also generally rather small, few being larger than a man's head, and not usually exceeding the size of the fist, although occasionally boulders weighing several hundred pounds may be observed. The body of the gravel is generally very nearly horizontally stratified. But near the bed-rock there are occasional appearances of cross-stratification, or "beach-structure." In one place an inclination of the layers as great as 15° was observed. The thin streaks of sand occurring here are generally of limited extent, irregular in shape, forming lenticular patches in the gravel, often bifurcating with it and entirely surrounded by it, and varying in thickness from a few inches to several feet.

At Independence Hill, a little northwest of Iowa Hill, in the Reno Claim, the thickness of the auriferous gravel is from 75 to 100 feet.

In the Morning Star Tunnel (near Iowa Hill) the gravel is what is called "blue cement," the pebbles being all metamorphic, often very hard, and consisting to a great extent of a dark-blue sandstone and slate, with less quartz than at Iowa Hill.

At the head of Refuge Cañon just west of the Wisconsin Hill school-house, some sluicing has been done, and banks of gravel twenty-five or thirty feet high are exposed. There are some very large quartz boulders here. This work is said to have yielded in a short time over $10,000.

A few hundred feet east of Mr. Teasland's house at Wisconsin Hill a shaft was sunk through gravel, said to be 190 feet deep, without reaching bed-rock.

In the Lebanon Tunnel, on the northeast side of New York Cañon, an air-shaft was raised to the surface giving the following section.

	Feet
Volcanic conglomerate	90
Auriferous gravel	4
Volcanic conglomerate	160
Auriferous gravel	60
Bed-rock	40

At Nahor's Claim, in Green Valley Gorge, the layer of gravel resting on the bed-rock is quite thin, more or less mixed with volcanic boulders, and covered over, first with more or less volcanic conglomerate, and then with a heavy mass of volcanic breccia.

At Sucker Flat, where a tunnel was driven on the surface of the bed-rock, N. 45° W. magnetic, some 1,300 or 1,400 feet into the hill, the quantity of gravel was very small, and it was immediately overlaid by heavy masses of volcanic débris. But very little gold was found.

At the Cement Knob Mine, in the spur west of Grizzly Cañon, and about a mile below Grizzly Flat, the layer of gravel is said to vary from six to eighteen inches or two feet in thickness, and it is overlain by volcanic materials.

The gravel in the Mountain Gate Tunnel, at Damascus, is all white in color, except that in some places it is stained slightly red with oxide of iron, the pebbles and boulders consisting entirely of quartz, which is remarkably uniform in character, compact, white and solid, and only occasionally a little crystalline. Some of the quartz boulders are very large, weighing, it is said, a hundred tons or more. But these large boulders seem to occur only near the bed-rock; and the gravel grows rapidly finer on going upward from it. The material between the boulders and pebbles consists largely also of quartz sand, mixed, however, to some extent with clay derived from the disintegration of the slates. The gravel is generally not very hard, except occasionally near the bed-rock, where it contains much iron pyrites.

On the east side of Damascus Cañon, and one eighth of a mile from the hotel, there is a hydraulic pit exposing a bank with from forty to fifty feet of gravel, covered by from twenty to twenty-five feet of water-washed volcanic gravel. The gravel here is very much like that in the Mountain Gate Mine, being white below, with plenty of large boulders near the bed-rock, growing finer above and assuming a yellowish tinge, sometimes deepened into red by oxide of iron. At the Cement Mill Claim on the east side of Damascus Cañon and one fourth of a mile from the hotel, the gravel in the channel, where worked, is composed of quartz and volcanic pebbles and boulders mixed together, with much volcanic sand ; the pay-stratum lies always close to the bed-rock, and is only a foot or two thick.   Toward the head of Indian Cañon, above the Mount Pleasant Flat, many tunnels have been driven into the hill, developing a system of shallow channels which seemed to come down from the central portion of the ridge obliquely to the present cañon.   The stratum of gravel here was thin, ranging from a few inches to two or three feet in thickness, and was covered with volcanic débris, while in places the volcanic matter closed down directly on to the surface of the bed-rock ; which, however, was everywhere water-worn, whether any gravel interposed between it and the volcanic, or not.

At the Red Point, near Damascus, the quartz gravel stratum is said to be twenty feet thick. The material is wholly quartz, but not so much rounded as in the Mountain Gate Channel.

At Sterrett's Claim, on the left bank of Sailor's Cañon, the gravel appears to be pretty deep, possibly 150 feet, and consists of a very heavy "wash," that is, of large boulders only partially rounded by water ; these boulders are generally very quartzose, but not pure quartz.   They seem rather to be a pebbly conglomerate metamorphosed into a jaspery mass.

In Yule's Claim, at Startown, the gravel is well washed and water-worn, and the boulders are of the same general character as the harder portions of the bed-rock, and do not appear to have come from any great distance.   The thickness of the gravel is from thirty to fifty feet, and it is immediately overlain by the volcanic débris, which forms all the crest of the ridge.

At Nick Anderson's Mine, just below Last Chance, the gravel is whitish and contains considerable quartz, but many of the boulders are of the same character as those in the two chief claims at Startown (Yule's and the Morning Star).

At the English Claim on the west side of the ridge fronting El Dorado Cañon, near Deadwood, the pebbles and boulders consist of a great variety of metamorphic rocks.   The bed-rock is well washed and the gravel is only a few feet deep, varying from one to seven feet, and overlain by volcanic débris, much of which is more or less clayey in texture and of somewhat the color of chocolate.

In the Reed Mine, near Deadwood, the gravel varies from an inch or two to six or eight feet in thickness, averaging probably about eighteen inches.   It contains some pretty large boulders, some of which will weigh several tons.   Some of these boulders are composed of white quartz ; but most of them are of metamorphic sandstone, very compact, and hard and bluish in color. Immediately above the gravel comes a layer of what the miners call "chocolate," which is from nothing to five feet in thickness, averaging about three feet.   This material looks like a mass of consolidated clay and has very nearly a chocolate-brown color.   It appears to be a mixture of clayey matter with almost impalpably fine sand, and is probably a volcanic mud-flow.   The order of succession, at a point where an air-shaft rises to the surface, in this mine is as follows : —

	Feet.
Cement and volcanic conglomerate	60 or 80 to surface.
Quartz gravel with fine scales of gold	7
"Cement," proper	40
Quartz gravel, auriferous	6
"Gray Cement"	30
"Chocolate"	4 – 5
Gravel, auriferous, six to eight feet in some places, averaging	1½
Bed-rock	

At the Rattlesnake Mine, near the Reed Mine, and near Deadwood, the gravel in the middle channel ranges from fifteen to twenty feet deep, and is similar in character to that in the Reed Mine, only it is much deeper, and contains large quantities of detached and more or less rounded fragments of bed-rock.   Many of the boulders in this gravel are of hard blue sandstone, and some are of quartz, weighing, occasionally, as much as two or three tons.   Above the gravel is a thin stratum of sand; and over this, again, the same "chocolate" as at the Reed Mine, resting horizontally on it. On the rim-rock the gravel disappears and the "chocolate" rests directly on the bed-rock. In both the Reed and the Rattlesnake claims the "chocolate" shows no indication of bedding or lamination.

In the Basin Channel, at the Devil's Basin, the average thickness of the gravel on the bed-rock was about two and a half feet, though it sometimes reached five or six feet. Immediately above the gravel was a stratum of the "chocolate" volcanic mud about two feet thick, and above this a bed of "gray cement" similar to that in the Reed Mine, which at one point where they raised up through it was found to be twenty-five feet thick, and to be overlain by another stratum of auriferous gravel of unknown thickness, resembling in character that upon the bed-rock itself, excepting that the boulders it contained were smaller and the gold was finer. Silas Griffett's Claim is located on the trail at the Devil's Basin, at the point where he believes the Basin Channel enters the Deadwood Ridge. The gravel in this claim is of about the same character as that in the Basin Channel, but is some ten or twelve feet thick on the bed-rock in the hydraulic face, and is overlain by about fifteen feet of exceedingly fine sand mixed with some clayey matter, and very thinly bedded in horizontal layers.

At Hornby's Tunnel, near Deadwood, the gravel runs all the way from a few inches to two feet in thickness. Over it lies the "chocolate," and above this as in the Reed Mine lies the "gray cement," which here, however, contains numerous smooth rounded volcanic pebbles and boulders, some of which will weigh over 200 pounds; these are usually very fine-grained, hard and tough.

At Weske's Claim, near Michigan Bluff, the layer of gravel is very thin, averaging probably not over a foot, although deeper in the depressions of the bed-rock. The quantity of quartz in the gravel is not large. It consists chiefly of metamorphic pebbles, with a few volcanic ones intermingled. Immediately above the gravel comes a heavy mass of "gray cement," similar in character to that of Reed's Mine, near Deadwood; the thickness of this bed of "cement" is unknown. In many places the gravel "shuts out" entirely, and the cement lies directly on the bed-rock. Yet even in such places the bed-rock sometimes pays; and, when there is only a little sand between the cement and the bed-rock, it is frequently rich. In Van Emmon's hydraulic claim, at Michigan Bluff, the gravel is from twelve to fifteen feet thick. It is overlain by a bed of sand, which at the point exposed when the locality was examined by Mr. Goodyear was about three feet thick. Above this there was another streak of gravel, of about a foot in thickness, and above that a material which they here call "cement," though it probably contains nothing volcanic, and consists of a very tenacious clay mixed with some sand, and containing a good deal of very fine gravel. The gravel in general is moderately hard and the majority of the pebbles and boulders of which it is made up are of quartz, though with these occur many of metamorphic rock. At the Specimen Claim, in Byrd's Valley, the gravel is very largely quartz, and many of the quartz boulders are of great size, weighing several tons; these are considerably water-worn, although not perfectly rounded.

At A. Bowen's Tunnel, near Michigan Bluff, the gravel averages a foot or two in thickness in the middle of the channel, and on the rims it runs out entirely. Over the gravel is, first, a mass of "chocolate" from one to four feet in thickness; above this the "gray cement," similar in character to that at the Reed Mine, near Deadwood. The chocolate also runs out on the rim-rock, and the gray cement closes down on the bed-rock. This gray cement is supposed to be from forty to fifty feet in thickness; the chocolate contains leaves of deciduous and coniferous trees in tolerably good preservation.

At Ayer's Claim, near Michigan Bluff, the gravel is well washed, but consists chiefly of volcanic boulders, which range from a few pounds to a ton in weight. At El Dorado Hill, near Michigan Bluff, the maximum height of the hydraulic bank is about 100 feet. There is very little quartz in the gravel here, the boulders being chiefly volcanic and not very large.

At the Dam and El Dorado claims, in the West Fork of El Dorado Cañon, between five and six miles north of Michigan Bluff, the gravel varies from nothing to seven or eight feet in thick ness; it thins out rapidly in going up on the rim to the west, and also in the opposite direction towards the deeper portion of the channel, which has not been worked so as to expose the rim-rock on that side. The gravel is chiefly made up of boulders of metamorphic rock, which are remarkably uniform in size, few of them being over 100 or 150 pounds in weight. Immediately over this gravel lies a heavy mass of the material called "gray cement" at Deadwood. It contains occasional large and hard boulders of volcanic material, and many small ones.

At the Paragon Mine, at Bath, or Sarahsville, near Forest Hill, a tunnel has been run first 2,200 feet N. 46½° W., then 1,250 feet N. 38° W., then 400 feet N. 7° W., all magnetic, making 3,850 feet in all, to a point in the gray mottled volcanic cement which here cuts off the gravel as elsewhere described. It is entirely in the gravel, except the last 100 feet, which is in the volcanic cement, and the distance down to the bed-rock is not accurately known, although it is probably between 75 and 100 feet. In this mine they are following a "pay-streak," which extends nearly horizontally through the gravel and varies from one to six or seven feet, but averages about three feet, in thickness; this they call the "lead," and the men working it can distinguish its limits at top and bottom, although the difference between the pay-streak and the material in which it is enclosed is so slight that it is hardly perceptible to one not trained to observe it. The "lead" appears to be a little less sandy, and of a rather more reddish color, and harder than the gravel above and below it. This gravel is made up chiefly of pebbles and boulders of hard metamorphic rock, with but little quartz and no perceptible volcanic rock. Below the "lead" is what the miners call "blue gravel"; it is grayish in color, and has somewhat more of a bluish tinge than the "lead" itself. Immediately above the paying stratum there is usually more or less gravel, the boulders and pebbles of which are perfectly similar to those of the "lead"; but whose finer portions consist of a gray granitic-looking sand, with much mica, and which appears to be of volcanic origin. Above this comes a heavy mass of what is now a soft rock, made up of a consolidated mottled gray and white sand, precisely like that in the gravel above the "lead," and containing much mica, in scales, and quartz in grains. This material is unquestionably a bed of volcanic ash, and belongs to the class of the so-called "white lavas" so common around Placerville and elsewhere; though its whole appearance is strongly suggestive of a granitic sand which has become consolidated and then partially decomposed, especially as it also has mixed with it considerable clay, apparently the result of the decomposition of feldspathic material. A vertical air-shaft was raised in this mine to the surface, and this furnished the following section. First 27 feet of gravel, including the "lead"; then 107 feet of the above-described gray volcanic sandy material; then 7 feet of quartz gravel, and above this to the surface the ordinary bouldery volcanic cement, 274 feet in thickness. The average breadth of the strip worked out in this mine is between 200 and 300 feet, and the "lead" has for a distance of 2,200 feet from the mouth of the tunnel a gently descending slope to the northwest. But at this point the grade changes, and from thence in to the end of the tunnel the "lead" has a gently ascending grade towards the northwest; in the drifts running northeast from the tunnel, however, there is always an ascending grade. In fact the form of this stratum is precisely that of a very wide and shallow V-shaped trough whose breadth is some 4,000 feet, and whose axis, running northeast and southwest, has a gently descending grade towards the southwest.

In the New Jersey Mine, at Forest Hill, the gravel on the bed-rock is generally but a few feet in thickness, varying from nothing up to seven or eight feet. The pebbles are very largely, although not exclusively, quartzose; those which are not quartz are of metamorphic rock, and there are none of volcanic origin. Immediately above this gravel there is, over large areas in the mine, a layer, varying from an inch to a foot or more in thickness, of a rather hard and pure white clayey material, containing some gritty particles. Over this again comes a heavy stratum of so-called "cement," which is of a very light granite-gray color, somewhat sandy, full of scales of mica, and containing some clayey matter. This cement is very similar to that which overlies the gravel at the Paragon Mine, but somewhat harder and generally of rather finer texture than that. In many places on the higher bed-rock this cement closes down on to the rock itself, the gravel being wanting. But even here the soft bed-rock is said to be rich and to have paid, over much of the area thus situated, almost as well as where there was gravel. There are still to be seen rich streaks in the gravel, "*lousy with gold*," to use the miners' phrase. The Back Channel* is believed to be rich, having been sufficiently prospected to prove this. The gravel appears to be deepest in the deeper portions of the channel; but it is not deep enough to cover the high bed-rock, nor does its upper surface lie level; but, to a considerable extent, it conforms to that of the bed-rock itself, especially in the larger channels. Numerous quartz boulders have been found here rich in gold, and some of them were remarkable. For instance, it is stated that one boulder weighing only twenty-nine pounds contained $3,700 in gold, which metal made up more than one third of its weight.

In the region lying between the heads of the three cañons, known as First, Second, and Third Brushy, directly north of Forest Hill, and from one to three miles distant from that place, there is a large amount of gravel. Some of this is uncovered; other portions have a deep covering of volcanic material. The gravel varies much in thickness, ranging from a few feet to as much, perhaps, as three hundred. A peculiar feature of this region is, the great depth to which much of the volcanic conglomerate has been decomposed. This decomposition has developed in a remarkable degree the concentric structure of the volcanic boulders. These are banks in which, almost without exception, every boulder will exhibit this structure, with sometimes a hard kernel in the centre, and sometimes without one. Another feature of this district is, the frequency of streaks or beds of auriferous gravel overlying the volcanic cement. Distinct channels can with difficulty be traced, in the region in question, on account of the wide diffusion of the gravel, the deep covering of volcanic material with which it is overlain, and the very limited extent to which the bed-rock itself has been exposed to view.

At Smith's Point, between First and Second Brushy cañons, in the banks exposed by the hydraulic operations, the gravel ranges from fifty to sixty feet in thickness. It is made up of pebbles and boulders of metamorphic rock chiefly, with some of quartz; they are generally under twenty-five pounds in weight. The gravel is in places more or less interstratified with beds and streaks of sand, which are not continuous for any great distance, and which are not exactly horizontal, but slightly inclined in various directions. Above the gravel lie alternating beds of ash, more or less clayey, and volcanic conglomerate, occasionally containing very large boulders. Farther back on the crest of the ridge the regular volcanic conglomerate occurs in heavy masses.

At Yankee Jim's the gravel is composed of all sorts of metamorphic rocks, and few boulders are met with which are too large to wash through the sluices with a good head of water. The proportion of quartz is small. Over the central axis of the Big Channel the gravel averaged, for most of the distance, over a hundred feet in thickness; and the average over the whole ground washed was, probably, forty feet. Over almost the whole of this ground the gravel was uncovered; but at the east end it is overlain by the volcanic cement, which here also shuts out a considerable portion of the thickness of the gravel.

At Georgia Hill on the south side of the Devil's Cañon, a little below Yankee Jim's, an area has been washed off 600 or 700 feet long by an average of 100 wide; in the western half of this ground the gravel was uncovered and reached a depth of 100 feet or more, but the eastern half was covered with volcanic cement which in going easterly from the middle of the ground worked gradually descends, and nearly shuts out the gravel.

At the Oro and Dardanelles claims, near Todd's Valley, the lower portions of the banks consist of blue gravel. Above this in the Oro and the eastern part of the Dardanelles is red gravel to the top; but in the central and southwestern portion of the last-mentioned claim the upper heavy bank of red gravel is overlain by a mass of volcanic cement, whose maximum thickness, so far as it is exposed, is not far from 100 feet. The gravel is generally very hard; and the banks, even where they have not been touched for five years, are usually very smooth, and often very nearly

* See page 106 for a description of the so-called Back Channel.

vertical from top to bottom. The gravel is a mixture of all sorts of metamorphic rocks, with but little quartz, and few large boulders. Many of the pebbles soften on exposure to the air.

At Todd's Valley the gravel ranged from a few feet to thirty or forty in thickness, and nine tenths of the pebbles and boulders of which it is made up are of quartz. They are not so much water-worn as is usually the case in deep gravel; indeed, the deposit has a more or less brecciated appearance, and there are streaks of the angular gravel so firmly cemented together by oxide of iron, that it will break through the centres of the quartz pebbles rather than crumble. Over the gravel lies a stratum from twenty to twenty-five feet in thickness of bluish-gray sand, and above this some thirty or forty feet of volcanic cement. This cement is soft and breaks into lumps, which pass easily through the sluices. The proprietor thinks that there is gold enough in the gravel to pay for handling 150 feet in thickness of this cement, if water were free.

At the Reed Claim, on the north side of Mameluke Hill, near Georgetown, the banks have a maximum height of fifty or sixty feet. On the bed-rock there was a thin layer of gravel ranging from a few inches to two or three feet in thickness. This has been all drifted out, and is said to have been extremely rich. At this point the surface of the bed-rock pitches southerly into the hill, and the southwest slope in the front of the claim was very rich, and is reported to have paid in spots as much as a thousand dollars per hour. From the area covered by a 4 × 4 foot shaft the sum of $7,000 was once taken. The gravel contains considerable quartz, which appears in general to have been but little washed. Immediately above the gravel comes a stratum of soft, gray, decomposed, volcanic bouldery cement, varying from ten to fifty feet in thickness. Over this there is another streak of quartz gravel, with small pebbles generally but little washed; this stratum is from one to four feet thick and contains a little fine gold. Above this there are a few feet of irregularly alternating layers of gravel, volcanic ash, and red dirt, more or less mixed together, and forming the top surface.

In the Roanoke Tunnel, about a quarter of a mile east of Bottle Hill, the channel is said to lie some 300 to 350 feet below the top of the crest of the ridge. The gravel is reported as not having averaged over two feet in depth, and as being covered with very hard volcanic bouldery cement. An upper stratum of gravel was found here, about four feet in thickness, which contained some fine scale gold, and ran nearly level from the upper bed-rock, or edge of the rim, through the cement over the deeper channel. The bed-rock is generally hard and dark-bluish in color; and the gravel consists of pebbles and boulders of dark-colored hard metamorphic rocks, the quantity of quartz being small.

At the Oak Grove Claim, on the north side of the ridge, a mile and a half east of Volcanoville, several tunnels have been driven, and irregular works extended from there for some distance into the hill. The gravel here consists of metamorphic materials, which are more than usually mixed with volcanic pebbles and boulders. The gravel is very hard in places; and the gold, which is mostly near the bed-rock, is not coarse.

At Flora's Mine, two miles west of Volcanoville, the bank exposed in the hydraulic workings is about 130 feet high, of which the lowest twenty-five feet consists of well-washed metamorphic gravel, which near the bed-rock is generally very hard, but higher up somewhat softer. The next twenty-five feet is a bed of volcanic ash cement, red in color, containing no large boulders, and rarely a pebble of any size, but full of little smooth rounded pellets of volcanic rock, which were once very hard but now considerably softened by decomposition. Over this, the whole bank, to the top, consists of well-worn, bouldery gravel, containing a little quartz and considerable metamorphic rock, in the form of boulders, as well as a little gold. This deposit, however, is more than half made up of volcanic materials.

At the Grizzly Flat Mine, half a mile a little east of south from Volcanoville, the character of the gravel is the same as that at Mameluke Hill, it being made up of small fragments of quartz, but little washed. The finer portion of it consists mostly of clayey matter, resulting from the disintegration of the bed-rock. This material appears in the hydraulic face to be some thirty to thirty-five feet deep, and above it lies red volcanic cement, containing a few hard boulders.

At the Buckeye-Sucker Claim, on the southeast side of the hill in Spring Tunnel Ravine, one third of a mile southeast of Mount Gregory, a hydraulic pit has been opened, exposing a bank of sixty to seventy feet in height. Of this, the first twenty or twenty-five feet above the bed-rock is a mass of angular fragments of the bed-rock mixed with earth, and containing but little gold. Over this comes a stratum of quartz gravel, from twelve to fifteen feet thick. This gravel is well washed, and contains many large quartz boulders. It resembles in general appearance the lower gravel at Michigan Bluff, the boulders and pebbles which it contains being almost without exception quartz.

At the Shoo Fly Claim, in Missouri Cañon, the gravel is generally some four or five feet thick on the soft bed-rock. It is of a dark-brownish color, and is made up of all sorts of slate and metamorphic rocks, with some quartz, and is not much rounded by water. The gold is generally coarse, heavy, and well-worn, and the rim is said to have paid very well. Over the gravel is gray volcanic cement.

At Kentucky Flat, two miles southeast of Mount Gregory, there is a bank exposed in the hydraulic washings, from twenty-five to thirty feet in height, and which is entirely of gravel. Nine tenths of the boulders in this gravel are quartz, like those at Michigan Bluff; some of them are very large, weighing as much as twenty-five or thirty tons. The deep bed of quartz gravel paid four dollars a day per hand, over all expenses, in coarse, heavy gold, which was smooth and well washed; the largest piece found weighed $94. Similar gravel is said to show in the bank on the opposite side of Otter Creek at a point perhaps half a mile distant, in a direction S. 35° — 40° E. (magnetic) from Kentucky Flat.

At West and Foster's Mine, about two miles above Auburn, a tunnel has been driven some 280 feet under an isolated gravel hill, in which a channel is said to run about in the same direction as the bed-rock, or southerly. The hill is about a hundred feet higher than the level of the tunnel. The gravel is all of metamorphic material, with but very little quartz. All the boulders and pebbles in the gravel are well washed and quite smooth. The hill is probably capped, at least in part, with volcanic cement. Above West and Foster's there are said to be scattered gravel hills all the way to Colfax, although the bed-rock is generally at the surface.

In Todd's Valley Ravine, near Peckham Hill, a shaft has been sunk between the rims of the Big Channel, to the depth of 103 feet, without reaching the bed-rock. After sinking forty feet through the volcanic material, a streak of gravel was passed through, three or four feet in thickness, and which is said to have prospected well. At ninety feet the top of another gravel bed was struck, which proved to be six feet thick. Below this was a stratum of exceedingly fine sedimentary matter, about seven feet thick, probably of volcanic origin, and below this, again, a stratum of boulders.

At Roach Hill, near Independence Hill, there is a clay bed resting directly on the bed-rock, and it is filled with impressions of leaves; this bed is from three to four feet in thickness. Above it is gravel from a hundred to a hundred and twenty-five feet thick, and over the gravel volcanic cement.

At Densmore's Claim, in Grouse Cañon, about one and a half miles a little west of south from Startown, there are generally but a few inches of gravel in the channel; this gravel is covered with volcanic cement, of a brecciated character.

At Wilcox's Claim, on the right bank of the North Fork of Long Cañon, the bed-rock is entirely granitic; beneath the gravel, however, it is soft and decomposed. The gravel is from fifteen to twenty-five feet thick, and is overlain by a light-gray volcanic ash cement, which contains occasional more or less rounded pebbles of metamorphic and other rocks. This cement forms quite a thin stratum at the Claim; but increases rapidly in thickness in going northwesterly into the ridge, where it attains a development of 200 feet or more. The channel seems to have come through the ridge in a direction of about S. 60° E. (magnetic), and is from 300 to 400 yards in width. The boulders in the gravel are in general not very large, although some attain a great size. They consist almost entirely of a very hard quartzose sandstone and of silicious slate, with some impure quartz, while those of granite are exceedingly rare. There is in the bank at this claim a great deal of opal, almost all of which is green, a part of it being of a bright emerald color. The boulders in the bank are almost entirely undecomposed, and the opal occurs in the interstices between them.

At Blacksmith Flat, on the south side of the ridge between Long Cañon and the Middle Fork of the Middle Fork of the American River, the maximum thickness of the gravel exposed in the hydraulic banks is fifteen feet. The gravel is here and there overlain by an irregular stratum of sand ranging from five to ten feet thick. Above this come several hundred feet in thickness of volcanic materials forming the crest of the ridge. The gravel is smoothly rounded and made up of a great variety of metamorphic rocks, with a large admixture of granite boulders, some of which are eight or ten tons in weight. For a distance of two or three miles above Blacksmith Flat the ditch, which runs 300 feet above the bed-rock, is cut through gravel, and there are indications at other localities that the quantity of gravel in this ridge between Long Cañon and the Middle Fork of the Middle Fork of the American River is very considerable, although at the same time the capping of volcanic materials along the central portion of the ridge is very heavy.

At Castle Hill, near Georgetown, the gravel varies, in the channel, from one or two inches to two or three feet in thickness, and contains some pretty well washed quartz pebbles, with many fragments of bed-rock. It is immediately overlain with volcanic cement, which is generally grayish in color. This volcanic capping is probably 125 feet deep on the crest of the ridge; and it contains many large boulders, which are equally plentiful towards the top or the bottom of the mass.

At Centerville, some ten or twelve miles below Georgetown, the material called gravel consists mostly of angular fragments of the bed-rock, of all sizes. Among the great variety of dioritic, hornblendic, and porphyritic rocks of which this gravel is made up, there are a few of granite, and these are well rounded. Numerous quartz boulders also occur, many of which are very large, some even weighing from eight to fifteen tons. The greater portion of these boulders, whether large or small, are but little rounded by water, although some of them are thoroughly so. One large boulder of compact white quartz found here yielded $8,000 in gold, and others have also proved valuable. The bed-rock at this place is chiefly slate, but there are dioritic rocks farther to the southwest, in the sides of Pilot Hill. The bed-rock slopes to the East and the West; and on the eastern slope there is a certain area covered with a few feet in depth of well-washed gravel, which is very firmly cemented together and very hard. This gravel is overlain by the brecciated mass ten to twenty feet thick, which is the ordinary "gravel" of the district, and which is in all probability only a local deposit, perhaps accumulated from the slopes of Pilot Hill.

The general character of Buffalo Hill, near Georgetown, is much like that of Mameluke Hill; the crest of the ridge runs for half a mile, or more, in a direction N. 18° W. (magnetic), between West and Illinois cañons. It is capped with volcanic bouldery cement, in places a good deal decomposed, and with a maximum depth of from seventy-five to eighty feet. There appears to be, in general, but little gravel on the bed-rock, beneath the cement; and, although numerous shafts have been sunk here, the results do not seem to have been pecuniarily satisfactory.

At Tipton Hill, two and a half miles a little west of south from Kentucky Flat, the gravel ranges from four to six feet in thickness, and is almost entirely made up of quartz, in fragments of moderate size, not much rounded. This gravel is covered with the ordinary bouldery cement, and the maximum height of the banks is about thirty-five feet.

At the Excelsior Claim, near Placerville,* the maximum height of the bank is 170 feet, of which the lower sixty or seventy feet are "pay gravel" consisting of quartz, metamorphic rocks, and sand; the upper hundred feet is made up of well-rounded and water-worn volcanic gravel. This is said also to contain a little fine gold. The line of demarcation between the pay gravel and the volcanic capping is pretty well defined, the former being yellow, and the latter bluish-gray. There are a few metamorphic pebbles in the volcanic beds; but no volcanic materials in the pay gravel. About twenty acres of ground have been washed off here, with an average thickness of something over fifty feet of "pay gravel." Along the northern side of Hangtown Hill, just south of Placer-

---

* For the position of the localities and the claims about Placerville, see diagram, Plate C.

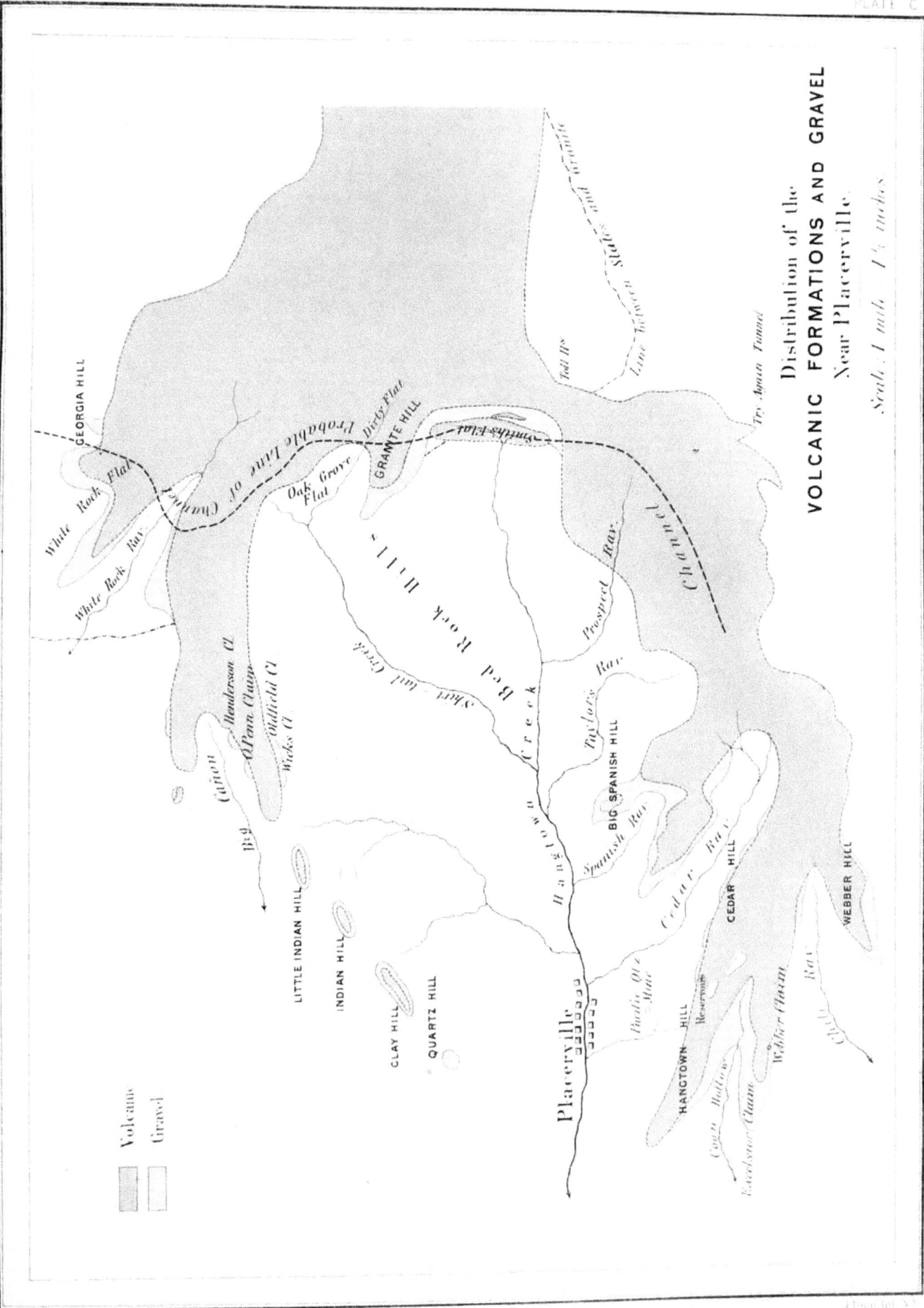

PLATE C

Distribution of the

VOLCANIC FORMATIONS AND GRAVEL
Near Placerville.

Scale: 1 mile = 1¼ inches

Volcanic

Gravel

GEORGIA HILL

White Rock Flat

White Rock Rav.

Probable line of Channel

Oak Grove Flat

Dirty Flat

GRANITE HILL

Smith's Flat

Line between Slates and Granite

Toll W.

Tr. Aqua Tunnel

Channel

Prospect Rav.

Bed Rock Hills

Shenland Creek

Hangtown Creek

Taylors Rav.

BIG SPANISH HILL

Spanish Rav.

Cedar Rav.

CEDAR HILL

WEBBER HILL

Cañon

Henderson Cl.

Penn Claim

Coldfield (?)

Wicks (?)

By.

LITTLE INDIAN HILL

INDIAN HILL

CLAY HILL

QUARTZ HILL

Placerville

Public Old Mine

HANGTOWN HILL

Reservoir

Coon Hollow

Webber Ravine

Rav.

Rav.

Bucksner Claim

ville, more than nine tenths of the whole mass of the gravel is of volcanic origin, and all the pebbles are very smoothly rounded.

At the Webber Claim, about a thousand feet east of the Excelsior, and on the south side of Coon Hollow Ridge, where an acre or more has been washed off, the bank is a little over a hundred feet in height. The upper twenty-five or thirty feet are of the so-called "black lava," and beneath are sixty or seventy feet of smoothly rounded "mountain gravel." The metamorphic gravel lies on the bed-rock, and varies from two or three to twenty feet in thickness, according to the inequalities of the surface of the bed-rock. The mountain gravel is said to contain gold enough to pay all expenses of wages and water. At Webber Hill the maximum thickness of the gravel is forty or fifty feet. It is overlain by white lava, which, in the eastern part of the hill, is from seventy-five to one hundred feet thick, and shows a decided tendency to assume columnar forms. Above this is "mountain gravel," forming the crest of Webber Hill.

At the Confidence Mine, 300 or 400 feet north of the flume at the head of Cedar Ravine, the slope goes down with an inclination of five feet in twelve and is 300 feet long to the bed-rock. On the bed-rock is a stratum of gravel from five to eight feet in thickness, overlain by the "white lava," which extends all the way to the surface. In Cedar Ravine, near Dickerhoff's Mill, the white lava shows something of a tendency to a columnar or prismatic structure.

At Dickerhoff's Mine in Cedar Hill the pay-gravel in the channel on the bed-rock is from three to four feet in thickness, and is made up almost entirely of pebbles of metamorphic rock, with much quartz, a good deal of which is but little rounded; it is immediately overlain by a body of exceedingly fine-grained, compact material, called by the miners "lava," and which, in fact, appears to be of volcanic origin. It is believed to be of great thickness, perhaps from 100 to 200 feet, and it is capped by the "black lava," which is also fully a hundred feet in thickness, in all probability. Just east of Dickerhoff's Mill in the north side of Cedar Hill, the hydraulic banks are from fifty to sixty feet high, the lower few feet only being a true gravel, and all the upper part of the banks consisting of sands and clays. In the hill immediately back of the banks these finer sediments are also overlain first by eight or ten feet of "white lava," then by a body of "mountain gravel," and finally the latter is capped with a mass of "black lava."

At the Hook and Ladder Claim in Big Spanish Hill, about one mile nearly east of Placerville, the gravel on the bed-rock averages from four to six feet in depth, and is covered by from fifteen to twenty-five feet of sand, above which come heavy beds of still finer sediment, containing occasional thin streaks of fine gravel; and over all a few feet of volcanic cement, the whole height of the bank being about a hundred feet.

On the north side of Little Spanish Hill, about the head of Spanish Ravine, the gravel has a total thickness of sixty to seventy feet. At the height of twenty-five or thirty feet above the bed-rock it is traversed by a stratum of mottled white and rose-colored "pipe-clay" six or eight feet thick. Below this "pipe-clay" is a mass of fine sand with some clay but no pebbles. But above it is pebbly gravel, which shades off again at the top into fine sand and clay, immediately beneath the final capping of "black lava."

The section of the bank, at this point, is as follows :—

	Feet.
Volcanic breccia	10 – 12
Very fine sand, irregularly bedded	4 – 6
Streak of white pipe-clay	0 – 1
Fine gray gravel and sand	8 – 10
"Pipe-clay," pinkish and rather hard	8 – 12
Sand, reddish below and bluish at top	15 – 20
Gravel, rather fine, with some sand	10 – 15
Total height of bank	75 to 80 feet.

There are places in Little Spanish Hill, where the volcanic capping is seventy-five feet thick.

The order of superposition of the different materials in the banks above the bed-rock, in Hangtown, Cedar and Spanish hills, near Placerville, appears to be as follows, beginning from the bed-rock :

1. Gravel proper, which is generally but a few feet in thickness.

2. Yellow sand, of varying thickness, generally rather fine, and free from pebbles, or nearly so. It has, occasionally, streaks of fine gravel running through it. Its thickness is sometimes as much as thirty or forty feet.

3. A mass of finer materials, consisting of extremely fine sand, with a large proportion of clayey matter. This is often very closely compacted, and is of a mottled yellow and white color. It is called "pipe-clay" by the miners ; thickness, up to twelve feet.

4. The so-called "white lava," a material which appears to be a consolidated and slightly silicified, or metamorphosed, volcanic ash. It is easily dressed by the hammer, and is in general use as a building stone, for which it is, in most respects, well adapted, although rather too easily crushed. It occasionally though rarely contains small and sometimes angular, but sometimes smoothly worn fragments of quartz and metamorphic rock.

5. The "mountain gravel," which is a well-washed or water-worn gravel chiefly of volcanic materials, although containing a few metamorphic and granitoid pebbles. This gravel is usually slightly auriferous in the neighborhood of Placerville, and is locally known as "mountain gravel." This with its sandy streaks is a perfect fac-simile, so far as structure is concerned, of the ordinary metamorphic gravel. There are occasional boulders of white lava all through the mountain gravel, along the north side of Hangtown Hill ; and there seems to be a much greater variety of texture among the volcanic rocks in the mountain gravel than there is in the black lava. This is very natural, since the former shows by its water-worn character that it has been transported from a great distance, and it would naturally contain a greater variety of materials. Probably fifty per cent of the whole mass of the mountain gravel consists of pebbles weighing less than five pounds apiece ; and there are comparatively few boulders in it too large to go through the sluices. All through the Placerville district this gravel is not so solidly compacted together that it cannot be "piped."

6. The "black lava," which is generally a strongly compacted volcanic breccia, containing, however, occasional pebbles and boulders of granitic and metamorphic rocks.

It is often the case that one or more of these layers is wanting at particular localities. But where they are all present, the above is the order of their arrangement.

In the Robinson Mine, on Prospect Flat, near the head of the ravine of that name, southwest of Smith's Flat, and east of Placerville, a pretty well marked channel, running about S. 42° W. has been worked for several hundred feet, by means of a shaft 100 feet deep to the bed-rock, through "white lava," which has several small streaks of gravel intercalated between its successive beds. The bed-rock gravel, at the shaft, is thirteen feet thick, and the mine pays well.

In one place at White Rock Point, northeast of Placerville, the banks exhibit a thin stratum of gravel lying on the bed-rock, and covered by twenty or thirty feet of compact "white lava," above which comes a stratum of fine gravel three or four feet thick, consisting entirely of metamorphic materials ; this last is itself overlain by a bed of horizontally and very delicately and thinly bedded clay, from five to eight feet in thickness, while over all comes the "mountain gravel." A short distance east of here, however, the "white lava" gradually thins out wedge-shaped towards the east, while the underlying gravel grows thicker till finally the "white lava" entirely disappears and the two streaks of gravel unite in one. At one locality near here also, the "white lava" was seen in mass resting immediately on the surface of the soft and decomposed granite bed-rock with no gravel or anything else between.

At Negro Hill, near Placerville, in the Oldfield Claim, where two or three acres have been worked, they have a bank of sixty to seventy feet in height. The gravel on the bed-rock is in general not very smoothly water-worn, and it ranges in thickness from twenty-five to thirty feet. It is immediately overlain by the "black lava," which is from fifteen to twenty-five feet thick, and extremely hard, so that it breaks up into enormous blocks, many hundreds of tons in weight, thus adding much to the expense of working the claim. One of these blocks was found, on measurement, to be fifty-five feet in length, and to be equal in dimensions to a cube of thirty feet, and was estimated to weigh 2,000 tons.

At the Henderson Claim, on the north side of Negro Hill, a little northeast of the Oldfield, there have been two or three acres washed off, the maximum height of the bank being 115 to 120 feet. The amount of metamorphic gravel here is very small, and it forms a stratum of a few feet in thickness upon the bed-rock. The great mass of the bank consists of beds of volcanic sands and gravels. These are beautifully stratified in horizontal layers, and much of the material is very thinly and delicately bedded. This ground is said to be very rich in gold.

The maximum depth of the gravel, including the "mountain gravel," at Indian Hill, is about sixty or seventy feet; and there are five or six acres over which the average depth is twenty feet. On the top is a thin capping of "black lava," perhaps from three to four feet in thickness, — not enough to give any serious trouble in hydraulicking.

At the southwest end of Clay Hill, a little southwest of Indian Hill, there has been a quarter of an acre of ground washed off, exposing a bank twenty-five to thirty feet high, of which the lower five or six feet consist of quartz and metamorphic gravel, generally rather fine: all the upper portion is a coarse "mountain gravel," occasionally containing very large and very smoothly rounded boulders; one of them was found to be not less than ten feet in diameter and nearly spherical in shape.

At the west end of Indian Hill, a little to the northeast of Clay Hill, a considerable extent of ground has been worked, showing the face of a bank from twenty to thirty feet high, for a couple of hundred feet in length. Only a thin layer on the bed-rock here could have been of metamorphic material, for the whole face now visible is volcanic. The lower portion consists, to a great extent, of gray volcanic sand, with occasional thin layers of fine "mountain gravel." The upper portion of the bank consists entirely of coarser mountain gravel; and the structure of the bank shows, that after the sand had been deposited, it was again here and there channelled out to a greater or less extent, before the mountain gravel was laid in heavier masses over it.

At the Sugar Loaf, a hill on the south side of Webber Creek and about a mile below Diamond Springs, an area about 800 feet long and 200 wide was first drifted out, and has since then been washed off, so as to have removed the whole original top of the hill. The gravel which averaged about fifty feet in thickness, and was not capped at all with volcanic materials, was metamorphic, well-washed, and contained a good many quartz pebbles, but no large boulders, stones of more than a hundred pounds' weight being rare. Most of the pebbles which were not of quartz were thoroughly decomposed, and the mass of tailings slacks quickly on exposure to the air. The total yield of this deposit was, it is said, not far from $ 3,000,000.

On the east side of Bean Hill, just northeast of the Sugar Loaf, at Diamond Springs, the bank exposed is about 1,200 feet long, and ranges from twenty to forty feet in height. About four or five acres of ground have been washed off here, with an average depth of about twenty feet. The gravel is mainly metamorphic, yet containing a good many boulders of "white lava" and occasionally of other volcanic materials. Nearly all the pebbles, except those of quartz, are much decomposed, so that the bank washes very easily.

In the Deadhead Claim, three quarters of a mile below Newtown, on the south side of Webber Creek, an area has been washed off estimated at about 800 feet long by 200 feet wide, with an average depth of 30 feet. The gravel here is chiefly metamorphic, but contains a good many volcanic boulders, a large proportion of which are "white lava," scattered throughout its mass, and also many boulders of a rock of granitoid texture, but consisting of quartz and feldspar without any appreciable quantity of mica.

At Brownsville, on the North Cedar Creek, the gravel in the front of the claims is said to have been seventy feet thick, at least, above the bottom of the workings, the bed-rock not being reached. The gravel grew thinner as they went into the hill, its upper surface sloping to the south. It was almost exclusively a quartz gravel, the pebbles being rarely very large.

In the Sugar Loaf Hill, at the north end of a short ridge or spur between Clapboard and Indian gulches, near Volcano, there has been considerable hydraulicking, exposing banks from seventy-five to eighty feet in height. This hill is capped with a thickness of from fifteen to twenty feet of black lava, which over the central portion of the hill lies nearly horizontal; but, at the sides, bends downwards, reaching thus sometimes thirty or forty feet lower. This bending, however, is not smooth and regular, but of such a character as to prove that there has been considerable sliding of the banks at some time in the past. So irregular is the surface of the limestone, that the thickness of the gravel cannot be made out; and there are large masses of decomposed slates imbedded in it, the whole having been crushed together by the slipping and sliding which has here taken place.

In the W. T. Jamison Mine, near Fiddletown, a tunnel was driven about 630 feet, in a southerly direction, under Loafer Hill. No well-defined channel seems to exist here, the broad, gently-undulating surface of the soft slate bed-rock being almost everywhere covered with from two to fifteen feet of gravel which is immediately overlain by "white lava." The gravel varies a good deal in color, is not very hard, and consists almost entirely of metamorphic material, though containing occasional boulders of "white lava." It also contains a good deal of clayey matter and furnishes very smooth casts of its pebbles. This gravel has paid well for drifting. The general surface of the bed-rock under it is said to have a gentle fall everywhere towards the southwest. The whole mass of Loafer Hill is "white lava."

At Michigan Bar the gravel is all metamorphic, and its pebbles are small. The gold also is generally fine and smoothly worn. The area of gravel here is estimated at three eighths of a square mile, with an average depth of 15 to 20 feet.

At Forest Home the gravel is generally pretty well washed, and many of the hills are covered with it, to depths ranging from ten to thirty feet; it contains no large boulders.

The gravel in the hills at Muletown ranges from a few feet to seventy-five, or possibly a hundred, feet in maximum depth. It is a ferruginous quartz gravel, but little washed, and containing very few, if any, large boulders. Nine tenths of the pebbles it contains are of quartz, and there are none of volcanic origin. These gravel beds form the southwestern termination of a series of spurs of the chaparral hills, and rise two or three hundred feet above the general level of the country immediately to the southwest. In the different spurs, immediately to the northeast of the gravel, the crests of the hills continue on perfectly smooth, but at the edge of the gravel, the bed-rock, consisting of hard clay-slates, rises abruptly to the surface and forms the crests continuously in that direction. The whole structure of the banks here is full of evidence of the action of shifting currents, on a small scale, and probably of no great force. None of the gravel is thoroughly water-worn; it consists chiefly of small angular fragments of quartz. In the soft bed-rock underlying this gravel just such quartz occurs in little crushed seams. At Irish Hill, two or three miles northwest of Muletown, the gravel is quite different in character, consisting chiefly of metamorphic slaty rock, with a good deal of dark-colored quartzose material, and occasional volcanic pebbles, but hardly any white quartz. The pebbles are more water-worn than those of the Muletown gravel.

Near Irish Hill the greatest part of the work has been done in a mass of rolling, chamisal-covered hills, scattered over a quarter of a section of ground, in which some eight or ten acres have been washed away, with an average depth of gravel of perhaps twenty feet. The surface is irregularly cut up by gulches, and the pits are equally irregular in their distribution. The maximum depth of the gravel is perhaps forty or fifty feet. The whole character of the gravel at Irish Hill is so different from that of Muletown, that it appears quite impossible that these deposits should have been formed by the same stream. The gravel of Irish Hill would appear to have been deposited by a stream draining nearly the same area of country as that now drained by Dry Creek.

At Tunnel Hill, near Jackson, an area has been washed off estimated at 1320′ × 440′ to an average depth of ten or twelve feet. The gravel is red, with plenty of moderate-sized boulders, but few large ones. The smaller pebbles are nearly all quartz, and but partially rounded; some of the quartz is jaspery. There are no volcanic pebbles.

§ 5. *The Volcanic Capping of the Gravel ; its Thickness and General Character.*

On the road, a short distance above Independence Hill, may be seen a small outcrop of "white lava," which material is very rare in this vicinity. It underlies the bouldery cement which forms the crest of the ridge. Half a mile northwest of Independence Hill the capping of volcanic débris, on the crest of the ridge, is probably 400 feet thick. On the northern slope of the first Sugar Loaf Hill, just southwest of Iowa Hill, a tunnel has been driven, at a point northwest of the Colfax road, N. 15° W. into the gravel. The crest of this hill is covered with an aggregation of earthy matter with fragments and boulders of all sizes, up to many tons in weight, and only partially weather or water worn. The thickness of this capping is uncertain ; it probably does not exceed one hundred feet, and may be less than fifty.

There are in the Lebanon Tunnel, on the northeast side of New York Cañon, many very large boulders of very hard and compact volcanic rock, thoroughly smoothed and rounded by water ; as much so, indeed, as any of the pebbles in the ordinary metamorphic and quartzose gravel.

In the ridge between Iowa Hill and Damascus the volcanic conglomerate is frequently overlain by heavy masses of breccia, the total thickness of volcanic matter often ranging from 300 to 500 feet. The bed-rock at the Hog's Bank, between Damascus and Secret Hill, is said to be from 600 to 800 feet below the crest of the ridge ; or, in other words, that is the supposed thickness, in that region, of the deposits of detrital and volcanic material.

On either side of Sailor's Cañon, four or five miles east of Canada Hill, and nearly parallel with it, are two small cañons, and the basin to which they belong is approximately semi-circular in form and has a radius of three or four miles. The whole of the ridge around this basin is capped with volcanic débris of all sorts, beds of ash alternating with masses of well-rounded and water-worn conglomerate, but with no solid lava. The depth of this capping seems to vary from a few hundred to a thousand feet in thickness. From here a sharp high peak, on the crest of the Sierra, bears N. 37° E. (magnetic), and from this peak around to the north the whole summit as far as visible, that is through an arc of some 15° or 20°, consists of volcanic rocks horizontally stratified, and the depth of this deposit must be very great, probably from 1,500 to 2,000 feet.

North of Deadwood, along the western brow of the ridge fronting the East Fork of El Dorado Cañon, the gravel, which is a few feet in thickness only, is overlain by volcanic deposits in horizontal strata, and which are made up of alternations of sandy and clayey materials with heavier masses of bouldery accumulations. This volcanic capping acquires a thickness of several hundred feet, as we go back towards the crest of the ridge. At the Reed Mine, near Deadwood, the whole mass of the gray cement contains large quantities of magnetic iron in fine grains ; but the white pumice-like spots which occur in the cement appear to be particularly rich in this mineral.

The volcanic "cement" occasionally contains fragments of quartz included in it, showing that it has been transported from a distance, as at Hornby's Tunnel near Deadwood, and other localities. On the crest of the ridge, just above Deadwood, two or three huge volcanic boulders were seen ; these were smoothly rounded, and one of them was estimated to weigh not less than twelve or fifteen tons.

In the sides of the ditch running from Deep Cañon to Last Chance, the volcanic formation is seen to consist chiefly of only partially rounded pebbles and boulders, it being rather a breccia than a conglomerate. The rocky fragments of which it consists are of all sizes, from small pebbles up to boulders and blocks of five or six tons, or even more in weight.

On the gravel in Jones's Hill, northwest of Georgetown, there lies a heavy stratum of micaceous volcanic sand ; then, a layer of gravel, about four feet thick, containing a little gold, and above that the ordinary bouldery volcanic cement.

Immediately north and northwest of Colfax the volcanic formations are seen in the crest of a ridge rising to 400 or 500 feet above the town. There is at least 200 feet in thickness of volcanic material here, and the crest is narrow and sharp and covered with boulders, some of which will weigh two or three tons. These boulders are rounded ; but not so much as the volcanic fragments generally are, when found resting immediately upon the bed-rock, or at no great distance from it. The course of this ridge is northeast and southwest, and a line of detached hills of similar character extends for several miles to the southwest, appearing to cross the railroad a mile or more below Colfax.

To the south of Forney's, on Pilot Creek, the country around, within a radius of two or three miles, spreads out into a broad, gently-undulating tract, where the gulches are shallow, and most of which seems to drain towards Pilot Creek, although its general surface is not very far from level. This whole area, so far as could be seen, is covered with volcanic cement. At a point not far from two miles, in a direction S. 75° E. (magnetic) from Forney's, is a high timbered peak, the summit of which is about 800 feet higher than the level of Forney's. The volcanic cement extends to the foot of this peak and a very short distance up its flank. But the top and the upper 300 or 400 feet are of bed-rock, which is here a quartzite, more or less stained with iron, and containing occasionally little seams of quartz. The law which generally holds good, in the region to the north, in the basin of the Middle Fork of the American, — that the highest crests are all capped with volcanic matter, — seems to be here reversed, the more elevated ridges being of bed-rock, and the volcanic deposits not extending above a certain zone. From this it appears probable that the present depth of the volcanic matter in the central portions of the broader crests in the basin of the Middle Fork is a fair indication of its maximum depth in the past, and that it has never been much deeper there than it now is. In that case, it can never have extended much farther up the sides of the highest projecting peaks of bed-rock than it now does ; or, at least, these peaks have never been covered by it.

Along Hangtown Hill and Cedar Hill, near Placerville, the whole crest of the ridge consists of " black lava," i. e., volcanic breccia, the thickness of which in places is perhaps a hundred feet, and which contains many fragments or boulders of enormous size, some weighing from fifteen to twenty tons. Many of these are very angular and unworn, although they are generally somewhat more rounded on the corners than is usually the case with the boulders seen on the surface of lava streams. This breccia is underlain by smoothly washed volcanic gravel, here known as " mountain gravel," and under this again is the ordinary metamorphic auriferous gravel.

In the Franklin Claim on the south side of Little Spanish Hill, as well as at Negro Hill, near Placerville and elsewhere, the " black lava " or volcanic breccia is traversed by horizontal planes of stratification, proving the occurrence of successive flows of brecciated matter over the same ground. In the Franklin Claim the lowest stratum of " black lava " is four or five feet thick. Above it is a layer of sand and fine gravel a foot to eighteen inches thick, said to contain some fine gold, and over this again are several successive layers of " black lava." Along the southern side of Little Spanish Hill the gravel is generally immediately overlain by " black lava." There is no " mountain gravel " in this hill.

At Smith's Flat, about three miles east of Placerville, the " white lava " occupies most of the surface. The bed ranges from twenty to thirty feet or more in thickness. The gravel beneath it is worked by inclines. At thirty feet below the bottom of the white lava, the gravel still contains boulders of this rock intermingled with the other materials. These boulders must have come from some considerably older deposit of the same kind of rock higher up in the mountains.

Near the Toll House, about one and a half miles N. 8° E. from the Try Again Tunnel, there is a high bluff of the " white lava," in which rude columnar forms are well developed ; the top of this bluff is not less than 200 feet above the Toll House, and the bed-rock at the house is buried beneath some sixty feet of the same material. In the hill back of the face of the bluff, therefore, there cannot be less than 250 feet in thickness of this material. It is overlain, on the narrow crest of the spur immediately east of Smith's Flat, by a shallow bed about half-way in character between the " black lava " and the " mountain gravel." The " white lava " here contains a good many small cavities filled with a substance which has much resemblance to little fragments of fossilized wood. The material does not show in its internal structure, at this point, any distinctly horizontal bedding ; but the weathered faces of the columns indicate it by a corrugation of their surfaces, the little furrows and ridges produced by weathering being often from half an inch to an inch deep, and running horizontally across the nearly vertical faces of the columns. Many of the columnar surfaces are, however, not corrugated; and some of them are as smooth as if polished; the external crust, in such cases, having its texture entirely changed to a depth varying from an eighth of an inch to three quarters of an inch in thickness. This change of texture, which on the whole seems most likely to have been produced simply by the chemical action of water, is one from the ordinary, sharp, granular condition of volcanic ash, to a peculiar compact, and almost opaline, or semi-vitreous state. The rock of this bluff is quite hard, rings sharply in large pieces under the hammer, dresses, splits, and chisels easily and well, and is said, probably with truth, to furnish the best quality of building-stone of any locality near Placerville.

The whole crest of the main ridge near and above the "Twelve Mile House," on the Carson road from Placerville, is made up of volcanic débris, and the thickness of this material must be, in places, at least, fully 700 or 800 feet.

Negro Hill, northeast of Placerville, is capped with "black lava" or volcanic breccia, which throughout the hill is generally very hard. The same material extends over all the highest portions of Cedar and Hangtown hills south of the town.

The crest of the main ridge between Jackson and Sutter Creek is capped with volcanic gravel. The same material frequently caps other ridges in this section of the country, and often has beneath it a little metamorphic gravel, which here and there has been worked to some extent by drifting.

§ 6. *The Channels: their Width.*

It is impossible to define with accuracy the width of the deep channel at Iowa Hill, since the hydraulic workings, extensive as they are, do not uncover the rims, and high bed-rock is known to exist at one or two points so located as to render it more than probable that the width is far from uniform. This difficulty is further increased by the great depth of the channel, which appears from the barometric observations to be about 200 feet in the bed-rock. But it is evident that at the northwest end of Wiessler's ground the channel was at least from 400 to 600 feet wide, while in the central portion of the ridge the strip of ground still standing beneath the town of Iowa Hill is so narrow, and the extent of the pits already worked on either side is so great as to render it extremely probable that the width of the deeper portion of the channel in its narrowest part is not less than 200 feet.

In the Lebanon Tunnel, on the northeast side of New York Cañon, they have followed a well-defined channel in the bed-rock which generally ranges from sixty to eighty feet wide, although in places much narrower; it curves somewhat, but its general course is southerly, and it has a decided descending grade in that direction.

At the Tunnel of the Eclipse Company, in Grizzly Cañon, the channel is about 300 feet wide, and some fifty or sixty feet deep.

At the Mountain Gate Tunnel, near Damascus, where the channel has been followed for 4,000 feet, it has been found to be from 175 to 200 feet in width, and the rise of the rock towards either side, so far as they have been worked, is very gentle, the extreme points reached at the sides being not generally eighteen or twenty feet higher than the central and lower portions of the channel. At the Cement Mill Tunnel, near Damascus, the channel has been worked out to an average width of nearly one hundred feet. The bed-rock rises considerably higher on each side of the channel.

In Yule's Claim, at Startown, the channel is very wide, but not deep. At Nick Anderson's Claim, just below Last Chance, near Startown, the average width of the outer channel is from seventy-five to one hundred feet; that of the back channel is about 300 feet.

At the Reed Mine, near Deadwood, the width of the channel is said to be sixty or eighty feet, and the Basin Channel at the Devil's Basin, where the trail from Deadwood to Last Chance begins to descend into the cañon of the North Fork of the Middle Fork of the American River, about eighty feet. This is the width actually worked in the Basin Channel and rich in gold; but the entire width of the real channel may be as much as 200 or 300 feet, as indications of a rim-rock are seen on each side, indicating about those dimensions. At Hornby's Tunnel, near Deadwood, the width of the channel averages forty feet, and the bed-rock rises high on each side.

At Weske's Claim, near Michigan Bluff, the rich channel has been followed for 600 to 700 feet, running in a direction nearly N. 45° W. (magnetic), with an average width of about sixty feet, and in places as much as eighty. A. Bowen's Tunnel, near Michigan Bluff, has developed a channel having a width of from twenty-five to thirty-five feet, with a high rim-rock on both sides, the direction of the channel being N. 68° W. (magnetic). The direction of the flow appears to have been southeasterly, as indicated by the position of the fossil trees, and the occurrence of richer patches of gold-bearing gravel under the lee of some of the boulders.

At Forest Hill the channel is very wide and deep; and the appearances are on such a grand scale, that it is desirable to give quite full particulars in regard to the workings at that locality. As an indication of the importance of this channel, it may be mentioned that between the southern edge of the Bath District and Todd's Valley, there are twenty-five extensive tunnel and hydraulic claims, along the north side of the Middle Fork of the American River.

At the New Jersey Claim the tunnel starts in on the bed-rock, and runs in a direction of N. 43° W. (magnetic), for a distance of 1,100 feet; it then forks, and one branch keeps on for about 500 feet farther in the same direction, while the other one turns and runs nearly west for about 400 feet, to the head of a slope going down into the "back channel," with a grade of twenty-seven inches to the rod. This back channel is very wide, and appears to run from northeast to southwest. In the New Jersey Mine it has never been worked, although prospected by the above-mentioned incline, which has gone down for a distance of 900 feet on the slope of the southeastern rim without reaching the bottom of the channel, the vertical depth attained in the incline being forty feet below the level of the tunnel, which is itself thirty feet or more below the level of the rim on the southeastern side of the channel. Nowhere in this vicinity has the rising slope of the other side, or northwestern rim, of the channel been seen. If the slope on the northwest side is as long and gentle as that on the southeast, the channel must be not less than 3,000 feet wide, and it is possible that it is considerably more. Immediately back of the highest southeast rim, where the general slope of the rock is northwesterly towards the bottom of the great "back channel," there are in the New Jersey Mine some five or six small channels or troughs, which are nearly parallel with each other, but at right angles with the great main or back channel, seeming to run down towards its bed. It is from these that the gold has been taken; and there has been obtained, mostly from a strip of ground 800 feet long and about 330 feet wide, not far from a million and a half of dollars, the bed-rock between these little channels themselves being everywhere found highly productive. These front channels are usually but a few feet deep. As a general thing the bed-rock gradually rises for a distance of about 900 feet from the front in to the southeastern edge of the rim of the back channel, whence it begins to fall off, and more rapidly, towards the northwest. An attempt was made to get to the bottom of the channel from the Devil's Cañon, on the northwest side. A slope was sunk down to the bed-rock, which it reached at a point ninety feet vertically below the level of the main tunnel, and at the bottom of this slope the rock was found to be still pitching to the northwest, and even more steeply than in the tunnel. From the foot of this slope a counter-slope was sunk, following down the bed-rock for a distance of 300 feet, and attaining an additional depth of thirty-five or forty feet, when the work had to be suspended in consequence of the machinery not being sufficiently powerful to keep down the water. The gravel is said to have been very rich, but extremely hard. It will be seen from the figures given above that the foot of this counter-slope was 155 or 160 feet below the top of the southeastern rim of the channel. From all that could be gathered at the New Jersey Mine, it appeared that the inner slope of the southeastern rim of the back channel had been prospected for a distance of 1,900 feet, so that the probable width of the channel is 4,000 feet, while it may be over a mile.

In Second Brushy Cañon, about a quarter of a mile below "Young America," a tunnel has been driven some 1,500 or 1,600 feet, in a direction a little east of south, and the bed-rock found here to pitch to the south. This is the only indication yet discovered of a northwestern rim to the great Forest Hill channel, and this is over a mile distant from the known position of the south-eastern rim.

Something over a mile northeast of Bath, the mining operations have shown that the bed-rock rises quite high in the immediate vicinity of a deep channel. In the bottom of Volcano Cañon, at the crossing of the upper road from Forest Hill to Michigan Bluff, a shaft was sunk 153 feet deep through volcanic cement. Before reaching the bed-rock, and on the crest of the ridge, a quarter of a mile west of this, the bed-rock rises to an elevation of about 400 feet above the mouth of the shaft. At the bottom of this shaft the bed-rock was found pitching to the northwest, show-ing that they were on the southeast side of the lowest part of the channel. They then followed down some ten or twelve feet lower on the rock, and found the gravel very hard, but rich in gold. About half-way between here and Bath, and on the northwest side of Volcano Cañon, is the Maine Boys' Tunnel, which was driven through very hard bed-rock, in a direction nearly north (magnetic) some 500 or 600 feet, and at its end broke through into what is believed to be the same channel as that reached by the shaft above described. They then went down with a slope on the surface of the bed-rock to a vertical depth of ninety feet below the tunnel before reaching the bottom of the channel, when the water drove them out.

Near the northwest end of the tunnel in the Paragon Mine, at Bath, the pay streak and all the accompanying strata, so far as here prospected, above and below it are cut sharply off by a mass of volcanic cement precisely similar in character to the so-called "gray cement" of the Deadwood Mines, which contains much partially carbonized wood and some metamorphic pebbles. On strik-ing this mass the tunnel was continued for about a hundred feet into it to the northwest, in the hope of passing through it and finding gravel again beyond it; but without success. The surface of demarcation between it and the elided gravel strata was however followed by drifts for a dis-tance of nearly 600 feet, and was found, though somewhat crooked, to have a general direction of very nearly east and west (true course) and a pitch or dip of about 45° to the north. There can be little doubt that this body of cement marks the position of a stream which, subsequent to the deposition of the gravel strata in the Paragon Mine, eroded a considerable portion of them, and then afterwards had its own channel filled with volcanic mud. And it is not at all improbable that it may have been the same stream whose channel was struck in the Maine Boys' Tunnel and in the shaft in Volcano Cañon. If this be true, the general course of the stream was southwesterly, and it came from somewhere in the Michigan Bluff divide.

The channel at Jones's Hill, four miles northwest of Georgetown, appears to run very nearly west (magnetic) in the Columbia Mine, with a grade falling in that direction, heavy enough to make brakes necessary on the cars. The channel passes under the southern slope of the hill, and is said to range from fifty or sixty feet in width to over two hundred, averaging perhaps a hundred. It is described as well-defined, with a high and steep rim-rock rising on the north side, while the southern rim is not more than ten or twelve feet high. The channel has been followed and worked here, by drifting, for over a quarter of a mile; it has generally paid well, and in spots has been very rich.

In the Roanoke Tunnel, near Bottle Hill, the channel varied from ten to a hundred feet or more in width. The whole length of the tunnel through the hill was rather more than a mile, and it was somewhat crooked, although its general course was N. 50° W. (magnetic). The channel is narrow, and the rim-rock high on both sides, although it is not known how high. It is said, how-ever, that an incline was once run up on the northeast rim to a height of fifty or sixty feet above the tunnel, and that here an upper stratum of gravel was struck, which was about four feet in thickness, the bed-rock then remaining nearly level in a northeast direction, so far as it was fol-lowed, or for a distance of fifteen or twenty feet, indicating that the top of the rim had been reached.

At a point where the channel in Roanoke Hill leaves the hill, a little side channel comes in from the south. This little channel, where seen by Mr. Goodyear, was not more than twelve or fifteen feet wide, and very shallow, yet it has been followed for three fourths of a mile, and is said to have been pretty rich. It runs directly through the ridge, a little to the west of Bottle Hill, its general course being about N. 10° W. magnetic, with a decided fall in that direction. There is also said to be, in Gravel Hill, a channel as well-defined as the one just described. It is described as coming into the hill from the south, and running northerly for a distance of 700 or 800 feet, then bending to the west and going about as far in that direction before making its exit from the hill. This channel has been distinctly traced for the distance mentioned, with a decided down grade to the north and west. It is supposed to be a branch of the Roanoke and Jones's Hill channel, which it may have joined in Jones's Hill at the Columbia Claim.

At Flora's Mine, two miles west of Volcanoville, there is said to be a well-defined channel, with high rim-rocks on each side, and which has been worked from 100 to 150 feet in width. The grade of this channel is descending towards the west, and it appears to be pretty heavy.

At Densmore's Claim, a mile and a half a little west of south from Startown, on Grouse Cañon, the channel is narrow, ranging from thirty to fifty feet in width, and it runs nearly parallel with the ridge, or in a direction a little south of west; it has been followed for about 100 feet in the tunnel at Densmore's. It is said that the same channel has been traced in a southwesterly direction along the right bank of Grouse Cañon, for some two miles.

At Peckham Hill there appears to be a broad and well-defined channel, with high rims on both sides, but which has not yet been much explored. There is considerable reason for believing that this is the continuation of the great channel at Forest Hill.

At Castle Hill, two or three miles above Georgetown, and near the Clipper Mill, a tunnel has been driven in Robbins's Mine, for 400 feet in a direction of S. 5° E., and then 250 feet farther S. 20° E., following a channel up stream. This channel is narrow, so far as followed in this mine, having been worked only about fifteen feet in width; it is well-defined, with distinct rim-rocks. It is stated that at some distance to the southeast this channel is wider, having been worked in places to a width of sixty or seventy feet.

At the Excelsior Mine, near Placerville, the channel is supposed by Mr. Alderson to run diagonally through the main ridge in a northeasterly direction, passing just southeast of the ancient village of White Rock, and coming out in the cañon of the South Fork of the American River just above that place. This channel at Coon Hollow seems to be very wide, exceeding 2,000 feet; above this, it is said to be still wider. A little below White Rock, there is another channel, called the Blue Lead, which is supposed to lie beneath the other broader channel. This blue lead channel, which runs through the main ridge, with a direction a little east of south, is said to be from 400 to 600 feet in width, and from thirty to forty feet deep, with a well-defined rim on each side, and the grade or fall is said to be towards the southeast.

§ 7.   *The Channels ; Recent Changes of their Position.*

The Golden Gate Cañon is a little branch of Damascus Cañon, on its east side. Following down this cañon, on its left bank, and running parallel with it for a distance of 1,500 feet or more, and from twenty-five to one hundred feet above the bed of the present cañon, is an older bed of the same stream, now buried in the bank, and with a rim of bed-rock, varying from a few to forty feet in height, separating it from the present cañon. This older channel is filled chiefly with fragments of bed-rock and soil, with more or less volcanic boulders, and a few of quartz. It has been drifted and worked considerably, and has paid well. It contains an abundance of trees and wood of various kinds, which seemed to be similar to those now growing in the region. One cedar log, in perfect preservation, measured three and a half feet in diameter.

In El Dorado Cañon, near Deadwood, there are many old channels parallel with the present cañon, and some of them several hundred feet above its bed, on the mountain sides.

§ 8. *The Channels; their Grade.*

At the Mountain Gate Tunnel, near Damascus, the channel has a grade of about two feet to the mile, for the first thousand feet beyond where the tunnel strikes the gravel, 2,600 feet from its mouth ; after that it is somewhat heavier.   The channel descends towards the southeast.

At Canada Hill the channel has a grade of about five feet in a hundred, descending towards the east.

At the Reed Claim, near Deadwood, there is a descent in the channel of about two and a half feet in the hundred towards the southwest.

In the Basin Channel, at the Devil's Basin, above Deadwood, the average grade of the channel is just about two and a half feet in a hundred, and the descent towards the west.

At Castle Hill, above Georgetown, the channel has a fall of about thirty feet in going 1,200 feet northwesterly from White & Co.'s Tunnel.

§ 9. *The Channels; their Varying Character.*

The Dick and Arkansas Claims are on the eastern branch of the West Fork of El Dorado Cañon ; the former is on the right bank, the latter on the left bank of the cañon.   In each of these claims tunnels have been run for some 2,000 feet.   In the Arkansas Tunnel a well-defined narrow channel with a high rim on both sides was followed ; in the Dick, no rim-rock was found, the channel having a basin-like character.   In both these claims the gravel and cement are perfectly similar to those in the Dam Claim, and the gold found in them was of exactly the same character also, so that there would appear to be little doubt that all three were in the same channel, which must however in that case have been extremely crooked and varying in character.

On the west side of the ridge fronting El Dorado Cañon for one and a half or two miles above Deadwood there has been considerable hydraulicking on the spurs, besides the tunnelling in the hills.   It is thought that there are three channels here ; one, called the "' red channel," occupying the outermost spurs, and is seventy or eighty feet lower than the next higher one, which lies a little farther back in the ridge, and is known as the "quartz channel."   In this latter the boulders are nearly all white quartz, pretty well worn, and much resembling those in the Mountain Gate Channel, at Damascus.   Still farther to the east, and still higher than the others, lies what is called the "cement channel."

§ 10. *The Channels; Occurrence of Basin-like Depressions in them.*

In the ground worked by Mr. Wiessler at Iowa Hill, there are two so-called channels, running in the direction of the strike of the rocks (N. 25° — 30° W.) and separated by a ridge of soft, light-yellowish shaly bed-rock, while their own beds are hard and dark-colored slates.   One of these channels is broad, and the other narrow and deeper.   Both appear to be simple basins, which are elongated in the strike of the bed-rock, and which are entirely surrounded by rims, higher than their central portions.   These basins were very rich in gold.

Mameluke Hill, near Georgetown, is said to cover a basin in the bed-rock, the rim on all sides being higher than the central portion.   All over the basin the gravel was very thin, ranging from only a few inches to five or six feet ; and this was capped with volcanic cement, which, though very soft around the edges and on the top, was pretty hard in the interior of the hill.   The gold is said to have been smoothly washed, coarse and heavy ; and some large nuggets are reported as having been found here.

§ 11. *The Channels; their Direction.*

At Iowa Hill, in Wiessler's Claim, the surface of the bed-rock, where hard, is water-worn in such a way as seems to indicate that the old stream flowed here approximately in the direction of the strike of the slates (N. 30° — 40° W., magnetic) i. e. N. W. or S. E.; but which way the water ran is not so clear; the probabilities are that it flowed southeast.

In the deep channel at Iowa Hill the direction of the flow was probably from the northwest towards the southeast, as the bed-rock in the North Star Claim, on the southeast side of the present ridge, appears to be somewhat lower than it is in Wiessler's Claim on the northwest side.

Northeast of Independence Hill the surface of the bed-rock in the interior of the ridge, beneath the gravel and the volcanic capping, is said to be furrowed by a system of nearly parallel channels which have a southwesterly course.

At the Mountain Gate Tunnel, near Damascus, the course of the pay-channel, which has been followed by a tunnel for a distance of 3,900 feet, is about S. 45° E. (magnetic).

The course of the Canada Hill Channel, so far as it has been worked, is S. 87½° E. (magnetic).

At Nick Anderson's Claim, near Startown, the course of the outer channel is S. 40° W. (magnetic), and the back channel runs nearly S. 23° E. (magnetic).

In the Morning Star Claim at Startown the tunnel runs S. 12° W. (magnetic) some 1,300 or 1,400 feet, being parallel with the tunnel in Yule's Mine, and only about 300 feet west of it. Near the mouth of this tunnel there is a so-called "outside channel," which runs nearly parallel with the present river cañon, and has been richer in spots than the interior of the hill, but has not paid so uniformly. This channel is said to have been traced along the mountain side towards the west and southwest as far as Last Chance, and it is generally believed by the miners at Last Chance to be identical with the so-called outside channel in the Nick Anderson Claim just below that town.

At the Reed Claim, near Deadwood, and on the southeast side of the ridge towards the North Fork of the Middle Fork of the American River, a tunnel has been driven about 200 feet in the bed-rock, then on its surface about 1,450 feet, N. 35° E., then about 2,000 feet S. 85° E., then 200 feet N. 23° E. (all magnetic); this tunnel follows the channel, and is driven on the surface of the bed-rock, and has a grade of about two and a half feet in the hundred.

The mouth of the Rattlesnake Mine is situated nearly east of Deadwood on the eastern side of the spur between Black Cañon and the North Fork of the Middle Fork of the American River, and not more than a fourth of a mile distant in a direction nearly magnetic east from the last bend of the long tunnel in the Reed Mine. The Rattlesnake Tunnel runs S. 75° W. (magnetic) about 300 feet, following the course of a well-defined channel from 30 to 40 feet wide between high rim rocks with a descending grade towards the west. The course of the Rattlesnake Tunnel if continued towards the west would carry it but very little to the south of the bend just referred to in the Reed Tunnel, so that it is extremely probable that the Rattlesnake Channel actually unites with the Reed Channel at or near this bend.

The general course of the Basin Channel at the Devil's Basin for a distance of 2,150 feet in the tunnel is S. 70° W. (magnetic).

The present face of the tunnel in Hornby's Mine near Deadwood, which has followed up the course of the channel for a distance of 1,500 feet, is not more than 400 or 500 feet distant from where the Reed Tunnel first strikes the channel, and the course of the Hornby Tunnel at the face is directly towards the latter point, showing that it is in all probability on the same channel.

At Wilcox's Claim, on the North Fork of Long Cañon, the course of the channel seems to have been from northwest to southeast.

Mr. Silas Griffett's Claim at the Devil's Basin, between Last Chance and Deadwood, follows the channel in a southwesterly direction. At Hornby's Tunnel, near Deadwood, the direction of the channel is from northeast to southwest. From a point near Robertson's Tunnel on the northwest side of the ridge, near Deadwood, a channel has been followed in a direction about N. 70° E. for a distance of 800 to 1,000 feet through the mountain to Kaylor's.

At Weske's Claim, near Michigan Bluff, they are following the channel in a northwesterly direction with a down grade.

At the Dam Claim, in the west branch of El Dorado Cañon, the tunnel runs nearly straight for

900 to 1,000 feet in a direction about N. 20° to 25° W. (magnetic) following, up stream, along the right bank of the channel, whose deepest portion lies, or appears to lie, east of the tunnel. At the northern extremity of the mine the channel seems to make rather a sharp bend to the west and southwest, and the tunnel, which is driven some 300 feet beyond the end of the straight portion, forms within this distance a curve from N. 25° or 30° W. to S. 75° W. (magnetic).

At Yankee Jim's the broad channel has been worked through the hill called "Cleveland Hill," just north of the village, for a distance of some 2,000 feet, in a direction nearly S. 75° W. magnetic; the width is varying, but it will average probably a thousand feet. The area which has been worked off here is estimated at fifty acres.

In the Gore Tunnel and Rough and Ready Claim, at Forest Hill, a channel runs in a direction of N. 87° W. (magnetic) and is narrow. It is perhaps the same as one of the outer channels in the New Jersey Mine.

At Jones's Hill, four miles northwest of Georgetown, the channel in all probability came in from the northeast, passed through the crest and then curved around to the west. The general course of the channel in Castle Hill, above Georgetown, is about N. 20° W. (magnetic).

About a quarter of a mile east of Bottle Hill a deep channel, on which is the Roanoke Tunnel, has been followed entirely through the ridge in a direction about N. 50° W. (magnetic), the course of the ridge being itself nearly east and west. This channel is said to have a heavy fall towards the northwest, so that a brake was needed on the mine-cars. The main channel runs from the mouth of the Nevada Tunnel, on the southeasterly side of the ridge, in a pretty straight general course, with but little curving, through the Nevada, Washington, Roanoke, Eureka, and Black Hawk Claims, a distance of about 7,000 feet; it then makes a sharp bend and goes some 600 or 700 feet in a direction S. 55° W., crossing Roanoke Gulch into the next spur, then bending again, and running about 700 feet in a northwesterly direction, then turning again to the southwest, and running for about a thousand feet to a point where it passes out of the last spur. From this it appears to have passed along over the southern portion of the present river cañon, and to have been carried away, nothing more being seen of it until it strikes Jones's Hill, at a point bearing S. 79° W., and distant about two miles.

There seems to be a deep channel crossing the Missouri Cañon, at the Shoo Fly Claim, about a quarter of a mile S. 50° E. from the Buckeye-Sucker Claim. This channel runs in a northwest-southeast direction, the bed-rock on each side of it being high.

The Blue Lead near Placerville is supposed to enter the hill on the southwest side of White Rock Cañon, and to pass diagonally through Negro Hill Ridge and beneath Oak Grove and Dirty Flat, thence to Smith's Flat, and through the ridge between that and Webber Creek, coming out at Try Again Tunnel. This would give it a very nearly north and south course.

At Dickerhoff's Mine in Cedar Ravine, near Placerville, a tunnel has been run into the hill for a distance of over 1,000 feet, and in a direction S. 10° — 12° W. (magnetic), the pay-gravel being from three to four feet thick. The channel followed is narrow, in front, being only from 200 to 300 feet wide, but it enlarges as the hill is penetrated, and at the back end of the mine is from 500 to 600 feet in width. It bent around to the southeast and southwest, being quite crooked, and in one place enclosing an island of high bed-rock. The rim-rock here rises on each side to the height of from forty to sixty feet.

At a point in the surface of the bed-rock, on Clay Hill, near Placerville, there are furrows worn by the current, their longer axes lying in a direction of about S. 52° W. (magnetic).

At the west end of Indian Hill, near Placerville, the strike of the bed-rock is about N. 35° W., and the direction of the furrows worn in its surface by the water ranges from S. 60° W. to S. 85° W., the average being probably not far from S. 70° W. (all magnetic).

In the hill between Smith's Flat and Dirty Flat, near Placerville, called Granite Hill, the general course of the deep channel, going "up-stream," is supposed to be about N. 40° W. (magnetic), which course it seems to hold nearly through the hill, and then to make another sharp bend to the north, running nearly north magnetic, across the head of Dirty Flat. It would appear from the statements of Mr. Carpenter, that the so-called "benches" of this "deep channel" are sometimes separate channels in the bed-rock, with their rims well defined on both sides, and running nearly parallel with the deep channel, only at a higher level.

It is thought by those acquainted with the region, that there are, in the hill south of Big Cañon, northeast of Placerville, several parallel channels, with ridges of bed-rock, from twenty-five to forty feet high, between them, running in a southwesterly direction through the hill. The bed-rock which shows near the crest of the ridge, immediately northwest of the Rocky Mountain Claim, is said to be an isolated point, the gravel being deep all around it.

§ 12.   *The Gold; its Distribution in the Gravel, and Position with Reference to the Channel and Rim-Rock.*

In Wiessler's Claim, at Iowa Hill, where there are two basin-like depressions in the bed-rock, surrounded by rims, the richest spots are said to have been invariably not in the central parts of their bottoms, but on the sloping sides of the bed-rock which surrounds them. The sides of the ridge which separates the two basins are said to have been extremely rich; but there did not appear to be any connection between the direction of the slopes and the amount of gold deposited on them. Both sides of the dividing ridge were rich.

At Independence Hill and elsewhere, on the ridge above Iowa Hill, as a general rule, the richest spots and coarsest gold are found immediately on the slopes of the rim-rock, and on the ridges in the interior of the hill, which sometimes rise as high or even higher than the rim. Streaks of coarse gold are occasionally found high up in the gravel above the bed-rock.

At Wisconsin Hill the rim-rock is said to have been rich in places, and to have yielded some pretty coarse gold; but generally the surface of the bed-rock here is not supposed to be rich.

At King's Hill the eastern rim-rock is said to have paid very richly, while the ground farther back in the hill did not pay expenses.

At the Morning Star Claim, at Startown, the bed-rock is very rough, and the richest spots are found in the depressions of its surface.

Many of the claims immediately about Deadwood are said to have been extremely rich at the outer edge, but poor farther back in the hill.

At Weske's Claim, near Michigan Bluff, the rim-rocks are not very well defined, but the high rock has generally paid the best; although the distribution of the gold is very irregular, or "spotted," as the miners say.

The richest portion of the pay-streak in the Paragon Mine, at Bath, was a strip from 200 to 250 feet wide, immediately northeast of the main tunnel.

In the Rough Gold Mine, southwest of the Paragon Mine, the pay-streak — which is the same stratum as that at the Paragon Mine — has, at one point between it and the bed-rock, as much as 140 or 145 feet of gravel.

At the Dam Claim, in the west branch of El Dorado Cañon, there is a great curve in the channel, which makes within 300 feet a bend from N. 25° W. to S. 75° W. (magnetic), and immediately below this bend, and on the right bank of the channel, an extensive bar was found, which, as is stated, would have paid, if worked by itself, from $15 to $20 a day per man employed. This bar was between 300 and 400 feet long, and from sixty to eighty feet wide, and it was separated from the main channel, on the east, by a narrow ridge of high rock, along which there was a large quartz vein, from five to eight feet in width. All along the straight portion of the tunnel, or for a distance of about a thousand feet, the surface of the bed-rock is pitching to the east, and there is a distance of some 1,300 feet or more in which they have never driven east across the channel, or seen anything of the east rim-rock. In the vicinity of the main or party tunnel, which forms the dividing line between the Dam and El Dorado claims, and runs for 250 feet in a direction N. 56° E. (magnetic), the channel, going down stream, seems to make a pretty sharp bend to the east, and a considerable area of ground has been worked out by the El Dorado

Company, which is said never to have paid much, with the exception of a certain portion on the northeastern or concave rim. The channel is well-defined, and the bar and the two curves are excellent illustration of the formation of bars and the accumulation of the gold in the channel on the concave side of the bends, just as has taken place so often in the more modern streams.

At the New Jersey Mine, near Forest Hill, the "great pay" came from immediately but entirely back of the highest rim-rock of the great channel.

At Smith's Point, between First and Second Brushy cañons, the pay-streak on the bed-rock has been found, at some points, to rise from the surface of the rock, and run nearly horizontally through the gravel, across the little basins and depressions in the rock.

At Yankee Jim's the best pay over the whole surface washed has been invariably found on the southwest slope of the bed-rock. The latter is not very uneven in the channel, yet there are, of course, some irregularities; and wherever the rock pitched off a little to the west or southwest, there it was rich; while wherever the slope was in the opposite direction, there it was very poor. At one locality near the central part of the ground worked there is a narrow depression cut some twelve or fifteen feet deeper than the average of the broad portion of the channel, and where the rock first pitched off into this depression, it was so rich that about $20,000 was taken from an area only twenty or twenty-five feet square. The true "pot-holes" were generally barren. In the upper or eastern portion of the Big Channel the richest pay was on the bed-rock; but towards the lower end, that is, near the head of Yankee Jim's Cañon, much of the bed-rock was not so rich, and the upper gravel paid much better. The rock here, too, was more uneven, and contained several large basins, which, however, are said to have been comparatively poor, the pay-gravel running nearly horizontally over them. The head of Yankee Jim's Gulch is said to have been enormously rich, single men having repeatedly taken out thousands of dollars in a day. At Georgia Hill the rim-rock was richer even than the spot at the head of Yankee Jim's Cañon.

In front of the Paragon and Rough Gold claims, at Bath, some parties, it is said, took out, chiefly from a thin streak in the gravel, not far above the bed-rock, and only an inch or two thick, the sum of $52,000.

The Roanoke Channel, near Bottle Hill, which was followed for about a mile with a width varying from twelve to a hundred feet, yielded over $500,000.

At Flora's Mine, west of Volcanoville, the lower portion of the gravel near the bed-rock is said to be tolerably rich, and occasionally very rich spots are found in it. One hundred dollars to the pan have been scraped up. The upper portion of the gravel is much poorer, and is said to pay no more than two dollars per day to the hand in drifting.

At the Slab Claim, on the trail three quarters of a mile below Last Chance, the pay was found almost exclusively on the bed-rock, the layer of gravel being very thin, and overlain by a stratum of chocolate cement; this again was covered by a bed of gravel about twelve feet thick, but containing very little gold.

§ 13. *The Gold; its Size and Character.*

At Wiessler's Claim, Iowa Hill, the gold near and upon the bed-rock is coarse and worn into elongated and flattened grains; it is said to be worth $18 per ounce. That from the upper part of the bank is scaly and exceedingly fine, or "flour gold," and is generally worth from fifty to seventy-five cents per ounce more than the coarse gold on the bed-rock.

At the Lebanon Tunnel the channel has proved rich in coarse and rather scraggy gold. One lump is said to have weighed ten ounces, and another forty-two.

At Nahor's Claim, in Green Valley Gorge, the gold is rather coarse and smooth, and it seems generally to grow coarser as we travel up the ridge from Iowa Hill.

The gold from the tunnels about the head of Indian Cañon was usually very coarse. Some of the claims paid well, and occasional spots were found that were very rich.

The gold in Grizzly Cañon is said to be coarse and generally rather smooth, although some of it was quite "scraggy."

At the Mountain Gate Tunnel, near Damascus, the gold is generally not very coarse, although it is sometimes so, a piece weighing nine ounces having been found here. The coarser gold is generally well-rounded and smooth, the finer is often flaky or scraggy, and sometimes pieces of considerable size are found very scraggy, with more or less quartz adhering to them. The gold from the Cement Mill Claim, at Damascus, is not so coarse as that from the Mountain Gate Tunnel, and is worth about $18.25 per ounce.

At Sterrett's Claim, in Sailor's Cañon east of Canada Hill, some coarse gold is found on the bed-rock, and finer gold in the gravel above. The gold is well washed.

Nearly all the gold in Weske's Claim, near Michigan Bluff, even when the pieces are quite large, is in thin, scaly, and tabular forms, which are generally pretty well water-worn and smooth.

In the Van Emmons Claim, at Michigan Bluff, the gold in the quartz gravel is not in such thin tabular forms as at Weske's Mine, but is thicker and rounder and more water-worn; it is also duller and darker in color, and is said to be worth $19.50 per ounce. The gold from the quartz gravel at this locality is said to be worth a little more per ounce than that from the so-called "black gravel," which contains less quartz and more metamorphic rocks.

Mr. Ströbel, who has been for many years a purchaser of gold-dust at Michigan Bluff, states that the fine gold is always richer than the coarse. For example, if coarse gold from a given locality be worth $17.25, the fine gold from the same locality will be worth $18.00 per ounce.

The gold from El Dorado Hill, near Michigan Bluff, is generally well water-worn, and about as coarse as that from Weske's Mine, but thicker and heavier, and often very black; it resembles that from the quartz gravel at Michigan Bluff.

At the Specimen Claim, in Byrd's Valley, many handsome specimens have been obtained of crystallized gold.

Mad Cañon, near Byrd's Valley, is said to have been very rich in scraggy gold, furnishing many beautiful specimens, with foliated and tabular crystalline forms. Ladies Cañon also paid well, but yielded little fine gold, the metal being mostly in large pieces, or "slugs," very rough and scraggy, and often associated with quartz.

In the Paragon Mine at Bath, the gold now obtained is always fine, a piece worth as much as five cents being very rarely seen. Yet the pay-streak is rich, and portions of it have yielded five dollars per car-load, the cars holding but little over a ton each. But Mr. Wheeler, one of the owners of this mine, states that while working the front portion of the mine, and for a distance of 2,400 or 2,500 feet back from the mouth of the tunnel, considerable coarse gold was found, and pieces worth from two to five dollars each were not uncommon. But absolutely no coarse gold at all has been found more than 2,500 feet back from the mouth of the tunnel.

In the New Jersey Mine, at Forest Hill, the gold from the back channel is not worth so much as that from the front. The gold from the quartz boulders contains considerably more silver than that from the gravel.

At Smith's Point, between First and Second Brushy cañons, a stratum worked by drifting, some forty or fifty feet above the bed-rock, was quite rich in rather coarse gold, while the gravel beneath it is far poorer, and the gold much finer. The gold here is generally fine and scaly, but the coarse gold is pretty well worn and heavy. Considerable "shot gold" is also obtained from points in this region.

The gold in Castle Hill, a short distance above Georgetown, is pretty coarse, well-rounded and heavy; it is often more or less covered with a kind of blackish incrustation.

At the Excelsior Claim, near Placerville, the gravel was drifted at two different levels, once on the bed-rock, and then again at twenty feet higher. The bed-rock gold is rather coarse, much of it being in pieces of from five cents to twenty dollars in value; it is also very smoothly rounded, and it averages about $19.10 per ounce in value. The gold from the upper streak is rather fine, not quite so well washed as that from the bed-rock, and is worth, on the average, one dollar more per ounce than that. This gold is said to contain but a mere trace of silver.

§ 14. Seam Diggings.

At Illinois Cañon, near Georgetown, are so-called "seam-diggings" which consist of decomposed bed-rock, filled with irregular seams of quartz containing gold. This ground is worked by the hydraulic method, and a spot having an area of fifty by a hundred feet has been washed off. This operation is said to have paid a little less than wages, on the average, although as high as $ 40 to the pan has been obtained here.

At Georgia Slide, near Georgetown, they have been working seam diggings in the bed-rock for some sixteen or seventeen years, and have reached a depth of 175 feet in the bed-rock, which is a rather soft, decomposed slate, generally more or less talcose, and full of decomposed crystals of pyrites ; much of the slate itself contains fine gold.

Young's Dry Diggings are shallow, surface excavations, covering two or three acres of sloping ground, on the southwestern side of the ridge which extends along the road from Georgetown to Spanish Dry Diggings. They are said to have paid well, whenever water could be obtained. It is stated that a little seam of quartz, some twenty feet long, was once struck here, from which in a short time between $ 25,000 and $ 30,000 was taken. The bed-rock at this locality consists partly of rather soft slates, and partly of very hard and tough metamorphic sandstones ; the latter occurs, in places, in the form of lenticular masses enclosed in the slates.

The so-called "Spanish Dry Diggings" belong entirely to the class of "seam diggings," although the gulches in this vicinity are said to have been rich in the early days. There are two principal claims, the "Grit" and the "Dam." The course of the main tunnels is very nearly north and south, and this appears to be the general course of the belt of rich ground. Immediately east of the Grit Claim there is a heavy mass of semi-serpentine, which appears to be a product of the metamorphism of a very fine-grained slate. The rocks in the mines are slates and sandstones, the latter very fine-grained ; these rocks are filled with crystals of iron pyrites, and traversed by numerous small irregular seams of quartz. It is said that, in the Grit Claim, the seams which have paid best run nearly north and south and dip to the east at an angle of 45°, or thereabouts. These seams traverse the decomposed rock and dip towards the serpentine ; but it is stated that they do not pass into it. The Grit Claim is said to have yielded not less than $ 300,000. It was from this mine that the magnificent specimen of gold was taken which was on exhibition for some time in San Francisco, and which afterwards was taken to Paris. Its intrinsic value was estimated at $ 4,000, and it was entirely made up of fine reticulations of imperfect arborescent crystallizations. It is said that some $ 6,000 or $ 8,000 more was taken out from the same place in the immediate vicinity of this remarkable specimen. The Dam Claim is said to have yielded $ 250,000 ; and the seam diggings, within a mile of the town of Spanish Dry Diggings, have yielded fully $ 1,000,000.

The small quartz seams in the Dam Claim run in every possible direction, without regularity, and the gold is very far from being uniformly distributed through them ; indeed, it occurs chiefly in "pockets," so that it is commonly said that a miner at these diggings either makes nothing or a fortune. It is stated that whenever the quartz seams run into hard rock they thin out or disappear altogether and the gold almost invariably gives out.

Immediately opposite Spanish Dry Diggings, on the other side of the river, at a place known as Shenanigan Hill, there is a spot where some work has been done, and where there are also seam diggings resembling those just described. There are other claims of a similar character in this vicinity, and from one — the "Short Handle" — it is said that $ 60,000 was taken, from a seam only twenty or twenty-five feet long, and worked to a depth of thirty feet only.

In the range of hills immediately west and southwest of the village of Greenwood, and which rise to the height of from 100 to 150 feet, there are seam diggings of some extent, in the so-called Franklin and French Claims. Here a piece of ground of an irregular shape has been worked out, covering an area of one or two acres in extent, and the maximum elevation of the banks is about sixty or seventy feet. The rocks exposed here are very fine-grained argillaceous slate and sandstones, and associated with these is a rock which was once a fine breccia or conglomerate, and which has since been so metamorphosed as almost to have lost its original character. There is also a porphyritic slate, which consists of a fine-grained silicious base, with large crystals of feldspar scattered through it. These crystals are so large and so peculiarly distributed through the rock as to give it an appearance slightly suggestive of its being covered with hieroglyphics. For this reason, probably, it is known to the miners as " China rock." All these rocks have been irregularly and extensively decomposed, and are traversed in every direction by little seams of quartz, but they are not numerous. These seams are occasionally rich in gold; but, to use the miners' phrase, they are very " spotted." It is said that about $81,000 has been taken from the French Claim, in ten or twelve years of working. There are several other claims of a similar character in this vicinity; but none have paid so well as the French Claim. In the early days of California, the gulches and ravines about here are said to have paid quite well; but water is not abundant in this region.

There are Seam Diggings between Johntown and Georgetown, on the hill at the head of Crane's Gulch. They are in slate rock, which is quite talcose in character, and considerably decomposed. The slates have the usual northwesterly strike and nearly vertical dip, and are a good deal contorted. The seams of quartz are irregularly distributed through the slates, curving more or less, but on the whole following pretty nearly their stratification. The whole area of ground which has been worked here is not far from a quarter of an acre, and from the flume or tailing sluice to the top of the bank at the highest point is about seventy-five feet. At one place a bunch of quartz is exposed at a depth of about forty feet below the top of the bank, and is there nearly three feet thick; but a few feet higher up it splits into eight or ten little irregular stringers which extend branching upwards for a short distance and then disappear, none of them reaching the surface of the rock. This is, indeed, the case with most of the quartz seams at this locality; they rarely extend to the surface. There has been a great deal of pyrites in the quartz as well as in the slates; but it is now decomposed and carried away. The quartz itself is considered of little value and is not crushed, the gold being all obtained from the sluices, for which purpose the rock has to be blasted and broken with the hammer, and then washed with a two-inch hydraulic jet.

§ 15. *Amount and Yield of Gravel Worked.*

At Wiessler's Claim, Iowa Hill, in 1871, the body of ground worked had a length of about 1,200 feet and width of 400 approximately; the maximum height of the bank was from 125 to 130 feet. Assuming an average of 1,200 feet in length, by 350 in width, and 60 in depth, the amount of gravel washed away will be 933,333 cubic yards. The amount of gold taken out could only be guessed at. It was put all the way from half a million to two millions.

From the southwest corner of the ground owned by Mr. Wiessler at Iowa Hill another large pit extends into the hill in a direction S. 30° E., magnetic, for a distance of 700 or 800 feet, with a maximum width of something over 200 feet, the height of the bank at the head of the pit ranging from sixty to eighty feet. The quantity of material removed from here was estimated at 200,000 cubic yards. The bed-rock was not reached, and the gravel immediately upon it is said to be too poor to pay for working. No information was obtained as to the amount of gold yielded by this work.

On the southeasterly side of the Iowa Hill Ridge, including the North Star and Sailors' Union claims, an area of about 1,000 by 250 has been worked off, with an average depth of 100 feet: this gives, as the total amount of ground washed, about one million cubic yards. In regard to the operations of the Sailors' Union Mine, the following statement was taken from the Company's books.

Ground worked: 580 feet long by 160 wide, and with an average depth of 116 feet.

Amount of gold taken from the bed-rock .	$124,598
Amount of gold from gravel above the bed-rock .	42,800
Total .	$167,398

This gives *as the yield of the gravel*, which equals in amount 398,700 cubic yards, the sum of 10.7 cents per cubic yard.

At the Morning Star Mine, near Iowa Hill, there has been about one quarter of an acre of the bed-rock stripped and cleaned, which is said to have yielded about $30,000, exclusive of the $40,000 which was taken from the crevice.

At King's Hill there have been four or five acres washed off to an average depth of 20 feet.

In the east pit at Elizabeth Hill an area 700' × 200' has been worked to an average depth of about 20 feet. In the middle pits at Elizabeth Hill an aggregate area of about three acres has been washed off, exposing banks whose higher portions range from 50 to 75 feet. In the south-west pit at Elizabeth Hill there has been about an acre of ground washed off to an average depth of seventy-five feet.

About 800 feet south of the Wisconsin Hill school-house, on both sides of a little branch of Refuge Cañon, hydraulic pits have been worked over an aggregate of five or six acres, exposing banks whose maximum height ranges from sixty to seventy-five feet.

In the claims worked by Mr. Vaughn, at Wisconsin Hill, about twelve acres of ground have been washed off, in five years' work to an average depth of fifty-five feet. The yield is estimated at about $30,000, which would give an average of about thirty-four cents per cubic yard.

It is stated that in one season's work the sum of $10,000 was taken from the Lebanon Tunnel at New York Cañon, and the total yield of the mine is estimated to have been not less than $75,000.

At the Clinton Claim, in Grizzly Cañon, an area of 600 feet by 100 was sluiced off to a depth of about twenty feet, yielding, it is said, $240,000.

The main channel at Canada Hill is said to have yielded about $100,000, and the adjacent gulches about $50,000 more.

The English Claim and the one adjoining it on the western side of the ridge, near Deadwood, are said to have yielded $100,000.

The Morning Star Mine, at Startown, has yielded over $300,000.

The Slab Claim, on the trail three quarters of a mile below Last Chance, is said to have yielded not less than $75,000.

The Basin Claim at the Devil's Basin is said to have yielded about $200,000.

The Dick and the Arkansas claims are said to have yielded over $70,000 each, and the Dam Claim some $40,000. All three of these claims are on the upper part of the West Fork of El Dorado Cañon, five or six miles a little east of north from Michigan Bluff.

At Weske's Mine, near Michigan Bluff, June 16, 1871, Mr. Goodyear saw washed out from eight car-loads of gravel, of about fourteen cubic feet each, the result of one day's work, with twenty men employed, $91\frac{1}{2}$ ounces of gold, worth $1,601.25. The total yield for the current week, at this claim, was $514\frac{1}{2}$ ounces, worth $8,939.44. At this claim the distance worked on the channel in five months was about 200 feet, and the yield during the first five days of the week of Mr. Goodyear's visit was 352 ounces. Weske's Mine had yielded something over $100,000 up to June 30, 1871.

El Dorado Hill, near Michigan Bluff, is said to have yielded over $150,000.

Fabulous stories are told of the richness of Dutch Gulch, which lies between the point of the Flat and Red Hill at Michigan Bluff. For instance, it is said that one man took out $1,100 from a single pan of dirt here. Another statement with respect to the Flat itself, and which was repeatedly made, was to the effect that in drifting here two men took out 1,200 ounces in one week; this was at the Empire Claim. Stickner's Gulch, which runs down on the opposite side of the point at Michigan Bluff into Skunk Gulch, is also said to have been extremely rich, although not comparable with Dutch Gulch.

At Ayer's Claim, on the east side of the ridge facing El Dorado Cañon, north of Michigan Bluff from a space, near the mouth of the tunnel, about a hundred feet long and from thirty to forty wide, about $30,000 was taken out, and after that nothing farther was found.

In the Franklin Claim, near Michigan Bluff, from a triangular piece of ground about eighty feet long and forty wide, or say 2,000 square feet, $37,000 was taken out.

The Express Agent at Forest Hill, Mr. K. B. Soulé, stated that the shipments of gold through the Express Office at that place in 1871, ranged a little over $30,000 per month; this comprised the yield of the Forest Hill, Brushy Cañon, and Yankee Jim claims. Mr. Soulé considered that the shipments from Michigan Bluff were a little greater in amount than those from Forest Hill, and that the total yield from the country between the North and Middle Forks of the American amounted (in 1871) to something like $1,500,000 per annum.

At Smith's Point, between First and Second Brushy cañons, a strip of ground some 2,000 feet long and having an average width of 200 feet approximately, has been washed off, with an average depth of fifty feet. It is estimated that more than a million of dollars has been taken from this ground. This was first drifted over, however, to a considerable extent and afterwards "hydraulicked," and a portion of the gold was obtained from drifts running from 200 to 600 feet beyond the present faces of the banks.

Mr. K. B. Soulé, Express Agent at Forest Hill, stated that the total yield of the Paragon Mine, at Bath, had been, up to 1871, certainly over $750,000; and that during five years from 1865 to 1870 its regular product was $100,000 per annum, no two years differing from each other by so much as $5,000, during that time.

The area which has been hydraulicked off at Todd's Valley is probably about a mile long with an average width of a quarter of a mile. The ground washed off here was in the form of a ridge, which along the middle line was some sixty to seventy-five feet in depth, and the average depth was about thirty-five feet. This gives a total amount of 9,000,000 cubic yards nearly, and the estimated yield from this ground is $4,000,000, or about forty-four cents per cubic yard. The estimate of the yield of this ground was given by Mr. Pond, who has lived at Todd's Valley since 1849, and is and has been largely interested in mining operations at that place.

A piece of ground in Pond's Claim, of the yield of which precise record had been kept by Mr. Pond, and which was hydraulicked in the five years beginning with 1866, was surveyed by Mr. Goodyear. It was found to contain 138,206 square feet, or 3.1728 acres, equal to 15,356 square yards. If the pay gravel averaged thirty-five feet deep, as estimated, this will give 179,153 cubic yards of auriferous gravel, and the total estimated depth of fifty feet will give 255,933 cubic yards, in all, washed away from this ground. Its total yield was $91,828.30. Considering, therefore, the auriferous gravel alone, the yield would average 51.257 cents per cubic yard; while, considering all the bank moved, it would be 35.88 cents per cubic yard. Since the water bill for the same time was $29,003.42, it also appears that the water cost about 11.332 cents per cubic yard of earth moved. The price of the water was ten cents per "inch" for ten hours, or one cent per "inch" per hour. But the method of measuring the water was an unusual one. The opening in the measuring box through which the water issued was ten inches high, its bottom being level with the bottom of the box, and the water standing four inches above the top of the opening, or fourteen inches deep. The amount of the discharge was regulated by a sliding gate which varied the horizontal length of the opening according to the quantity of water required. For example, if 400 "inches" were required, the gate was set to leave the opening forty inches long.

The former Express Agent at Georgetown, Mr. Murphy, stated that Mameluke Hill had yielded not less than two millions of dollars.

The Roanoke Channel at Bottle Hill, near Georgetown, was found to pay well wherever there was any gravel, and many spots were extremely rich. From a single candle-box full of dirt as much as seventy ounces are said to have been taken. The total yield of the channel has been, it is thought, not less than a million of dollars. Almost all the ground has been worked over three times, and a part of it four times, each time paying satisfactorily. In the first spur reached by the channel after crossing Roanoke Gulch, a spot about a hundred feet long, in the Buffalo Claim, yielded, it is stated, about $60,000. In the Bottle Hill Claim, immediately below the last bend in the channel, before it passes out of the hill, there was a spot about 200 feet long, where the whole width of the channel did not exceed seventy-five feet, but which yielded as much as $200,000, which would be over thirteen dollars per square foot of surface of the bed-rock. In this claim it is said that there was in 1851, or 1852, a face of gravel five or six feet in thickness, stretching entirely across the channel, and, to use the miner's expression, "perfectly yellow" with gold. Yet rich as this gravel was, it has not generally paid expenses to hydraulic the top off, after the bottom had been drifted out. In all the spurs crossed by the channel, after leaving Roanoke Gulch, the cement is decomposed and soft, and easily piped, and the banks would not anywhere be more than about a hundred feet in height, if the whole bed of this portion of the channel were hydraulicked. Consequently several pits have been opened, and the attempt made to wash the whole off; but, as a whole, the work has not even paid "water-money."

At Wilcox's Claim, on the north fork of Long Cañon, the gold is distributed entirely through the gravel from top to bottom. It is both coarse and fine, and generally very smoothly washed, although flaky. The bed-rock surface is not rich, and would hardly pay for drifting. About two acres have been washed off here, and $100,000 obtained.

The proprietor of Flora's Mine, west of Volcanoville, states that his gravel averages at least two dollars a car-load, and that his claim has yielded about $20,000.

At the Excelsior Claim, near Placerville, about twenty acres of ground have been washed off, with an average thickness of from fifty to sixty feet of pay gravel. The yield from this ground is estimated at over $5,000,000, about half of which was taken out by drifting, and the other half by hydraulicking. The whole ridge has been drifted out at two different levels; first on the bed-rock, and then along a second pay streak which was about twenty feet above this. The latter streak paid about $12 per day to the hand employed. Mr. Alderson, the principal proprietor, estimates the yield of the Excelsior gravel at one dollar per cubic yard of dirt washed away; this includes also the top, which is volcanic gravel.

It is estimated that the yield of the ground in Coon Hollow, near Placerville, has been at the rate of five dollars per cubic yard, and that the total amount obtained there is $5,000,000.

§ 16. *Amount of Gravel still remaining.*

The distance in a straight line from the outcropping bed-rock a little southwest of the Parker House, at Iowa Hill, to the base of the first Sugar Loaf is some 2,300 or 2,400 feet, and within this distance nothing but gravel is to be seen on the crest of the hill. The Odd Fellows Hall is not far from midway between these two extremities, and the drifting never extended so far southwest as this Hall. As good a rough estimate as can now be made of the total quantity of gravel originally existing, and capable of being hydraulicked, in this ridge, between the first Sugar Loaf and Independence Hill, would be half a mile square, or 160 acres, with an average depth of one hundred feet over the whole of it. The little flat on which stands Independence Hill is about a quarter of a mile long, and it is covered, though not very deeply, with volcanic matter, and may possibly yet be all hydraulicked. This estimate would give, in round numbers, about 26,000,000 cubic yards of auriferous gravel in the Iowa Hill Ridge, between the first Sugar Loaf and Independence Hill. Of this quantity it may be estimated that about 20,000,000 cubic yards were concentrated in the deep channel which crosses the crest at Iowa Hill itself, and the remainder scattered along the sides of the ridge at various points, but chiefly along the southeastern side near Indian Cañon, and in the shallow patch which caps the ridge near the Catholic church, a little northeast of Iowa Hill. The total quantity already washed away is probably between two and three million cubic yards, of which about nine tenths may be assumed as having come from the deep channel, and the remainder from smaller pits at other points. This would leave between twenty-three and twenty-four million cubic yards yet capable of being hydraulicked. How much of this it will ever pay to wash is a difficult question to answer; but probably the greater portion of it will be worked with profit, whenever the supply of water shall be plentiful, reliable, and cheap. Such estimates as the above must, however, be considered only as approximations. In this case, the area was calculated from measurements made by pacing, the heights of the banks being determined by the aid of the barometer. There was also as much information obtained as possible in regard to the probable lay of the bed-rock under the gravel, both from personal observation and from the statements of men who had worked in the various shafts and tunnels in the region in question.

Mr. Goodyear makes the following estimate of the amount of ground remaining to be worked by the hydraulic process, in the vicinity of Placerville : —

Locality.	Dimensions.	Cubic Yards
Hangtown Hill to Reservoir	$2640' \times 1000' \times 75'$	7,333,333
Excelsior Claim, Coon Hill	40 acres 135' deep	8,712,000
Webber Claim, Coon Hill	$900' \times 1200' \times 100'$	4,000,000
Webber Hill	40 acres 75' deep	4,840,000
Little Spanish Hill	20 acres 100' deep	3,226,666
Big Spanish Hill	10 acres 100' deep	1,613,333
Smith's Flat	$1200' \times 200' \times 60'$	533,333
Clay Hill	3 acres 15' deep	72,200
Indian Hill	6 acres 20' deep	193,600
Negro Hill, below Reservoir	50 acres 75' deep	6,050,000
Steven's Claim, spur	15 acres 40' deep	968,000
White Rock Cañon, southwest side	$1200' \times 1000' \times 40'$	177,778
White Rock Point, spur	$1500' \times 400' \times 90'$	2,000,000
Cedar Flat	15 acres 45' deep *	1,089,000
	Total,	40,809,743

If now to the above estimate for special localities we add 25 per cent, which is probably a liberal allowance for the quantity which remains distributed along the sides and spurs of various hills, and not taken into account above, we may say that as fair an estimate as can now be made for the total quantity of gravel capable of being hydraulicked in the vicinity of Placerville will amount, in round numbers, to fifty millions of cubic yards.

SUBDIVISION II. — THE REGION SOUTH OF THE MOKELUMNE RIVER.

The region between the Mokelumne and the North Fork of the American River having been described in the preceding pages, chiefly from the notes of Mr. Goodyear, we turn next to that portion of the western slope of the Sierra which lies to the south of the Mokelumne. The region which is now about to be described is extremely interesting from many points of view; but it is not a great hydraulic mining district. It has been the scene of extremely rich placer, river, and tunnel mining operations, and many quartz mines have been worked in the three counties of Calaveras, Tuolumne, and Mariposa; but no one of these interests can be said to be, at the present time, in a very flourishing condition. Few, if any, of the quartz mines have proved to be permanently productive and capable of being worked to great depths like those in Amador County; the river beds have long since been thoroughly cleaned out. The rich channels under the various lava flows, like the Sonora Table Mountain, have been worked out by tunnels, and

* This item was added by Mr. Alderson, who examined and approved of Mr. Goodyear's estimates as given above.

the gravel deposits, although numerous and often rich, are on too small a scale and too "spotted" to be capable of being attacked with success by hydraulic mining operations on a large scale.

In describing the geology of the region south of the Stanislaus we may follow an order somewhat similar to that already indicated for the region farther north.

§ 1. *The Bed-Rock in Calaveras, Tuolumne, and Mariposa.*

So much has already been stated in regard to the bed-rock in Tuolumne and Mariposa[*] that this branch of the subject may be passed over with rapidity.

The most interesting orographical feature of this region in connection with the bed-rock series is the existence of a subordinate range in the foot-hills, parallel with the Sierra, but having more of an isolated and independent character than any other of the more or less continuous foot-hill ranges which lie along the base of the great mass of the Sierra itself. The range in question is so distinctly marked as to have received a name, and the group is known as the "Bear Mountains." It is, in fact, a double range, having two quite well-marked divisions; the more southwesterly, which is parallel with the other but quite subordinate to it, is known as the "Gopher Hills." A glance at the map of Central California will show the position of these elevated ranges. They extend between the Calaveras and the Stanislaus rivers, and govern the distribution of the drainage within that region, giving it a character quite unlike what it usually has in this part of the Sierra. The various streams which rise in the upper portion of Calaveras County flow with the usual southwesterly course, until they reach the northeast edge of the Bear Mountains, when they turn at right angles, and flow northwest and southeast to the Calaveras or the Stanislaus. The smaller creeks, which rise in the rather broad and regular valley between the Gopher Hills and the Bear Mountains, also have courses parallel to that of the ranges themselves.

The rocks of the Bear and Gopher ranges are all of the bed-rock series, and highly metamorphic. The slates are well developed in the depression between the two ranges, and are there associated with irregular bands of serpentine which are sometimes of great length and width. This region was formerly the seat of the most active copper-mining operations in the State, Copperopolis being the principal mining centre. The only important

* See *ante*, pp. 43 – 47.

vein at this place, the so-called "Reed Lode," was enclosed, for a portion of its course at least, between walls of argillaceous slate on one side and serpentine on the other. In Salt Spring Valley, a continuation of the depression between the two ranges, previously mentioned, ten or twelve miles northwest of Copperopolis, the rocks consist almost entirely of slates, generally thin-bedded and fine-grained; the prevailing character is argillaceous, although portions are talcose or chloritic. These slates, with the associated serpentine, appear to occupy the whole width of the valley between the two ranges, which themselves are made up of much harder rocks, evidently highly metamorphosed and probably, as farther south, composed in good part of volcanic materials. The chemical changes in these rocks have been carried so far that the stratification is almost or quite obliterated, and could hardly be made out at all unless from the peculiar form of the elongated outcrops on the surface, which everywhere on these ranges is but sparsely covered by detritus. At the western base of the Gopher Hills the granite appears at the surface and occupies a considerable width, where crossed by the stage-road from Stockton to Copperopolis.

Not far from Telegraph City is the locality called Quail Hill, which at one time attracted much attention on account of its supposed richness in gold. There is here a belt of several hundred feet in width, through which the rock has been acted on by chemical influences so as to have become more or less irregularly decomposed, the resulting material, which is of a clayey and ochery character, having assumed a variety of brilliant colors. The local name for this decomposed formation is "calico rock." The same peculiar belt may be traced at intervals for several miles along the edge of the foot-hills, and being near the granite, which, as already mentioned, occurs in this position, it is not unlikely that the proximity of the eruptive mass is connected with the peculiar decomposition of the adjacent slaty rock. This belt of calico rock was at one time believed by some persons to be very rich in gold, and Quail Hill was the scene of considerable excitement in consequence of this.*

On the north side of the Bear Mountain Range, and in close proximity to it, is the belt of rock through which the Great Quartz Vein passes, and which presents many phenomena of interest with reference to the occurrence of the various formations characteristic of the western slope of the Sierra, although

---

* See Proceedings of the California Academy of Sciences, Vol. III. p. 349, in which article the whole mass of rock at Quail Hill is stated to be worth $50.22 per ton in gold and silver.

by no means a region of important hydraulic mining operations. A large part of Calaveras County above the Bear Mountains is covered by volcanic materials, which in places come down within a few miles of the foot of that range, although nowhere invading it. The Bear Mountains seem to have acted as a barrier during the volcanic period; neither gravel nor lava are found in this range, or in the Gopher Hills. The volcanic, however, in the central belt of Calaveras, on the line between Angel's Camp and San Andreas, occurs in patches, covering only the higher portions of the country; the bed-rock is well exposed in the lower regions, and especially in the valleys of the rivers.

In the vicinity of Mokelumne Hill the slates are very quartzose and micaceous. Their dip varies much more than their strike. At French Hill, one fourth of a mile from Mokelumne Hill, the strike of the mica slates is N. 67° W., and their dip 45° to the north-northeast. In Burns's Gulch, on the opposite side of the Mokelumne River, the strike is nearly the same as in the last-mentioned locality; the dip is a little higher, namely, 65°. There is also a large amount of hornblende rock, or amphibolite, in this region. This rock, which is made up of hornblende, associated with a little quartz and mica, is of dark-greenish color, and is occasionally traversed by veins of epidote, with which are associated masses of crystallized brown garnet.

The slates in the vicinity of Angel's Camp are gray or light-green in color, and fine grained. Near the Great Quartz Vein they become porphyritic. The bed-rock all along the line of outcrop of this vein is very similar in character to that described as occurring in similar relations in Tuolumne County.

The most interesting feature of the bed-rock geology in Calaveras County is the great development and peculiar character of the surface of the limestone belt, the occurrence of which farther north has already been noticed. This belt of rock is continuous from Kincaid Flat, a little east of Sullivan's Creek, in Tuolumne County, as far as the Stanislaus River, and its average course is nearly uniform, although its width and the space it occupies on the surface are quite variable. At Kincaid Flat its development hardly exceeds a few hundred feet in width, but beyond Sonora it expands rapidly, and just before reaching Springfield a line drawn across it at right angles to its trend would be fully a mile in length. Just beyond Springfield its northeastern boundary bends sharply around at a right angle, and runs for nearly two miles towards the northeast, then turns again as sharply, at Shanghai Mountain, and assumes its ordinary course of northwest. By this change of trend the limestone belt acquires a width at Columbia, and from there west to the Stanislaus, of from one and a half to two miles. Great numbers of dykes of dioritic rock cross it nearly at right-angles to its strike; they vary in thickness from a few inches to many feet. It would appear that these dykes are a peculiar feature of the limestone belts, since they occur not unfrequently in this connection.* It is not easy to see why there should be any special reason for their occurrence in the limestone belt in preference to other portions of the bed-rock series. It may be that they are more conspicuous in that position, on account of the marked difference in color of the two rocks and the complete exposure of the surface of the limestone. This rock, which between Kincaid Flat and the Stanislaus was formerly covered by a heavy mass of detritus, has been washed clean, and exhibits a most curious appearance, as already noticed in describing the bed-rock in the vicinity of Volcano. Its surface is everywhere worn into cavities, sometimes as much as fifty feet deep, but oftener ten to twenty, the inclination and direction of which coincide with the dip and strike of the rock itself, while their sides are smooth in appearance, looking as if the material had been removed by some corroding, rather than eroding, agency. These cavities were once filled with auriferous materials, which also appear to have extended over the whole surface in distinctly stratified beds, rising several feet above the general level of the higher projecting portions of the limestone.

Beyond the Stanislaus the limestone belt is quite irregular in its development; but it is, as before remarked, a prominent feature in the bed-rock geology of Calaveras County. At Abby's Ferry, where it is intersected by the river, it is more than a mile wide, and rises in highly picturesque cliffs on either side. The strike of the limestone here is nearly east and west, magnetic, and the rock underlying it is almost exclusively made up of hornblende, which occurs in very coarsely crystallized masses. On the west side of the Stanislaus, away from the immediate vicinity of the river, the limestone is covered by the lava flow of Table Mountain; but it is well exposed again in the neighborhood of Douglass Flat and Murphy's. The strike of the beds in this county is very variable. At the Ferry, as already remarked, it is nearly east and west; and while the general trend of the formation from Vallecito to Murphy's is about north and south, yet the strata appear to run in a direction nearly at right-angles to this. At the Blue Wing Mill, for instance, three fourths of a mile north of Mur-

---

* See ante, p. 87.

phy's, the strike of the limestone is N. 80° E., and its dip is to the southeast at an angle of 75° or 80°. The outcrop of this rock at Murphy's, where the belt is a little over a mile wide, is in part similar to that described as occurring on the other side of the Stanislaus, and the cavities have been extensively washed for gold. On the northern side of the town, however, the limestone rises in high hills, very conspicuous from their peculiar blue-gray color, so different from the usual tinge of the bed-rock in the mining districts. This character of the limestone continues between Murphy's and San Domingo Creek, on the trail to Cave City. Just beyond the last-named creek it disappears again, and is not seen until we arrive near Cave City, about five miles, in a direct line, northwest of Murphy's. At this point the limestone belt is very conspicuously exposed, resembling in the character of its outcrop the hills north of Murphy's. The strike of the beds at Cave City is about N. 30° W., and the dip to the northeast at an angle of 60°, the width of the belt being here about one fourth of a mile. From this point the limestone can be seen trending to the northwest and presenting, for a distance of two miles, a series of precipitous and almost bare outcrops. Beyond this it has not been seen by any member of the Geological Survey Corps until the neighborhood of Volcano is reached (see page 87).

The bed-rock on the northeast side of the limestone belt in Calaveras County is mostly slate of a very quartzose character, or quartzite. It is much less auriferous than the rock on the other side of the limestone, there being hardly any mining, whether quartz or placer, to the northeast of that belt of rock.

§ 2. *Gravel and Volcanic Formations in Calaveras, Tuolumne, and Mariposa Counties.*

The gravels are comparatively of so little importance in the region south of the Mokelumne, that in a rapid description of the geology it is not desirable to describe them separately from the volcanic formations. The latter are, however, of great extent and interest; but the region requires a much more detailed investigation than has yet been given to it, based on an accurate map; and this field work would have to be followed by a microscopic examination of the specimens collected. All that can be done, at present, is to point out roughly the extent of the region covered by the volcanic beds, and to give the character of the gravels at the principal points where they have been worked.

Unlike the counties to the north, of which the geology has just been described, Calaveras does not extend so far up on the range as to take in any portion of the summit. Alpine County has been formed out of the territory on both sides of the dividing ridge at the expense of Amador, Calaveras, and Tuolumne. This was done at the time the mines near Silver Mountain were expected to be very productive, which they have not proved to be.* This peak is the centre of a grand eruptive region, from which the lava flows have spread themselves far down the slope of the range. Nearly all the crest of the Sierra, through Alpine County, is covered with volcanic materials, which are sometimes accumulated to a great thickness, as in the mass of Silver Mountain itself; but it is everywhere evident that granite forms the basis on which the volcanic materials have been piled up, the lava occupying only the crests of the spurs. One of the grandest, and perhaps the most picturesque, of all the volcanic ranges in the State is the one about six miles northwest of Silver Mountain, called on Holt's map the "Cliffs of Nonatore."† From here seem to have originated great flows of lava, which have spread down the slope of the Sierra into Calaveras County, and which are so conspicuous to one looking from Murphy's towards the north, and over which the traveller passes in going from the Calaveras Big Tree Grove to West Point or Railroad Flat. The last-named place is quite surrounded by volcanic ridges and tables, which are elevated from 600 to 800 feet above the Flat, and capped by basalt which weathers into rounded, boulder-like masses. The volcanic materials continue almost or quite uninterruptedly from Railroad Flat to Mokelumne Hill and far down on the divide between the Calaveras and the Mokelumne, into the very foot-hills. From the appearance of the low flat-topped elevations which occur all along the base of the Sierra between Dry and Little John Creeks, a very large amount of volcanic material has been carried down the range by currents of water, and deposited at its very base in an almost or quite continuous mass of considerable extent and thickness. This, however, has been so extensively and irregularly eroded away, that, without maps on a very large scale and a long course of detailed exploration, it will be impossible to lay them down with accuracy. The materials of the different flows have, in all probability, been much mixed together by shifting currents, so that it would, under any circumstances, be

---

* See Geology I. p. 449. Hardly anywhere in the country have more persistent efforts been made to work worthless mines than here.

† See Geology of California, I. p. 446.

extremely difficult to refer any portion of it to its original source in the High Sierra. The region near Q. Ranch, a few miles northwest of Ione City, shows a number of these isolated mound-like outliers of volcanic sedimentary deposits. One of these elevations, called Pratt's Hill, four or five miles west of Ione City, is made up of a series of strata of clays of different colors and textures. There are five of these beds, which seem to have been formed, in large part, from the trituration of the débris of volcanic materials, resting horizontally on the bed-rock, which is here a hard, compact, fine-grained silicious slate, somewhat ferruginous in character. On the clay beds is a deposit of coarse gravel, made up almost exclusively of well-rolled pebbles of lava. This is succeeded by a fine gravel cemented by volcanic sand, and capping the whole a solid (andesitic?) lava.

The volcanic deposits cover a large amount of surface in the vicinity of Jackson, which town is situated upon the slates, but is surrounded by an almost continuous line of elevations capped with volcanic materials. Large areas to the south and southwest of the town have been washed off and found rich in gold. North of Jackson and extending along the south side of Sutter Creek is a continuous ridge of lava, known as Humbug Hill. The thickness of the detrital deposits here is about 250 feet; of this, about 200 feet consist of finely stratified materials, and the upper fifty of imperfectly rounded masses of lava (andesite?) which grow larger towards the top of the stratum. The thickness of the pay-streak is quite small, in places not over one or two feet, and it is made up chiefly of pebbles and slightly rolled fragments of quartz, including, however, some very large boulders of the same. The channel is said to be about 125 feet in width.

The "Butte," about four miles east of the town of Jackson, is an isolated knob, from which a fine view of the adjacent region may be had. The summit is elevated about 1,200 feet above the town of Jackson, and 800 feet above its base. At the base of the Butte, on the west side, an excavation made by prospecters shows horizontal strata of white clay, sand, and finely powdered pumice-stone, the thickness of which could not be ascertained. Above this apparently horizontally stratified and undisturbed sedimentary volcanic deposit is at least 500 feet in thickness of a hard eruptive material of a reddish color, supposed to be basaltic in character.* This mass is distinctly laminated or stratified, and the layers have a variable strike and inclination. The prevailing direction, however, is northeast and southwest, and the dip, which is very irregular, is in places as much as 50° or 60°.

* The specimens obtained in this vicinity were all destroyed by fire.

The cañon of the Mokelumne River, between Jackson and Mokelumne Hill, in Calaveras County, is about 1,500 feet deep, and the rock exposed here is of a gneissoidal character. The river, along this portion of its course, has been "flumed" again and again; it was long since abandoned to the Chinese, and probably has ceased to be worked even by them. The detrital formations at Mokelumne Hill were, in the early days, the scene of the most active mining operations, and they seem to have been, in places at least, extremely rich. Mr. Hittell says, "Every variety of mining operations has been carried on successfully in the neighborhood of Mokelumne City..... There is scarcely one of the numerous ravines and gulches in the vicinity the bed of which has not been overturned, year after year, since 1849; first, with the butcher-knife and pan or *batea*; then with the pick and shovel and the rocker; next, with the long-tom; and finally, with the sluices. At last they were abandoned to the Chinamen with their rockers, and the Digger Indian women with their little crowbars, horn scrapers, and tin pans."* It is said that the claims at Mokelumne Hill were limited to fifteen square feet, and that some of them produced as much as 250 pounds of gold each.† The gravels in this vicinity were thin, but extremely rich. They were covered by a mass of sedimentary volcanic materials as much as 200 feet in thickness. This deposit was somewhat variable in character; but generally very fine-grained and homogeneous, having a pinkish-red color, and breaking with a conchoidal fracture. Over this, again, was a mass of boulders of lava (andesite?) not polished and smooth as if formed by the action of water, but roughly rounded in such a way as to lead to the inference that their form was due to dry friction — if the expression may be allowed — of a mass of lava broken up on the surface and carried downwards by the aid of gravity alone. Large masses of this lava detritus were closely examined by the writer, without the discovery of a single pebble of quartz or any other form of metamorphic rock. The Mokelumne Hill channel seems to be continued for considerable distance to the southwest. Along the borders of Chili Gulch, for a distance of six miles from Mokelumne Hill, active mining operations were being carried on from 1860 to 1864, and perhaps later. Here there were several beds of gravel intercalated in the lava, the "pay-streak" being at the bottom, however, and about eight feet thick. Much of it was so compacted together that it had to be stamped before it

---

* Mining in the Pacific States, p. 90.

† P. Laur, Métaux Précieux en Californie, p. 27.

could be run through the sluices. That this could be done with profit,—as must have been the case from the length of time the operation was carried on,—shows that the gravel must have been quite rich in gold. The continuation of the Mokelumne Hill channel below Chili Gulch, towards Double Springs, has, apparently, not been found worth working to any considerable extent.

In that portion of Calaveras County which lies to the southeast of the river of that name there are numerous localities where gravel deposits occur under the volcanic materials; but none of these are of great extent, or have been worked very continuously. The localities where these gravels occur are chiefly in the vicinity of San Andreas, Altaville, and Vallecito.

The gravel deposits at San Andreas have been worked in former years to considerable extent. The formation appears to be from 100 to 150 feet in thickness. A section, as given by two of the miners at this place, will be found farther on, in the chapter devoted to the remains and works of man in the gravel under the volcanic beds. The eruptive masses here would appear from their position to be connected with the flow which has come down from the Sierra behind the Big Tree Grove, and which passes about two miles north of Cave City, and is there known as Table Mountain, a name very commonly given in California to lava flows, and especially to isolated patches of such flat-topped masses.

In the vicinity of Douglass Flat and Altaville there is a large development of volcanic materials, and some gravels which have been worked from time to time. There is a nearly level tract of land extending north from Vallecito to Douglass Flat, known as Vallecito Flat, which is an area depressed below the surrounding region, and filled to a considerable depth by gravel, of which the exact limits seem to be not well known. The bottom of this depression is too deep to be drained, except by a long tunnel, and such a one was commenced many years ago, but never completed. A tunnel from the forks of Coyote Creek, about a mile in length, would, it is supposed, open this ground at a sufficient depth to permit its being worked by the hydraulic method. There is an elevated ridge of volcanic breccia extending along the northwest side of Vallecito Flat, which rises to a height of nearly 600 feet above the lower portions of the surface of the Flat, and the Sonora Table Mountain extends, with a nearly equal elevation, along the east side, occupying most of the space enclosed between Murphy's, Douglass Flat, and Vallecito Flat on the west, and the Stanislaus River on the east, as far as

Abby's Ferry. The surface of the depression on the west side of Table Mountain is drained by Coyote Creek, a branch of the Stanislaus, both streams running nearly south in this part of their course. The depth of the gravel which partially fills this depression is stated by those who have worked there to be in places considerably over a hundred feet. The position of the rim-rock is also said to indicate that the course of the channel is nearly west. Only a small portion of the Flat seems to have been worked by drifting down to the surface of the bed-rock. Mr. Goodyear, who examined this locality, in 1877, for a private company, says that no definite information can now be obtained in regard to the yield of the gravel of Vallecito Flat, in the early days, except that it was "extremely rich," and that many large nuggets were found there. He adds: "That it was indeed rich may be inferred, however, from the extent of the drifting which was done, and the conditions under which it was done. From the best information which I could obtain, it appears that about eight or ten acres of this ground at the lower end of the Flat were drifted out. . . . . It was done by the sinking of vertical shafts from fifty to over a hundred feet in depth, through which all the water had to be pumped and all the gravel hoisted. Moreover, as the individual 'claims' in those days were very small, and as only a few claimants would unite in the sinking of any single shaft, the shafts were very numerous, and often very close together; and their mouths, together with the heaps of dirt around them, still remain to attest the extent of the work which was done."* Of late years a portion of the ground which had been thus previously exhausted, as far as practicable, by the earlier drifting has been washed away down to the bed-rock. This operation was still going on in 1877, and up to that time two or three acres of ground had been worked over in this way to an average depth of thirty-five or forty feet, and in places to as much as sixty or seventy, but under great disadvantages, for want of the necessary deep drainage, without which the gravel has had to be hoisted in cars on inclined tracks and the water pumped by machinery. Immediately north of this drifted ground there is, as Mr. Goodyear thinks, an area of from forty to fifty acres in the bottom of the valley which has not been touched, and over which the gravel may average from seventy-five to eighty feet in depth, as appears from prospecting shafts sunk over it to a depth, some of them, of over a hundred feet, and from which the miners were driven out by water. Altogether this is a very remarkable locality, espe-

* From a pamphlet report by Mr. Goodyear, to private parties, made in April, 1877.

cially on account of its great depression below the adjacent channel in Table Mountain.   Until it has been more extensively worked, but little can be made out in regard to the connection of this deposit of gravel with any other in the vicinity.

The volcanic formations described as occurring to the west of Vallecito Flat extend down nearly to Angel's Camp, and may be observed in low flat-topped elevations on the north of that place and northeast of Altaville. There are several alternations of lava and gravel; but the hills are too deeply covered with soil to allow the geology to be well made out from an inspection of the surface.   There have been several shafts sunk and drifts run to strike the gravel in this vicinity; but they have not been accessible when visited by the writer.   What has been ascertained in regard to the details of the occurrence of the gravel and lava near Altaville will be noticed in the section devoted to the occurrence of human remains at Bald Mountain, the famous locality of the so-called " Calaveras skull."

The Sonora, or Tuolumne County, Table Mountain, already several times alluded to in the preceding pages, is a lava flow of so much interest, from a variety of aspects, that it demands a somewhat detailed description.   It is perhaps the best marked of all the flows of volcanic materials down the slope of the Sierra, not so much on account of its great length, but because it is more continuous and more isolated than such flows usually are, as will be apparent from the description.   It also has a special interest, due to the fact that the stratified materials underlying it have been found to contain a considerable variety of plants and of animal remains, including the bones as well as the works of man.   The Tuolumne Table Mountain begins in Calaveras County at some point high up in the Sierra, which has not yet been positively identified.   It seems, however, the same flow which is crossed on the way from the Calaveras Big Tree Grove to the South Grove, so called, and this appears to connect with the lava ridge which is so prominent south of Hermit Valley, on the Big Tree and Carson road.   At all events, this flow is well ascertained to be continuous on the west side of the Stanislaus for at least ten or twelve miles above the point where it crosses that river between Abby's and Pendola ferries.   On that side of the river, however, there has been but little mining done, although the channel seems to have been prospected, without success, at various points near the river.   From that crossing, which is about on the line connecting Sonora with Angel's Camp, the Table Mountain has been cut into by tunnels at quite a number of points and the channel opened, so that a very good idea can be had of its structure.

This lava flow has long been known and has excited much attention, not only on account of its peculiar features, but because of its richness in gold. It has already been referred to as having been described by Dr. Trask (page 68) in his Report for 1856. Mining was commenced under this flow in November, 1854. the first discovery having been made, it is said, near Shaw's Flat, by the Brown boys, who in cleaning up an old shaft struck gravel, which although only one foot in thickness was very rich in gold. Two years later, judging from Dr. Trask's description, there must have been a good deal of activity displayed here. The region was examined by the writer in 1861, when work was going on at several places; when revisited, ten years later, there seemed to be nothing more doing. All the tunnels, between Shaw's Flat and Jamestown at least, had been abandoned. Mr. Rémond also in the course of his examinations near Sonora, in 1866, made two sections across Table Mountain. A description of the principal features of this lava flow will be found in Geology, Vol. I., with some sections illustrating its structure and peculiarities.

The fact that the Sonora Table Mountain formerly extended over what is now the cañon of the Stanislaus was early recognized by the miners, and is so evident that it could not escape their attention. To one standing on the summit of the mountain at the point of crossing, where it first passes from the west to the east side of the river, and looking from either side over to the other, the evidence of a former junction of the dissevered portions seems to be very clear. There is no mistaking the fact that the portion on the west once connected with that on the east. After crossing here, the flow continued on for about two miles in an easterly direction to a point near where the present town of Springfield is situated, then turned and ran a little east of south in a line nearly parallel with the present Stanislaus, for about ten miles, to a point a little west of Montezuma; here it divided into two portions. of which one ran west for four or five miles and again crossed what is now the cañon of the river and continued down on that side, while the remainder seems to have kept to the east of the present river. From Montezuma downwards, however, the current spread out over such a wide area, and has been so much eroded away, that the original form of the flow could not be restored without much more labor than we have been able to bestow upon it. From the diagram which is here given (Plate D), however, a good

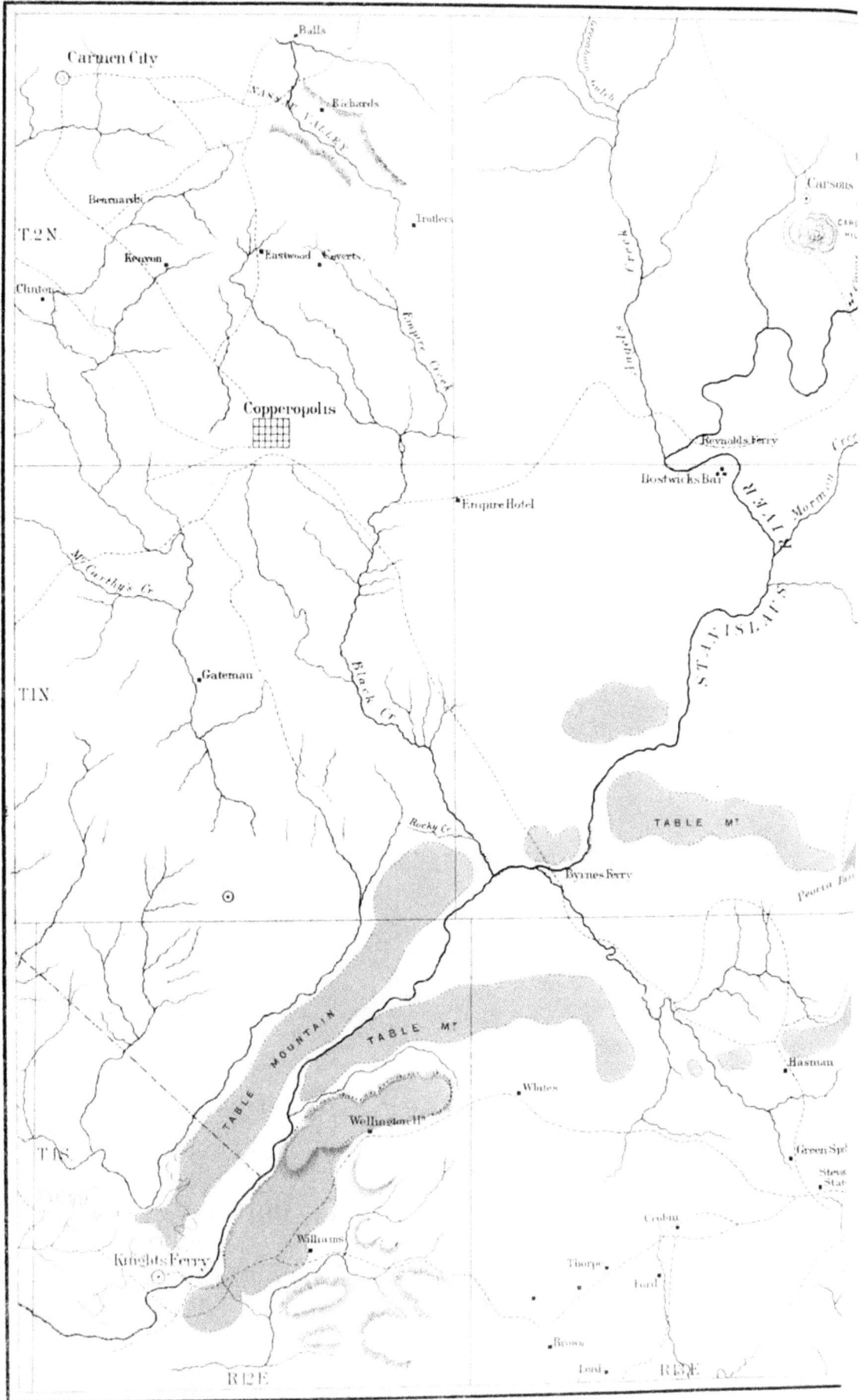

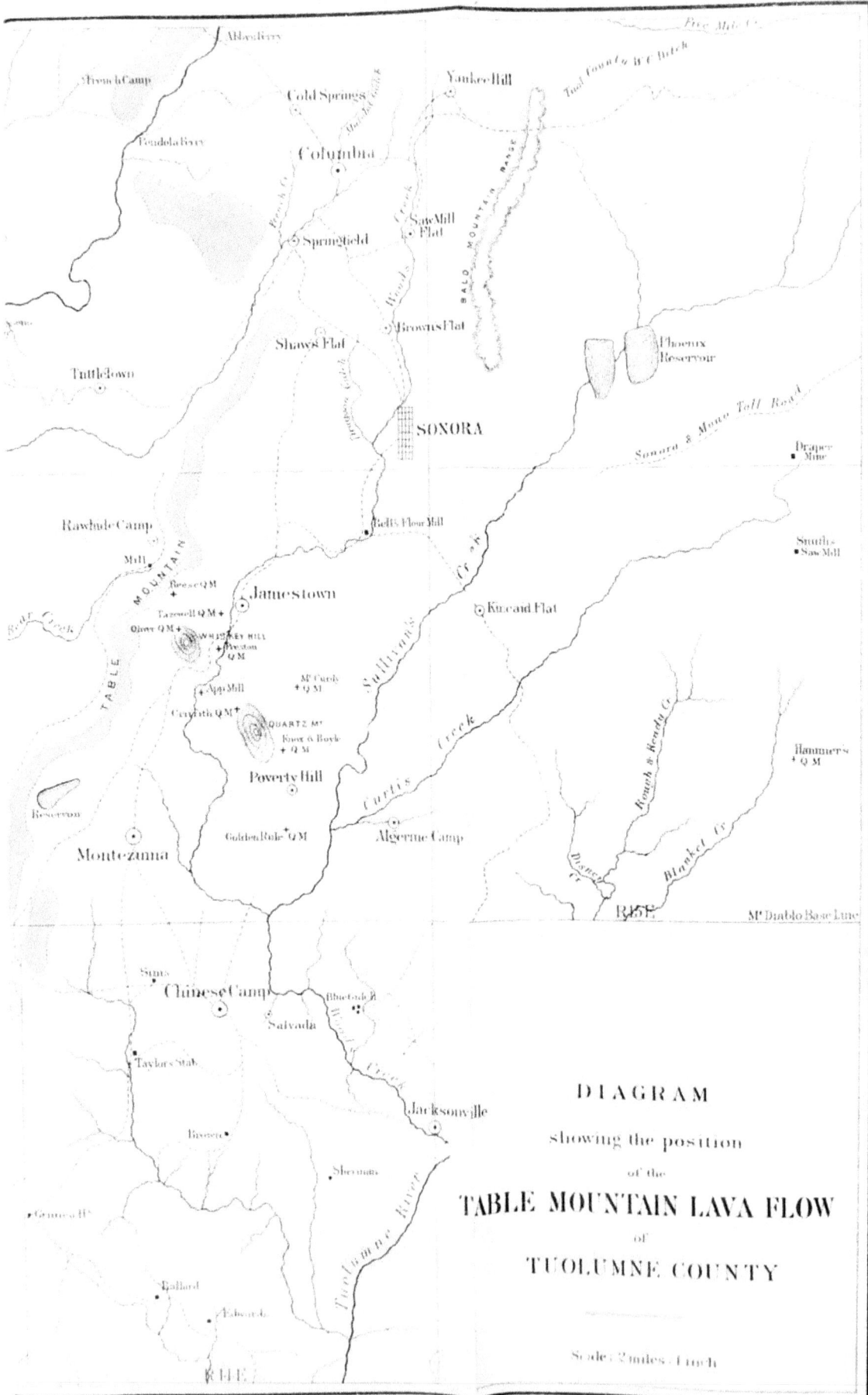

DIAGRAM

showing the position

of the

TABLE MOUNTAIN LAVA FLOW

of

TUOLUMNE COUNTY

Scale: 2 miles = 1 inch

SONORA

Columbia

Yankee Hill

Cold Springs

Shaws Flat

Browns Flat

Saw Mill Flat

Springfield

Tuttletown

French Camp

Abbeys Ferry

Pendola Ferry

Phoenix Reservoir

Draper Mine

Rawhide Camp

Mill

Reese Q M

Tazewell Q M

Oliver Q M

WHISKEY HILL

Preston Q M

App Mill

Corinth Q M

QUARTZ M^t

Knox & Boyd Q M

Mc Curdy Q M

Jamestown

Poverty Hill

Golden Rule Q M

Algerine Camp

Kincaid Flat

Smiths Saw Mill

Hammers Q M

Montezuma

Reservoir

Chinese Camp

Salvado

Bluetub R

Taylors Stab

Brown

Sherman

Genoa H^o

Ballard

Jacksonville

Sonora & Mono Toll Road

Mt Diablo Base Line

TABLE MOUNTAIN

Sullivans Creek

Curtis Creek

Tuolumne River

Woods Creek

Reed Creek

BALD MOUNTAIN RANGE

general idea of the position of the flow may be obtained. It breaks off pretty abruptly close at Knight's Ferry, where the basaltic masses at its terminus are very conspicuous. The total distance from the first crossing of the river to the foot of the flow is about twenty miles in a straight line. The exact amount of fall in this distance is not known; but it appears to be on the average about seventy-five feet to the mile. On the trail which crosses the top of Table Mountain, on the way from Murphy's to Pine Log, the elevation of the summit, as determined barometrically, is 2,515 feet; southwest of, and at the nearest point to, Columbia it is 2,247 feet; over the Buckeye Tunnel west of Sonora 2,033 feet, which would give an average fall of about eighty feet to the mile in this part of its course.

Between Shaw's Flat and Montezuma, a distance of eight miles, the flow of Table Mountain is very regular and uniform, and it is here that the channel has been reached in quite a number of places by tunnels driven in through the rim-rock, and mining has been carried on by drifting in gravel, which here, however, forms but a narrow and thin belt. There is perhaps no place in California where all the peculiar features of one of these lava flows covering an ancient river channel are so well seen as here. To one standing at a distance and looking in the direction of this flow, from a point affording an unobstructed view of it, the horizontal upper edge of the dark mass of lava, looking perfectly straight, but with an evident gentle descent towards the valley, offers a marked contrast to the usual curving lines of the eroded hillsides of the Sierra. The absence of vegetation on this table-like elevation also helps to render it very conspicuous. It is seen at once that the top of this flow is higher than the surrounding country, and that there must have been a large amount of rock removed, by denudation, from each side of it, in order that it should stand out above the adjacent country in such isolation. It is evident that being, as is clearly seen from the nature of the rock and its position, a flow of volcanic material down the slope of the Sierra, it must have occupied a pre-existing depression or valley; by no possibility can such a flow be conceived of as having followed the summit of a ridge or spur. But there are other features which also furnish corroborative evidence as to the character of the surface when this erupted material assumed its present position. Under the lava is found an extensive formation of sedimentary origin, distinctly stratified, in a horizontal position, which could only have been deposited from water standing still, or flowing with a very gentle current. This stratified material varies in character; much of it is a fine white clay made up, probably, in considerable part of volcanic mud and ashes, often in very fine layers; other portions are more like ordinary river sand, not very closely compacted together, and usually disintegrating rapidly on exposure to the air. The entire thickness of the sandstone and clay under the centre of the lava flow is, in one locality at least, fully two hundred feet. Under this stratified deposit is a distinctly marked river channel, containing pebbles and boulders of metamorphic rock, intermixed with more or less of fragments of trees, and presenting exactly the same appearances as any of our present river channels.

The peculiar features of the Sonora Table Mountain will be best made out by examining the sections given on Plates E and F. Of these the one given on Plate E, Fig. 1, by Mr. Rémond, was made at a point east of Mormon Creek, about one and a half miles below Springfield, and in the vicinity of Shaw's Flat. In this section only one of the "rims" of the channel has been preserved, the other having been eroded away, together with a considerable portion of the lava flow itself. This partial denudation of the region on one side of the Table Mountain is what led originally to the discovery of the channel at this point, or in its immediate vicinity, as already mentioned. On the east side the basaltic lava forms a perpendicular wall a little over a hundred feet high, under which a few feet in thickness of sands and clays are exposed, the position of the channel under these sediments being indicated on the diagram. On the western edge of the flow, at this point, the rim-rock rises nearly as high as the edge of the lava itself. There is under the basalt a greater or less thickness of brecciated material, called by the miners "cement," and which is probably chiefly made up of andesitic lava. The thickness of this deposit seems quite variable, according as more or less of it had been eroded away previous to the flowing down of the basaltic mass which everywhere covers it, and in most places conceals it entirely.

At the Buckeye Tunnel (Plate F, Fig. 1), a couple of miles farther down the flow than the locality of the section just described, the following data were obtained, by the writer, from observation and from information given by the miners then at work there. The whole flow, so far as visible on the surface, seems here to be of solid basaltic lava, which is very smooth and flat on the summit and almost entirely destitute of vegetation. The width of the mass was here found by measurement to be 1,700 feet, and its thickness varies from 40 to 140 feet. Under the basalt is more or less "cement" or brecciated (andesitic?) material; but this rock could not be well exam-

Fig. 1.

Kaw Hah Ranch

Metamorphic Slates, dipping Northeast

Scale about 300 feet to 1 Inch

Eureka Tunnel

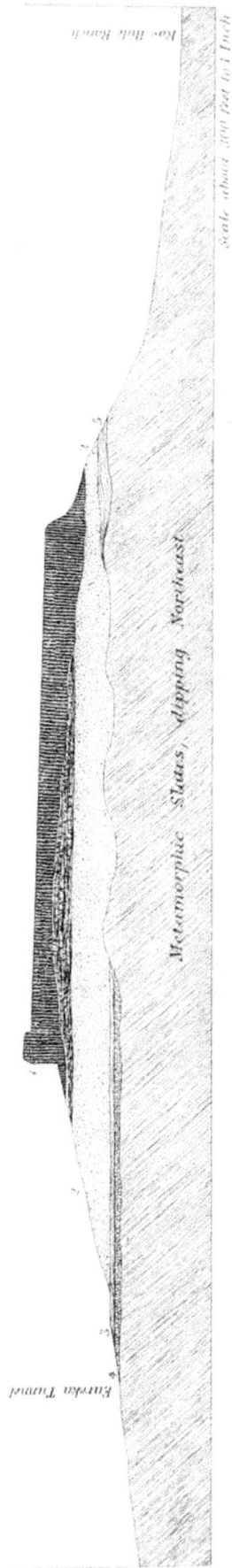

SECTION ACROSS TABLE MOUNTAIN, NEAR JAMESTOWN, TUOLUMNE COUNTY

1 Basalt, overlying Andesitic Cement (Bluff 60 feet high)   2 Bluish soft Sandstone, 70 feet ;   3 Whitish Clays, 10 feet ;   4 Auriferous Gravel, 2 feet ;   5 Yellow Clay

Fig. 3.

Andesitic Breccia

Yellow Sands & Clays

Fine white Tufa

Granitic

Granite Sands

Gravel

SECTION NEAR SOULSBYVILLE

Fig. 2.

Mica Schists, dipping Northeast

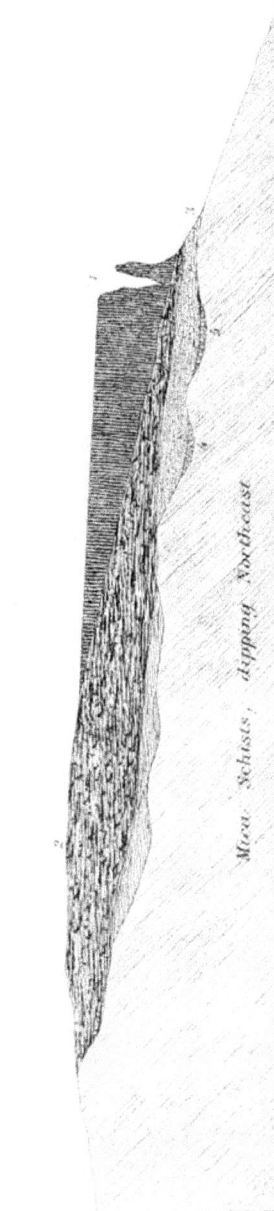

SECTION ACROSS TABLE MOUNTAIN, EAST OF MORMON CREEK

1 Basalt (Bluff 105 feet high ;   2 Cement ;   3 Whitish Clays ;   4 Yellow Sands & Clays ;   5 Gravel

ined for want of good exposures at the time the locality was visited; neither could the exact thickness of the detrital beds be made out. On the north side a tunnel was run in through the rim-rock, and at a distance of 700 feet from the entrance the first channel was struck, the tunnel passing just under it, so that by rising a little (excavating upwards) it could be drifted upon, which was done and some gold obtained, but not in paying quantity. From this channel, which was found to be eighty feet in width, the tunnel was continued in the bed-rock for 700 feet farther, which point it had reached when examined by the writer; it was then said to be 285 feet below the surface. This tunnel runs S. 80° E., the silico-argillaceous slates which it traverses having a dip of 70° to the northeast; they were of a dark color, containing much disseminated pyrites. On the opposite side a tunnel had been run in for a distance of 300 feet, following the surface of the rim-rock, which here descends with a steep pitch, or at the rate of thirty-six feet in a hundred. Here the bed-rock was overlain by "cement," and a channel had not been reached when the place was examined by the writer.

At the Down East Tunnel, about a third of a mile below the Buckeye, a tunnel had been run in 875 feet and the main channel reached, on which drifting was being done at the time of the writer's visit. At this place they passed under the small channel struck by the Buckeye Company, but did not rise to it.

At the Boston Tunnel the condition of things was more as represented in the section at the Eureka Tunnel. The tunnel was run on the bed-rock with sandstone just over it. After striking the main channel the drift was continued alongside of it, with cross-drifts, so as to take out the pay gravel and leave the tunnel securely standing. The thickness of the stratum of gravel at this place was from one to two and a half feet; it consisted of rounded boulders mixed with smaller pebbles, exactly like the material in the bed of a modern stream from which gold is now being taken out.

At the Maine Boys' Tunnel (Plate F, Fig. 2), on the west side of Table Mountain, near those already described, the excavations made at the time the locality was visited permitted the relations of the rim-rock to the channel and the overlying beds of sedimentary and volcanic materials to be well made out. The lava capping, which is mostly of basalt, with some cement underlying it, is 152 feet in thickness. The rock between the volcanic and the bed-rock is sandstone, and is well exposed by an inclined shaft sunk in it, for some distance, from a point just at the edge of the junction of the sedimentary and the volcanic. The vertical thickness of the sandstone on the top of the outer edge of the rim-rock is somewhat over forty feet. The tunnel was run 1,000 feet before the channel was struck, and the rim-rock rises 142 feet above its floor. The channel is 100 feet wide at this point, and in it there is a sharp projecting ridge of slate three feet high, the gravel being four or five feet thick. Work was begun in this tunnel in October, 1855, and the pay dirt struck in March, 1860. The cost of the work, up to the time of striking the channel, had been about $38,000, according to the statements of the owners.

Still farther down, nearly opposite Jamestown, Mr. Rémond made another section at the Eureka Tunnel, which is here appended (Plate E, Fig. 2). The rim-rock is eroded on the northwest side, so that the tunnel is run in on the gravel, the channel appearing here to be very wide. The basalt capping is about sixty feet thick, and 700 feet wide. The underlying sedimentary material consists of bluish soft sandstone, resting on whitish clays, the whole stratified deposit having a thickness of about eighty feet, while the underlying gravel is only two feet thick.

According to information obtained from the miners, the channel under Table Mountain is of irregular width, sometimes dividing into two parts, but averaging perhaps sixty or seventy feet. The gravel is shallow, and the rim-rock exceedingly well marked, when not eroded off on one side or the other. The course of the channel is also very easily distinguished, since the flow is so isolated in its position. The main channel is said to have occasional depressions, or "sinks," as they are termed, which are sometimes as much as forty feet deep. The gold is pretty coarse, usually in the form of what is called "shot gold"; that is, in small rounded pieces like shot. The largest nugget the writer heard of as having been found under Table Mountain was of $40 in value.

The large quantity of silicified wood and of impressions of leaves found in the sedimentary beds under Table Mountain has already been noticed. Bones of animals and works of men's hands are also among the materials obtained in the tunnels which have been described or in other workings under the lava. These will be noticed at length in a future chapter of this volume.

When Table Mountain was last visited by the writer, in 1870, all the above-described tunnels seemed to be entirely abandoned. That mining operations lower down on the channel were not quite given up seems apparent from Mr. Skidmore's remarks in his report to the United States Commissioner of

Fig 1

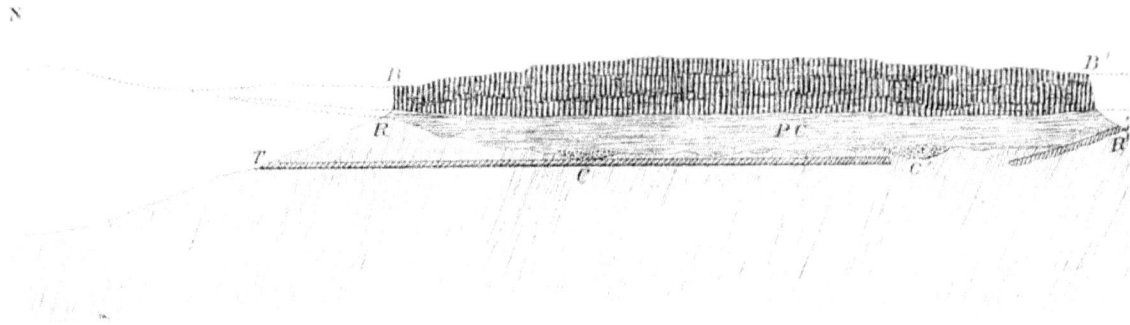

SECTION AT THE BUCK-EYE TUNNEL, TABLE MOUNTAIN

T T Tunnels   C C Channels   B B Basaltic Capping   R R Rim-rocks   P C

The dotted lines indicate the probable former position of the surface

(Scale about 450 feet to the Inch)

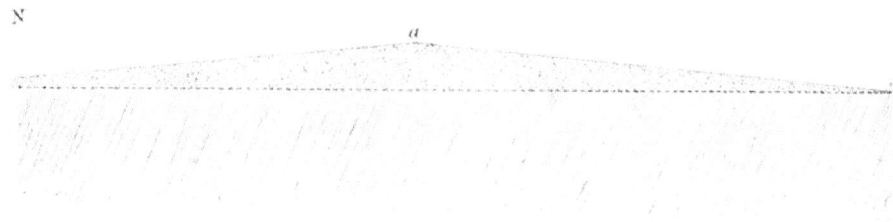

SECTION ACROSS

a Christmas Hill   c Manzanita Hill   d Centra

(Scale

Fig 4.

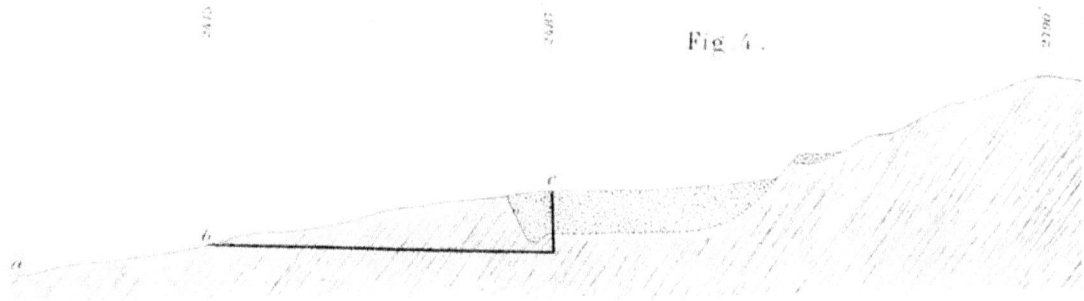

SECTION OF THE GRAVEL AT THE CARIBOO CLAIM.

a Mouth of Cariboo Ravine   b Mouth of Cariboo Tunnel   c Shaft   d Tunnel from Nigg

The figures are the heights of the bed-rock at the points below them, above the Sea-level

(Scale 300 feet to the Inch)

S

Fig. 2.

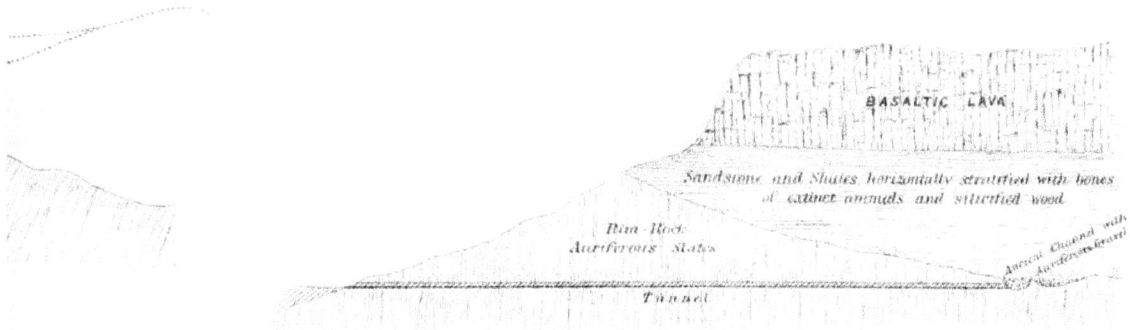

BASALTIC LAVA.

Sandstone and Shales horizontally stratified with bones of extinct animals and silicified wood.

Rim Rock
Auriferous Slates

Ancient Channel with Auriferous Gravel

Tunnel

SECTION AT MAINE BOYS TUNNEL, TABLE MT.

showing the Rim rock.

(Scale 500 feet to the Inch.)

.3.

S

RAVEL AT LITTLE YORK.

hannel: b. Rim-rock on S end Christmas Hill.

o the Inch)

Fig. 5.

SECTION ACROSS GRAVEL AT HUSSEYS

(Scale 200 feet to the inch.)

Mining Statistics for the year 1874.* He says: "Drift mining on the old channels underlying the basaltic capping of Table Mountain still continues, and, in the main, with excellent results. The Alpha Company, near Jamestown, has demonstrated the existence of two channels; these are known as the "front" and "back" channels, and are separated by a high rim. That on the east side is the most ancient, and is known as the Caldwell or Saratoga channel, and carries heavy, black, coarse gold, wherever it has been struck in Table Mountain. The west channel is of later formation, and carries finer and brighter gold; no black gold is found in this, and no pipe-clay, as is the case in the other." That the Table Mountain gravel channel is divided into two or more parts, for some portion of its course, has been already stated; but there is not sufficient evidence to justify the generalization reported by Mr. Skidmore. It is not likely that two channels running so near each other and so nearly on the same level should be very different from each other in age and character.

Of the pecuniary results of mining under Table Mountain but little is known to the writer. In 1867 Mr. Hittell wrote as follows in regard to this point:† "Table Mountain has been an unfortunate locality for miners. It is estimated that at least $1,000,000 more have been put into the mountain, counting the regular wages, than were ever taken out. Nine tenths of the miners who undertook to work claims there were the losers. There was enough gold to pay well, but the miners did not know how to get it." The reasons given for this want of success are chiefly ignorance of the position of the channel, so that many tunnels were too high for drainage; and dividing the channel up into claims of insufficient length, so that many more tunnels had to be run than would have been necessary to effect drainage and to open the ground for working. The principal difficulty, however, was undoubtedly the fact that the amount of gravel was too small to pay for operations not conducted with skill and economy; both of which elements seem to have been wanting, in a very high degree. A long list of the companies at work under Table Mountain in 1867, and in previous years, will be found in the report from which the above extract has been made, and to that the reader may be referred.

The following are some of the principal localities in Tuolumne County

* Statistics of Mines and Mining. Seventh Annual Report, Washington, 1875, p. 61.

† J. Ross Browne's (second) Report on the Mineral Resources of the States and Territories west of the Rocky Mountains. Washington, 1868, p. 39.

where placer gravel mining has been carried on, off the line of the Table Mountain channel just described. At Chinese Camp there is a flat, or level, area some three miles in length and covered with detrital material to the depth of from three to five feet, which has been mostly worked out; although as late as 1872, according to Mr. Skidmore, furnishing employment to a considerable number of Chinese, and likely to do so for several years to come. Some spots were found to be very rich; but these have long since been exhausted. In this vicinity a number of patches of cemented gravel occur on the summits of the higher spurs. One of these, situated immediately to the east of Chinese Camp and about 180 feet above the level of the Flat, and covering an area of about ten acres, is said to have been exceedingly rich, the pay gravel, which was of a blue color, varying from one to twenty inches in thickness and lying immediately above the bed-rock.*

Two miles northeast of Chinese Camp is the town of Montezuma, which was once noted for its rich placers, the gravel covering several small irregular patches to the north and east of the town and a larger one between it and Table Mountain, which runs about a mile to the west. The thickness of the gravelly material was here, in places, as much as fifty feet. Between the Stanislaus River and Wood's Creek there are various gravel areas, but no one so extensive as that to the west of Montezuma. Jamestown, on Wood's Creek, was once the centre of an active placer mining population, the bed of that creek having been exceedingly rich, and many small patches of gravel yielded largely, although not usually over two feet in thickness. It is said that the sales of gold at Jamestown for several years averaged over a thousand dollars a day.

The whole surface of the limestone belt between Kincaid Flat and the Stanislaus, already described, has been worked over and nearly or quite exhausted, in all the localities where the deeply eroded cavities in this rock could be got at, without too much expenditure for drainage. Columbia, three miles northwest of Sonora, was for some years the centre of activity in this mining region. It is situated on the limestone belt, in the midst of what was once a beautiful valley; but which now presents a most extraordinary appearance, the soil and detrital accumulations, which once covered over the underlying rocks, having been entirely removed, over the whole extent of the belt, so that nothing is to be seen except the projecting tops of the ragged edges of the eroded limestone, rising sometimes fifteen or

* W. A. Skidmore, in Fourth Annual Report of the Commissioner of Mining Statistics, p. 60.

twenty feet above the general level of the surface, and glistening white in the sunlight. It is said that the crevices, or cavities, between the "boulders"—as the projecting portions of the limestone are called by the miners—have in some places been worked down to a depth of a hundred feet; the usual depth, however, was much less. Many attempts have been made to drain this region by deep tunnels; but they have not, it is believed, been in any case pecuniarily successful, owing to their great cost, the peculiar nature of the ground, and especially the very irregular form of the bottom of the eroded rock. The usual method of mining has been to hoist the pay dirt into a dump-box, placed high enough to allow the refuse to be carried away in sluices, into which the material was washed by water thrown upon it from the hydraulic pipes. The hoisting was done by power obtained from the water brought to the spot in pipes, under sufficient pressure to rise to the top of the overshot wheels employed. Many stories are told by the miners of deep holes in the limestone, resembling caves in their extent and ramifications. The material which filled the cavities between the "boulders," in the neighborhood of Columbia, was largely made up of clay, and much of it was evidently the result of the trituration of volcanic rocks. These deposits have proved to be extremely rich in the bones of various extinct animals, especially of the mastodon, as will be noticed farther on. The most extensive and interesting locality of mining in the limestone cavities is near Columbia, from which place as much as $100,000 per week is said to have been shipped in the years between 1853 and 1858. Work was still going on there in 1870 and 1871, but on a very limited scale, as compared with its former activity. Some claims are said to have paid, for a time, very large profits,—as much as $600 per day for each man employed. The scene at Knapp's Ranch, just adjoining Columbia on the east, where the "boulders" are very large, was a most extraordinary one, at the time when operations were being actively carried on, and could hardly be compared with anything else in the way of mining, unless it might be the scene at the diamond fields in South Africa, known to the writer, however, only through the medium of photographs. The ragged projecting masses of limestone, between and among which ran innumerable lines of sluices supported on trestles, the lofty wheels mounted on tall scaffolds, with the accompanying ladders, ropes, derricks, and other appliances, formed a most striking picture of mining activity, displayed in a very peculiar field.

Kincaid Flat, two miles southeast of Sonora, is also on the limestone belt, and in its mining features resembles those of the localities just described. It is said that the area of mining ground here was originally 200 acres in extent, and that it has been worked continuously since 1850, in places to a depth of seventy-five feet. The total yield of the Flat, up to 1868, is stated at $2,000,000.* The attempts to get rid of the water, which has been very troublesome here, making it impossible to mine to the bottom of some of the cavities, have not been pecuniarily successful, for the reasons which have already been given in connection with the account of the localities on the limestone farther west. Sonora was one of the principal mining centres of California during the time of the great productiveness of the placer diggings. The town is situated on the junction of the limestome and the slates, in which latter there is in the vicinity a large amount of volcanic material, — a diabase. The limestone itself is intersected by numerous dykes of diorite, as already, mentioned. The bed of Wood's Creek, on which the town is situated, has been most thoroughly worked over, yielding largely.*

The Soulsbyville district, about five miles east of Sonora, on Curtis Creek, is a region of considerable interest, not only on account of its gravels and volcanic rocks, but because there are several well-developed veins, which have produced considerable gold, and which are enclosed in the granite. This rock occupies a large area, between the Tuolumne and the Stanislaus, extending almost down to Sonora. Upon it rests a lava flow which has descended from the Sierra, spreading out in its downward course into several branches, and covering considerable deposits of sedimentary materials. This flow follows Curtis Creek, and occupies a belt on both sides of it. The surface of the granite, under the volcanic masses, is somewhat irregularly rolling, and in the depressions rests a small quantity of gravel, which does not seem to have been found to contain a workable quantity of gold. Over this is a stratum of pipe-clay, succeeded by a mass of rhyolitic tufa or ash, fine-grained and white in color, enclosing fragments of pumice ; and over this, again, a deposit of yellow clays and sands, the whole capped by a mass of volcanic materials, of variable thickness, which in places is as much as 140 feet. The larger portion of this mass of lava is of a brecciated character, consisting of fragments of andesite, cemented together by gray volcanic sands, the whole

---

* J. S. Hittell, in Report of J. Ross Browne on the Mineral Resources of the States and Territories west of the Rocky Mountains. Washington, 1868, p. 38.

† For an account, compiled by Mr. Skidmore, of the doings (social and other) at Sonora in the early days, see Report of United States Commissioner of Mining Statistics, for 1872, p. 65.

being hardened into a solid mass. The andesite is of a dark gray color, and contains long, prismatic crystals of black hornblende. As is often the case, the masses of which the breccia is made up grow larger toward the top of the deposit, and there attain sometimes a diameter of three feet or more. The unequal wearing away of the cementing material often causes the volcanic mass to assume a very rough and irregular surface. Over the volcanic table-land in the vicinity of Soulsbyville are scattered large blocks of a more compact and finer grained andesite than that occurring in the breccia beneath. It is of a grayish-brown color, and contains minute crystals of feldspar. These fragments appear to be the remnants of a bed of lava which was once continuous over a considerable area of the surface in this region. The gravels under the volcanic deposits in this vicinity do not seem ever to have been extensively worked for gold; they appear to be shallow and irregular, but the surface is so covered by the lava, that but little can be ascertained in regard to the real position of the channel. For an illustration of the character of the table-land near Soulsbyville reference may be made to the section given on Plate E, Fig. 3.

Volcanic rocks also occur over a considerable surface between Soulsbyville and the Bower Cave; they are andesitic in character and are associated with gravels; but the latter do not appear to contain a paying amount of gold; at least, although they seemed to have been prospected in many places, they have not been worked to any extent.

Farther south in Mariposa County there have been, in former times, very productive placer diggings, but all shallow; and these have been long since worked out or abandoned to the Chinese. The shallowness of the gravel in this region has been the chief reason why water has not been brought down from the higher parts of the Sierra, for which undertaking long and costly ditches would have been necessary.

As we proceed south from Mariposa County we find the gravels becoming less and less productive and important. Gold has been found in some quantity in some of the gulches in Fresno County, and the detrital accumulations in the beds of the lower tributaries of the Chowchilla, San Joaquin, and Fresno rivers have been worked to some extent; but we have no particulars of importance in regard to them. Their production has been so small that, so far as known to the writer, the limits of the gold placer mining region of the Sierra may be said to terminate with the southern limit of Mariposa County.

Subdivision III. — The Region between the North Fork of the American River and the South Yuba.

Having, in the preceding pages, given an account of that portion of the gravel region of the Sierra Nevada which lies to the south of the North Fork of the American River, we now pass on to the consideration of the very important hydraulic mining districts which are found on the north of that stream. In describing this region, as far north as the South Yuba, the work of Professor Pettee will be chiefly relied on for details, as already mentioned.* The large map, also previously referred to, of the country between the American and North Yuba rivers may be consulted in connection with this subdivision of the detailed geology of the auriferous belt of the Sierra. An inspection of this map will show at once the position of the gravel deposits, where uncovered by volcanic materials; and it will be easily recognized that there are quite extensive areas which are not thus overlain. Of course where the gravels are covered their position cannot be indicated; but that under the volcanic there may be almost anywhere more or less detrital material, containing some gold, will be readily inferred from what has been stated in the previous pages of this volume. Following the system already indicated, however, we shall defer general remarks on the occurrence of the gravels to another chapter, and for the present confine our observations to the details of their occurrence in the region under review. It is believed that by following this order it will be possible to give a better and more connected idea of the whole series of phenomena than could be had from a description in which the general should precede the special.

An examination of the map will show that directly north of the North Fork of the American River, and in immediate proximity to the Central Pacific Railroad, there is a very large deposit of uncovered gravel, on which a number of mining claims are indicated. This area lies in very close proximity to that lying north of Shirt Tail Cañon, and including the mining centres of Iowa Hill, Elizabeth Town, and other camps already described from Mr. Goodyear's notes. Indeed, the southern extremity of the gravel deposits north of the American is hardly two miles distant from the nearest gravel on the south side of that river, while there is good reason for believing, as will be noticed farther on, that these deposits, now separated by a deep cañon, may once have been connected.

* See *ante*, p. 80.

The Central Pacific Railroad runs on the divide between the North Fork of the American and Bear River, and the gravels which come first in geographical order lie chiefly on this divide, a small portion of them, however, being to the northwest of the last-named stream, and between it and its tributary, Steep Hollow. The main mass of the gravel extends uninterruptedly from Elmore Hill, just south of Bear River, for a distance of about five and a half miles to Indiana Hill on the American. The towns of Dutch Flat and Gold Run are situated on this range of gravel, the former near its northern extremity, the latter about two miles north of its southern end, the railroad passing through the latter and a little to the east of the former locality. A wide branch of this gravel also extends off from the main mass in a southwesterly direction, although intersected by Bear River, on the west side of which is the third important mining town of this district, Little York.

This main mass of gravel, as is evident from an inspection of the map, is prolonged in the deposit which, with only a slight break at Bear River, extends continuously from Liberty Hill to Lowell Hill, on the divide between Bear River and Hollow Creek. Beyond Lowell Hill are still three isolated patches of gravel, near each other and in the same general line of direction, upward towards the summit of the Sierra. Besides these, there are a few small areas of gravel to the south of and near the line of the railroad; the mining camps dependent on these are known as Blue Bluffs and Lost Camp. These different deposits, at all of which somewhat active mining operations were going on in 1870, will now be described in the order indicated, and principally from Mr. Pettee's notes.

§ 1. *Dutch Flat, Gold Run, and Little York.*

As before remarked, the gravel deposits are almost continuous from Indiana Hill, on the North Fork of the American River, to Elmore Hill on Bear River above Dutch Flat. In fact, with the exception of the gap at the cañon of Little Bear River, there is no absolute break of continuity between the two places mentioned, for none of the intermediate ravines cut through to the bed-rock at all points. In endeavoring to trace the connection of the channel, or channels, in this great mass of gravel, the problems presented are, as usual in such cases, not without difficulty. Above Elmore Hill there can be no question as to the direction of the ancient stream; for from Elmore the bank at Liberty Hill is plainly in sight at a distance of less than two miles, and so much higher that the tops of the trees on the summit of Elmore Hill are about on a level with the Liberty Hill bed-rock. This gives a good and sufficient grade, with no intervening obstacles. Since the time when the river flowed from Liberty to Elmore Hill, the present channel of Bear River has been cut several hundred feet deeper, although not in a materially different direction, the angle between the old and the new channels not exceeding ten or twelve degrees. Elmore Hill is on the left, and Liberty Hill on the right bank of a stream whose general course is nearly straight. If any additional evidence were needed, to show that there was once a connection between these two points, it is found in a deposit of gravel on one of the projecting spurs just below Liberty Hill, nearly on a line between the points in question, and at the proper altitude. From the other end of this gravel deposit, that is from Indiana Hill, the Iowa Hill gravel is plainly in sight, and with such a position and relative elevation that there is no difficulty in imagining that there was formerly a connection in this direction, as will be noticed farther on.

The presence of gravel in a continuous mass between Elmore and Indiana Hills seems sufficient proof of the existence of a continuous channel between those two points. But to actually trace this channel through the various workings along its line, so as not only to prove its existence but to recognize its exact position and grade from point to point, is by no means an easy matter. Among the miners in this channel there were, in 1870, and probably still are, various conflicting views as to the direction of the current, and these will be alluded to and discussed in a future chapter. After a long series of patiently-conducted observations for level on the bed-rock, at all points where this was visible, under the gravel, or where gravel had previously existed, Professor Pettee came to the conclusion that the evidence, although contradictory in some points, was, on the whole, in favor of the theory that the channel from Liberty and Elmore hills at one time found an outlet by way of Indiana Hill.

Taking that portion of the gravel range between the line of the railroad and the Cement Mill, near Indiana Hill, the eastern and western rim of the channel is pretty well defined, although it is not always easy to tell exactly where the line between the slate and the gravel shall be drawn. On the west is Cold Spring Mountain, near Gold Run Station, 3,679 feet above the sea-level. The main body of this elevation is slate, capped by a broad, flat, volcanic table, covering an area of as much as a hundred acres and about 200 feet in thickness. The height of the bed-rock is such here that there is no possibility that the channel had any exit in this direction. Near Gold Run Station, and north of the railroad, is another quite well marked hill of slate, which may be considered as the continuation of the slope of Cold Spring Mountain towards Bear River. At the southern end of this last-named elevation there is also a prominent spur of bed-rock, near Betton Ravine, which rises considerably above the line of the rim-rock, until the steep descent into the cañon of the North Fork of the American is reached, which is here 2,270 feet below the summit of Cold Spring Mountain. On the east, the rim-rock is seen on the right bank of Cañon Creek, for nearly the whole distance from the railroad to the neighborhood of the Cement Mill. On the left bank of the Creek there is a high ridge, the top of which has all the characteristics of a lava flow, and seems to be a portion of the same one to which Cold Spring Mountain belongs. There are one or two small patches of gravel on the left bank of the Creek, near its junction with the North Fork, and about on the same level with the main deposits of the same material on the opposite side; but they are not of much extent or value.

The width of the gravel will average, for the whole distance between the railroad and Indiana Hill, nearly or quite half a mile; although there are many places where it has not been washed off to that width. There were still, in 1870, a number of claims staked out and held by their owners, on which no work had been done, nor could be until the banks in front had been washed away, or some outlet provided. In the spring of 1868 Mr. E. C. Uren was employed to make the necessary surveys for a deep tunnel, through which all the claims — north of Potato Ravine, at least — could discharge low down into Cañon Creek. The total length of the proposed tunnel would be over 9,500 feet, and then its head would still be 125 feet below the present surface at a point quite near the railroad. In reference to this map Professor Pettee remarks as follows: "Mr. Uren's map affords, by the way, a good illustration of the difficulty of getting any full or correct list of claims or locations of mining ground. Even with his facilities for obtaining information, and considerable expenditure of time, some few inaccuracies — mostly slight, it is true — could not be prevented from creeping in."

An attempt was made to collect full details of everything connected with the hydraulic mining interests in this district, Messrs. Pettee and Bowman having been instructed by the writer to make a thorough survey of all the claims, for the purpose of ascertaining with accuracy the amount of gravel washed away, it being hoped that an approximation to the yield of the district might be obtained, and thus the average yield of the gravel per cubic yard be ascertained; a problem which, so far as known to the writer, had, up to that time, never been solved for any portion of the hydraulic mining region of California. It was found, however, entirely impracticable, with any, even unlimited, expenditure of time to collect an *accurate* statement of the location of the companies, the length of the channel held by them, and the depth of the gravel worked; and it was also found still more difficult to obtain an accurate account of the amount of gold washed out. The following statement of the detailed examinations made in the vicinity of Dutch Flat and Gold Run is condensed and arranged from Professor Pettee's notes.

I found it impossible, with the time at my disposal, to collect full details of location of companies, length of channel held by each company, width and depth of gravel, or height of bank. And, even if the time had been unlimited, such a tabular statement for Gold Run could have had little or no value, for the reason that the course of the channel and the position of the bedrock are entirely unknown. The claims are mostly of irregular shape, and seem to have been located entirely at random, or at least without any reference to the main deep channel; although probably some of the irregularities of boundary are owing to subsequent consolidations, and modifications after the original location.

In the accompanying list of claims between the railroad and Indiana Hill, the measurements are given as roughly determined by putting a tracing of Mr. Uren's map over a sheet of paper ruled in squares, each representing an acre, and estimating as nearly as possible the fractional parts of those squares not entirely included within the boundary lines of the claims.*

[Here follows, in Professor Pettee's notes, a list of forty companies, holding 737½ acres, the names of which it seems unnecessary to give at length, and they may be omitted.]

Since Mr. Uren's map was made there have been some few changes in the boundaries of the different claims, and some consolidations and consequent disappearance of old designations, and introduction of new ones. The measurements given are intended to include only the gravel contained within the boundary lines of the companies. In addition there is as much as seventy or eighty acres within the gravel limits, mostly in the ravines, but not included within the lines. Several rough measurements of the area included on Mr. Uren's map, not excepting the ravines, were made, and results obtained varying between 825 and 875 acres.

One of the objects in making the plane-table survey of these mines was to obtain some accurate data as a foundation for computing the average amount of gold contained in each cubic yard of gravel. I accordingly mapped the outline of the present banks, without any special regard to the lines of the individual claims, and determined the area of the ground actually worked over, as nearly as I could make it without actual tape measurements, to be : —

Between the railroad and Goosling Ravine	.	.	168 acres or 7,318,080 square feet.
Between Goosling Ravine and Gold Run Ravine	.	77½	"	3,375,900 " "
Between Gold Run Ravine and Potato Ravine	.	37	"	1,611,720 " "
South of Potato Ravine	. . . .	144½	"	6,294,420 " "
Total	. . .	427	"	18,600,120 " "

---

* In Mr. J. Ross Browne's (second) Report all the names on Mr. Uren's map will not be found; but there will be found as many as twenty additional or different ones. It is clear, in some cases, that the same places are referred to under different names, Mr. Uren giving the name the claim goes by, and Mr. Browne the names of the owners at the date of his Report; in other cases, it is not so easy to reconcile the map with the Report.

It was of course impossible to tell what the depth of bank had been at all points over this extent of ground, but relying upon the statement that the slope of the country had been pretty gradual from west to east, and upon my measurements and estimates of heights at a great number of points, I obtained the results given below. On the Cañon Creek side and near the ravines the banks were seldom more than thirty or forty feet high on the average for any considerable distance, while on the western and northwestern sides of the diggings we find as a rule banks of from ninety to 120 feet or more in height. At the Cedar No. 2 Claim, where the highest bank of all is to be seen, it is as much as 240 feet from the bottom of the flume to the top of the bank, but this depth is reached only over a comparatively small area. It will hardly be necessary to give in detail all my notes concerning the heights of banks, the results obtained from their use being all included in what follows. The three prominent ravines crossing the Gold Run deposit of gravel and emptying into Cañon Creek furnish such good lines of division that it is most convenient to subdivide the whole mass into four portions, measuring and calculating each one by itself. It is not claimed that absolute accuracy is attained in this way, but I think as good and trustworthy a result is here offered as could be expected from the small amount of time available for making the measurements. I am convinced that I am not far out of the way, and really cannot tell whether any of my figures are too low or too high, probably the former, if either. My endeavor was to strike as near a mean value as possible in all cases.

My estimate of the average thickness of gravel which has been removed between the railroad and Goosling Ravine is sixty feet, — based upon observations of height of bank every few rods around the whole region. One or two small ravines had their heads within this region, but I have made such allowances as seemed to me most nearly correct. The same estimate I made for the portion between Goosling Ravine and Gold Run Ravine. Between Gold Run Ravine and Potato Ravine the average is a little lower, — say fifty-seven feet. But to the south of Potato Ravine (even without the deep part of the excavation at Cedar No. 2) the average would be notably higher. For this portion I adopted seventy feet as the average thickness. My calculations to determine the number of cubic yards of gravel washed away gave the following results : —

Between the railroad and Goosling Ravine	.	.	.	.	439,084,800 cubic feet.
Between Goosling Ravine and Gold Run Ravine	.	.	.	202,554,000	"	"
Between Gold Run Ravine and Potato Ravine	.	.	.	91,868,040	"	"
South of Potato Ravine (excepting a part of Cedar No. 2)	.	.	440,609,400	"	"
Add for deep part for Cedar No. 2, say	.	.	.	.	50,000,000	"	"
Total	.	.	.	1,224,116,240	"	"

These calculations are made on the supposition that the bottom of the old workings is always smooth and level, which is far from being the case. In many places there are sluices of considerable size which have been cut deeper than the surrounding gravel, — and in many other cases considerable masses of gravel have been left standing. This is particularly the case where different claims have been worked from opposite directions, and a wedge or pyramidal shaped mass has been left as a landmark. To allow for what has been left in this way I estimated that a deduction of about five per cent from the grand total would be nearly correct. Making this deduction, I get 1,162,910,428 cubic feet as the amount of gravel which has been removed since the commencement of the washings between the railroad and the Cement Mill. Or, in round numbers, 43,000,000 cubic yards. This number I shall refer to again.

It is pretty safe to say that there remains within the limits referred to in the above calculation an average depth of as much as a hundred or a hundred and twenty feet more of gravel before bed-rock will be reached. On this supposition, then, we can calculate upon as much as 86,000,000 cubic yards, left to be washed, to say nothing about the outside claims which have not yet been touched to any great extent. These will add considerably more to the grand total, and make in all, say, 125,000,000 cubic yards ; much of this will undoubtedly prove richer than the portions already worked out, though some, with equal certainty, will be poorer.

That portion of the gravel lying to the north of the railroad to and including Dutch Flat and Elmore Hill was not surveyed with the same minuteness as the portion which has been previously referred to. This was owing partly to the fact that the rainy season seemed to be approaching and our time was limited. It was, furthermore, not possible to get such trustworthy statistics of the production of the district as were obtainable at Gold Run; and, as a consequence, there was less need of minuteness of detail in the measurements. There has been no map made, so far as I know, of the different claims in this district, — and in my own work I confined myself mainly to the outlines of the portions actually worked out. The following statement, however, may be taken as a pretty fair approximation to the actual gravel area which has been already washed over : —

Between the railroad and Squire's Cañon .	52½ acres or	2,286,900 square feet.
Between Squire's Cañon and Dutch Flat Cañon .	40 "	1,172,400 " "
In the main Dutch Flat Diggings between Dutch Flat Cañon and Bear River .	144 "	6,272,640 " "
Total .	236½ "	10,301,940 " "

The means for determining the average thickness of bank over this district are not so complete and satisfactory as they were south of the railroad. The original country was more uneven, and the outlines of the openings are not so regular as a whole. For the portion next north of the railroad I have adopted an average height of bank of forty-five feet. It is considerably higher than that on the side next the track, but as I have included in my estimate of the area the long shallow tongue through which an outlet is found into Squire's Cañon, the calculated number of cubic yards will be pretty nearly correct.

Between Squire's Cañon and Dutch Flat Cañon there was originally a high hill of gravel which has been so washed away that while we have on the southeast side of the excavations banks of a hundred, a hundred and twenty-five and more feet, the northwest banks are almost nothing. If I take an average of fifty feet for the whole area I think I am fully high enough.

On Gray's Hill and the other hills which go to make up the gravel deposit between Dutch Flat Cañon and Bear River, it is even more difficult to decide upon a fair average depth. For the western quarter of the area, where the slope is more distinct from the top of Thompson's Hill towards Dutch Flat Cañon, or Bear River, and the banks are pretty high on the northwestern side, it seems not far out of the way to take sixty-five feet for the average depth of gravel already washed away. For the rest of the district I cannot speak with so much confidence. The few facts upon which a judgment is to be based are as follows. The average height of the numerous projecting points on the Bear River side of the diggings is about fifty feet above the present general level of the gravel. Near the town of Dutch Flat there is hardly any bank at all, — certainly not more than fifteen feet on the average. On the northeastern end, near the small reservoir, the bank is a little higher, — say thirty feet; and on the southwest, near Thompson's Hill, it will average as much as sixty feet. The hill, however, rose considerably higher than the present level, — excepting perhaps on the southwestern extremity. At the Dutch Flat Tunnel (referred to on page 151) Mr. Colgrove said the top of the hill was from eighty to ninety feet above the present average gravel surface. At Mr. Teaff's shaft (in Teaff's Diggings) there is a discharging tunnel, which was 206 feet below the original surface of the ground, according to Mr. Julian's survey. This tunnel is still considerably below the general level of the gravel. From such facts as these it is clear that the highest part of the ground must have been somewhere between the site of the tunnel and Teaff's Shaft. For the southwestern half of Gray's Hill I am inclined to accept an average depth of 100 feet, and for the northeastern half an average depth of fifty feet, — or of seventy-five feet for the whole. This would give us for the cubical contents the following results : —

Between the railroad and Squire's Cañon .	102,910,500	cubic feet.
Between Squire's Cañon and Dutch Flat Cañon .	87,120,000	" "
Between Dutch Flat Cañon and Bear River (west ¼) .	101,930,400	" "
Between Dutch Flat Cañon and Bear River (east ¾) .	352,836,000	" "
Total .	644,796,900	" "

From this total I make a deduction of ten per cent to allow for the numerous blocks of gravel left standing within the limits comprised. This will leave 580,317,210 cubic feet, or approximately 21,500,000 cubic yards, just half the amount for the district south of the railroad. In regard to the amount left still to be washed I feel unable to make any estimate, on account of the uncertainty attending the position of the bed-rock.

The gravel is not uniform in character throughout the whole distance from Indiana Hill northward. Near the Cement Mill the lower stratum of the gravel is a compactly cemented mass about sixty feet thick, of which the lowest eighteen feet was so hard as to require drifting and blasting.

The method of excavation was to run a number of drifts and loosen, by blasting, a large quantity at a time. Some of it could then be treated by the hydraulic process, but the largest part had to be run under the stamps. This lower cement contains (as has been said before) a great quantity of large boulders. At Iowa Hill there is said to be the same kind of cement as at Indiana Hill, but I had no opportunity to visit the place. Going northward from the Cement Mill we soon lose sight of the cement, because none of the excavations are deep enough to have reached it yet, even if we are willing to admit that it really continues for a long distance. The great mass of the gravel, as far as Squire's Cañon at any rate, is fine and easy to wash, by no means firmly cemented together, and also without any noticeable amount of clay or sand.

At the Cedar No. 2 Claim, where the highest bank is to be seen, the top, to the depth of (say) 115 or 120 feet, is a red gravel. Next follows about the same depth of a blue gravel, though not cemented together. What there is below the blue remains to be seen at some future day, probably a hard cement. The line of demarcation between the red and blue was not very distinct at the time I was there, because the bank had been exposed to the air for some months and was very dusty. But the difference of color seems to point to some radical difference in the gravel itself, its origin or time of deposition. Mr. Brogan told me, for instance, that all the petrified wood seen in the claim came from the red gravel, while all the wood in the blue gravel is thoroughly charred, there being no charred wood in the red gravel at all. The gold, too, in the red gravel at this claim has a fineness of $\frac{961}{1000}$, while the gold of the blue gravel has a fineness of only $\frac{800}{1000}$, and is also a little coarser than the gold found above. The difference, however, is not great; the whole being classed with fine, or flour-gold. In the cement near the mill the gold is coarse, rounded and smoothed by washing. Some pieces, I was told, have been found worth as much as ten or fifteen dollars.

In the claim next southwest from Cedar No. 2, — the Brink Claim, — the gravel has been washed away to the bed rock, which here pitches rapidly to the east. There was also a much larger quantity of clay met with in and upon the gravel than elsewhere in this neighborhood. It would appear that when the channel had been filled up to that depth the quantity of water diminished or was spread out over a so much wider bed that, particularly on the banks, there was a good opportunity for finer sediment to lodge.

To the north of Cedar No. 2 the distinction between red and blue gravel is not so marked; or it may be that the upper claims have not yet exhausted the upper stratum of red gravel, and that blue gravel will soon be struck, when work is recommenced. Concerning the gravel sunk through at the '49 shaft in Potato Ravine (previously referred to) I have no information excepting that given in Mr. Browne's Second Report, viz.: "Pay gravel was found all the way down, and it was soft until within six or eight feet of the bottom." What the last few feet of gravel was like is not mentioned, though probably a cement similar to that near the mill was struck.

The general character of the gravel remains, then, essentially unchanged as far as Squire's Cañon. In the Jehoshaphat Claim, however, between Squire's Cañon and Dutch Flat Cañon, there are a few peculiar features to which my particular attention was called by Mr. Kelsey, and which may or may not be thought to have any great significance. Starting from the outlet of the mine, in Dutch Flat Cañon, there is a distinct line of demarcation, between red gravel above and blue gravel beneath, which rises in a southeasterly direction for a distance of about 175 feet at such a rate that it follows the line of the sluice (of which the grade is 10 inches in 12 feet).

From that point on the rise is more rapid for a short distance, — perhaps 100 feet, — so that the sluice is entirely in blue gravel, and then the blue pitches off suddenly and disappears. Commencing at the bottom of the present bank, then, the order of deposit in this claim is, first, the blue gravel and then a fine red gravel, above which is a coarser streak of about sixty feet in thickness in which the pebbles are mostly angular and unwashed. Mr. Kelsey says this streak can be traced across the Dutch Flat Cañon to the upper part of Thompson's Hill, and is remarkable for containing a noticeable quantity of copper. I had no opportunity of testing the correctness of this latter statement for myself, but have full confidence in Mr. Kelsey's honesty in making it. This cupreous streak is not wide, — only about 100 or 125 feet. It looks as if it lay lengthwise in a stream, and gives additional reason for thinking that a current once ran in a southerly direction at this point. Above the cupreous streak there lies about twenty-five feet more of red gravel.

A short distance beyond the point where the blue gravel disappears there are sandy deposits dipping at angles varying from ten to twenty degrees (mostly about twelve degrees) to the south and west of south. There are also extensive deposits, in alternating layers, of soft yellow and blue clay, the blue being made up largely of bark of trees and other unmistakable vegetable matter. These beds of clay dip to the south, or nearly south, at a considerable angle, as much as 10° or 15° on the average. These different clays are not, moreover, in all cases superposed quietly one upon the other. In one or two instances it seems as if a pipe of blue clay had managed to effect a connection through the yellow clay between two otherwise distinct blue beds. A similar phenomenon was also observed at the line between red and blue gravel. It looks as if a stream of blue coloring matter had made its way up through some open channel and soaked into the red until it was completely absorbed. The effect on the face of the red gravel was as if some blue gerrymander-like figures had been drawn there. I looked at this phenomenon pretty closely. The line between the red and blue was always distinct and well defined; there was no shading off of tint. In many cases, if a pebble as large as a man's fist was in the line of division, one half of it would be red and the other half blue, showing that the cause of the coloring is to be sought for in some agency acting after the gravel had been deposited.

The boundary line of the blue gravel presents also another peculiarity in this claim. At a point a few hundred feet from the cañon Mr. Kelsey has a shaft sunk forty-five feet in a red sandy material without striking any blue gravel at all; but twenty feet to the west of the shaft blue gravel sets in and rises abruptly. It is seen at the top of the shaft, but was not met with below within the distance of forty-five feet.

The yellow clays in this mine abound also in little nodules of rather irregular shape, but still so uniform as to have given the impression that they are "fossil oysters." There is indeed a certain resemblance to small oysters, and the nodules are all found lying on their flat sides in the clay. When broken open the central portion is, in almost every instance, a black powder differing entirely in appearance from anything else found in the neighborhood. I saved a few specimens in the hope of being able to determine the character of the interior black powder at some future time. Up to the date of this writing, however, nothing has been done about it. I am hardly of the opinion that these nodules had any organic origin, still the fact of their all lying upon the flat side is rather striking.

Some cause undoubtedly acted at this point of the gravel deposit to introduce irregularities which are rare or less marked at other localities near by. The dipping of the line between the blue and the red gravel in opposite directions, the deposits of vegetable mould, the coarse streak with angular pebbles, all taken in connection with the form of the bed-rock as indicated in the three old shafts near the outlet of the Jehoshaphat Claim, seem to corroborate to some extent the idea previously advanced that at some time in the history of the old river there was a sudden change in the direction of the current; that it perhaps found itself choked up, and then broke through or over some small divide and continued its way in quite a different direction. As has been before remarked, there is evidence of a channel's having also once gone from Dutch Flat to Little York, — and if there ever was any such change of channel as I have supposed, the Jehoshaphat Claim is just on the spot where the change must have taken place. It is true, I cannot see just why any such change of channel should have produced just the effects observed at this point, and it is possible that some far better explanation may be found. If so I shall be glad to adopt it.

South of Dutch Flat Cañon, it has been observed, we meet with no big boulders excepting in the cement at the Indiana Hill extremity of the gravel; but as soon as we cross to the Gray's Hill Diggings and the district which lies to the north and west of the town of Dutch Flat we strike at once gravel of a different character. Almost the whole extent of ground that has been worked over is now thickly covered with boulders varying from one or two up to five or six feet in diameter. Whether the gravel all the way up to the original surface was of the same nature or not I cannot tell positively, but I think the boulders were smaller and not so frequent towards the top. Mr. Ross Browne, for instance, speaks in his report of the soft gravel at Dutch Flat above the "blue lead."

There were not many places where good information could be gathered concerning the nature of the Dutch Flat gravel below the present surface. At Teaff's Shaft (already referred to in another connection) I could see that the mouth of the shaft was quite near the line between red and blue gravel. A few strokes of the pick sufficed to break through the weathered outside and bring the unmistakable blue color to light. The gravel at this point was very coarse and many of the boulders were as much as five or six feet in diameter. The largest ones were apt not to be perfectly rounded, though very sharp angles were rare. The blue color of this gravel is owing in part to the colors of the big boulders themselves, many of them being of metamorphic slate with a decidedly bluish tint, though there is also a good percentage of quartz boulders to be seen. In the red gravel, which overlies the blue, the boulders and the smaller pebbles as well are almost exclusively of white or reddish quartz. Southward from Teaff's Shaft big boulders begin to be rare. The shaft itself is sunk in blue gravel for 128 feet before reaching bed-rock, and discloses signs of gold pretty uniformly distributed all the way down. It was shown, too, that the gold in the blue gravel is, as a rule, coarser than that in the red, which is scaly, flat, and fine. According to Mr. Teaff, some of the gold in the blue gravel looked like pin-heads, — resembling the gold which has been found in Bear River. With respect to fineness of gold, this is about the same distinction which was made between the two kinds of gravel at the Cedar Claim No. 2. At Teaff's, however, I was told that charred as well as petrified wood was found in the red gravel, which was not the case at the Cedar No. 2.

Another place where I had a chance to learn something about the character of the gravel below the present surface was at the Dutch Flat Tunnel. The information was obtained from Mr. Colgrove, and is, briefly, as follows. The tunnel was run for about 500 feet in a general southeasterly direction without reaching gravel. It was then decided to rise, and gravel was struck on the northwestern rim at a distance of forty-two feet. Between this point and the present surface of the gravel is about sixty-five or seventy-five feet of blue cement covered by about fifteen feet of red gravel: above which there was still as much as eighty or ninety feet of gravel to the original surface. From the point where the channel was struck at the top of the forty-two feet rise, a horizontal drift was run for 200 feet in "blue cement." At the extremity of the drift sinking was commenced again. The first six feet was in the same kind of blue cement as in the 200 feet drift, and then followed ninety feet of clean blue gravel to bed-rock. From this point a new drift of forty feet in length was run in blue gravel, at the end of which a sink of seven feet was necessary to reach bed-rock. The last two feet were in a hard cement.

At the northeastern end of Gray's Hill, on the ground belonging to the Buckeye Company, the boulders are also very large and numerous. As would be expected where big boulders are so common, clay is scarce. In Mr. Teaff's claim the thickest body of clay was only ten feet in thickness, and that only over a small area.

At the upper end of Elmore Hill, where the gravel first makes its appearance on the left bank of Bear River, it is said to look like a deposit of rotten rock, but as we get nearer to the Little Bear River it takes on more the character of a blue cement similar to that in Gray's Hill.

At Liberty Hill there is a bank of thirty feet in height of bluish gravel, covered by about the same thickness of red quartz gravel. The boulders are very large and not much worn, as if they were lying near the head of a stream, and had neither been carried far, nor had much gravel washed over them. On account of the size of the boulders, it has not been very profitable to work the gravel at Liberty Hill.

The gold product of the Gold Run district has passed largely through the banking-house of Messrs. Moore & Miner, at Gold Run, who kindly prepared for me a statement of their gold purchases from March, 1865, to October, 1870, from which the following summary has been prepared :

Year.	Weight.		Value.
	oz.	dwt.	
1865 (March to December)	7,554	19	$ 140,182.05
1866	12,723		237,909.62
1867	16,542	13	311,812.65
1868	13,587	10	259,208.05
1869			189,968.73
1870 (January to September)			183,384.86
From Church Co., in addition,			62,814.60
Total,			$ 1,385,280.56

The gold from some of the claims south of the railroad never got into Moore & Miner's hands, and occasionally small lots from other claims would be sent directly to San Francisco, or sold at Dutch Flat. On the other hand, only a small portion of what they did buy came from claims other than those between the railroad and Indiana Hill. They were thus justified in saying that their statement comprises certainly two thirds, and probably three quarters, of the total production of the district. The quantity taken out at Gold Run previous to March, 1865, would not be sufficient to vary the grand total materially. In round numbers, then, we can assume two million dollars as the amount of gold saved from the Gold Run gravel banks. The amount of gravel washed away, in the same district, has been previously fixed at 43,000,000 cubic yards.* This makes the average amount of gold per cubic yard only four and three quarter cents. I have examined the data carefully on which this calculation is based and can find no material source of error, although the amount of gold per cubic yard falls much below the estimates of miners and claim owners.

At Dutch Flat it was not possible to get statistics similar to those at Gold Run. It would seem from Mr. Browne's report of the shipments of bullion from Dutch Flat, for the first six months of 1867, as compared with the statement of Messrs. Moore & Miner for the same period, that the Dutch Flat production was considerably greater than that of Gold Run ; but it must be noticed that a large part of the gold shipped from Dutch Flat is brought in from other districts, namely, Little York, You Bet, and other places, even from Gold Run itself, and no one can tell now just what amount was saved from the 21,000,000 cubic yards of gravel washed away in the immediate neighborhood of the town. I am convinced that most of the estimates of the yield per cubic yard have been too high ; but investigations similar to those made at Gold Run should be carried on in other districts, before any decided opinion could be given as to the amount of such over-estimates. If, as I believe, my calculations are free from material error, it appears that a poorer gravel can be worked with profit than has hitherto seemed possible.

The Nary Red Mine lies about half a mile east and a little south of Elmore Hill, and it has all the appearance of being a tributary to the main Dutch Flat channel. Between Elmore and the Nary Red there seems never to have been any gravel connection ; and, indeed, the remains of a spur of slate, which may have separated the two localities, are still to be seen. A similar remark may be made concerning Nary Red and Gray's Hill. If we take the Elmore Hill and Dutch Flat channel as the representative of the original Bear River, the Nary Red stands, with equal

---

* See *ante,* p. 147.

certainty, in the place of the Little Bear River, the junction of the two having been somewhere near the northeast end of the Dutch Flat Diggings. The Nary Red is a narrow channel, in which the gravel is mostly a small, clear, red quartz, thus differing almost entirely from the gravel on Elmore or Gray's Hill. The description of the region above Nary Red is, in the words of Mr. Colgrove, substantially this: "The Nary Red is a narrow channel which crosses Cañon Creek near Alta, and then spreads into a broader lake-like expansion, terminating by a slide into the North Fork of the American at Green River." On the opposite side of the North Fork is a lava flow following nearly the same course as the present stream. To the southeast of the Nary Red the ground rises rapidly.

Where the gravel range is crossed by Squire's Cañon, the country rock is seen on each side, with a width of about 500 feet of gravel and tailings, in the bottom of the cañon. How much more slate was visible, before the accumulation of the tailings began, it is not easy to determine; but it appears as if the slate did not extend entirely across the pitch being quite rapid both on the east and on the west side of the narrow place.

"Plug Ugly" is the name of the high hill between Squire's Cañon and Bear River; it is just south of the Dutch Flat Diggings. The western rim in the cañon appears about 200 feet in a westerly direction from the outlet of the old Juniata Claim, on the upper end of Plug Ugly Hill. It appeared, on first inspection, as if the river here must have made a very sharp turn in order to include the gravel on this elevation; but it afterwards became pretty clear that this deposit is one which was formed at a time when a large extent of surface, not included in the main deep channel, was covered. If the bed-rock is ever fully exposed throughout this district, a comparatively flat table will probably be found on the top of Plug Ugly, having a rapid descent on the northeast side, into the bed of the main channel. But how far back one would have to look for the commencement of the pitch it is not easy to decide, for it is clear that a considerable depth of the rim has been cut away at the point where the line of bed-rock was seen in the cañon. It is not so probable that Plug Ugly is the relic of a tributary to the main stream, for the general slope of the country is rapidly to the west from this point, and the drainage must have come mainly from the east.

The general course of Squire's Cañon is about southwest, and for nearly half a mile it has a breadth of 150 feet or more on the tailings, with a pretty steep grade toward Bear River. The course of this cañon was followed until the gorge became narrow, and the descent so steep that tailings would not collect. Here was no gravel, but a narrow gorge, with slates on both sides. On climbing, however, by the most direct path, to the top of the hill between the cañon and Bear River, it was found to be capped with gravel, which could still be traced for some little distance down the spur. The lower end of the deposit was not very distinctly marked, small patches being found here and there for a considerable distance below the main mass. A little way up the hill from the lowest gravel seen was what appeared to be a genuine rim, inclining N. 40° E. (magnetic), at the outlet of a small excavation about 150 feet long by seventy-five wide, and twenty-five deep, in fine red quartz gravel. This point was found to be about 175 feet below the top of Plug Ugly Hill, although it is probably 350 feet higher than the bed-rock at Missouri Hill, on the opposite side of Bear River. Separated from this opening by a narrow ridge of gravel is another large one, 400 feet long by 125 feet wide and about thirty deep, the quality of the material being the same in the two. The direction of the longer axis of the opening is N. 30° E. (magnetic), which is also that of the top of the hill itself. No well-defined rim was found on the Squire's Cañon side of the deposit. The projecting spurs are covered with slides of gravel, making the total amount of this material seem, at first sight, larger than it really is; although the main mass, on the summit of the hill, must be nearly or quite a thousand feet wide. There is evidently a large number of cubic yards of gravel in this hill, but the amount could not be estimated. At the time this examination was made parties were engaged in making preparations to work it extensively, by means of water brought from a ditch near Dutch Flat Station.

The following is a synopsis of Professor Pettee's observations made in examining the region around Little York and along the spur between Bear River and its branch called Steep Hollow, below the Camel's Hump.

The Dutch Flat gravel was found chiefly on three hills, — Gray's Hill, Thompson Hill, and Ellis Hill; the latter two being at the southwestern extremity, below the point where the change of channel in the direction of Gold Run is supposed to have taken place. The altitude of the bed-rock at the bottom of Teaff's Shaft, at Dutch Flat, was ascertained to be 3,004 feet, which is a trifle too high, on account of the shaft's not having struck the lowest point. The rim-rock which is visible at the Iowa Claim rises to a height considerably greater than 3,000 feet. What the form of the bed-rock will be when uncovered is purely a matter of conjecture, but there must be a sudden fall, as if there had been rapids, or a cascade, in the original river, as we go to the southwest from the Iowa Claim. The next point in that direction where there was an opportunity to take an observation was nearly or quite at the end of the diggings on Thompson Hill; its altitude was determined to be only 2,838 feet; unexpectedly lower than the bottom of Teaff's Shaft. But, though this is the lowest rock measured north of Dutch Flat Cañon, it will be seen to be decidedly higher than the bed-rock at the Cement Mill below Indiana Hill. In no way can we find other than an up-hill course for the river if it flowed northward from Indiana Hill.

In regard to the course of the channel between Dutch Flat and Little York, the following facts are to be noticed.

Directly opposite Thompson Hill, on the other side of Dutch Flat Cañon, at an elevation of between 350 and 400 feet above Bear River, was a spur of gravel known as Eastman Hill. This has been washed away pretty cleanly to the bed-rock, which, however, is almost hidden from view by the piles of boulders which have been left. The gravel was, in 1870, not entirely gone; for there were still two or three men at work at the upper end of the hill. Eastman Hill is one of the spurs leading down from Plug Ugly, but there is a long distance of slate bed-rock between the two deposits of gravel, so that no intimate and immediate connection can be traced between them, the distance across the cañon being hardly a quarter of a mile.

As we descend into the cañon of Bear River, and ascend on the other side by the wagon-road from Dutch Flat to Little York, we find, at a distance of about half a mile from the latter place, at a point where the road bears more to the west to follow up the line of Scott's Ravine, a spur of gravel known as Missouri Hill. This point was not in sight from Thompson or Eastman Hill, and no direct comparison with the hand level was possible. The barometer shows, however (by a mean of two closely agreeing observations taken on different days), an altitude of 2,753 feet for the Missouri Hill bed-rock.

Standing at Missouri Hill, it was seen that a level line would strike part way up on the face of the gravel banks on Independence Hill, below Little York, which proved that the bed-rock in Scott's Ravine was, at all events, a little lower than that on Missouri Hill. This observation was confirmed by subsequent barometric measurements, by means of which the altitude of the deep bed-rock in Scott's Ravine, east of the town of Little York, was shown to be only 2,704 feet. To the west of the town no bed-rock is found uncovered, until the claims on Empire Hill are reached. Between the last-named place and Scott's Ravine a high ridge of gravel has been left standing, along which the road to You Bet runs. From this ridge it is not quite easy to decide on which side the deepest bed-rock lies. A number of rough measurements have been made, from time to time, with discordant results, some indicating that the Empire Hill bed-rock is a few feet higher than that in Scott's Ravine, while the majority point to a slope in the opposite direction, but differ among themselves as to the amount. At Empire Hill there is a considerable body of bed-rock exposed, over a length of nearly a thousand feet, along which there is a narrow channel of blue gravel, between one and two hundred feet wide, which is a few feet deeper than the average bed-rock of the rest of the claim. The observation with the barometer was intended to be taken at the bottom of the deep channel, and the calculated altitude was 2,652 feet. This series of numbers shows conclusively the possibility, indeed the almost absolute certainty, of there once having been a channel from Thompson Hill, by way of Eastman Hill, Missouri Hill, and Little York to Empire Hill. As to its continuation from the last-named place, it may be noticed that from there across Steep Hollow the banks of Waloupa and of Squirrel Point are plainly in sight, with no obstacle in the way to force us to believe that connection between them and Empire Hill was an impossibility. The hand level and the barometer also agree in giving a suitable grade from Empire Hill (2,652 feet) to Squirrel Hill (2,639 feet) and Waloupa (2,590) feet. Concerning these last altitudes, it must be observed that they are liable to some slight alterations, but not enough to change essentially the general deduction to be drawn from them. Put in tabular form, these elevations are : —

Name of Station.	Approximate Distance from last Station.	Elevation.
Thompson Hill		2,838
Eastman Hill	¼ mile	2,8??
Missouri Hill	¾ mile	2,753
Little York	½ mile	2,704
Empire Hill	½ mile	2,652
Squirrel Point	¾ mile	2,639
Waloupa	¼ mile	2,590

The average grade would be, therefore, about eighty-three feet per mile between Thompson Hill and Waloupa.

Corroborative testimony, in regard to the course of the channel and the direction of the current, was also obtained by examination of the surface of the bed-rock, although a large part of it is hidden from view by the piles of refuse. In spite of this difficulty, a sufficient number of places were found on Thompson Hill, where the course of the stream was indicated, beyond possibility of doubt, as being toward Eastman Hill. Similar evidence was obtained that the direction of the flow, from the last-named locality, was toward Missouri Hill ; although, in this case, the condition of the bed-rock did not furnish quite as satisfactory indications as had been observed higher up on the channel.

After mapping the course of the old channel from Liberty Hill to Waloupa, and comparing it with the present course of Bear River, it is impossible to resist the conclusion that we are dealing with two different beds of one and the same stream. Even the sharp bend at Missouri Hill ceases to be an object of surprise, for we find that the present stream makes a similar curve near that point, and now runs, at the present day, in a parallel course about a mile farther south. The Bear River of to-day, instead of joining Steep Hollow, or its representative, at or near Waloupa, merely swings around the next hill to the south, and enters it a little farther down. What caused Bear River to change its course in this way it may be difficult to determine ; but that it did so seems almost demonstrated. Or it may have been that the present channel is the original one, which became choked up, and changed its course, for a time, so as to flow across the low divide between Christmas and Manzanita hills, returning afterwards to its original position. This supposition, however, seems not a very probable one, because the grade from Missouri Hill to Waloupa is so regular. If there had been an overflow of the kind suggested, we should have been likely to find more unevenness in the bed of the stream.

On Ellis Hill the top gravel is quite fine and of a reddish color ; but, as we approach Dutch Flat Cañon, the percentage of large boulders increases, and the bottom of the gravel becomes decidedly blue ; so that, at the outlet at Thompson Hill, the bed-rock is now pretty well hidden from sight by the piles of refuse boulders. What the character of the gravel at Eastman Hill was can only be inferred from the piles of boulders left on the bed-rock, or from the statements of those who were familiar with the ground when the washing was going on. There is no reason for thinking that the Eastman Hill gravel differed, in any essential particular, from that on the opposite side of the cañon.

At Missouri Hill the surface of the bed-rock exposed is not large, neither was the amount of gravel considerable. All that is left is an abundance of boulders, of such form and dimensions as to remind one strongly of the upper end of the Dutch Flat claims. There are many sharp corners and edges visible, which may be owing to the fact that a good proportion of the boulders have been broken since they were uncovered.

At Little York proper — that is, east of the bridge of ground left standing, along which the road to You Bet crosses — there is a large extent of gravel. At the deepest part is a channel, from one to two hundred feet in width, of hard, blue gravel, — in some places almost " cement," — and in which are numerous boulders of considerable size. The depth of this blue gravel it is not easy to give with precision. At the lower end of Scott's Ravine the stratum between the blue and the overlying red gravel consists of a sort of clayey sandstone, which is six or eight feet in thickness, and appears to pitch to the west, at an angle of 8° or 10°. Above the blue channel is a depth of from 125 to 150 feet of a finer red gravel, and this is spread out over a broader area, the bed-rock rising pretty rapidly on the south side, to a point about twenty-five feet above the level of the town at the flag-staff, so as to form a well-defined trough of red gravel at least 600 or 700 feet wide at the top. From that point the bed-rock still rises, but more gently, toward the south, for a quarter of a mile or more, in the direction of Manzanita Hill. The rock, especially in its upper portion, is a clayey slate, which decomposes very rapidly when exposed to the action of air and water. At some points it was difficult to tell whether one was standing on the bed-rock, or on a mass of clay belonging to the gravel series. As a rule, the uppermost gravel of Manzanita Hill is quite fine, of a reddish color, and easily washed away. There are also strata of clay to be seen in it, which attain, on the eastern side of the hill, a thickness of six feet or more. These strata are seen to dip toward the southwest, at an angle of from 12° to 15°, which is about the position they might have been expected to assume if they were deposited on the sloping side of a bay or indentation between two hills, to one side of the main channel of the river. A nearly north and south section across the Little York gravel is represented on Plate F, Fig. 3. This figure is drawn so as to include the gravel on Christmas Hill, which will be described a little farther on. The position of the small narrow and deep channel, at d in the centre of the section, will be observed, as well as the broader channel of red gravel, extending from c to e. The summit of Manzanita Hill is at f, and that of Christmas Hill at a, the position of the bed-rock surface under the last-named elevation being unknown, but probably, as represented in the section, sloping to the south. The upper line representing the original surface of the gravel is, of course, only an approximation.

At Empire Hill there is, at the bottom and middle of the channel, a mass of blue gravel thirty or forty feet in depth. Its general direction is east and west, but the drifting under the easterly bank shows that the course of the channel was not exactly straight from Scott's Ravine to Empire Hill, but that there was a slight bend around to the north. The southern bank of Empire Hill is nearly washed away; but on the northern one there is, above the blue gravel, a thickness of thirty or forty feet of red, the succession being the same as at Scott's Ravine. Toward the west end of Empire Hill boulders of from one to three feet are common. They are found mostly in the upper thirty feet of the gravel, and many of them are quite angular, the corners being hardly rounded at all.

The main mass of the Christmas Hill gravel covers a surface about three quarters of a mile long and an eighth of a mile wide; or, in round numbers, about sixty acres, the longer axis having a nearly north and south (magnetic) direction. It is on the top of the ridge which separates Steep Hollow from Bear River, and is about midway between the two streams. The county road from Little York to Liberty Hill follows this ridge, and for half a mile is quite near the bank of gravel; for some portions of the way, indeed, there are openings on both sides of the road, with just width enough between to allow teams to meet in safety. The top of the Christmas Hill is pretty flat, although sloping gradually from the highest point toward Bear River, Steep Hollow, and Little York, as far as the gravel extends. Then, on both sides, the pitch into the two cañons is quite steep, there being a descent of fully 700 feet within a distance of little over a quarter of a mile. Toward Little York the fall is not quite so rapid. On the northern end of the hill, near Mellor's house, there is a divide a trifle lower than the summit of the hill; and in following the road beyond that the gravel grows thinner, and disappears before the sharp ascent of the Camel's Hump commences. To the northeast of this elevation there is a small patch of washed gravel; but it seems not to belong to the main deposit at all, but to be, rather, the relic of some small feeder of Bear River or Steep Hollow.

The western face of the bank on Christmas Hill, for nearly the whole distance, will measure from thirty to sixty feet in height. The section thus offered shows mostly a fine red gravel, intermixed to some extent with strata of pipe-clay and sand, which latter is so hardened as to be almost a proper sandstone. There is here hardly a boulder as large as a man's head, although on the western side such small ones are rather frequent. The eastern bank, which is much cut up by the sluices and by ravines which empty into Bear River, will average about twenty feet in height, and its summit is from twenty to thirty feet lower than that of the western bank. All along the eastern side of the hill, at distances ranging from fifty to 200 feet from the edge of the bank, the line between the gravel and the slate is easily traced; but it is not easy to decide whether this is to be regarded as the rim of a channel. It is more probable that this is not the case; this high gravel appearing rather to indicate that there was once an extensive lake at this place, or else that the river choked up to such an extent that the lateral ravines were filled with detritus for a long distance back from the main channel. Possibly, however, there was here once a tributary of the main stream. In the case of Christmas Hill, this latter supposition seems hardly tenable; because, to have deposited so much gravel, there must have been a great deal of water, and a longer stream than we have any evidence of higher up. The second of the above theories seems, on the whole, the most reasonable, as was the case in regard to Plug Ugly Hill. Indeed, these two hills resemble each other in many particulars. Plug Ugly has its capping of fine red gravel, with Eastman Hill on its flank, just as Christmas Hill has its capping of similar material, and is separated from Missouri Hill by an extensive stretch of slate. The parallelism is complete; and it would be desirable to have an opportunity to examine the bed-rock under these hills, to see whether it shows such signs of wear as would be left by running water.

On the southeastern side of Christmas Hill are two or three small openings, which are eight or ten feet lower than the main mass of gravel, and separated from it by a rim of slate bed-rock. The largest of these openings is about 350 feet in diameter, with a bank of gravel about thirty feet high on the west, and one of six or eight feet on the east. There is also a considerable quantity of clay, and of boulders measuring more than a foot in diameter. The union of notable quantities of clay and of boulders of such large dimensions in the same restricted deposit points, not so much to a channel of running water, as to a hole in which the boulders have been stopped, and where the clay has been packed around them, at a later period, when the force of the water had abated. If this whole deposit of gravel were in a ravine, which had been filled up by the back water from the main stream, it would be only natural to find small boulders, which have been rolled down from above, and not carried far from their original place of lodgement.

Manzanita Hill is the highest point south of Christmas Hill, on the ridge between Bear River and Steep Hollow. Toward the junction of the streams the slope is gradual, until the steep descent into the cañon is reached, as is common in this region, wherever there is a ridge between two cañons. Near and around the summit of Manzanita Hill are the heads of a number of ravines, some emptying into Steep Hollow and some into Bear River. The most important one on the Bear River side is known as Nigger Ravine, and that on the Steep Hollow side as Cariboo Ravine. Between the heads of these two ravines is a low divide, or saddle, which connects the top of Manzanita Hill with the subordinate ridge between Nigger Ravine and Steep Hollow. On the highest point of this latter ridge, not far from the saddle just mentioned, there is a small patch of rolled quartz gravel, covering an area of about a quarter of an acre, and not appearing to be of any considerable depth. It is several feet higher than the bed-rock, although not as high as the top of the gravel at Empire Hill. This indicates that the old river possibly spread out to this extent at one time; in which case Manzanita Hill must have stood out like an island in the broad stream. Southward and eastward from this gravel-patch there is nothing but slate bed-rock to be seen. But at the saddle between the heads of Nigger and Cariboo ravines there is a mining claim, called the Cariboo, which presents some curious and interesting features.

The course of the small ravine or gulch, in which the Cariboo Claim is situated, is nearly due north, for the distance of about a quarter of a mile, the fall in that distance being a little over 300 feet. At the upper end of the ravine the slate has its ordinary, nearly vertical, dip to the east. To the east and west the bed-rock is seen in its usual position, within two or three hundred feet of the ravine; while, to the north it is hardly necessary to go beyond the mouth of the ravine to find the slate with its ordinary dip. But in the ravine itself the slate bed-rock lies nearly horizontally, or with an inclination to the west not exceeding 15°. Near the shaft, in the ravine, there is an easterly dip to the slate on the east side of the ravine, and a dip in the opposite direction on the west side. Accompanying this horizontality of the slate are other anomalous features. The deposit is frequently spoken of as a "channel" by the miners at Little York; but if it be a channel, it is one *lying on its side*. On the east side of the deposit, and resting against the horizontal slate, is a vertical stratum of heavy boulders; while the fine gravel, which ought to be on the top, forms a vertical stratum on the west. On the side where the heavy boulders lie the crevices in the slates in which the nuggets are found are also horizontal, like shelves cut out of the rock. From one such crevice in the eastern wall, eighteen inches long and eight or ten deep, there were taken (as is stated on good authority) three pans of dirt, which yielded $ 29.00. Relatively to each other, then, we have bed-rock, boulders, and fine gravel in the position they would have occupied, had there been a true channel. The width, which corresponds to the thickness in ordinary channels, is not much more than thirty feet at the top, and diminishes as we go down.

The cross-section of the slate, observed going up the ravine, is peculiar. This rock occurs all the way up to a point about fifty feet north of where the shaft was sunk. Previously to reaching that point, however, there was a crevice met with, which was about twenty feet deep, and which had been worked out, by sinking several small shafts and refilling them with rock. The slate which is seen to be nearly on a level with the top of the shaft, at the point mentioned as fifty feet to the north of it, suddenly pitches almost vertically for sixty-nine feet. Beyond the shaft, as we go southward up the ravine, the slate appears again, at a point about 350 feet distant, and ten or twelve feet higher than it was when last seen to the north. Above this point rises a vertical wall of slate thirty feet or more in height, beyond which there is, again, a little gravel to be seen. A north and south section on the line of this ravine is represented on Plate F (Fig. 4), the whole distance included in it being about three eighths of a mile. The section also includes a tunnel run in from Nigger Ravine, to which reference was made on the preceding page. In opening this deposit of gravel, a tunnel 600 feet long (*b* in the figure) was run in the slate-rock at the west side of the ravine. From the head of this tunnel it was necessary to rise twenty feet before gravel was struck, and the deposit was found to be forty-eight feet thick at this point. To the west of the shaft, and as low, or lower, than its bottom, there is said to be gravel still, although the amount appeared not likely to be large. As nearly as could be ascertained, the gravel near the bottom of the shaft was quite narrow, probably not exceeding eighteen inches in width, and, in fact, presenting, in some respects, the appearance of a vein.

The quality of the gold found at the Cariboo Diggings is poorer than that of the product of any of the other mines in the neighborhood; it sells for from $ 16 to $ 16.50 per ounce, the average price paid by Messrs. Moore & Miner for Gold Run gold being nearly $ 19.00. At the Cariboo, also, there was both coarse and fine gold; some pieces being smooth and rounded, others "like tacks," to use the expression of those who worked in these diggings. The nearest gold of a similar kind being found, as was said, at Secrettown, has led to a belief, on the part of some persons, that there was once a connection between these two places; a theory which is supported by the fact that the country between You Bet and Secrettown is generally low. Above the shaft, at the

Cariboo Diggings, to the west, the top dirt is made up partly of small rounded stones and pebbles, partly of decomposing slate, together with the usual red dirt, so common in connection with the gravel deposits. Among the small stones in this ravine were some of a porphyritic character, quite different from any of the pebbles found in the Little York or Empire gravel, and also different from any rock in place in the neighborhood.

Another feature of the Cariboo Ravine is the magnificent boulder of white quartz. Its maximum dimensions are fourteen feet in length, seven in width, and four in thickness. Its average length would be fully twelve feet, and its width five and a half. The thickness of this block is pretty uniformly four feet, the sides being parallel, as if the mass had come directly from a quartz lode. This boulder shows, it is true, signs of wear from water; the surfaces are all smooth and the angles rounded. But still there are, quite close to the surface, numerous vugs and cavities containing quartz crystals; one of the largest of these crystals, which is more than half an inch long, projects within a quarter of an inch of the surface of the boulder, and is perfectly sharp on all its edges and angles, showing no sign of wear. It seems impossible that this particular piece of rock can have been carried to any great distance, or subjected to any considerable amount of erosive action. A hundred and fifty feet farther up the ravine there is another large boulder, of bluish quartz; which, however, is inferior in size to the one just mentioned. These two masses of quartz are different from anything else found in the neighborhood, and add to the general mystery which hangs over the Cariboo gravel deposit.

The same reasons which prevented our getting full and complete details of the management of the mines at Dutch Flat appeared in even greater force at Little York. The best which could be done was, to get measurements for an approximate calculation of the amount of gravel which has been washed away in that district. The original surface, however, must have been far from even, and my estimates as to the thickness of gravel may be considered too high by some, and too low by others. The results, such as they are, here follow: On Christmas Hill, forty-four acres worked out to a depth of thirty-five feet; in and around Scott's Ravine and Little York, as far as the You Bet road (excepting Manzanita Hill), thirty-two and a half acres, to an average depth of sixty-five feet; on the northerly slope of Manzanita Hill, thirty-five and a half acres, to a depth of forty-five feet; and at Empire Hill, eighteen acres, to a depth of thirty feet; or, in cubic yards, in round numbers:—

Christmas Hill	2,500,000
Little York and Scott's Ravine	3,500,000
Manzanita Hill	2,500,000
Empire Hill	875,000
Total,	10,375,000

The number of cubic yards of gravel still remaining to be washed it is impossible to estimate, with even a tolerable degree of exactness. On Manzanita Hill, on the gently-sloping bed-rock, there is still considerable gravel, but not depth enough to make a profitable bank for hydraulic working, until the upper end of the hill is reached. Here there are still good banks in sight; but it will be necessary to clear away some of the old remnants in front, before they can be worked with profit. In Scott's Ravine, the accumulation of boulders has been so rapid that the sluices are full, and a considerable outlay will be required to get things in running order again. But there is gravel enough between Little York and the Empire Hill bank to last several seasons yet, even with the best facilities for piping and sluicing. The gravel left on Christmas Hill is so high, that it seems hardly a profitable undertaking to bring the water to that point, as long as there is so good a demand for all that can be furnished at the lower claims. The old flumes have been partially destroyed, and will probably not be rebuilt for the next two or three years, at any rate. Negotiations were said to be pending for the sale of the whole ditch and gravel property at Little York to an English company.*

* See farther on, page 161.

While meandering the beds of Steep Hollow and Bear River, an attempt was made to gather data for calculating the number of cubic yards of tailings in the head of those two streams. Such a calculation can, however, at best give but a rude approximation to the reality. The width of the mass of tailings at a number of points along the stream could be, and was, obtained with a good degree of exactness, and the length of the portions meandered is known with sufficient accuracy. But in regard to thickness of the deposit, reliance must be placed entirely on estimates. The opinion of those familiar with the regions was, that the average depth of the tailings was between fifty and seventy-five feet. Some persons were confident that Bear River had been filled up as much as seventy-five feet, between Dutch Flat Cañon and its junction with Steep Hollow; a few even set the number as high as one hundred. In using these estimates it is clear that there is great room for differences of opinion. And furthermore, the amount to be allowed for the sloping banks, under the edges of the tailings, will be almost entirely a matter of conjecture. Where the sides of the cañon rise very steeply on both sides, it is not so difficult to get a fair average; but, in the broader parts of the stream and where side ravines have brought down their additions of tailings, it is almost certain that, for a considerable portion of the surface covered, the depth is much less than the average in the middle of the stream. The portions of the water-courses included in the following estimate are: on Steep Hollow, from the outlet of Wilcox Ravine to the junction with Bear River; and on the latter from Steep Hollow to Scott's Ravine. The stretch on Bear River is a little the longer of the two; but the average width of the stream is less, so that the quantity of the tailings in the two cases is almost identical. The sources of the Bear River tailings are principally Dutch Flat and Little York, including Christmas Hill; while Steep Hollow has been supplied mostly from You Bet, with a small addition from the Empire Hill side of the Little York gravel. Below the junction, Bear River is also filled to a considerable distance with tailings; but the amount has not been estimated. In round numbers, the number of cubic yards of tailings in each stream, within the limits above indicated, may be placed at 5,000,000. The distance from Scott's Ravine to Wilcox's Ravine, by the stream, is not far from four and a half miles. In the absence of any data, it is not possible to make an estimate of the amount of gold and quicksilver lying buried in these 10,000,000 cubic yards of gravelly pebbles.

In the course of the meandering of Steep Hollow and Bear River, little of interest was found connected with the geological bearings of the gravel question. At one spot on Bear River, however, about a mile above its junction with Steep Hollow, where the stream changes from a southerly to a westerly course, a small quantity of gravel was found, at an elevation of a few feet above the present tailings, probably about seventy-five feet higher than the bed of the river before washing began. This must have been left where it now is, while the river bed was at or near that elevation. There was not enough of the gravel to pay for working; but the bank had been opened by some prospecters.

The condition of things in the region about Dutch Flat and Gold Run has been considerably changed since the above description by Professor Pettee was written. Consolidations of property seem to have been made, and sales effected, which have had for a result the opening, on a very extensive scale, of the deeper gravels previously not reached by the hydraulic workings. There is no recent information from this district in the possession of the writer*; but some idea of the extent of the preparations making, for the purpose of reaching the bottom of the Dutch Flat and Gold Run channel, may be got from Mr. Skidmore's report to the Commissioner of Mining Statistics, published in his last (seventh) volume, which embodies the work of the

---

* Should it be possible to obtain such information, it will be added in the Appendix to this volume.

year 1874.*   At the time of Mr. Skidmore's examination, the "Indiana Hill Blue Gravel Mining Company" was running a tunnel, in the channel itself, from the extreme southerly end of their claim, on the cañon of the North Fork of the American River. No rim-rock tunnel being necessary, the channel was followed northerly on the bed-rock by means of a main tunnel, 1,600 feet in length, with side-drifts toward the west, where the richest ground was discovered. A width of about 110 feet was carried forward in this way, and a thickness of six or seven feet of gravel excavated. The material taken out is "cement," and has to be stamped in a mill run by water-power. In spite of this, the material appears to be so rich as to have paid handsomely, the company having made a profit, from the time of their incorporation (April, 1872) up to the close of the season of 1874, of $34,853.47 on a total yield of $75,422.47, the cost of mining and milling having been $40,569. According to this official statement, the average yield of gravel was $5.29, the expense $2.90, and the profit $2.39, per cubic yard. It is said that, in one instance, $1,000 worth of gold was obtained from two car-loads (or thirty-nine cubic feet) of dirt. Mr. Skidmore adds: "The Indiana Hill is the only notable successful enterprise in operation in California of mining gravel by the crushing or mill process."

From the above-cited authority we also learn that an extensive consolidation had taken place, immediately after Professor Pettee's examination, in the region about Gold Run, where the "Gold Run Ditch and Mining Company" had acquired an area of 328 acres, together with an extensive system of ditches.† To reach the deep gravel in this consolidated claim, the position of which will be understood from the description given in the preceding pages, the new company was engaged, in 1874, in running a bed-rock tunnel from Cañon Creek towards the "'49-Shaft," in the supposed centre of the channel in the deep gravel. This tunnel was intended to be 2,200 feet long, and twelve feet wide and nine high, and it was begun in July, 1873. The total cost was estimated at $125,000; and it was expected to be completed in about two years. The Cedar and Sherman claims, near Gold Run, were also sold to an English company, and preparations were making, in 1874‡

* Seventh Report of the Commissioner of Mining Statistics, pp. 99–102.

† See the large map, appended to this volume, on which it appears that a patent had been issued under the name of "Church and Golden Gate," for an extensive gravel claim near Gold Run.

‡ The most astonishing stories of the richness of the surface of the bed-rock struck in a prospecting shaft sunk by this company, were reported by the Superintendent. The shaft is said to have reached the bed-rock at the depth of 181 feet, and the dirt at the bottom to have "prospected" at the rate of $2 per pan, or $232 per cubic yard.

for opening them on an extensive scale. The same may be said of the Cedar Creek Claim, near Dutch Flat, which according to Mr. Skidmore has been opened by a bed-rock tunnel 3,000 feet long, and eight by eight feet in dimensions, the property having been sold to an English company.*

Before passing on to a description of the You Bet and Red Dog gravel deposits, next adjacent on the northwest to those just described, a few small claims south of the line of the Central Pacific Railroad and at a considerably higher elevation than those at Dutch Flat, may be described.

About half a mile south of Blue Cañon Station there is a cut of about twenty-five feet in depth; this is in a mass of gravel made up of rounded pebbles and rocks of different sizes, up to a foot in diameter. These pebbles were almost without exception of volcanic origin; and at first sight it seemed as if they must have been transported from a considerable distance; but a second look showed that all of them, almost without exception, were weathering from the outside and scaling off in such a way as to gradually assume a more and more perfectly spherical form. Directly opposite Blue Cañon Station is also some of this cemented gravel, but with more angular pebbles. It is said that "the color" can be obtained from the cuts south of the Station (elevation a little over 5,000 feet).

The mining settlement of Lost Camp gives its name to Lost Camp Spur, which is about a mile from Blue Cañon Station in a nearly southerly course. In this vicinity is the so-called Slumgullion Claim, at an elevation of 4,386 feet, and having its outlet into Texas Cañon. The area of ground which had been washed here in 1870 was of an irregular shape, about 500 feet long by 250 wide. The bed-rock is slate, with a dip of from 80° to 85° to the northeast. Overlying the bed-rock is a coarse quartz gravel, none of the boulders being much rounded; the thickness of the deposit is about sixteen feet. Above the gravel is a stratum of pipe-clay, about fourteen feet thick, and containing a few lenticular masses of gravel; the altitude of the bed-rock here is 4,386 feet.

Three eighths of a mile nearly due west of the Slumgullion is the Boston Claim. The bed-rock in these two claims is of the same character, but its surface is about 175 feet lower in the Boston than in the other. The height of the bank varies from twenty to sixty feet, and there is an almost entire

---

* The boulders too large to pass through the sluices are now got rid of, without the necessity of drilling holes in them, by exploding giant powder on their surfaces, which breaks them up sufficiently, at slight expense. The masses of pipe-clay, formerly a great annoyance in hydraulic mining, are now bored into by an auger, and torn to pieces by the aid of a cartridge of giant powder.

absence of pipe-clay.   Between these two claims there is a pretty continuous body of gravel ; at least, rounded pebbles were seen, on the surface, all the way along the ridge between Blue Cañon and the North Fork of the American.   In both the Boston and Slumgullion claims, as was stated by the Superintendent, Mr. Coyn, there is no fine gold ; but nuggets of considerable size have been met with.

In regard to the origin of this rather isolated and very elevated mass of gravel Professor Pettee remarks as follows : —

The question now arises, whence came this gravel? My first impression was, that its source was to be sought in the ridge above the railroad.   But the gravel differed, in almost all respects, from anything met with in the railroad excavations ; it was not easy to trace any connection between the two.   There is a considerable mass of slate intervening between these two gravel mines and the railroad, on which no gold has been found.   The conclusion to which I was finally led was, that this old current came independently from some point farther east.   The fact that the bed-rock on Lost Camp Spur is some 350 to 400 feet below the station, was also a ground for thinking the gravel in question has no connection with the channel above the railroad.   Standing at Mr. Coyn's house, near the Slumgullion Mine, it was clear to the eye that there was plenty of room for this gravel to come down the valley of the present North Fork of the American ; and Mr. Coyn informed me that there was, or had been, a claim on the left bank of the North Fork, known as the Texas Claim, from which gold of the same quality and of similar character had been taken as from the Slumgullion.   The altitude of the Texas Claim would be considerably higher than that of the Slumgullion, and would admit of a reasonable grade between the two places.   If this channel came in the direction supposed, it probably continued farther down to the west.   In support of this view, Mr. Coyn said he had traced a similar gravel along all the spurs between the Boston Claim and Blue Bluffs.   If this theory proves to be substantiated, it would seem as if this was a relic of an old channel in which the North Fork of the American used to run before cutting out its present deep cañon.

A small patch of gravel will be noticed on the map at Blue Bluffs, a short distance northeast of Shady Run Station on the Central Pacific Railroad.   At the locality called Blue Bluffs Point, there is said to be a hard mass of cement, of a dark-brown color, which cannot be worked without the aid of blasting.   The Blue Bluff Diggings proper are about a thousand or fifteen hundred feet from the Point, on the south side.   A shaft is said to have been sunk there seventy-five feet deep to bed-rock, and all the way in gravel, with the exception of a stratum of clay, three feet thick, met with at a depth of thirty-five feet below the surface.   The bed-rock at this place is estimated to be 400 feet lower than the railroad track, at a point a quarter of a mile above Shady Run Station.

Leaving, for the present, the consideration of the gravel claims lying on the ridge between Bear and Steep Hollow creeks, but considerably higher up than those already described, we pass to the very extensive deposits which cover the region between the last-named creek and the Greenhorn. The towns of You Bet and Red Dog are the most important mining centres of this district, the description of which here follows, as drawn up by Professor Pettee.

§ 2.   *You Bet and Red Dog.*

The road from Little York after crossing Steep Hollow follows up the line of Wilcox Ravine toward You Bet. The country rock is the ordinary slate of the region. On the east of the ravine is Chicken Point, with high gravel extending nearly to the edge of Steep Hollow Cañon. The road rises rapidly, and, at a point a little less than half a mile from the town of You Bet, strikes the main mass of gravel, which covers without any material exception an area of twelve or fifteen hundred acres. A line drawn in a southwesterly (magnetic) direction from the end of Chicken Point, for a distance of a mile and three quarters, would mark nearly the southeastern boundary of this gravel. Starting from the same point and running for the same distance in a nearly northwest (magnetic) direction along the base of the Sugar Loaf and Chalk Bluffs gives us, approximately, the northeast line of the gravel as far as Boston Hill, near the cañon of the Little Greenhorn. Beyond this point lies the gravel of Buckeye Hill, Hunt's Hill, and other places, which will be noticed farther on. The northwestern boundary of the gravel in question is an irregular line running to the northwest of You Bet, and including Red Dog and the mines of Independence and Bunker hills. It must not be supposed that there are absolutely no points within these limits which are not covered with gravel; but they are not many in number.

To the northeast, the Sugar Loaf and Chalk Bluffs form as it were a high, steep wall, at the foot of which the highest gravel and bed-rock are found. How far the gravel extends underneath the volcanic material of which the bluffs are composed, if at all, is a question for future consideration. From the foot of the bluff towards the west the general slope of the country is more moderate, though decided. The surface, however, is by no means level and smooth. The town of You Bet, for example, near the head of Wilcox Ravine, is built upon a sharp ridge of gravel, along the comb of which there is room for only one street. To the south, the slope is rapid towards Steep Hollow; and to the north there is also a considerable depression — Missouri Cañon — between You Bet and Red Dog, while to the west, towards Pine Hill, the slope is quite gradual. The town of Red Dog is also built upon the gravel between Missouri Cañon and Arkansas Cañon.

At Pine Hill, the western end of the district, the gravel is spread over a nearly level plateau, of several acres in extent, fifty feet or more above the gravel at Waloupa, which lies on the opposite side of a ravine. The principal points of interest on Pine Hill are at Hubbard's Tunnel and near Cahel's house. The Waloupa * Diggings are at the southeastern extremity of a spur of slate between two ravines which unite below to form Birdseye Cañon. Across the ravine from Waloupa, and to the south of You Bet, the principal mines — or those to which reference will have to be made in the following pages — are those of Niece & West, Williams, Heydliff, Brown and Mallory, and Hyatt or Haight. To the east of the mines just mentioned, and in part separated from them by Sardine Flat, lies the high gravel of Chicken Point. At the head of the ravine the gravel is continuous from You Bet to Chicken Point. From Chicken Point northwesterly the gravel is continuous along the base of the bluff at the head of the ravines which unite to make Missouri Cañon. These mines have been worked mostly by Williams, Timmens, and Brockmann. Beyond these, and separated by only a small interval, come Hussey's claims, and the openings on Darling's and Boston hills. At Red Dog there is gravel on both sides of Arkansas Cañon; the mine first to the north of the cañon being called Independence Hill. The continuation of Independence Hill at the point where Williams has another claim is known as Bunker Hill.

In attempting to trace the geological relations of this mass of gravel, difficulties upon difficulties arise so rapidly that it seems almost impossible ever to hit upon any explanation of the facts which will satisfy all conditions. In what follows, therefore, I shall try, not so much to establish

---

* The correct orthography of the name Waloupa is involved in considerable doubt. Most persons have supposed the name to be of Indian origin; but there are some who say that the claim was first opened by a party of Spaniards who gave to it the name of Guadaloupe, which in time became corrupted into Waloupa, or Wauloopa, or any one of a number of other ways of writing it.

a theory or announce the solution of a difficult problem, as to state as clearly as I can the facts upon which a theory will have to be based and the conditions which must be satisfied.

Proceeding upon the assumption that the gravel had been deposited from the waters of some river, the first natural inquiry was as to the original course of the channel : and, as a key to the solution of the problem, barometric measurements of the altitude of the bed-rock, wherever it could be found exposed, were made. The necessary observations were taken with great care. The barometers were frequently compared with each other ; and in almost every instance synchronous observations at the hotel in You Bet, and at the different stations in the mines, were obtained. None of the stations visited were more than a mile and a half from the hotel nor more than 400 feet above or below it. Under these circumstances we are warranted in saying that in no case can there be a relative error of more than a few feet. The altitude adopted for the hotel at You Bet was fixed by comparing the observations of several days with those of the station barometer at Colfax, and, at the worst, cannot be far out of the way. There have been, furthermore, from time to time, partial surveys of the claims in this district ; and, in some instances, differences of altitude have been determined by the spirit-level. Wherever this has been the case the results have been almost identical with those obtained by us in our barometric series.

On Pine Hill no observation was taken at the top of the gravel ; but there is a descent of fully 100 feet from the average level of the summit of the gravel plateau to the point where Mr. Hubbard has run in his tunnel. My observation was taken near the mouth of the tunnel, and showed an altitude of 2,652 feet. According to Mr. Hubbard's statement this was about five feet below the point on " the rim " where gravel was first struck in the tunnel. Farther in, the bed-rock was nearly on the same level, — possibly one or two feet deeper in some places ; thus 2,655 feet may be taken as the altitude of the Pine Hill bed-rock. The length of Hubbard's Tunnel is between five and six hundred feet in all, and its course at the mouth is nearly N. 40° W. (magnetic). Beyond Pine Hill there is no chance for an outlet unless by way of Mule Cañon, which empties into Greenhorn. This cañon having been rich in gold, it is probable that there was once an overflow from Pine Hill, if nothing more, in this direction. The most rational supposition is that the deposition of gravel gradually filled up the original channel, until it caused the water to spread over the Pine Hill region and then overflow into the adjacent low country to the northwest. That the main channel, however, could not have run at first in this direction is shown by the elevation of the bed-rock. And this idea is strengthened by the character of the Pine Hill gravel. At the bottom are found large boulders of blue and white quartz, — said to be eight or ten feet in diameter, — together with more or less decomposed material, while the main mass of the gravel all the way to the top is a rather fine washed quartz. Such a combination points rather to the filling up of a lateral ravine after the main channel has outgrown its original bed, — the large boulders not showing signs of any great amount of wear. To the south and west of Hubbard's Tunnel the slate bed-rock rises to such a height as to preclude the possibility of the channel's having had an outlet in that direction ; there is, to be sure, a small ravine leading down from the mouth of the tunnel, but no signs that it can ever have been the path of the stream from which the gravel was deposited. A quarter of a mile to the southeast there was a small extent of bed-rock exposed, the altitude of which I made to be 2,677 feet ; this was near Cahel's house. The amount of bed-rock uncovered was so small that it was impossible to decide whether I was in the centre or only on a piece of the rim of a channel. The country slopes very rapidly towards Birdseye Cañon. From where I stood, the banks at Empire Hill and Little York were plainly visible across the cañon of Steep Hollow, showing that there was plenty of room for the old streams to have formed a junction anywhere within quite an extended area of country. The low divide to the southwest of Cold Spring Mountain and the Secrettown gap were easily distinguishable, and it was not hard to see the grounds on which the theory was based of there once having been a channel from Waloupa or Pine Hill by way of Cariboo to Secrettown and beyond. The probability of the correctness of any such theory, however, diminishes very rapidly in the presence of others which are so much more worthy of credence.

The chief point decided by these measurements on Pine Hill is that the slate rock is altogether too high for the deep channel to have found its way through here. The details of the form of the bed rock on the hill itself cannot be known with certainty until further explorations have been made, but this is a matter of only secondary importance so far as the deep gravel is concerned. The key to the solution of the gravel question is to be sought on the lowest, not on the highest bed-rock. It is possible that a channel once flowed in at one side of the present Pine Hill and out at the other, so that the well-defined rim of a large bend in the channel may some time be brought to light; but I hardly think so. I prefer to adhere to the theory of the filling up of the lateral ravines.

To the northeast of the Pine Hill gravel, and across a ravine cut in bed-rock, comes the frequently mentioned Waloupa. Here there is a considerable exposure of bed-rock. The general course of the long axis of the diggings, across the spur between the two ravines which unite below to make Birdseye Cañon, is north and south. The point where my observation was taken was at the north end of the diggings, directly opposite Niece and West's. The altitude I made to be 2,590 feet. Two hundred feet farther to the south there was a place in the bed-rock at least ten feet deeper than where I measured; but the rock rose again on the opposite side, so that the deep place was probably only a hole. On the east side of the claim the gravel is nearly all removed, and on the west there is a bank of about thirty-five or forty feet in height. At the bottom there is a blue gravel for ten or fifteen feet in thickness, while the upper twenty-five feet, more or less, is made up of layers of red gravel alternating with red or gray sand. This arrangement of the strata — a substratum of blue gravel covered with red — corresponds in its main feature with what is seen at Niece and West's on the opposite side of the cañon, and points unmistakably to the former connection of the two places before the wearing away of the cañon between. Above Waloupa there is gravel on the ridge to the north and northwest for several hundred feet, and there are several small mines or openings where work has been commenced; but there are none at a higher level than the top of the high bank on the opposite side of the ravine, and slate is struck again before we reach the main divide along which the road from You Bet to Colfax and Grass Valley runs. The bed-rock at Waloupa, it will be observed, is the lowest that has yet been found anywhere throughout this whole gravel district.

Across the cañon from Waloupa in a northeasterly direction is the mine of Messrs. Niece and West, where a great deal of bed-rock has been exposed. Here, as at Waloupa, the bottom of the gravel is blue. The thickness of the deposit, however, is greater. The blue gravel is so firmly cemented together that it is more profitable to run powder drifts and loosen large quantities at a time, than to try to cut it away by the force of the hydraulic stream alone. The upper strata are lighter in color and easier to work.

The deepest part of the centre of the old channel, as nearly as could be made out, was selected as the point whose altitude was to be determined. The barometer hung at one corner of the old mill, which is built upon the bed-rock, and indicated an altitude of 2,617 feet, a result which agrees well with what was shown by the hand-level from Waloupa.

To the north of Niece and West's there is a large gravel claim belonging principally, or entirely, to Mr. Williams, but as only the upper gravel has been removed and no bed-rock has been struck in the deep part of the channel, no observation was taken for altitude.

The next important points, where deep bed-rock is to be seen, are in a northeasterly direction, and on the opposite side of the hill from Niece and West's. Following up the line of Wilcox Ravine, to the east of the Little York and You Bet road, there is bed-rock seen at Heydliff's ground, at Brown's, and at Mallory's. There has been considerable drifting done under these banks, and there cannot be much, if any, doubt that the deepest parts of the bed-rock have been cleaned off. There is not much rock to be seen without entering the drifts, however. Observations were taken at the mouth of the tunnel at Heydliff's and at the bottom of the old incline at Mallory's. Between these two points lies Brown's Mine, where it was not convenient to take any observation. Our results agree very well with those of the spirit-level surveys in giving a fall of about twenty er twenty-five feet between Brown's and Niece and West's. And the high knob of slate to the south shows that the channel must have had a nearly southwest course between the two places. As to the difference of level, however, between Mallory's and Heydliff's our results are not quite so satisfactory, — the reductions giving 2,641 feet for each place. But the three mines in question have been so far connected underground as to show, from the running of the water, that the grade of the bed-rock is from Mallory's towards Heydliff's. No great difference was expected, owing to the proximity of the two stations; but the bed-rock at Mallory's is probably five or six feet higher than at Heydliff's. The calculation of the altitude of the bed-rock at the foot of the incline is probably less trustworthy than the others, partly on account of the difficulty of getting an accurate reading underground, and partly from the slight uncertainty resting upon the temperature term. For places so close together and so nearly on the same level, some other method of determining the difference of altitude is to be preferred, when the closest accuracy is desired. In the present case, however, there seems to be evidence enough that the bed-rock slopes at an easy river grade from Mallory's to Waloupa.

On the opposite side of Wilcox Ravine and a little to the east of Brown's house there is slate-rock decidedly higher than the bed of the ravine, which may be taken as the east rim of the channel or as a part of an island in the stream. The latter supposition seems preferable, because farther to the east we find other gravel similar in character, to say nothing about the high deposits on Chicken Point.

So far, then, there seems to be no difficulty in tracing the channel by the grade of its bed. But beyond this point trouble arises. For the next half-mile in a northerly direction there is no deep bed-rock to be seen. In the mine just to the east of where the Red Dog road crosses the ravine between Savage's and Williams' there is a little bed-rock dipping rapidly to the east. Its altitude we made to be 2,828 feet. How much deeper the rock extends in an easterly direction there is no means of knowing, for it is not seen again until it comes out high up on the Sugar Loaf, three quarters of a mile distant. Somewhere between these points, it is natural to suppose, a deep channel lies. Still farther to the north, at a point on the south side of Missouri Cañon, — opposite the Cozzens and Garber shaft, — bed-rock is seen at the outlet of a mine at an altitude of 2,762 feet. Its dip is nearly northwest; that is to say, under the cañon. But as the cañon has been filled to a considerable depth with tailings, this observation may have been deceptive. The course of the sluice, up which bed-rock could be followed for a distance of four or five hundred feet, was S. 20° E. (magnetic). Following up the sluice for about 350 feet I found the slate continually and regularly rising. Farther up the mine the rock rose on the west side while it disappeared from view under the gravel on the east. The highest point of slate-rock seen in this mine was where it disappeared under the gravel of the west bank which has not yet been washed away. This was at least forty feet above the mouth of the sluice, or, in round numbers, at an altitude of 2,800 feet. Farther to the west there was a still higher slate hill, so that there can be no doubt of my having been on the west rim again, — as at the rock opposite Savage's. So far it seems that the observations cannot lead to any erroneous conclusions, for we have found two points on the west rim with plenty of room for the channel to the east and no chance for it at the west.

The Cozzens and Garber shaft, mentioned above, is on the north side of Missouri Cañon, and was sunk through gravel for about ninety feet before reaching bed-rock. The depth of the shaft I cannot give with precision, on account of the diversity of statements made by different persons. Ninety feet is not far from correct. The altitude of the top timbers we made to be 2,747 feet; which would give for the bed-rock at the bottom about 2,657 feet. The prevalent idea among those acquainted with the region was that the deepest part of the channel was found at or near the bottom of this shaft. If this is so there would seem to be nothing to prevent the original channel's having passed from the Cozzens and Garber shaft to the south and southwest by Mallory's and Waloupa. But by going to Red Dog, bed-rock is found exposed in abundance at an altitude of only 2,621 feet; only four feet higher than at Niece and West's (though the distance in a straight line is a mile and a half), and actually lower than at Mallory's or Cozzens and

Garber's, which is only a few hundred feet distant up the cañon. If there ever was a deep channel from Red Dog to Waloupa, there must have been subsequent important changes of level, unless some such explanation as the one which will be proposed farther on (page 172) be adopted ; for, as the rock now lies, the water must have run up hill at some part of its course. That the rock at Red Dog is not exceptionally low (as might possibly be supposed) is clear from the fact that the exposures at Independence and Bunker hills are only a few feet higher, — enough, barely, to give an easy grade from the north. The discussion of the question of probable change of level I will not enter upon ; for it will lead too far, and involve too many considerations with which I am only imperfectly acquainted. As far as our observations extended, however, the weight of evidence seems to be against any theory of great change of level ; and for the present it will be assumed that the position of the slate-rock is essentially what it was at the time of the deposition of the gravel.

On this supposition it seems more natural that the course of the — or a — deep channel was from Cozzens and Garber's towards Red Dog. At a later day, no doubt, the whole of the region between Red Dog and You Bet was covered with water, and channels may have been cut in a great many directions. Having met this difficulty in our attempt to trace the course of the old channel, two questions arise : first, if the stream which flowed to the southwest from You Bet is not a continuation of the Red Dog channel, where *did* it come from ? and, second, if a channel flowed from Cozzens and Garber's to Red Dog, where was its outlet, and what was its source ?

It may be thought possible that the lowest points at Cozzens and Garber's and at Mallory's have not yet been reached, and that there is still a chance for a nearly level bed-rock all the way from Red Dog to Niece and West's. Such a supposition is, to be sure, possible, but I can see no independent evidence in its favor.

In attempting to find the continuation of the You Bet channel at some other point than at Red Dog, attention is naturally turned first to the east and northeast, to see if there is any chance for the channel's having come down either under or to one side of the present Sugar Loaf and Chalk Bluffs. A careful exploration was made of the whole extent of gravel from Chicken Point to Boston Hill, which led to the conclusion that the main channel could not have come through this way. The observations on which this conclusion was based were substantially as follows.

At the extreme southern end of Chicken Point the gravel reaches out on a narrow spur for about a quarter of a mile beyond the main body. The banks are not very high, and there was never any great depth to bed-rock. I took no special observation at this part of the mines, but the altitude of the hill at the end of the gravel will not be far from 3,050 feet. A spot near the middle of the Chicken Point mines, where bed-rock was to be seen, was pointed out to us from the bank by Mr. Heydliff. When I went to the spot, I first saw the rock in a pit twenty-five or thirty feet across and twenty-five feet deep, through which a tunnel had been run as an outlet for a claim higher up. The bed-rock had been cut into about fifteen feet. Following up the line of this tunnel in a direction about N. 10° E. (magnetic), the bed-rock was seen to rise regularly for a distance of two or three hundred feet. Farther east I saw no bed-rock in the Chicken Point mines ; but the sandy strata dipped to the north and east, as if the rock might possibly slope off in that direction again, — though of this there can be no certainty. The altitude of the highest point of rock seen I determined to be 3,066 feet.

Previously to this I had taken an observation with the barometer on the rim-rock at the head of Sardine Ravine, at a point 350 feet south of the high flume. Its altitude I made 2,907 feet. It dipped to the north, — that is, under the gravel, — and shortly disappeared from sight ; but, notwithstanding, the dip of the strata of clay and sand near by was mainly to the south and southwest. The gravel in this neighborhood contained considerable sand and clay, — some black and indurated, and some more yellowish in color. The presence of large and rather angular boulders was also noticeable.

These were the only two places on Chicken Point south of the ridge of gravel along which the Chalk Bluffs road runs, where the altitude of bed-rock could be measured, though there was plenty of verbal evidence that it had been struck only a few feet below the present surface of the gravel at a number of other points. The character of the gravel — the clay, sand, and angular boulders — and the dip of the bed-rock where seen seem to point to the filling up of a large basin after the main stream had been choked by its accumulations. At any rate, high bed-rock has been found in so many places along Chicken Point that it is hardly possible for room to have remained for a deep channel from the east or northeast south of or under the Sugar Loaf. It is true, there are some few places where bed-rock has not been actually reached; but if a deep channel exists at any of these points, we should expect to see some signs of it in the character and stratification of the gravel as well as at points farther up the ridge beyond the Sugar Loaf. It will also be noticed that between the bed-rock on the Point and that in Wilcox Ravine there is a difference of altitude of over four hundred feet, within a distance of less than three quarters of a mile, which would be rather too much for the quiet deposition of gravel.

In the mines at the base of the Sugar Loaf — those north of the Chalk Bluffs road and having an outlet into the most southern fork of Missouri Cañon — I saw no bed-rock at all, but determined the altitude of the top of the present gravel, near the corner of the tool-house, to be 3,068 feet. There was a depth of at least twenty feet of gravel here, as could be seen in the cuts and gullies worn by the hydraulic streams, but how much more it was impossible to tell.

In Missouri Cañon, which is now filled with tailings to a considerable depth, a spot was pointed out to us, a few hundred feet below where the last observation was taken, at or near which bed-rock used to be seen in the original bed of the cañon. We, of course, had no means of determining its altitude with precision, but may adopt, as a rough estimate based upon the observations at Cozzens and Garber's shaft and in the Sugar Loaf mines, an altitude of 2,825 feet.

In the Sugar Loaf mines it was noticeable that there was relatively much more sand and clay in the gravel than at points nearer You Bet, reminding one rather of the Chicken Point gravel. There was also a considerable number of rounded lava boulders, increasing rapidly in size and frequency as the base of the bluff was approached. These had not been met with, or at least not noticed, on the south side of the road. From such indications it seems fair to conclude that the present high gravel at the head of Missouri Cañon is not part and parcel of the main deep You Bet channel, but belongs rather (the top, at any rate) to the later period when the whole region was overflowed.

Crossing from the Sugar Loaf mines in a northwesterly direction to the next outlet (where there were a few square rods of standing water with rushes growing, on which account we designated the spot, for transient purposes, as the "Rush Swamp"), I determined the altitude of a point, where the bed-rock, we were told, was covered by only ten or twelve feet of tailings, to be 2,989 feet. This would assure us of the existence of bed-rock at an altitude of about 2,975 feet. There was no rock in sight, however, and no evidence as to the direction of its dip. At a point three or four hundred feet north there were indications in the stratification of the sand, clay, and gravel, as if the underlying bed-rock dipped both in an easterly and westerly direction, — but nothing conclusive.

An eighth of a mile northwesterly from the rushes bed-rock was plainly seen at the outlet of Timmens's and Brockmann's mines. As exposed, it slopes rapidly down the cañon in a southwesterly direction. But in the mine it was seen in patches here and there within a distance of three hundred feet of the outlet, — apparently nearly level or with a possible pitch to the northeast under the bank. Some of the strata of sand near the bed-rock had a dip of as much as ten or twenty degrees to the northeast; though, as I have said before, I do not think too much stress ought to be laid on this appearance. The altitude of the highest point of bed-rock seen in this outlet was determined to be 3,051 feet.

The next point where the altitude of bed-rock was measured was at the outlet of Hussey's mine (a quarter of a mile, more or less, to the northwest of the last point of observation), where the rock exposed at the upper end of the sluice, near Mr. Hussey's cabin, was found to have an altitude of only 2,913 feet. Following up the line of the sluice, — the direction of which is N. 10° E. (magnetic), — the rock is hidden by the wash of the gravel, but was said by Mr. Hussey to rise six or eight feet in the course of 150 feet and then to pitch off suddenly to the northeast. And for the next three or four hundred feet no bed-rock has been found, because, there being no chance for a cheap drainage into Missouri Cañon, there has been no attempt made to reach bottom by shaft sinking. On the farther side of the mine the bed rock rises almost precipitously, so that its highest point (near the northeastern bank of the mine) has an altitude, as estimated by the hand-level with reference to known objects, of fully 3,040 feet. In the central part of the mine, according to Mr. Hussey, "blue gravel" was uncovered in digging to connect with the sluices and boxes at the outlet. A section across this part of the gravel is given on Plate F (Fig. 5), — the vertical as well as the horizontal scale being two hundred feet to the inch. Beyond this high rock Mr. Hussey thinks there is still another channel, with a southerly or southeasterly course, but of this I was not fully convinced. There is, however, in the next small excavation to the north, a comb of bed-rock, dipping both to the east and west and apparently a continuation of the high rock in Hussey's Mine. To the east of this comb the bed-rock has not yet been struck, which circumstance adds, it is true, some weight to the theory of the additional channel. Still, there may be no real connection between the two exposed masses of rock, or the latter may be nothing more than a high place, like an island in the stream.

Keeping along in nearly the same direction, the next point of bed-rock measured, beyond Hussey's, was in the bottom of the mines on Darling's Hill, near the western bank. Its altitude we determined to be 3,064 feet. Other rock was seen at about the same elevation at a number of points on Darling's and Boston hills, but it was by no means clear that there was any dip to the east. The general slope of the country from Darling's Hill is to the west, and considerable slate is seen between the mines on the hill and those at Red Dog, as is seen by following the line of the Bunker Hill ditch from near the outlet of Hussey's Mine. For one or two hundred feet from Hussey's slate-rock is seen; but it soon disappears, and the ditch is dug in gravel for nearly a quarter of a mile. At the westerly end of the hill, about a quarter of a mile east of the town of Red Dog, slate-rock was met again in the bottom of the ditch, and its altitude determined to be 2,908 feet, — almost identical with that at the head of Hussey's sluices. From this point on (with the exception of perhaps a hundred feet of gravel near by) the ditch runs entirely in slate around the north side of Darling's Hill and the head of Arkansas Cañon, to where the water is taken down to the Bunker Hill Mine.

The altitude of the bed-rock at Red Dog has been already given as 2,621 feet. The same result was also obtained for the bed-rock at Independence Hill on the opposite side of Arkansas Cañon from Red Dog. The rock not being very even at either place, an error of two or three feet could easily be made in fixing upon the point to be measured, so that these two results are to be regarded not so much as deciding anything about the direction of the channel between the two places, as corroborating the general statement that there is a fall of about thirty feet from the bottom of the Cozzens and Garber shaft to the bed-rock at Red Dog.

The highest point of the Independence Hill gravel is nearly 200 feet above the bed-rock, being about on a level with a small gravelly knoll, at the east end of the high flume, the altitude of which was determined to be 2,821 feet. On the northern side of Independence Hill I also determined the altitude of a point on the rim — at the outlet of a small claim discharging into Greenhorn — to be 2,723 feet.

The observation at the bed-rock of Williams' mine on Bunker Hill — by which the altitude was made to be 2,632 feet — was taken at a later date than the others, when the nearest station barometer was at Colfax, and is consequently not to be trusted quite so implicitly as the others in the series, though there is no probability of any great errors. Both at Independence and Bunker hills the gravel shows the same succession of blue and red as is seen at Red Dog and below You Bet.

These are the principal points where observations for altitude of bed-rock were taken within the You Bet and Red Dog districts. For convenience of comparison, the results, so far as they shed light upon the question of direction of channel, may be brought together as follows : —

[I.] *Points supposed to be on or near the main central line of the deep channel.*

Waloupa. Bed-rock	2,590 feet.
Niece and West's Mine. Bed-rock	2,617 "
Heydliff's Claim. Bed-rock	2,640 "
Brown's Claim. Bed-rock	[2,640½] "
Mallory's Claim. Bottom of Incline	2,641 "
Cozzens and Garber's shaft. Bottom	2,657 "
Red Dog. Bed-rock	2,621 "
Independence Hill. Bed-rock	2,621 "
Bunker Hill. Bed-rock	2,632 "

[II.] *Points on the westerly rim.*

Pine Hill. Hubbard's Tunnel	2,655 "
Opposite Savage's house. You Bet	2,828 "
Opposite Garber's shaft. Missouri Cañon	2,800 "

[III.] *High bed-rock to the east of You Bet.*

On Chicken Point	3,066 "
Outlet of Timmens's Mine	3,051 "
High Rock at Hussey's Mine	3,040 "
Bed-rock on Darling's Hill (in the mine)	3,064 "

[IV.] *The lower line of high bed-rock to the east of You Bet.*

Sardine Ravine	2,907 "
Missouri Cañon. At outlet of Sugar Loaf mines (estimated)	2,825 "
Rush Swamp. Partly estimated	2,975 "
Outlet of Hussey's Mine	2,921 "
On Darling's Hill. In the ditch	2,908 "

Before attempting to draw any conclusion from this in some respects inconsistent testimony, there may be a few words added in regard to the other kinds of evidence which we tried to get. The form and position of the pot-holes or excavations worn in the bed-rock attracted our particular attention, though not until a late day. The opportunities for studying them were, unfortunately, few in number. At Niece and West's Mine we examined the bed-rock carefully in company, and found what seemed to us incontrovertible proof that the rock had been worn by a stream running in a direction from Niece and West's to Waloupa. At Dutch Flat we met persons who were strong in the opinion that the original course of the stream at this point was from southwest to northeast, in spite of the grade of the bed-rock ; but we found in the pot-holes absolutely nothing which even suggested such a direction. The argument for the up-hill direction of the channel is substantially as follows : " Whenever, in the present river beds, there is an unusually high place in the rock, there will be no gold found on the slope towards the head of the stream, but it will all be found on the lower side. In the old channels the reverse is the case, — if the channel flowed to the south. Ergo, the old channel must have flowed to the north, and owes its present position to a slow upheaval or relative change." Inquiries among the miners in the gravel claims failed, however, to establish as a rule that the gold was in reality more likely to be found on the northern than on the southern slopes of inequalities in the bed-rock. In isolated cases it may have been observed, but the rule is by no means universally recognized. It is only just to add that those persons who defended the theory of the northerly course of the channel, also claimed to have seen in the position of the pebbles with respect to each other, and in the accumulations of sand around the larger boulders, evidence in corroboration of their theory. But evidence of this nature it seems to me very unsafe to trust to any great extent.

At Red Dog I failed to find the conclusive proof I had hoped for as to the direction of the current. The rock was so much weathered, that I could find no hole sufficiently well preserved to convince me beyond a doubt. On a different occasion, however, Mr. Bowman assured me that he found positive evidence that the channel once came across from Independence Hill, and made a bend to the east where the Red Dog mines now are. *A priori*, such a bend in the stream is not to be expected, for there is almost no instance in the present streams or ravines in the neighborhood where there is an east course for even a short distance. The only case of the kind which came to our notice was where the present Greenhorn, a little above Quaker Hill, runs for a couple of hundred feet in an east and northeast direction in order to get around a high spur of rock.

I have thus completed the arrangement of my notes on the You Bet and Red Dog gravel deposit, and shown where the main difficulties of the problem lie. There seems to be no one answer to the questions proposed which will satisfy all the conditions. If we suppose the deep channel below You Bet to be the continuation of the stream which flowed by Red Dog, we are confronted by a rise in the bed-rock, which obliges the water to run up hill, or, at the best, with an almost absolutely level river bed for a distance of more than a mile and a half. Low places and holes may occur, it is true, in the beds of running streams, and there may be long stretches where the fall is only slight; but if that had been the case between Red Dog and the site of Niece and West's claim, it seems to me that we should find a finer gravel than we do find, or perhaps only a sandy mud. And, furthermore, wherever the deep bed-rock is exposed, the indications are strong that there was once a current of considerable rapidity. It is indeed possible, in spite of these objections, that the connection of the channels was as has been supposed, but the most that can be claimed for the theory is that there has been as yet no insuperable obstacle found in the way.

If, on the other hand, we attempt to find any different source for the You Bet channel we are soon driven to the wall by the high slate-rock whose existence is proven at so many places between Chicken Point and Boston Hill. The nearly uniform elevation of the highest bed-rock, at the four points where measurements were taken, is rather remarkable, though the coincidence may have been only accidental. The general slope from the highest bed-rock towards the west is in all cases very rapid, as is seen by reference to the observations at different points lower down. These circumstances point rather to a long easterly "rim," than to a district through which a channel from the east is to be looked for. To be sure, there is a possibility that the upper part of the You Bet channel may have skirted along the base of the bluff; and to such a theory a degree of probability is given by the shape of the bed-rock at Hussey's Mine, of which an east and west section has been already given (see Plate F, Fig. 5), though there is no other positive evidence to adduce, and there are some reasons which may be urged in opposition.

And, again, it is possible that the key to the solution of the problem still lies hidden underneath the lava flow which terminates so abruptly at Chalk Bluffs. There may have been a small watercourse — or even more than one — which had its origin at some point considerably higher up the mountains, and down the steep bed of which the lava found an easy path. Indeed, the presence of the lava itself indicates that there was some shallow trough in which it could flow. On this supposition we should not look upon the You Bet channel (in its deepest portion) as a part of a long and deep stream, but rather as the lower end of a large ravine or small creek, originally distinct from the Red Dog River, but connected with it after the partial filling up of the channels with gravel. As a part and parcel of this same theory we can suppose that the original Missouri Cañon is represented in part by the gravel which now extends from the Cozzens and Garber shaft to Red Dog. In these suppositions there is nothing incompatible with the final covering over of a large extent of country, including not only the ravines but the intervening ridges. In opposition, chiefly to the latter part of the theory, we have Mr. Bowman's observations of the Red Dog pot-holes, and his conviction that the stream made a bend to the east at that point. And, furthermore, in supposing the original drainage to have been almost identical with that of the present day and that there was a cañon or ravine from the east, corresponding to the present Missouri Cañon and emptying into the main channel at Red Dog, we meet with difficulty in fixing both the outlet below Red Dog and the source of the cañon to the east. If we could have seen the ground before the accumulation of tailings in Missouri Cañon it is barely possible that a more satisfactory conclusion might have been reached. The fact that there was slate-rock to be seen in the bed of the cañon, between the high gravel at the base of the bluffs and the deep gravel at Red Dog, taken in connection with the depth of the gravel at the Cozzens and Garber shaft, and the general appearance of the country, may have been an important element in the settling of our doubts.

In looking for an outlet to the west and southwest of Red Dog, we find ourselves restricted to the lines of Arkansas Cañon and Missouri Cañon. To the south of the latter rises the high ground of Pine Hill; and between the two there is an elevated spur of slate which bars all progress. That the channel cannot have found an outlet by way of Missouri Cañon itself is clear from the rim of rock which is still to be seen between the cañon and the Red Dog gravel. There is no room to change the direction of the channel from east to west, without requiring an almost impossible bend in the stream; and, further, the narrowness of the cañon adds to the improbability of the idea. The notion, however, of an outlet by way of Arkansas Cañon into the present Greenhorn is favored by the slope of the bed-rock at Red Dog, so far as it can be estimated, and would be the most probable one of all were it not for the evidence of the pot-holes that the stream made a curve to the east. It is true that I know of no additional evidence lower down the stream that the valley or cañon of Greenhorn was the natural outlet from Red Dog (unless possibly something may be found at the "Buena Vista Slide," * of which I heard frequent mention, but which I never had an opportunity to examine), and, according to all reports, this particular stream was never remarkably rich in gold. This is, to be sure, only a very feeble piece of negative evidence, and is not to be allowed much weight.

Supposing the outlet to have been by Arkansas Cañon amounts to about the same as saying that the gravel at Red Dog and vicinity was deposited by essentially the same stream as the present Greenhorn, though following a slightly different course. And, indeed, when we trace the gravel mines from above Quaker Hill along the banks of the present Greenhorn by Hunt's Hill to Red Dog, it seems almost impossible to resist the conclusion — similar to the one reached at Dutch Flat and Little York with reference to Bear River — that there has been no essential change in the directions of the water-courses since the gravel era, and that the confusion and difficulty of tracing channels arise from the chokings up of the old beds and the overflows upon the surrounding country.

There is a theory held by some of the miners in this region, and of which mention will be made farther on, that there were formerly two large rivers analogous to the Sacramento and San Joaquin, which united near Red Dog and found an outlet towards the sea somewhere in the neighborhood of Grass Valley or Nevada City. In accordance with the theory, while we were at You Bet, work was begun by some men from Dutch Flat on a tunnel in Rocky Ravine, — the first large ravine emptying into Greenhorn from the north below the Nevada City road crossing. The tunnel as commenced lies about in the middle of the northwest quarter of Section 25. The projectors were so full of confidence that they had found the hidden outlet of the main stream formed by the junction of the other two that they called the location "Eureka Gate." The altitude of the point selected for the commencement of operations proves, however, to be 2,692 feet, — or seventy-one feet above the bed-rock at Red Dog. The rock at the head of the ravine — or of one of its branches — near McLeod's house is also slate, in which I could find no place low enough for any channel to have come through. To the north and northwest of "Eureka Gate" there has been considerable exploring and sinking of shafts in past years in the hope of striking some hidden channel, but, so far as I could learn, with no favorable results. The crest of the main ridge below Quaker Hill is made up, for a considerable depth, of volcanic material, but the spurs which make down from the top of the ridge between the ravines to the Greenhorn Creek are mainly slate, though there have been slides from time to time of volcanic material which hide in some places the original bed-rock. And it is underneath some of this lava covering that the hoped-for channel is expected to be found, or the low place in the slate through which the channel might have come. Since leaving You Bet I have learned that the prosecution of the Eureka Gate project has been stopped for the present, — and from all that I could see, in my short forenoon's explorations, I am well satisfied that the sooner the work was abandoned the better for the prospectors. From the house of McLeod, Senior, I climbed up the ridge to a point a half or three quarters of a mile dis-

* The "Buena Vista Slide" was said to be in the N. E. ¼ of Sec. 31, T. 16 N.; R. 9 E., which would bring it within two or three miles of Grass Valley, and as much as that from the line of Greenhorn.

tant to examine a white deposit which looked like that which is so prominent a feature on the Sugar Loaf and Chalk Bluffs. The two deposits were so nearly on the same level that I at first supposed they marked the water-line of an ancient lake. The one above McLeod's, however, appeared to be made up of volcanic material, which had not been deposited by water. While near the top of the ridge I saw small patches of washed pebbles, which were clearly too high to have belonged to the main deposits by Red Dog, and may have been the remnants of some smaller stream down the bed of which the lava flowed.

To the southeast of McLeod's house there is a ravine emptying into Greenhorn about three eighths of a mile above the Nevada City road crossing. On the right bank of the ravine there is nothing but slate, while on the left there are two or three gravel claims. Strangely enough, as I was assured by Mr. McLeod, that ravine has never been found rich in gold. It would seem as if the rim within which the channel or the gravel was confined lay just to the northeast of the present line of the ravine.

The only other point in this neighborhood where I saw gravel was half a mile south of McLeod's, at the divide between Rocky Ravine and the small ravine which empties into Greenhorn quite near the Nevada road crossing. The amount was not large, and the elevation was not so great as to shut out the probability of these few pebbles having once belonged to the main Red Dog gravel, at the time when the greatest extent of country was covered. At least, they can hardly be taken as evidence that the grand outlet was at or near that place.

If time had allowed I should have made a more extended examination of the Greenhorn slope of the ridge between that creek and Deer Creek, but, as it was, shall be obliged to stop with this imperfect sketch of only a small portion of the ground.

A very few inquiries showed that it would be impracticable for us to ascertain with even a decent degree of approximation the total gold production of the district, and no steps were taken to estimate with any degree of precision the number of cubic yards of gravel which have been removed, or the amount left. On those points I have no estimates to offer. Neither will it be worth while now to attempt an explanation of the reason of so many claims lying idle, or to discuss the general economic relations of the gravel industry at You Bet. Such a discussion would require too much time and space, and the data for making it really valuable are not at hand in proper amount and shape.

§ 3. *Lowell Hill and Remington Hill.*

The most direct road from You Bet to Liberty Hill is by way of Little York, and thence up the ridge between Bear River and Steep Hollow. The road is good, and the ascent from Little York gradual, excepting for a short distance at the Camel's Hump. The ordinary slate bedrock is visible nearly all the way up. The small patch of gravel on the upper side of the Camel's Hump has already been referred to (*ante*, page 157). Near Liberty Hill the character of the bedrock changes somewhat ; all that is necessary to be said in reference to the gravel at that locality has been given in connection with the description of the adjacent Elmore Hill, on the opposite of Bear River.

There is a prominent elevation about a mile and a half northeast of Liberty Hill, called on the map Maguire's Mountain. Its summit is capped with lava, which has a thickness of several hundred feet. The altitude of this mountain was found to be 4,460 feet, which is 1,110 feet higher than the bed rock at Liberty Hill. The descent on its eastern side into the cañon of Bear River is almost precipitous. From its summit the ridge above Dutch Flat and Alta, on the opposite side of the cañon, was in full view for a number of miles, and its smoothness and regular grade pointed unmistakably to a lava flow. Directly opposite our point of view the Dutch Flat ridge appeared decidedly higher than the one on which we were standing, and there seemed to be a considerable extent of table land on its summit. The point of the ridge on the same level with us was three or four miles distant, and bearing about south-southeast.

The connections of the gravel above Liberty Hill were not made out clearly; but from Mr. Bowman's investigations it appears that it forms a continuous deposit, as far as Lowell Hill, and it is so represented on the map. The barometrical observations at the last-named locality were made under rather unfavorable conditions; but it is clear that the bed-rock at that point is from five to six hundred feet, at least, higher than at Liberty Hill. This gives ample fall between the two places, which are a little less than three miles apart.

Higher up Bear River, a mile and a half east of Lowell Hill, is a small gravel area, known as Kinder's Diggings. It was stated, by those acquainted with this region, that this gravel is on the Bear River side of the divide, across which the old channel to the South Fork of Steep Hollow went. It was also said that the character and mode of occurrence of the gravel and gold at Lowell Hill indicated a connection with Kinder's rather than with Liberty Hill. Such evidence may perhaps not be considered sufficient for basing a decided opinion as to the original course of the streams; but, as the heads of the present rivers are so near together at Bear Valley, it would not be surprising to find the upper portions of the ancient representatives of the same streams in such close proximity to each other, that they might be with difficulty distinguished, and connections supposed to exist where there really were none.

The town of Lowell Hill, consisting of perhaps half a dozen houses, stands, as seen from the opposite side of Steep Hollow, at the end of a rather flat ridge, which runs parallel with the general course of the Creek for about half a mile. From Lowell the descent into the cañon is very steep. South of this small ridge is a ravine nearly a mile long, emptying into Steep Hollow, just below Lowell Hill. The gravel is at the westerly end of the ridge, just were it begins to pitch into the cañon. The extent of the diggings is not very great. Their total length in a north and south direction would be, perhaps, a little less than 800 feet, and the average width may be taken roughly at 300. This gives an area of a little over five acres. The only unmistakable bed-rock seen in the diggings was near the eastern bank, and was pitching rapidly in a direction N. 80° W. (magnetic). Whatever bed-rock has been laid bare in former years has since been covered again by the large quantities of clay which have slidden down from the upper banks. The rock seen was in the line of a sluice through which there had been an outlet from the diggings into Steep Hollow. Everywhere along the eastern bank of the mine the indications pointed to a bed-rock sloping to the west; and nowhere were any signs of a rim with an eastern inclination. This may be explained on the supposition that the western rim has been worn away during the formation of Steep Hollow Cañon. The height of the eastern bank of gravel will average about thirty or forty feet. At the point where the altitude of the bed-rock was measured there was a covering of eight or ten feet of red dirt over the gravel proper. Then came seven feet of fine gravel and sand, followed by six feet of coarser gravel and ten feet of clay. At the bottom there was a coarse gravel, with rather large boulders. The presence of so much clay has made the working of the Lowell Hill bank unprofitable, and it is not probable that there will ever be much more money taken out there than will be needed to pay expenses.

On the opposite side of Steep Hollow is the Remington Hill gravel, bearing from the Lowell Hill Mine, N. 22° W. (magnetic). A little lower down the creek, and on the same side with Remington, is a small gravel deposit, known as Melbourne Hill. The observations with the hand-level showed Remington Hill to be a little the higher of the two, while the altitudes of Melbourne and Lowell were nearly equal. To get from Lowell to Remington Hill it is necessary to cross the cañon of Steep Hollow, the depth of which, as determined barometrically, is about 500 feet. The trail from Lowell follows a narrow spur between two ravines, in a direction N. 40° W. (magnetic), to the bottom of the cañon, and then up on the other side, on a spur between Snake Creek on the east and a ravine which heads near the Remington gravel on the west, ascending with a general course of N. 22° W. (magnetic). The Remington Hill gravel is hardly half way up the side of the main ridge to the northwest of Steep Hollow; it is on the spur which lies between Snake and Dry Creeks, the general course of which is S. 10° E. (magnetic). Snake Creek is hardly anything more than a large ravine, heading just above the gravel deposit at Remington Hill, and having a length of about a mile before entering Steep Hollow. The extent of the gravel opened on the end of the spur at Remington is about five or six acres, over a large portion of which the bed-rock has been exposed. The character of the gravel is much the same as of that at Lowell Hill, although the lower stratum has fewer large-sized boulders. As at the last-named place, there are heavy masses of clay interstratified with the deposit. The bed-rock is slate, which in many places is jet-black in color. This peculiar discoloration seems to be due to some local causes, as the patches of black are irregularly scattered all over the exposed surface. The bed-rock is very flat, there being no rim discovered at all, or indications of a channel such as are expected to be ordinarily found under the gravel deposits. Besides the open gravel diggings, there has been considerable drifting under the lava with which the main ridge is capped. The two principal tunnels are known as Frank's and Joyce's. The mouth of the former is a little more than an eighth of a mile east of the principal house at Remington, or about a quarter of a mile east of the main open gravel, and it discharges into Snake Creek. This tunnel runs nearly due north for a couple of hundred feet, then north-northwest for four hundred farther, to the gravel, on which it has been run for a distance of about two hundred feet. Joyce's Tunnel is a short distance farther north. In neither of these excavations have any signs of a rim been discovered.

The direction of the channel at Remington seems to be a matter in regard to which the miners themselves have no theory. The finding of gravel under the lava in the tunnels has suggested to some the idea that the main channel passed directly across the ridge. But this seems hardly probable; and it is more likely that this deposit will be found to have been connected with that at Klipstein's, striking along on the southeastern side of the present ridge, as will be noticed when describing that locality. The altitude of the Remington Hill bed-rock was made, by barometrical observation, 3,870 feet; this is higher than the eastern rim at Lowell, and the bed-rock at Melbourne Hill. It is possible that the channel crossed to Lowell, and went thence to Liberty Hill; or it may be that Lowell and Melbourne are only relics of the time when Steep Hollow spread over that whole width of country; with our present knowledge it is impossible to come to a positive decision in regard to this point.

Klipstein's Claim is about three quarters of a mile, in an easterly direction, from Remington Hill. This gravel deposit covers several acres, and lies on a spur between Democrat and Lucky John's ravines, both of which empty into the North Fork of Steep Hollow, half or three quarters of a mile above its junction with the main creek. The general course of Democrat Ravine at Klipstein's is S. 25° E. (magnetic). On the west side of the ravine there is also a small patch of gravel. These deposits have been well prospected by means of tunnels. One of them, run in on the bed-rock on the Democrat Ravine side, was found to have an altitude of 4,020 feet, which is 150 feet above the bed-rock at Remington Hill. This tunnel runs in an easterly direction, is 180 feet long, and is nearly level: at its end the bottom is fully ten feet below the gravel. At a distance of 140 feet from the mouth of this tunnel a branch was started in a north-northeasterly direction, soon curving round, however, so as to run parallel with the main tunnel and about fifty feet distant from it. From the form of the surface of the bed-rock, as shown in these excavations, it would appear that there is here a basin-shaped depression, rather than a regular channel. The gravel is, however, well rounded, and some portions are well cemented together, and the larger boulders, which are quite numerous, are smoothed and polished; thus presenting all the peculiarities of the ordinary channel deposits.

There are other reasons for supposing that Klipstein's gravel forms a portion of an old channel. On the opposite bank of the North Fork of Steep Hollow, and bearing from about the middle of the gravel at Klipstein's in a direction N. 38° E. (magnetic), is the deposit known as Excelsior, or Secret Hill. This is on the spur which starts from the main lava ridge and descends between Steep Hollow and its northern fork. At a point as nearly as could be estimated on the same level with the Excelsior bed-rock, the height of the latter was found to be 4,090 feet, or seventy feet higher than Klipstein's Tunnel. Between the two places, it is true, there could not have been a perfectly straight channel, on account of a little spur of high bed-rock between Lucky John's Ravine and the North Fork ; but there was plenty of room for a channel, if we suppose it to have made a slight curve to the south. And that such a connection did once really exist seems established by the following facts, for which Mr. Klipstein is the authority. Directly through the bed-rock of the Excelsior Mine there is to be seen, striking north and south, a quartz lode with peculiar greenish streaks, and specimens of this easily recognized veinstone are met with both at Klipstein's and at Remington Hill. It was also stated that pebbles of distinctly gold-bearing quartz had been frequently found at the last-named place; although it is not certain that these were from this particular lode at the Excelsior Claim. There is also said to be a point of gravel on a spur between the Middle and South Forks of Steep Hollow, about a mile and a half above Excelsior, and similar in character to that at this locality. The level of this gravel was said to be about 100 feet higher than that at Excelsior. No work had been done there, and the quantity of the deposit seemed small. It was thought by those acquainted with the region that there was no high bed-rock offering any obstacle to a connection between Excelsior and the point mentioned as being higher up.

Between Klipstein's and Remington Hill there does not appear to have been any connection of channel actually proved. On the opposite side of Democrat Ravine — as has already been mentioned — there is a small quantity of gravel ; but, beyond that, the trace is lost. Three eighths of a mile westerly from Klipstein's cabin is a long spur, up which the trail leads to the summit of the ridge. Between this spur and Klipstein's is a pleasant open flat, a quarter of a mile wide, and sloping gradually for about the same distance toward Steep Hollow and the North Fork. The spur itself, from where the hill commences, is lava, and has a general course of N. 40° W. (magnetic) along the ascent of the trail. Lower down, it bends more to the south, so as to form one of the walls of Snake Creek, and is composed entirely of slate. The continuation of the channel between Klipstein's and Remington is probably to be sought for under this spur of lava. There seems to be no obstacle in the way, in the form of high bed-rock ; and if we were to attempt to seek a channel lower down, it would probably be too low to reach Remington Hill. On the whole, the evidence seems strong that the old channel is to be found hugging the base of the present lava flow, and following essentially the course of the existing cañon.

The so-called " Bald Eagle Diggings" are located at a point where the North Fork of Steep Hollow has cut through the lava capping of the ridge, and thus exposed gravel under the volcanic. The elevation was said to be considerably lower than that of Excelsior. The method of working was by drifting under the lava ; but whether gold has been found in paying quantities was not ascertained. It is possible that the Remington Hill gravel may be connected with that under this ridge ; but the opportunities for observation in this vicinity were not sufficiently extensive to furnish the data required for a settlement of the question.

§ 4. *Quaker Hill and Vicinity.*

On examining the map it will be seen that north of the You Bet and Red Dog district, which lies between Steep Hollow and Greenhorn creeks, and which has been already described, there is an extensive gravel area, occupying a belt of country north of the last-mentioned creek. This deposit is somewhat over three miles in length, and from half a mile to a mile in breadth, its longer axis lying nearly north and south and extending from the Greenhorn as far as Deer Creek. Into the last-named stream the northern end of the gravel area in question is drained ; but much the larger portion of it is tributary to the Greenhorn and its numerous small branches, the main stream having a nearly north and south course, from its junction with Steep Hollow nearly to its source. This stream heads in the lava ridge which lies to the west and northwest of Remington Hill, and the other adjacent mining claims described in the preceding sub-section. From these diggings Professor Pettee descended on the divide between Greenhorn and Deer creeks to Quaker Hill, the principal mining camp of the district, now to be described from his notes. This description may be prefaced by some remarks as to the character of the Mount Oro Ridge and the country traversed in descending to Quaker Hill: —

The main features of the head waters of Greenhorn, as observed on this hasty ride down the ridge and subsequently from the top of Quaker Hill are, then, summarily as follows: After descending the steep hill to the north of Mount Oro, my road followed what is known as the Quaker Hill ridge between Deer Creek and the North Fork of Greenhorn nearly to Osborn Ravine; the junction of the North Fork with the main creek being only a quarter or a half a mile above the outlet of the ravine. A half or three quarters of a mile above the junction of the North Fork, Greenhorn forks again into a Middle and South Fork, of which the Middle is the longer. The whole of the high bluff between the North and Middle forks is known as Mount Oro. From this elevation two prominent spurs appear to diverge; the more northerly running to the North Fork and the more southerly — called Jossy Point (or Yossi Point, the correct orthography I cannot be sure of) — extending to the Middle Fork above its junction with the South Fork. Between these two spurs the country is rather flat and cut up irregularly with ravines. Above Jossy Point the Middle Fork appears to wind around Mount Oro and to have near its head two main branches, the hill between them (which is plainly visible from the top of Quaker Hill up the cañon to the south of Mount Oro) being Kilbury Point. The Buckeye Hill Ridge, below the South Fork, shows a couple of prominent ravines, to which no names have been given and in which no mining has been done to amount to anything.

The flat top of the ridge is lava throughout. Upon Mount Oro I was told that shafts had been sunk to a depth of 200 feet and that gravel had been found under that depth of lava, but there was no one at work at the time of my visit, and I could get no details of value. The first slate on the road, that I have noted in my book, was at the beginning of the long curve which swings around the head of Osborn Ravine. From that point down to Quaker Hill the road follows not far from the line between lava and slate. As well as I could judge from the top of Quaker Hill, the top of the Buckeye Hill Ridge is also capped with lava. I have heard of the existence of slate-rock ridges and slate-rock country of some considerable extent above Buckeye Hill and Chalk Bluffs, but my track did not lead me positively in sight of any. Whether all the country at the base of Mount Oro and around the heads of Greenhorn is covered with lava or not I cannot tell, but think it probable that it is.

The hasty sketches that I made of the mines at and near Quaker Hill in the intervals between the rains have been transferred to the map. As I only had the aid of a pocket compass and was obliged to estimate a great many distances by the eye, it is very likely that more or less errors of detail will be found in them, if there is ever an accurate survey made; though, judging from the close agreements of the result of my meanderings of Greenhorn with the sketchings from the banks of the mines, I feel confident that the main features will be found essentially correct. I was not able, either, to make a thorough examination of all the important points near Quaker Hill as regards course of channel, dip of bed-rock, thickness of gravel, and possibility of outlet in different directions; the notes which I took are necessarily incomplete and imperfect, but I will arrange them in as clear a form as I can.

In my description I will begin at the northeasterly end of the main series of mines, near Osborn Ravine. I have already mentioned meeting with slate rock on the road at the point where the curve is made to pass around the head of this ravine. The spur between the ravine and the next one east, or the North Fork of Greenhorn, has a course of S. 7° W. (magnetic), and is composed of slate, which — in the prolongation of the line of the gravel mines — is fully 300 or 400 feet higher than the bed-rock in the first mine to the west. And it is by no means clear on which side of that spur the original stream flowed. It may have made a sharp bend around the southerly end of that spur, or there may prove still to be a low channel to the north now covered with lava. My information is too limited to hazard an opinion on this point. Both upon the high spur of slate rock, and high up in the head branches of Osborn Ravine, I found a little quartz gravel, which may have come down from under the lava. And as far as this goes, it is an indication that that particular flow of lava followed an old creek channel or ravine. But if so, it would seem as if it must have been anterior to the wearing away of the channel in which the present gravel has been deposited. The question is still involved in considerable doubt.

The "Dutch Diggings" lie between Osborn and Sapsucker ravines, and extend over an area of ten or twelve acres. In no place did there appear to be more than fifty feet in depth of gravel. There was an extensive amount of bed-rock uncovered, and a decided rim to the channel, both on the north and south. In the mines lower down the northern rim has nowhere been uncovered, I believe.

Around the head of Sapsucker Ravine is a small claim, which has been worked mostly by drifting. Between the Sapsucker and Knickerbocker ravines is an open gravel mine, comprising eight or ten acres, known as the Hazelgreen Mine. The Knickerbocker Ravine heads high up in the east gap of Quaker Hill.

Still farther to the southwest, which is the general course of the mines from the Dutch Diggings to the Railroad Mine on Prior Ravine, comes a large opening which includes the Knickerbocker, Aurora, and Newton claims. At the northern end of this opening there are large masses of clay, which have stood in the way of any search for the upper rim. On the southeastern side, in the outlets of the mine down the steep ravines into Greenhorn, there is a plenty of slate rock to be seen. And, indeed, slate rock is visible on the southern and southeastern sides of all the claims as far down as Prior and Green Mountain ravines. Next westerly from this large opening is the smaller Chollar (or Aurora) Claim, beyond which comes the main and largest claim of all, which lies just to the south of the town of Quaker Hill, and is owned by Messrs. Jacobs and Sargent. Including the Railroad Mine on the opposite side of Prior Ravine there must be as much as forty acres of gravel which have been worked over. And it is just at this point that the main interest of the problem centres. The grade, the succession of gravel almost unbroken from Osborn to Prior Ravine, the rim rock on the southern side, and all the conditions unite to establish the fact that there was once a channel following essentially the same course as the present Greenhorn for certainly more than a mile and a half. One evening, indeed, when the moon was shining brightly, the occasional glimpses which I got (from a high point of the road) of the brilliant white gravel contrasted with the deep green or black of the trees and bushes, seemed like the flashes of reflected light from the living waters of a winding stream. The deception was complete in all its details, — excepting a certain ghastliness and pallor which showed me at once that it was, as it were, the ghost of a stream long since departed.

But besides this stream from the northeast, there is also supposed to be a part of the famous "blue lead" making its way under the lava of the West Gap of Quaker Hill and connected directly with the gravel at Scott's Flat, on the Deer Creek side of the ridge. Whether this is so or not I am not fully prepared to say, not having had the time to pay any attention to the northern side of the ridge, though my present opinion is adverse to any such connection.

In the Quaker Hill mines proper (those directly south of the town and including the Railroad Claim) there has been hardly any deep bed-rock ever struck. Toward the western end of the main mine east of Prior Ravine I was shown the spot where a shaft had been sunk seventy-one feet to bed-rock below the present level of the gravel, — the last twenty feet of the sinking having been in " blue cement, good and solid." At a point about four hundred feet to the east of this shaft there was bed-rock ten feet higher than the top of the shaft and pitching rapidly to the west. There was no drifting at the bottom of that shaft, but the bed-rock was still pitching rapidly to the west. This circumstance in itself, however, has but little weight, for the shaft might have struck upon the edge of some hole in the bed of the stream.

Prior Ravine has its head in the volcanic rock of the west gap of Quaker Hill and flows in a general S. 33° E. (magnetic) direction for about a mile to Greenhorn. In this ravine, I understand, there has been no northerly rim of slate-rock found. The bottom of the ravine itself, where it crosses the mines, is still in gravel; and the depth to bed-rock is not known. I am inclined to think, however, that the depth will not prove to be very great. On the western side of the Railroad Mine the bank of gravel is nearly or quite 150 feet high. The bottom of the gravel as exposed is a bluish cement for about thirty feet. And nowhere else in this district, at Red Dog, or You Bet, has there been a much greater depth than thirty feet of blue cement found at the bottom. I have no observation for altitude at this point, nor indeed at any other in the neighborhood. Owing to the storm and other unfavorable circumstances I made no use of the barometer. In July, however, Mr. Bowman took a few observations, of which I can make some use here. The altitude of the blacksmith's shop in the Railroad Mine he makes to be 2,974 feet; but whether that was near the bottom of the mine or not I do not know. The main mass of the gravel in the Railroad Mine is fine and of a reddish or grayish color. In both the red and the blue gravel there is also considerable clay and sand. Charred wood is also plenty. One peculiarity of this gravel deposit is the presence of a large number of rather angular volcanic pebbles and boulders (up to a foot or more in diameter) found thickly distributed in the upper eight or ten feet only, indicating that a portion of this channel was filled after the main flow of lava. It is possible, however, that we shall have to seek no farther than the west gap of Quaker Hill for the origin of these particular pebbles, and that they were brought down by the waters of the original Prior Ravine.

About an eighth of a mile below the lower outlet of the Railroad Mine slate-rock is seen on both sides of Prior Ravine. On the left (east) bank there was an overflow of gravel upon the bed-rock to a point a little farther south than on the right bank. On the right bank of the ravine, not far from the outlet of the Railroad Mine, I was shown the top of "Hotellen's Incline." This was meant to be sunk on the bed-rock, which was here pitching to the west. The length of the incline is 600 feet, and it has a fall of one foot in ten. It followed pretty nearly the line of the rim for some distance and was then run through gravel until it struck the bed-rock, or western rim, rising on the opposite side of the channel. This fact is much depended upon by the advocates of the theory of a deep north and south "blue lead," as showing the presence of a deep channel, with an unmistakably north and south direction. It needs but a glance at the map, however, to see that the present Greenhorn, which had been following a nearly southwest course for some distance, bends to the south, not far from Quaker Hill, and flows in that direction for nearly two miles. It is not strange that a similar curve should be made by the old stream if, as I have supposed, the ancient and recent drainages are essentially the same.

West of Prior Ravine is the Green Mountain Cañon, the ridge between having the general course S. 30° E. (magnetic); though half a mile distant from Greenhorn the general course of Green Mountain Cañon is S. 57° E. (magnetic). The ridge between the two ravines terminates in a prominent hill of slate rock. To the west of Green Mountain Cañon there is slate rock at an altitude of 3,009 feet, which would be sufficient to turn the channel to the south. In Green Mountain Cañon there is a shaft sunk about 130 feet in gravel to bed-rock. From the bottom of the shaft there has been drifting, mainly in a direction a little west of south. I find the altitude of the mouth of the shaft at the Green Mountain Mine to be 2,810 feet, from which 130 is to be taken to give us the altitude of the bed rock, viz. 2,680 feet. My informant could not say positively as to the depth of the shaft, and so this result is not to be taken as the most accurate determination possible. Higher up the same cañon, probably between forty and fifty feet above the top of the Green Mountain shaft, and bearing N. 78° W. (magnetic) from it, is another shaft, in which bed-rock rising to the west was found at a depth of only forty feet. This seems almost conclusive as to the position of the west rim of the channel at this point. The course of the channel being pretty nearly south, it will be seen that it crosses the present Green Mountain Cañon at a rather acute angle. From near the mouth of Gas Cañon, which joins Green Mountain Cañon a quarter of a mile from Greenhorn, the left bank of Green Mountain Cañon is bed-rock for a distance of six or eight hundred feet, while on the right (west) bank the gravel is seen in or near the cañon, and crosses Gas Cañon quite near its mouth. On the right bank of Gas Cañon near its junction with Green Mountain Cañon is the Empire Mine and Mill. On the opposite side are Fisher's Diggings. At the Empire Mill there is a tunnel, on a course S. 40° W. (magnetic), which was started in bed-rock, and which, for 400 feet, had bed-rock for its bottom, though the top was usually in gravel. The grade of the tunnel was "water grade," — say one fourth of an inch to twelve feet. At the end of the tunnel the bed-rock was still pitching slightly in a westerly direction. Higher up Gas Cañon, above the boarding-house, there is high bed-rock again, so that the eastern and western limits of the channel can be established with a tolerable degree of approximation.

The country between the Empire Mill and Red Dog I did not visit at all. I will only add that, taking our altitude determinations as approximately correct, there is a fall of about fifty feet between the Empire Mill and Bunker Hill bed-rock; which gives a very convenient grade, the distance being only a little over a mile. At the Gouge-Eye Mine, which lies between these two places, we determined an intermediate value for the altitude of the bed-rock. The course of the channel, then, may be taken as pretty well established from Quaker Hill southward.

There was another small extent of country around the heads of Green Mountain Cañon and Gas Canon, which I wished to explore for the sake of finding high gravel or overflows, if any existed, but was obliged to leave that part of the work unfinished.

§ 5. *Grass Valley and Nevada City.*

The next group of hydraulic mines to be described comprehends those situated on both sides of Deer Creek, in the neighborhood of Grass Valley and Nevada City. A reference to the map will show that extensive lava flows have made their way from the higher portions of the Sierra far down toward the foot-hills on both sides of Deer Creek. The flow which has already been described as forming the high and conspicuous ridge known as Mount Oro, on the upper waters of the Greenhorn, seems to be connected with the gravel deposits at Quaker Hill, being there interrupted for a distance of somewhat less than a mile, but again resuming its regular course, and continuing, not southwardly parallel with the present Greenhorn, but in what may be called the normal southwesterly direction of the Sierra slope. This flow forks again just beyond Banner Mountain, one portion continuing southwest, and soon terminating at a point three miles east of Grass Valley. From this termination a mass of gravel of considerable area extends off to the west, nearly reaching the last-mentioned town. On this area are the Town Talk mines; and there are also shallow patches of gravel, south of these, and between those mines and Osborn Hill.

The other branch of the lava flow, mentioned above as dividing near Banner Mountain, runs for nearly ten miles almost due west, passing a little to the north of Grass Valley, and terminating just before reaching Rough and Ready. There are numerous hydraulic mining claims at various points along the edge of this branch; but they are not now, neither have they ever been, of very considerable importance.

The group of gravel mines near Nevada City, which is only three miles northwest of Grass Valley, is connected with another flow than that just indicated as having come down from the direction of Mount Oro. This flow, one point of which, just north of Nevada City, is quite a conspicuous elevation, and, as usual, known by the name of the "Sugar Loaf," seems to have come down on what is now the divide between the South Yuba and Deer Creek, having headed, however, not far from the great volcanic centre near Mount Stanford, or very near where the other flow originated, which has made its way along to the north of the North Fork of the American, and which appears to be connected with the occurrence of the Dutch Flat and Gold Run gravels. In fact the whole appearance of these volcanic masses which occupy so much of the space between the North Fork of the American and the South Yuba rivers, is that of a connected series of flows, originating nearly at the same central focus, spreading fan-like in their descent, and terminating at the various points already indicated, which are from four to five thousand feet lower than the starting-place.

There are a large number of gravel deposits connected with these lava flows, the position of which can, however, be better made out from a consultation of the map than from a written description. Some of these were examined at various times, both by the writer and by Messrs. Pettee and Bowman, and such particulars as are on record in regard to them will here be given, beginning with the localities near Grass Valley and following chiefly the notes of Professor Pettee.

On Alta Hill, just northwest of the town of Grass Valley, there is a capping of lava overlying gravel, into which several shafts have been sunk, the sill of the shaft-house at "Alta No. 2" being 2,758 feet above the sea-level, those of "Alta No. 1" ten and a half lower, and those of the Hope Company ten higher. In regard to these mines, which were not very extensive and which were examined by Mr. Bowman, the writer has no information of importance. In the Grass Valley Ravine there is a locality known as the Grass Valley slide, the opposite side of the ravine being called the Eureka slide. The elevation of the bed-rock here is 2,490 feet. It seems clear, according to Professor Pettee, that this is really a "slide," and not a part of any main channel. This conclusion seems justified, partly because the measurements show the bed-rock to be so much below the bottom of the shafts on Alta Hill, and partly because the ravine seems to be filled with alternations of gravel, clay, and volcanic material, in endless confusion. At a point near the upper end of the ravine, sixty-two feet above the bed-rock just referred to as having been determined by the barometer, there is an irregular stratum of gravel varying from seven to fifteen feet in thickness, and into which four tunnels or drifts had been driven. These were all on the same level, and not very far apart. At the time of the examination they were abandoned; although, as appeared from the condition of the plant, not permanently.

A little farther west, north of the road to Rough and Ready, is the Rock Tunnel, whose mouth has an elevation of 2,482 feet. This tunnel runs in a direction a little west of south for a distance of 1,200 feet, and then turning to the left continues on a course a little east of south. The total fall in the tunnel was said to be twenty-three feet. An eighth of a mile east of this are the Lola Montez Diggings, worked by the proprietor, Mr. Weed, in the intervals of ranching, and said to be quite rich in spots. From the Rock Tunnel there is no ravine leading down to Slate Creek, the mouth of the excavation being on the bare hillside. Along this slope, something over a quarter of a mile farther west, are the Jenny Lind Diggings, which are open excavations on the side of the hill, which must have been worked for a considerable time. These diggings were also deserted, although tools were lying about, as if work had been done within a short time. Tunnels had been run in, and small shafts sunk. The bed-rock exposed was slate. Around the mouth of the tunnel were masses of bluish clay and small boulders of volcanic material.

At the head of a ravine, about a quarter of a mile in a westerly direction from the Jenny Lind Diggings, is the North Star Tunnel, said to have been run in for a distance of about 2,000 feet. The rock at the mouth of this tunnel was slate, having a dip of about 45° to the southeast. There were lava boulders in abundance lying about, which appeared to have rolled down from above. An eighth of a mile still farther west is the Virginia Tunnel, also abandoned, and in 1870 used as a milk cellar, but said to be nearly a mile in length. The measurements at the North Star and Virginia tunnels show that they are both lower than the Rock Tunnel; but, of the two first mentioned, the latter is about thirty-five feet higher than the former.

There are gravel mines near Rough and Ready, three or four miles west of Grass Valley, on the continuation of the channel which passes to the north of that place. These localities have, however, not been examined in detail by us, a plan to make Rough and Ready headquarters for some days having been frustrated by want of time.

At Goshen Hill, at the head of Hawk Ravine, about a mile west of Rough and Ready, a few scattering rounded masses of lava were found, and there is a deposit of gravel, consisting mostly of massive boulders, in which a shaft had been sunk to the depth of forty-three feet to the bed-rock. There appeared to be no indications of a regular channel. The altitude of the mouth of the shaft is 1,838 feet.

East of Grass Valley, and probably on the same channel-system which passes that place, but higher up toward the summit of the Sierra, there are other localities where hydraulic mining is, or has been, carried on. One of these is Bannerville, where a tunnel has been run in, under the lava, on the slate bed-rock. The altitude of the mouth of this tunnel is 2,986 feet, and that of the summit of the hill above it 3,231 feet.

The prominent point just north of Nevada City, known as the Sugar Loaf (as already mentioned), is 3,111 feet above the sea level, and 670 feet above the bed of Deer Creek at the Suspension Bridge. The extension westerly of the volcanic ridge, on which the Sugar Loaf forms a partly isolated and therefore rather conspicuous mass, is called Cement Ridge. Its general direction just beyond the gap which separates the portions of the ridge, is S. 21° W. (magnetic), but a mile or so farther on it becomes more nearly west. Cement Hill is a few feet lower than the Sugar Loaf, directly opposite which it seems to form almost a flat table of lava, sloping off gradually toward the west. Dean's Tunnel is on the northwest quarter of Section 1 (T. 9 N.; R. 16 E.), about two miles northwest of the centre of Nevada City. The course of this tunnel is a little west of south, and it is run in a much decomposed syenitic granite, resembling the usual bed-rock in this vicinity. This granite has undergone decomposition in a very irregular manner; so that, on the line of the tunnel, there are still "boulders," — as the undecomposed portions of the rock are called, — which cannot be removed by the use of pick and shovel, as the rest of the material can be. This tunnel had been driven 400 feet at the time when visited by us; but another one, forty-three feet higher up, had been carried in for a distance of 900 feet on the bed-rock, before striking the gravel. This higher tunnel was not low enough to strike the channel, but was run 575 feet in the gravel without having got entirely across it. The question of the connection of the channel here with that of the mines east of the Sugar Loaf, known as the Manzanita Diggings, is one of very considerable difficulty. The following extract from Professor Pettee's notes gives the results to which he was led by his examinations in this region: —

In exploring the Cement Hill Ridge, in the direction of Peck's Diggings, nothing but volcanic was seen until I began to descend the spur between Long Tom and Native American ravines; and here, about 700 or 800 feet south of Peck's house, the end of the lava was reached, this altitude being about 2,700 feet. The measurement of the lower edge of lava on Cement Hill, opposite the Sugar Loaf, had given me as a result 2,835 feet. Peck's house is on the unsurveyed land, but he said the corner stake between sections 2, 3 (and what would have been and will be when sectionized), 34 and 35, was between a quarter and half a mile pretty nearly due south of his house. Long Tom Ravine is on the east, and Native American Ravine on the west of Peck's; into the last-named of which his old diggings, which are mostly open excavations, discharge. He selected a spot for me which was nearly an average bed-rock, and the barometer indicated an altitude of 2,632 feet, which is only about twenty feet lower than the bed-rock in the Manzanita Diggings. A difference so small as this was unexpected (though there were some who declared that the bed-rock at Peck's was even higher than at the Manzanita Diggings), and I am sorry that arrangements were not made to have a short series of synchronous observations at the two places, extending over three or four hours. The distance between the two places is not far from three miles, and it is not easy to believe that the deposit at Peck's is in the direct continuance of the Manzanita gravel; but it is even more difficult to suppose that the stream went in the opposite direction. There are also difficulties in the way of supposing that a tributary to the main Manzanita channel passed through here, in a direction from north to south, which do not exist at Dean's Tunnel, where I made a supposition of that kind. Another explanation, which seems in many respects the most plausible of all, is that the gravel under Cement Hill at Dean's and at Peck's has no connection at all with the Manzanita channel; at least no direct connection. It may have been that the Manzanita channel choked up and overflowed into some depression at the north and then flowed in its new channel down towards Peck's Diggings; but I am at present more inclined to the opinion that there was a small, nearly parallel, channel, perhaps only a large ravine, which came from somewhere near Round Mountain, and passing along where now we have the north side of Cement Hill, emptied into the Yuba. The tracing of hidden channels is always a matter of great difficulty, because we can scarcely ever be sure that the particular piece of bed-rock which we see exposed is not exceptionally high or exceptionally low, and so I advance the above solution of the problem, not as a final settler of the disputed question, but as a possible and plausible explanation. I said, above, that there were difficulties in the way of supposing that Peck's Diggings were on a tributary which helped feed the main Manzanita channel. One of the chief of these is that the present ravines and creeks, across which any such tributary must have gone, do not show, and have not shown, any such amount of gold as would have been probable, if the tributary had ever existed. Concerning the country to the south of Cement Hill, and along Rush Creek, I shall have more to say when I review the notes of my second visit to Nevada, made toward the end of September. At Peck's Diggings I saw some of the largest granite boulders that I found anywhere in the gravel region; the bed-rock was also granite. The pay gravel was mostly or entirely quartz sand and boulders, and not very thick, say from one to four feet. A considerable part of the gravel on the bed-rock to the east has been worked out by means of a tunnel and drifting. The mouth of this tunnel is on the Long Tom Ravine side of the spur, between an eighth and a quarter of a mile about south-southeast from Peck's house. The tunnel itself was said to be 600 feet long, running about west-southwest. I took an observation for altitude at its mouth, and, allowing six feet for the rise in the tunnel, made the bed-rock under the hill twelve feet higher than the exposed rock in Peck's Diggings. Mr. Peck supposed the difference would be fourteen feet, which was an excellent corroboration of my measurement. Later in the season, October 1, I visited the place again and repeated my observation at the mouth of the tunnel, obtaining a result agreeing very closely with the first one. This higher bed-rock on the east would seem to indicate, if anything, that the course of the old channel was directly towards the present bed of the South Yuba. When the washing away of the whole bank has been accomplished, — as is contemplated by Messrs. Rolfe and Stranahan, who had arrangements nearly completed for commencing work as soon as a supply of water could be had, — and more bed-rock is exposed, we shall be in a condition to judge with considerable certainty as to the old channel's course. A peculiar feature of the end rim at this point is that it does not rise above the general level of the bed-rock (the only place in the county, according to Mr. I. N. Rolfe, of Nevada City, where this is the case). Why the rim should so universally bend upwards on all sides of the gravel deposits in the Sierra is a question into the detailed discussion of which I will not now go. Possibly in this case the fact that the bed-rock is granite may have something to do with the solution, especially if, as I have supposed, the old channel went down nearly where Native American Ravine now is to the South Yuba. I am not prepared to say whether or not there are any other gravel deposits on a granite bed-rock where the position of the rim has been noticed. In the great majority of cases the bed-rock is slate, which may possibly have swollen and expanded (as some think) when exposed to the action of the air, though I have no fact nor experiments to adduce in corroboration.

The following is a *résumé* of the observations made in the vicinity of Nevada City, at various times during the season of 1870, by Professor Pettee, for the special purpose of making out, more satisfactorily than had been previously done, the position and direction of the channel in that vicinity: —

Among those interested in hydraulic mining at Nevada City there has been considerable difference of opinion concerning the probable original course of the old gravel channel. It is certain that there is a deposit of gravel on the southern side of the Sugar Loaf, extending from the Manzanita Diggings to and beyond American Hill, for a distance of a mile and a half or more, in a general west by south direction. It is also known that there is gravel under and crossing the Washington Ridge (the top of which is clearly an old lava flow); for between the Kansas shaft, on the northern side of the ridge, and the Manzanita Diggings there has been underground connection made for a distance of a mile or so. To the north of the Sugar Loaf there is about a square mile of country where there have been extensive surface diggings and where there are still one or two larger isolated banks of gravel or pipe-clay. To the west and northwest of Sugar Loaf the same lava flow continues, and is known as Cement Hill for a distance of two and a half or three miles, to a point between Native American and Know Nothing ravines, where it comes to a sudden end or has been cut off by the South Yuba. Under this Cement Hill lava there is also gravel, as is proven by the prospecting at Dean's Tunnel and at Peck's Diggings (Rolfe and Stranahan), to which reference has already been made. So much is positive; in the shape of negative data we have also the following: To the west of American Hill — or Red Hill, a point a little farther west than American Hill — down the present valley of Deer Creek we find but little if any trace of old channels; beyond Peck's Diggings in Know Nothing Ravine there was never much gold found, and to the west it is entirely a bed-rock country with no traces of gravel for some miles at least, — not, at any rate, until Rush Creek is crossed. Whether the main mass of lava of the Washington Ridge covers a gravel deposit or not I cannot tell with certainty; my explorations did not lead me any great distance up that ridge.

The problem to be solved is, how are these different deposits of gravel to be connected? One theory advanced was that the Manzanita Diggings are a part of an old channel which came down from the north — from the neighborhood of Montezuma Hill; another theory, based upon the supposed higher altitude of the bed-rock at Peck's Diggings, supposed that the north and south channel came around by Peck's and thence to the Manzanita ground; while others think that the old channel came from a nearly easterly direction (under the lava of the Washington Ridge, or not, as the case may be) and continued to the west under the lava of Cement Hill. Neither of these theories seems to be entirely satisfactory. Against the general theory of a channel flowing south through this part of the country may be urged the general east and west course of the lava flows, — and that is the main argument that I can urge at present, for I had no opportunity of visiting the region to the north of the South Yuba at all, and have no direct means of knowing what positive or negative evidence could be found there. Against the theory of the channel's having come *from* Peck's towards the Sugar Loaf, I find the difficulty of the grade's being the wrong way, to say nothing of the course of the old lava flow. Against the third theory, that the Manzanita channel extended under Cement Hill towards Peck's, there is less objection to make. To be sure the grade is almost too small, and it seems clear that a part of the channel at any rate must have found its outlet by way of American Hill. Still, it is not impossible that there may have been a forking of the stream from some cause or other, or a change of channel; but if so, I cannot believe that the two forks or channels ever united again near Nevada City, for to do this one of them must have made a very sharp turn and returned almost along its original course. Without meaning to say that the problem is solved beyond all cavil, I am more and more inclined to the opinion, the more I look at the subject, that there was an old stream or ravine on the northern side of what is now Cement Hill, entirely independent of the one which came down on or nearly on the line of the present Washington Ridge and discharged down the line of the present Deer Creek.

Being alone, and having only one barometer, I was not able to make such a close measurement of the differences of level between the bed-rock at the Manzanita Diggings and the Kansas shaft as would have been possible under more favorable circumstances. But by repeating my observations on successive days, and by choosing my hours in such a way as to best correspond with the observations at Colfax, I think I attained results sufficiently near to answer all practical purposes. Concerning the difference between the Manzanita Diggings and the bed-rock at Peck's I have already spoken.

At the Manzanita Diggings the deepest bed-rock that has been exposed has been covered up again by slides from the bank or washings from the sluices. I hung the barometer on the building in the mine near the centre of the channel, and had the authority of Messrs. Maltman and Marcelus for saying that the barometer cistern was five feet above the bed-rock. The surface of the loose material covering the bed-rock is very uneven, and this estimate as to its thickness at this particular point may easily have been a foot or two out of the way. If there was any such error, I think the correction would increase rather than diminish this estimate of five feet. At the Kansas shaft (Allen's) I hung the barometer at the level of the top sill. The location of this shaft was given on the spot as about 300 feet south of the line between Sections 6 and 31, and about a third of a mile from the corner of Sections 5, 6, 31, and 32. The road meanderings of Mr. Bradley corroborate this location quite closely. In regard to the depth of this shaft to bed-rock the accounts were at variance with each other,—some giving 220 feet as the depth, and others 228 feet. It seemed probable that the latter statement included a sump in the bed-rock. But allowing a depth of only five feet to bed-rock at the Manzanita Diggings, and giving to the Kansas shaft the greatest depth claimed, I still make the Manzanita bed-rock to be fifteen feet below the bottom of the Kansas shaft,—and by adopting the other suppositions I could easily increase this number by ten feet at least. This is enough to allow a fair grade between the two places, to say the least. I also attempted to get additional evidence by means of measurements at the Live Oak (or *White Oak*,—I find the shaft referred to under both names in my notes) shaft and at the Nebraska incline. The Live Oak shaft has been abandoned and covered up, so that I could not tell with any degree of certainty where the proper point to measure from was, and there was also uncertainty as to its depth, though Mr. Maltman said he believed it was 220 feet deep. The location of the Live Oak shaft is just above the present bank of the Manzanita Diggings, and near Hitchcock's vineyard. The Nebraska incline is on the northern side of the ridge, and was said to be 400 feet to bed-rock, the angle of slope being 34°. This would correspond to a vertical depth of 224 feet. But in this case, also, the initial point of measurement was destroyed. The mouth of the incline had been allowed to cave in. The evidence obtained from these two points will then have little or no real value ; but as far as it goes it seems to corroborate the idea that the rock rises between the Manzanita Diggings and the bottom of the Kansas shaft. The course of the channel between the Manzanita Diggings and the Kansas shaft I could not trace in person, because the old drifts are not now accessible. I do not think that a continuous line of bed-rock has been exposed between the two places, but I was assured that connection had been made through. According to Mr. Maltman. the centre of the channel follows a line running nearly due north from the building in the middle of the diggings, where I hung the barometer, for a distance of three or four hundred feet ; then it turns sharply to the east for about the same distance, and then to the north again directly under the Live Oak shaft, and from there on towards the Kansas shaft. At the Live Oak shaft, however, the most of the gold was not found in the deepest part of the bed-rock, but on a plateau a hundred or a hundred and fifty feet to the west and about fifteen feet above the bottom of the shaft. From that point on I have only the general information that the ground has been thoroughly prospected by drifting, so that there is no hesitancy in proceeding with the hydraulic washing of the whole bank. The information concerning the course of the channel which I obtained at the Kansas shaft is as follows : At a distance of about 600 feet and bearing S. 48° W. (magnetic) from the Kansas shaft (bringing us into the Northeast quarter of Section 6) is an old shaft which struck at a point a hundred feet to the southeast of the south rim of the channel, — the Kansas shaft being pretty nearly in the centre. It would require only a slight additional curving to the south from this point to bring us into the known line of the channel at the Live Oak shaft. All attempts to trace in detail the course of the channel to the west of the Manzanita Diggings were unsatisfactory, on account of the accumulation of boulders and other material on any bed-rock that may have been exposed at any previous time.

What I have thus far given is substantially all the information I obtained concerning the probable course of the old channel. In regard to its width the statements are, as might be expected, anything but precise. At the Kansas shaft the width of channel was given at about 200 feet. At the Manzanita Diggings Mr. Maltman said the main channel seems to find its deepest bed-rock in a narrow gorge of, say, thirty feet in width. Near the building in the diggings (previously referred to) I was told that the width of the channel was about ninety feet, while farther to the east and north it spread out in such a way that there is a width of six or eight hundred feet which it will pay to wash. From such data it is clear that we can get only a rude approximation to the reality.

The total thickness of the river deposit in this neighborhood is considerable, but there is a large percentage of clay and much that contains little or no gold. The top of the Kansas Shaft is about fifty feet (vertically) below the lower line of the lava on the Washington Ridge, and is sunk for the whole distance of 220 or 228 feet through alternating strata of pipe-clay and poor gravel; the layer of pay gravel on the bed-rock being not much if any over fifteen inches in thickness. At the Manzanita Diggings the thickness of the blue gravel (substantially, according to Mr. Maltman's statement) varies from a few inches up to nearly or quite twenty-five feet in the bed of the channel, — the variation arising mostly from the irregularities of the bed-rock. Above the blue gravel comes pipe-clay for a considerable thickness, say, in round numbers, twenty feet, and above this is a newer gravel channel made up mostly of a mass of whitish gray pebbles. The thickness of this bed may be taken at twenty feet. Above this comes still another deposit of less pure clay reaching to within a few inches of the grass roots. This would be an approximate description of the eastern end of the bank — where the principal washing was going on — and where the height of bank is not far from a hundred feet. At another part of the bank I observed the section and estimated the thickness of the different strata something as follows : Above the top of the face of the bank was a considerable thickness of gravel and clay reaching up to the lower line of the Sugar Loaf lava, amounting perhaps in all to fifty feet. The first stratum exposed at the top of the bank was about four feet of gravel. Below this was a stratum of pipe-clay sixteen feet in thickness, followed by five or six feet of gravel. For the next fifty feet the material was mostly clay, with only occasional lenticular masses of gravel. Whether there was any gravel below this or not could not be seen, on account of the accumulations which had slidden down from the top. Mr. Maltman claims a total height of bank of 170 feet. The face I observed was about 300 feet in length. Beyond those limits, in either direction, the section would have varied in more or less particulars. The outlet of the Manzanita Diggings is nearly on the line of the old Coyote Diggings (to which reference will again be made). The next hill to the west — distant about a quarter of a mile, and lying to the east of the present road to Blue Tent — was known as Pontiac Hill, and next to this, on the west of the present road, came Buckeye Hill. Next west of Buckeye Hill came Oregon Hill. These hills have all been washed away, and no means were available for finding out what the peculiarities of their gravel were, if any, excepting that Oregon Hill was said to be the last which carried any "fine quartz gravel." Next west of Oregon Hill is American Hill, where there is still considerable coarse gravel left to work over; but the prevailing impression is that it will not quite pay. Between American Hill and the Sugar Loaf a large mass of gravel has been removed so as to expose considerable bed-rock. At and near where the Newtown road (or one of them) crosses these old workings the bed-rock slopes gently to the south. The house and part of the vineyard of Mr. Rogers stand directly on the American Hill gravel. On three sides of his house the perpendicular bank is only a few rods off, while to the south the slope to Deer Creek is more gradual, excepting where the Plymouth Rock cut has been excavated. This empties directly into Deer Creek without having any connection with the other banks on American Hill. A little north of the house Mr. Rogers has a well sunk to the depth of ninety-three feet. At the bottom bed-rock was struck, rising rapidly towards the south. The gravel on Red Hill I did not visit; there was only a small bank opened on one side of the hill, and it was of no great importance either theoretically or economically.

The following items of interest concerning the Manzanita Diggings were kindly communicated by Messrs. Maltman and Marcelus, and are given pretty nearly in their own words. The water for washing these banks is obtained partly from the South Yuba Canal Company and partly from sources controlled by the owners of the diggings. The Canal Company charge ten cents per inch for furnishing water ten or eleven hours a day, but if taken for the whole twenty-four hours the charge is seventeen cents per inch. Messrs. Maltman and Marcelus have a reservoir into which the water is allowed to run. During the day the supply will be nearly exhausted, but in the night the reservoir will fill again. In this way, by buying 300 inches for $51 they, have during the working hours the equivalent of nearly 700 inches, which without the reservoir would cost $70. At the bank where they are now working they have an average head of 125 feet. The sluices from these diggings are nearly a mile long, discharging into Deer Creek near the Union Hotel. The sluices are from forty-four to forty-eight inches wide, and have the unusually low grade of only four inches to the box. This grade would not be sufficient, if the gravel were not easy to work and the supply of water good. If the grade were increased, too, in this case, they would have to shorten their sluices materially and discharge into the creek at a higher point. At the lower end of the sluices they have also an undercurrent with a grade of only one inch to the box. This is, of course, only for the very finest stuff. The sluice bottom is made entirely of blocks of wood instead of stone. The men at Smartsville prefer rock sluices because their gravel is more tightly cemented together and would wear out a wooden sluice bottom too rapidly. For a 48-inch sluice, the blocks are made eleven inches square and set so as to leave a space for the accumulation of gold and amalgam all around each block. Messrs. Maltman and Marcelus have been driving a bed-rock tunnel through which to sluice away the remainder of their claim, — the claim extending back to the summit of the Washington Ridge, above Hitchcock's. If their present sluices were continued up to that point, they would be considerably above bed-rock. This tunnel is to be 2,400 feet long, four feet three inches wide, and six feet three inches high in the clear inside the timbering, and have a grade of four inches to twelve feet. The cost of the first half of the work has been a little under $10,000. It is expected that this tunnel will be completed in 1871, at a total cost of about $20,000. Possibly the death of Mr. Maltman may have changed all these plans.

The length of a run at the Manzanita Diggings varies. The upper part of the sluice is cleaned up as often as once in twenty or thirty days, while the main clean-up of the whole length occurs only once in a season. The old custom was to use two or more small nozzles; now a single six-inch nozzle is found to be the most economical and effective. Some Chinamen are employed in connection with the white laborers. The latter receive three or three and a half dollars per day, while the former are obtained for $1.75, which Mr. Marcelus said was "too much." The loss of mercury is estimated as "small, perhaps five per cent." The loss of gold is thought to be as much as twenty per cent, and very likely more. The most of the gold would be classed as "fine," though some of it is tolerably coarse.

In regard to the annual or total production of these gravel deposits I never could get any very trustworthy statements. Mr. Maltman, the last time I saw him, said he would run over the old books and give me a detailed statement of production for a number of years back, but I think he never did it. The common idea was that from a million to a million and a half of dollars had been taken out since the first working of the Coyote Diggings. It was here that the first discovery of the old channels was made. The Coyoters continued their burrowing under the hill, following the bed-rock, — and intending to do so as long as gold enough was found to pay, — until they unexpectedly found the rock pitching the other way, and on that inclined surface the gold appeared in surprisingly large quantities. On the other side of the ridge — Shelby (or Selby) Flat — the prospecters worked up in the same way until they found the rock pitching away from them, and then *stopped*. The Coyote Diggings were said to have been the richest diggings in this immediate neighborhood, and it is a pity that no accurate record of the amount taken out was kept. Not knowing the yield per year nor the amount of bank washed away per year, of course I can make no estimate as to the amount of gold found in each cubic yard of gravel.

The petrified and charred trunks of trees which have been found in the Manzanita channel may just be noticed. Some of them have been as much as five or six feet in diameter, — sometimes so much petrified on the inside as to require blasting to remove them, while the outside is easily broken and cut away. I could not hear of any instance where the position of the tree-trunks would serve as a key to the course of the channel.

There are several isolated patches of gravel, sometimes covered by a volcanic capping, and sometimes uncovered, to the west of Nevada City, along on the divide between Deer Creek and the South Yuba. The first of these is Connor Hill, which is on the divide between Rush and Deer creeks and about four miles west of Nevada City. There is a considerable amount of gravel, made up of rounded quartz pebbles, on the northern end of this hill. It had not, however, been attacked, in 1870, on account of the difficulty of getting water to so high and isolated a locality. According to the single barometrical observation taken at this point, the top of Connor Hill is several feet higher than the office of the South Yuba Water Company, at Nevada City. Where this gravel came from, or what the direction followed here by the channel, is, at present, only a matter of conjecture. To the east and northeast is the valley of Rush Creek, and it is possible that the original channel of this hill was once as high as the present top of Connor Hill. That it could have crossed from Cement Hill seems, according to Professor Pettee, hardly probable. To the northwest and west lies Illinois Ravine, once the scene of rich shallow placer diggings. This ravine heads near Newtown, and follows a nearly north course to its junction with Rush Creek, where it contracts to a narrow gorge. The gold of this ravine had evidently the same origin as that on Connor Hill.

In attempting to solve the question of the probable continuation of the lava flow of Cement Hill, the valley of Rush Creek was examined in considerable detail. No indications were observed of a probable passage of the channel in that direction.

At the Empire Shaft, which was sunk at the western end of Cement Hill, the thickness of the detrital formation is about 140 feet; of this seventy-five feet was lava cement, then a stratum of pipe-clay, and under it gravel. From a point not far from the bottom of the shaft, the bed-rock, which here is granite, rose to the south at an angle of nearly 45°; and other excavations in the opposite direction showed a similar condition of things, there being here a deep and narrow ravine in the granite, filled with gravel and pipe-clay. In this gravel the drift-wood was found on the west side of the boulders, near which the gold was accumulated on the north and south edges, as would have been the result, if the course of the channel had been from east to west.

Near the Key-Stone saw-mill on Owl Creek,* six miles west of Nevada City, there is a small outlier of volcanic materials covering gravel. A short tunnel has been run under the lava and a few shafts sunk; the gravel, apparently, has not proved to be very rich. The elevation of this lower edge of the lava at this place is 2,133 feet, and its thickness about thirty feet. About three eighths of a mile east-northeast from the saw-mill are some old surface diggings, the work of Portuguese miners, which are said not to have been rich.

§ 6. *Smartsville and Timbuctoo.*

The gravel deposits in the vicinity of Smartsville and Timbuctoo are interesting on account of their great extent, and because they are at such a low altitude. They are also comparatively isolated; as a reference to the map will show, there being only occasional patches of lava along a line connecting the Smartsville gravels with those great deposits between the South and Middle Yuba. In describing this district the notes furnished by Professor Pettee, of his examination of the region, in 1870, will be chiefly relied on.†

In attempting to give a list of the different companies in this neighborhood a difficulty is met, arising from the changes of name and ownership, as well as the consolidations of title which have taken place at various times. In some instances, the new name has not entirely superseded the old ones, and in others an old name has been made to cover more ground than formerly. In either case, it was not always clear that two persons had the same claim in mind when using the same term, or that they meant different grounds when using different names.

Beginning at the most westerly end of the Timbuctoo deposit, we find the names of the companies, in 1870, to have been nearly as follows: *Warren* Company; Gallagher Claim; *Bourgogne*; Wolf; *Antone*; Union; *Michigan*; *Hyde*; Estner; Greenhorn; Live Yankee; *Babb*; Kentuck; Pennsy; Marlow; Maple; O'Brien (probably the same as the *Pactolus*); *Rose's Bar*; *Pittsburg and Yuba River*; *Blue Gravel*; *Blue Point*; *Smartsville Consolidated*; and perhaps one or two more, such as Campbell's, around Sucker Flat. The names given in italics are those which are most in use. As near as could be ascertained in 1870 the length of channel owned by the different companies then at work was as follows: Antone, 205 feet; Babb, 1,800; Pactolus, 900; Rose's Bar, 1,000; Pittsburg and Yuba River, 1,500; Blue Gravel, 1,200; Blue Point, 1,200; Smartsville

* In the southeast quarter of Section 1, T. 16. N.; R. 7. E.
† Professor Pettee mentions Mr. Ackley and Mr. McAllis of Smartsville and Mr. Redfield of Timbuctoo as having furnished valuable information concerning the claims in their neighborhood.

consolidated, 1,490. This makes, in all, for the seven companies following each other in order from the Babb to the Smartsville consolidated inclusive, 9,000 feet, or nearly one and three fourths miles; so, the total length of a line drawn through the centre of the whole gravel deposit would be a little over three miles.

The width of the channel, or pay gravel, at each company's workings it is not easy to give with accuracy. At Timbuctoo, the Antone Company has worked down to the bed-rock, so as to show a deep channel — the old original water-course, as it were — now filled with blue gravel, or blue cement, a material so closely cemented or compacted together as not to be capable of being moved by water without the aid of previous blasting. This lower channel has a width of seventy-five or a hundred feet. But higher up the gravel was deposited over a much greater width, so that the Antone Company's workings were 700 to 800 feet wide; and in one place, it is said, as much as 1,200. At the Babb Company's claim the width of the channel worked is said to vary from fifty to 300 feet. At one point in the Rose's Bar workings the channel was said to be 600 feet wide; in the Blue Gravel ground the estimated width is from 150 to 300 feet. It is evident that the width of ground which it pays to work is not a fixed quantity, the breadth of the channel varying considerably within short distances.

The thickness of the gravel at this locality is considerable, but variable. Near the Warren and Antone Company's claims the top of the bank was found by barometric measurement to be 172 feet above the bed-rock at the nearest accessible point. The top of the bank in the Antone Company was stated by Mr. McAllis to be 150 feet above the "white cement," — this being a local term for the blue gravel filling the deepest portion of the channel, given probably on account of the preponderance of quartz pebbles in it at that particular place. Eastward of the Antone Company the following data were obtained in reference to the depth of the gravel worked: At the Babb, average height of the bank, to the level of the old river channel, 100 feet, and the old channel itself eighty feet, and here some of the lowest portion of the bed-rock has been exposed by washing. At the Pactolus, a bank of 130 feet in height was worked off before reaching the "old channel"; a shaft was then sunk ninety-six feet — for the last thirty feet in blue gravel — without striking the bed-rock. The Rose's Bar Company were, in 1870, working off their upper stratum of gravel to a depth of from 100 to 130 feet, a deeper tunnel being required to work the remainder, the exact depth to the bed-rock at this point not being known. The Blue Point Company were working a bank of from forty to 120 feet in height; but required a deeper tunnel for washing their deeper gravel. Thus it will be seen that there is in this region a channel of variable width, the upper portion of which in places expands out to as much as 1,200 feet; that ordinarily the higher portion of this channel varies from 300 to 600 feet in width; that the bottom of the gravel lies within comparatively narrow limits, and that in this "old channel," which is less than a hundred feet wide, the gravel is more solidly compacted together than it is in the higher and wider portions of the deposit. The general course and position of the channel will be easily recognized on the map. The thickness of the gravel is in places over 200 feet, and over a large portion of the deposit nearly as much as that. Owing to the great depth of the channel, very long and expensive bed-rock tunnels have been required here to reach the bottom of the gravel. That of the Blue Point Company occupied four years in its construction, and cost, it is said, $146,000; it is 2,270 feet in length. The first tunnel of the Blue Gravel Company was eight years in building, and cost over $100,000; but does not drain the lower portion of their ground. Another and a deeper one was in progress in 1870, to be 1,500 feet in length and to occupy five years in its construction.

In the early days of hydraulic mining in this district the sluices were comparatively short, that of the Michigan Claim, for instance, being from 400 to 800 feet in length only, and of course with a higher grade, perhaps as much as ten or eleven inches to the box.* In those earlier days the

---

* A "box" is always twelve feet in length; it is one section, or piece, of the sluice, and by putting more or fewer "boxes" together the sluice is made of any length desired.

tailings were often rewashed, with greater profit than that obtained from the original washing, and the operation has in some cases been repeated several times on the same material.* The improved sluices differ from those formerly in use principally in being much longer and less steep.

Both "rock bottom" and "block bottom" sluices are in use in this district. When quarried rock is used, it is laid in the box with a thickness of from fifteen to eighteen inches. Boulders, selected from the coarse gravel as it comes from the bank are also used, and it is said that one man can lay eight boxes a day (ninety-six feet) with stone fifteen inches deep, the width of the box being not less than three, nor more than four, feet. When wood is preferred at Smartsville, that of the "digger pine" (*P. Sabiniana*) is taken, and the blocks are cut seven inches thick. They are set with the fibres of the wood vertical, and can be turned and used on the other side, when the first surface gets too rough. A "run" with wooden blocks lasts from twenty to thirty days; at the end of that time the bottom gets so uneven that a "clean-up" and repairs are necessary. With a rock bottom, on the other hand, no change need be made for 100 or 150 days.

In regard to the yield of the gravel in the Timbuctoo and Smartsville district, Mr. Pettee was not able to procure any information upon which an average per cubic yard could be obtained. From all that could be learned, it appeared that the yield of the gravel varied so much at different times and in different places that nothing like a fair and trustworthy average was obtainable. And a similar remark may be made with reference to the total production of the mines as a whole or individually. The most satisfactory item obtained was this: That the Blue Gravel Company's ground had been worked since 1864 with great success, over a million of dollars having been taken out, with a supposed expenditure of thirty-five per cent of the yield; but whether this included the cost of the expensive bed-rock tunnel was not ascertained. The gross yield of the Smartsville Consolidated Company was given at $225,000 per year, with an expenditure of sixty per cent of the yield. A few items may be added in regard to the use of powder, in this district, for loosening the gravel as a preliminary for sluicing it. There are two principal methods in use. One is, to run a drift near the bottom of the bank for a distance of twenty or thirty feet, and then side drifts of sufficient length to insure the throwing out of the whole bottom front, after which the upper portion must fall of its own weight, or when subjected to the action of the water. The other method is to run in a drift to a distance of 100 or 200 feet and then excavate side chambers and side drifts to such an extent that a large quantity of powder can be exploded at once (by electricity), so far from the surface that none of the gravel will be thrown to any considerable distance, but the whole mass shattered. From 200 to 450 kegs (each keg weighing twenty-five pounds) of powder are commonly used in a moderate blast; but sometimes much larger quantities are employed. In the spring of 1870 the Smartsville Consolidated Company set off a blast of 1,500 kegs.

As to the loss of gold in the operations carried on in the Smartsville district, the information which could be obtained was, as everywhere else, exceedingly vague, there being no means of determining it accurately. It was the opinion of some well-informed miners that not over twenty per cent was lost, while others set the amount as high as fifty per cent, and expressed the opinion that the bars of the Yuba, into which the tailings are run, would be found to be as well worth working as they were in 1849.

In regard to the quality of the gold obtained in this district, and its position in the gravel, Mr. Pettee obtained the following information. The fineness of the gold is such that its value ranges from $17.50 to $19.25 per ounce. Mr. Ackley's statement of the returns of the fineness of the bullion from the Smartsville Consolidated were in four cases, respectively, .919 to .936; .919 to .964; .917 to .961; .918 to .934. The highest figures in the last three cases belong to gold obtained from the distillation of the mercury which passed through the meshes of the strainer, during the straining of the amalgam. This seems to contain the smallest and purest grains of all, and pays for a separate distillation. In one case this gold ran as high as .976. The average fine-

---

* This is called "tail-sluicing."

ness of the Smartsville Consolidated Company's gold may be taken at .927, and its general mode of occurrence is in fine particles ; there are no nuggets. Mr. Redfield stated that the gold of the highest percentage near Timbuctoo came from the upper gravel, — the "red gravel," as it is called in that particular locality. The gold from the channel nearest the bed-rock is more in amount, but has a larger admixture of silver. In the Michigan ground — in the upper end, at least — the richest dirt was near the top of the bank, according to Mr. McAllis, while there was nothing but a rather fine sand on the white cement. Farther down, toward Timbuctoo, the richest stratum lay on the white cement. The best ground of the Smartsville Consolidated is in a red streak. In the Smartsville region, throughout, the coarse gravel or cement is generally rich and the fine poor. It seems as if the gold came when the stream had power enough to bring a coarse material and boulders of considerable size ; but when the fine gravel and sand were deposited, the gold was left higher up the stream.

The richest spots in the present channel of the Yuba, in the neighborhood of Smartsville, — namely, Rose's Bar, Barton's Bar, and Park's Bar — Barton's Bar being not quite so rich as the other two, — seem to owe their productiveness to the erosion of portions of the old river channel.

In the region bordering the Yuba on both sides of that stream, and between Timbuctoo and Marysville, there are extensive areas covered with shallow detrital deposits, with more or less volcanic material resting on them. A similar condition of things has already been mentioned as occurring at various points in the Sierra along the edge of the foot-hills, as, for instance, in the vicinity of the Merced River. These shallow gravel beds do not usually appear to contain gold enough to be worth working ; and those below Smartsville and Timbuctoo are, so far as known to the writer, of no value. The area which they cover is designated on the map, and no further information can be given at present in regard to them.

There being such an extensive area near Smartsville, and one so deeply covered with highly auriferous gravel, it would naturally be expected that this channel could be followed almost continuously upward, so that its connection with the immense gravel deposits higher up in the Sierra along the Yuba could be easily made out. This, however, appears not to be the case. From Smartsville upward there is a distance of eight miles, in a line due east, to the west end of the Rough and Ready gravel, without any intervening deep gravel deposits, unless it be a very small area just south of Deer Creek, opposite the Anthony House. This would seem to cut off all idea of a connection, on the south side of that stream, between the Smartsville and Grass Valley channels, the shallow placer deposits found for two or three miles below Rough and Ready not being of importance enough to be more than mentioned in this connection.

On the north side of Deer Creek, however, there are quite a number of detached elevations capped with lava and accompanied by gravel beds, occurring at intervals of from one to two miles along the divide between Deer Creek and the South Yuba. Those between Owl and Rush creeks, about a mile south of the Yuba, are about equally distant from the western terminus of the Nevada City channel, that at Montezuma Hill, and the great line of deposits which occupies so much of the area between the South and Middle Yuba and seems to terminate at French Corral. Before describing these last-mentioned gravels, which are the most interesting and important of any in the State, the isolated patches mentioned as occurring just north of Deer Creek will be noticed. The question of their probable connection with the other deposits mentioned as occurring higher up may be perhaps discussed farther on in this volume.

Tracing up the gravel from Smartsville towards the summit of the Sierra, the valley of Deer Creek is crossed. This stream has in the lower part of its course a very rapid fall; in the last mile before it joins the Yuba its descent is as much as 300 feet. On the north side of Deer Creek, a little northeast of Fiene's, which is about a mile north of Mooney's Flat, there is an isolated patch of gravel, at an altitude of about 1,200 feet. This deposit is thin, and of its productiveness nothing is known. Of this Mr. Pettee says: "There were no other high points to the west to interfere with the course of this channel, and it is impossible to say in just what direction it went. It may very likely have been a part of the original Mooney Flat and Smartsville channel, or one of its tributaries."

On the summit of Pearl's Hill, about a mile northwest of the Anthony House, nearly on the line connecting the last-mentioned deposit with those of Smartsville, there is a bed of gravel, at an altitude of 1,549 feet, capped with a mass of lava about a hundred feet in thickness. The bed-rock in this region is slate. At Stark's Tunnel, on Stark's Hill, a little northeast of Pearl's, there is a tunnel, run in a white clayey material, not seeming to be connected with any old gravel channel, but rather the result of the decomposition of the bed-rock.

Beckman Hill, about a mile north of Deer Creek, has an elevation of 1,950 feet. This is capped with volcanic material, which lies chiefly or entirely on the western slope of the elevation; but which has, in many places, rolled or slid down into the ravines and gulches, so that the line of demarcation between lava and bed-rock could only with difficulty be made out. The body of lava extends in a northeasterly direction for a distance of somewhat less than a mile. The material capping the hill is made up of rolled fragments of lava, intermingled with others of granite and metamorphic rocks. The thickness of the mass is nearly 200 feet. Tunnels run under it, and shafts sunk from the upper surface — to but little depth, however — seem to have developed nothing of value.

Subdivision IV. — The Region North of the South Yuba River.

§ 1. *The Divide between the South and Middle Yuba Rivers.*

The intention of the writer of this volume was to have the hydraulic mining region north of the South Yuba River surveyed with great care; but circumstances, to which allusion has already been made in the Preface of this work, rendered this impossible. Several of the most important localities in the district in question were visited at various times during the continuance of the Geological Survey, either by the writer himself or by some one of his assistants; but most of these examinations, which were not detailed and systematic, were made before the hydraulic mines in this extremely interesting and important region had attained anything like their present development and importance. Mr. C. F. Hoffmann, as already mentioned, made special surveys in this district, for private parties, which were utilized in preparing the large map appended to this volume, and to which reference may be made in studying any portion of the gravel region from the Middle Yuba, as far south as Michigan Bluffs.

Parties not connected with the Geological Survey have published important documents in regard to the gravel deposits between the South and Middle Yuba, and their work will be made use of to some extent in preparing the following pages.*

* The most important of these is a Report of James D. Hague, entitled "The Water and Gravel Mining Properties belonging to the Eureka Lake and Yuba Canal Company and to M. Zellerback, Esq.," and dated December 22, 1876. Mr. A. J. Bowie, Jr., also gives valuable information, statistical and otherwise, in regard to the region in question, in his "Hydraulic Mining in California," published in Volume VI. of the Transactions of the American Institute of Mining Engineers, dated 1878. In the Fifth, Sixth, and Seventh Reports of the United States Commissioner of Mining Statistics, for the years 1873 – 1875, will be found various statements of value in regard to the mines of Nevada County, collected and compiled by Mr. W. A. Skidmore. It may be stated, however, that it is the intention of the author of this volume to procure additional information in regard to the region north of the South Yuba, which will be published either in the form of an appendix or supplementary chapter, as a portion of the present work; or in some other form, so that the important results which have been attained in the development of the hydraulic mining interests in Nevada and Sierra counties during the years since the stoppage of the Geological Survey shall be placed before the public, in connection with these observations of earlier date.

The Middle and South Yuba rivers run nearly parallel with each other, through the gravel region, maintaining a pretty uniform distance of from five to six miles apart, but converging rapidly just below North San Juan, and uniting about three miles below French Corral, which is the terminal point of the gravel range on this divide. The two branches of the Yuba head near each other, about thirty-six miles above their junction, on the crest of the Sierra, — a region of granitic rocks, consisting of ridges of moderate height, the depressions between which are often occupied by small lakes, or lake-like expansions of the streams, where the very numerous branches of the Yuba head, offering every possible facility for the collection and storage of water on the large scale required for the working of the hydraulic claims situated lower down on the divide. The system of ditches and storage reservoirs belonging to the "Eureka Lake and Yuba Canal Company," probably the most extensive organization of the kind in California, is fully described in Mr. Hague's pamphlet, reference to which has been made in the note on page 196. This company has a system of ditches aggregating about 200 miles in length, and three or four storage reservoirs, at an elevation of from 5,600 to 6,600 feet, which reservoirs are on the head waters of Cañon Creek, one of the main branches of the South Yuba. Connected with the Eureka Lake ditch, which is supplied from these reservoirs, there is another one, called the Miner's ditch, taking its water high up on the South Yuba, and these two ditches are so connected that whenever the supply from the river is insufficient, it can be supplemented from the reservoirs. The position of these ditches and that of the mining localities supplied by them can be seen on the map. In addition to the foregoing, there are also several smaller branches or tributary ditches, all belonging to the same system, and centring finally at Columbia Hill, the principal point of operations of the company owning this water; they working mines on their own account, as well as selling water to other parties. From Columbia Hill, the Eureka Lake ditch is continued down the divide to French Corral, or to the lowest point occupied by the gravel.

The possible delivery of water by the ditches centring at Columbia Hill is stated by Mr. Hague at 7,500 inches in wet weather, and 3,300 in dry. Adding the San Juan ditch, which takes its water from the Middle Yuba, but at too low a point to be available except for mines lower than Columbia Hill, the total is 8,800 inches in wet weather, and 4,600 in dry. The average possible delivery during the year would be 6,700 inches for 300 days; or, in round numbers, 2,000,000 inches annually, or about 15,000,000 cubic feet per day.* Of the reservoirs of this company the most important is French Lake, which was originally a mountain lake of great depth, and which has been increased in size by means of a granite dam, sixty-eight and a half feet high and 250 long, giving a reservoir of 337.32 acres' surface with a capacity of 661,000,000 cubic feet. The Faucherie reservoir is also a natural lake, raised by a dam to an area of ninety acres and of 58,800,000 cubic feet.†

Besides the French and Faucherie reservoirs there is another extensive one in the Weaver Lakes, having a capacity of over 100,000,000 cubic feet. Practically, the supply of these reservoirs, if full at the commencement of the season, is counted on as sufficient for five months' run.

The Milton ditch extends from high up on the Middle Yuba to French Corral, and has a capacity of 2,500 or 3,000 inches. The reservoir of this company, called the Rudyard or English reservoir, is formed by means of three dams, of which the central one is said to be 114 feet high.‡ It is built of timber crib-work, filled with stone; and of its character and appearance some idea may be obtained from the accompanying plate, copied from a photograph by Watkins (Plate II). The reservoir is said to hold 535,000,000 cubic feet of water.

The North Bloomfield Company has also a complete system of ditches, and a large storage reservoir, called the Bowman reservoir, of which the dam, originally sixty-five feet high, and in 1875 and 1876 raised to eighty-five feet, is intended to be eventually one hundred feet in height, and to hold back about 1,000,000,000 cubic feet of water.§

The actual delivery of water through all the ditches in the divide between the Middle and South Yuba rivers, by the three companies — the Eureka Lake, the Milton, and the North Bloomfield — does not exceed (according to Mr. Hague) 3,000,000 inches annually; while the total possible delivery might be 4,000,000 inches, or, say, 13,500 inches a day for 300 working days in the year.‖ At a rough estimate, this amount of water would be sufficient to wash about 100,000 cubic yards of gravel per day.

---

* This is about equal to the average discharge of the Ohio River in a minute and a half.

† The total catchment area of the French and Faucherie reservoirs is, according to the Surveys of Mr. Hamilton Smith, 8,432 acres. The average rain-fall (including snow converted into water) is about seventy-eight inches; and of this it is estimated that about three quarters runs into the reservoirs, so that these are supplied with sufficient water to fill them.

‡ The height of this dam is given as 131 feet "from the deepest portion of its foundation to its summit," by Mr. Bowie.

§ This magnificent work is fully described, with a section, in Mr. Bowie's article.

‖ The cost to the Eureka Lake Company of their water may be stated at from 4¼ to 5⅝ cents per inch,

With the above rapid sketch of the facilities for procuring water on the divide between the Middle and South Yuba rivers, we may proceed to a statement of some facts in regard to the gravel deposits.

A reference to the map will show that there is a continuous capping of lava on the ridge from Eureka down as far, nearly, as North Columbia. Below here is only one isolated patch, at Montezuma, about a mile and a half long and pointing directly toward the isolated lava-capped hills mentioned as lying near the Key-Stone saw-mill. Of the thickness and character of the volcanic capping the writer knows almost nothing. Neither are there any details of altitude for this region, the elaborate series of barometrical observations of the Geological Survey, of which the tabulated results are given in the Appendix, having been discontinued on reaching the South Yuba. In general terms, however, it may be stated that the summit of the ridge, at its lower extremity near French Corral, is a little over 1,000 feet above the surface of the water in the Yuba; also that the height of the flat portion of the ridge, above the river valleys north and south, increases gradually as we go up the divide; but that the depth of the cañons in this region is not so great as it is farther south among the branches of the American River, or farther north in Sierra County. Still, there is ample room for tailings in the various steep ravines which furrow the sides of the central high divide, whose table-like top rises on the whole with rapid and regular grade towards the granitic High Sierra region, where lie the various reservoirs which have been mentioned, and which appear to be at an elevation of about 6,000 feet.

There is on this ridge, and especially towards its lower portion, a large area of gravel uncovered by any volcanic capping, of very remarkable thickness, and most of it not so compacted together that it cannot be moved by the hydraulic jet. It is also apparently auriferous throughout, so that with large supply of water available, as already described, this portion of the State offers a most favorable combination of conditions for the development of the hydraulic mining industry on a grand scale. A brief account of the operations going on in this region will be given here; but with the expectation that a fuller one will follow in a supplementary chapter or appendix to this volume.

The general position of the uncovered gravels in the region between the South and Middle Yuba rivers will be seen on the map, as well as the names

according to the amount utilized. This includes repairs and all expenses, but not interest on the cost of the property and improvements. The selling price varies from sixteen to thirty cents per 24-hour inch, according to the locality where delivered.

and locations of the principal mining towns. The highest gravel is at Snow Point, on the north side of the lava flow, close by the Middle Yuba, and there are three other large areas of the same material in such close proximity to that at Snow Point as to be almost continuous with it. These are known as Orleans, Moore's, and Woolsey's flats, and they with the Snow Point gravel are the only uncovered deposits of this kind on the north side of the lava flow. On the south side there is a small gravel area at Relief Point, which appears to be a detached deposit; at least it is so represented on Mr. Hoffmann's map.* From North Bloomfield off to the southwest the gravel occupies a broad area, about a mile in width and two miles long, but it narrows down to much smaller dimensions at Lake City, then extends for five miles farther in a westerly direction, with a maximum width of something over a mile, turning to the north towards its western end at Cherokee and approaching close to the Middle Yuba at Badger Hill. A long branch extends also to the south, from Columbia Hill, and terminates on the south close to the South Yuba.

From Cherokee, for a distance of a little over three miles, as far as San Juan North, there is a break in the gravel, granite bed-rock occupying the surface. But from San Juan to French Corral, a distance of about six miles, following the winding course of the channel, the gravel is almost continuous. At French Corral we are, as already explained, about equally distant from the detached lava-capped elevations near the Key-Stone saw-mill, and the one called Beckman Hill, the former two miles to the southeast, on the south side of the South Yuba; the latter about the same distance to the southwest.

In regard to the general position of the main channel in this region, there can be but little doubt. The lava flow and the masses of gravel are so connected together and isolated from other such deposits, as is evident from inspection of the map, that it would require very strong evidence of an opposite character, to overcome the weight of these proofs of a descent of the volcanic and detrital materials in what may be called the normal direction, or somewhat nearly at right angles to the crest of the Sierra. Mr. Hague gives, in the following words, a statement of the course of the channel on this divide, as recognized by those who have the best opportunities to become acquainted with the facts. He says: "Commencing at the eastern

---

* On the map accompanying Mr. Hague's Report, and which he refers to as Hoffmann's map, but which bears on its face the name of J. P. Wilson as its compiler, the gravel area at Relief Point is represented as extending around so as to connect with that at North Bloomfield, and as occupying a belt a mile or more in length along the volcanic belt in the opposite direction, or in the direction of Mount Zion.

or upper end, the first and highest appearance of the gravel is at Snow Point, on the north slope of the ridge overlooking the Middle Yuba, and the localities lying west of it, known as Orleans, Moore's, and Woolsey's Flats. These are obviously portions of the main channel, now separated by modern ravines. At Woolsey's Flat the gravel channel evidently turns to the south, and with a clearly descending grade passes under a high ridge of lava and disappears from view. At North Bloomfield, three or four miles southwesterly from Woolsey's, it reappears on the south slope of the main ridge, and continues uncovered by lava with a southwesterly course toward Lake City, two or two and a half miles distant. Somewhere under the lava covering before referred to it is believed that the channel is joined by a branch coming in from the southeast, the only uncovered portion of which on this ridge appears at Relief Hill near the South Yuba.

" The main channel at Lake City is again partly covered by lava, but only for a comparatively short distance; beyond which it continues with a westerly course and on the south slope of the crest of the ridge, toward Columbia Hill and Cherokee (the last-named place about five miles distant), being joined near Spring Creek, two miles below Lake City, by another branch coming from the southeast (Grizzly Hill).

" At Cherokee the channel turns to the north, reaching the Middle Yuba River in the distance of a mile or a mile and a half, where it is abruptly cut away by that stream at Badger Hill. Thence the modern channel is supposed to have followed generally the course of the older for five or six miles to San Juan, sweeping the gravel entirely away along that part of its course. At San Juan the ancient channel reappears upon the ridge, and is an almost continuous body of gravel for six miles or more to French Corral, its continuity being only broken by two or three modern ravines."

The point of gravel at Relief Hill, mentioned by Mr. Hague as being probably an indication of a tributary channel coming in from the southeast, appears to be continued on the other side of the South Yuba in a series of gravel areas, extending in the line of Relief Hill, through Gold Hill and other smaller patches to the important mining localities known as Alpha and Omega. In these various localities the channel appears distinctly indicated, the gravel masses being very closely connected with each other, and the grade suitable.*

---

* According to Mr. Skidmore (in the Seventh Report of the Commissioner of Mining Statistics) the gravel at Omega Hill has been found very rich, having yielded up to 1874 at least $1,500,000, and there being a large area of ground left to be washed; enough, as is supposed, for twenty seasons' work.

Of the grade of the supposed main channel, on this divide, there are few detailed measurements. The difference between the height of the bed-rock at Snow Point and at French Corral is stated by Mr. Hague at 2,500 feet, which would give a grade of about 100 feet per mile on the average. It is said that the height of the bed-rock at Woolsey's Flat, as compared with that at North Bloomfield, indicates that if, as generally supposed, the channel crosses under the lava so as to be continuous between those points, its grade must be very steep; as much as 180 feet to the mile. Between Grizzly Hill and Badger Hill the bed-rock has but a very slight inclination, or only nine feet to the mile. Between San Juan and French Corral the descent is 450 feet, giving a grade of between eighty and eighty-five feet per mile.

The width of the channel and the form and character of the rim-rock in the region between the South and Middle Yuba rivers are as yet but very imperfectly known. Enough has been revealed by the various shafts, tunnels, and other excavations made by the different companies at work here, either in their ordinary mining operations or for prospecting purposes, to show the existence of an astonishing amount of gravel, which cannot be washed away, even with the utmost possible activity, until many years have elapsed.

From San Juan to French Corral, the channel is comparatively regular in its form, and its width is estimated by Mr. Hague at something over 1,000 feet. Above this it is much more irregular, and, on the whole, much wider, being in places, as at Columbia Hill, fully a mile and a half wide. In fact, this channel far surpasses in its dimensions the "big channel" at Forest Hill, besides having the advantage of being much less extensively covered by the volcanic formations.

In the element of depth, also, the gravel between the South and Middle Yuba shows itself pre-eminent. In the lower portion of the channel below San Juan, the depth of the gravel is not much over 150 feet;[*] but above that place this figure is greatly exceeded. The ground of the North Bloomfield Company is said to have been extensively prospected by shafts, next to the bed-rock, revealing a total depth of gravel amounting to as much as 300 feet, and in places (according to Mr. Skidmore) of fully 600 feet. At Columbia Hill the depth of the gravel is also stated as being from 300 to 600 feet.

In its general character the gravel of this region resembles that already

* Mr. Hague calls the depth of the gravel below San Juan 140 feet, probably intended as an average.

described as occurring at Smartsville. The bulk of the material consists of water-worn pebbles and small boulders, of moderate size and made of all the usual varieties of metamorphic and granitic rocks occurring in this portion of the Sierra ; there are also many pebbles of quartz. Among the boulders are some of great dimensions ; but, so far as the writer has observed, it is remarkable how near an approach to uniformity of size there is in the mass of detrital material taken as a whole. As hardly ever fails to be the case, there are occasional beds of tough clay and others of fine sand inter-stratified with the gravel, and there is no mistaking the general stratified character of the deposit, although portions seem to have been rapidly heaped together under the influence of shifting currents. The heliotype reproduc-tions of Watkins's admirable photographs (Plates A — frontispiece — and I) show most distinctly the lines of stratification, and the general homogeneous character of the boulders, at the North Bloomfield Company's Malakoff Diggings.

The upper portion of the gravel in this district, as is usually the case in the hydraulic mining region of the Sierra, is less compacted together than the lower, and can be washed away without previous loosening, or shaking up, by means of powder. The top gravel is also of a lighter and more red-dish color than the lower portions of the mass, as would naturally be expected. the easier access of the atmospheric agencies to the higher parts of the banks allowing more or less disintegration of the material and oxidation of the iron. The lower strata are, as usual, of a more or less decidedly blue color, and are often so solidly compacted as to require a liberal use of powder before the hydraulic jet can be brought to bear successfully on the mass. This "blue gravel" is of irregular thickness, varying from a few feet to over one hun-dred. In shaft No. 1 of the Malakoff Diggings, according to the section given by Mr. Bowie, there is mostly blue gravel and "cement" indicated, for the whole depth below seventy feet, to the bottom of the shaft, which is about 220 feet deep.

To reach the deep gravels in the region between the South and Middle Yuba rivers, a considerable number of long and very expensive bed-rock tunnels have been run. Mr. Hague gives a tabular statement of the length and cost of nine of these. That of the North Bloomfield Company, which was sunk with the aid of eight auxiliary shafts, is the longest and most expen-sive of them ; in fact, with the exception of the Sutro Tunnel, it is the lar-gest work of the kind in the country. It commences in the deep cañon of

Humbug Creek, a short distance from the South Yuba River, and follows nearly the course of the cañon, some 200 feet below its bed. This tunnel, which is 7.874 feet in length, has a grade of four and a half feet in a hundred, and is eight feet by eight in dimensions above shaft No. 6. Its total cost is stated at half a million of dollars. Other tunnels in this region are : the Boston, leading to Woolsey's Flat, length 1,600 feet, and cost $40,000 ; Farrell, to Columbia Hill, length 2.200 feet ; English to Badger Hill, length 1,400 feet ; American, below San Juan, 3,900 feet long, cost about $140,000 ; Manzanita to Sweetland, length 1,740, cost $62,000 ; Sweetland Creek, also to Sweetland gravel, length 2,200 feet, cost $90,000 ; Bed-Rock, below Sweetland, length 2,600 feet ; French Corral, length 3,500 feet, cost $165,000.*

The fact that most of the gravel between the Middle and South Yuba rivers belongs to large companies renders it probable that we shall eventually obtain from this region much fuller and more accurate statistics than it has been possible to procure from other parts of the hydraulic mining region ; where, usually, as will have been recognized from an examination of the preceding pages, trustworthy information on most of the important points can hardly ever be obtained. Some data of this kind are already at hand, and have been published in the papers of Messrs. Hague and Bowie, which have already been cited several times. The investigations made at the North Bloomfield Company's works quite harmonize with the general result already made evident in the preceding pages of this volume, that the particles of gold are in most cases scattered through the whole body of the gravel ; but that they are more numerous and of larger size in the lower portions of the channel, and especially immediately over and on the bed-rock surface. The results obtained by Professor Pettee,† in regard to the small yield of the top-gravel near Gold Run, are quite similar to those published by Mr. H. Smith, Jr., the able Engineer of the North Bloomfield Gravel Mining Company, of the amount obtained from the upper portions of their deposit. It appears from his statement,‡ that this company in

---

* Mr. Skidmore, in the Sixth Report of the Commissioner of Mining Statistics, states that the cost of the improvements made by three of the principal mining companies in this district, the "North Bloomfield Gravel Mining Company," the "Union Gravel Mining Company" and the "Milton Mining and Water Company," had up to 1874 amounted to $3,500,000 ; and that nearly $1,000,000 more would be required to complete the works then in progress, so as to fully open up the gravel deposits belonging to these various companies.

† See ante, p. 152.

‡ Quoted by Mr. Bowie in his paper previously referred to.

washing, from 1870 to 1874, three and a quarter million cubic yards of *top-gravel*, made a profit of $ 2,232.84, the yield per cubic yard having been only two and nine tenths cents; that of the Gold Run top-gravel having been four and seventy-five hundredths cents. This result led to some investigations in regard to the comparative value of the upper and lower portions of these gravel deposits; such investigations being considered desirable, in view of the large expenditures contemplated by the company, and, indeed, necessary, in case the lower gravels were to be worked. The following results are quoted from Mr. Smith's report: "To test the comparative values of ground developed by the shaft-workings and top-gravel, two hundred and forty samples, weighing in all two and one-half tons, were taken at even distances from the sides of the drifts, and the same quantity sampled from different layers of the upper bank. These samples were carefully panned out, and yielded, the blue $ 1.10 per ton, the white a large number of colors, but an inconsiderable weight of gold. The gold from the blue dirt was from fifty to one hundred times heavier than that from the white gravel." It appears, also, that from every one of the 240 pans one or more colors of gold were obtained.*

The following general summary of the economical results of hydraulic mining, in the region between the South and Middle Yuba, is extracted from Mr. Hague's Report: —

"The hydraulic method has now been in use on the ridge for over twenty years, and the experience of this period affords some means of judging of the value of the gravel and the profit in working it. The general results have been very satisfactory. Wherever the richer blue gravel has been accessible, as at the Flats, Badger Hill, and below San Juan, it has, with very rare exceptions, paid profits, and sometimes large profits, to its owners. The top-gravel, though much poorer than the blue, has often been found very rich in streaks (due to concentration by surface streams), and has, in general, paid large sums of money to the ditch companies furnishing the water, leaving something besides for the owners of the ground. There are few, if any, trustworthy records of operations showing in detail the costs and profits or losses of the business in the earlier years of hydraulic mining; but so far as the top-gravel is considered, the price paid for the water used in mining it is some indication of the result obtained.

"In early years the price of water was twenty-five cents per inch, for ten hours' flow. This price has fallen, by gradual reductions, to twenty, sixteen and two thirds, twelve and a half, ten, and eight cents per inch, for ten hours, the price at the present time varying from eight to twelve and a half cents per inch for ten hours, or twice that price for twenty-four hours. In many claims in which top-gravel only was washed, the water was paid for at twenty to twenty-five cents per inch for ten hours, and instances are reported in which, after paying these charges, the owners retained handsome profits; such cases were, however, exceptional. On the other hand, it is well known that under the high rates charged for water in early days, many attempts to wash the top-gravel resulted in loss.

* Mr. Bowie gives a section of shaft No. 1 in the Malakoff Diggings, in which the number of colors obtained at various points beneath the surface is stated.

"At the present day, where the top-gravel only is washed, it is thought to do very well if it yields from ten to fifteen cents per inch of water for ten hours. In the North Bloomfield Company's mine, in 1870–71, the yield of surface gravel was sixteen cents per 24-hour inch, or six and six tenths per 10-hour inch. From 1870 to 1874 the yield was only thirteen and six tenths cents per 24-hour inch, equal to five and sixty-six hundredths cents per 10-hour inch. In 1875 the top-gravel, including a little blue gravel, but nothing within forty feet of the bed-rock, yielded nineteen and two tenths cents per 24-hour inch, or eight cents per 10-hour inch. It is quite probable, however, that the water in this mine, furnished as it is from the Company's own ditch, is much more lavishly used than in mines where it is purchased, and the relation of water to product would, on that account, be an unfair criterion for other mines. At Columbia Hill, where only top-gravel has been washed, its yield in several instances, of which more details are given farther on, has varied from about twenty cents to fifty-eight cents per 24-hour inch, affording in the instance last referred to exceptionally good profits.

"Some of the mines below San Juan afford the most satisfactory examples of the results of washing the entire bank, including together the top and bottom gravel. The record of their observations during the past four years, where they can be obtained, furnish the best data for judging the value of the ground. Of this portion of the channel some two or three miles have been already washed out. The American Mine has been worked for a length of about 3,000 feet along the channel. The width of the cut from rim to rim is probably 1,000 feet, and the workings about 140 or 150 feet. From some data furnished from the Company's books a few years since, it appears that the gross product of the mine from December 19, 1860, to August 6, 1872, was $1,241,240.30. The water used in this period was 1,454,174 inches of ten hours, and the yield per 10-hour inch would accordingly be eighty-six cents. The price paid for the water varied from sixteen and two thirds to twelve and a half cents per 10-hour inch, amounting in the aggregate to $218,749.58, or 30.6 per cent of the whole working expense. The last-named item was $714,771.04, and the net proceeds $526,469.27, or 42.41 per cent of the product.*

"These examples might be supplemented by others, but enough has been said to warrant the conclusion that the top-gravel alone usually contains gold enough to pay the expenses of mining, and leave a profit for the owner of the water ; and, further, that where the blue or bottom gravel has been washed, it has, with very rare exceptions, made satisfactory profits for the owner ; finally, in order to reach the blue gravel, the top-gravel must be removed."

In regard to the quantity of gravel yet remaining on the divide between the South and Middle Yuba rivers, and not too deeply covered to be washed, it is impossible, at present, to make any statement which could have any claim to accuracy. It is safe to say, however, that with the present available supply of water this gravel will not be exhausted during the present century. Mr. Hague estimates the linear extent of the main channel and its branches, on this ridge, at not less than twenty-five miles, and considers that, deducting liberally for the portion already worked, and for that too deeply covered by lava to be available for hydraulic mining, there are probably not less than fourteen miles of channel still available for washing.† Roughly estimating the average width of the remaining gravel range at 400 yards, and the average

---

* In 1871, Mr. Hamilton Smith estimated that the yield of this ground amounted to twenty-four cents per cubic yard, and that each linear foot of channel worked had paid at least $750.

† This estimate was made two years ago.

depth of the portion yet left to be washed at seventy yards, the total amount of gravel still available for hydraulic mining purposes will be, in round numbers, 700,000,000 cubic yards. Estimating seven cubic yards as the amount of gravel that may be washed by a 24-hour inch of water, on ordinary grades, and knowing the total possible delivery of water by the three companies owning the ditches by which this region is supplied, which delivery may be stated, in round numbers, at 4,000,000 inches annually, it follows, according to Mr Hague, that the whole length of fourteen miles of channel, if opened and rendered available for mining, may be washed in from twenty-five to thirty years.

§ 2.   *Cherokee Flat, and the Region near Oroville.*

Low down on the foot-hills, about on the same elevation as the Smartsville gravel, there is in Butte County, between Oroville and Dogtown, an important placer and hydraulic mining region which properly comes up at this place for description. Unfortunately the data in possession of the writer in regard to this district at the present time are exceedingly meagre. The geological features of the vicinity of Oroville and Pence's Ranch, ten miles farther north, near Cherokee, have been described with some fulness in the "Geology of California"; but the hydraulic mining operations, which have only rather recently assumed great importance, were never examined by the Geological Survey with any detail. The general character of the region is similar to that of the other districts farther south already described. There is, namely, a volcanic capping overlying heavy beds of sandy material, with gravel at the bottom, and the latter is rich in gold.

The Table Mountain of Butte County, which extends from Oroville to Cherokee, a distance of about eight miles, resembles in many respects the Table Mountain of Tuolumne, being like that capped with basaltic lava, and forming a very conspicuous feature of the topography of the region. The quantity of gravel which underlies the Oroville basaltic table is, however, as it appears, much larger than that under the Sonora flow. There are numerous isolated table-capped elevations north of the main mass, which is itself cut into and eroded away at various points, especially at Cherokee and Morris's Ravine, exposing large areas of gravel. It is said that the channel extends as far north as Dogtown, a distance of fully twenty miles. This north and south direction of the channel corresponds with that of the present drainage, the North Fork of Feather River running almost exactly parallel with this channel for several miles, and the west branch of the same taking up and continuing the parallelism for many miles, keeping almost the same direction as far as the great volcanic flow which extends to the south-southwest from Lassen's Peak, in the region where the Butte County Table Mountain probably heads.

According to Mr. Waldeyer's notes,* the gravel under this lava flow is extremely thick, being nearly equal to its greatest development on the divide between the Yubas. The bottom gravel is, as usual, of a bluish color not cemented together. The pay streak is said to be only a few feet in thickness, and the finer material above, mostly quartzose sand, is poor, although not destitute of gold. The basaltic capping is from eighty to a hundred feet in thickness. The gold in the gravel is often found in nuggets of considerable size.

The peculiar position of this locality — so low down in the valley and so isolated as it is — has rendered it a matter of difficulty and great expense to procure the necessary amount of water for carrying on hydraulic mining here on a scale suitable to the magnitude of the deposits. The principal works are at Cherokee Flat and Morris's Ravine. At the first-named locality the Spring Valley Canal and Mining Company is engaged in operations on a scale almost or quite unprecedented, even in California. The water used by this company is brought by two ditches, sixty miles in length, from Butte Creek and the head waters of the West Fork of the Feather.† The product of gold here is said to be very large, and the profits considerable. At Morris's Ravine, a few miles south of Cherokee, there has been a heavy expenditure required to bring in water and for other necessary preparations for washing on a grand scale.

§ 3. *The Region North of the Middle Yuba.*

The Middle Yuba River forms the boundary between Nevada and Sierra counties, and the hydraulic mines described in the first section of this subdivision of the detailed geology of the gravel districts belong to the first-named of those counties; those between Oroville and Dogtown are in Butte, and the mines now to be briefly noticed extend across the west portion of Sierra into the centre of Plumas County.

---

* Published in the Fifth Report of the Commissioner of Mining Statistics, p. 72.

† Mr. Waldeyer puts down the expenditure at Cherokee in improvements, up to 1873, at $ 1,305,000.

In addition to the information gathered during repeated visits to Sierra and Plumas counties by the writer and other members of the Geological Corps, we have some valuable materials from the pen of Mr. C. W. Hendel, who prepared an elaborate account of the mining operations in the district in question, in 1872, which was published in the Fifth Report of the Commissioner of Mining Statistics, and to which frequent reference will be made in the course of the following pages.* A much more detailed and thorough investigation will have to be carried on in this region than has yet been before all the important questions can be solved which are here presented. Some most interesting and suggestive facts can, however, be brought forward at the present time, and it is to be hoped that these will be speedily supplemented by further information, to be gathered and utilized in the manner already suggested in a former page of this volume.

Sierra County is principally drained by the branches of the North Fork of the Yuba, the main stream having a nearly due west course for most of the way from its source, but turning suddenly just before reaching the foot-hills and running south for about twenty miles to its junction with the Middle Yuba, just below San Juan. It is with the head waters of the North and Middle Yuba that the peculiar course of the rivers of the northern portion of the Sierra first becomes noticeable. These branches of the Yuba run for some distance in a northwest direction parallel with the crest of the Sierra, before turning to assume their normal direct descent down the slope of the range. The Feather River exhibits this peculiarity in a much more marked degree, as already mentioned.† The branches both of the Middle and the North Fork of that river run for a long distance toward the northwest before turning to cross the main belt of the Sierra, and after reaching the foot-hills the course of the united stream is nearly south, and parallel with that of the Sacramento, which it does not join until it arrives within fifteen miles of the junction of the American River.

In view of these peculiarities of drainage, it is not surprising that the channels of the ancient rivers exhibit corresponding features, differing from those of the Sierra slope to the south of the Yuba. In Plumas and Sierra counties the general direction of the channels seems to be nearly north and south. The data at the writer's command, at present, are far from sufficient to enable

---

* An extensive series of barometrical observations were made in Plumas County by the writer and Mr. Wackenreuder, chiefly in 1866, but they have not yet been computed and reduced.

† See *ante*, p. 11.

him to speak with any degree of certainty in regard to the number and connection of the various channels in Sierra and Plumas. That there are several of them, and that their general course is in a southerly direction oblique to that of the streams in the region, seems not to be doubted. That they are very extensive and as yet but imperfectly developed is also a matter of certainty.

An abstract of Mr. Hendel's views (taken from the account referred to on the preceding page) in regard to the course of the channels will be given here,* as it contains the latest views of those practically acquainted with the district, and appears trustworthy; although, perhaps, the resources of the county are depicted in somewhat too glowing colors. The difficulty and expense of getting water to the high gravels of Sierra and Plumas counties must be taken into consideration. The form and drainage of the summits of the range are not so favorable for the collection of the large bodies of water necessary to keep up a supply during the dry season as they are farther south. Some of the gravel deposits, indeed, are, as will be seen farther on, on the very summits of isolated peaks, where an abundant supply of water would be an entire impossibility. Still, even with these drawbacks, there is undoubtedly an immense field for the development of the hydraulic mining interest in the region in question. In regard to this point Mr. Hendel remarks that, although there are fifty mining ditches in the county, with an aggregate length of 220 miles, and having cost about $ 750,000, there is still demand for more water, as the present supply lasts only from four to eight months.

According to Mr. Hendel's views, there are three distinct lines of gravel, or channels, running across Sierra County with a general north and south course. One of these, however, is divided into two parallel branches for a considerable distance, as will be seen from the following description given in Mr. Hendel's own words : † —

"The most eastern of these channels appears to come from Plumas County, in the north. It crosses Feather River near Beckwith's Pass,‡ continues thence, in conjunction with a channel coming from the northwest (passing Gold Lake), as one grand river in a southwesterly course, passes the Key-Stone quartz mine and Milton's Ranch, crosses the old Henness Pass wagon-road near the 'middle waters,' — having been so far but very little opened, — thence along the mining camps of

* See Fifth Report of the United States Commissioner of Mining Statistics, pp. 77 – 90.
† A diagram of the position of the Plumas and Sierra County channels will be given farther on in this volume.
‡ Probably a misprint for Beckworth's Pass.

Nebraska and American Hill, both of which have proved its richness, and finally on toward Eureka south, in Nevada County, where it is extensively opened and worked with great success and profit by many mining companies. Here the celebrated North Bloomfield Company (at an expenditure said to have been, for some time past, $60,000 per month for labor and material alone) is now opening this channel with a bed-rock tunnel one and a half miles long,* in order to work out the lower rich strata of the company's extensive mining ground, located on the above-described grand gravel-channel or dead river. This channel is, in Sierra County, about twenty-five miles in length, and averages one mile in width.

"The channel lying next west enters Sierra County near its northeast corner, on the dividing ridge of the heads of Hopkins and Nelson creeks, in Plumas County, and Cañon Creek, in Sierra County; runs a southerly course, and is covered to a great depth by heavy layers of lava and volcanic sand (conglomerate), or 'mountain cement,' as it is generally termed by the miners. The main channel has been tested by partial working in the following rich places through which it passes, namely: Cañon Creek, Poker Flat, Deadwood, Sebastopol, Excelsior, Fir Cap, Monte Cristo, City of Six, Rock Creek, Forest City, West Ravine, Alleghany, Chip's Flat, and Minnesota, all in Sierra County; and Orleans and Moore's Flat, &c., in Nevada County on the south. In all these places, with the exception of the four last named, the deposit has been worked by means of shafts or tunnels, by drifting, and in most instances the front of it only has been hydraulicked, where water could be obtained, with a satisfactory result. The four last-named places are still worked with great success by hydraulic process. All these mining camps paid richly in early days, producing many millions; and this channel has of late proved as rich as formerly wherever followed and opened low enough into the centre of the overlying hills. This has been demonstrated at the mining ground of the Bald Mountain Company, at Forest City, and the Highland and Masonic Company, situated between West Ravine and Alleghany.

"This channel has several branches, which have proved equally rich in several places, as, for instance, the celebrated 'blue lead' or 'blue banks' near Downieville, situated on the left bank of the North Fork of the North Yuba River, which is apparently a different gold-bearing channel from that of the ancient-river beds before described. Without its branches the above-mentioned channel is over twenty miles long in Sierra County, and more than one mile wide, containing an area of over twenty square miles, having a grade of seventy feet per mile in average.

"Farther west comes the celebrated and more extensively developed so-called 'Slate Creek Basin,' on each side of which is a well-proved and very rich auriferous gravel range or dead river channel. These run nearly parallel with each other in a southwesterly course from the northeast, uniting, however, again near Bald Mountain, in the neighborhood of Scales Diggings and Poverty Hill. The eastern of these two, lying between Cañon Creek on the east and Slate Creek on the west, apparently enters Sierra County in its northwest corner, under Pilot Peak. This isolated mountain is over 7,000 feet high,† of volcanic origin, its northeastern slope heaved up and walled with basaltic columns, while its lofty summit, commanding a sublime panoramic view of the Sierra Nevada mountains for hundreds of miles distance, Sacramento Valley, and the Coast Range, is capped with a bed of lava 600 to 700 feet thick. This channel has an average fall from the base of Pilot Peak (where it is worked by the North American Mining Company) down to Scales Diggings, near the junction of the western channels, of eighty-four feet per mile, reckoning these two points as twelve miles apart, as they are in an air-line."

This channel on the eastern side of Slate Creek is most extensively worked all along between Pilot Peak and Scales, passing through Howland Flat, Pine Grove, Chandlerville, Saint Louis, to Portwine, where numerous companies

---

* See *ante*, p. 203.

† The elevation of Pilot Peak is 7,605 feet above the sea, and 1,216 above Onion Valley at its base, according to the Geological Survey measurements. See Geol. I. p. 306.

are at work and where many millions have been produced. The most westerly channel of the Slate Creek Basin also enters Sierra County under Pilot Peak, according to Mr. Hendel, and passes through Gibsonville, where it has been extensively worked. From there on, it has not been much developed until it reaches Laporte, in the vicinity of which place placer mining is very active. A little below this the channel crosses Slate Creek to Poverty Hill, near which it unites with the easterly branch previously mentioned; and the main channel then passes on to Brandy City and Camptonville, the latter being on the south side of the North Yuba, connecting (as Mr. Hendel thinks) with the channel at San Juan. The identity of the various ancient rivers in Plumas and Sierra counties with the channels on the south side of the Middle Yuba is a matter in regard to which our data are far from sufficient to justify any decided expression of opinion at the present time. The subject is one which must be reserved for discussion after further investigation.

Mr. Hendel has published various statements in regard to the very large yield of the mines on the Slate Creek Basin channels.* According to his authority, the total shipments of gold dust and bullion from Laporte, in the eighteen years previous to 1873, were equal in value to $ 60,000,000, of which nine tenths are believed to have come from the Slate Creek Basin mines.† Four companies at Howland Flat and Potosi — the Down East, Union, Hawkeye, and Pittsburgh — took from 2,365,000 square feet of surface $ 2,251,653.95, the pay gravel being estimated at four and a half feet in thickness. This would give an average of ninety-five cents per square foot of surface, or $ 5.70 per cubic yard of gravel washed. This material was mined, it is stated, at a cost of forty-seven cents, leaving a profit of forty-eight cents per square foot. At Grass Flat, in the Pioneer Company's ground, the yield per cubic yard of gravel is said to have been $ 1.59. The mines in the Slate Creek Basin are mostly hydraulic claims; but a portion of them appear to be on regular channels, and a part are of the nature of placer mines. The bed of Slate Creek itself has been the reservoir for more than twenty years of the tailings poured from the rich diggings on all sides. This deposit of tailings is more than twenty miles in length, extending as it does to the Yuba River, and is from 150 to 800 feet wide. Two companies, at

---

* See Fifth and Seventh Reports of the United States Mining Commissioner.

† These statistics are very vague; but those who endeavor to collect information in regard to the produce of our mines in the United States soon find that accurate figures in regard to these matters are of extremely rare occurrence.

least, have purchased extensive claims on the channel of the creek, with the intention of washing over this enormous mass of material, which is thought by many to contain a very large amount of gold.

The other great channel, which has already been mentioned as running almost due south through the county, forms the central one of the three systems indicated, and is next in importance to the one following the course of Slate Creek, as just described. It is at a very high elevation, Forest City, about six miles north of the Middle Yuba, and near its southern terminus, being, it is said, 4,350 feet above the sea-level. This channel has been traced from Plumas the entire distance through Sierra County to Minnesota on the Middle Yuba. Being for most of the distance deeply covered by volcanic materials, it has been chiefly worked by tunnels and drifts; which in places have been run for long distances in a connecting line, as between Forest City and Alleghany. It is difficult to make out from the published statements what the width of the channel and the character of the gravel are. Indeed, the explorations seem, thus far, to have been quite insufficient to settle the important points of the exact position of the main channel and of its branches. One of the most extensive workings on this line of gravel was that of the Live Yankee Company, at Forest City. This claim was opened in 1855 and worked until 1863, at which time the whole length of the channel in the claim, 2,600 feet, had been nearly drifted out. The yield during this time is said * to have been $ 698,534, of which $ 328,368 was profit. For a period of seven consecutive years after the opening of this mine the dividends averaged nearly $ 50,000 a year.

The Bald Mountain Company commenced operations about the time of the stoppage of the Live Yankee, on ground near that of this company, but to the east of what was popularly supposed to be the course of the channel. A shaft was sunk to the bed-rock at the depth of 269 feet and pay gravel struck at 260. The prospecting of this gravel having proved satisfactory, it was opened by a tunnel begun in June, 1870, and completed in twenty-two months. This tunnel was 1,800 feet in length, 400 of the distance being in serpentine, and the cost of the work (including that of the prospecting shaft) was $ 20,000. According to Mr. Skidmore's published statement, the books of this company show a remarkably successful career from the time of the completion of the tunnel up to July, 1874. At this time 292,200 square feet of ground had been worked on the channel, which had paid at the average

* Seventh Report of the Commissioner of Mining Statistics, p. 152.

rate of $1.09 per square foot, the net profit being just half the yield. Up to October, 1874, the bullion produced had been $345,079.22 from a piece of ground about 7,000 feet in length and 500 in width. The gold is coarse, one piece having been found weighing sixty-five ounces.

The hydraulic mines which have been described in the preceding pages of this section devoted to the region north of the Middle Yuba are partly in Sierra and partly in Plumas County, the line between which now passes across the central channel, previously spoken of, in a diagonal direction near Alturas. The county line also divides the Slate Creek Basin mines in such a way as to throw a portion of them into one county and the remainder into another. Pilot Peak, previously referred to, seems to be on the boundary between Plumas and Sierra.

North of Pilot Peak there is a distance of about six miles, between that point and the Middle Feather, through which the channels just described must have extended, and where they have probably been traced by the miners with more or less continuity. But there are no details in regard to this region in the possession of the writer. North of the Middle Feather is an extensive district, within Plumas County, which is drained by the numerous branches of the North Fork of the same river. The region in question is very rough, being intersected by numerous ranges whose highest points are from 7,000 to 8,000 feet in elevation: their crests have a general northwest and southeast trend, and there is no one which can properly be called the dominating range.

As we go north of the Middle Feather we find the bed-rock more and more covered by the volcanic formations, which beyond the North Fork occupy nearly the whole surface. From Pilot Peak north, through Quincy, Elizabethtown, and Greenville, there is a belt of argillaceous and talcose slates, very thinly laminated, and closely resembling the more slaty portion of the auriferous belt farther south; as, for instance, at and near Placerville. It is on, or immediately in the neighborhood of, these slaty rocks that the mines of Plumas County chiefly occur. The American Valley, in which Quincy is situated, is entirely surrounded by rocks of this character. Both east and west of the slates are large areas of granite, which forms the summit of Spanish Peak nine miles west of Quincy. The cañon of the North Fork of the Feather is deeply cut through a great variety of metamorphic rocks, which also occupy most of the surface between the Middle and North Forks, to the southwest of the Spanish Peak granitic mass. The very highest portion of the divide between these two Forks is occupied by a mass of basaltic lava, some three or four miles in length, in the neighborhood of the Buckeye House ; and below that, as far as Oroville, nothing but bed-rock was seen.

To the east of Quincy the geology is very difficult to decipher. The ranges are mostly capped with lava, the main portion of their mass being made up of granitic, granitoid, and various highly metamorphic rocks, with intercalated porphyries and altered volcanic beds, the whole forming a series of the greatest possible complication, as may have been inferred from what has been said in a previous chapter in regard to the geological age of the fossiliferous strata in this country.

The distribution of the gravel in Plumas north of the Middle Feather is as hard to make out as is the geology of the bed-rock. So far as the writer could discover from a rapid reconnaissance of the region, the gravel interests were of very secondary importance ; at least from an economical point of view. The scientific features of the volcanic and detrital deposits in this region, however, are as interesting as anywhere in the State, and their careful study will undoubtedly throw light on some important points which are yet obscure.

The results of the writer's investigations in Plumas County would seem to indicate that, north of the Middle Yuba, gravel occurs chiefly in isolated patches, at very high elevations ; that these patches are the remains of a former extensive system of channels ; and that the amount of erosion which has taken place since the gravel was deposited in these channels has been astonishingly great.

These inferences will be, in some degree at least, substantiated from the description of two of the localities of gravel examined in 1866 by the writer ; one of these is on the summit of Clermont, the other on Spanish Peak.

Clermont is a lofty mass, the highest point of the range between Quincy and the Middle Fork of the Feather, and having an elevation of 6,844 feet above the sea.* It seemed to have no point of equal altitude anywhere in the immediate vicinity, but is slightly overtopped by Pilot Peak, eight miles farther south, by Spanish Peak, ten miles distant on the west, and perhaps by some points in the range to the northeast which lies between Indian and American valleys, and of which Mount Taylor and Mount Hough are two of the dominating peaks.

---

* At the time this observation was taken there was no station barometer nearer than Copperopolis, so that the result can be received only as an approximation.

The mass of Clermont is made up of talcose, silicious, and argillaceous slate, having the usual northwest strike and dipping to the northeast at an angle of 50 — 70°. Near the summit of the mountain there is a very large outcrop of quartz, and indeed the number of seams or veins of this mineral scattered through the rock in this mountain is very great. The bed-rock is very little covered by detrital material except just at the top, and here is a mass of lava, having a length of about two miles in an easterly and westerly direction, and comparatively narrow north and south. It forms a capping of perhaps 200 to 300 feet in thickness. This has been worked under, in the Excelsior Claim, by a tunnel driven in, on the south side of the flow, for a distance of 662 feet. Two men were occupied for five years in this work. The material passed through, near the end of the tunnel, was a mixture of pipe-clay and gravel irregularly interstratified. At the extreme end of the workings, when examined by the writer, there was a thickness of about seventeen feet of gravel, chiefly quartzose in character. A similar tunnel was run in under the lava from the opposite or north side. Of the auriferous character of the gravel nothing definite was learned. Of course, the difficulty of getting water to this isolated and elevated point, on the very summit of Clermont, will be apparent. A fact of interest observed here was the occurrence of large logs of wood in the lava, partly carbonized, and which had evidently been enveloped and borne along by the flowing mass.

Remarkable as is the above occurrence, that on the summit of Spanish Peak is still more so, as the mountain is more isolated and higher, and does not seem to be in the line of any known channel. With the exception of the patch of gravel on Clermont, and a possible deposit under the lava at the Buckeye House, previously referred to, the writer was not able to discover indications of any channel which could possibly connect with the one on Spanish Peak, the summit of which is 7,058 feet above the sea-level.*

Spanish Peak was examined by the writer in 1866, when the existence of a deposit of gravel, with a volcanic capping on the summit of the mountain, was distinctly recognized. A tunnel had been run here to develop the possible channel, but the workings had been abandoned and were not then accessible. Some years later Mr. J. A. Edman, of Meadow Valley, kindly furnished the writer with a section of the formations noticed in running

* According to barometrical measurements made, as in the case of Clermont, with a station barometer no nearer than Copperopolis.

Volcanic tufaceous
a Aglomerates & Conglomerates

b Soft Sandstone *Sand Cement*

c Coarse gravel *Auriferous*

d Pipe clay *with fossils*

e Loose gravel *Auriferous*

Syenite

SECTION AND PLAN
OF
SPANISH PEAK GRAVEL DEPOSIT

SECTION
on Line of *A C F* of Plan

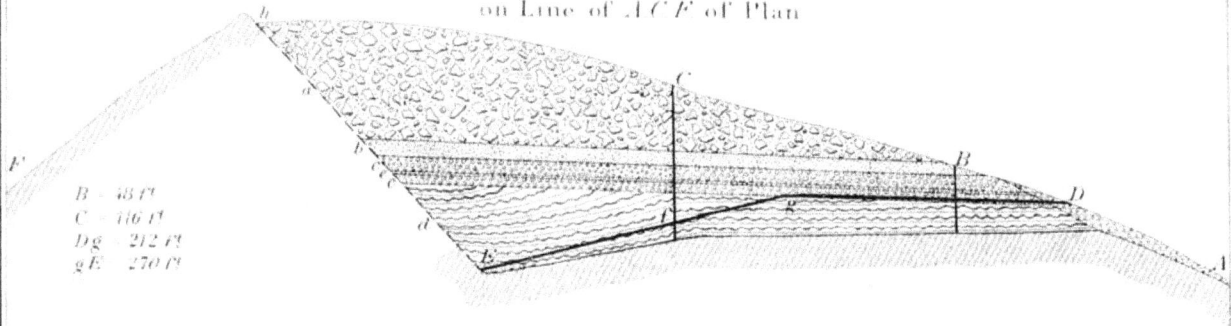

B   48 Ft
C   116 Ft
Dg   212 Ft
gE   270 Ft

Scale 1 inch 160 feet

PLAN

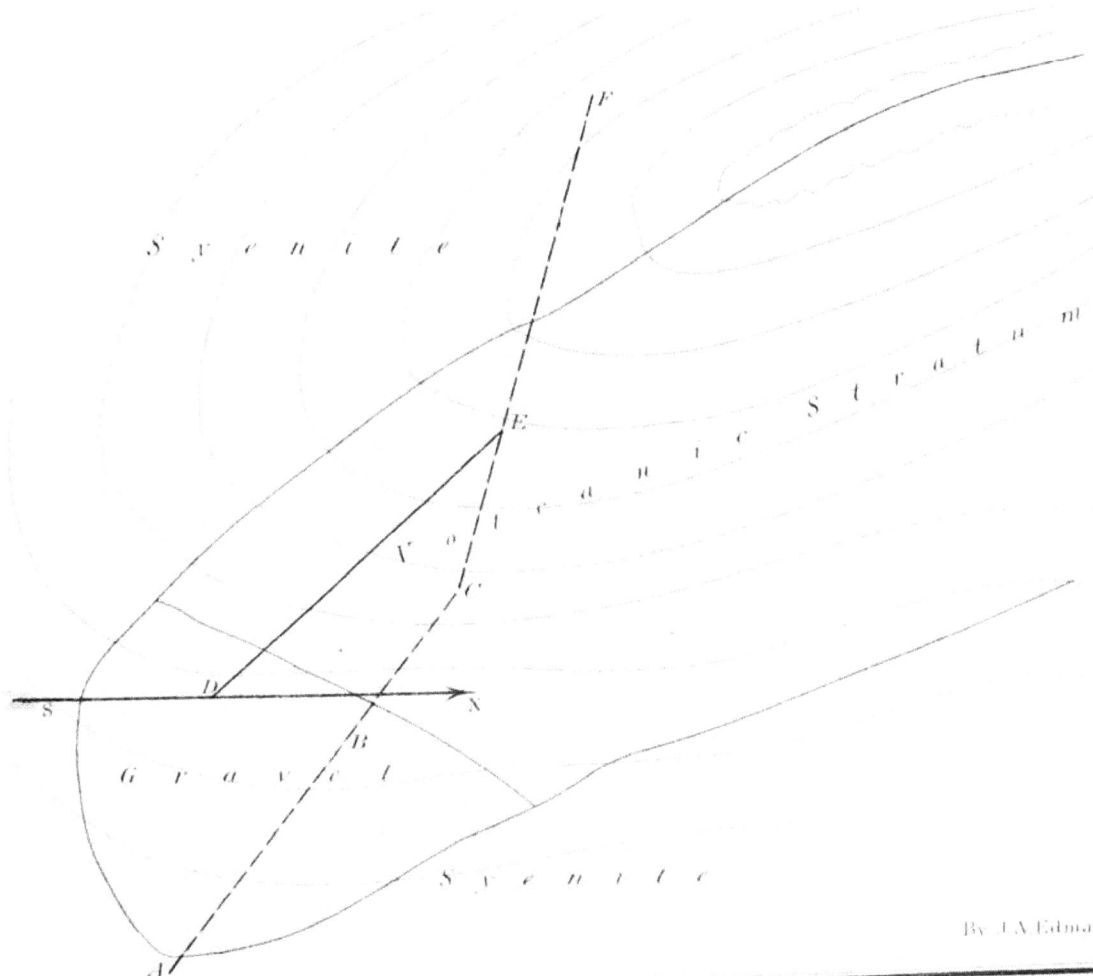

Syenite

Volcanic Stratum

Volcanic Stratum

Gravel

Syenite

By J A Edman

the tunnel, as well as a ground-plan of the locality, which are reproduced on Plate K. There is a thickness of from sixty to eighty feet of detrital material, covered by a mass of volcanic "cement," breccia, and tufa, which where least eroded away is about a hundred feet in thickness. At the bottom of the detritus is a thick bed of pipe-clay with impressions of leaves, and over this several alternations of gravel and sand (as shown on the section), one of the strata of coarse gravel containing some gold. It is not likely, however, that mining ever was, or could be, successfully carried on at this locality. Of the theoretical bearings of the fact of the existence of these deposits on Spanish Peak and Clermont more will be said in a future chapter.

There is a gravel deposit of considerable extent at a point on Spanish Creek, above four miles west of Quincy, where considerable work has been done, although nothing could be ascertained with regard to the productiveness of the material. From the large amount of washing which had been done here previous to 1866, it would appear that the undertaking had been profitable. The thickness of the gravel was about seventy-five feet, and it was quite homogeneous in character from the surface down to the bedrock. The locality has an elevation of about 400 feet above Quincy, which itself is 3,270 feet above the sea-level. Consequently this gravel is about 3,400 feet below that on Spanish Peak.

No extensive gravel deposits were being worked in 1866 in Plumas County, north of the Middle Fork of Feather River, so far as the writer's observation extended. In later years, according to Mr. Skidmore's reports to the Commissioner of Mining Statistics, there have been some undertakings of considerable magnitude commenced at various points. One of them is said to be situated "near Quincy," where the Hungarian Hill Company was at work in 1873, preparing to introduce all the modern improvements in hydraulic mining. This locality, the exact position of which is not specified, may be the one mentioned just previously. At Ohio Creek, a small tributary of the North Fork of the Feather, which enters that stream about six miles below the lower end of Big Meadows, a dam has been built, for the purpose of getting the necessary power to pump water up to an elevation of 328 feet, into reservoirs, for the purpose of washing a (placer?) deposit of gravel said to be very rich in gold.

Fourteen miles east of Quincy there is said to be an important placer-mining claim, which from the description given must resemble in character the so-called "seam diggings," previously mentioned as being extensively worked near Georgetown.* The claim is said to belong to E. A. Heath, and the gold to be exceedingly fine, and seeming to come directly from decomposed auriferous quartz seams which abound in the clay slates on the eastern side of the excavation.

It is also reported that, in 1872, an extensive body of gravel was found at a point, on the very summit of the Sierra, twelve miles south of Susanville, which would be somewhere near the peak designated by the Geological Survey as Red Butte, at the head of Light's Cañon.

It is also stated that a large gravel deposit has been discovered on the Susan River, a few miles west of Susanville, near Big Spring. This deposit is said to rest upon the granite, to have a thickness of about sixty feet, and to be capped with volcanic rock. The gravel, as is stated, is auriferous from top to bottom of the mass. If these statements and those in the last paragraph be true,—and, from what has been mentioned in regard to the deposits on Spanish Peak and Clermont, there is no reason why they should not be,—the difficulty of making out a connection between the Plumas County gravel channels is not by any means lessened. The elevation of the gravel on the range south of Susanville must be one or two thousand feet greater than of the deposit on Susan River. At all events, it would appear that gravels exist in Plumas County in localities of very different altitude, and that all attempts to trace a continuity in the channels in that region must be given up until more detailed explorations shall have been made to the north of the Middle Feather. Nothing is known to the writer of the existence of gravels, sufficiently extensive to be worked by the hydraulic process, at any point north of Honey Lake Valley.

* See *ante*, p. 115.

CHAPTER III.

THE FOSSILS OF THE AURIFEROUS GRAVEL SERIES.

SECTION I. — *Introductory Remarks.*

NOTHING has been said in the preceding pages of the geological age of the series of deposits included under the comprehensive term of "gravels." As will have been seen from the detailed description of the mode of occurrence of these accumulations of detrital material, there is a mass of sedimentary material often one or two hundred feet in thickness, and sometimes three or four times that, piled up on the upturned edges of the bed-rock series, and one of the important questions to be answered in regard to this formation is: What is its geological age? To answer this we must turn to the investigation of the fossils which it contains, and which — as has already been hinted several times in the preceding pages — are quite numerous and varied in character.

The fossiliferous contents of the strata belonging to the gravel series group themselves naturally in three divisions: 1. *The microscopic organisms;* 2. *The plants;* 3. *The animal remains.* Each of these will be considered separately in a special section of this chapter.

The division between the microscopic organisms and the plants proper is, of course, more or less an arbitrary one. Many of the bodies found in these strata, and too small to be seen by the naked eye, are in fact comminuted fragments of undoubted vegetable forms, such as the carices and grasses proper; while other, and in many localities almost the whole of them, belong to more obscure forms, — so obscure, indeed, that there is not an entire unanimity of opinion among naturalists as to whether they are really to be classed in the animal or in the vegetable kingdom.

While the infusorial deposits of the gravel series, or those containing microscopic organisms, are both abundant and important in the Sierra Nevada, they are of little value as aiding in the determination of the geological age of the formation. The remains of plants, on the other hand, which are found abundantly in the formation in question, are of great value in that respect. Their abundance and importance made it desirable that they should be specially investigated, and such specimens as the writer could procure were submitted to Mr. Lesquereux, the eminent authority in fossil botany, for examination and description. His work constitutes Part II. of this volume, and will of course be referred to by those who are interested in the various questions involved in the discussion of the gravels of the Sierra.

The animal remains found in the auriferous detrital series are also of great interest and importance. They naturally divide themselves into two classes, for the purposes of such a work as the present, which does not profess to be palæontological in character. Each of these classes will be considered in a separate section, one of which will be devoted to human remains and works of human hands, the other to remains of animal origin and not human. The reasons for the different treatment which these various subdivisions of this chapter here receive at the writer's hand will be apparent after a perusal of the following pages.

SECTION II. — *The Infusorial Rocks of the Auriferous Gravel Series.*

Probably the most suitable, and certainly the most convenient, term to use for speaking of the strata containing microscopic organisms is to call them by the general term "infusorial," which does not demand or imply an adherence to any particular theory of their origin and biological affinities. Under the term "infusorial rocks," therefore, we include all those deposits which are made up in part of bodies or fragments of bodies of organic origin, but so small as only to be visible under the microscope. As the bodies in question are chiefly of a silicious nature, it is very common to speak of the aggregation of such as "infusorial silica."

It is chiefly owing to the labors of Ehrenberg that we have become acquainted with the geological importance of the infusorial element in rocks of various kinds and ages in extensive and widely separated regions. The various publications of this eminent naturalist on this subject will be found in part in the Proceedings and Memoirs of the Berlin Academy of Sciences. The magnificent folio entitled "Die Mikrogeologie," published in 1854, contains a complete *résumé* of all the results attained by Ehrenberg in the examination of rocks, soils, dust, ashes, and other masses or accumulations of matter from every part of the globe. Among the vast variety of materials here described were some collected by Fremont in Oregon, in the region of the Des Chutes River, which were obtained from a deposit of great thickness, associated with volcanic rocks and found to be very rich in organic forms. The frequent occurrence of infusoria in connection with volcanic overflows and formation of various kinds — as, for instance, in the *moya* or volcanic mud of the Andes — had already excited considerable attention, and there was not a little discussion as to the meaning of this, at first sight, seemingly strange association. The attention of Ehrenberg, which had been repeatedly called to this remarkable association by specimens sent from various parts of the world, was especially aroused by an extensive collection of material furnished by Castillo from Mexico. These were chiefly obtained by means of Artesian borings, which had been executed in and about the city of Mexico. The results of the investigation of this material were published in the Memoirs of the Berlin Academy.*

The examination by the California Geological Survey of various portions of the Pacific coast from Mexico to British Columbia brought together a considerable number of facts and a good deal of material from various localities of infusorial rocks. Those specimens collected which seemed from their appearance likely to contain microscopic organic forms underwent a preliminary examination at the hands of Professor Brewer and the writer, but most of them were forwarded to Mr. A. M. Edwards, at that time residing in New York and professing to make a special study of this class of subjects. Several years having elapsed without any report having been received from him, a duplicate series was sent to Ehrenberg, — this was in 1870, — and by the aid of this series and the specimens sent by the Fortieth Parallel Survey he was placed in a position to acquire a quite thorough knowledge of the organic forms enclosed in these deposits.

The writer of this volume, perceiving that the nature of the connection between the volcanic formations and the infusorial strata which they enclose had not been fully comprehended, published an article on this subject in 1867.† This article was in considerable part translated and made the basis of a communication by Ehrenberg in the Memoirs of the Berlin Academy.‡

* Under the title of "Ueber mächtige Gebirgs-Schichten vorherrschend aus mikroskopischen Bacillarien bestehend, unter und bei der Stadt Mexiko." Abhandlungen der phys. Klasse der Königl. Akademie der Wissenschaften, 1869.

† On the Fresh Water Infusorial Deposits of the Pacific Coast, and their Connection with the Volcanic Rocks. Proceedings of the California Academy of Science, Vol. III. p. 319.

‡ Abhandlungen der Königl Akademie der Wissenschaften, 1870. The title of the communication is "Ueber die wachsende Kenntniss des unsichtbaren Lebens als felsbildende Bacillarien in Californien."

who also included in his paper a general review of the progress of microscopical discovery in this department over the region of the Cordilleras, with a description, accompanied by plates of forms observed at various localities in the Humboldt Valley, on the Truckee River, and in the neighborhood of Great Salt Lake. In this paper of Ehrenberg's a number of questions were proposed relating to topics connected with the investigation of the infusorial deposits. These questions were answered by the writer, so far as it was in his power to do so, in a manuscript communication to the eminent Berlin microscopist, and by him published, with comments, in the Proceedings of the Academy.*

From all the various papers and communications above cited, a clear idea can be obtained of the relations of the infusorial strata to the volcanic formations with which they are associated, and of the great variety and interest of the organic forms which are found therein. All that it seems necessary to do, in connection with the present volume, is to give a brief *résumé* of the facts, and to state a little more clearly and connectedly what has been especially observed in the region of the auriferous gravels, — that is, in the central portions of the Sierra Nevada.

From the various local details incorporated in the preceding chapter, it cannot fail to have been noticed that among the volcanic materials associated with the gravels there are large quantities of fine-grained rock, usually white or grayish-white in color, and known to the miners by various names ; as, for instance, " volcanic ash," " cement," " white lava," " chocolate," " pipe-clay." The bulk of the material composing these various deposits is almost invariably of volcanic origin, as is readily discovered by microscopic investigation. Sometimes the different mineral ingredients have been so finely triturated that the result is simply a mud, which has become consolidated, and in which little of the original structure can be discerned. Other portions of these deposits retain their distinctly crystalline character, and the component minerals can be easily recognized. Most of the white fine-grained lavas seem to be of a rhyolitic character. It is not unreasonable to assume that the original form of much of this material was that of an ash, having been thrown in that condition from the various volcanic vents which must have crowned the summit of the Sierra during the later Tertiary times. This ash was carried down the slope of the Sierra by the streams which were the agents in the

---

* Monatsbericht der Konigl. Akademie der Wissenschaften zu Berlin ; Sitzung der phys. — math. Klasse, 19 Feb. 1872.

formation and deposition of the gravel beds. Much of it being ground up into fine powder, and becoming mixed with foreign material, assumed on deposition and consolidation the form of clay, more or less homogeneous according to circumstances. In other cases the ashes seem to have fallen to considerable depth and to have become consolidated, without having undergone much movement from their original position, and to have been but little, if at all, acted on by water.

The thickness of this deposit, like that of the other members of the volcanic capping of the gravel series, is very irregular. At Webber Hill, near Placerville, the " white lava " is from seventy-five to a hundred feet in thickness. On the divide between the South and North Webber creeks, near Burns's Ranch, this material has perhaps a greater development than has been observed anywhere else in the mining region of the Sierra; it there attains a thickness of from 200 to 300 feet.

It is in connection with the " white lava " that the infusorial beds, — so far at least as the writer's observations extend — are almost exclusively found. The portions which contain organic forms are interstratified with the other non-fossiliferous materials, and seem not to be always capable of being distinguished from them by the unaided eye. Usually, however, the layers which are rich in microscopic forms are not only extremely fine-grained, but are also light and porous, resembling commercial magnesia in appearance; indeed, the name magnesia is frequently given to this substance by the miners.* It is fine enough to make an excellent polishing powder, for which purpose more or less of it has been brought to San Francisco and sold. It does not seem, however, for some reason or other, to have ever come into general use, either in California or elsewhere.†

The most interesting locality within the mining region where infusorial material has been quarried for economical purposes is one near Newtown, which was examined by Mr. Goodyear, and from whose notes the following description is taken.

About a mile above the house of Mr. Samuel Fleming, which is two miles

* To the writer's knowledge the chemists and assayers of San Francisco have frequently called this infusorial silica "magnesia," when it has been submitted to them by miners and others for examination. Since it has become somewhat extensively used as a polishing powder it has received the name of " electrosilicon."

† The " rotten stone " of commerce, so much used for polishing, is a soft material resulting from the decomposition of impure silicious limestone. " Tripoli " and " Bath brick " seem generally to contain more or less infusorial silica.

above Newtown on the South Webber Creek, is the locality of the so-called " photograph rocks " or " picture rocks," which were exhibited in San Francisco some years ago, and which were supposed by some to be natural photographs of the surrounding scenery of the locality where they were obtained. These " photographs " were dendritic markings on a large scale, and they seemed to penetrate the whole mass of the rock, which was itself a pretty fine sandy and clayey material, horizontally stratified and in places very delicately and thinly bedded. The materials of which this rock is composed seem to be partly at least of volcanic origin, and the hill in which they occur is capped with a heavy mass of white lava. Infusorial material has also been obtained here and sold in San Francisco, under the name of "El Dorado polish." The greater portion of the dendritic markings occur in horizontal bands parallel with the bedding of the material and from a fourth of an inch to two inches wide. These bands usually have a dense black border on the upper and lower margin, and between these borders is a narrow medial band of white ground mottled with the dendritic markings, which are often extremely varied and elaborate, forming two lines of " landscape-scenery," as these markings are called, the one being upright and the other inverted. These curious markings are sometimes continuous for fifteen or twenty feet. Between the bands of dark variegated material the rock is white. These markings are not merely superficial, but penetrate the mass of the rock. The bands do not always run horizontally, but occasionally rise and fall in wavy lines, or are developed in rather complex convolutions, although the bedding of the rock itself is everywhere nearly horizontal. In one portion of the bank the color of the mass of the material is a pretty deep and brilliant pink, instead of white as usual. This portion also is traversed by markings, which, however, do not so often assume the form of bands, but are rather irregular although often extremely beautiful.

There are two points here, on about the same level, and within a hundred feet or less of each other, at which some little quarrying has been done. One of these openings yields fine specimens of the dendrites, but none of the " polish " of good quality ; the other opening yields both. The passage from one material into the other is gradual, and there is no well-defined line of demarcation between them. These openings are small and do not offer facilities for thoroughly investigating the geology of the locality. The hill rises to a height of several hundred feet above the quarries, and appears to be made up of " white lava," — that is, of white, consolidated, volcanic ash. These are enormous masses of this "white lava" in other hills of nearly equal height in the immediate vicinity, and there are also very numerous boulders of it in the gulches running down the hill above the "picture-rock" quarries. What the material is which lies below the level of these openings is not known; neither has the depth to the bed-rock surface been ascertained, but it is probably in the neighborhood of a hundred feet. It is very likely that a considerable portion of this thickness is made up of "white lava," and that there is a small amount of gravel at the bottom.

Another locality of a very white, fine-grained, and probably infusorial material is the Clydesdale Claim, in the "Long Cañon Country," about a mile from Blacksmith Flat.* This claim is on the south side of Long Cañon, on a spur between it and Wallace Cañon, a little branch coming in here from the southeast. Here there have been some two or three acres washed off by the hydraulic process, the general appearance of which ground is very similar to that at Michigan Bluff, there being a great preponderance of quartz boulders in the gravel, and some of them being very large. In the western portion of the area washed off the gravel is covered by some twenty-five or thirty feet of gray, micaceous, sand "cement," precisely similar to that which forms the roof of the mines at Bath and Forest Hill. Here, at one point, a large detached block of snow-white material, appearing to be infusorial silica, was found by Mr. Goodyear, which undoubtedly came from somewhere in the volcanic ash "cement" overlying the gravel, which itself was overlain by a thickness of eight or ten feet of clayey matter, mixed with small fragments of the bed-rock, some of which was exceedingly delicately and finely stratified. Over all, and forming all the higher portion of the ridge, came the ordinary, bouldery, dark-colored volcanic "cement."

At Prospect Flat, three fourths of a mile southwest of Smith's Flat, which is about three miles east of Placerville, on the Carson Road, is a locality of infusorial silica, from which material seems to have been obtained for commercial purposes. There is a shaft on the Flat, about a hundred feet deep to the bed-rock, and the whole of which is in "white lava," with the exception of three or four little streaks of metamorphic gravel intercalated in the mass at different depths. The locality from which the infusorial material has been obtained is in the side-hill, just southwest of the Flat, and probably

---

* For the locality of Blacksmith Flat, see *ante*, p. 98; consult also the diagram illustrating the region covered by Mr. Goodyear's explorations.

150 feet above it. It is known as the "silicon lead."* The thickness of the bed of infusorial matter could not be ascertained, as the workings did not expose its full extent. It appeared, however, to be underlain by "white lava," and immediately above it was a stratum of quartzose and metamorphic gravel some four or five feet in thickness, and over this the dark volcanic breccia which forms the crest of the hill.

Still another locality of infusorial silica is that in Chalk Ravine, at the head of Shirt Tail Cañon. This place was not visited by any person connected with the Geological Survey. As described by Dr. J. F. Smith, " Chalk Ravine" is a little branch of Shirt Tail Cañon, situated about three miles southwest of Damascus. The bed is said to have a thickness of fifty feet or more, and the material to be an excellent polishing powder. Fossil leaves have also been found here, it is said.

It must have been apparent, from the detailed description of the gravels and the associated volcanic deposits, that the epoch of their deposition was on the whole a turbulent one. Hence it will not be surprising to the reader to learn that the infusorial strata are considerably more extensively developed in some other portions of the Cordilleras than they are in that region with which this volume is specially occupied. In the northeastern counties of California, on the high volcanic plateau which occupies that portion of the State, and where — so far as we know — no gravels occur, the infusorial rocks occasionally acquire a great thickness. The same is the case all along on the eastern slope of the Sierra Nevada, in the State of Nevada, and also in Oregon, where volcanic rocks are abundant and gravels of rare occurrence. Hence it is that the most prolific localities — those furnishing the most abundant material for description to the Berlin microscopist — have been those outside of the gravel region. Some of the most interesting localities are in Nevada, in quite close proximity to the boundary of California. Our parties at various times obtained abundant specimens from the Silver Peak mining district, from the vicinity of Aurora, and from points north of Virginia City and along the Truckee River. Mr. King indicates the localities of infusorial silica observed by the Fortieth Parallel Survey in this region, as follows: " The infusorial silica overlying the palagonite has its most important outcrop at Fossil Hill, and along the whole northeastern edge of the Kawsoh

---

* Whether it was to the material obtained at this locality that the term " electro silicon " was first applied the writer does not know. He has seen infusorial silica, labelled with this name, from various places in California and Nevada, offered for sale in San Francisco.

Hills, and striking their northern base nearly as far west as Warm Spring Valley; also near the site of Sam's Station, northwest of Mirage Station, and on the banks of Little Truckee River, between Pyramid and Winnemucca lakes; also west of Reno Station, on the Central Pacific Railroad, near the boundary of California. The deposits of Warm Spring Valley are obscure, and show no very great thickness of beds That near Hunter's Station, west of Reno, is an extensive exposure on the right of the railway-cut in approaching California, and consists of several hundred feet (certainly as many as 300) of pure-white, pale-buff, and canary-yellow beds of remarkably pure infusorial earth." *

The locality on the northeast point of the Kawsoh Mountains was found by Ehrenberg to be very prolific in microscopic forms. The total thickness of the series of grits, fresh-water limestones, and infusorial silica exposed at the northern end of the Kawsoh Mountains is, according to Messrs. Emmons & Hague, about 450 feet. Above this is a heavy deposit of basalt, and under it one of palagonite tufa of over 250 feet in thickness. The infusorial beds are said to have a development of 200 feet.†

Another interesting locality is that on Little Truckee River, a few miles above its mouth, also very rich in microscopic organisms, forty-six distinct species of diatoms having been observed in the examination of twenty-five different slides; of these, Ehrenberg classed twenty-eight as *Polygastera* and eighteen as *Phytolitharia*. The thickness of the infusorial mass at this locality could not be made out on account of the absence of sufficient exposures.

The plateaux of the Des Chutes basin, in Oregon, first visited by Fremont, and afterwards by Newberry, are covered by a smooth sheet of basalt, under which is a great thickness of volcanic conglomerate, tufas, and marls, with more or less of infusorial material intercalated. These tufaceous strata are, as Dr. Newberry remarks, cut by the Des Chutes and its tributaries to the depth of more than a thousand feet, without exposing the floor on which they rest. There is no reason to suppose, however, that — as has been often stated — there is a thickness of 1,200 feet of infusorial rock here. The tufas do indeed exhibit that amount of development, and with horizontal, undisturbed stratification. Dr. Newberry, however, does not say that the whole mass is infusorial, but that some of the finer varieties are. It would appear highly probable

---

* Report of Fortieth Parallel Survey, Vol. I. p. 419.

† Ehrenberg got, from the information sent him with the specimens, by the Fortieth Parallel Survey, the impression that the thickness of the infusorial strata was a thousand feet.

that here, as farther south on the Sierra Nevada, by far the larger portion of the mass is made up more or less of finely triturated volcanic material.

The number of localities of infusorial silica along the flanks of the Sierra, as well as in the Cascade Range and other portions of the Cordilleras, is so great that it would be impossible to enumerate them all. It is sufficient to have mentioned some of the most important, and to have called attention to their mode of occurrence.

In view of what has been said in the preceding pages, in regard to the former exaggerations of the thickness of the infusorial deposits at various localities in the Cordilleras, it will only be necessary to briefly refer to Ehrenberg's endeavor to account for this supposed enormous development of rocks made up of such extremely minute organisms. Observations made by him on the island of Ischia showed that, in that locality at least, an appearance of great thickness of the infusorial masses had been given by the carrying down from above and deposition, at the base of a cliff of volcanic rock, of material originating in a hot spring, the water of which fell from the edge of the rock to the bottom, thus giving rise to a conical deposit; which, although really of comparatively small dimensions, might possibly be mistaken for an interstratified mass of considerable extent. In the case of a very high mural face of rock, like those in the Des Chutes Basin, it would no doubt be possible that such deposits should have been formed on terraced edges or shelf-like projections which are often observed in lofty cliffs. This, however, does not seem to have been the case in any of the localities of infusorial material examined by the writer, or in those described by others as occurring in the Cordilleras. The appearances everywhere in that region are such as to decidedly justify the inference that the microscopic organisms were developed in the position in which they were found, once occupying the surface, and afterwards becoming covered by other layers of similar material, the result of successive growths, with intercalated non-fossiliferous deposits of greater or less number and thickness, according to the varying conditions prevailing at the time the mass was in process of accumulation.

In concluding this section, a few paragraphs may be added with reference to the cause of that constant association of the infusorial deposits with the volcanic formations, which observation has so clearly established. These remarks are quoted from the article previously referred to as having been published, in 1867, by the writer of this volume in the Proceedings of the California Academy of Sciences: —

The mode of occurrence of these fresh water infusorial deposits in California, and on the Pacific coast in general, is very simple. They are accumulations of organisms which have been collected at the bottom of the lakes, or in the lake-like shallow expansions of rivers, in which they grew. This growth took place at a time when volcanic agencies were busily at work, giving rise to accumulations of ashes, pumice, and other materials. The rapidity with which these infusorial deposits form, at the present time even, the vast extent over which they are distributed, and the general importance in the geological history of the earth of the masses thus accumulated and in regard to which the store of facts has been rapidly growing in magnitude during the past few years, are now matters which are well understood. The mud deposits and deltas of rivers, the bottoms of lakes and swamps, and the bed of the ocean itself, are the repositories of these forms. Heat and stagnant water seem to be what is required for their rapid reproduction and the consequent rapid accumulation of their remains.

The infusorial deposits of Central California — I refer now to those of fresh water origin, and connected with volcanic masses — are all situated in such positions as to show, that they were formed and deposited in shallow water; that, through the various alternations of calm and convulsion in the Sierra, they were at one time allowed to accumulate in quiet, then swept over by masses of gravel and sand, indicating a furious rush of water, then covered with a shower of ashes and pumice from the neighboring volcanoes of the Sierra then in active operation; and finally, at the grand finale of the basaltic lava overflow of the chain, capped with this indestructible material, which has effectually prevented the washing away of the otherwise easily removed infusorial deposits. This is the connection between the volcanic and the infusorial masses; by their absolute indestructibility the former have protected the latter from denudation, and consequently we see them always accompanying each other: for where the cover did not exist there the denuding forces have swept away every vestige of the soft and easily yielding material, or else it remains concealed under the water. To form an idea of the extent of the erosion which has taken place since these infusorial beds were deposited, and the consequent change in the configuration of the country, we must bear in mind that the whole of the present river cañons on the west slope of the Sierra have been excavated since that time, and that, in many places, the strata have been removed to a vertical depth of between two and three thousand feet.

Everything shows that the surface covered by fresh water in the region east of the crest of the Sierra was, at a not very distant epoch, much greater in extent than it now is. There existed, probably during or immediately after the glacial epoch, a chain of great lakes occupying a large portion of the country from Walker's Lake to the Des Chutes River, a distance of about four hundred miles, and extending over a breadth of not less than one hundred. A large portion of this region is now a volcanic plateau; and, where cut into by the force of running water, the deposits of infusorial strata may be seen, sometimes thin and unimportant, but often of great thickness. Observations and measurements of terraces and determination of the altitude of all these old lake deposits will enable us at some future time to indicate on the map the area once occupied by this great chain of inland seas. The vast extent of the lacustrine infusorial formations on the east side of the Sierra is thus accounted for, as well as the comparatively small area which they cover on the western slope.

In addition to the stratigraphical reason given above why the infusorial strata should occur connected with eruptive masses, there may be a chemical one which shall, in part, account for the apparent great development of the *diatomaceæ* in volcanic regions. These organisms require an amount of silica, infinitesimally small for each individual, but in reality enormous for the number of organisms required to develop themselves over the vast area and with the thickness which they occupy. That a volcanic region should supply a larger amount of silica in the state in which it can be appropriated by the *diatomaceæ* is extremely probable. We know that silicification of all organic matters occurring in these volcanic regions of our coast proceeds with the greatest rapidity, and has taken place on an extensive scale. The thermal springs contain a great amount of free silica, and it is in the vicinity of such springs that large infusorial deposits are frequently found. It seems that it could only be in regions particularly favorable for the secretion of their silicious coverings, that these infusoria could be accumulated with such rapidity as to form what may be called, without exaggeration, mountain masses. It is also possible that temperature may have something to do with this rapid development, and that volcanic regions may on this account be favorable to it.

To my apprehension, the phenomena of infusorial deposits in connection with volcanic masses admit of an easy explanation on this coast, at least; and I can hardly believe that any of the localities of *diatomaceæ*, if closely examined, would present any such difficulties as to make the assumption necessary that they have been ejected from the interior of the earth. In cases where infusoria seem to have been actually ejected from craters, as is said to have been the case in some of the South American volcanoes, it is not difficult to understand that an ancient crater may have become filled up and temporarily converted into a lake; and that, after the growth and deposition of an infusorial deposit at the bottom, a new eruption may have broken out in the same place as a previous one, or in its immediate neighborhood. In such a case, among the ejected material a large quantity of the infusoria would be found, mingled with the ashes, which must pass through the material collected in the bottom of the crater as they rise from the interior of the earth. The bursting of lakes at the bases of volcanic cones, caused by the rapid melting of the snows above them, have often given rise to torrents of volcanic mud, called "moya" in South America, in which both animal and vegetable remains are often enclosed in great quantity; but the connection between the organic and inorganic phenomena, in such cases, is perfectly evident.

SECTION III. — *The Fossil Plants of the Auriferous Gravel Series.*

As the fossil plants found in connection with the auriferous gravels is the subject of a special report, prepared by Mr. Lesquereux, and forming a portion of the present volume, it will not be necessary to devote any considerable part of this chapter to that branch of the subject. Some additional information, however, may with propriety be given in regard to certain points connected with the mode of occurrence, the distribution and the relative abundance of the remains of a former vegetation imbedded in the detrital deposits which have been described in the preceding chapter. Such general considerations as properly connect themselves with the climatic conditions prevailing at the time of the growth of the plants in question may, in accordance with the plan of this volume, be reserved for a future chapter.

From what has been stated in the preceding pages it will have been inferred that the remains of vegetable life, in the form of trunks of trees, impressions of leaves, and the like, are of common occurrence in the strata worked by the hydraulic mining process. This is indeed the case; for from the most southern to the extreme northern localities mentioned, in the detailed description of the gravel region, there seem to be but few districts where such remains have not been noticed. The material which could be placed in the hands of Mr. Lesquereux for description was, however, but meagre, in comparison with what might have been obtained, had it been possible to give the necessary time and labor to the collection and preservation of specimens in this department. In the first place, however, it should be mentioned that a very considerable number of excellent specimens were collected during the early years of the Geological Survey, and that most of them were destroyed by the fire which in 1865 consumed a large part of the material which had been gathered up to that time. There were other difficulties, however, in the way of procuring satisfactory suites of specimens even at localities where plant remains were abundant in the rocks. Everything passing through the sluices is, of course, broken up and destroyed; so that, in places where the hydraulic process is exclusively employed, it could not be expected that anything should be preserved except fragments of wood and trunks of trees too large to be handled by that method. Hence, most of our fine specimens were obtained at specially favorable localities, where tunnel mining was chiefly practised, and where the material excavated was brought out in cars and dumped on the surface before being thrown into the sluices. But even in such places it was usually necessary to be on the spot at the time the mining work was actually going on in order to procure valuable material, because the fine clays in which the delicate and perfectly preserved leaves are almost exclusively imbedded are liable to swell up and disintegrate rapidly on being exposed to the air. It was generally necessary to soak the specimens in glue or gum and cover them with varnish, without any delay, or they would soon be destroyed. This was the case particularly at the very prolific localities under the Tuolumne Table Mountain, and this was done with a large number of specimens, which unfortunately were destroyed, as mentioned above. Thanks to the zeal and patience of Mr. C. D. Voy, a fine suite was obtained some years later from the Chalk Bluffs, near Red Dog and You Bet, and these form the bulk of the material examined.

There is no limit to the amount of fossil wood which can be obtained in the various hydraulic mines of the gravel region. But it was not supposed that much valuable information could be obtained from the examination of the wood itself, especially when leaves, and even fruit, were comparatively abundant; hence no attempt was made to collect fossilized wood for any purpose, except to throw light on the manner in which such material has been preserved and the changes it has gone through subsequent to its being imbedded in the detrital masses.

It having been stated that remains of fossil plants are abundant in the gravel deposits, one of the first questions to be examined in this connection is: In what way are they distributed? Are they found in such positions as to indicate an effect produced by difference of altitude or of latitude? We know that the present vegetation of the Sierra Nevada, as is always the case in mountain ranges of considerable elevation, shows in the most marked degree the effect of change of altitude, and that the element of latitude also plays a part which is especially conspicuous in the distribution of the forest growth of that chain.

There are at the present epoch four pretty well marked belts of vegetation on the western slope of the Sierra. These belts, however, pass gradually into each other, and are not so defined that lines can be drawn exactly limiting their range. Still, in the central portion of the State the succession of the different groups of species as we rise in altitude is very easily recognized. The great predominance of coniferous trees is the most conspicuous feature of the arboreal growth of the Sierra. Even in the lowest belt, — that of the foot-hills, — which extends up to an altitude of between 2,000 and 3,000 feet, the pines are decidedly superior in number to the oaks, the digger-pine (*P. Sabiniana*) and the black oak (*Q. Sonomensis*) being the predominant species. With these two trees are mingled some quite conspicuous shrubs, some of which attain to considerable height. Of these the California buckeye (*Æsculus Californica*), the Manzanita (*Arctostaphylos glauca*), and the *Ceanothus* are the most striking, as helping to give some variety to what, on the whole, is a decidedly monotonous vegetation. The next succeeding belt, which ranges up to 4,000 or 5,000 feet, is peculiarly the forest zone of the Sierra, distinguished by the great size and beauty of the individual trees, but still without sufficient variety and intermixture of deciduous foliage to be beautiful as well as grand. The pitch pine (*P. ponderosa*), the sugar pine (*P. Lambertiana*), the white cedar (*Libocedrus decurrens*), and the Douglas spruce (*Abies Douglasii*) are the principal trees of this belt, in which also the Big Tree (*Sequoia gigantea*) belongs, although the latter is quite limited in its range. Above the zone of pines come the firs, *Picea grandis* and *amabilis*, with a considerable number of the pine characteristic of high altitudes, the so-called tamarack pine (*P. contorta*). This belt of firs ranges, in the Central Sierra, from 7,000 to 9,000 feet. *P. monticola*, in many localities, rather usurping the place of the firs at very high elevations, while a variety (as considered by some botanists) of *P. ponderosa*, *P. Jeffreyi*, is also of somewhat common occurrence. Above all the other trees comes *P. albicaulis*, or *flexilis* as it has been often called, which however occurs chiefly in scattered clumps, marking the upward limit of vegetation, but hardly forming a distinct zone.

If now we inquire whether any arrangement or order of succession, of a character similar to that just indicated for the present vegetation, can be traced in that of the auriferous gravel series, we have, in the first place, to confess to poverty of material necessary for giving a definite answer. Still, what evidence has thus far been collected is decidedly in favor of a uniformity of vegetation, rather than of variety or separation into zones, during the epoch of the gravels.

In illustration of this statement, the entire absence of any proof of the existence of coniferous trees during the gravel period may be brought forward. In none of the collections made, so far as the same have been examined by Mr. Lesquereux, have any traces of the existence of conifers been observed. Neither has the writer any recollection of having observed such in his investigations in the gravel region. Although further examinations may bring to light proofs of the former presence of coniferous trees in some portions of the gravel deposits, it seems almost certain that, if existing at all, they must have been very subordinate to the deciduous vegetation.

We have now to inquire in regard to the vertical range of the fossil plants collected and examined by Mr. Lesquereux. Although leaves have been observed in many localities, as has already been explained, extremely few of these have yielded material in sufficiently perfect state for description. By far the most prolific locality is that of Chalk Bluffs, and the one next to it in importance is the Tuolumne Table Mountain. These two places are separated from each other by a distance of somewhat over eighty miles; the former is in latitude 39° 12′, the latter in 38°. Bowen's Claim, the next locality in importance, is fifteen miles south-southeast of Chalk Bluffs. The difference in altitude between the two principal localities is, however, considerable, the plant-bearing beds of Table Mountain being about 2,000 feet in elevation, and those of Chalk Bluffs from 3,500 to 3,600 feet, the difference between the two being fully 1,500 feet. The fossil flora of the localities, thus seen to be considerably different both in latitude and altitude, would seem — from the investigations of Mr. Lesquereux — to be essentially the same. That is, nearly the same genera are represented, although the species are, in most cases, considered as differing. From the annexed tabular view of the species described by Mr. Lesquereux, in which the localities are designated, it will be seen that there are only two species common to both Chalk Bluffs and Table Mountain, but that ten genera are represented at both places.

Name of Species.	Chalk Bluffs.	Bowens'.	Table Mountain.
Sabalites Californicus			
Betula æqualis	*		
Fagus Antipofi			*
" pseudo-ferruginea	*		
Quercus elænoides			*
" convexa			*
" Nevadensis	*		
" distincta	*		
" Goepperti	*		
" Voyana	*		
" pseudo-lyrata	?		?
Castaneopsis chrysophylloides	*		
Salix Californica			*
" elliptica	*		
Populus Zaddachi	*		
Platanus appendiculata	*		
" dissecta	*		*
Liquidambar Californicum	*		
Ulmus Californica	*		*
" pseudo-fulva	*		
" affinis			*
Ficus sordida	*		
" tiliæfolia	*		
" microphylla			*
Persea pseudo-Carolinensis			*
Aralia Whitneyi	*		
" Zaddachi ?			*
" angustiloba	*		
Cornus ovalis			*
" Kelloggii	*		
Magnolia lanceolata	*		
" Californica	*		
Acer æquidentatum	*	.	
" Bolanderi			*
Ilex prunifolia			*
Zizyphus microphyllus	*		
" piperoides	*		
Rhus typhinoides			*
" Boweniana	?	*	
" mixta	*		
" myricæfolia	*		
" metopioides			*
" dispersa			*
Zanthoxylon diversifolium	*		
Juglans Californica	*		
" Oregoniana	?		?
" laurinea	*		
" egregia	*		
Cercocarpus antiquus			*

In regard to the geological age indicated by the plants in question, it is not necessary to do more than refer to what Mr. Lesquereux has said in Part II. of this volume. This subject will come up to some extent again in a future chapter; and the supposed influence of the glacial epoch in effecting changes in the flora of the Atlantic and Pacific slopes will also be discussed in its proper place, after the facts relating to the former presence of ice in the Sierra Nevada have been set forth. It is sufficient, at this time, to note the facts that the fossil flora of the gravel deposits is entirely different from that now prevailing in that region; and that it is considered by Mr. Lesquereux as Pliocene, while he admits at the same time that it is related by some identical or closely allied forms to the Miocene.

It will be proper to add some of the most important facts gathered during the investigation of the gravel deposits in regard to the mode of occurrence of the fossil plants of the Pliocene epoch. The vertical range of these has been alluded to, and it may be more distinctly stated that either fossil wood or leaves have been found at every elevation, from the lowest to the highest, where gravels occur. Even as high as Silver Mountain City, at 7,000 feet of elevation, large masses of fossil wood are found in the volcanic deposits; and in Plumas County the same occurrence has been noticed on several of the highest mountains in the region, as Penman's Peak and Clermont, peaks from 7,000 to 8,000 feet high. The impressions of leaves, of course, are chiefly limited to the finer clays and tufaceous beds, and in these are often preserved in the greatest perfection; except that while small fragments are abundant, whole leaves are very difficult to procure. Fragments and often large masses of wood are found, both in the gravels and the associated clayey and tufaceous beds. In the gravel they frequently bear the marks of transportation from a distance, as would be expected.

Much the larger portion of the wood found in the volcanic and gravel deposits is almost or quite completely silicified. This process has usually had, as its result, the conversion of the woody fibre into what is called by mineralogists, opal, — that is, quartz in the amorphous condition; for, although water appears to be almost or quite invariably present, yet it is regarded as not essential to the composition of the mineral. Specimens of opal, on analysis, are usually found to contain from six to ten per cent of water; rarely does the percentage fall as low as three. A piece of fossil wood from the Sierra Nevada, which was placed in the hands of Mr. S. P. Sharples for analysis by the writer, was found to have the following composition: —

Silica .	90.00
Oxide of iron .	3.27
Water, and organic matter .	6.29
	99.56

The opalized wood is often translucent, and sometimes almost transparent. It has usually a vitreous or almost resinous lustre, and is of various shades of gray or whitish-gray. Not unfrequently, however, it is light-green or even yellow, or of varying tints, the colored portions contrasting beautifully with darker bands of brilliant lustre. These latter seem to be the result of an incipient carbonization of the wood previous to silicification. This carbonization was evidently the beginning of a series of changes which, carried far enough, would have converted the mass into lignite or even coal, as has happened on so grand a scale in the older formations, but hardly at all in rocks so new as the Pliocene. Sometimes, however, in the gravel region, the wood is black throughout, and would appear to a casual observer to be considerably carbonized. As far as the writer's observations extend, however, most of these specimens are really silicified, and contain but little organic matter. Specimens of the almost colorless opalized wood do sometimes, it is true, blacken on exposure to heat, from the presence of organic matter, no doubt, and the naturally black portions do not burn white when exposed to the blowpipe flame. The carbon seems to be too thoroughly enveloped in the silicious material. In one remarkable instance the body of the dark-colored silicified wood was observed to be penetrated by large and well-formed quartz crystals, from a third to half an inch in diameter. These crystals, on examination with the microscope, proved to be partly chalcedonic in structure, the exterior or border, which was sharply defined on the outer edge, being chiefly quartz. The process of silicification has gone on, in the gravel and accompanying volcanic beds, on the most extraordinary scale. In some localities — as, for instance, at Chalk Bluffs — one walks over the hydraulicked area among the fallen silicified logs, which lie scattered pell-mell, as if they had been prostrated by a tornado. The well-known " petrified forest," in the Coast Ranges, near Calistoga, is another good instance of the same thing. In the latter case the silicified trees are imbedded in a dark andesitic ash; in the former, the enveloping material is mostly a fine rhyolitic tufaceous mass. At a locality in Nevada, near the California border, not far from Black Rock Cañon, a large silicified log was observed by Mr. C. F. Hoffmann, resting on the surface, in a region where at present there is no forest growth at all. This log was four or five feet in diameter, and more than a hundred feet of it was exposed to view, — a remarkable witness of the great climatic and geological changes which have gone on in this region during the latest epoch.

The most striking kind of fossil wood occurring in the mining region is the fibrous variety, of which beautiful specimens are occasionally found. Some of these have the most delicately fibrous structure, somewhat resembling raw silk both in texture and in color. The fibres are often very long and straight, like those of the finest asbestus. This peculiarity evidently has nothing to do with the original structure of the wood. The fibres have been formed, in the process of the replacement of the ligneous particles by silica, in a manner analogous to that which is not unfrequently observed in minerals and rocks which have undergone or are undergoing metamorphic changes. Chrysotile, the fibrous variety of serpentine, and asbestus, a similar form of hornblende, are good instances of this kind of structure in simple minerals, where there is no reason to suppose the preëxistence of anything organic to bring about this peculiar arrangement of the particles.

It is not rare to find that portions of the wood occurring in the gravel deposits are mineralized with sulphuret of iron. This substance may form the whole mass of the fragment, but is sometimes deposited in crystalline plates in the fissures, or as a crust on the exterior.

The following particulars in regard to the occurrence of fossil wood and leaves in the region explored by Mr. Goodyear are extracted from his notes.

At the Reed Mine, near Deadwood, at the elevation of about 3,600 feet the gray cement, as well as the so-called " chocolate," — a sort of indurated volcanic mud, — often contains more or less apparently half carbonized wood. And, besides these fragments of wood, numerous trees have been found here, still in their upright position, with their roots upon and ramifying into the bed-rock, and their stems and tops projecting through the "chocolate" and up into the "gray cement." The trees thus found standing are not very large, none of them being much over a foot in diameter. All had their roots on and in the bed-rock, and none in the "chocolate." This last-named material contained also occasional impressions of leaves, very perfectly preserved. At Weske's Claim, near Michigan Bluff, where fragments of trees and wood are common in the cement, some trees are also said to have been found in an upright position, with their roots in place on the bed-rock. In the Basin Channel, at the Devil's Basin, Mr. Goodyear saw, at one locality, a fossil tree standing vertically in the "gray cement."

Fossil wood is of frequent occurrence in the finer material of volcanic origin which overlies the gravel, in the claims along the western side of the ridge near Deadwood. At Kentucky Flat, also, there is considerable wood in the volcanic ash. In the Oldfield and adjacent claims at Negro Hill, near Placerville, the whole mass of the "black lava" contains frequent casts of sticks and broken bits of wood, and sometimes, also, delicately preserved impressions of leaves, which latter are generally more or less bent or twisted.

The village of Fairplay is just south of Perry's Creek, a tributary of the Middle Fork of the Cosumnes River. There is a large amount of volcanic gravel in this region, and at Fairplay considerable mining has been done. It is said that in the volcanic gravel on Fairplay Hill there was once found a "natural shaft" three or four feet in diameter, with its sides covered with a substance resembling soot. This was probably the cast of an erect fossil tree.

At the Roanoke Mine, near Georgetown, large quantities of wood were found in the cement; and it is said that it was a common thing to observe here, along the sides of the channel, at about the level of the top of the gravel, which was about two feet thick, stems of bushes and grass-roots, forming bogs, all standing in the place where they grew, the stems of the bushes being sometimes two or three feet high, although the tops were generally gone. The grass is said to have been well preserved, and the form and outlines of the whole thing unmistakable.

In a hydraulic pit opened at Kentucky Flat, two or three miles southeast of Mount Gregory, there is a stratum exposed which is very full of fossil wood, a part of which is carbonized and a part petrified by iron pyrites. The sticks are very numerous, and the texture of the wood is generally well preserved.

In Bear Hill, at Diamond Springs, fossil wood has been very plenty, and is generally converted into semi-opal.

SECTION IV. — *The Animal Remains, not Human, of the Auriferous Gravel Series.*

The reasons which have been given in the preceding section for finding the fossil plants of the gravel beds in a more or less fragmentary and imperfect condition apply with equal force to the animal remains which have been imbedded in these deposits. The collecting of fossil bones has been, in some respects, an exceedingly unsatisfactory business. No locality has, as yet, — so far as the writer has observed, — ever been discovered in the hydraulic mining region, where animal remains seemed to occur in the position in which they had been left at the time of the death of the individual. The only entire skeletons which have been observed were found quite low down in the foot-hills, where conditions were more favorable to their preservation than was the case higher up among the gravelly beds. Still, a considerable amount of material has been obtained, and enough, with the aid of the plants, to throw considerable light on the age of the formations in which they occur. The presence of the works of man and of human bones at various points, however, has made the geological relations of the strata in question an object of very great interest, so that it becomes desirable to lay before the reader as complete an account as possible of what has been observed, and what has been learned from the miners themselves, in regard to their discoveries. The testimony of uneducated men, not accustomed to take into consideration all the conditions necessary for insuring accuracy of observation, must, of course, be taken with caution. The statements of one person must be weighed against those of another, and from the whole body of evidence thus brought together and compared inferences may be drawn, and even definite conclusions arrived at, which shall have all the weight of authority; since it is not likely that a considerable number of persons, at different times and places, each knowing nothing of the other's doings, should make exactly the same misrepresentation.

The miners, in general, do notice the occurrence of animal remains in their workings with some attention, and this for several reasons; chiefly, however, because the bones found are often of great size, and so make a strong impression on their imaginations. It may easily be conceived that the skull of a mastodon, or even a single tooth of this animal, is not an object likely to be passed by without notice, even by the most ignorant miner. Indeed, so far as our observation goes, these men almost always set a wonderful value on such specimens, and are often unwilling to part with them at any price. Hence it is that we, in some instances, have been obliged to be content with a cast or even drawing of some unique bone or tooth, while in other cases the object could only be borrowed for a short time, — just long enough for study and comparison.

Nearly all the animal remains obtained from the gravel deposits belonging to the vertebrates, our specimens have, almost without exception, been referred to Dr. Joseph Leidy for examination and description. This was done from time to time during the continuance of the Geological Survey, and a portion of his results have been given in the volume of Geology published in 1865. Among the material submitted to this eminent authority in vertebrate palæontology there was nothing of human origin. In regard to objects belonging to this latter class, advice and information was sought from Dr. Jeffries Wyman, who was made acquainted by the writer with every step in the progress of the discovery of evidence bearing on the antiquity of man in California. None of this, however, has been published, up to the present time, and it will be given in a body in the next section of this chapter. Dr. Leidy's results have been in part communicated to the scientific world in Geology Vol. I., as already mentioned, and the remainder will be found included in the elaborate volumes published by him in 1869 and 1873.* All that it will be necessary to do, at the present time, in regard

* The Extinct Mammalian Fauna of Dakota and Nebraska, including an Account of some Allied Forms from other Localities. By Joseph Leidy, M. D., LL. D. Philadelphia. 1869. Contributions to the Extinct Vertebrate Fauna of the Western Territories. By Professor Joseph Leidy; the same being Vol. I. of the Report of the U. S. Geological Survey of the Territories. Washington. 1873.

to Dr. Leidy's work, will be to bring this scattered material together, accompanying it with such purely geological comments as shall seem required for enabling the reader to get the most light possible on the question of the geological age of the formations rendered so interesting not only by their richness in gold, but by the presence in them of human remains.

From what has been said in the preceding chapters, it will be apparent that one of the first and most difficult questions to be settled in the discussion of the age of the auriferous gravel series will be, Where is the line to be drawn between the epoch of the present and that of the past? The whole body of material with which we have to do is, geologically speaking, of very recent age. There is evidently a passing, as we trace the progress of events backwards, from circumstances and conditions actually existing to quite different ones. Yet the striking analogy between former conditions and those now prevailing cannot but have been impressed on the mind of the reader of this volume. The gravels were then as now the result of fluvatile action: the rivers which did the work of rounding and polishing the innumerable boulders and pebbles which those older deposits contain are doing the same thing now, although with diminished power. The very channels in which those former currents ran have, in most cases, their representatives now; at a lower level, and on a diminished scale, it is true, but still essentially the same, since modern and ancient streams do not probably differ very much in their areas of drainage.

What does essentially distinguish the present epoch from a former one is the cessation of volcanic energy. For it has been abundantly shown in the preceding chapters that, during the formation and deposition of a portion of the auriferous detritus, the gravel region was the scene of powerful and persistent eruptive action, the seat of which extended through the whole of the Sierra. The volcanic vents were undoubtedly high up in the range, in the gravel region at least, as will be evident to any one examining the large map appended to this volume, as well as from the description of the position of the various detrital masses which has been given in the preceding pages. This peculiar geological phase of events has now entirely passed by, and that a very considerable period of time has elapsed since volcanic agencies ceased their activity, at least over all the mining region of the Sierra, cannot be doubted. The fact is demonstrated by the large amount of erosion which the most recently erupted masses have undergone, and which has been repeatedly mentioned and commented on while describing the gravels and associated volcanic rocks. The extent and nature of this erosion, and the character of the agencies by which it has been brought about, will be still further discussed in a future chapter, and the subject is only alluded to here, on account of its intimate connection with the determination of the position of the various fossils occurring in the formations described.

It so happens that the close of volcanic action in the part of the Sierra now under discussion seems, almost beyond a doubt, to have been marked by the eruption of a kind of lava differing from those which had been previously emitted from the interior. The material referred to as, in a measure, closing the volcanic epoch, is the basalt; which, wherever the series is complete, is found overlying the rhyolitic and andesitic flows, and over which no extensive deposits of gravel, so far as known to the writer, have ever been found.

A peculiar character of the basaltic masses is their solidity and unbroken condition. The underlying volcanic masses are very largely — indeed, almost exclusively — brecciated and fragmentary; and it is clear that they have been in large part brought into their present position by the agency of water. The basaltic flows, on the other hand, are exclusively igneous: they rest where they descended in a molten condition. This solidity of the basalt connects itself, also, with its indestructibility. The material is very slowly acted on by the ordinary erosive agencies, as has already been explained in describing the way in which it has protected the underlying infusorial strata. Hence it forms the capping of the various "Table Mountains," of which class of flat-topped elevations the one in Tuolumne County is so remarkable an example.

As a consequence of the conditions above described, we are justified in feeling confident that deposits found under the basaltic masses remain in their original position; or, at least, that they must have been accumulated in the place where we find them previous to the cessation of the period of volcanic activity. The gravels which have not been protected by a capping of basalt, if only thinly or not at all covered by erupted materials, may in some places have been overlain by recent deposits in such a way that the line between volcanic and post-volcanic cannot be distinctly drawn. Mention has already been made, in the preceding pages, of cases where slides have taken place on the steep walls of cañons, mixing older and more recent formations in inextricable confusion. And it must not unfrequently have happened that fossils have been washed out of the less coherent detrital beds, belonging to the volcanic series, carried far from their original resting-place, and deposited in such a position that they seem to belong to the present epoch.

In view of the above considerations, it seems reasonable, in endeavoring to find a line of demarcation, as indicated by fossil remains, somewhere in the gravel and volcanic series, to inquire what is positively known to have been found in strata lying undisturbed under the basalt; or, at least, so far down in the volcanic formations as to preclude any possibility that the object in question could have found its way down from a superficial deposit of post-volcanic age. It will be well, therefore, in the first place, to take up the material collected by the Geological Survey, and investigated by Dr. Leidy, and ascertain what has been discovered unquestionably older than the epoch of the eruption of the basalt.

The number of species in regard to which the evidence is clear that they are prior in age to the eruption of the basalt is not large. The most important localities are those of Douglass Flat, Chili Gulch, and the Tuolumne Table Mountain, the position of which has already been sufficiently explained. From Douglass Flat and Chili Gulch remains of a species of rhinoceros were obtained, which have been described by Leidy under the name of *R. hesperius*.[*] In both cases the specimens exhibited a considerable portion of the jaw with a number of teeth. That from Chili Gulch consisted of the right side of the lower jaw, without the ascending portion, and with the symphysial portion of the opposite side. It contained the true molars, the fangs of four premolars, one lateral incisor, and the fang of the other, and the alveoli of the internal incisors. The form of the jaw is nearly like that of the corresponding portion of the Indian rhinoceros, and the formula of dentition the same as in that species. The size of the species, as indicated by the jaw, is nearly that of *R. occidentalis*, a common species of the Mauvaises Terres, described by Leidy, and considered by him to be from half to three fourths as large as the living Indian species. Of the Chili Gulch jaw Dr. Leidy remarks: "The specimen so closely resembles in its general aspect and state of petrifaction the Mauvaises Terres fossils of White River, Dakota, that it would have been viewed as one, if the locality from which it was obtained were not known."

Of the other specimen, from Douglass Flat, Dr. Leidy's notes, furnished the writer, give the following account: "This second specimen is remarkable on account of its condition of preservation, so totally different from the other. It consists of a portion of the left ramus of a lower jaw of a young

[*] Extinct Mammalia of Dakota and Nebraska, p. 230.

animal containing molar teeth. These are the last pair of temporary molars, and the succeeding pair of permanent true molars in functional position, and the crown of the last true molar visible through a fracture within the jaw. The teeth are tolerably well preserved, though generally much fissured, and appear completely fossilized. The specimen is partially imbedded in a mass of coherent auriferous gravel. All distinct appearance of ossific texture in the jaw fragment is obliterated, and it looks as if it had been a layer of plastic clay rudely and very roughly modelled into the form of the portion of the jaw containing the teeth, and mingled with some of the gravel in which it is imbedded. The size and form of the corresponding parts agree with the upper jaw first described [that from Chili Gulch], so as to render it probable that the specimen belongs to the same species."

A portion of the Chili Gulch jawbone, from one of the condyles, was analyzed at the request of the writer by Mr. Sharples, with the following results:

Phosphate of lime	49.40
Carbonate of lime	18.33
Fluoride of calcium	4.77
Silica	22.70
Oxide of iron	4.58
Magnesia	Trace
	99.78

The Douglass Flat locality also furnished a unique specimen, a single tooth, which is thus described in Leidy's notes: "Not less remarkable than the discovery of the former remains [the rhinoceros jaws] is the finding of an incisor tooth, apparently indicating a species of the extraordinary pachyderm *Elotherium*. The incisor appears to be the right lateral one of the upper jaw, and may perhaps belong to one of the species detected in the Mauvaises Terres of White River, Dakota, though from its size I suspect it belongs to a larger species than either of them. The crown of the tooth is conical, compressed from within outwardly, and subacute laterally. The apex is rounded; the base somewhat expanded, and at its fore part produced in a short embracing ridge. The fang is conical and curved. The tooth is twenty-nine and a half lines long; the crown thirteen lines; its breadth nine lines, and its thickness six and a half lines."

This was considered by Leidy as probably indicating a new species, and named by him *Elotherium superbum*,* and of it he says, "Perhaps the same as

---

* Extinct Mammalia, &c., p. 388. Also Proc. Philadelphia Academy of Natural Sciences, 1868, p. 177.

E. ingens, of the Mauvaises Terres of White River, Dakota, although it would appear to belong to a larger individual than the remains referred to the latter, if not to a yet larger species." No trace of this animal has been found in the gravel deposits except the above-described tooth. A number of teeth, however, were obtained by Rev. T. Condon in the valley of Bridge Creek, a tributary of the Columbia River, which were referred by Dr. Leidy to the genus Elotherium, and described under the name of *E. imperator*.* Of these specimens it is said that it "is not improbable that part or the whole pertain to the species named *Elotherium superbum* from an isolated incisor tooth found in Calaveras County, California."

Another species is represented by a single fragment of a tooth or tusk belonging to some large animal, which is said to have come from the Buckeye Tunnel, under Table Mountain, the position of which has already been described. From its appearance and partially fossilized character it would seem probable that it was obtained from this or some other position under the lava. The fragment is about two inches long and consists of a portion of a tooth split through the centre, its diameter being about an inch and a half. Of this specimen Dr. Leidy only says in his notes: "Uncertain. Apparently the fragment of an incisor or canine of some large pachyderm, not the mastodon or elephant, and probably allied to the hippopotamus."

The analysis of this specimen, by Mr. Sharples, gave the following results: —

Phosphate of lime	67.89
Carbonate of lime	8.74
Fluoride of calcium	16.97
Oxide of iron	2.34
Water and organic matter	3.03
	98.97

The amount of fluorine given above is quite remarkable, as it considerably exceeds that previously obtained in the analyses which have been made by chemists of the teeth and tusks of various animals.

The only other animal remains which have come into the possession of the writer, appearing without doubt to have come from deposits positively prior in age to the basalt, are such as have also been found in an uncertain position: that is to say, where it is impossible to tell whether they were or were not deposited prior to the cessation of eruptive action in the Central Sierra. It will be proper, in the first place, therefore, to enumerate all the species

* In Contributions to the Extinct Vertebrate Fauna of the Western Territories, p. 247.

which have been found, or which have come into the writer's possession with the statements of others as to their being found, in the various detrital deposits in and adjacent to the mining region of the Sierra. The names of the species will be given in each case; then such remarks as have been furnished to the writer by Dr. Leidy, in the form of notes on the specimens submitted to him at various times during the progress of the Geological Survey. Some additional information obtained from other sources will occasionally be added; and, especially, such as may be found in the "Extinct Mammalia of North America" and in the Proceedings of the Philadelphia and California Academies. The order followed will be that of the first-mentioned work of Dr. Leidy. Under each species indicated will be given such facts as have been obtained with regard to its locality and stratigraphical position. And, after all have been mentioned, the various occurrences will be classified, as far as possible, and the attempt made to ascertain what geological age should be assigned to the beds in which these remains occur.

Of the *Carnivora*, none of the *Felidæ*, so far as known to the writer, have been met with in the Sierra Nevada, but the remains of one species at least have been observed in the Coast Ranges. They were discovered by Dr. L. G. Yates, at a locality near Livermore Valley. The specimens consist of fragments of the jaw, and have been described by Leidy under the name of *Felis imperialis*.* Of the *Canidæ*, a portion of the lower jaw of a wolf was obtained from the locality last mentioned, and considered by Leidy to be probably identical with *C. Indianensis*,† a fossil species described as occurring with *Megalonyx* on the banks of the Ohio River; although possibly not different from the existing *C. occidentalis*. A tibia obtained by the Geological Survey, probably from Murphy's and certainly from the auriferous gravel, in a perfect condition and somewhat fossilized, was referred by Leidy to *C. latrans*, as being of about the size and form of the corresponding bone in the prairie wolf.‡

The remains of the *Bovidæ* are of somewhat frequent occurrence in the detrital deposits of the mining region, and they have also been found in the Coast Ranges. Dr. Leidy says:§ "Remains of large oxen which were contemporaneous with the American mastodon have been discovered in several

---

* Contributions to the Extinct Vertebrate Fauna of the Western Territories, p. 228.

† l. c., p. 230.

‡ Dr. Leidy considers the formation in which these fossils (the tiger and wolf) were found to be "quaternary." There is abundant reason for classing them as Pliocene.

§ Contributions, &c., p. 253.

parts of North America. They have been referred to several extinct species, but the materials have been too incomplete to determine the question with any degree of satisfaction whether they pertain to more than one. The fossils indicate individuals very greatly differing in size, but the difference is perhaps sexual rather than specific."

The finest specimen ever discovered in California was found in Pilarcitos Valley, a few miles south of San Francisco, imbedded in blue clay, twenty-one feet beneath the surface. It has been figured and described by Leidy, under the head of *B. latifrons*.* It consisted of the cranial portion of a skull, with both horn-cores. Two fragments of the lower jaw of *B. latifrons* were obtained in connection with remains of elephant and mastodon, near Millerton, Fresno County.† Of these specimens Dr. Leidy's notes give the following description: " One of the fragments is a portion of the left ramus of the lower jaw containing part of the second and the last true molars. The teeth have advanced to that position that the crowns have wholly protruded. They agree in their proportions with those represented of *B. latifrons*, which probably is the male of the same species [*B. antiquus*]. The other fragment is a portion of the right ramus, and contains parts of the hinder pair of premolars and the succeeding pair of true molars. Measurements derived from the specimens are as follows:—

Length of the true molar series	56 lines.
Fore and aft diameter of last true molar	24 "
Depth of jaw below middle of last true molar	36 "
Depth of jaw below middle of first true molar	28 "

Another specimen, which came from Kincaid Flat, near Sonora, consisted of two inferior molars, with the dentine decomposed and in a fragmentary condition, and was considered by Leidy as belonging to *B. latifrons*. It was stated to have been found at a depth of eighteen feet in the auriferous detritus of that locality, which has already been noticed.

Still another specimen, referred to the same species, consisted of a lower molar, in a similar condition to the last. It was obtained at Saw-mill Flat, not far from the last-named locality; and was stated to have been found at twenty feet in depth beneath the surface. Other specimens representing the same animal have been obtained at various points in the Coast Ranges.

An interesting group of fossils was discovered by Mr. C. D. Voy, when ex-

* Contributions, &c., p. 253, and Plate XXVIII.

† The locality will be more exactly designated under the head of the Elephant.

ploring for the Geological Survey, on a nameless dry creek tributary to Bear Creek, in Merced County, near the line of Mariposa, about six miles southwest of Indian Gulch. The rocks at this place consist of a coarse, friable, light-colored volcanic ash, which envelopes a large quantity of bones, and also contains the remains of vegetation, and especially the casts of some small fruit, or seed vessel, the relations of which have not been made out.*

The most striking of the bones found here were those of an extinct lama, much larger than the ordinary camel, of which several fragments were obtained. With these were associated bones of the deer, and those of one or more species of horse, together with others which could not be determined. The following statement in regard to these interesting relics was furnished by Dr. Leidy: †

1. A metacarpal bone of a ruminant of large size. In form and construction it bears more resemblance to that of the lama and camel than of other ruminants with which I have the means of comparing it. As in the lama and camel the lower articular extremities are divergent, and the articular surfaces are provided with a median ridge only at the back part. In ordinary ruminants, as in ox, deer, sheep, &c., the median ridge is produced the entire extent fore and aft of the articular surfaces. The peculiar arrangement in the extinct animal, as in the lama and camel, allowed a greater spread or divergence of the toes in the extended condition. The fossil bone is nineteen inches long; the breadth of its proximal end is three and a half inches, of its distal end four inches. In the skeleton of a camel in our Museum the corresponding bone is thirteen inches long.

2. The distal extremity of another metacarpal of the same animal.

3. The proximal end of a femur, probably of the same animal, with the head of the bone three inches in diameter. An acetabulum of corresponding size appears to have belonged to the same individual.

4. Two fragments of a tibia of probably the same animal.

The bones mentioned probably represent a large extinct species of lama, which may be distinguished with the name of *Auchenia Californica*. Perhaps the fossils represent a distinct genus, allied to the lama, but this is a question that can only be determined by the discovery of other and more characteristic remains of the animal.

5. A first phalanx, in the collection, resembles in form that of a represent-

* This locality is of limited extent; but deserves a thorough exploration.
† See Proceedings Phila. Acad. Nat. Sci., 1870, p. 125.

ation of the same bone in the lama, and is about the size of that in the camel. Perhaps it belonged to a small individual of the preceding extinct form; probably to a smaller species. It is three and a half inches long, one and a half wide at the proximal end, and one and a quarter wide at the distal end. The articulation of the latter is not expanded beneath, as in the camel, for the apposition of the sesamoid bones.

6. The proximal three fourths of a metacarpal, probably of a deer. It is of rather more robust proportions than the corresponding bone of the Virginia deer.

7. An incisor tooth of a small horse, partially imbedded in a coherent mass of gravel, which also contains the impress of a nut-like fruit.

8. Portion of the tibia of a small horse, probably pertaining to the same individual as the tooth just mentioned.

9. The lower extremity of a metacarpal, probably of the same horse. It is proportionately thicker and less wide than in the corresponding bone of the domestic horse. The articulation is one and a half inches wide, and sixteen lines fore and aft at the median ridge. The equine remains perhaps belong to a *Hipparion*.

10. A few fragments of undetermined bones of other animals.

Among the collection of fossils in the cabinet of Wabash College, purchased of Dr. Yates, and submitted to Dr. Leidy for examination, there is a well-preserved series of lower molar teeth,* which from their size and constitution would appear to belong to a species of lama exceeding in size not only the existing lama, but also the camel and *Palauchenia*. The question then arises, as Dr. Leidy remarks:† Whether these teeth belong to *Auchenia Californica*, *Palauchenia magna*, or to a third species? He adds: "The proportions of the bones upon which the former was founded indicate an animal one third larger than the camel, but the teeth above noticed might belong to an animal but little exceeding a large camel or the *P. magna*. If the characters assigned to the latter as a genus are correct, it is clear that the series of teeth from California do not belong to the same animal, and they then could only pertain to a small individual of *Auchenia Californica*, or to another species rather larger than the existing camel. Under the circumstances, until further light is thrown on the subject by the discovery of additional material, we may suppose that two large species of lama, perhaps exclusive of *Palauchenia magna*, were once inhabitants of the western portion of the North American conti-

---

* Figured in Contributions, &c., Plate XXXVII. Figs. 1, 2.	† l. c., p. 256.

nent, contemporaneously with the *Mastodon Americanus*. One of these species, a third larger than the existing camel, is the *Auchenia Californica*; the second, intermediate in size to the two latter, may be named *A. hesterna*."

The locality of Dr. Yates's specimens of *A. hesterna* is Alameda County.

A single tooth, in excellent preservation, was seen by the writer in the collection of Dr. E. S. Snell of Sonora, and by him stated to have been found in the auriferous gravel in the vicinity of that place. It proved, on examination, to be the last inferior molar of a large ruminant probably allied to the camel. It resembles a tooth of corresponding size, but a penultimate molar, from the Niobrara River, for which Dr. Leidy proposed the name of *Megalomeryx*.[*] Of this genus Dr. Leidy remarks: "The genus to which the above name was applied has not been determined by positive characters, and may prove not to be distinct from *Procamelus*."[†] This is a remarkable instance of the proof of the existence of an interesting species — once probably abundant on the Pacific Coast — dependent on the accidental preservation of a single tooth; for this seems to be the only specimen of the kind found east of Nebraska.

Of the *Capridæ* it is not known to the writer that any remains have been found in the Sierra. A single specimen of a large sheep-like molar, of uncertain reference, was among the specimens submitted by the writer to Dr. Leidy for examination. The tooth was in a very broken condition. It came from near Centerville, Alameda County, and was collected by Dr. Yates.

Of the *Cervidæ*, the only fossil remains known to the writer as having been found in the Sierra are those mentioned in connection with the notice of the bones of the lama discovered in Mariposa County and a single metatarsal bone, of a species of *Cervus*, smaller than the Virginia deer. This latter was found in the gravel near Murphy's at an unknown depth.

Two lower molar teeth of an animal not differing from the living American tapir were also seen in Dr. Snell's collection; and were said to have come from the gravels in the vicinity of Sonora.[‡] These also are unique specimens, no other evidences of the existence of the fossil tapir anywhere west of the Mississippi Valley, although the remains of this animal have been found in many localities in the Eastern States. As Dr. Leidy remarks: "Seeing that different known species of tapirs exhibit little or no difference in the

[*] Proceedings Phila. Acad. Nat. Sci., 1858, p. 24.

[†] Contributions, &c., p. 260.

[‡] Casts of these teeth were taken and submitted to Dr. Leidy for examination. They were the same specimens afterwards examined by Professor Owen of London. See Am. Jour. Sci. (2), XLV., 381.

parts corresponding to the fossil specimens just indicated [teeth and jaw fragments], it is not improbable that these really belong to an extinct species."

We come next to the most abundant and widely disseminated of all the animal remains found in California, namely, those of the *Proboscidea*, including both the mastodon and the elephant. The widespread occurrence of these animals on the American continent during the later Tertiary times is a fact well known to all naturalists, and the literature relating to this subject is already very extensive.* It will be impossible to enumerate all the localities in California where the remains of the mastodon and elephant have been observed; the most that can be done will be to mention some of the most important facts connected with their occurrence.

In the first place, as to the mastodon, the remains of which animal in California are decidedly more abundant and more widely distributed than those of the elephant. On the map given in Murray's "Geographical Distribution of Mammals," to illustrate the range of the fossil proboscideans during the Pliocene period, the whole of North America is colored to indicate their existence, excepting the extreme northeast, along the coast of Labrador and farther north. It would not be correct, however, for California, to extend the range of either the mastodon or the elephant over the whole of the Sierra. So far as known to the writer, the bones of these animals have not been found high up on the mountains. Their habitat seems to have been in the Coast Ranges, and along the foot-hills of the Sierra, up to an elevation little, if at all, exceeding 3,000 feet. By far the larger number of the remains which have come under the writer's notice have been at elevations little exceeding 2,000 feet; while, as before remarked, the skeletons which have been found under such circumstances as to indicate that they had not been moved since the individual's death are limited to the region at the base of the foot-hills. The portion of the State where mastodon remains have been found in the greatest abundance is that in the vicinity of Sonora and Columbia, along the limestone belt, so often alluded to in the preceding pages. Cart-loads of mastodon bones, as has been repeatedly stated to the writer on good authority, have been accumulated at various places between Sonora and the Stanislaus River at the workings in the limestone crevices. Most of these have been destroyed by the inevitable fires which periodically consume the mining towns, and those not burned have generally crumbled to pieces for

* In Dr. Leidy's article on the Mastodon, in the Extinct Mammalia of North America, there are four solid quarto pages, in fine print, of references to authors who have written on the *Mastodon Americanus*.

want of the necessary care. Some fine skulls, nearly perfect, were still in existence in the miners' camps, near Sonora, when that region was first visited by the writer; but they were not to be bought at any reasonable price. Occasionally such specimens are carried to San Francisco and placed on exhibition, always at a great pecuniary loss to the exhibitors. Two fine heads from Horseshoe Bend, on the Merced River, were thus exhibited some years since, the owner believing them to be of fabulous value. So far as known to the writer, they did not find their way into any public museum, where they would be likely to be cared for and preserved.

By far the larger number of specimens collected in California or elsewhere in the United States belong to the common, well-known species, *M. Americanus*. But besides this, there appear, according to the latest expressed opinion of Dr. Leidy,* to have been at least three others which inhabited this continent: these are *M. mirificus*, *M. Andium*, and *M. obscurus*. Of the first of these, remains have been found by Dr. Hayden, in association with an abundance of those of other species of extinct animals, in the Pliocene formation of the Loup Fork of Platte River. Remains apparently identical with the South American *M. Andium* have been discovered in Central America, and the fourth species, *M. obscurus* of Leidy, is known only by specimens collected in California and New Mexico. For a full account of the relations of these different species, their mode of occurrence, and their distribution, many different works must be consulted. The species which interests us particularly, next to *M. Americanus*, is *M. obscurus*, and the specimens of this which have been found in California are described and figured in Dr. Leidy's "Contributions, &c."† There are two localities, so far as known to the writer, where the remains of *M. obscurus* have been found, one in the Coast Ranges and one in the foot-hills of the Sierra. The Coast Range locality is Oak Springs, Contra Costa County; the other is at Dry Creek, in Stanislaus County. Both of these were discovered by Dr. Yates. There is also a cast in plaster of a mastodon tooth in the Museum of the Philadelphia Academy, the original of which is supposed to have been found in the Miocene of Maryland. The specimen from the foot-hills of the Sierra was at first considered by Dr. Leidy as distinct from the one obtained in Contra Costa County, and was described by him under the name of *M. Shepardi*; but he afterwards included it with *M. obscurus*. He says, in reference to this matter and to the

* Contributions, &c., p. 231.

† l. c., pp. 231–237, and Plates XXI. and XXII.

relationship of the species: "I think it probable, without being positive in the matter, that the mastodon remains above described [those discovered by Dr. Yates] which have been referred to species under the names of *Mastodon obscurus* and *M. Shepardi*, including those from New Mexico, belong to one and the same species. This, from the form of the molar teeth, the constitution of the upper tusks, and the prolonged symphysis of the lower jaw, was clearly a near relation of the *Mastodon angustidens* of Europe."

The fossil remains of *M. Americanus* are widely scattered over the State, as already mentioned, and appear to be almost as common on the west side of the Great Valley as on the east, although there seems to have been no one locality discovered as yet in the Coast Ranges where such a quantity of bones were heaped together as were found at and near Gold Springs, a little west of Sonora. It would be natural to expect, however, that the finds would be more numerous — other things being equal — where there was the greatest activity in making artificial excavations. The seeming relative abundance of fossil remains in the mining region may be simply due to the fact that the soil and gravel have there been so thoroughly worked over with pick and shovel.

The geological range of the mastodon seems — so far as present evidence goes — to have been greater than that of any of the extinct mammals found in California. While far more abundant in the gravels which are not covered by volcanic deposits, and which therefore, as already shown, may be of somewhat uncertain age, they have also been found in deposits which are covered by the basalt. In the ordinary gravels, not so covered, the mastodon has been found at all depths, from a few feet up to a hundred or more. There seems also to be abundant evidence that the remains of this animal have been met with in the excavations under the Tuolumne Table Mountain. A tooth of *M. Americanus* was also found at a depth of forty-eight feet beneath the surface, at Douglass Flat, according to Mr. A. Jaquith, a careful and trustworthy observer. This tooth, however, did not appear as thoroughly fossilized as the rhinoceros jaw from the same locality, which was said to have come from a great depth in the gravel, there in places probably over 200 feet in thickness. The weight of the evidence, thus far collected in California, is certainly in favor of the mastodon's having been more persistent than any other of the animals of the gravel period. But, as more will have to be said in regard to the occurrence of this animal in connection with the discovery of human remains and works of art in the same association, further remarks on this point may be deferred until the next section of this chapter.

Passing next to the elephant, it may be said, in general, that its remains, like those of the mastodon, are widely distributed over the State, occurring both in the Sierra and in the Coast Ranges. In point of relative abundance, the mastodon seems, however, to have the advantage. A number of specimens, collected by the Geological Survey, were submitted to Dr. Leidy for examination. Among them was an upper last molar tooth of the right side, obtained at Murphy's, in Calaveras County, at a depth of about thirty feet in the auriferous detritus overlying the limestone. Several other fragments of teeth were also obtained, all from a moderate depth in the gravel. Of these specimens Dr. Leidy remarks in his notes: "All the elephant remains from California, consisting of molar teeth and fragments, belong to the coarse-plated variety, referred by Falconer to a species with the name of *E. Columbi*. While I do not deny the probability of the latter being distinct from *E. Americanus*, a number of teeth in the Museum of the Academy exhibiting transition forms lead me to view it only as a variety."

The finest specimen of the fossil elephant ever discovered in California — so far as the writer has learned — was one found near the Fresno River, the locality of which was visited and carefully examined by Dr. E. C. Winchell, of Millerton, in 1866, who kindly furnished the writer with full notes of his observations. According to this gentleman, the elephant was found reposing on the top of a bed of hard "cement," consisting of sand and yellow clay, eight or ten feet above the solid granite ledge, or bed-rock. The locality is on the south bank of the Fresno, about a hundred yards from its present channel, and three miles above the crossing of the stage road from Hornitos to Visalia, and twenty northwest of Millerton, the county-seat of Fresno. The remains were covered by only three or four feet of sandy alluvium, mingled with disintegrated granite washed from the adjacent hillsides.

As this locality is of importance, as being one of the very few instances in which such remains have been found in the position in which they were left by the death of the animal, and as indicating also its recent geological age, Mr. Winchell's careful description of its appearance, as it lay exposed in the excavation, will be given in his own words, as follows: —

"The vertebræ lay, without disarrangement, in their natural position, relatively, in almost a straight line from the tail to the skull, each separated from its fellow by a space of an inch to an inch and a half. The first joint only of the tail was found.   By actual measurement, the length of the vertebral column was twenty feet.   The great bones and joints of the fore and hind legs lay on either side of the column in their appropriate places, but in disorderly array.   The skull rested apparently upon its face, the lower jaw uppermost; it was about four feet in length and two in width.   From the upper jaw sprang two massive, curved, black tusks, diverging as they lay, until the tips were four or five feet asunder.   Their size and beauty induced Mr. Bailey [the miner by whom the skeletons were discovered] to spend two days of labor in the attempt to disengage the right-hand one for preservation.   With great care and patience he removed the cement and soil from above and around it, until it rested only on a thin narrow ridge of clay curving from base to point; but, on applying a slight degree of force, it crumbled at once to fragments.   Its diameter at the base was six inches, and it retained this size with great uniformity for four feet; thence it tapered to the point, two feet two inches farther, making its entire length six feet two inches.   Its surface was hard, smooth, black enamel.   This color penetrated one fourth of an inch in depth.   It does not appear to have been colored by any ingredient in the soil.   The mass of the interior was white.   All the bones crumbled rapidly on exposure."

A mile below the locality where the above-mentioned elephant was found, a large quantity of bones were discovered by Mr. Bailey in the course of his mining operations.   The "cement" — clay and sand — was there twenty-two feet thick, and the remains were mostly found in a stratum about two feet above the granite bed-rock.   Here, as Mr. Winchell remarks, hundreds of decaying bones, as large as those of the elephant, were found lying together in great confusion, with some bones of smaller animals among them.*

The most important point regarding the occurrence of the fossil remains of the elephant in the mining region of the Sierra is, that they have not thus far been found as low down — geologically speaking — as those of the mastodon.   No instance has as yet reached the writer's notice of their occurrence under the basaltic capping of the gravel.   And, in general, the appearance of the various specimens seen in the possession of the miners, or collected together in museums and saloons,† was never that of partial fossilization, such

---

* It is much to be regretted that it was impossible for any member of the Geological Survey to visit this locality at the time these explorations were going on.

† "Saloons," i. e., places where whiskey is dispensed, are excellent places to see what of interest in the way of fossils has been found in their respective neighborhoods.   A large portion of the most interesting specimens used to find their way to saloons, where they were carefully preserved and highly valued.

as has been repeatedly noticed to be the case with the mastodon remains. Further remarks on this point and on the probable relative geological age of the mastodon and elephant may be found in the following section.

Next to those of the mastodon and elephant, the most abundant fossil remains found in the gravel region are those of the horse, and of this animal several species have been recognized by Dr. Leidy among the specimens collected from different parts of the State and submitted to him by the Geological Survey. It seems proper to publish in full the notes furnished by him in regard to the equine remains examined, although some years have elapsed since they were written. These notes do not appear to have been rendered any the less valuable by what has since been published on this subject. They will be followed by some general remarks on the probable geological position of the different species mentioned, which will be introduced when discussing the age of the human remains discovered in the same deposits with the various animals which have been described.

Equus excelsus. Leidy : Proc. Ac. Nat. Sci., Phila. 1858, 26.
Equus occidentalis. Leidy : Ibidem, 1862, 94.

An extinct species supposed to be different from other North American species, and about the size of ordinary varieties of the recent domestic horse. It is indicated by the following specimens :

a. The greater portions of both jaws of an individual of mature age. The specimens are thoroughly imbued with bitumen, and were obtained near Buena Vista Lake.*

The portions of the upper jaw contain the incisors, canines, and the anterior four large molars of the left side. The teeth resemble in their size and relationship those of the living horse. The course of the enamel lines on the triturating surfaces of the molars presents an extreme condition of simplicity. The bottom of the principal internal valley is devoid of the inflection of enamel seen in the recent horse, and in this respect the fossil teeth resemble more those of the ass.

The portions of the lower jaw contain all the teeth of both sides except one last molar. The teeth present nothing peculiar distinguishing them from those of the recent horse.

b. Portion of the left ramus of a lower jaw of another individual from the same locality. It belonged to an older but somewhat smaller animal, and like the preceding specimens is imbued with bitumen. It contains the anterior five molars.

c. An isolated upper molar, from the same locality and in the same condition. It exhibits the peculiarities already mentioned in the series of upper molars.

The collection of the Academy of Natural Science also contains a portion of an upper jaw with the anterior five molars, from the same locality and in the same condition, presented by Dr. George H. Horn. The teeth likewise exhibit the peculiarities above mentioned, distinguishing them from those of the recent horse.

d. An isolated upper molar, from auriferous clay, twenty feet below the surface, Columbia, Tuolumne County. It is a second of the series, is hardly changed in texture, and is only partially stained with iron. The tooth exhibits the same peculiarities as the upper molars above indicated.

e. Two upper molars, from different individuals, with no locality indicated [but probably from Sonora]. They both exhibit the peculiarities above mentioned.

* The exact locality is believed to be the "Sta. Maria Oil Springs," about twelve miles northwest of the north end of Buena Vista Lake.

f. Two lower molars, one marked Matlock Gulch, Tuolumne County, the other with no locality indicated. Probably belonging to the same species as the preceding.

Equus caballus.

a. Skull, of the size of that of the recent mustang or Indian horse of Western America. Specimen entire, unchanged, and not differing in any respect from that of the living domestic horse. Labelled "From auriferous gravel, thirty-five or forty feet below the surface." * Length of skull from oc. condyles to incisive alveoli, 19¾ inches. Length of upper molar series, 6½ inches.

b. An isolated upper lateral incisor differing in no point from that of the recent horse. Unchanged in texture. From calcareous tufa, overlying auriferous gravel, fifteen feet in depth. From Texas Flat, Tuolumne County.

c. Two superior molars, apparently a third and a fourth, but from different individuals. Not fossilized and not differing from those of the recent horse. Twenty-five and twenty-nine feet below the surface. From Kincaid Flat, Tuolumne County. A last inferior molar from the same locality, and of the same general character.

d. An inferior molar, not differing from those of the recent horse. From five or six feet of depth of vegetable. Columbia, Tuolumne County.

e. Four upper molars and the fragment of another, without locality indicated. Clearly of recent origin. Another specimen of the same character is marked Mojave Valley.

Fragment of the left ramus of the lower jaw containing three molars. Labelled "From the Post-pliocene of Oregon." It belonged to a young animal, and does not differ from the corresponding part in the recent horse. The teeth consist of the last temporary molar and the succeeding pair of permanent molars, of which the fifth is so far protruded as to be worn at its fore part.

Equus pacificus, n. s.

a. An extinct species, apparently larger than any ever before indicated. The most characteristic specimen upon which it is founded consists of a second upper molar tooth nearly half worn. It is from Martinez, Contra Costa County, from a formation viewed by Professor Whitney as of Pliocene age. The tooth is well preserved, retaining its outer cementum, and appears but slightly changed in texture. The triturating surface in the arrangement of its enamel lines presents nothing strikingly different from that in the corresponding tooth of the recent horse, and differs from *Equus excelsus* in the same manner as the latter. Its measurements are as follows:

Length externally, independent of the fangs	26½ lines.
Fore and aft diameter of triturating surface	16¼ "
Transverse diameter of triturating surface	15 "
Transverse diameter with cementum	16 "

b. Three water-rolled fragments, one of an upper molar, and two of lower molars, marked from near Martinez, Contra Costa County, probably belong to the same species.

c. Another fragment of a large inferior molar, marked "Murphy's Diggings," perhaps also belongs to the same.

d. Another specimen, the greater portion of a fourth or fifth lower molar, from the "Elephant bed, Centerville, Alameda County," may belong also to the same.

Equus.

a. A left metacarpal, of small size compared with that of the ordinary domestic horse. Length eight and a half inches; breadth of extremities twenty lines; circumference at middle forty-four lines. Unchanged except having lost some of the bone cartilage. Labelled "Southern California, sixty feet below the surface."

b. The lower extremity of the humerus of a small equine animal. Diameter at condyles twenty

* This specimen was found at Brandy City, Sierra County, at an elevation of about 2,000 feet above the North Yuba River.

six lines. The specimen is firm in texture and appears partially fossilized by the substitution perhaps of calcareous matter for some of the bone cartilage. "Found under the lava at a depth of 210 feet from the surface. From Table Mountain, Tuolumne County."

c. An inferior molar, apparently a fourth, resembling in constitution the corresponding tooth of the recent horse, but much smaller. It has lost its outer cementum and the dentine is chalky, but the enamel is unchanged in texture. It is about half worn, and measures internally, without the fangs, ten lines in length. Its fore and aft diameter at the triturating surface is eight and a half lines; its transverse diameter four and three fourths lines. "Found beneath the volcanic rocks, in gravel resting upon granite. Soulsbyville, Tuolumne County."

An upper molar, from an intermediate position of the series, with its outer part broken away. The tooth in size and constitution bears a near resemblance to some of the small equine teeth from the Niobrara River, Nebraska. It had just commenced to be worn. In the perfect state, it has measured about fourteen lines in length externally; nine lines fore and aft; and about eight lines transversely. The specimen appears partially fossilized, that is to say, some of its bone cartilage appears to have been substituted for earthy matter. The tooth looks as if it might belong to *Merychippus insignis*, which also has been found in Texas, as well as in Nebraska. No locality is marked on the specimen.*

SECTION V. — *Human Remains and Works of Art in the Auriferous Gravel Series.*

We come now to a most interesting portion of this volume, namely, the occurrence of human remains and the works of human hands in the various deposits which have been described in the preceding pages. And, in commencing this section, it is necessary to say a few words in addition to what has been previously stated in regard to the nature of the evidence which is here to be brought forward bearing on the question of the antiquity of the human race. It is true that there was a time, but few years ago, when no attention whatever was paid to any statements which seemed, if not to prove, at least to lend probability to the theory that the epoch of the appearance of man on earth must be carried back into the shadowy past of geological time. The writer well remembers how, a little more than twenty years ago, the first statements of Boucher de Perthes in regard to the finding of flint implements in the "drift" at various localities near Abbeville, in France, began to be circulated in the country, and with what incredulity they were received.† Gradually, however, the evidence has accumulated from widely separated regions, until the idea of prehistoric man has become

---

* But it is probably from near Sonora.

† The first discoveries of M. Boucher de Perthes were made in 1841, and a full account of his investigations published in 1847. In 1854 Dr. Rigollot obtained similar results, and they were made public in a paper, accompanied by careful sections and drawings, in the succeeding year. With few exceptions, however, these discoveries remained unknown to, or were considered inconclusive by, antiquaries and geologists, both in France and England, until, in 1858, Dr. Falconer, and a little later Mr. Prestwich, went over the ground thoroughly, and fully confirmed what had been previously published.

familiar to geologists, and it may now be considered as generally admitted by them that the human race was living in Europe during the later Pleistocene age. The great question now is, How far back can man and his works be traced? and in regard to this point a large store of reputed facts are gradually being brought together, and most geologists are ready and willing to examine them and discuss their authenticity, the time having gone by when they were contemptuously thrown aside as conflicting with one of the fundamental ideas of the science. The following pages are offered, therefore, as a contribution to the history of prehistoric man, as setting forth the results obtained during several years of geological work in California, — years in which this subject was not made a special object of research, but when facts which came under our observation were examined into, so far as time and circumstances admitted, and which, taken together, form a considerable body of material.

Before entering upon the setting forth of what has been collected relating to the antiquity of man in California, it will be desirable to say a few words in regard to the nature of the evidence presented and the manner in which it has been collected.

It will have become abundantly evident, from what has been stated in the preceding pages, that most of the material collected bearing on such a question as the one now under discussion must be expected to be of a very fragmentary character. The nature of the auriferous deposits is such as to preclude, except in a few specially favored situations, any hope of finding a large number of fossil remains in an undisturbed position. Objects found in the gravel must, of course, have been subjected to the same long-continued course of abrasion and transportation as the material in which they are imbedded has undergone. It has already been mentioned how of several species of animals found in the auriferous detrital deposits only a single tooth has as yet been obtained, while others are represented by two or three fragments of teeth or bones at most. The finer-grained beds, such as pipe-clay and sand, are usually unproductive in gold, and therefore rarely worked except in cases of necessity. But it is in these that one would expect to find remains in the most perfect condition and in the largest quantity, other circumstances being favorable.

There is another point which must be mentioned in this connection, and especially as bearing on the question why so much of the evidence presented in relation to the antiquity of man dates back quite a number of years. The hydraulic method of attacking the gravel deposits has become more and more universal of late years, and is now almost the only one employed, ordinary placer mining with portable implements for washing having been almost exclusively relegated to the Chinese, while tunnel mining is also practised on a much less extensive scale than formerly. But in hydraulic mining, with sluices many hundred feet in length, nothing can be seen of what the gravel may contain of human or other remains, unless under very exceptional circumstances. The coarser, larger implements, if sufficiently strong to endure the wear and tear of the sluice, would be carried down and deposited far away in the tailings and then speedily covered by other materials. The finer and more delicate portions, such as bones, are ground to powder between the cobble-stones and are lost forever. By the tunnel method some portion of the material excavated stands a chance of being examined; or, at least, of being for a while in such a position that it could be inspected if there were any one present who had the curiosity to do it. Hence most of our evidence of the former existence of man during the gravel period comes from a region of former tunnel mining on a large scale, and is, much of it, of ancient date, because this region is pretty nearly worked out and abandoned. That evidence greatly exceeding in fulness and value any that has heretofore been obtained may yet be secured is certainly possible, if not probable. It is not necessary, however, to speculate on what may happen in the future, but rather to set forth what is already known.

It will be observed by the reader of the following pages that the bulk of the evidence presented is that furnished by the miners themselves, sometimes supported by the actual presence of the objects found, at other times without such support, there being only the bare statement of a former find. That the miners themselves should be, in most cases, the persons furnishing the evidence is too natural a circumstance to need comment. Their statements have to be taken — as already suggested when speaking of the occurrence of animal remains in the gravel — for what they are worth, balanced against each other, and the most probable result accepted. A long chain of circumstantial evidence is frequently more convincing than a single statement of an eye-witness. It might be asked, Why should not the writer, or one of his assistants, have been on the spot when some of these discoveries were made, or have taken with his own hands some one of the objects mentioned from its original resting-place? The answer to this is, that such finds have not, thus far, proved to be sufficiently common to make it worth while to give up to such an undertaking the amount of time which would probably be necessary to insure success. It was believed that a better result would be secured by collecting a large body of reputed facts, even if observed by others, than by watching at any one locality in the hope that something of interest would turn up. The principal difficulties in the way of such a course have already been sufficiently hinted at. When very important discoveries have been made, as in the case of the "Calaveras skull," the locality has been visited and carefully examined in order that the geological conditions of the occurrence might be fully comprehended.

With these preliminary remarks we proceed to lay before the reader the facts which have been collected, mentioning them in geographical order. While thus going as a general rule rapidly over the ground, certain special occurrences deemed to be of more importance than the others, for reasons which will be given, will be taken up and enlarged upon, as full details as possible being presented. Finally, the probable bearing of the whole mass of evidence with regard to the occurrence of remains of man, as well as of other animals, will be briefly summed up, leaving a more complete general discussion for a future chapter of this volume.

At this point it will be proper to mention the valuable services rendered in the collection of evidence bearing on the antiquity of man in California, by Mr. C. D. Voy, who partly in the employ of the Geological Survey, but chiefly as a volunteer, travelled over the State at intervals during several years, investigating and gathering materials in this department. His ample collections at present belong to the University of California, where they may be examined by those interested, and where quite a number of the most important objects mentioned in the succeeding pages will be found.

MARIPOSA COUNTY.

At Horse-Shoe Bend, of the Merced River, in October, 1869, mastodon bones were found at a depth of twelve feet beneath the surface. In the immediate vicinity were numerous human bones, none of which were preserved. Stone implements, however, were obtained, and one of them is preserved in Mr. Voy's collection. It is a spear or lance head of obsidian, five inches long and one and a half broad, quite regularly formed.

At Hornitos and in No. 1 Gulch, five miles northeast of Hornitos, stone implements are reported to have been found at various times, in considerable number, and particularly in the year 1864. Mortars were the articles chiefly found. Accompanying these were bones of the elephant and horse; and also, as is supposed, of some species allied to the camel; but they have not been preserved. The depth at which these various articles were found is reported at fifteen feet.

About the year 1863, in some mining claims near Princeton, a considerable number of interesting implements of stone were discovered, imbedded in the auriferous gravel, at a depth of about ten feet below the surface. The implements consisted of large stone mortars, with pestles, and spear and arrow heads made of obsidian. One of these spear heads, in Mr. Voy's collection, is about six inches long and three wide, and rather roughly made. One of the largest stone mortars ever met with in the State was found here. It was made of granite, eighteen inches high and thirty-six in circumference, weighing over fifty pounds.

Other localities in Mariposa County where stone implements are reported to have been found are : Three miles north of Mariposa (town), in various mining claims ; at Indian Gulch, a few miles south of Mariposa ; near O'Neal's quartz mill, two miles south of the Buckeye Ravine.

MERCED COUNTY.

In the vicinity of Snelling numerous stone implements are said to have been found in the gravel at various times.

STANISLAUS COUNTY.

A tusk and some of the molar teeth of the elephant were found, in 1870, at Dry Creek, nine miles south of Knight's Ferry, and are now preserved in the Voy Collection. The tusk is ten feet long and thirty-six inches in diameter at the base. These remains are said to have been taken from a depth of thirty-seven feet below the surface, where they were found imbedded in hard "cement gravel." It is also stated that numerous stone implements have been discovered in this county, at various points, in the gravel workings. There are no particulars in the writer's possession, however, in regard to these occurrences.

TUOLUMNE COUNTY.

We come now to a region which has been more prolific in human remains and works of art than any other in California. Some of the more important localities where these things have been found will be noticed on the map of the Table Mountain Flow (Plate D). Dr. Perez Snell, of Sonora, made a large collection of fossils obtained in that vicinity, for many years the scene of the greatest activity in placer and tunnel mining. This collection, now dispersed or destroyed by fire, it is believed, was several times examined by the writer, and valuable items of information obtained in regard to some of the objects it contained, as will be mentioned farther on. Mr. Voy also spent considerable time in this region, and secured a large amount of evidence bearing on the question of the truth of the supposed occurrence of human remains in the uncovered gravels, as well as beneath the basaltic capping of Table Mountain, so often referred to in the previous pages.

The first information of importance obtained by the writer in regard to the antiquity of man in California and his coexistence with extinct mammals was procured, in 1863, at Gold Springs, a little west of Columbia. Mr. Lot Cannell, a native of Bangor, Maine, stated that he had in the course of his mining operations, in that vicinity, found stone mortars and " platters " in the same stratum with the bones and teeth of the mastodon, an animal with which some of the miners in that locality had been quite familiar, on account of the repeated finding of its remains as already mentioned. The objects discovered had all been destroyed by a fire ; but there was no reason to doubt the correctness of Mr. Cannell's statement, and it became evident to the writer, from repeated conversations with him, that he was a man of intelligence and a careful observer. This was the beginning of a long chain of evidence which has been gradually accumulating during the past sixteen years, and for publishing which the writer has been patiently waiting the proper occasion. Already, however, in 1865, there had been a sufficient number of facts collected of a similar kind to that mentioned above, to justify the statement made in Geology I.,* " that it is

---

* See Geology I. p. 252.

hardly possible to escape the inference that the human race existed before the disappearance of these animals [the mastodon and elephant] *from the region which was once so thickly inhabited by them.*" The bones and relics found by Mr. Cannell were from a depth of about ninety feet below the surface.

Mr. Voy, several years later, obtained abundant corroborative evidence of the above statement from various parties in the vicinity of Gold Springs. The following is an extract from his notes on the subject furnished to the writer: "A wagon-load of bones of mastodon and other large animals was destroyed by a fire here, some years ago. In close proximity to these were found numerous stone implements at various times, and at different depths. A very interesting mortar which was discovered here is made of hard granite, and is ornamented on the outside with diagonal markings, about three quarters of an inch deep and half an inch wide. This mortar is thirty-seven and a half inches in circumference, weighing some thirty pounds. It was found in the year 1863, near other relics and animal remains, imbedded in auriferous gravel mixed with calcareous tufa, at a depth of about sixteen feet beneath the surface." Some distance below this — in Gold Springs Gulch — other relics were found, one of which is in Mr. Voy's collection. It is a large oval, shallow stone dish, fifty-three inches in circumference and weighing about forty pounds. This was found, in 1862, in auriferous gravel beneath an accumulation of about twenty feet of calcareous tufa. Among the relics obtained at this locality are certain discoidal stones, from three to four inches in diameter, and about an inch and a half thick, concave on both sides, with perforated centre. Some of these implements are made of granite, others of sandstone. The purpose for which they were used it seems not easy to make out.

At Kincaid Flat, famous as a locality of animal remains,* stone mortars and pestles have been found in the auriferous detritus at various depths below the surface, up to twenty feet, as is stated. Some of these are in Mr. Voy's collection, as also certain implements supposed by him to have been intended to be used in connection with a bow, for enabling the hand to get a better grasp of the weapon.

On Woods' Creek, in various years from 1862 to 1865, numerous fossil remains of the mastodon, elephant, and other animals were found, and with them stone implements of different kinds. Among these was a large stone dish, or platter, as well as mortars. The depth at which these objects were found varied from twenty to forty feet. The same statement may be repeated with regard to other localities in this vicinity; as, for instance, Springfield and Columbia. On Mormon Creek similar discoveries were made during the years from 1851 to 1865. It is not necessary to delay longer on this class of occurrences. It may be stated, in general, that all about Sonora the auriferous gravels which have been worked as placer mines, and the material filling the crevices in the limestone belt, already described, have in a great number of localities been found to be filled with the bones of animals of extinct species; and that with these many relics of the works of human hands have also been discovered, at various depths, down to about a hundred feet.†

We come now to a more interesting branch of the subject, namely, those facts which indicate the existence of man in this region previous to the basaltic overflow forming the capping of Table Mountain, so often referred to in the previous pages. There are a number of occurrences of this kind, and they will be discussed in the order in which they have been brought to the writer's notice.

* A photograph sent by Mr. Voy, and examined since page 252 was in type, shows — as the writer believes — that *Mastodon obscurus* has also been found at Kincaid Flat. The photograph is labelled "Two fossil mastodon teeth found in 1861 with numerous others; also elephant, camel, and horse remains, and stone mortars and other stone relics, imbedded in auriferous gravel, about sixteen feet below the surface, at Kincaid Flat, Tuolumne County, California. In Voy's Cabinet."

† A large number of these relics are now preserved in the Museum of the University of California, which contains not only the Voy Collection, but the materials of the Geological Survey. They have all been examined by the writer, and there can be no possible doubt as to their being the work of human hands.

In Dr. Snell's collection, previously mentioned, there were several objects which were marked as having come "*from under Table Mountain.*" Among these was a human jaw, which was repeatedly examined by Mr. Voy and the writer. It was five and a half inches across from condyle to condyle. "Near this," according to Mr. Voy, "were found several curious stone implements. Among them was a piece of stone apparently designed as a handle for a bow. It was made of silicious slate and had little notches at the end, which appear to have been formed for tying the stone to the bow. There were also one or two spear-heads, from six to eight inches long, and several scoops or ladles, with well-shaped handles." In regard to most of these specimens it can only be said that they were given to Dr. Snell by the miners as having been taken from under Table Mountain. There was one object, however, of the locality of which more certain proof could be furnished. This was a stone muller, or some kind of utensil which had apparently been used for grinding. It was carefully examined by the writer, and recognized as unquestionably of artificial origin. In regard to this implement Dr. Snell informed the writer that he took it with his own hands from a car-load of "dirt" coming out from under Table Mountain. Dr. Snell also, a short time before his death, namely, in November, 1869, called the attention of Mr. G. A. Treadwell, of Big Oak Flat, to this specimen, and informed him that "he took it from a tunnel under Table Mountain, and that it was the only specimen he had of which he could say positively that it came from under that formation." *

Soon after the writer's communication to the American Association for the Advancement of Science at the Chicago meeting, in 1868, of some of the principal facts connected with the discovery of the "Calaveras skull," and when that remarkable relic had begun to be talked about a good deal, he was informed by Dr. J. Wyman that there was in the Museum of the Natural History Society of Boston a small fragment of a skull, which he had identified as being human, and which bore the following label : "From a shaft in Table Mountain, 180 feet below the surface, in gold drift, among rolled stones and near mastodon débris. Overlying strata of basaltic compactness and hardness. Found July, 1857. Given to Rev. C. F. Winslow by Hon. Paul K. Hubbs, August, 1857." It was also soon ascertained that a similar fragment of a human skull existed in the Museum of the Philadelphia Academy of Natural Sciences, with a similar label. And it appeared that Rev. Mr. Winslow divided the specimen given him by Mr. Hubbs between the two societies in the manner indicated. In Volume VI. of the Proceedings of the Boston Natural History Society, page 278, under the head of October 7, 1857, is a communication from Rev. Mr. Winslow sent with the skull fragment, from which the following extract is made (date not given) : "I sent by a friend, who was going to Boston this morning, a precious relic of the human race of earlier times, found recently in California, 180 feet below the surface of Table Mountain. . . . My friend Colonel Hubbs, whose gold claims in the mountains seem to have given him much knowledge of this singular locality, writes that the fragment was brought up in pay-dirt (the miner's name for the placer gold-drift) of the Columbia Claim, and that the various strata passed through in sinking the shaft consisted of volcanic formations exclusively."

This find evidently excited no attention at all, as was natural at the time, especially as the locality was so far removed ; indeed, the well-authenticated discoveries of Boucher de Perthes had, in 1857, gained no credence in this country, as has already been mentioned. The writer, however, on receiving the above information from Dr. Wyman, proceeded at once to investigate the matter. Hon. Paul K. Hubbs was soon discovered to be a well-known citizen of Vallejo, California, and a former State Superintendent of Public Instruction. He was kind enough to furnish to Gorham Blake, Esq., at the writer's request, a full written statement of all the circumstances of the find, the principal points of which he remembered perfectly. In a letter dated "Vallejo, November 23, 1868," he described the locality at considerable length ; and from the writer's own knowledge of the place, added to that of Mr. Blake, who was familiar with the region and took great pains to make himself acquainted with the mining operations formerly carried on there, we were able to

---

* Quoted from a letter from Mr. G. A. Treadwell to the writer, dated Stockton, February 16, 1870.

understand perfectly the geological features of the point at which the discovery was made. It was in the Valentine Shaft, which was sunk on the side of Table Mountain, a little south of Shaw's Flat. There were several shafts ranged nearly in a line between the Flat and the summit of the mountain. The Valentine Company's claim lay between the Sampson and the Columbia claims, all of them working through vertical shafts. The Sampson shaft passed through six feet of surface soil; then forty-two of pipe-clay; then four of hard "sand cement" containing numerous impressions of leaves, and bones towards the bottom of the stratum; then under that the pay gravel, nine feet in thickness. The section in the Valentine Shaft was almost identically the same as that in Sampson, except that, the former being higher up on the mountain, its depth was necessarily greater, and more of the pipe-clay had to be passed through. The essential facts are, that the Valentine Shaft was vertical, that it was boarded up to the top, so that nothing could have fallen in from the surface during the working under ground, which was carried on in the gravel channel exclusively, after the shaft had been sunk. There can be no doubt that the specimen came from the drift in the channel under Table Mountain, as affirmed by Mr. Hubbs. This gentleman was on the ground himself, at the time the fragment was found, and he says: "I saw the portion of skull immediately after its being taken out of the sluice into which it had been shovelled, and some of the marine dirt was sticking to it." By "marine dirt" is meant the pay gravel; Mr. Hubbs, like many other miners* at that time, having the idea that the auriferous gravels were of marine origin. It is clear from Mr. Hubbs's statements that the fragment was raised from the stratum of pay gravel, and that it was noticed when the contents of the bucket were dumped into the head of the sluice, and either picked up by Mr. Hubbs, or by some one else who happened to be standing by, and who handed it to him on the spot. The evidence seems very clear, in all respects, so far as the fact of the occurrence of human remains in the strata underlying the Table Mountain basalt is concerned. Unfortunately the piece of skull preserved is too small to be made the basis of any craniological investigations.

Additional and entirely independent information in regard to the discovery of proofs of the former existence of man at this locality was also obtained from Mr. Albert G. Walton, who was one of the owners of the Valentine Claim, and the carpenter under whose direction the shaft was timbered, as already mentioned. According to his statement the section of the formation passed through in sinking this shaft was as follows: soil, six to ten feet; pipe-clay, seventy feet; cement, with fossil leaves and small branches of trees, three to four feet; pay gravel, five to nine feet, making the total depth of the shaft from ninety to ninety-five feet. The depth of 180 feet, previously stated by Mr. Hubbs, means the vertical distance from the surface to the workings at the end of the drift leading from the shaft to the point where the fragment of a human skull was found. According to Mr. Walton, who remembers nothing of the finding of this piece of bone, there was a mortar found in these workings in the gravel. It was about fifteen inches in diameter, and, as he says, resembled those so often found in the diggings in California. Mr. Valentine, on the other hand, who was one of the owners of the claim, corroborated the main facts about the position of the workings, as being in the channel under the basaltic capping, but remembered nothing about any discovery of either bones or implements. It is clear that, had it not been for the accidental presence of Mr. Hubbs on the spot, at the time the piece of skull was found, we should never have heard anything of it. And if Mr. Hubbs had not given it to an enthusiastic observer, like Dr. Winslow, it would probably never have come to the notice of scientific men. One should bear in mind how few of the discoveries of human relics or remains which are made are likely ever to be heard of beyond a very limited area, even under the most favorable circumstances, as is well illustrated by the facts in this case.

Mr. Voy was able to procure still further evidence bearing on the question of the occurrence of implements under Table Mountain. This evidence is given as it came into the writer's hands, in the form of an affidavit, duly sworn to before a magistrate:—

---

* And some scientific men. See *ante*, pp. 70, 71.

SHAW'S FLAT, TUOLUMNE COUNTY, CALIFORNIA, May 23, 1870.

This is to certify that I, the undersigned, did about the year 1853, visit the Sonora Tunnel, situated at and in Table Mountain, about one half a mile north and west of Shaw's Flat, and at that time there was a car-load of auriferous gravel coming out of said Sonora Tunnel. And I, the undersigned, did pick out of said gravel (which came from under the basalt and out of the tunnel about two hundred feet in, at the depth of about one hundred and twenty-five feet) a mastodon tooth in a good state of preservation, which afterwards was partly broken, in the hollow of which was sulphuret of iron. And at the same time I found with it some relic that resembled a large stone bead, made perhaps of alabaster, about one and a half inches long, and about one and one fourth inches in diameter, with a hole through it one fourth of an inch in size, which no doubt had been used, some time, to put a string through. I also certify that I gave the above specimens to C. D. Voy, about the year 1864, to put in his collection of some similar fossils, which had been found in this county at various depths and in various localities.

[Signed]   OLIVER W. STEVENS.

Here follows the affidavit : —

STATE OF CALIFORNIA, COUNTY OF TUOLUMNE, SS.

On this twenty-third day of May, A. D. 1870, personally appeared before me, A. Bullerdieck, a Notary Public, the within named Oliver W. Stevens ; who, being duly sworn, deposes and says, that the within statement is true and correct.

In witness whereof, I have hereunto set my hand and affixed my official seal the day and year last above written.

L. S.   [Signed]   A. BULLERDIECK, *Notary Public.*

Mr. Voy adds to the affidavit the statement, that he visited the locality, in company with Mr. Stevens, and found it to have the geological position indicated in the affidavit.

The bead was carefully examined by the writer. It is correctly described above, except that the material of which it is made is white marble, not alabaster. It had evidently been much handled, and unfortunately cleaned of the incrusting material ; but quite distinct traces of a former filling of the hole with sulphuret of iron were still visible. The mastodon tooth bore, also, as stated by Mr. Stevens, evident marks of an incrustation of the same mineral ; and it may be added that several of the bones, which are said to have come from under Table Mountain, have been found to have more or less abundant crystallizations of pyrites in the cellular portions. There can be no question as to the artificial character of the so-called bead. It is regularly and symmetrically shaped, and looks as if intended for an ornament.

In the Proceedings of the Boston Society of Natural History, Vol. XV. p. 257, under date of January 1, 1873, will be found a communication of Dr. Winslow's relating to another discovery of human remains in Tuolumne County. Captain David B. Akey is the authority for the statement that a complete human skeleton was found in a tunnel under Table Mountain ; the name of the tunnel, however, he did not remember. He saw the bones after they had been brought out from the excavation by the miners. This occurrence the writer has had no opportunity to inquire into or verify, as it did not come to his notice until after he had left California. The date of this find seems to have been 1855 or 1856.

We pass to another occurrence of a similar character vouched for by Mr. Llewellyn Pierce, in the form of a written statement which is given here in full as furnished to Mr. Voy : —

SONORA, TUOLUMNE COUNTY, CALIFORNIA, December 28, 1870.

This is to certify that I, the undersigned, have this day given to Mr. C. D. Voy, to be preserved in his collection of ancient stone relics, a certain stone mortar, which has evidently been made by human hands, which was dug up by me, about the year 1862, under Table Mountain, in gravel, at a depth of about 200 feet from the surface, under the basalt, which was over sixty feet deep, and about 1,800 feet in from the mouth of the tunnel. Found in the claim known as the Boston Tunnel Company. In these claims, at various times, there have also been found numerous bones of different animals.

[Signed]   LLEWELLYN PIERCE. (L. S.)

The mortar is two feet seven and a half inches in circumference.

I hereby certify that I have, this day, visited the above claim, and was shown about where the above relic was found ; and I believe the above statement to be about correct.

[Signed]   C. D. Voy.

(Here follows an affidavit to the above facts, sworn to by Mr. Pierce before James Letford, Notary Public of Tuolumne County, which it is not necessary to repeat in full. It is dated December 29, 1870.)

CALAVERAS COUNTY.

We come now to a county where occurrences of human remains do not seem to have been as frequent as they were in the adjacent Tuolumne ; but where one specimen has been obtained which has excited more interest than all the others put together, and which is popularly believed to be the only instance of the kind which has been met with in California. A perusal of the previous and of the following pages will, however, it is thought, satisfy the reader that the belief of the existence of man in that region previous to the cessation of volcanic activity there does not, by any means, backed up by one item of evidence alone. The peculiar interest of the so-called "Calaveras skull" depends, in good part, on the fact that it is, thus far, the only relic of the skeleton of prehistoric man in California, which has come into the hands of scientific authorities in such a condition of completeness as to give some basis for ethnological conclusions, or, at least, for craniological measurements. Public attention has been so much attracted to the Calaveras skull,* and so much has been said and written about it, that it will be well for the writer to state what he knows with some detail in regard to this find, in order that those who are interested in the subject may have all the facts which are at hand on which to base their opinions.

The manner in which the skull in question came into the writer's possession is as follows : June 18, 1866, Dr. William Jones, of Murphy's, Calaveras County, a physician of extensive practice in that part of the mining region, and who had been long known to the writer, and for whose veracity and scientific tastes he can personally vouch, wrote to the office of the Geological Survey, at San Francisco, stating that he had in his possession "a human skull of Indian type, in a good state of preservation, with the exception of the parietal and occipital portions, — the frontal, facial, and temporal being complete, — which was recently found by Messrs. Mattison & Co., in their claim on Bald Mountain, near Altaville and Angel's, one hundred and thirty feet from the surface, and beneath the lava, in the cement, and in close proximity to a completely petrified oak."

The State Geologist being absent from the city at that time, Mr. Gabb, the Palæontologist of the Survey, answered Dr. Jones's letter, and requested that the skull might be sent to the office of the Survey for examination, which request was immediately complied with, and the skull forwarded on the 29th of June.†

On his return to San Francisco, a few days later, the writer examined the skull, and at once proceeded to visit the locality. He saw Mr. Mattison, the principal owner of the claim from which the relic was taken, and heard from his lips the same statement which Dr. Jones had communicated in his letter, with several additional items of information, some of which are of importance as bearing on the question of the authenticity of the supposed find. And here it may be remarked, that nothing is known unfavorable to the credibility of any of the witnesses to the facts in this case ; and, were it a question of only ordinary importance and interest, the statements made by them would have been received as being, without doubt, the exact truth. The extreme care, however, with which all the facts, in a case like this, should be weighed, must be my excuse

---

* Calaveras means skulls, and was the Mexican-Spanish name of the river which gave its name to the county, Rio de las Calaveras. Skulls and bones of dead animals are common enough in the Western country, in the vicinity of small streams. A larger collection than usual of such remains at some point on this particular stream may have been the origin of the name.

† This skull was temporarily intrusted to the writer ; and, after the discontinuance of the Survey, given to him by Dr. Jones.

before these gentlemen for seeking to sift the evidence, and endeavoring to ascertain, by comparison and putting together of various circumstances, whether there were any flaws in their statements, or whether any reason could be found for doubting the exactness of the information given by them.

Mr. Mattison, on being questioned, stated that he took the skull from his shaft in February, 1866, with some pieces of wood found near it, and, supposing that it might be something of interest, carried it in a bag to the office of Wells, Fargo & Co.'s Express, at Angel's, and gave it to Mr. Scribner, the agent, also well known to the writer as a man of intelligence and veracity. He stated, on being questioned in regard to the appearance of the skull when it was brought to him, that it was so imbedded in and encrusted with earthy and stony material that he did not recognize what it was. Mr. Mattison had previously made a similar statement, saying that when he found the object he thought it to be a piece of the root of a tree, only a portion of the frontal bone being visible. Mr. Scribner's clerk cleaned off a portion of the encrusting material, discovered that the article in question was a human skull, and, shortly after, gave it to Dr. Jones, who was well known in that region as an enthusiastic collector of objects of natural history, and in his possession it remained for some months before it was placed in the writer's hands.

The skull is by no means a perfect one, as the whole of the parietal and nearly the whole of the occipital, as well as a large part of the right half of the base, are missing. The line of fracture through the base is from the right temporal fossa through the opening for the spinal cord, leaving its fore part, and ending about an inch and a half behind the left ear. The frontal bone is nearly entire. A fracture extends across the upper jaw, a short distance below the orbits, otherwise the bones of the face are in most respects complete. This describes the appearance of the skull as it now exists. When delivered into the writer's hands its base was imbedded in a conglomerate mass of ferruginous earth, water-worn pebbles of much altered volcanic rock, calcareous tufa, and fragments of bones. This mixed material covered the whole base of the skull and filled the left temporal fossa, concealing the whole of the jaw. A thin calcareous incrustation appears to have covered the whole skull when found; portions of it had been scaled off, probably in cleaning away the other material attached to the base.

Nothing was done to the skull to alter its condition in any way, after it came into the writer's hands, until it had been examined by Dr. Wyman, when we together carefully chiselled off the foreign matter adhering to its base, so as to fully expose the natural surface of the skull, leaving it in its present state as figured on Plate L.*

On exposing the jaw, the skull was found to be that of a very old person; as the teeth, with the exception of a single root of a molar on the right side, have disappeared. All the alveoli in front have been wholly, and those on the sides partly, absorbed; in consequence of this, if any peculiarity of the jaws existed, it is no longer to be recognized.

In cutting away the mixed tufa and gravel which covered the face and base, several fragments of human bones were removed; namely, one whole and one broken metatarsal; the lower end of a left fibula, and fragments of an ulna, as well as a piece of a sternum. These bones and fragments of bone might have belonged to the same individual to whom the skull had appertained; but, besides these, there was a portion of a human tibia of too small size to be referred to the same person. There were also some fragments of the bones of a small mammal. Under the malar bone of the left side a small snail shell was lodged, partially concealed by one of the small human bones which was wedged into the cavity. This shell was recognized by Dr. J. G. Cooper as *Helix mormonum*, a species now existing in the Sierra Nevada. Cemented to the fore part of the roof of the mouth was found a circular piece of shell four tenths of an inch in diameter, with a hole drilled through the centre, which had probably served as an ornament. Several very small pieces of charcoal were also found in the matter adhering to the base of the skull.

On chemical examination of a portion of the skull by Mr. Sharples, it was found that it had lost

---

* The upper jaw is detached, and had to be temporarily attached by wax when the drawing was made.

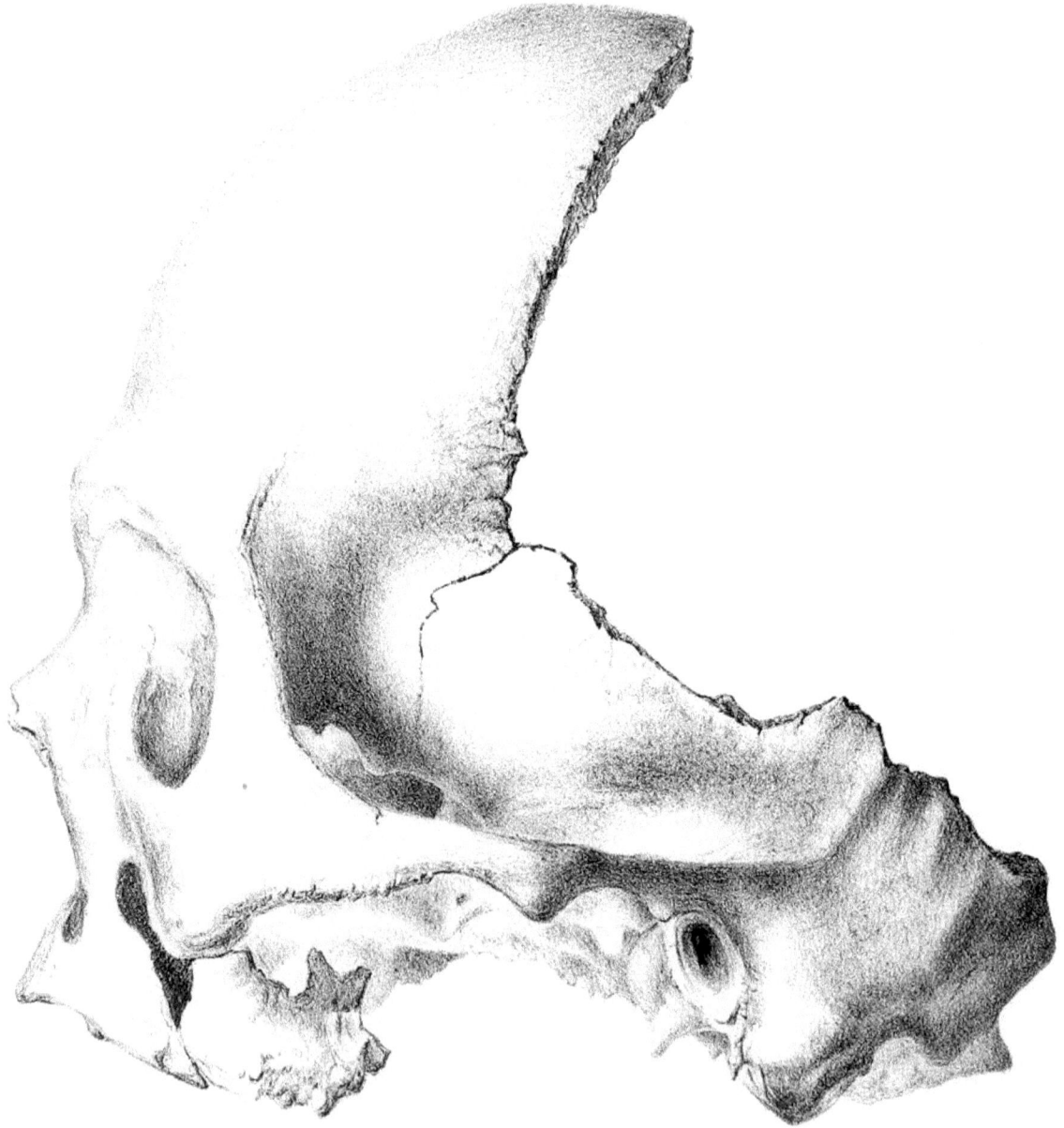

THE " CALA'

PLATE L.

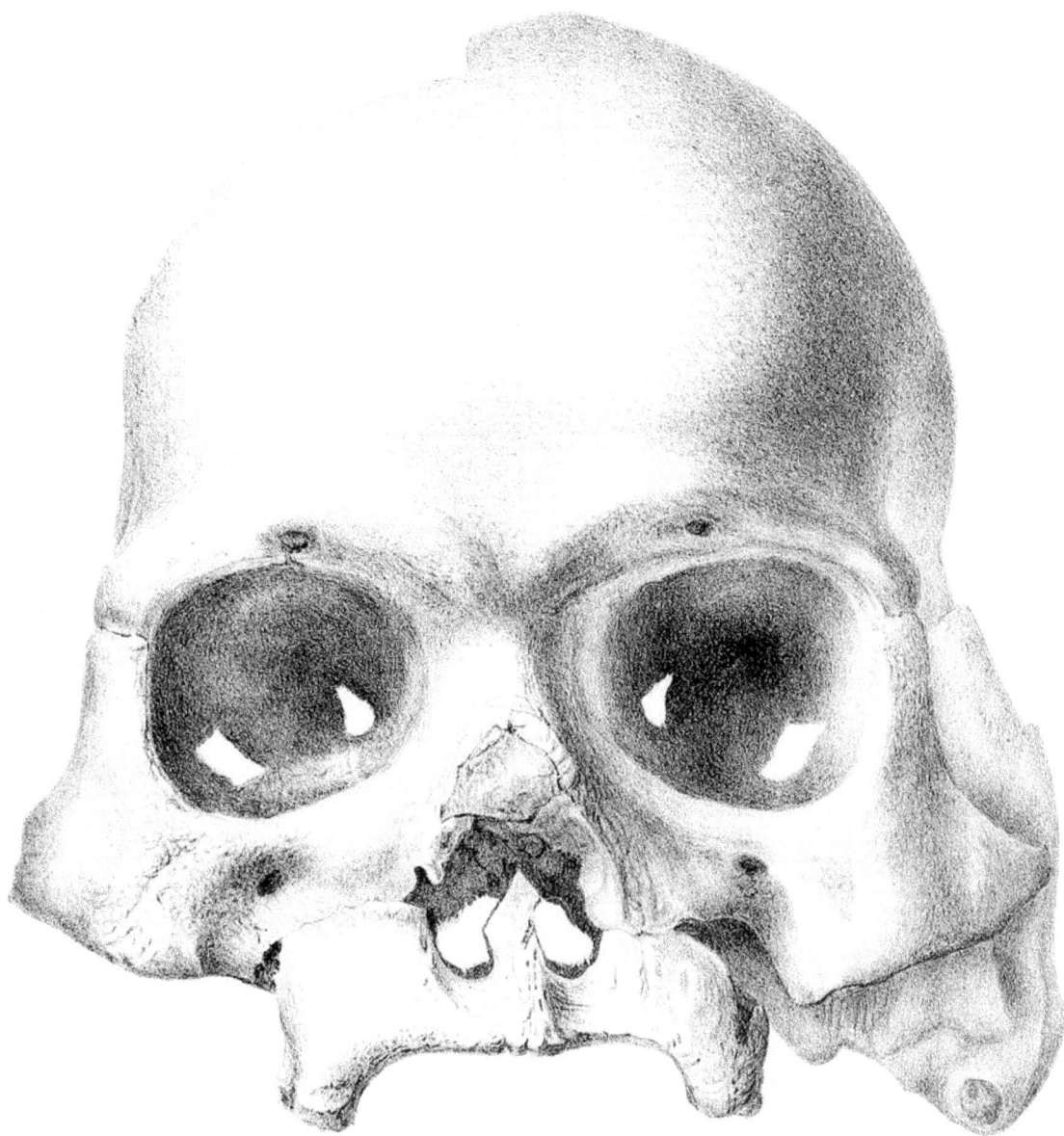

J. Bien print

S SKULL"

nearly all its organic matter, and that a large portion of the phosphate of lime had been replaced by the carbonate. In other words, it was in a fossilized condition. The following are the results of the analysis : —

Phosphate of lime	33.79
Carbonate of lime	62.03
Silica	1.44
Oxide of iron	.81
Carbonate of magnesia	1.86
Water and organic matter	Trace
	99.93

The skull having been coated with wax to protect it, the analysis was made on a piece which had been treated with ether to remove any still adhering particles of this substance : this also took up any organic matter remaining over from what originally existed there. A separate examination of another piece of the skull showed, however, that only a trace of this existed.

Such are the facts concerning the Calaveras skull, in regard to which there can be no dispute. It remains to describe the geological position in which it was said by Mr. Mattison to have been found, and then to see how the facts observed agree with his statements, and especially how far the condition and appearance of the skull itself lend plausibility to the theory — widely circulated and believed in — that the find is " a hoax."

According to Mr. Mattison's statement, the skull was taken from a shaft which he himself had sunk on the northwest slope of Bald Hill, at a distance of just half a mile northeast of Altaville, which is about a mile and a quarter northwest of Angel's, the outcrop of the Great Quartz Vein intersecting the road between the two places, and running close to and nearly parallel with it. All along the region adjacent to this road, on the northwest, is a series of moderately high table-topped elevations, which rise from 300 to 400 feet above the bed of Angel's Creek, just below the town of that name. Bald Hill is the most conspicuous of these hills, and it runs in a northeast-erly direction for half a mile, its culminating point being 388 feet above the creek at Angel's, which itself is 1,380 feet above the sea-level. The higher portion of all these elevations is made up of volcanic and detrital materials, consisting, so far as could be seen, of alternations of more or less consolidated volcanic ashes and of gravel beds. The volcanic deposits are either white or dark bluish-gray in color ; the whiter varieties being very fine-grained and pretty solidly compacted together. It is what is usually called " white lava " by the miners, and is probably rhyolitic in character. The deep gravels in this vicinity do not seem to have been profitable to work, and all the excavations made about Bald Hill had been abandoned at the time of the writer's visit ; and, so far as known, have not been resumed. The opportunities for getting an exact section of all the beds as in this elevation have therefore not been satisfactory. Mr. Mattison, however, gave the following as the section of the strata penetrated in sinking his shaft, which he said was 153 feet deep to the bed-rock : —

	Feet.
1. Black lava	40
2. Gravel	3
3. Light lava	30
4. Gravel	5
5. Light lava	15
6. Gravel	25
7. Dark brown lava	9
8. Gravel	5
9. Red lava	4
10. Red gravel	17
Total	153

The skull, on Mr. Mattison's authority, was found in bed No. 8, just above the lowest stratum of lava.

Immediately after visiting the locality a notice of the discovery was presented to the California Academy of Sciences, at a meeting held July 16, 1866.[*] The débris in which the skull was imbedded were not, however, removed, nor any chemical examination made of it, until some two years later, after it had been taken to Cambridge. The supposed discovery created considerable interest, and was generally looked upon with great suspicion. It has hardly received any notice in Europe. The religious press in this country took the matter up, however, and were quite unanimous in declaring the Calaveras skull to be "a hoax." One of the editors of "The Pacific," a so-called religious newspaper published at San Francisco, hastened to visit the locality, and "interviewed" the different persons through whose hands the skull passed, received — apparently — almost, if not quite, exactly the same statements from them which had previously been given to the writer by the same men, and then returned to San Francisco and wrote and published the following: "Strange memory this, we thought, to retain such minute particulars of such a supposed unimportant discovery, two years before. Then, to have it [by *it*, the skull appears to be meant] pass through all these hands, and varied operations, till it came out a skull of most ancient days. Certainly, it looks laughably suspicious of a regular hoax, got up as a fine California joke upon our State Geologist. California miners have always, as a class, had a low estimate of Eastern geologists who have attempted to show their wisdom by the raid on our mining regions, and they would much enjoy a joke like this on one in high position. We believe the whole story worthy of no scientific credence, and are also more fully established in this belief by the declaration of an able Congregationalist minister, who has preached some time in the region, and who told us that the miners freely told him that they purposely got up the whole affair as a joke on Professor Whitney." [†]

By the time this valuable information reached Boston it had become much more precise and authentic; for, — as may be learned from the "Congregationalist," [‡] — "Once more, we have the 'highly developed' Calaveras skull, found in a shaft 150 feet under the volcanic deposits of California, where it had been placed by some mischievous miners as a hoax upon one of their own number, who was of an anti-Scriptural and geologic turn of mind. He swallowed the hoax, and carried the news to Professor Whitney, who thereupon secured the skull for the State Museum, and introduced the world to the oldest man yet known, a man of the Pleiocene period. The writer has the facts of this case from a clergyman of the Pacific coast, whose brother was a party in the affair by which Professor Whitney was led to a conclusion to which, as far as we are aware, he still adheres. The facts, we believe, are capable of being proved beyond all question. We do not want to press these scientific sores unduly. But when we remember that they are specimens from almost innumerable conclusions that a certain class of scientists have always been eager to draw, if they could be made to bear against the Bible, we have a right to hesitate before accepting their present equally confident conclusions; possibly to suggest a reasonable modesty in the future assertions of those who have made such egregious mistakes in the past." [§] Similar articles were published in many other religious newspapers, and the cause of the unwillingness to believe in Mr. Mattison's statements is, in such cases, easily to be understood.

A decidedly more amusing solution of the problem presented by the Calaveras skull was offered soon after its discovery by Bret Harte, whose poem, entitled "The Pliocene Skull," went the rounds of all the newspapers and made the name of Calaveras classic. It remained, however, for a French writer to put a finishing touch on the absurdities to which this discovery has given rise. M. L. Simonin, in the Revue des Deux Mondes, [||] accepting Bret Harte's "But I'd take it kindly

---

* Proceedings of the California Academy of Sciences, Vol. III. p. 277.

† The Pacific, Vol. XVIII. No. 48.

‡ See number of September 27, 1876.

§ This article stands under the name of "Professor T. S. Childs, D. D., Hartford, Ct."

|| Vol. XII. (Troisième Série) p. 288. This gentleman is the same one previously mentioned (see *ante*, p. 35) as having done his best to bemuddle the question of the geological age of the auriferous slates of the Sierra Nevada.

if you 'd send the pieces home to old Missouri," as scientific authority, says, in describing the gold mines of California : " C'est au fond d'un de ces puits que fut écrasé un jour un mineur missourien, qui resta saisi dans l'éboulement. Plus tard le savant Whitney, qui devait attacher son nom à la géologie californienne, rencontra dans le même lieu, un crâne humain fossile. . . . . Disons nous bien vite que l'homme préhistorique de la Californie ne semble pas plus authentique que celui d'Abbeville," &c.

Having in the preceding pages set forth somewhat in detail the condition of the skull when received from Dr. Jones, and described its appearance after being freed from the debris in which it was imbedded, and having shown what the materials were thus found associated with it, it remains to add a few remarks in elucidation of the question how far the condition and appearance of this skull and all the facts connected with it justify us in believing that the statement of Mr. Mattison in regard to the place in which it was found may be accepted as true.

The skull, being as nearly deprived of its organic matter as fossil bones found in the Tertiary usually are, and having had a large portion of its phosphate replaced by carbonate of lime, is undoubtedly *a fossil*. Chemical analysis proves that it was not taken from the surface, but that it was dug up somewhere, from some place where it had been long deposited, and where it had undergone those chemical changes which, so far as known, do not take place in objects buried near the surface. In view of these undoubted facts, the absurdity of the statement previously quoted in regard to the placing of the skull in the shaft "as a hoax" to be played off on the "anti-Scriptural" miner becomes apparent. The miners who are supposed to have done this clever trick must themselves have obtained from somewhere the object thus used ; and as all the diggings in that vicinity are in the gravels intercalated between the volcanic strata, it becomes, really, a matter of but little consequence, from a geological point of view, from whose shaft the skull was taken. The following are the considerations, then, which lead us to put confidence in Mr. Mattison's assertions as to his having taken the skull out of his own shaft, and from the position already indicated.

In the first place, the locality and the neighborhood were several times visited by the State Geologist, and also by three of his assistants, and by several of his personal friends not connected with the Geological Survey, and to all these Messrs. Mattison and Scribner have given exactly the same statement in regard to the finding of the skull. All of these gentlemen have returned from the place strongly impressed with the idea that there was no mistake, and certainly no intentional misstatement, on the part of either of the principal parties whose names are associated with the find. Messrs. Mattison and Scribner have uniformly told to all the same story, and nothing has developed itself as offering a motive to either of these gentlemen to enter into a combination for the purpose of deceiving individuals or the public. The skull remained on and near the place where it was obtained for several months after it was discovered ; and no doubts were expressed by any one as to the good faith of the parties concerned, until after it had been sent to San Francisco and had been much written about in the newspapers. At the time of the writer's first visit to the region, after the discovery had been made, the miners were evidently entirely unaware of the geological significance of the find. Similar ones had been made before, in repeated instances, and in various districts in the neighborhood, without exciting, so far as it appears, any special interest. It is giving the miners far too much credit for geological knowledge to believe that they would recognize the importance of such a discovery. Much less would it have occurred to them to see anything "anti-Scriptural" in it.

Again, evidence in regard to the skull obtained from an inspection of the ground in the shaft about the spot whence it was said to have been taken is, as yet, wanting, as the excavation has remained filled with water during the whole time since the skull came into the writer's possession, and it has never been in his power to visit the place at a time when the shaft could conveniently, without considerable expense, be emptied of its water. It was his intention to have this done ; but circumstances have rendered it impossible that the desired end should be attained. Mr. Mattison has always said that he expected to resume work in the shaft at some future, not distant, time, and that he would give notice whenever he did so, and that a full opportunity should be afforded of making a careful inspection of the vicinity.

The appearance of the skull when it came into the writer's hands, and especially its appearance when obtained by Mr. Scribner, whose statements may be considered as beyond suspicion, shows that this is not an ordinary skull picked up at random in order that it might be palmed off as a curiosity on an unsuspecting "Eastern geologist," or even an "anti-Scriptural miner." The skull was unquestionably dug up somewhere, and had unquestionably been subjected to quite a series of peculiar conditions. In the first place it had been broken, and broken in such a manner as to indicate great violence, as the fractures go through the thickest and heaviest parts of the skull; again, the evidence of violent and protracted motion, as seen in the manner in which the various bones were wedged into the hollow and internal parts of the skull, as, for instance, the bones of the foot under the malar bone. The appearance of the skull was something such as would be expected to result from its having been swept, with many other bones, from the place where it was originally deposited down the shallow but violent current of a stream, where it would be exposed to violent blows against the boulders lying in its bed. During this passage it was smashed, and fragments of the bones occurring with it were thrust into all the cavities where they could lodge. It then came to rest somewhere, in a position where water charged with lime salts had access to it, and on a bed of auriferous gravel. While it lay there the mass on which it rested was cemented to it by the calcareous matter deposited around the skull, and thus the base of hard mixed tufa and pebbles which was attached to it when it was placed in the writer's hands was formed. At this time, too, the snail crept in under the malar bone, and there died. Subsequently to this the whole was enveloped by a deposit of gravel, which did not afterwards become thoroughly consolidated, and which, therefore, was easily removed by the gentlemen who first cleaned up the specimen in question, they only removing the looser gravel which surrounded it.

Now, such is the condition of things and the chain of events through which the skull passed, as vouched for by its own appearance when it left Dr. Jones's hands, and by the perfectly reliable statements of Messrs. Jones and Scribner.

How does this compare with Mr. Mattison's statements as to the position of the skull? And this is a question of great importance, as, if this gentleman told one story and the skull another, we should not doubt which authority to accept. If, on the other hand, there is no discrepancy in the evidence thus furnished by the dead and the living, then we have here a very strong corroborative link in the chain of testimony, going to show the genuineness of the find.

Mr. Mattison told me that he with his own hands took the skull from near the bottom of bed No. 8, in the section given on page 269, and that it was found lying on the side of the channel with a mass of drift-wood, as if it had been deposited there by an eddy of the stream, and afterwards covered over in the deposit of gravel by which bed No. 8 was formed. Now here seems to be a very satisfactory coincidence between the statements of Mr. Mattison and the facts revealed by the condition of the skull itself. Indeed, the coincidence is as complete as could be desired, and in view of these facts it seems very difficult not to accept the statements made by the gentleman in question as authentic.

We have the independent testimony of three witnesses, two of whom were previously known to the writer as men of intelligence and veracity, while in regard to the third there is no reason for doubting his truthfulness. Each one of these gentlemen testifies to some points in the chain of circumstantial evidence going to prove the genuineness of the find. No motive for deception on the part of Mr. Mattison can be discovered, while the appearance of the skull itself bears strong, though silent, testimony to the correctness of his story.

The following is Dr. Wyman's notice of the craniological peculiarities of the Calaveras skull : —

"The volume of the frontal region is large, so that if the skull were viewed from above, the zygomatic arches would be nearly concealed. As a large part of the occiput is destroyed, it is uncertain whether the head was long or broad. The face is somewhat deformed, the left orbit being smaller, and the left cheek higher than the right, thus giving the whole an unsymmetrical appearance. The ridges over the orbits are strongly marked, and the lower border of the opening of the nostrils is not sharp; but, as in some of the crania of many savage races, is rounded, and the malar bones are prominent. The strongly marked borders of the orbits are the most striking features of the fragment.

"Extended comparisons of crania clearly show that the range of variation in the individual characters of a given race is quite large. This is well illustrated in the results obtained by the eminent American craniologist, Dr. Meigs, in regard to the ratio between the breadth and the length among the American aborigines, in one and the same race, some having the long and others the broad head. In a series of skulls from one of the Hawaian Islands we have found the long and broad heads in nearly equal proportions, the breadth varying from 0.72 to 0.94 of the length. As other features offer similar differences, any conclusions based upon a single skull are liable to prove erroneous, unless we have sufficient grounds for the belief that such a skull is a representative one of the race to which it belongs. If this consideration had been kept in view, much useless discussion in regard to the celebrated Neanderthal skull might have been avoided. We have no sufficient reason for assuming in the present instance that the skull is a representative one; and, in view of this circumstance, the results given in the following table must be considered as applicable only to an individual, not to a race. Future discoveries can alone decide its real value.

	Breadth of Cranium.	Breadth of Frontal.*	Frontal Arch.	Length of Frontal.	Height of Cranium	Zygomatic Diameter
22 Esquimaux	134.5	94	296.5	126.6	135	137.6
5 From Alaska	133.5	92.8	285.5	121.8	129.5	132
11 from different parts of California .	150.5	93.5	260	117	120.8	134
3 Digger Indians	136.6	88.3	280	119	120.3	141.5
Fossil skull	150	101	300	128	134 †	145

(The measurements are in millimeters.)

"The above table of measurements shows:—

"1. That the skull presents no signs of having belonged to an inferior race. In its breadth it agrees with the other crania from California, except those of the Diggers, but surpasses them in the other particulars in which comparisons have been made. This is especially obvious in the greater prominence of the forehead and the capacity of its chamber.

"2. In so far as it differs in dimensions from the other crania from California, it approaches the Esquimaux."

The above seems all that is necessary to be said, in this place, in regard to the famous Calaveras skull. Further discussion of the facts may be reserved for the close of this section; and it is intended that the general *résumé* of all the results of the gravel investigations, in a future chapter, shall also include some additional remarks on the question of the antiquity of the human race on the Pacific coast, and the catastrophes which it has survived. We may pass now to some other occurrences of a similar character, in the same county, but which have not as yet been laid before the public.

The fact that human implements had been found in some of the mining claims near San Andreas, in gravel under the volcanic strata, was repeatedly mentioned to the writer by persons living in that vicinity, and Mr. Voy was successful in finding some of the parties personally concerned in these finds, and getting their written testimony in regard to them. The geological conditions in the vicinity of San Andreas closely resemble those of the locality just described. The

---

* This is the breadth of the frontal at its narrowest part when the skull is viewed from above.

† Measured from the anterior edge of the foramen magnum to the level of the top of the frontal, and an inch behind it on the inside.*

* These measurements can, of course, be only considered as approximations; the fragmentary condition of the skull must be taken into consideration in this connection.

distance between Angel's and San Andreas is about twelve miles, the latter place being on the divide between Calaveritas River and Murray's Creek, a small branch of the Calaveras. Through all the higher southeastern portion of this county the streams run in deep parallel cañons, quite close to each other, and having the ridges between them capped with volcanic overflows, all seeming to form part of the grand lava system which has spread far down the Sierra slope from the vicinity of Silver Mountain. In the vicinity of San Andreas the volcanic accumulations consist of alternating layers of sand, gravel, and volcanic ashes and conglomerates, overlying, as usual in the Sierra, gravel deposits more or less auriferous, the pay gravel being usually quite thin, and the whole series of detrital and volcanic materials reaching a thickness, in places, of from 150 to 200 feet. In fact, the geological conditions are very much like those described as existing near Altaville and Angel's.

The most important detailed evidence of the occurrence of human remains in and under the volcanic in this neighborhood is contained in the following affidavits, which are given in the form in which they were obtained by Mr. Voy : —

SAN ANDREAS, CALAVERAS COUNTY, CALIFORNIA, January 3, 1871.

This is to certify that we, the undersigned, proprietors of the Gravel claims known as Marshall & Company's, situated near the town of San Andreas, do know of stone mortars and other stone relics, which had evidently been made by human hands, being found in these claims, about the years 1860 and 1869, under about these different formations : —

		Feet.
1. Coarse gravel		5
2. Sand and gravel		100
3. Brown gravel		20
4. "Cement" sand		4
5. Bluish volcanic sand		15
6. Pay gravel		6
	Total	150

The above [mentioned relics] were found in bed No. 6.

[Signed]   R. D. HUBBARD.
                                                           JOHN SHOWALTER.

Subscribed and sworn to before me this 3d day of January, 1871.

W. O. SWENSON,
Justice of the Peace, Township No. 5, Calaveras County, California.

To which Mr. Swenson also adds his own personal testimony, as follows : —

I certify that I have seen one of the above described mortars, taken from said claims, and know the above to be true.

WM. O. SWENSON,
Justice of the Peace, Calaveras County, California.

In Smilow & Company's claim, on Gold Hill, about one mile west of Marshall & Company's, stone mortars were found at a depth of about one hundred feet in the pay gravel, under the volcanic, the formation being closely similar to that of the last-mentioned locality. This find is vouched for by Mr. Smilow himself.

Once more, in the way of evidence from Tuolumne, the following is submitted : —

SAN ANDREAS, CALAVERAS COUNTY, CALIFORNIA, January 3, 1871.

This is to certify that I, the undersigned, did, about the year 1858, dig out of some mining claims known as the Stanislaus Company, situated in Table Mountain, Tuolumne County, opposite O'Byrn's Ferry, on the Stanislaus River, a stone hatchet similar in shape to this [here is inserted a rough drawing of a cutting implement of a triangular shape] with a hole through it for a handle, near the middle. Its size was four inches across the edge, and length about six inches. It had evidently been made by human hands. The above relic was found from sixty to seventy-five feet from the surface in gravel, under the basalt, and about 300 feet in from the mouth of the tunnel. There were also some stone mortars found at about the same time and place; and at various times there were also found numerous fossil bones of different animals, and fossil wood.

[Signed]　　　　　　　　　　　　　　　JAMES CARVIN.

Subscribed and sworn to before me,

WM. O. SWENSON,
Justice of the Peace, Calaveras County, California.

At Murphy's, in the detritus accumulated in the deep crevices intersecting the limestone belt at that point, a large number of teeth and bones of various animals, including the elephant and mastodon, have been found. It is reported that with these various relics of the human race were obtained, at depths of from ten to twenty feet beneath the surface. No further particulars, however, have been received. The same is true with regard to several other localities in this region, especially Railroad Flat and San Domingo. At the latter locality a large number of mortars, pestles, and stone dishes have been found in the auriferous gravels at various depths.

AMADOR COUNTY.

The list of localities in Amador County at which human remains have been found is short, as would naturally be expected, since there are comparatively few deposits of deep gravels there.

Near Jackson it is reported that various stone implements were found in the gravels between the year 1852 and 1857. The locality was about two miles south of the town. Some mortars were dug up of large size, weighing from twenty to forty pounds.

Other localities where similar discoveries have been made, but where no special investigations have been made either by Mr. Voy or the writer, are: Little Grass Valley, near Volcano; Pokerville, where discoveries of implements were made in 1858; Fiddletown; Forest Home. At the last-named locality some stone mortars were found, about the year 1864, at a depth of forty feet below the surface, in some hydraulic claims. One of the mortars is now in Mr. Voy's collection. Mr. Goodyear heard of similar occurrences in this county, and at Randall's Claim, near Aqueduct City, was informed by the owner that he had found mortars (of which one was still kept) in the auriferous gravel at depths of eight or ten feet beneath the surface.

EL DORADO COUNTY.

El Dorado has been prolific in its yield of human relics and implements. Indeed, the number of localities where such things have been found is so great, that most of them must be passed over with mere mention of the name, and a few additional particulars in relation to finds of special importance.

Between 1852 and 1865 numerous stone mortars, and other interesting relics, as also some mastodon remains, were found imbedded in the auriferous gravel at Shingle Springs, about ten feet below the surface. One of these mortars, which is about ten inches high and six in diameter, is preserved in Mr. Voy's collection.

At Diamond Springs, in 1855, stone mortars and other stone relics were found in auriferous gravel, at a depth of one hundred feet.

In the vicinity of Placerville quite a number of discoveries of a similar kind have been made, at various times. One of the most interesting of these is described in the following letter from a physician, once a Californian, but who in 1870 was residing in Geneva, Wisconsin. This is given in full, as containing very detailed and valuable testimony in regard to a discovery of human remains at a locality at which he was formerly engaged in mining. This letter was written in answer to one addressed to him by the State Geologist, asking for information in regard to a find reported to have been made in former years.

GENEVA, WAL. COUNTY, WISCONSIN, November 2, 1870.

DEAR SIR, — In accordance with a wish expressed in your letter of June last, I would state the particulars of the discovery of human bones on Clay Hill,* El Dorado County, California. While engaged in the business of mining, in the spring of 1853, I purchased an interest in a claim on this hill, on condition that it prospected sufficiently well to warrant working it. The owner and myself accordingly proceeded to sink a shaft for the purpose of working it. It was while doing so that we discovered the bones to which you refer.* Clay Hill is one of a series of elevations which constitute the water-shed between Placerville Creek and Big Cañon, and is capped with a stratum of basaltic lava, some eight feet thick. Beneath this there are some thirty feet of sand, gravel, and clay. The country-rock is slightly capped on this, as on most of the elevations, the slope being towards the centre of the hill. Resting on the rock and extending about two feet above it, was a dense stratum of clay. It was in this clay that we came across the bones. While emptying the tub, I saw some pieces of material which on examination I discovered were pieces of bones ; and, on further search, I found the scapula, clavicle, and parts of the first, second, and third ribs of the right side of a human skeleton. They were quite firmly cemented together ; but on exposure to the air began to crumble. We made no further discoveries, and indeed I did not then take sufficient interest in such subjects to make any efforts in that direction. I preserved this part, and presented it to Dr. Harvey of Placerville. I did think it of interest, as showing the great age of the human race. No one who should give the slightest attention to the physical configuration of the country, but must be struck with the almost infinity of years which have elapsed since the being whose skeleton I had found was animated with life. It must be evident to every unprejudiced mind that the auriferous drift was deposited long anterior to the present drainage system. And then, when we look at the great depth which the rivers have worn through the drift, and into the rocky sides of the mountains, in some places nearly two thousand feet, the mind is lost in wonder in contemplating the time which has elapsed since man was first born into the world.

Begging pardon for the length of this, I remain very truly yours,

    [Signed]   H. H. BOYCE, M. D.

In reply to further inquiries made by the writer, Dr. Boyce stated that there could be no mistake about the character of the bones, and that he had made a special study of human anatomy. His description of the geology of Clay Hill agrees, in the main, with that given by Mr. Goodyear, who states that the deposit on the bed-rock was from twenty-five to thirty feet thick, all but the lower five feet consisting of "mountain gravel," a local name for the volcanic material capping the hills in that vicinity.† Dr. Boyce's statement carries with it the weight necessarily attaching to the observations of an educated man, evidently accustomed to observe and reflect on what he saw in the remarkable region in which he was temporarily engaged in mining. The bones themselves seem to have disappeared, as no trace of them could be found at Placerville, although there was still some remembrance of them in the minds of persons who had seen them after they had been taken out from the diggings.

Similar occurrences are reported from a place bearing the somewhat characteristic name of Soapweed, and situated near Placerville.

Spanish Flat is reported as having been a prolific locality of stone implements, which were dug up at various points and at a depth of from ten to sixteen feet beneath the surface. These relics are described by Mr. Voy as consisting of "tools, kitchen utensils, and other indestructible traces of man's presence and activity." Among them were oval stones with continuous grooves cut around them, not transversely, as in the case of the stone hammers so common on Lake Superior, but around the outer edge of the flat ovals. The use to which they were put it is not easy to make out. Besides these, there were other curious implements of which the use is unknown to the writer, unless possibly they may have served as ornaments. Mr. Voy thinks it more likely that they were intended as handles for bows, being hollow on one side, as if to fit the weapon, and convex on the other to give the hand a better grasp. Two of these are nicely worked out of a very fine-grained stone which resembles diorite. These are five or six inches long, and one inch thick in the middle. Specimens of all these implements are in the Voy Collection.

* See diagram of localities near Placerville, Plate C.
† See ante, p. 101.

Other localities in El Dorado County where similar discoveries have been made and which may simply be referred to without giving particulars are : Kelsey's Diggings, a few miles north of Spanish Flat ; Dry Creek, three miles west of Georgetown, where a large stone dish, or platter, was found ; Coloma ; Georgetown. At the last-named place Mr. Gabb, the Palæontologist of the Survey, saw a human femur imbedded in "cement"; it was in the possession of Mr. A. Hyatt, the express agent. For some good and sufficient reason, the nature of which has escaped the writer's memory, this find was not investigated further.

Mr. Goodyear learned at Brownsville, from Mr. Ford, that near the head of Spanish Creek a perfect mortar and pestle were once found in the gravel beneath the volcanic matter. According to Mr. Ford, these implements were taken to Placerville, and were for a long time in the possession of a gentleman named Douglass. Farther investigations into this find were not made.

PLACER COUNTY.

At various times, between the years 1851 and 1864, quite a number of stone relics were found at and near Gold Hill, in the different auriferous gravel deposits. The depths from which they were taken varied from ten to twenty feet. The objects found were mortars of different forms and sizes, large stone platters or dishes, and "other interesting things." One of these mortars, which is made of granite, is in Mr. Voy's collection. It is sixteen inches high and twelve in circumference ; its weight is about twenty pounds.

According to Mr. Voy, who is also the authority for the statements in the preceding paragraph, stone implements were found, in 1864, about a mile south of the town of Forest Hill, at a depth of about ten feet. One of the most interesting of these was a flat dish, or platter, worked out of hard granite, and about eighteen inches in diameter. Similar discoveries have been made, as is stated, at Byrd's Valley, a short distance below Michigan Bluff.

The Missouri Tunnel runs from the Devil's Cañon southerly into the ridge between it and the Middle Fork of the American River, a little above Yankee Jim's. This region has been described in the preceding pages as deeply covered with volcanic materials. In this tunnel, under the lava, two bones had been found, some thirteen years before Mr. Goodyear's visit to the locality, which were pronounced by Dr. Fagan to be human. One was said to be a leg bone ; of the character of the other nothing was remembered. The above information was obtained by Mr. Goodyear from Mr. Samuel Bowman, of whose intelligence and truthfulness the writer has received good accounts from a personal friend well acquainted with him. Dr. Fagan was at that time one of the best known physicians of the region.

NEVADA COUNTY.

Mr. Voy collected a large amount of information in regard to the occurrence of implements in the gravel diggings of Nevada County.

At Grass Valley, between the years 1853 and 1864, numerous stone relics, such as mortars, pestles, and grooved oval disks were found at various depths, ranging from fifteen to thirty feet beneath the surface. Some of these implements are in the Voy Collection.

In Myers's Ravine, about two miles northwest of Nevada City, stone mortars and other implements were found, in 1866, at depths of from ten to sixteen feet in the auriferous gravels.

On Brush Creek, about two miles north of Nevada City, similar discoveries were made about the year 1866. The same is reported from Sweetlands, about eight miles north of the last-mentioned locality.

BUTTE COUNTY.

There seems to have been a great variety of stone implements found in the gravel mines of this county at different times during the past twenty years. It is unnecessary to give the details, which have, in most cases, not been carefully inquired into, partly for want of time and partly because the mass of evidence already accumulated seemed to be so large already. The principal localities seem to have been : the neighborhood of Oroville, Bidwell's Bar, Junction Bar, Saw-mill Ravine, and, in general, the whole region of the foot-hills described in the preceding pages as being covered by deep gravels overlain with heavy deposits of volcanic materials.

Mr. Amos Bowman, for a time one of the assistants on the Geological Survey, collected considerable information in regard to the occurrence of implements at Cherokee, a few miles north of Oroville, in a locality briefly described in the preceding pages.* The facts reported by Mr. Bowman will be found in the Overland Monthly (Vol. XV. p. 34, 1875). As this now defunct periodical will probably be accessible to but few of those who have occasion to consult the present volume, a portion of Mr. Bowman's statements (omitting the accompanying fanciful and somewhat absurd speculations) may here be quoted : —

"I presented the California Academy . . . . along with a stone mortar . . . . which I obtained recently at Cherokee, Butte County. . . . . The mortar is from the hydraulic mines, where from half a dozen to a dozen or two have been found, — enough to establish the presence of a large population in the vicinity, taken in connection with all the surrounding facts and circumstances. Several of these mortars I was able to trace through the finders to the particular spot where they were found. . . . . One of the mortars, found by Mr. R. C. Pulham, of the Spring Valley Mining Company, was taken out of a shaft which he dug himself in 1853, and was found, according to his testimony, twelve feet underneath undisturbed strata, the character of which is still visible in the bank adjacent. He is certain that the mortar was placed there before the overlying gravel.

"This mortar was found standing upright, and the pestle was in it, in its proper place, apparently just as it had been left by the owner. The material around and above it was fine quartz gravel intermixed with a large proportion of sand; in short, just the material of an ordinary sea-beach. This was forty or fifty feet above the bed-rock, and about thirty feet above the blue gravel.

"About 300 feet east of this shaft Mr. Frederic Eaholtz took out in 1853 a similar mortar at a greater depth. I visited both places with Mr. Pulham, and found several mortars still lying around on the top of the blue-gravel bench which is not yet mined away. The locality is about seventy yards east of Charles Waldeyer's house. Mr. Eaholtz was sent for, and he told me further that, in 1858, while engaged with Wilson and Abbott in mining in the southwesterly part of the Sugar Loaf, he found in place, forty feet under the surface, a mortar of the same sort in unbroken blue gravel. This blue gravel nowhere comes to the surface, and it extends with the before-mentioned white and yellow gravel, under the Sugar Loaf, and under the Oroville volcanic mesa. It appeared only on the bottom of this claim. He was picking the blue gravel to pieces with a pick, when he found the mortar, which was a portion of the mass of cemented boulders and sand. He picked it out with his own hands. Both these witnesses are trustworthy men, widely acquainted in the county, and they are willing to appear before a notary to certify to the above.

"The fossils are from two different gravel beds immediately underlying the auriferous gravel formation and the volcanic outflows, at a distance of about one and a half and two and a half miles from Cherokee, in a southwesterly and northwesterly direction respectively. The latter is only about thirty feet underneath the volcanic capping of the Dogtown and Mesilla Valley table-land, in a ravine immediately back of Van Ness's house on Dry Creek."

Cases similar to the above-cited are reported from a number of localities in both Siskiyou and Trinity Counties. It would be tedious to enumerate all the instances in these counties in which mastodon and other bones of extinct animals are said to have been found in connection with human implements.

* See *ante*, pp. 207, 208.

A sufficient number of these have already been given. We may now pass, therefore, to a brief discussion of the facts which have been set forth; and this discussion will have to do chiefly with the probable authenticity of the mass of evidence taken as a whole, and the question what geological age is to be assigned to these various remains. A more general treatment of the whole subject, especially from the point of view of the physical conditions which have been influential during the later Tertiary times in the Sierra Nevada, may be expected in a future chapter.

In considering all the evidence which has been offered above, a most striking thing is its coherence. It all points in one direction. There has never been an attempt made — so far as known to the writer, and he has kept his eyes pretty widely opened in this direction — to pass off on any member of the Survey, when engaged in seeking for proofs of the antiquity of man in the mining region, anything out of keeping, or, so to speak, out of harmony, with what had been already found, or might be expected to be found. No hieroglyphics, ornaments, or implements of any kind, the workmanship of which would indicate a high degree of civilization, have ever been offered as finds in the auriferous gravel region. It has been always the same kind of implements which have been exhibited to us, namely, the coarsest and the least finished which one would suppose could be made and still be implements at all. Stone mortars are by far the most common and the most striking utensils which have been unearthed from the gravel. This may be partly due to the size and strength of these articles, which are usually such as would enable them to withstand a great amount of hard wear without going to pieces. Still, it is difficult to avoid the conclusion that the race inhabiting this region in prehistoric times did, for some purpose or other, make great use of these utensils. The most natural supposition in regard to the mortars is, that they were used for grinding food. They are not in use at present among the Indians of that part of California where the implement in question is so abundantly found. The Digger Indians seem now, for some unknown reason, to prefer cavities worn in the rock in place, and in these the writer has often seen them crushing their acorns; but never once has he found them using the portable mortar.

It is not intended, however, at the present time, to go into any discussion of the ethnological relations of the prehistoric race as evidenced by the implements which have been discovered. All that is desired is to impress upon the reader the homogeneous character of the evidence from the point of view of the nature of the objects which have been discovered. It shows, as the writer thinks, very decidedly that no portion of the evidence has been gotten up to deceive. Persons anxious to impose on the "unsuspecting geologist" would have been likely, in some cases at least, to endeavor to palm off on him some specimen of writing in an unknown tongue, some article of finished workmanship, or perhaps an imitation of some portion of the human figure; or possibly even something as preposterous as the "Cardiff giant" in its make and style.

One reason why this has not been done, perhaps, is that the miners, as a general rule, have very little appreciation of the great geological age of the formations in which they are working. They see that the "high gravels" look very much like those often occurring in the present river beds, and they do not ordinarily take into consideration the immense amount of erosion and denudation which the region must have undergone since these gravels were deposited. This inclination to see in all the objects found in excavating in the older detrital formations a resemblance, or identity even, with species now existing in the region is well illustrated by the fact that the miners almost invariably declare the impressions of leaves and the fragments of wood occurring in the gravel to belong to trees and shrubs exactly like those now growing in the Sierra. Yet the reader will have seen, from what has been published in the preceding pages, how different most of these are.

Another consideration may be introduced. Some persons are inclined to think that the implements found in the gravels are simply works of the present Indians, which, lying scattered over the surface, have been washed or carried down into the gravels, through fissures or in slides, or in some other not exactly defined manner. Some would suppose that these things might have been intentionally buried, either in funeral ceremonies or for concealment. Had this been the case, it would have been expected that finds similar to those announced as of such frequent occurrence in the mining region would have turned up — occasionally, at least — in the Coast Ranges. No instance of this, however, so far as known to the writer, has ever been heard of. The soil and detritus of the region about the Bay of San Francisco have been excavated for all sorts of purposes, and in a great many localities bones and teeth of extinct animals have been found in abundance. Never, so far as known, have any human bones or works of human hands been met with in connection with these remains, while they are common enough on the surface. This non-occurrence of proof of the existence of prehistoric man in the Coast Ranges has been regarded by some as being strongly indicative of the improbability of his existence in the Sierra. But this appears to the writer, on the other hand, confirmatory evidence of the genuineness of the discoveries in the mining region. It will be made apparent, in a future chapter, that there are strong reasons — if the past geological history of the Pacific Coast has been made out with some approximation to correctness — for believing that man could not have maintained an existence on the Coast Ranges any earlier than towards the very close of the Pliocene epoch.

The considerations here adduced, taken in connection with the large body of facts which have been presented in the preceding pages, seem to justify the belief that it is not possible for any candid mind to deny that the human race has existed, during a prehistoric period of great length, contemporaneously with a fauna which has now entirely passed away. That man co-existed with the elephant and mastodon and other extinct animals has been abundantly proved by the — so to speak — every-day occurrence of proofs of his former presence in connection with the remains of those animals, at all depths up to a hundred feet or more. To go one step farther, and admit that there is ample evidence of the existence of man in California previous to the cessation of volcanic activity and to the erosion of the deep cañons of the Sierra, will probably be to many persons a more difficult matter. And yet, incredible as the facts may appear, it is difficult to see how they can be set aside. Leaving out of consideration the Calaveras skull, we have such evidence as that of Dr. Boyce and Mr. Hubbs, which appears as clear and direct as possible, while it is also amply supported by other reported facts from various quarters which it seems in no way reasonable to ignore. That a deposit of human bones should be opened in some position clearly anterior to the period of the basalt, and remain accessible to geologists for an indefinite period, so that they might all go there and dig for themselves, is hardly to be expected, after all that has been stated with regard to the gravel deposits in the preceding pages. It is much more likely that general belief in the existence of man in California during the Tertiary epoch will be brought about by the gradual accumulation of similar facts from other portions of the world. The discoveries made in Europe, which have already obtained general credence, carry man back close to the verge of the Tertiary ; if not, indeed, a little the other side of the line.

Other facts, which seem equally well vouched for, but which have not met with general credence, would place man in Europe in the Tertiary as decidedly as does the Calaveras skull in America. Among these the brilliant investigations of Ribeiro and the Geological Survey of Portugal ought to be mentioned, as an excellent illustration of the way important discoveries — backed up by an abundance of evidence — are ignored, when not fitting in with the preconceived ideas of the majority.*

It will be expected by many that the epoch of the Tertiary at which man made his appearance in California, or, at least, in which he can earliest be recognized as having existed, should be stated as definitely as possible. This is by no means an easy thing to do. The divisions of the Tertiary into Pliocene, Miocene, etc., as originally instituted by Lyell, were based on the relative abundance in the different groups of the shell-bearing mollusca, and far more on those of marine than on those of fresh-water origin. To coordinate strata thus classified with others which contain only the remains of land animals is, of course, a very unsatisfactory proceeding. Indeed, the divisions of the Tertiary in any region ought not, in the nature of things, to be strictly comparable to those in another somewhat distant one. In the upper part of the geological series animal life begins to be differentiated according to zonal position and climatic conditions, hence general classifications of the Tertiary become vague and often even misleading; yet so convenient are they, that it will probably be long before they drop out of use. As one of the consequences of the use of Lyell's names for the Tertiary there is no representation of the Eocene in California; or, at least, none which can be distinctly recognized as such. Yet, no doubt, while Eocene rocks were being deposited in other parts of this country, and in other more distant regions, some accumulations of sediment were being formed, and some kind of animal or vegetable life existing on the Pacific slope.

---

* See Descripção de alguns Silex e Quartzites lascados encontrados nas Camadas dos Terrenos Terciario e Quaternario das Bacias do Tejo e Sado. Lisboa, 1871. In this memoir Ribeiro shows that cut flints, evidently the work of human hands, have been found in abundance in the Pliocene and Miocene, even, of Portugal; that there are more than 1,200 feet of strata piled over the beds containing these implements; that these rocks have been upheaved and turned up at an angle in places quite vertical, since man's appearance; and that this appearance was prior to the cessation of volcanic activity in that region, so that the human race must have witnessed those grand dislocations of the earth's crust to which the present topographical features of that part of Portugal are due. Verneuil himself was obliged to admit that there was every reason for calling these strata Tertiary, except that they contained proofs of the existence of man at the time of their deposition. Lyell does not even mention the Portuguese discoveries in the last edition of his " Antiquity of Man," published in 1873, two years after the appearance of Ribeiro's memoir.

The important investigations made in India, by the officers of the Geological Survey, have been equally ignored by Lyell; they seem to show, beyond doubt, the immense antiquity of the human race in that country. These results were published several years before Lyell's last edition was issued.

It appears probable on stratigraphical grounds, as will be set forth in a future chapter, that the detrital beds overlying the "bed-rock" of the Sierra Nevada represent the whole Tertiary period; that is, that they have been forming since the beginning of that epoch. It also seems likely that there has been no break in the series of events; none, at least, of sufficient magnitude to justify the drawing of a sharp line anywhere, so as to say, above this is Pliocene, and below it is Miocene, or Eocene.

The evidence as to the geological age of the gravel deposits afforded by the plants found in the sedimentary beds underlying the latest eruptive masses in the mining region of the Sierra has already been discussed by Mr. Lesquereux. He distinctly recognizes the presence in this flora of forms identical with, or closely allied to, those of the Miocene; but still calls the age of group Pliocene. Something of the same kind seems to be legitimately inferred from the animal remains of the same deposits. There are certain fossils which have been found only in deep-lying gravels, like those of Douglass Flat and Chili Gulch. No traces of the rhinoceros, the elotherium, or the small equine animal referred, with doubt, by Leidy to *Merychippus*, have ever been found in deposits which could by any possibility be proved to be more recent than the basaltic overflow. It is true that the evidence, thus far collected, is but fragmentary; still, taking it for what it is worth, it may be said that the affinities of the animals found in these lower deposits would indicate a Miocene rather than a Pliocene age. There are also, it is believed, stratigraphical reasons for admitting that some, at least, of the deposits containing these older fossils may be proved, by other than palæontological evidence, to belong to an older series than those strata which, though anterior to the basalts, yet contain a fauna decidedly more Pliocene than Miocene in character. This subject may, however, be discussed more at length, from the stratigraphical point of view, in a succeeding chapter.

To draw the line between Pliocene and Recent, or to say whether between these two groups any distinct series of beds of an intermediate age exists in the fresh-water Tertiary of the Sierra, is a matter of even greater difficulty than to separate the Pliocene from a possible lower member of the series. It seems that the mastodon lived through a portion of the volcanic era, and flourished exceedingly down to a geologically recent period. The elephant, on the other hand, has not been detected in the beds below the basalt, so far as the writer can learn. That the fauna associated with these animals was entirely different from that now living in the same region is beyond doubt.

Whether any of the species occurring in the gravels are not at present extinct is not to be decided from the material thus far collected. The tapir found in the gravel, for instance, may or may not be identical with a living species. But, taking the whole evidence together, it is certain that the fauna of the gravel deposits is almost exclusively made up of extinct species, and we are justified in saying that it is far more Pliocene in its aspects than Postpliocene or Recent.

On the whole, then, it seems reasonable, in the light of the evidence thus far collected, to assert that the conclusions drawn by Mr. Lesquereux from the study of the fossil plants of the gravel series are strengthened by the indications afforded by the fossil animal remains. The Miocene may be represented by some of the lowest deposits, while a portion of the series lying under the basalt is probably of Pliocene age. The passage from Pliocene through Post-pliocene, if such a division can hereafter be maintained in this region, has been a gradual one, and some of the Pliocene animals have certainly lived on close up to the Recent period. That a portion of the human remains and implements described in the preceding pages are as old at least as Pliocene, it seems hardly possible to doubt.

It may be proper to add a few remarks in regard to the amount of time which has elapsed since the human race began to exist in California, as shown by the evidence which has been brought forward on the preceding pages.

In the first place, the total change in the fauna and flora of the region, and we may say, in fact, the succession of changes, will always be admitted by geologists as having required an immense amount of time for their completion. Nothing that we know of the history of the past will justify the assumption of a sudden extinction of the animal and vegetable creation within an extensive region and the immediate introduction of a new creation. The more zoölogists and palæontologists have studied the record of the past, the more they have become convinced that such changes have taken place step by step, and that time has been an essential factor in their occurrence.

The necessity of the grant of time demanded for the changes in the organic life of the region is vouched for by the phenomena of erosion which the region presents, and which has already been sufficiently insisted on. We know, for instance, that after the fossiliferous strata under the various basalt-capped mountains in California were deposited there were repeated phases of volcanic activity and of aqueous invasion, during which the alternations of gravel, sandstone, and volcanic materials were piled upon each other to the height of from 200 to 300 or more feet. The time required for this accumulation of strata — hurrying up the operation as much as we possibly can — will have been considerable. Some of the finer deposits of clayey materials are of great thickness, and must have occupied a long period in their deposition. Even the more turbulent torrents bearing down the coarser gravels from the more elevated regions must have persisted during long periods of time, in order so effectually to spread the débris they conveyed over such an extensive area. Neither is the transition from a period of volcanic activity to one of repose likely to have been always abrupt. The analogy of present volcanic action shows clearly enough that considerable time is required for thick accumulations of erupted materials of various texture and composition to be formed.

But if time is needed for the accumulation of the strata in question, how much more is imperatively demanded for the accomplishment of the immense denudation and erosion which the region has undergone, and of which some idea has been given in describing the phenomena of Table Mountain! And the farther north we go, the more extensive this erosion proves to have been.

The discoveries in California, even admitting all that seems to be indicated by the truly wonderful occurrences of human remains and works of art under Table Mountain, are by no means unprecedented or unsupported by similar testimony from other countries. Almost every month brings us from some part of the world proofs of man's having existed at a remote epoch, and for an indefinite period, in what we may call his primitive condition, that is to say, in the lowest stage of development in which he could continue to exist as man. It is only necessary to mention in this connection * the discovery of stone implements by the officers of the Geological Survey of India, scattered over a considerable portion of the Indian Peninsula, and occurring under circumstances which prove very clearly that great changes have taken place in the physical geography of that country, since the people lived by whom these implements were fashioned. The evidences to this effect have been accumulating for many years, and are as unmistakable in their nature as is possible for any geological evidence to be. The whole

---

* In addition to what has been already said of similar and perhaps even more extraordinary discoveries in Portugal.

of Southern India has been raised 600 feet vertically, and then subjected to immense denudation, after which the coast was depressed again ; and all since the appearance of man, indicating a lapse of time which may have been even greater than that required for the denudation of Table Mountain. There is hardly any portion of the globe, however remote, which has not afforded some evidence of the long-continued existence of man in his primitive condition ; even Japan and Australia have their testimony to offer.

What, then, is this primitive condition of man ? Nothing, it may be replied, in any important respect different from what we see exhibited by the lowest types of the human race now existing on the earth. All the investigations of geologists and ethnologists thus far have failed to obtain satisfactory evidence of the existence at a previous epoch of any type of being connecting man with the inferior animals, or decidedly lower in grade than races now inhabiting portions of the earth, or anything that we fail to recognize instantly as man. In fact, we cannot conceive of man's existence at all at any lower stage. The implements of the palæolithic age are simply the implements necessary for the support of life to a being having the requirements of man, but destitute of strong claws and powerful teeth. They consist of the rudest possible implements of attack and defence, fashioned in the simplest manner, and of the only material which nature everywhere offers suitable for the purpose, namely, certain varieties of rock which are tough and heavy, and which can be fashioned into shape without the aid of fire, and with the smallest conceivable amount of intelligence, certainly not more if as much as we in repeated instances see displayed by animals we are accustomed to call inferior to man. Next to weapons and instruments of defence, or perhaps even before them, would come implements for procuring and preparing food, and possibly for cooking it. No doubt man for an immense period of time has been in the habit of cooking his food ; but how long the human race existed before the use of fire we have no means of knowing or of even guessing. Probably the desire for some kind of personal ornament was early implanted in the human breast ; and man may have had a ring in his nose, or women a similar appendage in their ears, before the art of cooking was introduced.

The discoveries in California, India, and elsewhere seem clearly to indicate that the human race must have existed, over a large portion of the world at least, for an immense period of time, in the primitive condition, that is, at the lowest possible stage of humanity, — civilization it cannot be called. So far as California is concerned, the evidence all points in this direction. The implements, tools, and works of art obtained are throughout in harmony with each other, all being the simplest and least artistic of which it is possible to conceive. Whether found in the strata under the basaltic lava, or above, at any point or depth in the detritus, we always recognize the same type.

There is nothing about either the remains of man or his works which indicates anything different from what we find in other parts of the world wherever the lowest stratum of humanity has been reached, or essentially different from what is now existing in California itself. Hence we may quote Huxley's statement in regard to the Engis and Neanderthal skulls, and, introducing the necessary modification demanded by the Californian discoveries, say that the advocates of progressive development must look farther back than the Pliocene (if that be the correct term to apply to the strata underlying the basalt) for traces of the primordial stock.

It is evident that there has been no unfolding of the intellectual faculties of the human race on this continent which can be parallelized with that which has taken place in Central Europe. We can recognize no palæolithic, neolithic, bronze, or iron ages. Over most of the continent man cannot, as it seems to the writer, be considered as having made any essential progress towards civilization. What the exact relations of the intellectual development of the Central and some of the South American peoples, to our civilization are, it is not easy to understand. At present our stock of information is so scanty that generalizations cannot be too cautiously drawn.

The steps of progress in Central Europe which are indicated by the successive use, first of more artistic stone implements, then of bronze and afterwards of iron, have no parallel on this continent. The use of copper tools in the Mississippi Valley does not, as the writer believes, indicate any considerable progress in intellectual development, for the copper at first may have been picked up from the surface in the native form, and to fashion it into rude tools required but little more skill than is indicated by the chipped obsidian implements which are now, and have been from all time, in use among the aborigines of this continent. The mining work executed in the search for copper, after the supply from the surface had been exhausted, proves the existence of an inexhaustible amount of patience, but nothing like the ingenuity and development of brain which the fabrication of bronze would imply.

Finally, as the summing up of the discoveries and investigations made by the Geological Survey in California, we have : —

1. The clear and unequivocal proof, beyond any possibility of doubt or cavil, of the contemporaneous existence of man with the mastodon, fossil elephant, and other extinct species, at a very remote epoch as compared with anything recorded in history.

2. That man, thus proved to be contemporaneous with a group of animals now extinct, did not essentially differ from what he now is in the same region and over the whole North American Continent.

3. That there is a large body of evidence, the strength of which it is impossible to deny, which seems to prove that man existed in California previous to the cessation of volcanic activity in the Sierra Nevada, to the epoch of the greatest extension of the glaciers in that region, and to the erosion of the present river cañons and valleys, at a time when the animal and vegetable creations differed entirely from what they now are, and when the topographical features of the State were extremely unlike those exhibited by the present surface.

4. That man existing even at that very remote epoch, which goes back at least as far as the Pliocene, was still the same as we now find him to be in that region, and the same that he was in the intermediate period after the cessation of volcanic activity, and while the erosion of the present river cañons was going on.

5. That the discoveries in California, and those in other parts of the world, notably in Portugal and India, present a strong body of evidence going to prove the existence, during an immensely long period, of the human race in its primitive condition, — that is to say, in the simplest and rudest condition in which man could exist and be man.

6. That, so far as we now know, there is no evidence of the existence of any primordial stock from which man may have been derived, as far back at least as the Pliocene. Man, thus far, is nothing but man, whether found in Pliocene, Post-pliocene, or Recent formations.

CHAPTER IV.

RÉSUMÉ AND THEORETICAL DISCUSSION OF THE PRINCIPAL FACTS CONNECTED
WITH THE GEOLOGICAL OCCURRENCE OF THE AURIFEROUS GRAVELS.

SECTION I. — *Introductory remarks.*

In a previous chapter of this work a brief general account of the mode of
occurrence of the Auriferous Gravels of the Sierra Nevada has been already
given, for the purpose of making that which was to follow intelligible to the
reader. It now remains to present a fuller discussion of a portion at least of
the problems involved in the gravel question, in which the object will be, to
throw some light on the physical conditions which have had as a result the
very remarkable deposits of which a somewhat detailed description has been
given in the pages of this volume.

It will be well to caution the reader, however, that an exhaustive discussion of the phenomena of the gravels is not to be expected. Indeed, such is
the nature of geological inquiries in general, that in no branch of that science
is there a final and absolute result to be attained. The geologist must content
himself with slow and often hardly perceptible advances towards the ideal
of a perfect work ; he must be satisfied if he throws here and there a little
light on doubtful points, removes a few erroneous ideas, and indicates to
some extent the direction which future investigations should take. Furthermore, it should be mentioned in this connection that the present volume is
not intended to answer as a technical guide to the practical side of hydraulic
mining. The processes employed in that kind of work are simple enough in
theory, and the machinery not complicated ; notwithstanding this, the business is one which requires great experience and much caution, since the
preliminary expenditures in such undertakings must of necessity be large,
and difficulties are wont to occur for which only the most far-sighted are
likely to be prepared.

An ideal report on the gravel region, from the practical point of view,
would demand a much larger amount of time and a vastly greater expenditure of money than have been at the disposal of the present writer. Such a
report is not to be looked for, indeed, if by an "ideal report" we mean one

which should be a trustworthy guide to the miner in determining before-
hand where work could be carried on, at what cost, and also with what profit.
The problems presented are too difficult and complicated, in most cases at
least, to be settled without actual experiment. Still, it is true that a good
deal of assistance could be rendered, in the way especially of preventing
wasteful expenditures, if a systematic examination of a sufficiently detailed
character could be made of the gravel region, accompanied by a careful inspection of all that was doing in each important district and a full record of
everything observed. Such an investigation should, however, have been
begun long ago, for already a great deal of valuable information has been
lost forever, there being no one on the spot sufficiently intelligent or interested in the matter to seize the opportunities presented for examination, and
which can never recur after the right moment has passed, and the locality
gone to ruin or been abandoned. As a preliminary to any such work, however, an accurate contour-line map, on a very large scale, would be absolutely
necessary, while the natural difficulties of such a survey are so great that it
could not be made without a large expenditure.

It is hardly necessary to say that the work of any geological survey, to be
of practical value, must be continuous and never-ending. The conditions of
every region of metalliferous or other deposits of economical value are constantly altering as exploitation goes on. Old mines are worked out, new
ones are discovered, new processes are invented, other more prolific regions
opened, forbidding competition on the part of the poorer ones, prices rise and
fall as the supply increases or diminishes : all these changes make the development of the mining resources of any region something to be always looked
after by the Government, if looked after at all. And it is in accordance with
such views as these that in civilized countries, where mining inspection exists
at all, its task is a never-ceasing one, its execution, moreover, being intrusted
to the competent hands of men specially and thoroughly educated for it.

The present chapter, then, must be looked upon as an attempt to throw
light on some of the scientific problems presented by the auriferous gravels
of California, rather than as a practical guide to the miner in his work.
Not that it may not interest those actually engaged in hydraulic mining.
On the contrary, it appears that the miners themselves are, in many instances, greatly exercised in their minds by the curious phenomena with
which they are brought in contact ; and it has already been noticed that,
while so-called scientific observers were utterly mistaken in their views

of the origin of the gravels, those practically engaged in working them had arrived at much more correct conclusions in regard to some of the important points involved. It is for the satisfaction, then, of both practical and scientific men, that the following discussion is entered upon; and it is hoped that at least it may have the result of causing, in some localities, a more careful and systematic observation of some of the facts than has been hitherto customary.

By consulting the Appendix to this volume,* it will be seen that after writing out his very voluminous notes on the region examined by him, which have been already presented in a condensed form in a previous chapter, Mr. Goodyear furnished a supplementary paper, containing a discussion of some of the difficult points presented by the gravels of the Sierra Nevada. This paper has been printed in full in the Appendix, and it should be read with care by all interested in these geological problems. This abstract, prepared by Mr. Goodyear, has been printed not only as an act of justice to him, but also as throwing light on some points of importance. The writer's opinions differ very considerably from those of Mr. Goodyear in regard to certain of the main questions involved, as those who take the pains to read the present chapter and Appendix B will readily perceive. The writer's views, however, are based on a long-continued series of investigations in every part of the gravel region, while Mr. Goodyear was naturally most strongly impressed by the phenomena which had been specially well exhibited in the district to which his work was mainly limited.

It should not be forgotten that the entire body of views and opinions in regard to the mode of occurrence of the gravel is something which has long been under discussion among the various members of the corps of the Geological Survey, carried on under the writer's direction. All that is definitely known to science in regard to these matters comes from the work of that survey, as has already been clearly set forth in a previous chapter, where indeed credit has been given to those of the practical miners whose views have in some important respects been much more nearly consonant with the truth than those of any of the scientific men who had visited the region previously to the appearance of the Geology of California, Vol. I.

What has here been stated in regard to the theories of the gravel, and the divergencies of opinion in regard to vital points connected therewith, will be better understood after a perusal of the following section, in which the

* See further on, Appendix B.

endeavor will be made to set forth that which may now be considered as quite clearly established in reference to the original mode of deposition and character of the auriferous gravel deposits of the Sierra. After that has been accomplished, the way will be clear to take up some of the more difficult and controverted points, the writer hoping that in thus doing he will be able in the discussion to throw light on some of them.

After these preliminaries, the remainder of this chapter will be devoted to a *résumé* of certain portions of the investigations embodied in the present work, in regard to which it seems desirable that the salient points should be brought together into a convenient form for a rapid review, for such persons as would not desire to make themselves acquainted with the mass of local details with which some of the essential facts are encumbered in the descriptive parts of this volume.

SECTION II. — *What may be assumed as having been clearly established in Regard to the Theory of the Gravel Deposits.*

In a previous chapter there has already been given a partial discussion of some of the more prominent points connected with the theory of the occurrence of the gravels.* This was done in order that the reader might have the preliminary information necessary for understanding the details which were to follow in regard to the nature and distribution of the various formations, to the special study of which this volume is devoted. It now remains to take up the more doubtful points suggested in the course of the investigation, and to devote especial attention to such of these as seem to require such treatment, and which the reader who has gone over the details embodied in the preceding chapters and Appendix A will now be better prepared to understand.

The present section is introductory to such a review and general discussion, its object being to clearly distinguish between those points in regard to the theory of the gravels, which are so well made out that it is proper to assume that there will be no difference of opinion about them, and, on the other hand, those which are more obscure and which therefore may suitably be laid before the reader at some length, and not without admitting that there are certain things which cannot yet be made so perfectly clear as to warrant the expectation of entire unanimity of opinion in regard to them.

* See *ante*, pp. 53–78.

From the historical summary presented in a previous chapter,* it will have been seen that, previous to the investigations of the California Survey, there were a variety of opinions current in regard to the nature and origin of the gravel deposits. Scientific men in general looked upon these detrital accumulations as having been formed by the agency of the sea, which spread them uniformly over the slope of the Sierra, while their present occurrence at various elevations and in detached areas was accounted for by subsequent supplementary upheavals of the range and the irregular erosion arising therefrom. That these views, based on very insufficient explorations, and quite at variance with almost all the prominent facts, can no longer be held by any one who is acquainted with the region in question, may be unhesitatingly asserted. That the gravels have been formed and deposited by the agency of fresh water may be set down as positively determined.

The idea that the gravels were of fluviatile origin having gained ground among the miners of the region even before it was accepted by scientific observers, it was quite generally held that it was essentially one stream which did the work; or, at least, that the most important and valuable gravel deposits, especially those represented by the " blue-lead," belonged to one system of drainage, and that this was parallel to the present crest of the Sierra. This was, probably, the most popular theory current among the miners at the time the writer began his work in California. Some persons even connected the sources of this north and south running stream with regions far beyond the limits of California, and diverted the waters of the Columbia so as to bring them into connection with the monster river by which the blue-lead was originated. What peculiarities there were in the position and mode of occurrence of the gravels which led to the adoption of such strange views will have been made sufficiently apparent from a perusal of the preceding pages.†

We may now proceed one step farther than we could at the time of the publication of Geology, Vol. I. It had then (1865) been clearly made out by the labors of the Geological Survey, that the high gravel-deposits were exclusively of fluviatile origin, or, in other words, the work of ancient rivers; but in regard to the relation of this old system of drainage to the present one but little has been definitely ascertained. It could be stated that it had been shown by our explorations that the materials of which the gravel deposits

* See *ante*, pp. 66–74.
† See also Mr. Goodyear's remarks in regard to the " blue-lead theory" in Appendix B.

are made up had "been brought down from the mountain-heights above, and deposited in pre-existing valleys"; but whether these valleys could be in any way correlated with those occupied by the present streams — whether the drainage areas of Tertiary times had any analogy with those into which the Sierra Nevada is now divided — could not be positively asserted. The most that could be said was, that the work was done "under the action of causes similar to those now existing, but probably of considerably greater intensity." This statement, as the writer believes, has been now shown to be strictly true, and it only remains to give the reasons why it is so, in a somewhat more systematic form than has been possible in the descriptive chapters of this volume, which portion has necessarily been arranged in a geographical order.

It can now be stated as having been clearly established that the great drainage systems of the Sierra were, during the gravel epoch, not materially different from what they are at present. Not only was there no river, or system of rivers, running in a direction parallel with that of the present crest of the range, but the divides between the principal streams descending the slope of the Sierra must, in several of the most prominent instances at least, have existed from the beginning of Tertiary times. This topic has been sufficiently enlarged upon by Mr. Goodyear, in his review (Appendix B), and his conclusions are also corroborated by the results of Professor Pettee's work, so far as regards the entire separation of the Feather River gravels from those of the Yuba. Farther north the gravel deposits partake of the now existing irregularity of the drainage system. Beyond the North Yuba the ranges are broken and irregular, and there is no dominating one; the courses of the streams are not only often parallel with the crest of the Sierra for long distances, but are sometimes just the reverse of what they usually are farther south. Add to this other disturbing conditions, the nature of which will be explained further on, and it appears quite reasonable that in the region in question it should be difficult to correlate the present streams with those of Tertiary age.

The main results which have been attained in the exploration of the high gravel deposits of the Sierra Nevada are these: that these detrital masses are the work of rivers which are of Tertiary age, as will be more fully set forth in a succeeding section of this chapter; and that the principal topographical features of the range existed when the gravels began to be deposited on its slope, possessing much the same character that they have now. That the whole of the mass of the upper portion of the chain was somewhat more elevated than it now is cannot, of course, be denied, since erosive agencies have been continually bringing down materials from higher and depositing them in lower regions. That volcanic agencies also have played an important part during the gravel epoch, and especially towards its close, is also perfectly clear, for the innumerable sections presented in the hydraulic mining region show the presence of eruptive materials almost everywhere, while an inspection of the various maps accompanying this volume would bear witness to the same fact. We may, however, with propriety discuss the bearing and importance of the volcanic deposits, as connected with the other occurrences of the gravel epoch, without its being necessary to enter upon any investigation of the nature of volcanic phenomena in general,—a subject, as is well known, of the greatest obscurity and difficulty. The effect of a heavy covering of lava over the gravels, acting as a mechanical protection of that which is beneath; the blocking up of old channels by masses of eruptive material borne down from above, either directly in a melted condition, or indirectly through the aid of water; the impossibility of getting at the gravel, except by means of drifting, where it has been too deeply covered,—all these conditions are of extreme simplicity, and without theoretical difficulty.

Again, it is perfectly clear that the shaping of the surface of the bed-rock and all the erosion which has taken place since the beginning of the gravel epoch have been exclusively the work of water. Of course it is not intended by this to exclude the idea that aqueous causes have acted independently of all others. On the contrary, the descending currents have taken advantage of every favorable condition to do their work more quietly and more thoroughly. Chemical decomposition of the bed-rock, under the influence of volcanic agencies; heavy grades, giving the necessary velocity to the down-moving waters; and, above all, copious precipitation, producing streams of large volume,—these are among the favorable conditions which would insure a rapid performance of the erosive work. It can be set down, however, as established beyond any possibility of doubt, that ice had nothing to do with any part of the erosion of the gravel period. It was not until the whole mass of gravels and other detrital deposits had been accumulated, and the volcanic strata piled over these, often to a thickness of several hundred feet, in short, not until the whole of Tertiary time had passed, that the higher portions of the Sierra Nevada were occupied by glaciers. More than this, it is clearly established that the present river cañons had been cut down into

their present form, and that the whole topography of the Sierra was, even
down to its minutest details, just what it now is, before ice had any existence in the range.[*]

The main features of the gravel epoch, therefore, as it can be unhesitatingly stated, have been recognized with clearness. There has been an
accumulation of water-worn detrital material in the beds and along the sides
of a system of rivers, which flowed down the slopes of the Sierra, the principal divisions of whose drainage-areas are essentially the same at the present
time as they were at the beginning of the gravel epoch. Upon, and to a
limited extent interstratified with, these detrital beds of purely aqueous
origin, there exist heavy deposits of volcanic materials, some solid, others
brecciated and fragmentary, and others water-worn, their whole aspect being
such as to warrant the belief that the epoch during which igneous agencies
prevailed was one of long duration. After the close of the period of volcanic
disturbances and overflows, water continued its work; and the result is visible in the form of the present river cañons and in the character of the
detritus deposited in their bottoms.

Such being the admitted facts, we now direct our attention to an inquiry
into the difficulties presented by the phenomena of the gravels, in the course
of which the whole series of events which took place during that period will
be passed in review, the principal object being to throw light on the physical conditions prevailing while this geological work was being done. And
in introducing this discussion it will be necessary, first, to state briefly what,
from a geological point of view, are the fundamental differences between the
present epoch and that of the gravels.

The difference in geological age between the high gravel deposits and
those in the beds of the present streams, in so far as it is a matter of palæontological evidence, presents no other essential difficulty than that of drawing the line between Tertiary and Recent. That the epoch of the gravels
represents a large portion of Tertiary time cannot be doubted, and the endeavor will be made in a subsequent section to show why, perhaps, there is
so imperfect a representation of the earlier stages of that epoch among the
fossils found in these deposits. The question of geological age, as determined
by fossil remains, is not one, however, which much interests the miners, who
almost invariably consider that they recognize the leaves and wood so frequently found imbedded in the gravels as being the same as those now living
in the Sierra.

* See the Climatic Changes of Later Geological Times, passim.

The first thing which does excite the curiosity of the practical man in regard to the Tertiary gravel deposits is their elevated position; hence the term which has often been applied to them, namely, "high." They are accumulations of detritus far above the level of the present larger streams, the difference of elevation being usually hundreds, and not unfrequently thousands of feet between the ordinary river-beds and the "high gravels." It having been already stated that the topographical features of the Sierra, so far as the outline of the main drainage-basins is concerned, is essentially the same now as it was during the gravel epoch, it becomes an interesting problem to inquire why the Tertiary detrital deposits occupy this peculiar position, for it would seem natural that the débris brought down by the streams should continue to be accumulated along the same lines of drainage so long as there was no orographic reason for a change in this respect.

This, then, is one of the most important questions to be investigated: namely, to show how it is that the gravel of the Tertiary period, as a general rule, occupies a position far above that which is now being deposited by those streams of the present day which are the representatives of the ancient rivers.

The next point claiming attention is the magnitude of the older deposits of detrital material as compared with those accumulations which we perceive to be directly connected with the present streams. This fact is the one which, next to altitude, impressed itself most strongly on the minds of the earlier placer miners; hence we find the deposits in question to have been called "in the early days," "deep" as well as "high" gravels; and, even at the present time, both these terms are used indiscriminately by those who do not wish to designate them as "hydraulic"; that is, as being so situated as to be suitable for exploitation by the hydraulic process,—a method of attack which is entirely inapplicable to detrital materials which are not raised above the ordinary valley levels, or which are of insignificant thickness.

Thus we are presented with another problem: namely, to account for the quantity of gravel of Tertiary age, which, as has been made abundantly evident from the detailed descriptions given on the preceding pages, is often several hundred feet in thickness and spread over wide areas. And in endeavoring to give a satisfactory explanation of the position and magnitude of the older gravel deposits, we shall have brought up for consideration a great number of minor problems, the solution of which is involved with that of the other main questions. The nature of these problems will become evident during a perusal of the following sections of the present chapter, and it will not be desirable to take the space which would be necessary for stating them, in advance of their presentation in the proper order as developed in the ensuing pages.

SECTION III. — *The General Distribution of the Detrital and Volcanic Materials on the West Slope of the Sierra Nevada.*

In proceeding to a general discussion of the phenomena of the gravel deposits, it will be desirable, in the first place, to give an account of their geographical distribution, and of the various materials of which they are made up.

In considering the position of the hydraulic mines of California, one could hardly fail to be struck at once with the fact that they are concentrated within quite a limited space as compared with the whole area of the State. It will also be apparent that gravels which are auriferous in character occur where no hydraulic mining has been attempted; and, furthermore, that large deposits of superficial detritus exist where no mining at all is carried on. Each of these conditions has its special and rather complicated set of causes, the nature of which has already been explained, to some extent, in the preceding pages,* but in regard to which a few words must be added in the present connection.

Supposing that gravel existed in large quantities in any region, the first question in regard to working it would be, Is it auriferous? If this were found to be the case, the next point would be, the quantity of water which could be obtained and the cost of the same. The existence of hydraulic mines in any region implies the presence of large bodies of auriferous gravel, and also of facilities for obtaining large quantities of water at a cost not too high in proportion to the yield of the gravel. There may be so much gold present that a high price can be paid for water, or a heavy expenditure incurred in bringing it to the spot; or, on the other hand, there may be so little gold in the gravel as to make the obtaining of it a remunerative enterprise only when water is abundant and cheap. Again, the deposits of detrital material may be fairly rich, or even very rich, and yet too thin to make it worth while to establish the necessary plant for hydraulic mining, which necessarily presupposes the occurrence of the material on a large scale. Another set of conditions connect themselves with the nature of the operations by which the gold-bearing detritus is handled and with the profit resulting

* See *ante*, pp. 74–78.

from the work. These have reference to the manner in which the gold is distributed in the detritus with which it is connected. The particles of the precious metal may be quite uniformly scattered through the mass, or they may be almost entirely concentrated within a narrow vertical range : in the latter case the pay-gravel or pay-streak is pretty sure to be on or near the surface of the bed-rock ; or the lower portions of the detrital mass may be much richer than the strata above, and yet the latter capable of being washed with profit by the hydraulic method, under favorable circumstances, even when the richer portions below have been previously worked out by drifting. Still another matter has often to be weighed in connection with gravel mining. The body of gravel may be heavy and rich in gold, but capped with volcanic materials which are of too solid a character to be capable of being washed away — or " cut," as it is usually called — by the jet from the pipes, even with the use of powder, in which case the hydraulic method cannot be used at all ; or, on the other hand, the capping, whether of volcanic material, which would yield to the force of the water, or of non-auriferous pipe-clay or gravel, may be too heavy to be capable of being handled without incurring more expense than the yield of the auriferous stratum beneath will justify.

Considering the above-mentioned circumstances influencing the development of the gravel-mining business, we may expect to find, and do find, a variety of methods employed, corresponding with the various conditions specified. Further on in this chapter some of the economical considerations connected with the hydraulic method, as well as the other kinds of mining employed in the gravel regions, will be briefly set forth. At present, it most concerns us to notice the fact that the gravel deposits of the west slope of the Sierra Nevada almost always contain some gold, and that there are comparatively few localities where they have not been " prospected " to some extent ; so that, on the whole, their distribution is pretty well understood.

The occurrence of bodies of gravel of considerable size in the Coast Ranges has already been alluded to.* Some of these, in the southern part of the State have, in former years, been worked for gold to some extent ; but it is not known to the writer that any of them are at the present time. By far the larger portion of the Coast Range gravels may without hesitation be set down as nearly or quite destitute of gold, as the activity with which prospecting has been carried on during the past thirty years is sufficient assurance that but little of value can possibly have escaped the notice of the

* See ante, p. 22, 23.

wandering miner. The position of the large, nearly isolated patches of gravel which occur in the Coast Ranges it seems not easy to account for. The probability is strong, however, that the localities where such masses exist were situated at the mouths of large streams, which drained considerable areas of the land forming the uplifted masses along the edge of the ocean. The material which is accumulated near Suñol Valley, for instance, must almost certainly have been brought down by fluviatile agencies from the great area of metamorphic rocks lying to the south, where the Monte Diablo Range is from thirty to forty miles in width. That these Coast Range gravels are not sufficiently auriferous to pay for working, it is not difficult to understand. The rocks themselves, out of which this detritus was formed, have not been mineralized in their present position. They may have come originally from the Sierra Nevada, which, as it seems reasonable to believe, was the source of much the larger portion of the material out of which the Coast Ranges have been built. But it is not certain that the impregnation of the Sierra rocks with gold had already taken place at the time when the Coast Range material was eroded from them; and this is a point which will be discussed somewhat further on. If, however, such impregnation had already occurred before the beginning of this erosion, it is still altogether probable that much the larger part of the gold which the abraded material might have contained would have been deposited before it had reached an area so distant from the place from which it started. Be this as it may, it is certain that the Coast Range formations have not been independently mineralized with the precious metals since their deposition. They are destitute of productive quartz veins, and contain among the valuable ores of the metals only those of quicksilver in sufficient quantity to pay for working.*

The distribution of the detrital formations in the Sierra Nevada is well indicated by the character and extent of the mining operations, which are at the present time, or have been formerly carried on. The geological structure, the topography, and the climatalogical peculiarities of different parts of the State are all reflected in the development of the mining interests.

In order to apprehend clearly the distribution of the detrital formations along the western slope of the Sierra, the nature of the different kinds of mining operations in use in California must be borne in mind, and may again be briefly noticed.† We have to distinguish, first: *river, bar* and *gulch mining,*

---

* Chromic iron is also considerably abundant in the Coast Range rocks, and may at some future time be worked with profit.

† See *note,* pp. 65, 66, and 74-78.

or washing the detritus either in or closely adjacent to the beds of the present rivers; and second. *high* or *deep gravel mining*. ordinarily carried on by the hydraulic process: third, *tunnel mining* or *drifting*, that is, working out the rich portions or "leads" on the bed-rock by means of drifts, as already explained;* and fourth, what may be called — for want of a better name — *surface mining*, which includes those shallow workings, or ordinary placer mines, which are carried on in the modern and usually thin detrital covering of the bed-rock, not immediately in the beds of the rivers, but on the rolling uplands between them, and chiefly in the foot-hills of the Sierra. The class of "diggings" designated as surface is intermediate in character between the river and the hydraulic mining operations. The latter require a large outlay for plant, and are expected to be quite permanent in character, the enormous quantity of material handled more than compensating for the moderate yield per cubic yard. River and bar mining, so extremely productive in the early days, are now almost entirely a thing of the past. Surface mining is of a transient character, the superficial deposits being too thin, usually, to justify heavy expenditures for bringing water from a distance, so that in such cases temporary reservoirs are mostly used, which are filled by the winter rains, the supply from which is soon exhausted, when the work is discontinued, to be resumed during the next season if sufficiently profitable.

As the more superficial kinds of mining operations — river and surface — have long since ceased to be of any great importance in the Sierra Nevada, it is not easy to designate the precise area over which such washings have been successfully carried on. From the point where the first discovery of gold was made by the Americans, on the South Fork of the American River, the gold-seekers spread themselves with rapidity far and wide; and but very few years had passed before every stream between the Kern and the Klamath had been "prospected." While it is true that some gold has been washed from detrital material lying on the granitic bed-rock, it is safe to say that. almost without exception, the area covered by the slaty rocks in the Sierra is that over which the detrital formations have been found sufficiently rich in gold to be worked with profit; and where, also, veins occur in the bed-rock which can be successfully mined.

Of course it is easy to see that gold may have been swept from its original place of occurrence in the slates downward on to rocks of a different character; and where the formations are much mixed together, as is the case from

* See *ante*, p. 78.

Nevada County northward to the extreme end of the gold region we may expect to find the auriferous gravels resting on rocks quite unlike those from which they were abraded.

The slate belt of the Sierra begins to appear as having some development a little south of the southern border of Mariposa County, and rapidly expands on entering it, so as to occupy a belt of over twenty miles in width, as already explained.* No sooner do the slates set in, in this manner, than powerful quartz veins begin to make their appearance; and those of Mariposa County are numerous, and have been extensively worked, although with varying fortunes. The conditions indicated for Mariposa are continued in Tuolumne County, near the northern boundary of which we have for the first time — proceeding from the south towards the north — a lava-flow, extending entirely across the auriferous belt, and covering an ancient river channel. To the south of this lava-flow (already in previous pages described as the Table Mountain of Tuolumne),† we find no proper deep gravels or hydraulic mines. The surface mines have, however, been quite important over various irregular areas, the localities of which have already been sufficiently designated.‡ These areas are chiefly in the vicinity of the Great Quartz Vein; and the affiliation of the deposits of paying surface gravel with the slaty portions of the bed-rock series, and with auriferous quartz veins, is a very marked feature of the region. It is in these counties and in the adjoining ones on the north, Calaveras and Amador, that those remarkable limestone belts, the surfaces of which have been corroded into such gigantic riffles, have their principal development, as previously described:§ these have been worked for many years, although probably by this time very nearly exhausted.

From the Tuolumne Table Mountain north, volcanic matter occupies, as a general rule, more and more of the surface on the higher portions of the divides between the streams; and, with the development of the volcanic, the gravel deposits also increase in importance. In Calaveras they have already acquired in places a considerable thickness, although by no means to be compared in this respect with those occurring farther north. The paying portion of the gravels in this county, however, is usually very thin, and the total development of pipe-clay, sand, and gravel only occasionally reaches as much as a hundred feet. The area occupied by the detrital deposits is also small;

---

* See *ante*, p. 43.   † See *ante*, pp. 131 – 137.
‡ See *ante*, pp. 137, 138   § See *ante*, pp. 138, 139.

yet Calaveras is prolific in quartz veins, some of which have been quite productive, although not usually holding out well in depth. But little progress was made in this county towards tracing out the connection of the channels. The most continuous one seems to have been that descending from the High Sierra in the vicinity of the Mokelumne River, and so extensively worked "in the early days" at Mokelumne City and afterwards near Chili Gulch. The channel was narrow and the gravel was thin, but exceedingly rich in places.* In Amador, the next county north of Calaveras, the association of gravels with volcanic materials becomes more and more a feature of the geology,† although the detrital masses have not yet attained anything like the importance which they have farther north. The quartz veins of Amador, belonging mostly to the system of the "Mother Lode," are large; and, although not as rich as in some other places, have been worked to greater depth, more continuously, and with perhaps larger profits in the long run, than anywhere else in the State.

The region drained by the different branches of the American River is one in which all the peculiar features of the gravel-mining portion of the Sierra are well displayed. From the neighborhood of Placerville north to the North Fork of the American, the central belt of the range, between the altitudes of 2,500 and 5,500 feet, is largely covered with volcanic materials; and between the Middle Fork of the Middle Fork and the North Fork all the higher portions of the divides are flat tables of lava, deeply cut into by erosive action, so as to leave many disconnected patches of this material, as will be noticed on the maps given in a previous chapter of this volume.‡ The total thickness of the detrital and volcanic formations lying on the bed-rock is, in portions of the region in question, very large, quite often reaching 400 or 500 feet, and occasionally much exceeding that. The larger part of this thickness, however, is made up of various forms of lava, the gravel proper rarely reaching as much as 100 feet. The heavier deposits of the latter lie usually pretty well down on the range, say at an altitude of from 2,500 to 3,500 feet. But a small portion of the gravel in this region has been uncovered by the erosion of the overlying volcanic materials, and an inspection of the maps referred to above will show how patches of it occur, in some places almost continuously along the edges of the lava, their position indicating how large a part of the non-volcanic sedimentary deposits must yet remain covered.

* See *ante*, p. 128.   † See *ante*, p. 102.   ‡ See Plates B and C, opposite pages 82 and 98.

When we pass the North Fork of the American we come into a region where not only is the gravel most heavily developed, but where considerable areas of it are uncovered by the volcanic capping. An inspection of the General Gravel Map, which embraces the region between the Middle Fork of the American and the Middle Yuba, will show this fact very clearly.[*] Relatively to the whole area of the country, the gravel occupies more space, and the volcanic formations less, to the north of the North Fork than they do to the south of that stream. The lava-flows reach from high up in the Sierra, in almost unbroken lines, along the summits of the flat-topped ridges which occur between the cañons on the head-waters of Bear River; but the workable deposits of gravel are accumulated chiefly along a line nearly parallel with the crest of the Sierra, in great, uncovered masses, which extend in irregular but almost unbroken succession from Indiana Hill to beyond Quaker Hill. These great areas of uncovered gravel, as described by Professor Pettee in the preceding pages [†] and in Appendix A, lie at very nearly the same elevation, namely, a little more or a little less than 3,000 feet, the bed-rock at Indiana Hill being about 2,800 feet, and that at Quaker Hill from 2,700 to 3,000 feet in elevation, while the height of the summit of the gravel banks at Plug Ugly Hill is 3,251 feet, and at Quaker Hill 3,130 feet. The thickness of these gravel deposits is also very considerable, — perhaps greater than anywhere else in the State, the fact being taken into consideration that most of it is unmixed with volcanic débris. Thus, at Indiana Hill there is a thickness of 400 feet of clean, almost uniform gravel, with hardly any admixture of either clay or sand.

Between Greenhorn Creek, the most northern branch of Bear River, and Deer Creek the gravels are continuous at a high altitude from Bunker Hill to Scott's Flat, forming the continuation of the great uncovered area just indicated as extending north from Indiana Hill to Quaker Hill. But below these there are also other gravels, of less importance, it is true, which lie around Grass Valley and Nevada City, and not at a much lower level than those just described. These deposits are of considerable thickness at a few points, and in places must have been fairly rich in gold; but they have never had anything like the importance of those farther north on the branches of the Yuba, or south on those of the American.

If the gravels at the head of Bear River present what may be called the type of the uncovered deposits, spread over broad areas, and with difficulty to be connected together into any one continuous channel, those between the South and Middle Yuba furnish, probably, the finest example to be found anywhere of a detrital mass possessing all those peculiar characteristics which mark the work of ancient rivers in the Sierra. The deposits of gravel are of great thickness, and may be traced almost continuously from high up in the range down almost to its very base. Large portions of it are uncovered, so that the banks can be washed by the hydraulic method; the auriferous particles are probably more uniformly scattered through the detrital masses than anywhere else on the slope of the Sierra; and the topographical conditions at and near the summit are favorable for securing large and permanent supplies of water. Hence we have in this region the largest and most important hydraulic mines in the world, which promise to hold their own for a considerable number of years to come. From Snow Point, at an elevation of 4,200 feet, to French Corral, 1,579 feet, the gravel has been traced not continuously, but sufficiently so to make it certain that it belongs essentially to one channel, although not without lateral branches, the position of which has not yet been clearly made out. This channel is extensively covered in its upper portion by volcanic materials, but below Columbia the gravel is entirely free from any lava capping; and although there is a break in its continuity, between Cherokee and North San Juan, of a little over three miles, yet there can be no doubt of the former existence of a connecting deposit, which has been washed away and disappeared. It is not necessary, however, to dwell on the peculiar features of this great deposit on the summit between the Yubas, because it has already been quite fully described in the preceding pages.* Certain questions connected with it will come up for consideration further on, after the present general sketch of the occurrence of gravels has been completed.

The divide between the Middle and North forks of the Yuba is very differently situated, with reference to the gravel deposits, from that between the Middle and South branches of the same river. There are two quite decidedly marked channels, and possibly a third; but we have them now for the first time exhibiting a direction not, in the main, coincident with that of the present lines of drainage; but, on the other hand, decidedly transverse to it. The gravel deposits occur on this divide in quite isolated patches, which are not of extensive area; but their position with regard to each other is such

* See *ante*, pp. 196–208, and Appendix A.

as to show clearly in what direction the former rivers flowed. The most westerly of these ancient streams can be traced by a series of gravel masses in a somewhat curving, but generally south-southwesterly direction, from Brandy City to Camptonville, at which point it is about six miles in a direct line from North San Juan. There seems to be but little doubt that these isolated deposits represent an old channel, and that there was formerly a northerly tributary coming into' the main channel of the present divide between the Middle and South Yubas from this direction. It must be noticed, in this connection, that the North Fork of the Yuba, which flows down the Sierra at first for a considerable distance in the normal southwesterly direction, makes a sudden bend about six miles west of Foster's Bar, and thence flows almost south until it enters the Middle Yuba a little below North San Juan, thus following a channel nearly parallel with that of the ancient river, which formerly ran from the north by way of Brandy City and Camptonville. There is another nearly parallel channel about twelve miles to the west, which passes by Downieville through Forest City and Minnesota, and which formerly joined the great stream of the divide between the Middle and South forks of the Yuba, at a point not far from Orleans Flat, as shown on Plate Q. Although there are higher deposits of gravel still farther east, which are considered by some as indicating a third channel as coming with a southwesterly direction down from the vicinity of Gold Lake towards the Middle Yuba, it seems more likely that these are only local deposits, swept down from higher up on the Sierra, and not belonging to a well-developed channel.

Between the North Yuba and the South Fork of the Feather River there are numerous gravel areas, some of which are of considerable extent, the region being one which has yielded heavily in former times, a large portion of the gold having, however, been obtained by drifting. Indeed, in crossing the North Yuba we have rather passed out of the region of the great hydraulic mining operations. Plates R and U show the position of the gravel areas near the creeks tributary to the North Yuba on the north side of that stream. These creeks have a general southwesterly course, and the divides between them are, almost without exception, covered in their highest portions by heavy masses of volcanic materials. The position of the gravel areas seems to indicate that the most important line of drainage for the ancient river system had a northerly and southerly direction, corresponding with what has been described as occurring between the North and Middle Yubas.

Indeed, it appears almost certain that during the gravel epoch, the ultimate connection of this whole region between the Middle Yuba and the South Feather with the Great Valley was by way of what is now the divide between the Middle and South Yubas; this being the case, it is not so difficult to understand how it is that the gravel deposits on this divide are so extensive and continuous. At the same time there seem to be other and probably subordinate channels coming in from the northeast, in accordance with the direction of the lava-spurs on the divides. It is naturally the inclination of the miners to seek for channels in this direction, as being in conformity with the general condition of things in the Sierra, and as promising new fields for discovery and profitable labor; and in the extreme northeastern portion of the district under consideration the gravels have indeed been traced to a great elevation in the range, namely, as high as 6,000 feet, — almost at the very summit, indeed. (See Plates R and T.) It will be noticed that, in leaving the gravels of the divide between the Middle and South Yuba behind, and following the general northerly direction of the channels, we have been rising to higher levels, while of course receding from the base of the Sierra. The low bed-rock at La Porte is over 4,800 feet above the sea-level; and to the south of Slate Creek, in the extreme southwest corner of the district shown on Plate R, the bed-rock under the gravels nowhere descends as low as 4,000 feet.

The thickness of the gravels also diminishes gradually as we recede from the Middle Yuba in a northerly direction. Still, there are localities where the amount of detrital material is very considerable, even far to the north and high up in the range; as, for instance, on the property of the Niagara Company, at the head of the North Fork of Slate Creek, where a total thickness of gravel, clay, and pipe-clay equal to 300 feet is reported, the gravel itself, however, making less than a hundred of this.

There are some deposits of gravel near Onion Valley, the stream heading there being a tributary of the Middle Fork of the Feather, the South Fork of this stream being much less important than the Middle Fork. Indeed, the former heads near Gibsonville and to the southwest of Onion Valley, while the latter rises much higher in the range, or, indeed, entirely to the east of it, and runs in a northwesterly direction for many miles before breaking across it, to unite with the North Fork, whose course is equally abnormal as compared with that of all the other rivers to the south. The Onion Valley gravels appear to be of but little importance, and not connected in any way with those to the south, which have been already noticed; but they have never received a careful examination at our hands.

The region between the Middle and North forks of the Feather contains some very interesting deposits of gravel, which, however, are by no means of as much economical importance as those farther south. That they are not connected with the gravel channels in the Yuba Basin seems clear to Professor Pettee, as it did to the writer after a hasty reconnaissance in 1866. Indeed, we have within the area of a hundred square miles in the vicinity of Quincy, and including those remarkable deposits on the summits of Clermont and Spanish Peak, a number of localities of gravel at very different elevations, and of very different lithological character, so that all attempt to co-ordinate them has thus far proved a failure. That there is in this region no system of ancient channels of Tertiary age comparable in magnitude with those farther south is evident, yet the geological problems presented by such occurrences as these on the two highest points of the region* are as interesting as those offering themselves in any portion of the Sierra.

The foregoing rapid survey of the distribution of the gravels along the western slope of the Sierra naturally opens the way for the discussion of a number of important questions therewith connected. As, however, the volcanic formations are so intimately associated with those of aqueous origin that it is impossible to separate the two classes from each other in any examination of the facts from the theoretical point of view, it will be well to give, before proceeding any further, a *résumé* of the most important conditions connected with the distribution of the masses of eruptive origin overlying the bed-rock on the west slope of the Sierra Nevada. After that, we may proceed to consider some of the more striking peculiarities of the gravels, as regards their lithological character and general mode of occurrence.

A large amount of detail has been given in the preceding pages, and in the Appendix, with regard to the character and position of the volcanic formations associated with the gravels in the Sierra Nevada. What remains to be done in this section is to present this branch of the subject in a more general way than has previously been possible, although want of space makes it necessary to do this in the briefest possible manner.

In the first place, as to the general fact that the gravels are associated with volcanic deposits. No one can doubt this for a moment who only inspects the various maps and diagrams accompanying the present volume.

* See *ante*, pp. 215, 216.

Almost the only large gravel deposits which are not intimately connected with and very largely covered by volcanic materials are those extensive areas between Indiana Hill and Quaker Hill; and even here the igneous rocks are present in force in the immediate vicinity, and do extend themselves in part over the detrital beds. In fact, we find the gravel and volcanic masses sometimes most intimately associated with each other; while, at other times, the connection of the two formations seems hardly anything more than an accidental juxtaposition. In the case of the Tuolumne Table Mountain, described in the preceding pages,* we have every feature marked with the utmost distinctness. The old channel of the river is there, and can be traced for many miles, perfectly continuous, in most respects closely resembling a river-bottom of the present time, but completely covered over and everywhere concealed by a protecting mass of hard, indestructible, basaltic rock, which evidently once flowed as a lava current down the river valley. Were all the localities as simple in their nature as this, there would be little difficulty in making out the order of succession and the character of the different geological events which followed each other. In other regions, the position of the volcanic masses is apparently independent of that of the gravels; as is well illustrated in the neighborhood of Forest City (see Plate Q), where the channel has a nearly north and south course, and the lava-flows cover all the spurs which descend from the summit between the southwesterly flowing streams. Still, on the whole, it is impossible not to admit the intimate association of the aqueous and the igneous masses. This, in itself, is a matter of great interest, especially as we find that almost exactly the same conditions are repeated in another and far distant country, as will be set forth in a subsequent section, these two gold-bearing regions — California and Australia — being by far the most important ones in the world.

The most obvious reason for the intimate association of the lava with the gravel is the simple mechanical one, that the former has acted as a protecting cover for the latter, the volcanic materials having, by their indestructible character or by their great thickness, either prevented the erosive agencies altogether from reaching the aqueous deposits beneath, or else greatly retarded their operations, so that portions at least of the underlying material are left behind. This view of the case is, indeed, to a considerable extent the correct one. The heavy capping of lava has preserved the gravels from being swept downwards, and has had a powerful

* See *ante*, pp. 131 – 137.

influence in directing the erosive agencies by which, since the gravel epoch, the present relief of the surface has been developed. But it is necessary also to inquire whether there is not some more intimate connection of volcanic activity with the auriferous character of the gravels than is implied in the very obvious one of a protection against erosion. It would not be safe to assume that where volcanic rocks do not now exist, there may have been in former times bodies of gravel which would have remained so as now to be worked for gold, had the volcanic vents on the summit of the Sierra above them emitted the material necessary to form the protecting cover.

The intimate connection of the occurrence of metalliferous deposits in general with the metamorphism of the associated rocks is something which is readily perceived, even if not thoroughly understood. That there is a more or less intimate relation between mineral veins and igneous rocks has also been recognized by some of those who have occupied themselves with the study of vein phenomena. For instance, it is a well-known fact that metalliferous deposits are often developed in the immediate vicinity of, if not directly in contact with, dikes of volcanic rock; also that portions of a sedimentary formation, which are found to be metalliferous over areas where igneous agencies have been in operation, may become entirely barren at a little distance from these regions of disturbance.*

It is not to be denied, however, that volcanic rocks are not usually metalliferous; it is only those which have been metamorphosed which enclose veins of importance; so that in this respect they hold the same relations to mineral veins that other formations do. The older any mass of rock is, the more likely it is to have undergone those chemical transformations to which the term "metamorphic" is applied. But the Tertiary volcanic rocks, which make up so large a part of the Cordilleras, or at least of the superficial portion of some of its principal ranges, are too recent to have been subjected to much chemical change since their eruption; it is only here and there, as near Virginia City, that they have been greatly metamorphosed, and are, at the same time, highly metalliferous.

That the débris of the volcanic rocks themselves in the Sierra are only very rarely found to contain gold, is a well-known fact. There are a few localities where it is said that washing gravel of this character has been

---

* As an instance of this, the relations of the Potsdam sandstone in the vicinity of the line of volcanic activity extending through Keweenaw Point and on westward, in a line parallel with the shore of Lake Superior, may be mentioned. The development of the lodes at Příbram is an excellent illustration of the occasional remarkable association and interdependence of mineral veins and eruptive dikes.

remunerative; but, even in such cases, it would be not safe to assume that the gold came from eruptive rock. At least, it would be difficult to prove that it did. Not a few of the miners, however, in the early days of gold-washing in California, were strongly inclined to believe that all the gold was thrown out from volcanic vents, and it was hoped that some lucky man would discover the central point from which the precious metal had been distributed. Mining has been carried on to a considerable extent, but without much pecuniary success, in the volcanic formations near Silver Mountain; but here, as in the Washoe Mines, there has been little or no native gold obtained. Ores of silver occur, which are auriferous, it is true, and a portion of the metalliferous contents of the Comstock lode consists of the native metals, silver and gold; but the former is present in much larger proportion than it is in the ordinary native gold of the Californian miner.

But if the volcanic rocks themselves on the west slope of the Sierra are not metalliferous, except in a few localities, may not the same agencies which gave rise to eruptive phenomena on so grand a scale along the axis of the range have had something to do with the formation of the quartz veins in the bed-rock and their impregnation with gold? This question may be taken up again for some discussion in the section devoted to the distribution of the gold in the gravel; but it may be admitted that a positive answer is not easily given.

In examining the geological features of the Sierra, we find quite different conditions prevailing in different portions of the range, with reference to the development of both volcanic formations and gravel. The southern portion, from Mariposa southward to the extremity of the chain, exhibits hardly any rocks which can properly be called slaty; there are, it is true, several areas of partly gneissoid and partly schistose rocks in the High Sierra, one of which passes through Red Slate Peak, while another traverses the summit of Mount Dana. It is not known that these bands, which are of somewhat doubtful character and origin, have ever yielded any perceptible quantity of gold. At all events, the Southern Sierra has never, when visited by any of our parties, exhibited any evidence of being sufficiently auriferous to make either surface or vein mining profitable. A large amount of work has been done and much money expended in the neighborhood of Kern River, whither, indeed, there was at one time a "rush" of excited miners; but it does not appear that anything in that region has been permanently profitable. Large deposits of gravel seem to be wanting; at least, none such have been observed by any of our parties, there are certainly no hydraulic mines in that part of the Sierra. Volcanic rocks are not entirely absent, but there are none of those long lava-flows like that of the Tuolumne Table Mountain anywhere south of the immediate vicinity of the San Joaquin River, where a flat, table-like mass of lava occupies a considerable area, extending off to the north for a dozen miles, and then disappearing, having apparently been eroded away in this direction, so that the place where it originated is unknown. The highest part of the range, from Kearsarge Mountain south, and all the western slope at the head of the Kern and of the South Fork of King's River, seems to be destitute of volcanic rocks. But north of Kearsarge, both on the crest of the range and on its eastern slope, lava-flows abound. All along near Owen's River, for six miles in both directions from Fish Springs, which is directly east of that part of the crest of the Sierra called the Palisades, there are large and finely preserved basaltic cones, from which great streams of lava have run down on the slope of the " wash," or detrital mass which lies piled up to a great thickness all along the eastern base of the range, from Owen's Lake north for a great distance. Indeed, from here to the north, through Inyo and Mono counties, on the east side of the Sierra, the evidences of former volcanic activity exhibit themselves on the grandest scale, as described in the Geology of California, Vol. I. None of the detrital materials on this side, however, contain any appreciable amount of gold ; neither are there any important or regular veins, containing the precious metals or any metalliferous ores, although occasional bunches of such are found a little to the north of Mono Lake, around Aurora and Bodie, none of which are of permanent value, while some of them have led to large expenditures with but little proportionate return.

All through the range on its western slope, as far as the Fresno River at least, all the conditions for favorable mining for gold have been wanting. There is no body of slaty rock there ; the volcanic flows have not descended on that side ; and most of the detrital materials eroded from the grand and rugged mass of mountains which here makes up the range have, as it appears, been carried entirely down into the Great Valley. This, it seems, is reasonable to suppose, since the Sierra is here nearly twice as high as it is farther north in the heart of the gold region, while the distance from the foot-hills to the crest is considerably less, so that the inclination of this part of the range is very rapid. And when we notice that farther north, where the slope is so much less steep, great masses of gravel have been swept down quite to the base of the mountains, it is not difficult to understand that when the streams had that greatly increased power which they would acquire in consequence of the higher grade, there would be a proportionately larger amount of detritus swept entirely away into the valley below.

Hence, we conclude that a remarkable combination of favoring circumstances has given rise to those peculiar conditions which make hydraulic mining a success in that part of the western slope of the Sierra which lies between the most southern tributaries of the American and the most northern of the Yuba River. The bed-rock itself has been mineralized, so that it is richly supplied with veins of quartz bearing gold. Orographic forces, acting through the geological ages, resulted in the formation of a mountain range of great extent, with a long, gently descending slope on the west side. Certain other conditions favored the rapid erosion of these mineralized rocks, raised in the way indicated; and the consequence was, that the abraded material was carried down the slope with just sufficient velocity, on the whole, not to spread itself out over too wide an area, and so as to have only its finer particles swept entirely into the Great Valley. In the course of this operation a large part of the gold which the abraded material originally contained found its way to the bottom of the detrital mass, where it was to lie until the activity and ingenuity of man should find the means of getting hold of it with profit. The bed-rock itself seems admirably adapted, in its lithological and stratigraphical peculiarities, to help in the operation; for its slaty laminæ are turned up in an almost vertical position, as if it was intended that, as they wear away irregularly, they should form natural " riffles " to arrest the gold in its downward progress. But this is not all. The auriferous gravel might have been spread so uniformly and so deeply over the surface of the country as to render gold at the bottom practically inaccessible. If the metallic particles all rested on the bed-rock, and were uniformly disseminated over its surface, and this again covered with a large mass of barren detritus, the chances for profitable mining would have been much less than they now are. If, however, in such a case the precious metal, instead of being uniformly spread over the bed-rock surface, were more or less concentrated in channels — or *gutters*, as they are called in Australia, — then shaft-sinking and drifting might be profitably carried on, but the conditions for successful hydraulic mining would not exist. For the application of that method, the material to be operated on must be raised to a considerable elevation above the adjacent lines of drainage, because otherwise there would be no room for the great body of tailings which accumulate as the work goes on. And still another favorable condition must be present: above the gravels themselves which are to be washed there must be a higher region, whence water may be obtained in large quantity, and under sufficient pressure. All these conditions are united in the region indicated as being *par excellence* the hydraulic mining district of California. Proceeding still farther north from the head of the Feather River, we find mining operations quite put an end to, because the covering of volcanic matter over the bed-rock becomes an unbroken one, so that anything of value, which might exist beneath it, would be entirely concealed. But there is reason to believe that the auriferous character of the bed-rock series is by no means so marked to the north of the Feather River as it is farther south. For, in the first place, there is a gradual falling-off in this respect already to be noticed, before the exclusively volcanic region has been reached. Plumas has fewer productive quartz-veins and smaller areas of gravel than Sierra, and the latter county is not equal in these respects to the region adjacent on the south. Again, the slaty formations of the Sierra Nevada system seem to emerge from under the lava covering in the extreme northwestern corner of the State, but here they are only quite moderately auriferous. And, still farther north, in Oregon, neither the Cascade nor the Blue Mountain Range — one or the other or both of which must be, as it would seem, the geological equivalent or continuation of the Sierra Nevada — can be compared in metallic wealth with the Californian division of the range.

SECTION IV. — *The Geological Age of the Gravels.*

In the preceding chapter will be found a condensed account of the principal fossil remains which came under the notice of the Geological Survey at various times, while engaged in the exploration of the gravel region. Not much need be added in the present section to what has been already said, in regard to the geological age of the detrital masses which form the subject of this volume; but a few points seem to demand further elucidation.

The age of the bed-rock series can be set down with certainty as being nowhere more recent than Jurassic, as has been already explained. It is not impossible that a portion of the metamorphic belt of the Sierra may prove to be older than the Carboniferous, although, in the light of the evidence collected up to the present time, it is quite improbable. But, on the other hand, the position and relations of the strata of Cretaceous age, all along the flanks of the range, are such as to demonstrate that there can be no metamorphic rocks of that age included within the crystalline formations underlying the gravels. Unmetamorphosed Cretaceous rocks, representing the later stages of that epoch, rest in an unaltered and almost undisturbed condition on the upturned edges of the bed-rock series, at a sufficient number of points along the base of the Sierra to prove that there was a great break at the close of the Jurassic, having as its result a complete change in the orography of the western side of the continent, as well as in the organic life of that region. This break between the Jurassic and the Cretaceous seems to have been the most important era in the geological history of North America west of the Wahsatch Range. It was pre-eminently the mountain-building epoch of that region.

It follows, therefore, that the detrital masses resting on the bed-rock may contain representatives of the various geological groups higher in the series than the Jurassic. The first question would naturally be whether the Cretaceous epoch was represented among these. There is, as has been stated in the preceding pages,* a very large development of the rocks of this period in the Coast Ranges, and a much smaller one along the western base of the Sierra Nevada. These, however, are exclusively of marine origin, and their position on the flanks of the Sierra is so low down as to show that the range had essentially its present elevation — so far, at least, as the effect of orographic causes is concerned — during the Cretaceous epoch. This statement, however, would not be true except for the mining region described in the present volume. Near Folsom, at the point where the American River issues from the foot-hills, the Cretaceous strata are but very little elevated above the level of the Great Valley. They rise, however, as we go north, and near Shasta City are more than a thousand feet above the sea-level. Farther on in the same direction, beyond Mount Shasta, in the Cottonwood Valley, they are found in considerable force at an elevation of over three thousand feet. It would seem, therefore, that there has been a decided uplifting of a region of large extent in the northern part of the State since the Cretaceous strata were deposited. This is in harmony with other known facts connected with the geology of that region, where the Coast Ranges, built up by orographic disturbances which have all taken place in Cænozoic time, come so close into contact with the Sierra Nevada, which is essentially a Mesozoic mountain-range, that it has not yet been possible to draw the line

---

* See *ante*, pp. 51, 52.

between the two systems. It does not appear, however, that this tilting of the land masses in the northern region produced any effect on the systems of drainage of the gravel region; at least, no important changes can be traced at present which might be referred to an uplift of the kind described.

There is no foundation in fact for the idea advocated by Richthofen,[*] in his masterly essay on the relation of the distribution of the volcanic rocks to the configuration of the surface of the globe, that, during the period of volcanic activity in that region, the crest of the Sierra was raised at a higher rate than its western foot, and that consequently those ancient rivers, which were flowing parallel to that crest, have been gradually turned from their channels and made to flow in the present direction of the drainage, that is, down the slope and at right angles to their former direction. At least, the most that could be said in favor of this view would be, that in the extreme northern portion of the mining region there have been disturbances since the deposition of the Cretaceous which have, to some extent, modified the courses of the channels of the ancient rivers; but in what direction, has not yet been made out. The natural course of things would seem to be, that elevating the region to the north would reverse the direction of the streams previously flowing towards the lower area. Such a change in the course of the ancient rivers as that supposed by Richthofen would not be likely to have been brought about by the simple more rapid rising of the crest than the base of the chain; this, as it would seem, might more naturally have only augmented the velocity of the streams previously flowing in a normal course, — that is, down the slope and at right angles to the crest. If we could suppose such a thing possible as a general southerly direction of the ancient rivers before the volcanic epoch, then it may be conceived that a rise of the chain, proceeding more rapidly at its southern extremity than at its northern, might have turned the streams down towards the Great Valley, one after the other, in the manner indicated by Richthofen. His views, however, are in fact tantamount to an assumption that the range of the Sierra was not elevated until after the old river channels had been in existence some time, and a large portion of the high gravels deposited; in short, it demands an entire reconstruction of the topography of the region during later Tertiary times. But this idea is entirely unsupported by the investigations, of which the results are recorded in the present volume, as has already

* Natural System of the Volcanic Rocks, p. 86.

been set forth, and as will be further explained in the course of the present chapter.

It appears to result, then, from a careful study of the mining region, that, for the practical purpose of the study of the gravels and the old river-channels, we may assume that orographic causes may pretty much be left out of consideration in the discussion of what has taken place since the gravel was deposited. The only exception to this would be, as already indicated, in the extreme northern part of the region, where confessedly the phenomena are too obscure for present explanation. There have, it is true, been slight disturbances even in the central and most important portion of the gravel district, as will be noticed further on; but these have certainly not been of such a nature as to give rise to any important changes in the general direction or grade of the ancient river systems.

It seems then, to be, if not absolutely proved, at least rendered highly probable, that, while the marine Cretaceous strata were accumulating at the base of the Sierra, along its central portion the range itself had at least its present elevation; and that from the material eroded from it these marine deposits were, in large part, built up. If the conditions were as thus indicated, why should not there be strata containing fresh-water or land animals, or plants of Cretaceous types, found above the line which marked the sea-level of that epoch? That there are not, seems a well-established fact. The animal remains are of distinctly Tertiary affinities, and there is nothing among them which is related to the characteristic forms of the Cretaceous, while the whole aspect of the flora is distinctly that of an epoch as recent as the Tertiary. The reason for this condition of things may be gathered, perhaps, in part at least, from what follows in regard to the absence or very imperfect representation of the lower Tertiaries in the formation under discussion.

That the western slope of the Sierra was above water, and in a position to be the recipient of detrital material, during the whole of the Tertiary epoch can hardly be doubted. All the geological conditions indicate this. At the same time marine strata were accumulating along the base of the range, the materials for which, as in the case of the Cretaceous, were derived, in large part, from the adjacent land to the east. There is no continuous belt of Tertiary, any more than of Cretaceous, along the foot-hills, because a considerable amount of erosion has taken place since the occurrence of the very moderate uplift which raised the formation to its present position.

There is, however, quite a large development of marine and brackish water Tertiary in the foot-hills at the base of the range toward the southern extremity of the Great Valley. Along the central portion of the Sierra, the Cretaceous occupies more area on the surface than the Tertiary does, but both formations are thin; and in some places the marine beds are so mixed with detrital and volcanic matter which has been washed down from above, bearing at the same time fragments of wood and bones of land animals, that the two formations cannot be separated from each other.

It being quite certain, then, that the Sierra was above the sea in Tertiary times, and has remained so, also that it had its system of rivers, with their lake-like expansions, it is to be taken for granted that this large area of land, with its bodies of fresh water, would become, in time, the home of various kinds of animals and plants suited to the conditions there presented. But our knowledge of the origin and development of life, and especially of the manner in which groups of species have spread themselves on the earth, is so imperfect that we have no right to be surprised if large areas occur, on which there are no organic remains to be found, as partial testimony of the lapse of geological ages. With regard to formations of marine origin we have less difficulty, because the ocean has at all times been a continuous body of water, extending over the larger part of the earth's surface; land, on the other hand, has existed in detached areas, of very small size as compared with the whole body of salt water, and of whose former connections with each other we can, at the present time, have but little idea: hence, when we find a gap in the series of geological formations in any particular region, we are wont to say that during the period represented by the wanting strata the land was raised above the sea. In doing this, we omit to notice that there must be also some reason why subaerial deposits, accompanied by a development of organic life, should not have been formed under such circumstances. It is true that we may take it for granted that erosive action is always modifying the surface of the land, and bringing about, on the whole, a more effective obliteration of the evidence of the lapse of past geological ages than may be expected to take place under the deep water; but even this consideration does not remove all the difficulties in the way of accounting for those great gaps in the formations, which present themselves in regions which we may assume to have been above water during certainly a large part of the time, for which there is nothing to show in the way of organic life. As an instance of this, we need only refer to the extensive area in northeastern North America, where there is no representation of any geological epoch between Palæozoic and post-Tertiary. It is impossible, in such a case as this, to suppose that the missing formations, or any considerable portion of them, have once existed there, and been removed by erosion. It is manifestly in the highest degree improbable that such a vast region could have been so thoroughly denuded of any considerable mass of overlying material that no vestige of it should remain behind. It seems, therefore, to be a legitimate inference that there were conditions unfavorable to the development of organic life on land during long periods of time. Such seems, at all events, to have been the case during the early portion of Tertiary times, on the western slope of the Sierra Nevada; for we have there no representation of the Eocene at all, and only a very imperfect one of the Miocene.

Indeed, while we find in the detrital beds accompanying the gravel a very considerable number of both plants and animal remains, there is far from being enough material to enable us to divide up the formation into groups. The reasons for the poverty of the collections from the gravel-mining region have been already given. In spite of these deficiencies, however, valuable inferences can be drawn with regard to the relations of the life of the gravel period and the more important questions which bear on the geological history of the region answered. The testimony of the fauna and the flora is essentially to the same effect. The general facies is decidedly later Tertiary, or Pliocene, while there are also both Miocene and Recent types present.

If there were no fossil remains in these gravels and the accompanying clayey and sandy strata, it would not have been possible to demonstrate that they did not belong to the present epoch. It might have been shown that during and since their deposition a very long period had elapsed, by pointing out the vast amount of time required for the erosion which has taken place since the end of the gravel period, not to speak of the thickness and complexity of the various masses, the accumulation of which must have occupied a still longer time. Still, in spite of this, it might have been said that we know little of how rapidly the work either of accumulation or erosion may have gone on in former times, especially in a region lifted up into high mountain ranges, and the centre of volcanic manifestations. Having the evidence of fossils, there can be no hesitation in allowing all the time necessary for the inorganic changes which have taken place in the gravel region, since we must have it in order to afford an opportunity for those successive developments of life, which we know could not have been accomplished without a corresponding lapse of time.

From the very nature of the case, there can be no such large and complete collections of fossil remains made in the gravel region as have been obtained from the lacustrine deposits farther east in the Cordilleras. The turbulent river channels, with their hundreds of feet in thickness of rolled gravels and boulders, are of quite a different character from those undisturbed deposits of fine sediment which have in so many places in the Cordilleras tranquilly filled up the gradually desiccated areas. All this, however, has been sufficiently explained in a previous chapter: it only remains here to emphasize the conclusions which have been previously drawn. We may say then, in brief, that the whole Tertiary period must be represented in the gravel accumulations of the Sierra; but only a small part of the life of that period. The nearer we come to the present epoch, the more complete the record; because, as will be shown hereafter, there was a gradual slackening of the forces inimical to life. The existence of the Eocene period can hardly be recognized at all in the gravel period; Miocene types occasionally present themselves; but forms which may most properly be referred to the Pliocene are by far the most numerous. There is no possibility of drawing any lines, however, in the field between one and the other of these divisions of the Tertiary. No section allows us to trace a succession of life in the various beds: it is only here and there a few fragments — a single tooth, perhaps — that we are lucky enough to secure.

That the mass of the gravels underlying the volcanic formations contains no remains of animals of existing species, is an established fact, so far as our present observations extend. The only exception to this would be in the case of man, the evidence of whose existence in the strata beneath the basalt has been laid before the reader in the preceding pages. It is, however, not to be forgotten that the mastodon, which also existed contemporaneously with man at an epoch anterior to the eruption of the basalt, continued to live until very recent times. There is a great body of evidence showing that this proboscidean was extremely abundant in California, as well as over a large part of the remainder of North America, during post-Tertiary times; and it was then contemporaneous with some other animals which have not yet disappeared. Thus the mastodon, which lived through a portion at least of the volcanic epoch in California, perished long afterwards under those subtle and but little understood influences which bring about changes in the nature of the organic life of various regions of the earth. It would be difficult to find a more striking example of the working of this mysterious cause than is presented in the entire disappearance of the mastodon and elephant, almost during the historic period, and very nearly at the same time, from over an area of some millions of square miles.

The question whether any part of the auriferous gravel series can ever be definitely separated from the overlying portions of the formation and distinguished as being clearly of Miocene age, is one which cannot at present be answered. Further light must be thrown on this subject, it would seem, if not by discoveries in the Sierra itself, at least by those made in adjacent regions. A full exploration of the fossiliferous Tertiary beds of Oregon would undoubtedly be of value in this direction, as giving some clew to the development of animal and vegetable life on the Pacific coast during the later geological epochs. The question would not be one of so much importance, were it not for the very considerable body of evidence which has been brought forward in a previous chapter, showing the existence of man during Tertiary times. Naturally it is asked, and especially by European geologists, What name shall we give to that division of the Tertiary in which these human remains are found? To this it may be answered, that the adoption of either the word Miocene or Pliocene in this case would not necessarily imply that the event designated by either of these terms precisely corresponded in time to the one similarly named in Europe. It will be safe to say, that the human race in America is shown by the evidence to be at least of as ancient a date as that of the European Pliocene; and to have an idea how far removed that epoch is from the present one, it is only necessary to recall the amount of erosion which has taken place since the cessation of volcanic activity in that part of the Sierra in which lie the formations which have been described in the present volume.

SECTION V. — *Lithological Character and Peculiarities of the Gravel.*

Certain peculiarities of the enormous mass of detrital material to which the comprehensive term " gravel " has been applied in this volume are worthy of special consideration. The main fact in regard to this material is a very simple one, namely, that it is *water-worn*. It is essentially the product of the action of water; and this action has been in many cases of such a nature as to give rise to masses of detritus of a surprisingly uniform and homogeneous character. Not unfrequently, however, we find the most abrupt transitions from one kind of deposit to another, or sudden changes in the size of the component materials, indicating corresponding variations in the physical conditions at the time of the deposition of the mass.

Perhaps it is more difficult to account for the remarkable uniformity which is displayed by the gravel deposits in certain localities, than it is to give satisfactory reasons for the ordinary variations in the size and nature of the component parts of the detrital masses. That in some places there should be a thickness of three or four hundred feet of gravel, almost homogeneous in character from top to bottom, seems to imply a persistent sameness in the work of the erosive agencies hardly to be expected. But it must be remembered that we usually see only a very small portion, longitudinally, of what once formed a continuous deposit resulting from erosion along one channel. It is only within very narrow limits that such homogeneous accumulations have taken place, and then only in very exceptional circumstances.

In general, the variations in the mass of the gravel deposits are frequent and abrupt. There is no difficulty in understanding that these would frequently be sorted out by currents shifting in direction and varying in force, so as to present themselves in a succession of strata of irregular thickness, made up of fragments of very different sizes and very unequally water-worn. There is also the additional series of complications arising from the presence, during the later period of the gravel epoch at least, of volcanic materials in all kinds of forms, which help to swell the thickness of the formation. Volcanic vents during the Tertiary epoch, as they do now, sent forth their products in a great variety of forms: lava, which, rendered fluid by heat, flowed down in the most accessible depressions of the surface, and then hardened to an almost indestructible rock; ashes, which may have been carried to a great distance from their place of origin either by the wind or by currents of water, which finally deposited them in the form of mud; fragmentary and brecciated masses, where the lava has been ejected in a solid form, but broken into large blocks and not pulverized to ashes, — all these varieties are common, but very irregularly distributed throughout the gravel region. That portions of the solid eruptive rock should be abraded and become water-worn, so as to form a volcanic gravel, is what would naturally be expected. One principal reason why such instances are rather rare is, that the flows of solid lava belong to the later part of the gravel epoch, when the eroding agencies had already slackened in their work.

Among the occurrences which arrest the attention of the investigator in the hydraulic mining region there are two which seem very noteworthy. One is, the fact that in some localities the gravel is almost entirely made up of quartz boulders and pebbles; the other, that some of the boulders are of such enormous and quite exceptional size. Each of these peculiarities demands a few words of comment.

The gravels, of course, consist essentially of the hardest portions of the rock masses, which have been eroded away during the time of their accumulation. A large part of the metamorphic crystalline schists of the auriferous belt are of a decidedly indestructible character. Hence the gravel deposits, as a general rule, do contain a very considerable proportion of this kind of rock, as has been mentioned so frequently in the preceding pages; but there are districts where the bed-rock is chiefly made up of finely laminated slates, which are sometimes very soft, or are easily rendered so by exposure to air and moisture. Such softer rocks are not unfrequently traversed by large quartz veins; and, in some cases, the slates adjacent to the veins seem to have undergone some chemical change, rendering them peculiarly liable to disintegration. It follows, therefore, that in the wearing away by water of a region where the bed-rock is of this character, it would be the natural result that the softer material should become almost entirely pulverized, and the resulting mud be carried off to a considerable distance, leaving behind only the indestructible portion, or the quartz. We know that, at the present time, the masses of quartz enclosed in the slates are sometimes of great width, occasionally exceeding a hundred feet. The breaking up of such masses would be much assisted by the occasional softer streaks of slaty rock which they are liable to contain, and of these only faint traces might remain after the abrasion, removal, and redeposition of the quartzose material had been effected.

There may have been localities, however, where the silicious deposit spread itself extensively over the entire surface, so that large masses were accumulated, quite free from any admixture of other rock. Something of this kind may be seen at the present time in process of formation at the well-known locality of Steamboat Springs, in Nevada. The breaking up and washing away of this enormous deposit might naturally give rise to an accumulation, somewhere along the line of direction in which it was carried, of quartzose material, quite free from any admixture of slaty or other kinds of metamorphic rock.

The immense size of some of the boulders in the gravel is a matter of interest, especially in view of the fact that they cannot, by any possibility, have been brought into their present positions by the agency of ice, which we well know to be capable of carrying the largest as well as the smallest of the fragments of rock which may be thrown down on to its surface from above. In some cases these enormous masses seem to be fragments from an adjacent vein, loosened from their original position by the decay of its walls, and thus allowed to slide down by gravity, the adjacent rock beneath having been eroded away in the ordinary manner. It is certain, however, that very large masses of rock can be transported by the agency of water alone. Instances of this kind have been witnessed, not unfrequently in recent times, where the giving way of the barrier confining a lake or reservoir, or one of those extraordinary rainfalls popularly known as a "cloud-burst," has allowed a large body of water to flow down a steep declivity.*

The chemical changes which have taken place in the gravel since its deposition are more difficult to explain than the chiefly mechanical agencies which resulted in bringing together the great detrital masses under consideration. In the preceding chapters, as well as in the Appendices A. and B., there will be found numerous references to phenomena observed, which indicate more or less chemical action as having taken place, and as being even yet continued, in the various strata opened to examination by the hydraulic mining operations. These changes are displayed in many cases by the gravel itself as well as by the other detrital beds associated with the gravel proper, and also by the organic bodies imbedded in the formation.

Decidedly the most important of these chemical reactions which present themselves for explanation in this connection are those indicated by the constant use of the terms "red" and "blue" gravel in the local descriptions given in the preceding pages. Indeed, the existence of these differences of color, which are accompanied by other practically important peculiarities, have sufficiently impressed themselves on the attention of the miner to lead to very curious theoretical views, to which reference has frequently been made in the course of the present volume. The famous "blue lead" is the result of a desire on the part of the miners to connect together the frequent occurrences of dark-colored gravel, as being a special formation, differing in

---

* It is stated — on good authority, as the writer believes — that during a cloud-burst in Mono County, California, a steam-boiler of considerable size was carried two miles down what but a few minutes before was a perfectly dry cañon.

origin as well as in lithological character, from the overlying softer and lighter colored materials.

The theory of the blue lead so commonly held by the miners, namely, that a special stream deposited this kind of gravel, — which idea is also closely connected with another favorite one, namely, that this stream came from the north, — is so contrary to the whole mass of observed facts that it is not necessary to spend any time in refuting it, especially as this has been satisfactorily done by Mr. Goodyear[*] in another part of this volume. What we have to do is, to set forth as far as possible what are the differences in character and position between the red and blue gravels, and to endeavor to account for the phenomena, which, however, as must be admitted, present a considerable number of points by no means easy of explanation.

The essential facts in regard to the blue gravel are these : Wherever there is a heavy deposit of gravel, the lower portion of it is likely to be solidly compacted together, often to such an extent as to require the use of powder to loosen it so that it can be acted on by the hydraulic jet ; indeed, it sometimes coheres so firmly as to require to be passed under the stamps, in order to break it apart. The writer cannot recall any instance where the blue gravel, if present at all, was not found underneath the red variety. But the relative thickness of the two varieties is by no means constant. At some localities a large part of the deposit is blue ; at others, only a thin stratum has that character. Often, especially, in cases where the gravel is thin, it is all soft and red ; but as a general rule, where there is a wide channel, with, as is often the case, a narrower, deeper trough or " gutter," — to use the Australian term, — the gravel filling this deeper portion is very apt to belong to the blue variety. Of course this kind of gravel is not limited to any special altitude above the sea-level ; its position is always related to that of the whole mass to which it belongs.

There is another important fact connected with the occurrence together of the blue and red varieties of gravel. The line of division between them is usually perfectly well-marked, but irregular in position, that is, not coinciding with the planes of stratification ; neither is it limited by any peculiarities in the lithological character of the gravel itself. Sometimes, indeed, a portion of one and the same boulder belongs to the red, and the other portion to the blue variety. In short, it is apparent that we have, in the phenomena in question, a case of chemical decomposition, proceeding from above down-

* See Appendix B.

wards, but with irregularity, the depth to which this change has penetrated varying with the varying character of the conditions.

That the red gravel is a more or less decomposed variety of the blue is evident not only from what has been stated above, but from the fact so frequently observed, namely, that the latter is rapidly changed on being uncovered so as to become exposed to air and moisture; and the presence of water in large quantity seems to favor rapid conversion of the blue into the red variety. When the blue gravel is uncovered so as to be exposed to the ordinary atmospheric changes, even without access of running water, it almost always becomes gradually softened and reddened. Indeed, it would appear to be true, as a general rule, that exposure converts the blue gravel into the red variety, although in some cases much more rapidly than in others. The inference, then, is that all or a large portion of the gravel has been originally blue, and that such portions as have become favorably situated, owing to mining operations or erosion produced by natural agencies, for the action of air and moisture, have been decomposed and oxidized.

This condition of things presupposes that all, or at least a large part of the gravel was originally in the condition indicated by the word "blue." Such is not usually the case, however, with detrital accumulations of a superficial character; and it becomes necessary to explain how it is that the gravel deposits on the west slope of the Sierra became so thoroughly consolidated as it is supposed they must have been, judging from the condition of certain portions still remaining unchanged in the deeper channels, as already described.

Detrital accumulations belonging to the older formations are usually very solidly compacted together, and there are not many exceptions to this. Occasionally, as in the case of portions of the St. Peter's sandstone in the Upper Mississippi Valley, the grains of which the rock is composed have so little coherence that they can be separated from each other by rubbing in the fingers. The causes of the compacting of the older rocks are apparently considerably varied in character, and, so far as the writer is aware, have not been very satisfactorily studied out. We know, however, that very durable and hard conglomerates and sandstones may be either of red or gray or bluish-gray color, the red varieties being supposed to contain a considerable quantity of iron in the state of peroxide, this forming a portion, at least, of the cement which holds the constituent fragments together. It is evident, however, that the newer the formation in which any detrital rock occurs, the more likely it is to be loosely held together. Thus most Tertiary conglomerates and sandstones are of very inferior value as building stones; while, in the case of some of the older pudding-stones, the forms of the originally water-worn fragments can be distinguished by the eye, but the whole mass has become so thoroughly welded together that it has no more tendency to separate into its original components than to break in any other direction. In such cases, that peculiar form of chemical action called metamorphism appears to have made over the mass so completely that the original surfaces of contact of the components have been entirely obliterated.

The case of the Tertiary gravels of the Sierra seems in some respects a peculiar one, as regards the original compacting of the mass and its subsequent local disintegration. To understand it more fully, it will be necessary to inquire what other evidences the formation presents of chemical action as having taken place at any period since the deposition of the mass. Proofs of such action present themselves in abundance, when the gravel and the organic bodies which it encloses at numerous localities come to be examined.

The most conspicuous of the chemical changes wrought in the gravel, as evidenced by the known change in substances imbedded in it, is silicification. As has already been stated in a previous chapter, the quantity of wood buried in the detrital masses of the Sierra is very large, and much the larger portion of it has become converted into opal, the amorphous form of silica.* Occasionally the fragments of trunks of trees have been slightly charred before being silicified, as is apparent from their color. This "charring" seems to have been the first step of a passage into coal, or, more properly, lignite. Indeed, there have been and are occasionally pieces of wood found of which the organic matter was so well preserved that they could be used as fuel.

So far as the writer's observations extend, the largest quantity of best preserved silicified wood is found in connection with deposits chiefly volcanic in character. At Chalk Bluffs, for instance, where the quantity of prostrate silicified trunks of trees which have been washed out is very large, the material in which the leaves are imbedded is a white pulverulent substance, apparently almost entirely made up of rhyolitic ash. The same is the case in the well-known "fossil forest" near Calistoga, in the Coast

---

* See ante, pp. 235–239.

Ranges. It would appear that the wood has been best preserved when surrounded by eruptive materials, in which the trees may perhaps have grown, or which may have been swept over the trunks, after they had been prostrated by age or other causes, so as entirely to envelop them. Many fragments of silicified wood are found in the coarse gravel; these, however, usually have a more or less water-worn appearance, and may have been carried by currents of water far from the place in which they grew.

The lava itself frequently exhibits signs of having been acted on by silicifying agents after its deposition. Some of the beds of white ashes have evidently undergone a change since they reached the place where they now lie; the whole mass, which, in all probability, was originally light and incoherent, has acquired more or less completely the texture and appearance of semi-opal, as if acted on by silicifying waters. The abundance of infusorial silica in portions of the volcanic rocks occurring in close connection with the gravels, as already described,* is another proof of the frequent presence of this element in such condition that it could be readily assimilated by the organisms requiring it for their development.

The legitimate inference, then, from what has been stated above, is that a large part if not the whole of the series of beds included within the gravel formation has been at some period since its deposition quite thoroughly permeated with water holding silica in solution in sufficient quantity and in the right condition to bring about extensive chemical changes of such a kind as to result in the replacement of such organic matter as was present by that element, while portions of the inorganic or mineral substances exposed to the same reaction exhibit a tendency to a similar replacement. There are, indeed, some proofs of the presence of fluorine as well as of lime in the vapors or waters which have permeated the formation, but the evidences of silicification are far more abundant.

The reactions thus indicated are decidedly those of volcanic regions. Hot springs, the water of which is charged with silica, are a very common form of volcanism, as may be seen on a grand scale in the geyser districts of Iceland, and on a still grander one in the neighborhood of the Yellowstone Park. We may assume, then, that the gravel deposits of the Sierra have been more or less subjected to those peculiar chemical influences which accompany volcanic action. This is by no means a matter to excite surprise, since, as has already been described, the Sierra Nevada was the theatre of

* See *ante*, pp. 220–231.

intense volcanic activity during the latter portion of the gravel epoch. Most of the chemical reactions indicated above can therefore be accounted for without difficulty by referring to the well-known results of that complex series of conditions embraced under the term " volcanism."

There are some points, however, in regard to which there is more difficulty, and these will now be taken up for a brief discussion. The first of these would be the question whether the chemical agencies connected with the volcanic displays were effected from above downward, or in the reverse direction. We find frequently large bodies of gravel overlain by equally heavy masses of volcanic materials; we almost never have the latter underlying the former. This, of course, is the natural result of the fact that the culmination of the volcanic era took place toward the close of the gravel epoch. Such being the case, the question arises whether the effect of an overlying mass of lava would make itself perceptible downward to the bottom of a deposit of gravel, perhaps several hundred feet thick. This seems, on general principles, hardly probable, and an examination of the conditions of the various beds which make up the whole mass of the formation, detrital and volcanic both being included, appears to confirm the idea that the chemical changes which have taken place in the body of the gravel cannot have been caused by any reaction originating in the overlying eruptive materials. Wherever the gravel is very deep and most confined within a limited space, instead of being spread out over a wide extent of surface, there it seems to have at the bottom most fully the *ensemble* of characters designated by the term " blue." Had the chemical influences producing this condition come from above, it would seem that, at least in some places, these deeper portions would have failed to be reached. The same thing seems to be indicated by the condition of the volcanic masses themselves. It is, as a general rule, the lower eruptive beds which exhibit the most marked indications of chemical alteration. The upper layers are usually but very little changed from their original condition, as far, at least, as can be made out without microscopical examination. One of the most striking facts connected with the mineralogy of the more solid upper portions of the volcanic formations of the Sierra is, the almost entire absence of the zeolitic and other minerals, which are so often found in connection with eruptive rocks in other parts of the world. Hardly a specimen of any zeolite has ever been detected in California, notwithstanding the large area there covered by formations in which such occurrences might reasonably be expected.

Such being the facts, it would seem that the influences by which the chemical changes in the gravels have been brought about, must have come from below ; and when we inquire more closely what these influences must have been, and in what way they were exercised, we find ourselves compelled to fall back on the volcanic epoch as that of the reactions in question, being obliged to admit, however, that it is through the bed-rock that they have been propagated, and not through the overlying volcanic masses. This may, at first thought, seem an unwarranted hypothesis; but the following explanations may perhaps remove a part of the difficulties.

It will be admitted by all, that the mass of slaty and other strata — designated for convenience by the term "bed-rock" — has been at some time the scene of very extensive chemical changes. Not to speak of the variety of those included in the commonly used term "metamorphic," we have exhibited, all through the formation, on the grandest scale those proofs of a former chemical activity, the result of which has been the development of the very numerous quartz veins by which it is everywhere intersected, and many of which are of great size, as already described. Here, then, are evidences of silicification on a grand scale, — the same phenomenon in another form which is presented to our notice so frequently in the gravels themselves. And the quartz veins are not exclusively confined to the bed-rock ; the writer has seen at least one cutting the gravel itself, which was well-defined, although of small dimensions *

The inquiry naturally suggests itself, then, At what epoch did the formation of the quartz veins in the bed-rock take place ? Was it contemporaneous with the grand exhibition of volcanic forces which took place in the Sierra, during the latter part of the gravel epoch ; and, if so, are the phenomena genetically connected? To the writer's mind it seems clear that both these questions can be answered in the affirmative, and the reasons for the belief will now be set forth, although space permits it to be done only in the briefest possible manner. Some further references to the same subject will, however, be found in the section of this chapter relating to the distribution of the gold in the gravel.

All the phenomena of occurrence in the quartz veins indicate most clearly that these were not formed until the chain of the Sierra had been uplifted into pretty nearly its present position, and that since their formation there have been but few, if any, important orographic disturbances of the region in

* See Geology of California, Vol. I. p. 276.

which they occur. Were this not the case, the lodes would be so faulted and disturbed the working them would be almost or quite an impossibility. Faulting of the lodes, however, even in a slight degree, is not common, and the existence of any very extensive breaks is something unknown to the writer; such must, at all events, be of rare occurrence There is also abundant evidence that the volcanic epoch was not inaugurated in the Sierra until the range had approximately its present form. This, however, must not be taken as excluding the possibility of the presence of eruptive masses in the bed-rock itself. These do exist, and probably in great quantity, although in a much metamorphosed condition, so that it is not easy to recognize their real nature, except by the aid of the microscope. But the era of volcanic action belonging to the bed-rock series is widely separated from that of the gravel formation, as shown by abundant facts: the one seems to have been the result of massive eruptions; the other was undoubtedly that of the ordinary crater ejections of the present day.

We have, then, the two sets of phenomena — the formation of the quartz veins in the bed-rock, and the covering of the gravel deposits with eruptive materials — brought very near each other in geological time. That these occurrences were absolutely synchronous cannot, of course, be maintained, because we have rolled fragments of quartz under the lava in many places, showing that veins or masses of this material existed before the opening of the volcanic epoch. The epoch of metamorphism proper seems to have been the first stage in the process which was to culminate in the pouring forth from the crest of the Sierra of the immense mass of eruptive material which now rests upon its flanks. That during or soon after the metamorphism of the slates the quartz veins began to be developed, is also a reasonable supposition, and we know that these phenomena were followed by the opening of a line of orifices from one end of the range to the other, through which vents came the eruptive matter in its various forms, as already indicated. That these various stages of chemical action are genetically connected cannot, in the present state of our knowledge of the mysteries of volcanic activity, be positively affirmed; but certainly the evidence pointing in that direction is very strong.

It might appear that what has been said above, in regard to the formation of the quartz veins after the range of the Sierra had assumed its present development, was in contradiction to what had been previously stated in this volume in reference to the peculiar form of the "Great Quartz Vein," as indicating powerful longitudinal compression of the whole body of the strata in which it is enclosed.* But this immense mass of quartzose, dolomitic, and magnesitic material, to which the name of Mother Lode, or Great Quartz Vein, is applied, is not by any means proved to be a fissure vein, or even an exclusively segregated one. It will require much more study than it has yet received before its real character can be stated with confidence. To the writer it seems, from present evidence, most likely that it is the result of metamorphic action on a belt of rock of peculiar composition, and perhaps originally largely dolomitic in character. That this belt should have undergone the same modifications of position suffered by the limestone formation of the Sierra at the time of the upheaval of the chain, would be, then, quite in accordance with what might be expected.†

We admit, then, as being on the whole the most reasonable theory, that the chemical changes in the gravel are the result, first, of volcanic impregnation from beneath, and of subsequent alteration by the ordinary meteorological agencies penetrating from above, and working downward as opportunity offered. It now remains to indicate a little more exactly the nature of these changes, which have resulted in the formation of the blue gravel, and in the development from this of the red variety. The latter stage of the process presents no difficulties, and may be rapidly passed over. Wherever air and moisture together can readily find their way down to some distance beneath the surface, there the rock becomes more or less completely disintegrated, this result being brought about, in considerable part at least, by the higher oxidation and hydratation of the oxides and sulphurets present in the mass. This is the phenomenon so often exhibited by metalliferous veins, which are, as a general rule, easily permeated by water, and which are in consequence decomposed down as far as the line of permanent water level. So far moisture and air easily penetrate; below that the air is almost excluded, and the ores remain in their original condition. The resulting oxidized material is the well-known "iron hat" of the miner.

It is a fact, the *rationale* of which, however, is not thoroughly understood, that reducing agencies predominate in all those chemical operations which are carried on at considerable depth, or which are propagated from below upward; while oxidation appears to be the invariable result of the access of

---

* See *ante*, pp. 46, 49.

† The Great Quartz Vein presents one of the most interesting fields for study offered to the chemical geologist in any country.

surface water and the atmosphere. This is particularly well illustrated in the case of metalliferous veins, which, almost without exception, tend to exhibit fewer indications of oxidized combination as greater depths are reached. The lower portions of the gravel deposits are in this condition. Overlying, as they do, regions of intense metamorphic action, preceding extensive volcanic manifestations in the adjacent region, they have been somewhat in the condition of the matter enclosed within the walls of veins. The finer materials between the pebbles and boulders of the gravels have been cemented together, partly by silicification and partly by reduction of the iron from the oxidized condition to that of sulphuret. In this latter operation the reducing agency of the organic substances enclosed in the formation has been, in places at least, very conspicuous in its effects. Fragments of wood penetrated by and incrusted with pyrites are of common occurrence, and the same is true to some extent of the bones and teeth of animals. Implements and works of human hands have also been found, under the volcanic masses, covered with a similar incrustation of sulphuret of iron.

SECTION VI. — *The Bed-rock Surface and the Channels.*

We have next to examine what is known in regard to the form and character of the bed-rock surface under the gravel deposits, for the purpose of ascertaining what light this class of facts is able to throw on the physical conditions prevailing during the period of the accumulation of the detrital masses. In this connection we have two somewhat distinct kinds of phenomena presented for our investigation: one of these relates to the form of the surface of the bed-rock and the character of the markings upon it, as indicative of the agency by which the erosion was performed; the other has to do with the connections of the channels or ancient river-courses with each other,— a question involving, to a considerable extent, a reconstruction of the former topography of the whole region embraced within the field of our investigations. This latter branch of the inquiry presents itself in a twofold aspect,— one distinctly theoretical, the other eminently practical. The geologist, looking at the question from the scientific point of view, desires to clearly establish the sequence of events which took place during the gravel epoch, and to be able to assign for each of these a satisfactory cause; the practical miner, on the other hand, seeks to ascertain how the old channels were connected, in order that he may be at as little expense as possible in making the necessary preparations for successful work. Many of the prominent points connected with this branch of the inquiry have already been up for consideration in the preceding pages; and some of the difficulties which present themselves have been discussed both by Professor Pettee and Mr. Goodyear in their contributions to this volume. The general mode of occurrence of the gravels may be considered as having been well made out: what remains to be done is to inquire more particularly into the causes by which these results have been brought about; and here we enter a field beset with difficulties, one in which but little work has yet been done, and where there seems to be but little harmony of opinion among the few careful observers who have entered it. The writer will, however, present, necessarily somewhat briefly, the theoretical results to which he has been led by a somewhat protracted study of the region in question.

We seem now to have arrived at the proper point for instituting a closer comparison than has hitherto been made between the high gravels and the detrital accumulations which we see forming at the present day; or, in other words, between Tertiary and Recent deposits. As already mentioned in a general way, quantity and elevation are the main characteristic features of the older gravel masses.

In the first place, we must notice the character of the present river channels, and the cañons of which they form the bottoms. These cañons, throughout the gold region, are very much of one type; they are deep,* their walls sloping steeply and being almost entirely free from débris. The amount of detritus in the beds of the streams is very small as compared with the size of the gorge or excavated V-shaped depressions, the lowest points of which they occupy, with rarely any considerable breadth of comparatively level ground on either side. Thus, a stream of only a few feet in width may often be seen at the bottom of a gorge the walls of which rise directly from the water on both sides, at an angle of 30° or 40°, and to a vertical height of two thousand feet and more.

The quantity of water carried by the different streams descending the slope of the Sierra, in the gold region, varies exceedingly from season to season, as well as from year to year. In the case of those which do not head in the very highest portion of the range, the amount becomes towards the close of the summer reduced almost to nothing. Even a river draining so large an area as the Yuba is but little more than a rivulet in August and

---

* See note, pp. 64, 65.

September. At times of unusually copious and long-continued rainfalls, — such as that of the winter of 1861–62, — or, still more, when very warm weather suddenly sets in after a heavy fall of snow, these streams are, for a short time, enormously swollen; but even then they do not rise so as to make any approach to entirely filling the cañons in which they run. Moreover, such heavy floods occur but rarely. The detrital material in the bottoms of these rivers, where they have not been filled with tailings, is apparently of quite moderate depth. It is true that the bed-rock is almost everywhere concealed, and the business of fluming the streams, so as to work the débris in their channels, — once so profitable, — is now entirely abandoned, so that opportunities for seeing the bed-rock under the present rivers are of extremely rare occurrence. Even as early as the time of the beginning of the Geological Survey (1860), this kind of mining had nearly come to an end.* The present natural regimen of the streams throughout the auriferous belt of the Sierra is so masked by the vast mass of tailings poured into them, that it is not easy to make out exactly what was their condition before mining began in that region.

After a careful review of all the circumstances and conditions connected with the occurrence of the gravel deposits, the writer has come to the conclusion that the essential differences between the epoch of the high gravels and the present one depend for their existence on one all-important condition; namely, that during the Tertiary or Gravel epoch there was a much larger precipitation than there is at the present time. How this greater rainfall of former ages can be shown to have existed, and how its results were combined with those produced by other but secondary agencies, such as the rise, culmination, and decline of the volcanic period, cannot be set forth in full detail within the limits of the present volume; but the attempt will now be made to elucidate a few of the more important points suggested by this theory, at the same time referring the reader to another work, in a measure supplementary to the present one, where some of the matters which must here be passed over in haste will meet with a fuller discussion.†

One of the most interesting circumstances connected with the gravel deposits of the Sierra is this: that we have offered to us in this region an exceptionally good opportunity of ascertaining what the work of water

* In such old photographs of river workings as have come under the writer's notice, the detrital material appeared to be thin and the surface of the bed-rock uneven.

† See Climatic Changes of Later Geological Times. Mem. Museum of Comparative Zoölogy. Vol. VII Part 2.

has been in the way of excavating valleys during the different geological ages. And, furthermore, we have here the action of the sea entirely excluded: all the erosion which has been done in the region in question has been the result of the combined action of rain and rivers. There can be no mistake about this; neither can there be as to the absence of ice during the entire period of the accumulation of the gravels, and of the wearing away of the channels which the detrital materials now fill.

That the ocean has not had anything to do with the erosion of the bed-rock or the accumulation of the gravels will at the present time be admitted by all. The labors of the California Geological Survey have established the main facts so clearly, that the days of the crude theories advocated before that work was begun * may now be said with truth to have entirely gone by. A few words, however, in this connection, in regard to the proofs of the absence of ice agencies during the whole of the gravel period may properly here find a place: for at the present time there seems to be no theory, however absurd, which does not find favor, provided we have the word " glacial " connected with it. The entire subject of the glacial epoch in the Sierra Nevada, and in the Cordilleras in general, has been discussed by the present writer in another work, to which reference has already been made.† It is therefore unnecessary here to enlarge on this subject, although it will be proper to give concisely the reasons why it appears to be unquestionably the fact that the glacial epoch in California did not occur until long after the accumulation of the gravels had ceased, and the topography of the country had assumed its present form, down to almost its minutest details.

In the first place, the reader will bear in mind that the operations of the hydraulic miner are constantly uncovering large areas of bed-rock all through the gravel region, so that we have far better opportunities for seeing the character of the surface under the detrital masses than we can usually have in drift-covered regions. For instance, a large part of the surface of New England is overlain by gravel deposits, but it is only here and there that we can see the surface of the underlying bed-rock well exposed. It is true that the slates of the bed-rock series in the auriferous belt of the Sierra are liable to decompose rapidly when uncovered, and this circumstance often renders the opportunities for inspection somewhat less satisfactory than they would otherwise be; but, on the whole, we have a

---

* See ante, pp. 66-72.   † See Climatic Changes, etc., Chap. 2.

very good idea of the bed-rock surface under the gravel, such as it has been fashioned by the passage over it of immense quantities of fresh water bearing detritus with it.

Nowhere does this surface present those markings which are essentially characteristic of ice-action. The rock is often furrowed with more or less deep and persistent longitudinal depressions; but these have nothing indicative of ice-work about them. There are none of those fine striæ, parallel for a considerable distance, and quite straight, such as we find to have been produced where a mass of rock has been very slowly moved under the pressure of a great body of overlying ice. It is true that to distinguish with certainty between the work of ice and that of running water requires some previous opportunities for studying both kinds of phenomena, under present glaciers as well as in the beds of running streams, aided by at least a modicum of natural ability for geological observation. Where, as seems to have been frequently the case, both these advantages have been lacking, it is no wonder that all kinds of mistakes have been made.

Besides the character of the bed-rock surface as an indication of the work of water rather than of ice, we have, as bearing testimony to the same fact, the form and distribution of the gravel itself. This never resembles morainic débris, or, at least, only in a few exceptional cases, and then only to a very limited extent. It is almost always more or less rounded and water-worn, and usually very considerably so. It is never deposited in the form of moraines, or in accumulations having any other character than that of materials laid in their present position by the action of water.

Again, the character of the fossils contained in the gravel deposits, especially that of the plants so abundantly distributed through the series, furnishes corroborative evidence that the climate has never been one of a glacial character. Everything, on the contrary, indicates a warmer epoch than that now prevailing in the same region.

It being, then, as the writer conceives, beyond doubt that all that we see as the work of eroding agencies in the Sierra has been effected by rain and rivers, we have now to examine, with some care, what the precise character of this erosion has been.

The circumstances which have contributed to bring about the present relief of the surface in any region of country form, in almost all cases, a very complicated series of operations. It is with difficulty that we can know anything absolutely certain in regard to any of them. It is only those

events which have happened at the very last that we can ever fully understand, and these only under specially favorable conditions. The beginning
of all erosion is, of course, the beginning of change of relative level as
between different parts of the earth's crust. A uniformly and perfectly level
surface would never suffer any change from any of the agencies which we
class under the term "erosive"; but let one area be raised above another,
and, unless this has taken place at a considerable depth beneath the surface
of the water, the work of the rain or of the running stream, or of the waves,
if the region affected be near the ocean's edge, will be at once begun. The
steeper the grade, that is, the greater the relative difference between the
higher and the lower region, the more will the eroding power of the water
be increased. This, then, will be the first condition to be considered, and the
power of running water to act is, as is well known, increased in an enormously
rapid ratio with its increase of velocity, while velocity depends primarily on
increase of grade.

But here another equally important factor comes in, — quantity of erosive
material, that is, of the water; this is important with regard to subaerial
erosion. In the case of the ocean and its work, the quantity of water remains
the same at all points, and the erosive effect which it is able to produce depends on the character and force of its movements, — whether tidal or other,
— and of the nature of the shore-line against which it acts; but in the case
of subaerial erosion, or the work of rain and rivers, the quantity of the rain
while falls is an element of prime importance. That under certain meteorological conditions, that is, with absence of rain and wind and with a uniform
temperature, there would be no erosion, even in a region having a surface
of the most diversified character, is quite evident. And without rain, even
with great changes of temperature, erosion could hardly be said to take
place; for although the higher portions would eventually crumble to pieces,
the fragments could not be carried away, but would be piled up as they fell,
gradually forming a protective cover to what was underneath. Such changes
we may with reason suppose to be taking place on the surface of the moon;
and that this is so rough at the present time is proof that it is not long since
volcanic activity ceased there, if, indeed, it has really come to an end.

The next, and by far the most obscure and difficult part of the problem
before us is the character of the surface as left by orographic forces, at the
time when these ceased to act, and the region was delivered over, so to
speak, to be worked into details by erosive agencies. We have abundant

grounds for believing that the uplift of great mountain-ranges has usually, if not always, been the result of forces which, from their very nature, could not have done otherwise than tear the surface of the region in which they operated into forms of utter raggedness. An inspection of the position of the stratified and eruptive masses as exhibited on the flanks and along the axis of any great mountain-chain, such as the Alps or the Himalayas, either on the spot or as represented in faithfully drawn geological sections, will show at once that such a crushing, folding, and overturning of the rock masses could not have taken place under the surface without leaving their impress upon it.

The study of the numerous sections with which we have been furnished by the great mining operations, especially for coal, which have been carried on in various parts of the world, shows us that in many instances great displacements of the rocky strata have taken place, where at present no corresponding effects are visible on the surface. It cannot be denied that, in such cases, erosive agencies must have smoothed down the previously much broken surface, and there can be little doubt that the sea has been the chief agent in such work as this. The larger the body of water which acts through its movements on the surface, the greater the tendency to reduce projecting masses to the common level. The sea smooths over the surface, rivers furrow it, — that is, in those portions of their courses where they flow with sufficient rapidity to act as erosive agents ; where, on the other hand, they flow over an almost level surface, they may cease to erode entirely, and merely deposit all of the material which they have brought down from higher regions, excepting perhaps the very finest.

With the excavation of mountain valleys the ocean has had little to do ; they are originated by orographic causes, and chiefly worked out in detail by the action of running water. The form of the cross-section of any eroded depression will, of course, vary at different points along its line ; and will, in all probability, have varied considerably, at any one point, at different periods.

To recapitulate : the following is a condensed statement of the circumstances and conditions connected with the work of the erosive agents.

First, the steeper the slope, the more rapid the erosion, provided this be subaerial. The same cause is effective in the case of marine erosion ; but the latter will soon be brought to a stoppage, unless effectively aided by other conditions. If the ocean waves beat against a high cliff, this will gradually become protected by a fringe of débris, unless there are ocean currents in the region of sufficient strength to sweep the material away after it has fallen and been sufficiently ground up by the action of the water. But, again, a sinking of the land will also bring fresh surfaces into contact with the waves, and thus allow the work of erosion to be continuously carried on, even where the abraded material accumulates at the base of the elevation from which the supply has been obtained. In this way, probably, the larger portion of the work of the ocean in levelling off the surface of the land has been performed. The tidal wave has beaten incessantly against an ever-sinking mass of land.

Again, second, in regions of great precipitation, there, other things being equal, the erosion will be most rapid. This statement is one of which the truth is so evident that it is not necessary to enlarge upon it. It needs only to recall what devastation streams swollen to very much enlarged dimensions have been capable of effecting.

Third, there are to be taken into account all those varied conditions, with regard to the character of the rocks themselves, by which rapidity of erosion is affected. As a general rule, the accumulation of débris is likely to be a somewhat slow process; the sweeping of such material away may, under favorable circumstances, be very rapidly accomplished. Much will depend, in regard to the amount of time required for detrital material to accumulate, on such circumstances as these: the character and position of the joints, cleavage planes, and lines of stratification by which the mass is intersected; the tendency of the rock to undergo decomposition on exposure to the atmosphere; the amount and rapidity of the variations of temperature; the violence of earthquake shocks, etc.

And, finally, the original form of the depression in which the water begins to run, as the result of the preceding orographic disturbances, will have a most marked effect on both the character and the rate of the erosion.

In applying the above considerations more specially to the resulting forms of valley sections, we arrive at the following conclusions.

As a general rule, large streams run in correspondingly broad valleys, with cross-sections closely approaching straight lines; the amount of the depression is but trifling compared with the linear extent included within the edges of the water-shed. The higher we ascend into the mountains, and consequently the smaller the stream, — because nearer its source, — the more the valley acquires a pronounced form, capable of being represented in cross-section. To this, there are, of course, striking exceptions; but these have usually special orographic causes for their existence.

Again, the character of the erosion effected by any stream depends, in a marked degree, on the permanence of the supply of water. Very different results must be produced, according as the stream is constant, subject to periodical fluctuations, or diminishing in volume. This is too obvious to need discussion. In point of fact, however, as the writer believes, there is abundant evidence that, throughout the world, rivers, although undoubtedly always more or less irregular in their regimen, have on the whole been diminishing in size at least during the later geological epochs.*

It is apparent, then, that every valley beginning in a high mountain-range, and terminating in a plain, must exhibit, in cross-section, a variety of forms, and that these result from an intricate combination of physical conditions, the nature of which can be made out only by a careful study of both the geology and topography of the whole adjacent region. If we wish to form an idea of the changes of form which various portions of any such valley may have undergone in past geological times, we must not only study present climatic conditions, but endeavor to get, from some quarter or other, light upon the difficult question of what changes have taken place in these conditions during former epochs.

It would appear that we have in the hydraulic mining-region of California a favorable opportunity for studying questions of the kind suggested in the preceding pages. It is doubtful if there has ever been anywhere so large an area of bed-rock surface underlying detrital material artificially exposed for study; and it is surprising how little could have been known about the most interesting points in the geology of the Sierra if mining operations had never been carried on there. It is not the intention of the writer to claim that the difficult questions presenting themselves in this connection can all be answered in full; on some of them, however, light can be thrown.

The first, and perhaps the most important result which can be drawn from the study of the mode of occurrence of the gravel deposits, as described in the preceding pages, is this: that the character of the channels indicates on the whole, beyond question, that a much larger quantity of water passed through them than that which runs at the present time from the same region. Thus the conclusion, which would be naturally drawn from a consideration of the character and quantity of the detrital material itself, is corroborated by an examination of the form and size of the depressions in the bed-rock in which the gravel has been deposited.

* See Climatic Changes, etc., Chap. II.

Before making any comparisons between the dimensions of the channels of the gravel period and the river beds of the present epoch, it will be necessary to revert to the fact, already clearly established, that the area drained by the Pliocene rivers was not, on the whole, essentially different in size from that which the present ones drain. Were this otherwise, it would be impossible to draw any trustworthy conclusions with reference to the special points before us for consideration; for if we could conceive the topography of the Pacific Coast so altered that the character of the drainage area should be essentially modified, we could institute no comparisons of value between the ancient and the present rivers. But it has been already abundantly shown that the chain of the Sierra was essentially, in Tertiary times, what it now is, so far as concerns its general elevation and the position and direction of the principal river basins into which its western slope is divided. Hence, the existence in former geological times of very much larger channels than any which the present streams now occupy is sufficient proof of the flowing of a much larger amount of water from the same area. To be sure, we cannot be certain that the number of streams reaching the Great Valley in Pliocene times was exactly the same as now; there may formerly have been a more general concentration of tributaries into main rivers than now exists. For instance, it is quite probable that the Pliocene equivalent of Bear River joined the Pliocene American at an elevation of 2,000 feet or more, instead of debouching in the Sacramento Valley, as it now does. This would account in part for the magnificent dimensions of the ancient river-course along the divide between the Middle and North Forks of the American River, the width of which channel Mr. Goodyear estimates at 4,000 feet, or possibly a mile.[*]

Two grand rivers flowed down the western slope of the Sierra in Pliocene times, one representing the present American with its tributaries, and with which Bear River was probably connected, as suggested above; the other was the equivalent of the present Yuba. Both of these streams were of such dimensions that their now existing representatives are but rivulets in comparison. Both the Pliocene Yuba and the Pliocene American rivers can be traced far down into the Great Valley by their broadly spread deposits of gravel and the usual accompanying volcanic masses.[†]   The width of the

---

[*] See *ante*, p. 106.

[†] The extension down into the Great Valley of the gravels and volcanic deposits belonging to the Yuba system is well seen in the General Gravel Map; they reach nearly as far as Marysville. The shape of the map does not admit of tracing the Pliocene American in the same way.

Pliocene Yuba must have been, in places, over a mile; at Columbia Hill the channel is estimated as fully a mile and a half across. While, therefore, there can be no doubt that the amount of water carried off from the basins of the Yuba and American by the Pliocene rivers was enormously larger than that conveyed away from the same areas by the present streams, yet numerical data are entirely wanting. It is impossible, for instance, to say whether the present flow of these rivers is five or fifty times less than that of Pliocene times, for we know too little of the form of the rock-bottom of actually existing large rivers. After weighing, however, the conditions favorable to the idea of a large flow of water which are apparent in the great width of the main channels of the gravel epoch, and the steepness of their grade, no one can hesitate to admit that the evidence is clear in favor of an enormous diminution of volume in recent times. This is particularly true with reference to the portion of the gold region included in the basins of the American and the Yuba. Similar conditions prevail both to the north and south, but not on so extensive a scale. The Pliocene representatives of the Stanislaus and Mokelumne, for instance, were undoubtedly larger than the present rivers; but there does not seem to be sufficient proof that the difference was as great as it was in the case of the other rivers mentioned.

The enormous thickness of the detrital deposits in the large channels of Pliocene age is a condition so intimately allied with and dependent on the size of the channels themselves, that similar conclusions may unhesitatingly be drawn from one as from the other set of facts. Indeed, it would seem hardly necessary to occupy space with insisting on the abundance of proof of the former greatly increased flow of water, as offered by the size of the deposits which this water has left behind as the evidence of its former presence. There is no other way of accounting for the existence of these masses of gravel than by admitting that the rivers which did the work of abrading and depositing this detritus were of a size corresponding to the magnitude of the results produced. To endeavor to substitute length of time in the place of energy of action, and to maintain that small streams operating for an indefinite period might bring about such results as are here manifested would be quite unreasonable, in view of the fact that the rivers of the present day are showing us exactly what kind of work they are able to do, and this, as already set forth, is something very different from that accomplished during the Pliocene epoch.

We may now take up the next most important feature of the gravels, namely, their elevated position. The fact that the large bodies of detrital material described in the preceding pages occupy the summits of the country or the divides between the present rivers in the hydraulic mining region seems at first view to be contrary to what ought to be expected; yet a careful consideration of the facts, in connection with what has been laid down in the preceding pages with reference to the nature of the erosion effected by gradually diminishing streams of water, will, as the writer believes, enable us to get over some of the difficulties which this branch of the subject presents. There are, however, points which seem involved in great obscurity; and even in regard to some of those for which an explanation is here offered there will, no doubt, be lack of harmony of opinion among geologists.

The reader may obtain an idea of the difference in relative height between the gravels and the present river valley by examining the sections on Plate G. Figure 1, to which reference has already been made,* illustrates by the relative grades of the channels in the most important portion of the hydraulic mining region, namely, that in the American and Yuba River basins. In preparing this diagram, the elevations, as determined by the Survey, of numerous points, both at the level of the present streams and on the bed-rock in the ancient channels, were laid down and then connected by lines, the broken ones designating the existing rivers, and the full ones those of Pliocene age. With regard to the South Yuba, there can be no difficulty in recognizing its former representative; in the region drained by the numerous branches of the American River, there was more room for fancy, in connecting the various localities so as to reproduce continuous channels which should be undoubted representatives of the present streams. For all purposes, however, of seeing at a glance the general difference of level between the Pliocene and the present rivers, the diagram is quite as valuable as if there could be no question, in any case, in regard to the proper manner of connecting the different gravel localities, so as to exhibit the relationship between ancient and modern rivers with exact truth.

Figure 3, Plate G, to which reference has been made in Appendix A,† shows the difference of level along a line drawn in a southeasterly direction, — that is, nearly parallel with the range of the Sierra,—from near La Porte, for a distance of about twelve miles, across four deep cañons, and indicates the position of the high gravels with reference to the depressions occupied by the present streams.

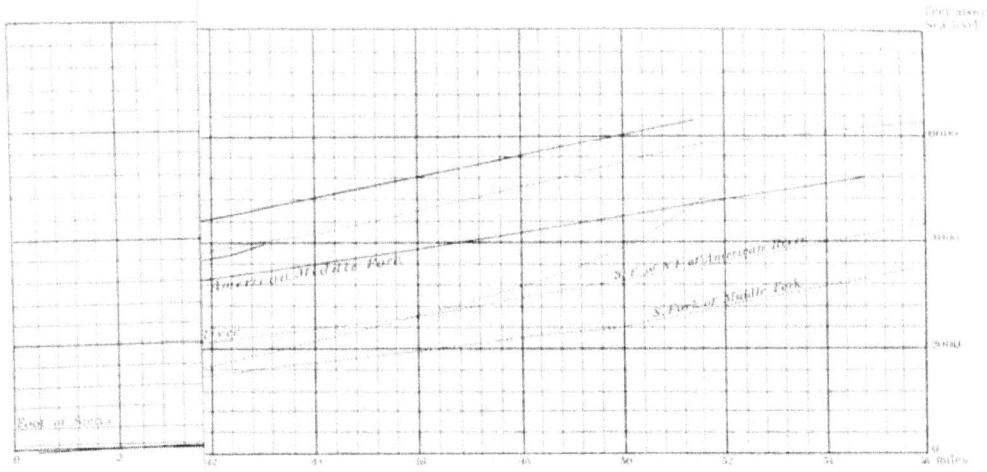

Feet above
Sea level

American of North Fork
Ravine
Bed rock at Suez
0 2

N Fork of American River
S Fork of Middle Fork

42 44 46 48 50 52 54 56 miles

Bed rock at Cement Hill 5793
Bed of N Fork of American River 108

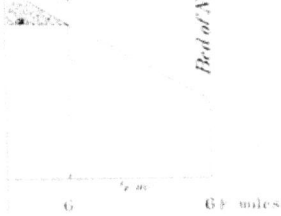

6 6½ miles

IVER.

Bed rock at La Porte 3677
Bed rock at Monte Cristo 5020
Summit on trail 5192
Bed rock at Excelsior 500
Bed of North Fork of North Fork of North Yuba River
Bed rock at Cincinnati

92 94 10
115 30

namely, their elevated position. The fact that the large bodies of detrital material described in the preceding pages occupy the summits of the country or the divides between the present rivers in the hydraulic mining region seems at first view to be contrary to what ought to be expected; yet a careful consideration of the facts, in connection with what has been laid down in the preceding pages with reference to the nature of the erosion effected by gradually diminishing streams of water, will, as the writer believes, enable us to get over some of the difficulties which this branch of the subject presents. There are, however, points which seem involved in great obscurity; and even in regard to some of those for which an explanation is here offered there will, no doubt, be lack of harmony of opinion among geologists.

The reader may obtain an idea of the difference in relative height between the gravels and the present river valley by examining the sections on Plate G. Figure 1, to which reference has already been made,* illustrates by the relative grades of the channels in the most important portion of the hydraulic mining region, namely, that in the American and Yuba River basins. In preparing this diagram, the elevations, as determined by the Survey, of numerous points, both at the level of the present streams and on the bed-rock in the ancient channels, were laid down and then connected by lines, the broken ones designating the existing rivers, and the full ones those of Pliocene age. With regard to the South Yuba, there can be no difficulty in recognizing its former representative; in the region drained by the numerous branches of the American River, there was more room for fancy, in connecting the various localities so as to reproduce continuous channels which should be undoubted representatives of the present streams. For all purposes, however, of seeing at a glance the general difference of level between the Pliocene and the present rivers, the diagram is quite as valuable as if there could be no question, in any case, in regard to the proper manner of connecting the different gravel localities, so as to exhibit the relationship between ancient and modern rivers with exact truth.

Figure 3, Plate G, to which reference has been made in Appendix A,† shows the difference of level along a line drawn in a southeasterly direction, — that is, nearly parallel with the range of the Sierra, — from near La Porte, for a distance of about twelve miles, across four deep cañons, and indicates the position of the high gravels with reference to the depressions occupied by the present streams.

Fig 1

SECTION
illustrating the relative grades
of the
PLIOCENE AND PRESENT CHANNELS
OF THE
AMERICAN AND YUBA RIVERS

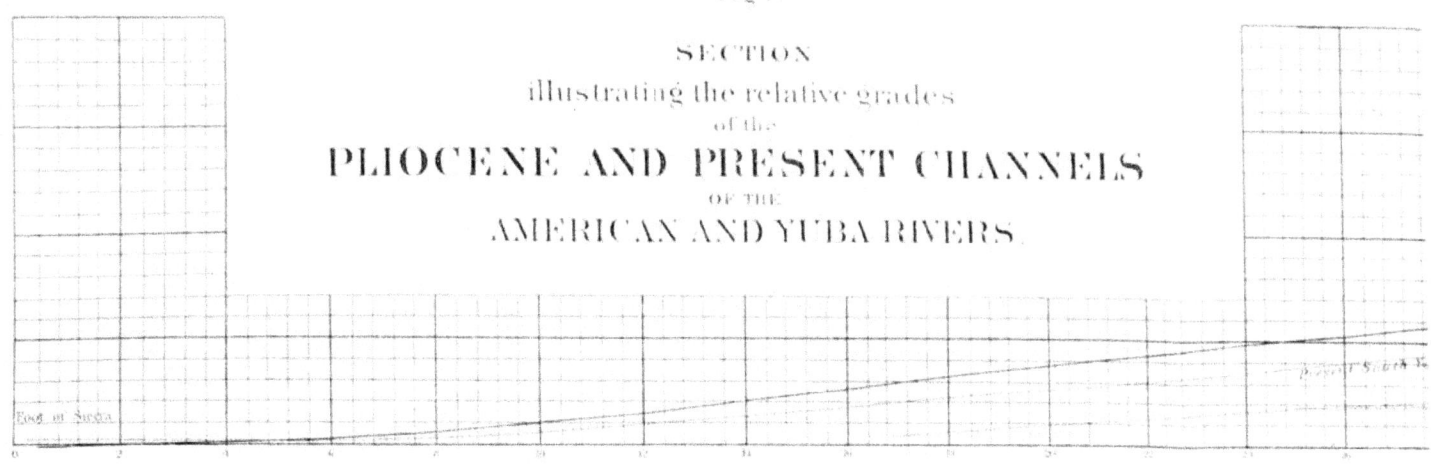

SECTION BETWEEN GREENHORN CREEK A

SECTION BEWTEEN LA

Feet above
Sea level

6000

4000

2000

0
26 miles

Profile of South Yuba River

Profile of North Fork of the American River

Profile of American Middle Fork

S. For s t of American River

S. Fork of Middle Fork

Profile of North Fork of American River

Profile of Middle Fork of the American River

Tailings in Bear River

Bed rock of Fug Ugly Hill

Top of Gravel at

Squires Cañon

Top of Gravel near Gold Run

Bed rock at Cement Hill

Bed of N Fork of American River

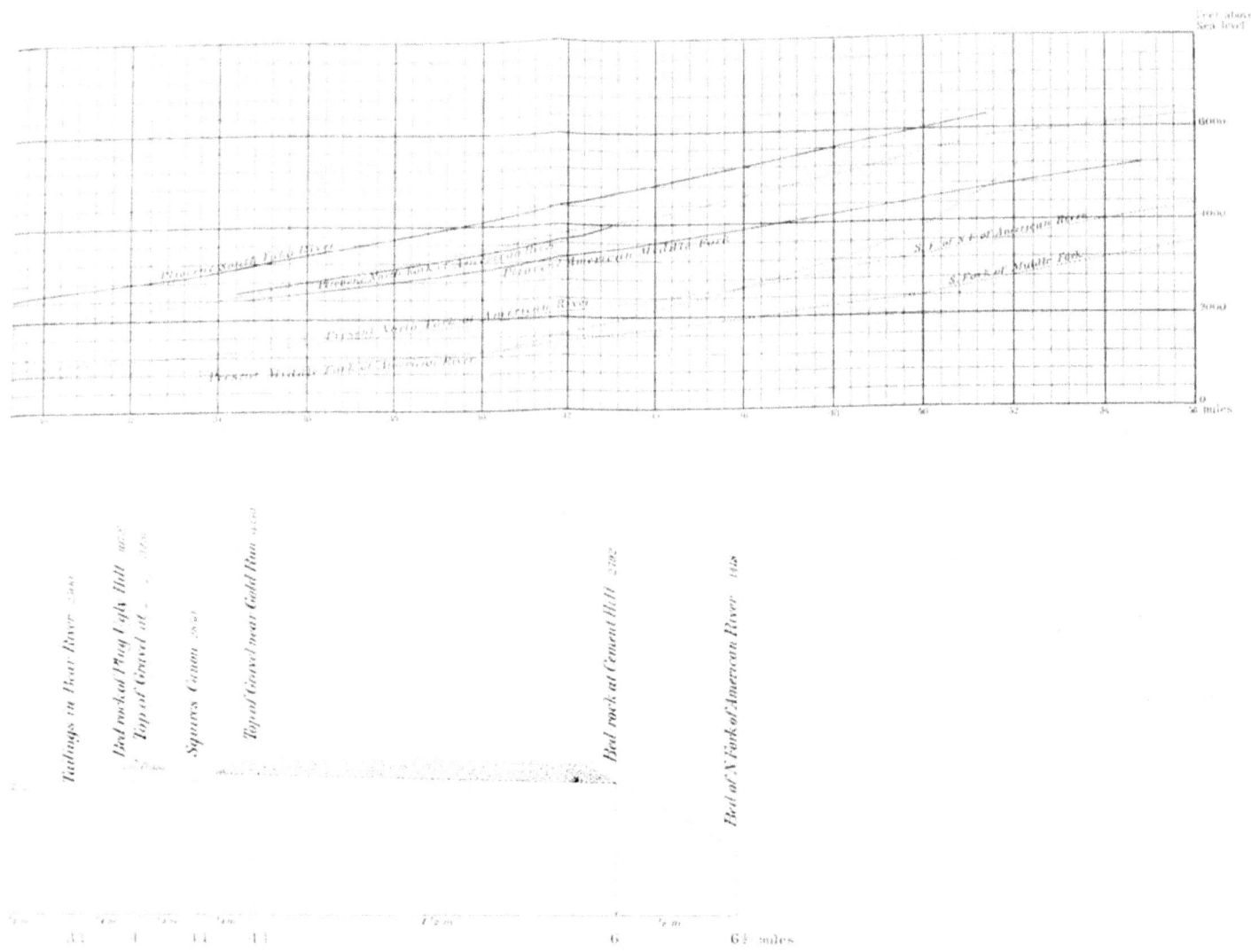

THE NORTH FORK OF THE AMERICAN RIVER

6½ miles

Bed of little Cañon Creek

Willis Gold Quartz

Magnusville bed rock
Top of Chapparal Hill

Bed of Woodmans Creek

Bed rock at Monte Cristo

Summit on trail

Bed rock at Excelsior

Bed of North Fork of North Fork
of North Yuba River

Bed rock at Craycrofts

TE AND CRAYCROFTS

11½ m

The third section on Plate G is introduced for the purpose of illustrating similar facts in the important district of Dutch Flat. Gold Run. and You Bet. This section. which is six and a half miles in length. is also drawn nearly parallel with the crest of the Sierra. and its position may easily be traced on the large map illustrating the position of the gravels in that district. The peculiarities of each of these regions illustrated in the three sections have been set forth with sufficient detail in the preceding pages.

That the detrital material resulting from the wear of a river upon its rocky bed, should occupy lower and lower positions as the work of abrasion goes on, seems to be in the ordinary course of events. It is true, however, only to a limited extent. As long as the stream has sufficient power to remove the detritus which is formed along its course, or brought into it by its tributaries, so long it must deepen its bed; but when the channel becomes covered with abraded material — gravel, sand, or mud — then, of course, all wear of the bottom ceases, and if the velocity of the stream slackens sufficiently. the height of its bottom will be raised. Thus the Lower Mississippi. like many other rivers. accumulates detritus along its course; and wherever it is artificially confined within fixed limits by dikes or levees, this takes place so rapidly as to become a practical question of serious importance. Thus, some regions are thought to have sunk, because the rivers draining them run at higher levels than they formerly did; when in reality it is only that their channels have become choked with their own débris, which the current has no longer the power to carry away.

A river remaining permanently of the same size, and not compelled artificially to heap up débris along one narrow line, will change its course so as to allow its deposits to retain about the same level over the whole width of the valley in which it is enclosed. Let the water diminish in quantity, and the stream will continually occupy narrower areas. terraces being often formed, as they have been along so many streams in New England.

The peculiarities which reveal themselves on comparison of the Pliocene and Recent drainage systems of the Sierra Nevada are, the great depth of the present cañons, the absence of débris from the slopes, and their extreme narrowness at the bottom. And the essential cause of these differences is, beyond question, the comparatively small quantity of water now being carried off from the region in question. Auxiliary conditions are. the steepness of the slope of the range and the peculiar climate prevailing there during the latest geological epoch.

During the gravel period the streams were broad, the slope of the Sierra being more uniform than it now is, and the valleys between the ridges decidedly shallow as compared with the present cañons. These streams, when choked with débris, had room to make for themselves new channels to one side or the other. They refused to be confined within fixed limits, and thus wear down one narrow channel, because the volume of water which they carried was too large.

Such were the conditions during the first portion of the gravel period. Vast quantities of débris must have been swept down to the very foot-hills and into the Great Valley itself. Even before the existence of this valley the western slope of the range was being eroded away; for, as already explained, it is necessary to look to this quarter for the source of supply for certainly a very considerable portion of the material out of which the Coast Ranges were built, these mountains having been uplifted while the gravels were depositing, and the detritus of which they are formed having been certainly accumulated, in large part, during the Cretaceous and early Tertiary periods. As the force of the eroding agents gradually slackened, more and more of the débris remained on the slope of the Sierra itself, where it accumulated, in favorable localities, to a depth of several hundred feet. The different river basins, however, were distinct from each other, the gravel not having filled them up so as to obliterate their boundaries, the erosive agencies having, previous to the deposition of the mass of the gravel on the flanks of the Sierra, worn out depressed areas more than large enough to hold all the débris afterwards accumulated.

Next follows in order the outbreak of the volcanic epoch, which piled on the previously accumulated gravels, in places, a thickness of several hundred feet more of various kinds of eruptive rock. This material, although, as it appears, mostly emitted from near the summit of the range, must rapidly have found its way to the most accessible depressions, carried there by gravity, with or without the aid of water. Thus the surface was still more nearly brought to a general level; and this fact is rendered very apparent when the observer stands on the surface of one of the old lava-tables, and looks across the country in a direction parallel with the crest of the Sierra, so that the immense cañons are concealed from view. The gentle, uniform slope to the west, and the smooth, flat surfaces formed by lava-flows on the divides between the streams, are most striking features of the scenery.*

* See *ante*, p. 65 ; and Geology of California, Vol. I. p. 244.

Portions of the erupted material were almost entirely indestructible, and remain to the present time very little affected by weathering, protecting the various underlying detrital materials from being swept away, while they themselves stand out in the well-known " table mountain " form, often rising high above the surrounding country. The cause of such topographical features is easily enough comprehended; and were all the volcanic materials as solid and indestructible as that of the Tuolumne Table Mountain, for instance, there would be no difficulty in understanding how it is that these lavas occupy so high a position. But, as Mr. Goodyear has shown in his review of his notes, much of the erupted material appears to be of a kind readily acted on by water, and more likely to be eroded away than the average bed-rock itself. This is particularly the case in the region which was the field of his special investigation; but both north and south of his district there is much solid basaltic lava, forming " table mountains," like that of which the one in Tuolumne County offers a typical example.*

The great difficulty in regard to the formation of the present river cañons is to explain how they were started in the position which they now occupy. Having once been begun, their present character can easily be shown to be the result of the peculiar climatological and geological character of the region. The gradually diminishing quantity of water which the streams carry prevents their rising above the channel in which they are confined; they must therefore continue to excavate along one line. But as the volume of water has decreased, so their erosive power has diminished; and the cañons, although gradually deepened, have become narrower, their whole form being that which must result from the action of a force continually diminishing in intensity. The sides of these cañons have been kept free from débris, as it appears, through the agency of the occasional sudden and heavy " freshets," which still take place, as previously described, although with, on the whole, gradually lessening intensity.

At the present time the excavating power of the streams seems to have diminished almost to nothing. Only the finer portions of the débris resulting from the hydraulic washings is carried away by the current; so that where the bottoms of the cañons have become filled to great depth with tailings, the streams, under ordinary circumstances, hardly disturb them. It is only during such winters as that of 1861–1862, when very extraordinary rises take place, that portions of this detrital material are swept down into the Great

* See Geology of California, Vol. I. pp. 219–211, for sections and descriptions of some of these.

Valley. On the whole, the streams in the hydraulic mining region are becoming rapidly choked with tailings, so that operations are already seriously interfered with; while the owners of farms at the edge of the foot-hills, where the larger streams debouch into the Great Valley, are much excited over the prospect of constant additions of mud and sand to the already extensive deposits of this kind on the adjacent land.

To return to the specially difficult point, — the cause of the beginning of the excavation of the present cañons along the lines which they now occupy. There seems to be no evidence to justify the falling back on any orographic causes in this case. There do not appear to have been any fissures or faults in the Sierra, since the gravel was deposited, of sufficient extent to materially influence the drainage, at least over by far the larger portion of the mining region.* Thus we are forced, not only to accept erosion as the essential agent in the formation of the existing river cañons, but to admit that the present streams were directed, from the beginning, into the positions which they now occupy through the agency of causes influencing the drainage of the region covered with gravel and volcanic materials.

The special mode of action of these causes seems, in many cases, to be involved in the greatest obscurity. Take, for instance, the Tuolumne Table Mountain, to which reference has been so often made. The preservation of the channel, with its well-defined rim-rock on both sides, and its softer pipe-clay deposits overlying the thin stratum of pay gravel, may be ascribed, with

* Disturbances of the gravel by faulting are somewhat numerous in the extreme northern portion of the mining region, as described in Appendix A. In the region so closely examined by Mr. Goodyear, however, the evidences of such movements of the bed-rock were but few, and the vertical displacement was never large. The points mentioned by him in his notes where such displacements were observed are the following: At the Missouri Tunnel, the bed-rock is traversed by a fault running N. 45° W. (magnetic), and the plane of which pitches to the southwest at an angle of 60°. The gravel on the southwest side of this fault has been raised from twelve to fifteen feet in vertical height; the facts at this locality are clear and unmistakable. Again, at Yankee Jim's, a fault in the bed-rock runs across the "Big Channel," in the form of an upward jump to the east of about fifteen feet, with a pitch in the same direction; the effect of this fault, however, could not be detected in the gravel. There are indications that this dislocation is continued with similar characters in Georgia Hill (see Map, Plate B, for the position of the localities specified). Mr. Goodyear was also informed of the existence of a disturbance of the bed-rock at King's Hill, showing itself in a displacement of the gravel to the amount of "twenty feet or more," the fault running northwesterly, and the upthrow being on its northeast side. This, however, he admits that he did not discover when at the locality in question. At Flora's Cañon the existence of a pair of parallel faults seemed to be strongly indicated by the sinking of the bed-rock so as to form a deep trough with a V-shaped section, the dimensions of which are not stated. Another fault was noticed at White & Co's. mine, at Castle Hill, near Georgetown. In this case, however, the displacement was only to the amount of about three feet, and there was some doubt about its existence at all. The above are all the instances of faults noticed by Mr. Goodyear in his investigations in the gravel region.

the greatest confidence, to the indestructible character of the basalt by which the lower portion of a pre-existing depression was partly filled as the result of a volcanic eruption. But only a very short distance above the place where all these phenomena are characteristically displayed, we find the Stanislaus River breaking directly across the ancient channel, and forming a gap or gorge of fully 1,500 feet in depth. From this intersection the Pliocene and the present rivers run nearly parallel with each other for several miles, when the basaltic mass is again cut through by the present river, and from here down to the foot-hills the Stanislaus winds between the old lava-flows in a manner for which it seems impossible to account.* In a similar manner the present Yuba cuts directly across the old channel in various places, where it appears extremely difficult to reconstruct the former topography in such a manner as to give the stream the necessary opportunities for beginning the erosive work, the final results of which are visible in the grand dimensions and depth of the cañon at the bottom of which the river now runs. That there was some reason, however, why the water selected the particular course which it followed, no one can doubt, however difficult it may be to point it out at the present time. The ways of water beneath the surface are enveloped in obscurity; so much so that the most contradictory views are held in regard to this subject by geologists of ability and experience. Portions of the lava and gravel deposits must have been very permeable to water, much more so than the average bed-rock. This might lead to the formation of subterranean currents, the result of which would be the diversion of a proportionally very large quantity of water along the lines of the former deep depressions. These would thus be deepened at a much more rapid rate than those portions of the surface where the bed-rock had the thinnest covering of volcanic and other débris upon it. Thus the streams finally came to occupy the old depressions, which they have ever since been deepening, for reasons which have already been given. In the case of a more rapid wearing away of the bed-rock, in consequence of the indestructibility of the lava, there would be usually well-defined rim-rocks on both sides of the channel. When the reverse took place, the rim might or might not be present, and this would depend, in great measure, on the form which the underlying bed-rock surface had before the channel became permanently filled up.

* See diagram of Table Mountain lava-flow, Plate D, opposite page 132.

SECTION VII. — *Source of the Gold, and its Distribution in the Gravel.*

In the first place, it may be unhesitatingly stated that the gold is almost exclusively limited to the quartz and metamorphic gravel. Volcanic gravel is of rare occurrence, and it still more rarely contains gold enough to be worth working. It may be set down as a general rule, that the volcanic rocks of the Sierra are not auriferous; and of course the detritus resulting from their disintegration is barren of the precious metal and of no value. This statement does not apply, however, to the older volcanic masses which are included in the bed-rock series, and which have themselves been subjected to such extensive metamorphism that they can no longer be distinguished from the metamorphic sedimentary rocks, except by the aid of the microscope. There is reason to believe that in the process of metamorphism gold and other metals have been introduced into the volcanic formations, as well as into those of sedimentary origin; and, indeed, we know of no reason why this should not be so. Volcanic rocks of Tertiary age on the eastern slope of the Sierra are traversed by veins of quartz and other vein-stones, richly impregnated with various ores, and especially with those which are both auriferous and argentiferous. These occurrences, however, seem limited to the oldest of the volcanic masses of the Tertiary epoch, and only to such as are highly metamorphic in character, — the "propylites" of Richthofen. In such cases we must conclude that the processes of vein formation and impregnation with the precious metals have been part and parcel of the chemical reactions which have resulted in the extensive metamorphism of the enclosing mass, or country-rock itself.

The whole series of phenomena observed in the Sierra Nevada gives ample grounds for the belief that the metamorphic rocks of the Range — the bed-rock, in short — were the original home of the gold now so generally distributed through the gravel. What has been already stated* in reference to the occurrence of the auriferous gravels themselves, shows clearly that their presence is entirely dependent on that of a body of slaty rocks, themselves enclosing great numbers of quartz veins, which in many cases are sufficiently well supplied with gold to repay the expense of mining. In short, the quartz-mining districts and the gravel region are essentially one and the same tract of country, as will be apparent on inspection of the various maps accompanying this volume. That a considerable amount of gravel

* See *ante*, pp. 302–314.

should in places have been swept downward beyond the limit of the area of productive quartz mining, is something not at all difficult to understand, when we consider the manner in which these gravels have been accumulated and moved from their original position, under the influence of powerful currents of water. Some very fine gold has been carried, no doubt, to a considerable distance from its native bed, and deposited upon strata which were not in the least auriferous; but the coarser portion of the gold remains either on or in close proximity to the place where it originated. Whether the large pieces of this ductile metal could be reduced by abrasion to particles so small as to be carried to a very great distance is not sufficiently well made out; but the gold, as it was originally formed in the quartz, seems, in very many localities, to have existed chiefly in the form of very minute particles, not visible to the naked eye. The abrasion of the quartz must set these free, and they might, as we may suppose, be easily transported with the aid of water.

If there is, as Mr. Goodyear thinks not unlikely, an immense amount of the precious metal buried in the detritus of the Great Valley, there is but little encouragement to look forward to a time when it can be profitably separated from the sand with which it is associated.

That the distribution of the gold in the gravel should be irregular and that it should be widespread are conditions which seem quite in harmony with its origin. Quartz veins are very generally distributed through the bed-rock series; a large portion of these contain some gold, many of them are rich enough — in places, at least — to pay for working. Whether the quartz veins in the bed-rock, as they at present exist, contain gold enough to furnish by their abrasion a quantity of the metal sufficient to correspond with what has formerly been accumulated in that way — equal quantities in each case being assumed, of course, as having been worn away — is a matter of considerable doubt. Yet the statement that the bed-rock at present would not furnish as rich a gravel under the same conditions, as to abrasion and accumulation, as those of the Tertiary gravels, is not one susceptible of positive proof. It can only be said that it appears — to the writer, at least — very clear that the older gravels have been formed from a material considerably richer in gold on the average than is the present bed-rock of the region. That this should be so, however, is not at all surprising, but, rather, in harmony with the general results of observations in mining regions. However unwilling those who have mining property for sale may be to admit it, there is no doubt in the writer's mind — and this statement is not here made without many years of careful examination of important mining regions in various countries — that the occurrence of metalliferous ores is rather a surface phenomenon than a deep-seated one. The conditions favorable to the formation of veins and vein-like or segregated masses are more likely to have existed near the surface than deep down below it. Such conditions would be — in part at least — diminution of temperature, relief from pressure, and, in the case of true veins, the existence of fissures. The history of mining operations shows beyond dispute that bodies of ore occurring in the segregated form are, on the whole, not to be depended on for persistence. And even true fissure veins must eventually give out in depth, if for no other reason than the change in the character of the enclosing rock brought about by intense heat. Neither would fissures be likely to continue to exist, nor the materials filling them to retain a distinct form, where the temperature was above the melting point. Thus, whatever theories we may adopt for the formation of mineral veins, we are led to the conclusion that they must, as a general rule, be better developed near the surface than at great depths. The fact that in some important mining regions the very upper portion of the veins has been oxidized, and then dissolved away by water, is very easy of comprehension, and not in conflict with what has been stated above.

The fact must be admitted that the quartz veins and masses were the chief source of the gold contained in the gravel, but at the same time it is true that gold does exist in the bed-rock where no quartz is found associated with it. As far as the writer's observations go, however, such occurrences are pretty much limited to the immediate vicinity of veins of quartz. The Great Quartz Vein, or Mother Lode of California, is in places associated with slaty rocks which carry considerable gold, although this appears to be very irregular in its occurrence. Such localities have never, so far as known to the writer, been found persistent enough in their yield to pay for working.* As far, however, as the special theory of the gravels is concerned, it is a matter of but little consequence whether the gold they contain came exclusively from the quartz veins, or in part also from the adjacent rock.

* In regard to such localities as those of Quail Hill and the Harpending Claim, where the slates are reported to have been highly productive in gold, but little can be said, except that they have proved, on working, to be conspicuous failures. How far the reported richness of the rock is to be accounted for by previous "salting" of the places from which the specimens were taken for assay, it is not easy to say. It is very clear, however, that there was more or less of fraud mixed up with the transactions at both of these localities.

The epoch of the formation of the quartz veins — that is, of their segregation and impregnation with gold — is a point of much interest, which has already been up for consideration,* and in regard to which a few words may here be added. There are four all-important occurrences in connection with the geological history of the bed-rock series, the co-ordination of which is particularly desirable; these are the upheaval of the chain of the Sierra; the metamorphism of the sedimentary rocks of which the range is composed; the formation of the quartz veins, which may be assumed to have been contemporaneous with their impregnation with their associated metals; and, lastly, the epoch of volcanic action, during which the flanks of the chain were so extensively covered with ejected materials.

With the upheaval of the range is most closely associated the appearance of its granitic axis, which it is impossible, in view of all the facts, not to regard as an intrusive mass, raised from beneath, and uplifting with itself the overlying sedimentary beds. That this took place at or soon after the close of the Jurassic epoch appears, as already stated, to be beyond doubt. But the granite itself is not the metalliferous rock of the Sierra; for, although it cannot be said to be entirely destitute of veins, yet, on the whole, as has been quite clearly shown in the preceding pages, when we pass above the slate belt, and enter the granite region, we leave the truly auriferous gravels behind.

If, however, the granite itself is not metalliferous, its appearance seems to be closely associated with the metamorphism of the adjacent sedimentary rocks; while this latter condition is, as a general rule, the concomitant of the occurrence of mineral or metalliferous veins. It may be stated with truth, that most great mountain ranges have an axis† of granitic rock, that this is flanked by metamorphic strata, and that these chemically altered masses are the home of the metalliferous ores.‡ Indeed, it has been said, and with truth, that mineral veins are only special forms of metamorphism: this is especially true of the rarer and more valuable metals, and particularly of those styled precious. Ores of iron are universally diffused through the formations; the ores of lead, zinc, and, to some extent, those of copper, are frequently found occurring in unaltered strata; but in such cases they are

---

* See *ante*, pp. 330, 331.

† This axis need not be one continuous mass, but may consist of several disconnected portions, as in the case of the Alps.

‡ In several of the most important European languages, "mine" and "mountain" are synonomous terms. A "mountain man" is a miner, a "mountain work" a mine.

rarely accompanied by the precious metals; while in the metamorphic rocks the baser metals are almost invariably argentiferous, or associated with ores of silver, which sometimes also contain gold.

The possible connection of volcanic activity with metamorphism, and with the impregnation of the veins with their metalliferous contents, has already been suggested in the preceding pages.* A few words may here be added with special reference to the occurrence of the precious metals as being more or less distinctly related to the presence of eruptive rocks. While the unmetamorphosed recent lavas and other volcanic ejections are, on the whole, very barren of metals, the rocks of this class which have been subjected to chemical reactions are often the repositories of extensive metalliferous deposits. This statement is not only true with reference to volcanic rocks older than the Tertiary, and which are often so much changed from their original character as to have — in many cases, at least — escaped recognition, but it applies especially to the Tertiary volcanic masses. These are believed to be the seat of some of the most important silver mining regions. In fact, the silver of the world comes chiefly from two sources: one is the association of this metal with the sulphurets of the baser metals, especially lead; the other is from the proper ores of silver. The first of these class of occurrences is now almost universally designated in the Cordilleras as "base-metal mines." These occur chiefly in metamorphic sedimentary rocks; the other class belongs essentially to the metamorphic volcanic, and includes such immensely productive regions as those of Mexico, Washoe, and Peru. The silver ores of these regions are always more or less auriferous, and sometimes, as in the case of the mines in the Comstock Lode, largely so.†

Again, the fact is not to be ignored that the two by far most important and productive gold-mining districts of the world are regions of former intense volcanic activity. Australia and California exhibit the same phenomena of rich auriferous detritus buried beneath masses of lava, as will be more fully noticed further on in this chapter. Although the eruptive rocks, in these cases, are not the direct repositories of the precious metal, there would seem to be strong reasons for believing that there is a genetic connection between the volcanic activity and the enrichment of the adjacent strata.

The ranges of the Andes and of the North American Cordilleras certainly

* See ante, pp. 310, 331.

† The average value of the ores from the Comstock Lode is about 47% gold and 53% silver.

surpass all other regions in the world in metallic wealth; for hundreds of years they have been supplying the world with a large part of its silver, and no inconsiderable proportion of its gold. The richness of the Peruvian mines long since became proverbial, and the word "bonanza" is now familiar to all from its association with the mines on the Comstock Lode and others of the Far West. These regions have been the seat of the most intense volcanic activity during Tertiary times, and in all probability in previous geological epochs. It is hardly possible, therefore, in view of all the facts, not to admit that there is likely to be some genetic connection between the manifestations of igneous forces and the impregnation of the rocks with the precious metals and metalliferous ores.

Quite a number of theories relating to the geological epochs at which the various metals have made their appearance, or been introduced into the formations where we now find them, have been put forth by chemists and geologists. Gold has been specially favored in this respect. No theoretical views regarding this metal have been so widely promulgated and generally accepted as those of Murchison. According to this eminent authority, gold in paying quantities is exclusively confined to the Palæozoic rocks, into which, however, it was introduced at a very late geological epoch. It was also a favorite dictum of this geologist that auriferous quartz veins are a superficial phenomenon, and that mines of this metal would not hold in depth as persistently as those of other metals. The discoveries of the California Survey in regard to the age of the gold-bearing formations of that State have entirely refuted the (until 1864) generally accepted theory of the exclusively Palæozoic age of rocks of this kind, although this fact was not admitted by Murchison in his latest publications.* The great depth to which some gold mines have been wrought, with profit, both in California and Australia, in connection with many other facts observed in various parts of the world, justifies us in asserting that auriferous quartz veins are as persistent, on the average, as those worked for the other metals. That the impregnation of the rock with gold took place at a comparatively recent geological epoch, at least in certain prominent and important mining regions, cannot be denied. These are regions of former intense volcanic activity, and the period to which that belongs is unquestionably Tertiary. In regard to mining districts where gold and other metals and metalliferous ores have been found in considerable quantity, and where there have been no striking

* See *ante*, p. 34; also Siluria, 4th Edition (1867), p. 469.

manifestations of volcanism, accompanied by ejections of lava, as, for instance, along the greater portion of the Appalachian Chain, and especially on its eastern border, no definite statement can be made in regard to the geological period of the metalliferous impregnation. It would appear, however, that the evidence is, on the whole, in favor of this having taken place at the time of, or shortly after, the upheaval of the ranges themselves. To prove that the rocks of such ranges as the Appalachian and Scandinavian, which are surrounded by entirely unaltered Cretaceous and Tertiary strata, have been the scene of extensive chemical reactions during those later periods would be a difficult task.* Under any circumstances, there is no basis for Murchison's idea that gold was — to use his own words — " the last formed of the metals"; for the impregnation of the quartz veins, or rather its segregation, at the same time with the quartz, into veins or vein-like masses, was merely a collecting together of particles previously existing in the rock, and not by any means a new creation of them.

On the purely chemical question, by the aid of what solvent the golden particles were carried into the position which they occupy, or have once occupied, in the veins, no light seems thus far to have been thrown by the investigators who, in various parts of the world, have occupied themselves with this problem. That the gold and the quartz were introduced into the vein-fissures or segregated into vein-like masses contemporaneously with each other, would seem, from the manner in which the two substances occur together, to be beyond doubt. Why they thus occur, or why this metal is so rarely found in any other gangue,—as, for instance, calcite,—is a question which cannot yet be answered. Neither has any chemist been able to advance the first step toward an explanation of the fact that native gold never occurs otherwise than as an alloy with native silver.

Some other problems have engaged the attention of those who have made a special study of the chemistry and mineralogy of gold regions. One of these is the apparently commoner occurrence of nuggets† of large size in the

---

* The most extraordinary of Murchison's theories is that held by him in regard to the relative value of silver and gold, which he thus sets forth in the last edition of "Siluria" (p. 475): "Before quitting this theme, I would simply say, as a geologist, that Providence seems to have adjusted the relative value of these two precious metals for the use of man, and that their relations, having remained the same for ages, will long survive all theories." One knows not which to wonder at most, the idea here expressed, or the fact that it is put forth as a *geological* statement.

† "Nugget" seems now to be the most generally accepted term for a "sizable" rounded piece of native gold. It appears to be a word of Australian origin, and is the equivalent of the Spanish "pepita" (seed, pip; *pépite* in French) or "chispa" (a spark or brilliant). "Scad" is the American for nugget, but, so far as the writer knows, of recent origin, and not in general use, even among the miners.

auriferous detritus than in the solid vein.   We say apparently, because it is not so easy to give satisfactory proof for this generally accepted article of belief. It does appear, however, as if there was some truth in the idea that the finding of large pieces of gold in the gravel is not justified by what we see of the occurrence of the metal in the quartz. It is certain, at all events, that the form of the ordinary nugget is something different from that which is offered by the gold as originally deposited. And before proceeding further in this discussion, a few words may be offered in regard to the shape and character of the metal, when in its natural condition, as associated with the quartz or with any other mineral or rock. The writer has had many opportunities to study the shape and appearance of the native gold enclosed in the quartz. The metal presents in such cases a great variety of forms, but it never occurs, so far as the writer is aware, in rounded, smooth pieces, such as used to be found not unfrequently in the placer mines along the course of the present streams.

The larger part of the gold contained in the quartz exists in the form of particles invisible to the naked eye ; and there are many mines, which are producing largely and paying handsomely, where free gold can hardly ever be seen at all in the rock going to the stamps. Indeed, there is a general belief among the miners that " specimen mines " — or those where the free gold is segregated from the quartz so as to form handsome specimens — are not likely to be persistent.

Where the gold is visible to the eye in the quartz, the predominating form which it exhibits is that which is best expressed by the term " scaly," — a word used by the placer and hydraulic miners in describing the small, rounded, flattened pieces with which they so frequently meet. When the gold is collected into continuous thread-like forms, these are usually found on examination to be made up of aggregations of very irregularly grouped filmy or scale-like particles. The metallic portion in such cases, if freed from the quartz without alteration of its form, would present itself in rough and extremely irregular shapes. A very large proportion of the gold, if visible to the naked eye at all, has this appearance. Occasionally, however, this metal occurs in thin plates or leaves, with rather smooth surfaces. Such forms appear, however, to be almost, if not quite, exclusively limited in their occurrence to veins in which the quartz is in combs, or parallel layers having their opposite faces lined with crystals ; it is between these crystalline plates that the leaf-gold usually occurs.

Crystallized gold is of extremely rare occurrence, and it appears to be limited to a few favored localities. No instance has ever come under the writer's eye of a perfect crystal with smooth sides and sharp edges; irregularity of outline, rounded faces, and imperfect edges are the rule. The octahedron is the predominating form, with various modifications; not a specimen in cubic form has ever been seen or heard of by the writer. Isolated crystals do occur; but groups are far more common, and they are almost always imperfectly developed, and aggregated in arborescent and filiform shapes, or implanted on broad leaves or plates. Some of the specimens of this kind are of great beauty. Spanish Dry Diggings is a famous locality for crystalline masses of gold. The specimen to which reference was made by Mr. Goodyear in his description of this locality* seems to have been one of the most strikingly beautiful ever discovered. It was very irregular in shape, fifteen inches in length, and about six and a half inches in its greatest, and two in its least width. While appearing, as described by Mr. Goodyear, to be a mass of "imperfect arborescent crystallizations," it exhibited few well-developed planes; but the writer had no opportunity of making a minute examination of this specimen. Others, much smaller, but of great beauty, from the same locality, and taken out at about the same time, are in irregularly branching, leaf-like forms, studded with triangular plates with bevelled edges, these plates are sometimes very sharply defined, but more generally they have a peculiar blunted look, as if they had been heated just to incipient fusion. Byrd's Valley † is another famous locality of crystallized gold. The specimens from that place, so far as the writer has had an opportunity to observe, are much more distinctly crystallized than those from Spanish Dry Diggings. They consist of groups of octahedra, of many of which the skeleton only exists, the crystal looking exactly as if it had been cast in a mould, and then the remainder of the metal allowed to run off as soon as the edges had hardened. Some forms are tabular and triangular in outline, with cavities of the same shape impressed in them. ‡

We may now proceed with a few remarks in regard to the occurrence of nuggets in the gravel. It seems to be the fact that the gold in the quartz

---

* See ante, p. 115.

† Mr. Goodyear gives "Mad Cañon near Byrd's Valley" as the precise locality of the crystallized specimens. See ante, p. 114.

‡ Professor W. P. Blake, in his Catalogue of California Minerals (in J. Ross Browne's First Report, 1867, p 204), gives several localities where crystallized gold had been found in the Sierra Nevada. All the crystals mentioned are described as having an octahedral form, sometimes slightly modified. Specimens from the Princeton Mine are spoken of as having "brilliant faces." Two localities are mentioned of gold in calcite or dolomite.

never has the proper "nuggety" character. The metal does not occur, as deposited from solution, in solid, smooth, and rounded masses, but in scaly, foliated, filamentary, arborescent, or crystalline forms. The question arises, then, How has this change been brought about? And connected with this is the inquiry whether there is really ground for believing that pieces of gold, after being separated from their original matrix, do increase in size, either by chemical or mechanical causes.

The finding of large nuggets in the hydraulic mines of California seems, so far as ascertained by the investigations of the Geological Survey, to be of very rare occurrence. There have been, however, occasional statements, in books and newspapers, of such lucky finds in the ordinary placer mines. Most of these, of course, took place many years ago. The largest nugget of which the writer has ever heard is one said to have been found at Vallecito, in 1852, which weighed twenty-five pounds.* Nothing is known of its form or of the character of its surface. Such finds seem to be much more common in Australia than they ever were in California, judging from the lists which have been given in the official publications of the Geological Survey of Victoria.†

Large masses of gold have occasionally been found in the quartz and in the bed-rock in California. The occurrence of such at Spanish Dry Diggings has been mentioned. Carson Hill is another locality from which similar facts have been reported. As far as the results of his own investigations in California are concerned, the writer is not able to find sufficient evidence to support the opinion that the large size of the nuggets in the gravel presents difficulties requiring the aid of chemistry for their solution. If it be true, as the writer believes, that quartz veins, as well as all others, as a general rule, have been richer near the surface than they are at great depths, then the occasional finding of large nuggets in the gravel would not be a matter of surprise. Heavy masses of gold are found, even now, in some of the quartz veins, and somewhat heavier ones may have existed nearer the surface.

---

* It is stated by Mr. W. Birkmyre, in his list of great nuggets found in various parts of the world, that there is in the collection of the Bank of England a nugget found in Carson Creek, in 1850, and weighing eighteen pounds three ounces. The above-mentioned find at Vallecito is given on the same authority.

† The largest Australian nugget — the "Welcome" — weighed 184 lb. 9 oz. 16 dwt., and contained, by assay, 99.2 per cent of gold, netting to the owners £9,325. This specimen was found at Ballarat. Many others have been found in Victoria weighing over fifty pounds. The writer saw in the collections of the Mining School at St. Petersburg the famous nugget from Miask, in the Ural Mountains, the weight of which is given by Tegoborski at 87 pounds 92 zolotniks, and by Humboldt at 36.025 kilogrammes; it is round and smooth, and free from quartz or other gangue.

With regard to the manner in which the gold in the quartz loses its characteristic forms, so as to become transformed into the smooth rounded masses occasionally found in the placer mines, there seems to be no theoretical difficulty. In the first place, however, it may be stated that by no means all the nuggets have this character. Many of them exhibit more of their original character than would be expected to be found remaining after ages of pounding between the boulders of the gravel. This is particularly true of specimens collected by Professor Pettee from the hydraulic mines during his last year's investigations, and which have been been carefully examined by Mr. Wadsworth and the present writer. The same fact has also been stated by Mr. Ulrich — who appears to be a close observer — in regard to the Australian nuggets.* There seems to be no doubt that a scraggy — to use a common miner's term — piece of gold can be transformed into a rounded smooth nugget by a sufficient amount of the right kind of rubbing and hammering, which must have taken place as these great piles of detritus were being shifted from place to place by currents of water. Some of the specimens collected exhibit in the most interesting and convincing manner the transitional form between the rough crystalline form and the smooth rounded one. One in particular, from an unknown locality, purchased by the writer in a shop at San Francisco, has one side almost perfectly smooth, and rounded edges turned over upon the back, which itself is covered with crystalline branchings, still retaining a large part of their original delicacy. It is evident, in this case, that the specimen has been protected on one side, while the other has been subjected to abrasion and pounding, the result being a nugget, presenting at the same time and in most remarkable perfection, the characteristic forms of quartz gold and placer gold.

That the masses of gold, when they have been released from the quartz veins and have begun to be rolled about in the gravel, could by any possibility be so situated as to become subjected to any chemical influences by which their mass could be enlarged, seems — to the writer, at least — highly improbable. That occasionally pieces of the metal may be united by pressure or by hammering between the gravel boulders, and that thus a larger mass may be formed by the union of two or more smaller ones, through purely mechanical agencies, seems not impossible; and some observations of Mr. Wadsworth appear to corroborate this view.

* See R. Brough Smyth's "Gold Fields and Mineral Districts of Victoria" (1869), p. 360.

The distribution of the gold in the gravel calls for a few remarks. Although the phenomena of this distribution are generally of a simple nature, and such as would naturally be expected to result from the disintegration and removal by water of a large mass of auriferous rock, yet there are some puzzling peculiarities, for which it is not easy to find an explanation. That a considerable part of the gold should be either lodged directly upon the bed-rock, or disseminated through the lowest layers of the gravel, nearly in contact with it, is easily understood. The tendency of the very heavy metal to sink, as the mass of detritus was being moved by the aqueous current, would be continually exerted; and it is only the extreme fineness of the particles which has kept even a small portion of it from finding its way as far down as to the solid rock.

In certain cases there is no gold at all upon the bed-rock; but such instances are rare. There is nothing in these exceptional instances which need excite surprise; for that in certain localities there should have been a covering of detritus spread over the surface of the bed-rock, coming from the disintegration of a quantity of rock which contained no gold, is very natural. It is not the case now, nor has it ever been, so far as we can judge, that all parts of the bed-rock have been impregnated with gold. Portions must have been entirely barren, and if the débris of such masses happened to find a permanent lodgment on the surface of the rock, and then become more or less consolidated, the particles of gold borne from other and richer regions would rest upon the underlying barren layers of gravel.

SECTION VIII. — *Ores and Minerals associated with the Gold.*

Some statements have been made in the preceding pages[*] in regard to the metalliferous ores which are most likely to be found accompanying the gold in the quartz veins. Something now remains to be said about the minerals and ores which have been detected in the gold-washings. These substances can, of course, be only such as are not too brittle or too easily oxidized; for otherwise they would have been ground to powder in the process of the formation of the gravel, and would then have rapidly disappeared, having been oxidized and dissolved away. If the metal gold is found so widely disseminated through the gravel and in river sands all over the world, it is because it is so very indestructible. It may be torn into very

---

[*] See *ante*, pp. 56, 57.

minute particles, as the fragments of rock containing it are swept hither and thither in the midst of the currents of water by which they are borne onward; but, however small these particles, they will not become oxidized, and thus be lost. Platinum is another metal which is almost indestructible, and one which is obtained solely from the washings of detrital material. Nowhere is it known to occur in sufficient quantity in the solid rock to repay the expense of its working.

By far the most common mineral found in connection with the gold, in the placer-mining operations, is the ordinary black iron sand, as it is commonly called. This is apparently usually magnetite; although, no doubt, some of it comes under the head of menaccanite, which is considered to be a hematite, or peroxide of iron, with part of the iron replaced by titanium. It seems, at all events, to be a very indestructible and not easily oxidized material. As magnetite is one of the most common minerals entering into the composition of volcanic rocks, and occurring also very frequently in granite and in slates, it is very natural that it should be found in considerable quantity and widely disseminated through the gold region. It is certain that it comes from the débris of the lava and bed-rock, and not from the abrasion of the quartzose vein-stone, for in the latter it never occurs.*

The most interesting mineral found associated with the gold in the gravel of California is the diamond. This gem, as it would appear, is usually, although not universally, found in gravels which are washed or searched with care. Such gravels are chiefly those which are auriferous in character. Hence it follows that diamonds have been found in the surface detritus of most gold regions, as in Siberia, Australia, along the flanks of the Appalachian chain, and in California. But not one of these localities can be classed among the productive ones. It appears that diamonds, like some other natural products, are widely disseminated over the surface of the earth, but only present in considerable quantity at a very few points. Formerly, India was the productive region; then Brazil had its day; and now South Africa furnishes by far the larger portion of the new diamonds put upon the market.

The mode of occurrence of the diamond is as yet but little understood, in spite of the amount of research which has been expended on the subject. That the diamond is " at home " in the older crystalline metamorphic rocks,

---

* The writer, in examining thousands of pieces of auriferous quartz, has never found one containing either hematite or magnetite.

and a product of metamorphic action, would seem to have been clearly established prior to the discovery of the South African localities, which appear to be not only more prolific than any yet discovered, especially in stones of large size, but to differ quite remarkably from all others, inasmuch as these are associated with and sometimes enclosed in a rock, which, although of a puzzling character, is generally admitted to be eruptive. But whether the diamonds were originally formed in this rock, or only introduced into it from some other source, is as yet undecided.

The diamond has not been found in sufficient quantity anywhere in North America, whether in the Appalachian or the Californian gold-fields, to be of much importance from an economical point of view. But from the methods pursued in mining in the latter region, it will be easily understood that but few of the diamonds actually contained in the detrital material washed would be likely ever to be seen. Still, the number of localities where this gem has been observed is considerable. The following is a list of those occurrences which have been brought to the notice of the members of the Geological Survey; these are mostly within the area surveyed by Mr. Goodyear, and the larger part of the detailed information about them has been extracted from his notes.

Mr. Goodyear says: " Mr. McConnell tells me that in the McConnell & Reed claim, which is the next one west of the one I visited on the south side of Webber Hill, he once found, in the sluices, a colorless and brilliant diamond. This was at a time when they were not touching the bed-rock, but working entirely above it, at a height of three or four feet. He says that the stone had twenty-four faces, and was ' about the size of a small white bean.' He also firmly believed that, previous to finding this one, he had thrown away one in the same claim ' as large as the end of his thumb.' He also says that there had been one or two found in still another claim, on the south side of Webber Hill, but farther west."

Mr. Goodyear further remarks: " I saw here [at Dirty Flat, near Placerville] to-day, in the possession of Mr. Robert Cruson, or rather of Mrs. Olmstead, his sister, a diamond measuring $\frac{9}{32}$ of an inch in maximum diameter, from apex to apex of opposite solid angles. This diamond is said by Mr. Cruson to have been found by Mr. Olmstead in clearing up the sluices in washing the gravel from the shaft connecting with the main Cruson tunnel at Dirty Flat. It was found not less than six years ago, and probably ten or twelve. It is a little longer in one direction than it is in another, and is a perfect crystal, though covered with a white coating which destroys its lustre. The surface is somewhat rough, under my pocket magnifying-glass, and the crystal in its present condition is but slightly translucent, not transparent, and has the common very slightly yellowish tinge of Californian diamonds, and probably one or two small flaws. It probably weighs not far from one and a half carats." In the same neighborhood, at Smith's Flat, Mr. Goodyear heard of two or three other probable diamonds as having been found there, and adds, "I have little doubt that a good many have been picked up here, and looked at and thrown away."

There are said to have been three or four diamonds found at the mines at and about White Rock. Mr. Goodyear purchased at that place, of Mr. Potts, a perfect crystal, having a very slight yellow tinge, and weighing half a carat. It was found in washing the gravel, which came from a tunnel driven into the White Rock Ridge. Near the same place three diamonds were found in the gravel, by the Ward Brothers, in 1867. The largest of them is said to have been valued at $ 50 by a dealer in San Francisco.

Other localities are: Jackass Gulch, near Volcano; Indian Gulch; Loafer Hill, near Fiddletown, from which vicinity quite a number have been reported at different times.

French Corral, in Nevada County, has also produced quite a number of diamonds; the largest Californian specimen which the present writer ever saw was said to have been from this locality; it weighed seven grains and a quarter. Others have been reported, one of which is said to have weighed a trifle over five grains.

At Cherokee Flat, in Butte County, Professor Pettee was informed that fifty-six diamonds had been picked out from the washings, at various times, and that one of these had an estimated value of $ 250.

Microscopic diamonds were discovered in 1869, by Wöhler, in gold washings from Oregon.* Associated with the diamond were platinum, laurite, iridosmine, chromic iron, and zircon. Similar finds have been made in northern California, in connection with black iron sands from the sea-beach. As in the case of the Oregon specimen examined by Wöhler, which was probably from a similar locality, the microscopic diamonds appear to be always associated with platinum.

Next to the diamond, perhaps platinum and the allied and associated minerals are of the greatest interest as accompaniments of the gold in California.

* American Journal of Science (2) XLVIII. p. 441.

The Geological Survey having, however, done little work in the platiniferous region, the writer has no original information of special value to offer on this subject. It is certain that the metals in question occur in the northern part of the State in considerable abundance, and it is not positively known that they have been found at all in the central and southern counties. The practically important ores of this group occurring in northern California are two: native platinum, which is an alloy of this metal with iridium, rhodium, palladium, etc.; and iridosmine, a mixture of the two metals, iridium and osmium, in varying proportions. Besides these, there are laurite, a sulphuret of osmium and ruthenium; and, according to Dr. Genth, probably platin-iridium, a combination of platinum and iridium in different proportions. Laurite (named in honor of Mrs. Laura Joy) was first discovered by Wöhler in the platinum washings of Borneo, and afterwards found by him, in connection with platinum, gold, chromic iron, zircon, quartz, and microscopic diamonds, in washings from the coast of Oregon; and the same mineral, in all probability, exists in northern California.

Native platinum and iridosmine occur together in the beach-sands from Cape Blanco to Cape Mendocino, as also at Cherokee Flat in Butte County. Of the quantity of these two mineral species obtained, or of their relative amount, no definite statement can be given.* According to Professor Pettee, the occurrence of platinum in the sands in the vicinity of Oroville, has attracted the attention of Mr. Edison, who last year was engaged in erecting works at that place for saving this valuable material.

The following analyses (by Sainte Claire Deville and H. Debray) seem to be the only complete analyses ever made of the California platinum.†

---

* The following information in reference to California iridosmine was communicated to Dr. Gibbs by the officers of the United States Assay Office, at New York: "For the first year or two after the establishment of the United States Assay Office, the proportion of osmiridium in the California gold did not exceed half an ounce to the million of dollars. Afterward, the proportion increased till the average was seven or eight ounces to the million of gold. Then for a year or more the quantity diminished, but for the last year it has been as large as ever. These differences depend on the variable composition of the native gold and the constant discovery of new diggings. The grains of osmiridium, suitable for pens, are roundish and solid, not liable to exfoliate when struck or heated. They seem to have a different composition from the compressed and tabular crystals. The proportion of them is usually more than a tenth of all the alloy, but it is sometimes as large as one fifth. The carefully selected grains used by the gold-pen makers are so minute that from 10,000 to 15,000 of them are contained in a single ounce. The very best are worth at least $250 an ounce, and a cubic inch, which would be equal to about eleven ounces, is worth $2,750. Am. Jour. Sci. (2) XXXI. p. 63.

† Ann. de Chimie et de Physique (3) LVI. p. 449.

	I	II	III
Platinum	85.50	79.85	76.50
Iridium	1.05	4.20	0.85
Rhodium	1.00	0.65	1.95
Palladium	.60	1.95	1.30
Gold	.80	.55	1.20
Copper	1.40	.75	1.25
Iron	6.75	4.45	6.10
Osmiuret of iridium	1.10	4.95	7.55
Sand	2.95	2.60	1.50
Lead			0.55
Osmium and loss		0.05	1.25
	101.15	100.00	100.00

Of other metals, the following have come under our notice as occurring in gold washings from California and the Pacific Coast region generally: copper, lead, iron, and nickel. Grains of native copper have repeatedly been noticed by the writer, and these were undoubtedly genuine. Small particles of lead were detected by Mr. Wadsworth in three specimens of gold washings from Rock Creek, Morris Ravine, and near Placerville.* He is unable to decide whether they are or are not really native metal. Small fragments of metallic iron, also observed in some of Professor Pettee's specimens, have almost certainly been introduced into the washings by accident.

The occurrence of native nickel, however, is one which is unquestionably authentic, and is due to Mr. J. A. Edman, by whom specimens have been presented to the writer. The locality is Trinity Bar, five miles below Fort Yale, on Fraser River. The nickel is in the form of minute rounded grains, associated with magnetite, garnet, gold, platinum, and (probably) iridosmine. Dr. Gibbs examined the nickel grains, at the request of the writer, and reported that they were nearly pure nickel, containing traces of iron and cobalt. This is an extremely interesting discovery, as being the first well-authenticated instance of the occurrence of native nickel.†

Gems of beauty or value are, with the exception of the diamond, rarely if ever found in the gold washings. Garnet is somewhat frequent, and is often mistaken for ruby. Topaz has been reported, and crystals supposed to be of

---

* The surface of the lead is much pitted, corroded, and coated with a whitish earthy substance (Pb CO_3?), and the grains are very irregular in form.

† The metallic iron found in the basaltic rocks of Greenland, at Ovifak, contains a little over two per cent of nickel. The question whether this iron is of meteoric or terrestrial origin has been considerably debated without a positive decision having been reached. The weight of evidence, as the writer believes, is strongly in favor of their having come from below rather than from above.

this mineral were noticed by Mr. Wadsworth in the washings from Morris Ravine. If present at all, it must be of quite rare occurrence. Zircon, on the other hand, has been frequently met with; the crystals are always minute. On the whole, the paucity of gems or other minerals in the gold washings of California is rather remarkable, as compared with their abundance in other parts of the world. Specimens of stream tin have been occasionally shown us, but under circumstances which threw much doubt on their authenticity.

The occurrence of cinnabar with native gold, which has been repeatedly noticed in the Coast Ranges, is also not unknown in the Sierra. A specimen of washings, described as having come from near Placerville, consisted of rounded grains of pure cinnabar, with crystals of magnetite, globules and grains of lead, and scales of gold. This need not excite surprise, as the writer has seen in the auriferous slates quartz veins containing well-defined crystalline masses of cinnabar, which were small in size, but extremely pure.

SECTION IX. — *Economical Considerations relating to the Working of the Gravels.*

Although this volume does not profess to concern itself especially with the economical aspect of the hydraulic mining business, yet a large amount of information in this department has been gathered in the course of the investigation, as will be seen by reference to the preceding chapters and the Appendices A and B. It is proper, therefore, to pass in review the principal facts, and to devote a few pages to setting forth such of the conclusions which can be drawn from them as may seem of special importance.

The first question would naturally be, What is the total yield of gold from the hydraulic mining operations? An answer to this question is desirable, as enabling the reader to form a correct idea of the magnitude of the business, the scientific side of which has been that most prominently presented in the course of this work. As is well-known, all statistics of the yield of the metals in this country are of the nature of guesses, more or less to be depended on in proportion as the parties guessing have had favorable opportunities for forming an opinion, and have been qualified by natural ability and education to make a good use of such opportunities.*

The yield of gold in the most productive years of gold mining in California, 1851 – 1853, was estimated by the present writer at from $ 62,000,000

---

* It is hardly necessary to add that disinterestedness is also a *sine qua non* in forming a valuable opinion in relation to the value and yield of mining property.

to \$63,000,000 in value, or in round numbers at from 250,000 to 260,000 pound, Troy in weight. The present yield — that is, the average yield of the last few years — has been much less than that. At the beginning of the present decennium, it was probably about \$20,000,000 in value, and since that time has slowly and irregularly decreased, the present production being about \$18,000,000. Considerably the larger portion of this comes from the hydraulic and tunnel mines, chiefly from the former. The best authorities estimate the present yield from this source at from \$12,000,000 to \$14,000,000.

It would be extremely desirable to be able to state how much has been expended in obtaining this amount of gold by the hydraulic method. This, however, is entirely impossible. The most that can be done is, to give approximate figures for a few localities where circumstances have been especially favorable for collecting information. An ideal report on the hydraulic mining business, from the economical side, would furnish information on the following points, for each district or special locality investigated: the total yield; the yield per cubic yard of material washed;* the cost of obtaining the gold, as divided between the various items of water, materials expended, labor, and interest of capital employed; the amount of water used and of gravel moved by this water. Such information as this would be of value for each locality, and as much as could be collected in the various districts visited in the course of our work has been presented in the present volume. It must, however, be impressed upon the reader that averages drawn from these statements have little value as a practical guide to those desirous of embarking in the business of hydraulic mining. The conditions are so complicated that each special district must be judged by itself, each circumstance influencing production being taken separately into account. Hence, it will be apparent that the figures professing to give the average yield of the gravel in the Sierra Nevada, and the average amount made per diem by the miners in working it, are entirely worthless.† This

---

* The cubic yard is universally employed as the unit of quantity in the hydraulic mining region, and the produce is stated in values not in weights. Taking gold at \$18.50 per ounce, the cent's worth would be equal to a little over a quarter of a grain in weight of the metal (0.2594 gr.).

† The estimates of average yield and profit in working the auriferous gravels of California, by various methods, which have been most widely circulated and credited, are those emanating from M. Laur, and are extracted from his report to the French Government, published in 1862. They are the merest guesses of a person of no experience, who had paid a flying visit to two or three localities of hydraulic mining, — a business at that time quite in its infancy. They are of a piece with his tabular statement of the daily average amount earned by each miner during the earlier years of placer mining, at a time when, in point of fact, the number of men at work is not

will be easily comprehended after a perusal of the facts set forth in the preceding pages.

In the first place, as to the yield of the gravel as worked by the hydraulic method. This question presents itself in two aspects: How much gold does the gravel contain? And how large a portion of it is obtained by the working? or, in other words, how efficacious are the means employed for saving the metal? It will not require much consideration to make it apparent that no satisfactory answer can be given to these questions; for the amount of gold in the gravel cannot be accurately determined by assay or any other method.* All that can be positively known is the amount obtained at the clean-up: of that which has escaped being caught in the sluices, no account can be had. Still some miners have their ideas of the proportion lost, although it is not easy to ascertain on what these ideas are based. That tail-sluices of great length have been put in below the ordinary ones, where the conditions were favorable, and that these have paid a profit on their cost and maintenance, seems to be proof that no inconsiderable portion of the gold escapes unless the sluices are very long.† That any other method of saving the gold will be invented to take the place of the sluice, and give better results, seems in no way probable. It may also be added, as will already have been made manifest to the reader, that by no other method of attack than the hydraulic could the high gravels of the Sierra be handled with profit.

If, then, the total amount of gold existing in the gravel at any point cannot be ascertained, we have to fall back on the quantity actually obtained, with special reference to the question how low a grade of material can be handled with profit. The total amount produced at any mine is not usually known with exactness, either for the whole time during which the workings have been carried on, or for any particular part of it for which the number of cubic yards washed away can be determined. When the more detailed survey of the gravel region was entered on by the writer and his assistants, there seemed to be no definite information to be had touching this important

known with any certainty whatever, different estimates (of those likely to have been best informed) varying by numbers as great as 10,000.

* The miner, by panning a sufficient number of samples, judiciously selected, can form a pretty good idea whether the gravel is likely to pay for working; but this is not by any means the same as ascertaining the exact amount of gold which it contains.

† Such tail-sluices are very long, and require but little expenditure, except the first cost of putting in, as they are usually cleaned up only after running for several months.

point. During that Survey, great efforts were made to obtain the necessary information, and since that time, at a considerable number of points, more or less systematic accounts have been kept, the chief results of which, so far as the same have been accessible, are laid before the reader in the present volume. From these data certain conclusions of value can be drawn; while in regard to other important points, we are left quite in the dark.

It will be evident to all that there is little account to be made of any statement of average value for the whole body of the high gravels. Neither is the highest value an element of much importance: it may be taken for granted that there are places, usually on or near the bed-rock surface, where the gravel is immensely rich.* Tunnel claims of course give higher results than the hydraulic mines do, because the former cannot be profitably worked, except when the ground is rich. It is not at all uncommon for the bed-rock surface to have been worked over by drifting, and the overlying gravel afterwards washed off by the hydraulic process.†

The most desirable practical result to be obtained in regard to the hydraulic mining operations is an answer to the question, How low a tenor of gravel can be profitably handled by this process? Here a difficulty arises, for the fact that a large amount of gravel has been worked, the average yield of which has been low, is not a proof that the operation has been conducted with profit or even without loss.‡

The following instances, however, of low rates of yield may be cited. At Blue Tent, where the Company owns its own ditch, and consequently gets its water at the lowest possible cost, a large quantity of top gravel was hydraulicked, yielding 2.6 cents per cubic yard. The gravel was loose and sandy, and easily moved. In this case the receipts " were barely sufficient to cover expenses." This, perhaps, may be set down as being the poorest

---

* The writer has seen $36 worth of gold panned out from as much dirt as could be conveniently carried away in a lady's handkerchief ; the auriferous material was scraped from the bed-rock surface under a heavy deposit of gravel. But Mr. Goodyear speaks of $1,100 in value having been obtained from a single pan of dirt. (See *ante*, p. 117.)

† While the gold, for reasons already given, is usually most concentrated on and near the surface of the bed-rock, there are some exceptions to this rule ; and there are also occasionally very curious irregularities in the distribution of the precious metal : sometimes it is all in the upper layers of the gravel ; sometimes concentrated in one or two pay-streaks, which may be very thin, while the intervening strata are entirely barren. But such exceptional cases are, on the whole, quite rare.

‡ Mining operations have been prosecuted, in repeated instances, for many years in succession, and on the largest scale, involving an expenditure of millions, without any return whatever having been received. This statement refers particularly to some of the mines on the Comstock Lode ; but what has been done there on a grand scale may easily have been repeated elsewhere, although perhaps on a considerably smaller one.

gravel which could by any possibility be handled by the hydraulic process without loss; and in this case it must be remembered that, in all probability, there was no account had of interest of capital invested in the dams and ditches. It is by no means to be understood that a company having no gravel richer than 2.6 cents per cubic yard could carry on their work without loss.

The average yield per cubic yard of the gravel washed by the North Bloomfield Gravel Mining Company from Jan. 1 to Oct. 14, 1875, was four cents; that of old washings of surface gravel near Malakoff, from 1870 to 1874, was approximately 2.9 cents. That in these instances the expense was met by the yield of gold is not likely. At least, Mr. Hamilton Smith, Jr., engineer of the Company, states the cost of working the gravel by that Company to have been, in 1877, 5.5 cents per cubic yard.*

The statistics relating to the yield per cubic yard of the gravel at Gold Run, obtained with so much labor by Professor Pettee,† are of great interest, but they do not throw the desired light on the question whether the operations were, on the whole, profitable. It is likely, however, that there may have been in this case some profit, if not a large one. Since the statistics extend over a period of five years and include so large an amount of gravel moved (43,000,000 cubic yards), it seems proper to assume that the work was done on the whole, with at least a small profit. If this be so, we should be able to say that, under the most favorable circumstances, gravel may be washed by the hydraulic method when it contains only 4.75 cents worth of gold to the cubic yard.‡

Here some more special definition of the term "favorable circumstances" is desirable. In the first place, abundance and cheapness of water are all-important. To show this, it will only be necessary to refer to a few facts illustrating the amount of water consumed in washing by the hydraulic method.

* This statement occurs in a printed report of the evidence given in the case of "James H. Keyes vs. Little York Gold Washing and Water Company," San Francisco, 1878.

† See ante, p. 152.

‡ When water has to be purchased, other parties owning the ditches, the exact cost of this item is, of course, well known to the miners; but the ratio of this cost to the yield has been but rarely ascertained, and still more rarely made known to the public. It is certain that gravel as poor as that indicated at Gold Run and some other places could not be worked without loss, if the water had to be bought at anything like the price ordinarily paid by private individuals to companies selling water. At Pond's Claim, for instance (see ante, p. 118), where the water was paid for at a moderate rate, namely ten cents per 10-hour inch, the cost of the water per cubic yard of gravel moved was 11.232 cents. The character of the gravel is not particularly given in Mr. Goodyear's notes, but, so far as the writer remembers, it might be called moderately easy to be hydraulicked.

At the " No. 8 " mine of the North Bloomfield Gravel Mining Company,* during the time from Jan. 1, 1875, to Oct. 13, 1877, 7,071,630 cubic yards of gravel were washed with an expenditure of 3,750,797,560 cubic feet of water. This gives an average of 534 cubic feet of water required to wash one cubic yard of gravel; or, in other words, the gravel at this locality required for moving it an expenditure of water nearly equal to twenty times its bulk. At the Blue Tent Company's mine, where careful record has been kept of the amount of gravel washed, water used, etc., for the past few years, the various kinds of gravel met with were moved at the rate of from 2.38 to 10.12 cubic yards per miner's 24-hour inch; † or, in other words, the gravel required, according to its condition, from eight to thirty-four times its volume of water to disintegrate it and carry it into the sluices. That which demanded the largest quantity of water specified is described as being " hard, indurated, and clayey."

Mr. Hague adopted seven cubic yards as the amount of gravel which, on the average, in the divide between the South and the Middle Yuba, could be moved by a 24-hour inch of water,‡ and this is said by Professor Pettee to be corroborated by the results obtained at Smartsville. Seven cubic yards to the 24-hour inch gives an amount of gravel not quite one twelfth of the volume of the water used.§

Mr. Ashburner considered that a 24-hour inch of water would move only about three and a half cubic yards of the lower portion of the gravel deposit in Bear River and its tributaries. This gravel may be considered as representing the hardest kind ordinarily worked by the hydraulic method.

It appears, therefore, that a 24-hour inch of water will disintegrate and

---

* The writer is indebted for these statistics to the kindness of Hamilton Smith, Jr., Esq.

† A miner's inch flowing for twenty-four hours is considered on the San Juan Divide equal to 2,230 cubic feet of water; in the Bear River mines it is a little less (about 2,200). The difference arises from the different forms of opening for the water to pass through, and the variable number of inches pressure allowed.

‡ See note, p. 207.

§ The following extract from Mr. Hague's report, to which reference has before been made, states clearly and concisely the main facts relating to the use of water by the hydraulic method : " At the present day the diameters of the nozzles vary from five to eight inches; the pressures under which they are used, at various places, from 150 to 400 feet; the velocity with which the water is discharged may vary from 75 to 150 feet per second, according to the pressure; and the quantity of water thus discharged through one nozzle, according to all these varying conditions, ranges from 300 or 400 to 1,200 or even 1,500 inches. A discharge of 1,000 inches in a single stream is not unusual. The volume of water thus discharged is 1,570 cubic feet per minute, weighing but little less than 100,000 pounds. The water used in the actively working mines on the ridge [the San Juan Divide, or the district between the South and the Middle Yuba] at present varies from 500 or 600 inches running ten hours, to 3,000 inches running twenty-four hours."

carry into the sluices from two to ten cubic yards of gravel, according to the character of the material. This, with the data previously given, will give an idea of the amount of water required in hydraulic mining operations. The use of powder as an auxiliary is not at all uncommon, and the loosening of the compacted mass by immense blasts previous to employing the water is, in some localities, a matter of absolute necessity.*

Another important matter must be kept in view, as one of the conditions on which the success of hydraulic mining, on a large scale, depends: this is the topography of the region where the work is to be carried on. There must be, in the first place, room for sluices of very considerable length, with a suitable grade; and, furthermore, a chance for the discharge of the tailings, which, of course, accumulate with a rapidity corresponding to the bulk of the gravel washed. From the descriptions of the topographical character of the Sierra Nevada, given in the preceding pages, it must have been clearly seen that the deep cañons, with which its western slope is furrowed, are an essential element in the hydraulic mining business: their existence is as important to its prosperity as is an abundance of water.

The use of a large quantity of water implies the handling of a correspondingly large amount of gravel, and this again connects itself with the low tenor in gold of the material. It is this which is the essential feature of the hydraulic system: a very large quantity of gravel can be handled, provided water is abundant and cheap, without any considerable increase of expense beyond what would be required for working a small amount of material. The poverty of the gravel makes the business unremunerative unless prosecuted on an immense scale; and this again is impossible without cheap and

* For information in regard to the use of powder in the hydraulic mines, see Mr. Bowie's paper in Vol. VI. of the Transactions of the American Institute of Mining Engineers, to which valuable contribution to the technical side of the hydraulic mining business reference has already been made. Reference may also be made to Mr. Waldeyer's paper in the Fifth Report of the United States Commissioner of Mining Statistics (1873). In this communication an excellent description is given of the hydraulic mining process, and of the various kinds of fixtures (they can hardly be called tools or machines) which go to make up the plant required for carrying on that kind of business. One of the most important recent improvements in the process is the employment of so-called "under-currents." These are boxes, from ten to twenty feet wide, and from thirty to fifty long, the bottoms of which are lined with riffles, and which are placed on one side of the sluices, and a little below them, at intervals varying according to the topographical and other conditions. These boxes receive from the bottom of the main sluices, through openings between steel bars, a certain portion of the finer washings, gold and gold amalgam, which then have a broad surface over which to spread themselves, and the velocity of the shallow stream being checked, these finer particles have an opportunity to become caught in the riffles; what is not thus detained goes back, lower down, into the main sluices again. This device is thought to be especially effective in saving the so-called "rusty gold," which will not readily amalgamate, on account of its coating of oxide of iron.

abundant water. But a low tenor of gravel, even if accompanied by abundant facilities for obtaining water, would be a fatal obstacle to success, were the detrital deposits thin. It is the enormous thickness of the gravel banks, in the hydraulic mining region proper, which gives permanency to the operations and makes them profitable. The plant cannot be rapidly shifted from place to place, as would be necessary if the gravel deposits were not very heavy, without great loss. This is certainly one of the principal reasons why hydraulic mining has rarely, if ever, been successful outside of California. Especially in the Southern Atlantic States, so often spoken of as an excellent field for the introduction of this process, the auriferous detritus lies too thinly scattered over the surface to make its handling the object of hydraulic mining operations on a large scale. What may be done there, in a small way, is what has been designated in the preceding pages as surface mining or sluicing. There are no deep placers, and no auriferous deposits, other than the ordinary surface detrital accumulations, on the eastern slope of the Appalachian Range.*

Similar statements may be made in regard to Australia,† — a region in so

* To this must probably be added the "spotted" character of the gravels in the Southern Atlantic States. Places appear to be very rich in gold; but there has been in that region no such general commingling of the detrital materials and consequent general diffusion of the metallic particles, neither was the original store of gold in the bed-rock by any means so large as it was in the Sierra Nevada. There are, perhaps, some localities at the South where sluicing may be done with the assistance of the hose; and this, in point of fact, is what properly constitutes hydraulic mining, the essential feature of which is the fact that the gravel is carried into the sluices by the aid of a jet of water through a hose or pipe, and not by shovelling.

† No mining region in the world, with the possible exception of some of the favored states of Central Europe, can boast a more thoroughly regulated and skilfully managed mining department than that of Victoria. The full and accurate descriptions of the mode of occurrence of the gold, of the distribution of the gravels, and of the methods of working, contained in the official documents published at Melbourne, enable us to form a very clear idea of the differences between the Australian and Californian auriferous deposits. In many of their most prominent features the two regions resemble each other to a degree that may justly be called most surprising. But the general type of the Victorian Tertiary gravels is that of the Tuolumne Table Mountain rather than of the San Juan Divide. Narrow channels, rich in gold, deeply covered by volcanic accumulations, and workable only through shafts and by drifting, are the rule in Australia. Heavy masses of gravel, uncovered by lava, and lying in a position sufficiently high to allow of the application of the hydraulic process, on any such grand scale as in California, appear to be entirely wanting. The sluice is extensively used in Australia for washing gold, and the gravel is, in some localities, moved by the hydraulic method, that is, with the use of the hose. These operations are, however, on quite a small scale as compared with those carried on in the Sierra Nevada. The largest quantity of water used at any one claim — that of the Yarra-Yarra Hydraulic Gold Mining Company — seems to be 510 gallons per minute, equal to about fifty-two miner's inches, the pressure employed being thirty feet. To quote the words of one of the best authorities (Mr. Peter Wright, Assistant Engineer for Water Supply), "Hydraulic mining [in Victoria] will be practicable only in a few places. The character of the earths which occur on our gold-fields, and the position of the auriferous alluvium, lying, as much of it does, at low levels, will prevent the general use of this method, but improved modes of sluicing on a large scale will certainly be invented when the miners are able to obtain water at a reasonable price."

many respects the most wonderful counterpart of California. That the gravel is on the whole richer, and that the number and value of the quartz lodes is greater in Victoria than in the Sierra Nevada seems to the writer a clearly established fact. But the Tertiary gravels of the former country are concentrated into narrow channels, and these usually covered with volcanic materials. There are no deposits which will compare in magnitude with many of the larger ones in California. Besides this, the facilities for procuring and storing water in sufficient quantity, and for bringing it into the gold fields under sufficient pressure, are not offered by either the climatological or topographical features of the Victorian mountains.

Whether any of the Russian gold-bearing districts will ever lend themselves to a development of the hydraulic mining business is a question which the writer finds it impossible to answer; for the careful examination of all the accessible material in regard to that country has not enabled him to form a clear idea of the mode of occurrence of the auriferous deposits through that vast region. The opinion of the Russian mining engineers in regard to the extent and future development of the gold-fields of their vast country seems to be of the most sanguine kind, as would appear, at least, from the recent work of Mr. Bogoliubsky on that subject.* This writer looks forward to the time as not far distant when the present number of gold washings on Russian soil will be multiplied by ten. That a portion of the gravels of that country are rich in gold would seem evident from the statistics of their yield; but how these richer deposits are distributed with reference to the poorer ones, and how much of the country they cover, cannot be made out from anything that has yet been published.

An all-important question, in regard to which the reader of the present volume might expect to find in its pages some information, is, How permanent are the high gravel deposits of the Sierra Nevada; or, in other words, how much longer can they be worked without exhaustion? No amount of labor in the field would have sufficed for procuring the information necessary for giving anything like a satisfactory answer to this question. In a few localities, as, for instance, near Placerville, where the conditions were especially favorable, the amount of gravel remaining unwashed and still available could be estimated with considerable approach to accuracy.†

In a few districts examined it has been plainly seen that the mass of the

---

* Zoloto, ego Zapasui i Dobuitcha. St. Petersburg, 1877.

† See Mr. Goodyear's estimates on p. 120, *ante*.

auriferous deposit is already almost entirely worked out; in others, the top gravel has been washed away, leaving the lower beds, which may be much richer than the upper ones, and, at the same time, more difficult of attack. Again, there are regions where a vast amount of gravel remains untouched, on account of the heavy capping of volcanic materials, either too hard to be hydraulicked, or too thick to be washed away with profit. In other districts there is still much gravel of low tenor to be washed, whenever the capital is forthcoming to supply water at a sufficiently low price. Again, there are apparently magnificent fields for enterprise, ready to hand, in laying open as yet almost untouched deposits, where the chief obstacle to development has hitherto been the difficulty of consolidating claims, by purchase or otherwise, without very large expenditures.

On the whole, it is evident from a consideration of the facts laid before the public in the present volume, that the hydraulic mining business may still be successfully prosecuted through many years before the field is exhausted. And it is probable that for quite a number of years to come the product of gold will not be materially diminished. The most serious obstacle likely to be met with is the filling of the valleys with tailings, thus hindering washing, and also exciting alarm and opposition on the part of the agricultural population of the belt of land adjacent to the foot-hills. Here complications may arise, putting a less favorable aspect on the hydraulic mining interest. Indeed, the possible condition of the western slope of the Sierra after the streams have become choked with tailings, the gold mines exhausted, and the principal part of the timber cut down, is not one satisfactory to contemplate. There is no part of the world where scientific oversight and judicious legislative interference is more desirable for the future welfare of the community than in the Sierra Nevada of California.

APPENDIX.

known as the Enterprise, which lies at the southern base of Mooney Flat Hill ; the opening of the Mooney Flat mines, on the northern side of the hill, near the head of Nebraska Ravine ; the consolidation of many of the water and mining interests ; and, in general, the improvements which have been made in order to increase the production or to diminish the expenses of the mines.

As mentioned in my previous report, the frequent consolidations of companies, or changes in ownership, have made it sometimes difficult to tell what piece of property was referred to by any given name. This difficulty is not yet entirely overcome, for many of the old names are still in common use, although at the present day nearly all the water and mining interests, by consolidations and sales, have come into the possession of the Excelsior Water and Mining Company, an organization formed, I think, in 1875. This company now holds, according to the statement of Mr. O'Brien, the superintendent, about seven hundred acres of gold-bearing gravel. The only other large owner of mining ground here, saying nothing of a few small claims, is the Nevada Reservoir Ditch Company, which holds the Blue Point claim, comprising about one hundred acres between the Blue Gravel and the Smartsville Consolidated ground. Including what is still to be "cleaned up" on the bed-rock, where the most of the top gravel has been removed, there must be nearly a thousand acres from which a greater or less yield of gold may be expected.

The Excelsior Company owns one of the oldest and most valuable water-rights in the State. One of its ditches was begun as long ago as 1851. The ditches belonging to this company, taken together, will measure more than one hundred miles in length. The water is drawn from the South Yuba River and from Deer Creek, at points so low down that there is no danger of injury from the accumulations of snow in winter. The capacity of these ditches cannot be given in detail from any data at my command, but, with the aid of storage reservoirs, they can be depended upon for a supply of three thousand miner's inches per day of ten hours during the dry season, which amount can easily be doubled in the winter. In Raymond's report for the year 1875, page 96, these amounts are given as twenty-five hundred, and seven to eight thousand for summer and winter respectively.

The Nevada Reservoir Ditch Company draws its water from Wolf Creek, near Grass Valley.

The accompanying map of the Smartsville district (Plate M), which I owe to the kindness of the Excelsior Company, is a tracing from a photographic copy of the original manuscript map prepared a few years ago by Mr. Bowman. I have copied only those parts which were essential for the work at present in hand. A lithographic copy of the same map may be found in Raymond's seventh annual report, for 1874, page 142. By reference to this map, where the dotted line represents the known outer edge of the rim-rock, or the line of junction, on the surface, of gravel and bed-rock, it will be seen that the gravel deposit extends, starting from a point above Mooney Flat near Deer Creek, for about a mile in a direction a little west of south, to a point half a mile above Smartsville, where it makes nearly a right angle, and continues to the westward for a distance of two miles farther, where it is cut off by Big Ravine.

The width of the gravel deposit, measured from rim to rim, varies from an eighth to something over half a mile. The thickness of the gravel has also been very variable, owing to the inequalities of the original surface. The highest banks at Smartsville have seldom, if ever, exceeded three hundred feet in height, though a bank of nearly four hundred feet may be looked for if a connection is made through from the Enterprise mine to the mines at Mooney Flat.

As before stated, the bed-rock in the deepest part of the channel is exposed for about a mile between Timbuctoo and the Blue Gravel claims. For the rest of the distance to Mooney Flat the position of the deep channel is not known, though inferences may be drawn from the appearance of the rock reached on the westerly rim at the Enterprise mine. The deepest bed-rock known is at a point about a quarter of a mile from the western end of the gravel. From this point the rock rises slowly, so far as known, in all directions, to the west as well as to the east. From observations taken at the Babb claim, at the Blue Gravel, at one point between the two just mentioned, and at Mooney Flat, the average grade of the bed-rock for the three miles is not far from one hundred and thirteen feet to the mile. In regard to the altitude of the bed of the channel at

YUBA RIVER

BIG RAVINE

To Marysville

TIMBUCTOO

Road to Timbuctoo

BIG RAVINE

Sucker Flat Road

SMARTSVI

MAP OF THE
SMARTSVILLE GRAVELS.

FROM A SURVEY MADE FOR THE

EXCELSIOR WATER AND MINING COMPANY
BY
A. BOWMAN.

Scale One Mile to Seven Inches

Volcanic

Gravel

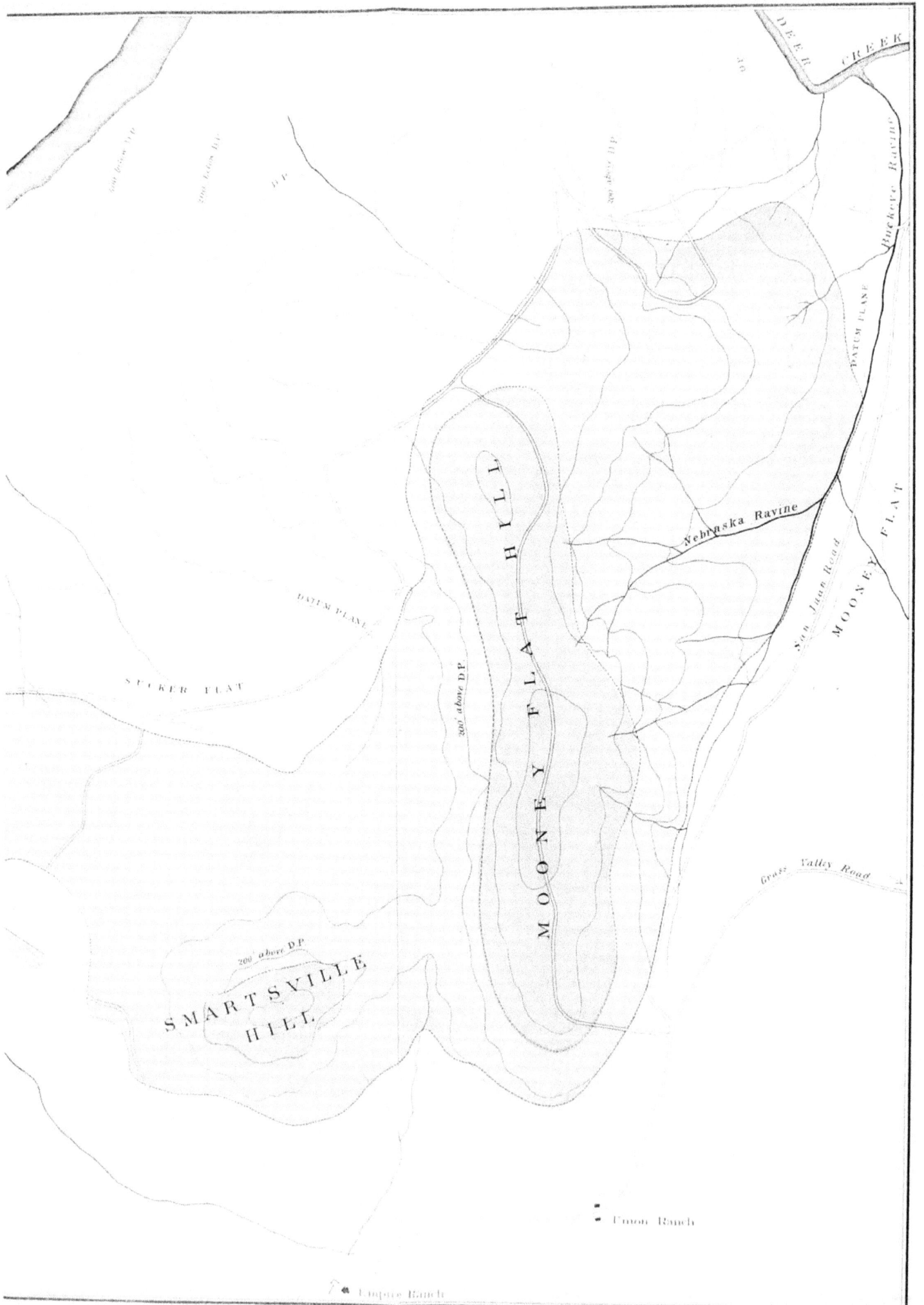

Mooney Flat there is some uncertainty, and, in the opinion of the captain at the mine, the difference of level between the blacksmith-shop in the mine and the mouth of the tunnel at Deer Creek is greater than I made it to be with the small aneroid. The error, however, cannot amount to more than a few feet, and it will not affect materially the estimate of grade above given.

A cross-section of the old gravel channel taken at any point between Smartsville and Timbuctoo would show that the deepest parts are a little to the north of the central line of the deposit, or that the rim on the northern side has on the average a steeper pitch than that on the southern side. The pitch is not uniform on either side, but increases very rapidly as the deepest portion of the channel is approached, which, taken by itself, is comparatively narrow. The diagram (Plate N, Fig. 1) is intended to represent a section across the gravel near the line between the old Pactolus and Cement claims. The top of the original gravel hill above the deep channel is reported to have been three hundred feet above bed-rock.

The bed-rock, where exposed, exhibits many irregularities of grade with high and low places, but nothing that can be called a fault, or that would cause cascades or waterfalls. It is worn into a great variety of fanciful shapes. The depressions are generally from three to six inches deep, and are sometimes nearly circular, though more frequently elliptical, oblong, or curved, and ranging from two to three feet in length. I did not hear of any deep pot-holes. The bed-rock is for the most part compact and fine-grained, changing frequently in appearance, but of a prevailing dark blue color. I thought, at first, that it was probably a highly metamorphosed clay slate, with the schistose characters nearly obliterated.*

The material of the Smartsville gravel is far from being homogeneous. As a rule the upper strata are reddish or grayish in color, and lighter than the lower, which are usually bluish, at least when first exposed to the air. But there is a very irregular distribution of the layers; the light and the dark gravels, the clays, sands, and cements being seen to be quite differently arranged, even when the faces of banks quite near together are compared. This points to frequent changes in the flow of the old current. The pebbles in the gravel, excluding for the present the volcanic capping, represent a great variety of rocks, pure white quartz being very rare, though not absolutely wanting. It is probable that rock in place corresponding to all the different pebbles could be found at points not far up in the mountains if there were favorable exposures. The faces of the banks which have stood exposed to the air for a few years give remarkable evidences of rapid change. In some of the banks almost every surface pebble could be easily broken, if, indeed, it were not so soft as to be crushed by the hand. This weathering or "slaking" will facilitate the working of the temporarily abandoned places. Some of the bed-rock also shows signs of easy decomposition. A rotten bed-rock is known here by the name of "callous."

With the finer gravel there occur also many heavy boulders. Even in the upper strata boulders have been met with, smooth on the surface, as if worn by the action of water, which are thirty, forty, or even as much as sixty feet in the longest diameter. As it does not seem possible that such large blocks of stone can have been brought down by the running stream, I am inclined to think that they had their origin upon the sides of the cañon in which the stream was flowing.

At Mooney Flat the top stratum of "white cement" is as much as one hundred and thirty feet thick in places. Under this there is a blue gravel, which contains heavy boulders, such as are not unfrequently found along the rim-rock. This is one of the reasons for believing that the deep channel has not yet been struck at this point.

At the Enterprise bank, below the volcanic capping, of which more presently, there is the ordinary light-colored quartzose gravel, similar to the top gravel in the adjoining mines. The total height of bank here, including the volcanic material, is not far from two hundred and fifty feet. The bed-rock, so far as exposed, is very irregular, and the gravel, particularly on the northwesterly

---

* These specimens were examined by Mr. Wadsworth. One taken from the surface of the bed-rock underneath the gravel, near the old Cement claim, proved to be melaphyr; the others, from near the extremity of the new tunnel in the Blue Gravel location, were diabase. Both might, however, have come from the same country-rock, the latter being more crystalline and altered than the former.

side of the claim, carries very large boulders. It is reasonable to suppose that the deepest part of the channel will be found farther to the south than the present limits of the workings.

The diagram (Plate N, Fig. 2) represents a portion of the face of the Enterprise bank. The upper fifteen feet and the forty feet below the intervening stratum are made up of a rolled volcanic gravel. The intervening stratum is a compact volcanic tufaceous rock; below this volcanic material comes the light-colored, fine, quartzose gravel.

A capping of volcanic material is found on Mooney Flat Hill, and on the hill called "Clark's" in my former report. This latter name I find is not in use at Smartsville, and in its place I should substitute "Smartsville." The thickness of this capping on Mooney Flat Hill may be taken as about 170 feet in all. On Smartsville Hill the thickness is less. This volcanic matter is mostly in the form of a rolled gravel, or a rounded gravel, the spherical form of which may be due in part to a concentric structure. I have brought a few specimens from this place, which fairly represent the form and character of the pebbles, but not the size.* The gravel of volcanic origin is interstratified to some extent with layers of solid and compact tufa or lava. The best exposures of this volcanic material are on the south side of the Mooney Flat Hill at the Enterprise bank, in the tunnel through the hill from the Enterprise to the Mooney Flat mines, and along the boundary ditch which separates the ground of the Excelsior from that of the Nevada Reservoir Company. This last-mentioned ditch is a shallow excavation extending from a point near the present Enterprise bank to the top of the hill. I went over this ground very carefully, and became satisfied that, above the level of the tunnel, which starts in the lower volcanic stratum shown in the section (See diagram, Plate N, Fig. 2) there is very little, if any, material, rolled or compact, which is not of volcanic origin. The tunnel is 972 feet long, and six feet by six in cross-section. Its course is N. 65° E. (magnetic). The lowest compact lava stratum is from five to ten feet thick, and crosses the face of the Enterprise bank about fifteen feet below the present top. Immediately below this compact stratum there comes a layer of rolled volcanic pebbles, thirty to forty feet in thickness, and of a pale bluish-gray color. The stratum above the compact lava is similar in character. The face of the bank was, of course, inaccessible, but I could determine the character of the layers by tracing them to the right and left of the bank.

Below Mooney Flat Hill and Smartsville Hill there is no lava capping in place over the gravel, but there have been found from time to time, near the surface and along the rim-rock, "float" boulders of lava with gold-bearing gravel adhering to them. In all probability the volcanic gravel, by itself, is not auriferous. I do not know that it has ever been tested at this point.

It is well known that there have been several bed-rock tunnels driven from time to time to open up different portions of the Smartsville ground. Those now owned by the Excelsior Company are as follows. The list was given me by Captain Flint, and includes portions of the flumes as well as the tunnels.

Name	Length.	Width.	Height.	Depth.	Grade.
1. Big Ravine Flume	4,800 feet	4 feet		34 inches	6 inches
2. Babb and Michigan Tunnel	2,000 "	6 "	8 feet		
" " " Flume	2,800 "	4 "		36 "	5 to 8 "
3. Pactolus Tunnel	1,600 "	6 "	8 "		
" Flume	3,400 "	3½ "		36 "	7 "
4. Cement Claim Tunnel	1,800 "	6 "	8 "		
" " Flume	3,100 "	4 "		34 "	4 to 7 "
5. Blue Gravel Tunnel	1,808 "	6 "	8 "		
" " Flume	3,410 "	{ 3½ "		32 "	?
		{ 4 "		32 "	
6. Enterprise Tunnel	3,092 "	? "	? "		
" Flume	6,156 "	4 "		34 "	6 "
7. Mooney Flat Tunnel	3,300 "	8 "	8 "		
" " Flume	3,480 "	5 "		34 "	5 "

* These pebbles, as well as the tufa mentioned further on, prove on examination to be andesitic.

Fig. 1.

Fig. 2

Rolled volcanic
gravel 15'

Compact Tufa 9'

Rolled
Volcanic
Gravel 40'

Gravel 185'

Sea level

1700'

SECTION ACROSS GRAVEL AT OLD PACTOLUS AND CEMENT CLAIMS.

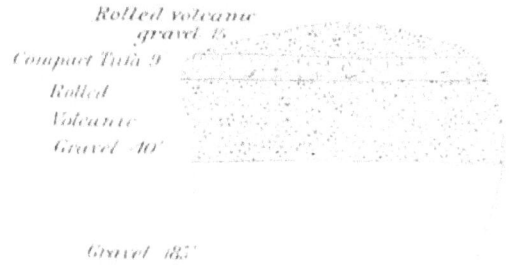

SECTION AT MOONEY FLAT.

SECTION AT MOONEY FLAT

Fig 3.

Lava

Gravel   150'

7'   160'

Bed Rock

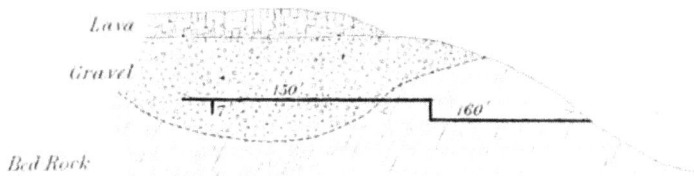

SECTION AT PEARL'S HILL.

Fig. 5.

Volcanic Tufa

Slide

Sand
and
Clay

Gravel

Bed Rock

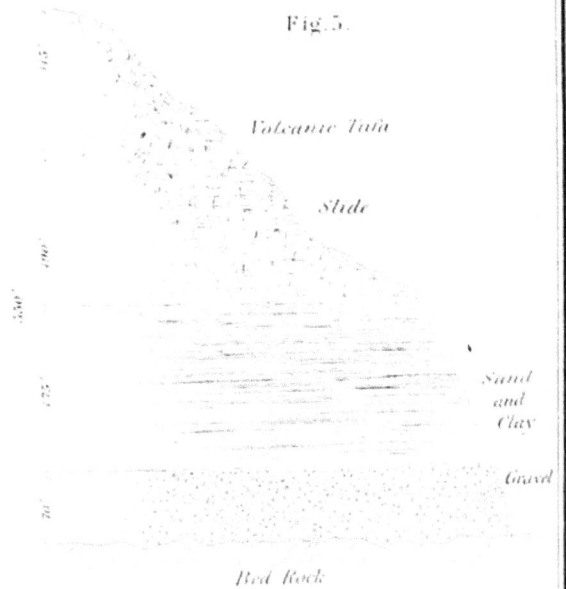

SECTION AT BOSTON MINE.

Fig. 4.

Middle Yuba

Summit

Old Channel

Summit

Shady Creek

Summit

South Yuba

1300'

2200'

2200'

1600'

South Yuba

Fig. 6.

Lava

Alpha

Bed
Rock

S. Yuba

SECTION AT ALPHA.

7 Miles

SECTION ACROSS GRAVEL FROM THE MIDDLE TO THE SOUTH YUBA.

The grade is given in inches for every twelve feet of length. The tunnels above mentioned all discharge directly into the South Yuba, except the last, which discharges into Deer Creek.

In addition to these the Blue Point tunnel should be mentioned, which is said to be 2,270 feet in length, and the branch tunnel which the Excelsior company is now driving in order to make an outlet for the Smartsville Consolidated gravel through the Blue Gravel tunnel. This list differs in many respects from that published in Raymond's report for 1873, page 131, but is probably a correct statement in regard to the tunnels now available for use.

The data in reference to the yield of gold at Smartsville, taking the district as a whole, or the yield per cubic yard of gravel, are too vague to be of much value; but there is no doubt that the gravel has been of unequal richness in different portions. The north rim, for instance, has paid better than the south. The ground of the Blue Gravel claim has been noted for its great yield of gold, while the claims lower down have been comparatively poor. In Raymond's report for 1875, page 97, it is stated that the gravel worked by the hydraulic process had yielded up to that time "$7,000,000, exclusive of $3,000,000 which was taken out in the early days, before the era of deep mining." This statement is in harmony with what I was told by Mr. O'Brien, who estimates the total yield up to the time of my visit at $13,000,000. Assuming these statements to be approximately correct, and adopting also the results of Mr. Bowman's measurements of the ground worked out, the average yield of the Smartsville gravel amounts to about twenty-three cents per cubic yard.

As to the amount left to be washed by the hydraulic process, I can only say that I have Mr. O'Brien's estimate of from $18,000,000 to $22,000,000 as the probable yield in the future.

The gold at Smartsville is fine, a nugget worth as much as five dollars being seldom found. In quality the finest grains are the best, reaching sometimes a fineness of .978. The average fineness of the Mooney Flat gold may be taken as .928.

I will add here a few notes taken at Smartsville, though not directly appertaining to what has gone before, or which are not sufficiently complete to justify any detailed statements.

The experiment of illuminating the banks at night by means of the electric light has been tried with considerable success at the Mooney Flat mines. The Brush machine is used, the power employed being water-wheels, driven by water taken from the company's ditches.

White labor is employed exclusively at Smartsville. Portions of bed-rock, partially cleaned, are, however, leased for limited periods of time to Chinamen, who pay a certain price for the privilege of working. The nozzles in use at Mooney Flat were $6\frac{1}{2}$ inches in diameter, two in number, each using 550 inches of water. The Smartsville inch of water is measured through an orifice four inches high, the level of the water being eight inches above the top of the orifice.

I made inquiry as to the existence of gravel beds to the north and west of Smartsville, and was told that there were none between Sicard Flat and Bangor, but that at the latter place and at Hansonville there were or had been gravel mines in operation. Time did not allow of my visiting the places, and I do not think them of much importance. Mr. O'Brien also mentioned Ohio Flat, near Forbestown. I was not able to go there. There is a short reference to the place in Raymond's report for the year 1872, page 71.

SECTION II. — *The Region between Smartsville and French Corral.*

No work of any importance has been done in this region, and no new discoveries have been made since the time of my visit in 1870, so far as I can learn. The tunnel at Pearl's Hill * has been driven in far enough to demonstrate the presence of gravel under the capping of lava, but the deposit cannot be worked without increased water facilities. Mr. Pearl told me that the gravel struck in the tunnel was reddish in color, and not cemented. He expects to be able to work a bank of a hundred feet in height at some future day. The diagram (Plate N, Fig. 3), from

* See *ante*, p. 195.

data given me by Mr. Pearl, shows all there is to be known about this deposit at present. The position of the bed-rock under the gravel is hypothetical.

It cannot be said with certainty that the old channel flowed directly from French Corral to Mooney Flat ; but the evidence is very much in favor of the hypothesis that those two places once lay upon the same stream. If this hypothesis is not accepted, two difficult questions arise for solution, namely: Where *was* the outlet for the French Corral channel? and, Where *were* the sources of the Mooney Flat channel? The two points are about nine miles apart in a direct line, and the bed-rock at French Corral is 822 feet higher than that at Mooney Flat. This allows a grade of ninety-one feet to the mile, a grade only four feet less than the average grade of the old channel between French Corral and Snow Point. The subject is complicated somewhat by the position of the gravel seen on the hill above Fiene's toll-house and of that at Pearl's tunnel. Both these masses of gravel are too high to belong to the old French Corral-Mooney Flat channel, unless there were irregularities of grade so extraordinary as to be almost inadmissible in theory, though they are very nearly in the line of the supposed stream. It is possible that they represent some tributary stream, or lateral ravine.

SECTION III. — *The Divide between the South and Middle Yuba Rivers.*

For the purposes of this report this region can be conveniently subdivided into five parts, which will be considered in the following order : —

A. FROM FRENCH CORRAL TO NORTH SAN JUAN,
B. LONE RIDGE AND MONTEZUMA HILL,
C. FROM NORTH SAN JUAN TO NORTH BLOOMFIELD,
D. FROM NORTH BLOOMFIELD TO EUREKA, AND
E. ABOVE EUREKA.

Before passing to the details, however, a few words may be introduced of a more general character.

Upon showing the Gravel Map to Mr. N. C. Miller, of French Corral, I learned that there were errors upon it which could not be corrected without great expenditures of time and money. Some of these errors may be due to inaccuracies of the original survey, and some of them are undoubtedly chargeable to the persons who ran the United States section lines in 1873. Certain it is that the relative positions of the section lines and the towns, as given on the map, are not the same as they are on the ground. The discrepancy amounts in some cases to as much as half a mile, saying nothing about those cases in which the error is even greater in amount but of such a character that it can be easily corrected.* Moreover, there is always room for considerable uncertainty as to the precise site of a town in the gravel mining districts. Take the town of Moore's Flat, for instance ; built originally upon the gravel, the rapid advance of the hydraulic washing has made at least one removal necessary, and, after the recent fire, the town was rebuilt in a still different place. At Woolsey Flat, also, two different towns have been washed away since hydraulic mining began. Columbia Hill has been moved from an uncertain gravel foundation to its present site upon the bed-rock. The town of Omega, on the Washington ridge, has been moved once, and will have to be moved again, before many years, if mining continues without interruption.

There is another class of errors, difficult to correct except at great expense. These relate to the positions of the mining ditches, and to the line of demarcation between the volcanic stratum and

---

* Some corrections in manuscript have been made upon the copy of the map left at Cambridge. Other errors may as well remain uncorrected until such time as more trustworthy and detailed surveys can be made. For example, it is known that the section corner, $\frac{16|15}{21|22}$, T. 18 N., R. 10 E., is a few rods only to the south of the hotel at Moore's Flat, and that the section line runs just in front of the hotel door. But to make that correction by itself, without making any change in the position of other places in the vicinity, would introduce confusion of a different kind.

the bed-rock. I place these in the same category for the reason that the ditches frequently give the best information, or the only information, in regard to the character of the rock beneath the surface. There are some errors — absolute errors — in the positions of the mining ditches, which arise from the incompleteness of the map; and there are other errors in the relative positions of ditches and lava, the true correction for which cannot be easily ascertained. In this connection I will call attention also to a fertile source of doubt as to where the true bounding line of the volcanic capping should be drawn. The sides of the cañons are so steep that much of the loosened lava of the crest rolls or slides down to a greater or less distance and covers the bed-rock to a thickness varying with the local conditions. If we decide to draw the line in all cases at the lower extremity of these slides, the space on the map to be colored for lava will be very much larger than it would be if we conclude to disregard the slides altogether and attempt only to give the boundaries of the solid and compact lava — the lava "in place."[*]

In regard to yield of gold on this divide I will give Mr. Miller's estimate. He places the production of the mines on the divide between French Corral and Snow Point for the year 1878 at $1,500,000. His facilities for gathering information are unusually good, and I place great confidence in his estimate. I made frequent inquiry for evidences of prehistoric man at several places on this divide, but could never hear of any evidence — even remote — that would lead one to suspect the existence of man during the gravel-forming period.

In speaking of the Smartsville mines I referred to the fact that Chinese labor was not employed. On the divide between the Yubas, however, I found Chinese employed in large numbers for certain kinds of work. Men who oppose any further immigration from China find their profit in employing the Chinese who are already in the country.

The system of "undercurrents," which has been pushed to a great extent upon this divide, would be worth a careful study from an economical point of view, as one of the most approved means of reducing the unavoidable waste of gold. I visited the undercurrents at some of the mines, but did not take the time to collect data about them, or their real efficiency. The subject has received a good treatment, however, by Mr. Bowie.

I will now take up the description of the different mining districts on the divide.

A. From French Corral to North San Juan.

At French Corral I learned from Mr. N. C. Miller, Secretary of the Milton Co.,[+] of the existence of a ridge of gravel at a considerably higher altitude than that of the main channel, to which reference is usually made when speaking of the French Corral mines. This high gravel has never been worked to any profit, so far as I could learn; but efforts have been made from time to time, by the sinking of shafts and the driving of tunnels, to open up a paying property. That there is some gold to be found at this higher elevation seems to be beyond doubt; but the difficulty of procuring water for washing and the probably small quantity of gravel make it very unlikely that the yield or profits of working will ever amount to much. I tried to find some of the men who had formerly worked at this point, but was not successful. All the information I have came from Mr. Miller, or was derived from a visit I made to the site of one of the old shafts one evening after sundown. Mr. Miller conducted me to a point known as the "coal pits," the altitude of which above the bed-rock at the French Corral mines is about 550 feet. We were here upon the top of the ridge, near the mouth of one of the old prospecting shafts. The ridge referred to lies

---

[*] The most serious errors of this class upon the divide between the Yubas are between Lake City and Eureka. I endeavored to find corrections for some of the errors, but now think it best to suggest no changes upon the lithographic stone, leaving them as they are, with a written explanation of the character of the inaccuracies, until further surveys of the territory in question can be made.

[+] I reached French Corral by stage from Smartsville, and presented my letter of introduction at once to Mr. N. C. Miller, the Secretary of the Milton Co. Mr. V. G. Bell, the superintendent, was not at home, and I did not see him at all. Mr. Miller was very kind and courteous to me during the time of my stay.

to the northwest of French Corral, and is distant from it about a mile.   According to Mr. Miller, a fine quartz gravel can be traced along the ridge for a mile or more.[*]

The old gravel channel, which follows the ridge between the south and middle forks of the Yuba River, is spoken of frequently as being continuous between French Corral and San Juan. This is not strictly true, as a glance at the map will show, — Woodpecker Ravine cutting quite across the old channel between Birchville and Buckeye Hill, Sweetland Creek between Buckeye Hill and Manzanita Hill, and Kent Ravine between American Hill and the mines at San Juan. These ravines, however, are all of modern origin, and there is no reason to doubt the former continuity of the gravel.

The distance between the upper end of San Juan Hill and the lowest point in the channel at French Corral, according to the map, is six miles and three quarters. I took observations for altitude at the extreme points, and at three points intermediate.   The grade seems to be pretty regular for the whole distance, and somewhat less than that previously given.[+]   I make the difference of level between the extreme points only 454 feet, which reduces the grade to sixty-seven feet to the mile.

The direction of the channel is southerly or southwesterly between the curves near the two ends of the portion now under consideration.   At French Corral there is a broad sweeping curve from the south to the west, and at San Juan a similar curve changed the course of the old stream from west to south.   When seen from some elevated point, as, for instance, from the top of the high hill which lies to the north of Woodpecker Ravine and west of Buckeye Hill, the old channel appears to follow a rather narrow trough or depression lying between continuous ridges or a series of high points of country rock, irregularly cut by ravines, which rise on either side to an elevation of three or four hundred feet above the gravel ; the ridge on the northwest separating the old channel from the present Middle Yuba, that on the southeast from the South Yuba or from Shady Creek. The diagram (Plate X. Fig. 4) represents a general section across the ridge in the neighborhood of Birchville.   No detailed measurements were made, the figures being only approximate.

The principal owners of the gravel claims between French Corral and San Juan are the Milton Company, and the company operating at American Hill. I did not make any attempt to get a list of names of all the owners, or to ascertain the precise boundaries of all the claims.   The principal town-sites are given upon the General Map.   The most important mining properties, to which allusion will be made in what follows, are the French Corral mines, comprising the portion extending from the outlet at the western extremity for a distance of about three quarters of a mile ; those at Empire Flat, the Kansas and Nebraska, and those at Kate Hayes Flat, comprising the ground opposite the upper town at French Corral, and extending for about half a mile ; the Esperance and Bed Rock, extending to the lower end of the Birchville diggings ; Buckeye Hill, between Birchville and Sweetland ; Manzanita Hill, the southern extremity of the large body of gravel lying between Sweetland Creek and Kent Ravine ; American Hill, at the upper end of the same body of gravel ; and the old mines at San Juan.

The bed-rock is exposed at several points between the lower end of the French Corral mines and Kate Hayes Flat, at Birchville, at Buckeye Hill, at Manzanita Hill, at American Hill, and at San Juan.   It changes its character several times in this distance.   I collected a few specimens, but could not include all the varieties, nor make out all the relations of bedding in the time at my disposal for the study of this section of country.   In general terms the bed-rock is either granitic (or syenitic) in character, or belongs to the series of metamorphic slates.[‡]   Below Kate Hayes Flat,

[*] I have indicated its approximate position upon the Gravel Map.

[+] See note, p. 202.

[‡] See the specimens numbered from 10 to 15.   Three varieties of the rock from (1) Kate Hayes Flat, (2) lower end of Birchville diggings, and (3) the upper end of Birchville diggings have been examined microscopically by Mr. Wadsworth, and are said by him to correspond to the so-called quartz-diorite of the German lithologists. The three specimens all came from near the line of junction between the granitic rocks which extend to the east and north and the slates which lie to the west, and do not differ so much from the former as to preclude the possibility of their being portions of the same mass, locally affected by the proximity of the slates. I have accordingly felt justified in speaking of them as granitic.

in the French Corral mines, the prevailing bed-rock is slaty or schistose in appearance, but it is not such a well-defined slate as is seen in Sweetland Creek, or at the upper end of American Hill. Further study will be necessary to make out its true relations.

Buckeye Hill I did not visit; the rock is said to be slate, as would naturally be expected from what is known of the rock above and below. In Sweetland Creek, not far below the stage road and near the outlet of the mines at Manzanita Hill, there is a line of junction between the slate and the granitic rocks. At Manzanita Hill the bed-rock is a gneissoid granite. It is softer and more scaly on the west side than it is on the east. At American Hill the bed-rock exposed is granite, excepting in part at the upper end of the claim, opposite San Juan. Here there is to be seen again the junction of granite and slate with an intervening thin layer of a soft, slaty, or gneissoid character. The slates are fine-grained, easily cleavable, and have a glistening surface, which makes them resemble to some extent ordinary roofing slate, but they are not strong enough to be used for that purpose. They stand with a vertical or very high westerly dip. The line of junction has a nearly north (magnetic) course, and along this line a tunnel was formerly driven with great ease. The position of this line of junction I have given approximately upon the map, but I did not trace it beyond the limits of the mines. The bed-rock at San Juan is also a granite, but of a variety which weathers very rapidly indeed, and crumbles to sand. About the middle of the San Juan mines the exposed bed-rock presented for a distance of five or six hundred feet along the channel a very peculiar appearance. From a little distance it looked like irregular heaps of a bluish-gray clay, with smoothly rounded surfaces, arranged like little knolls or hummocks. The surface looked as smooth and clean as if it had been brushed by Chinamen. A closer inspection showed that the rock had crumbled to a sand, so little compacted together that I could remove it with the finger alone to a depth of five or six inches. Below this, the rock became gradually harder, and appeared to be a true granite. Near the eastern end of the mines the rock in its decomposition showed a decidedly marked concentric structure.

At Manzanita Hill the granite was worn into large, broad, smooth surfaces, quite unlike the wear previously described as seen at Smartsville. There was also a high point, like an island, near the centre of the workings. There were some long furrows, and an occasional crevice. At American Hill the most marked irregularity in the surface of the bed-rock was the occurrence of "pot-holes." One of these was from six to seven feet across at the top, about three feet at the bottom, and twelve feet in depth. Its surface was smoothly polished.

It may here be added that the rim-rock at the outlet of the French Corral mines is on all sides higher than the lowest rock a few rods to the east. Prospecting shafts and tunnels have been made in many places in the hope of finding some deeper trough. The conclusion reached is that there is a bowl-like ending to the channel at French Corral, as at Timbuctoo. Probably the continuation of the old channel was in the direction of the ravine, which is now followed by the stage road from French Corral to Bridgeport.

There is observable throughout this district the usual division of the gravel into a red above and a blue below, but the thickness of the blue stratum is by no means uniform. The term "channel" is sometimes used in different senses, here as well as elsewhere, either to include the whole body of gravel from rim to rim, or only that deeper portion which "pays" the best, whatever be the color of the gravel or the conformation of the bed-rock. For example, the "deep channel" is said to be in some places only from 125 to 175 feet in width, while blue gravel is met with over a width of twelve or fifteen hundred feet. As a rule, however, the blue gravel is confined within narrower limits than the red, and there is usually a narrow trough, or "gutter," which can be traced pretty well where the bed-rock is exposed. At Manzanita Hill the bed-rock is nearly level for a width of 900 feet, while at American Hill and at San Juan there is a deep trough nearly midway between the rims from which the bed-rock rises on both sides. The average width of channel, from rim to rim, between French Corral and San Juan is certainly not less than one thousand feet, and probably as much as twelve hundred. I found it impossible to get accurate data upon this point without going to the expense of new surveys. At Manzanita Hill the rims have not been uncovered, but the whole width of gravel may be taken as at least from fifteen to eighteen hundred feet. At American Hill the width is nearly the same as at Manzanita Hill. At San Juan I estimated the width from bank to bank as not far from one thousand or twelve hundred feet.

The highest banks worked in this district are at Manzanita and American Hills, where the thickness of the gravel amounts to as much as 220 or 240 feet. Of this thickness nearly one half is frequently blue in color. A thickness of from fifty to seventy-five feet of blue gravel is quite common at these two mines. At Manzanita Hill there are no banks worked at present lower than about 125 feet, excepting near the rim or where the surface has been worn down by ravines. Mr. Hague's estimate of 140 feet for the thickness of the gravel below San Juan is a very good fair average.*

The pebbles which are met with in the gravel, or which go to make up its bulk, represent a great variety of rocks. At Kate Hayes Flat, the top gravel seemed to be, on the average, finer than the corresponding gravel at Smartsville, and to contain fewer independent strata or masses of sand and clay. At Manzanita Hill there was a very small percentage of white quartz in the gravel, while at San Juan, in the upper strata of red gravel, the pebbles seemed to be almost exclusively quartz. At this last-mentioned bank work has been stopped for several years, and the whole face, from 125 to 150 feet in height, is of a reddish or brownish color. At the lower end of San Juan Hill there is still some gravel left upon the bed-rock, which shows the characters of the ordinary blue cement. Chinamen are at work in organized gangs stripping and cleaning this portion of the mines.

The larger boulders, which are met with in connection with the finer gravel, present some features of interest, particularly at American Hill. The bed-rock has been uncovered at this mine for a distance of nearly three quarters of a mile, measuring from the upper or northern extremity. On the eastern rim large granitic boulders, similar to the bed-rock, are to be seen in abundance, while on the western rim, the side towards the slate rock, boulders of quite a different character have been found. In the upper part of the mine these boulders were not very frequent, but near the present face of the bank there is a remarkable accumulation of them, resting immediately upon or lying very near the granite bed-rock. They range in size from three or four feet in diameter up to at least twenty. I did not see any very small or very large ones. They lie very close together, and occupy a space, — say 500 feet in length by 200 in width and nearly 50 in height. It is not impossible that they may be found beyond these limits, when the banks to the south and west are washed away. The chief peculiarity of these boulders is their lithological character, which is widely different from that of any bed-rock known to exist in the district, or of any boulders met with in the other mines. I was told that similar boulders had been seen in the cañon of the Middle Yuba, near American Hill. Seen from a little distance they present a mottled or weathered appearance, a dark-colored base being marked with light spots of different sizes and shapes, some of them being as much as three or four inches across. The white surfaces are not distributed with any regularity, but make up, on the whole, about one half the whole surface of the boulders. A broken boulder shows on a fresh surface no obvious signs of being a mixture, but a closer inspection brings to light evidences that the rock is a conglomerate of some kind. The rock is exceedingly hard, and rings like an anvil when struck with a sledge.†

The boulders at the eastern end of San Juan Hill comprised some large granitic, with other smaller ones, largely quartz or quartzose in character. Possibly one third of what are left lying on the bed-rock at this point are of metamorphic rocks not distinctly quartzose.

The bed-rock tunnels for the mines in this district have been run from the cañon of the Middle Yuba, excepting at French Corral, at which place the tailings are emptied into the South Yuba. I have no complete list of the tunnels which have been driven, but will give such data as I have in respect to them, and to the methods of working.

---

* See note, p. 262.

† This rock appears from microscopic examination of a thin section to be a much altered volcanic material, — an amygdaloidal melaphyr tufa, or *porodite*.

The main flume at French Corral is 4,500 feet in length, below which there are ten undercurrents of varying length and either fourteen or twenty-one feet in width. The normal amount of water used is 2,400 miner's inches, under a pressure of 130 feet. The nozzles are seven or seven and a quarter inches in diameter. The top gravel has been pretty much removed as far as Kate Hayes Flat, and the principal work now doing is on the lower bench.

The tunnel at the Bed-Rock claim, between French Corral and Birchville, is 2,900 feet in length to the first shaft. Work is still going on.

The Birchville mines are practically exhausted, though Chinamen are at work cleaning up leased portions of bed-rock. The exposed banks of gravel on the rim are mostly reddish in color, with an occasional appearance of blue gravel in the deeper cuts and near the bed-rock. The material of the gravel seems less liable to decomposition than that at Smartsville.

Buckeye Hill I did not visit. There was not much doing at the time I was there beyond keeping the tunnel and flumes in order and preparing for the future. There is said to be some deep gravel left to work, besides the banks on the rim-rock. According to Raymond's report for the year 1874, pages 126 and 127, the property is owned by an English company, and is worked through a tunnel 5,000 feet in length.

At Manzanita Hill there is a tunnel 2,360 feet long, followed by 4,200 feet of flume and eleven undercurrents. The grade is six inches to twelve feet. The pressure-box is at an elevation of 430 feet above the bed-rock, and the water is brought through a pipe of 4,000 feet in length. The diameter of the pipe varies from twenty-two to thirty inches, and the iron used in its construction is in part No. 14 and in part No. 12. The nozzles are seven and a half inches in diameter, and deliver 1,500 inches of water. According to Mr. Miller the pipe would not stand the pressure if the nozzles were reduced to six inches, even if the quantity of water were reduced to one thousand inches. The principal excavation at the present time is on the lower bench of gravel, the upper bench having been removed for as much as a quarter of a mile to the northeast of the lower bank. The blue gravel has to be loosened with powder and broken up with sledges. The explosive used is the Judson powder, which works more rapidly than common black powder, though not so rapidly as the giant powder. According to Mr. Thomas, the captain of the mine, the richest gravel is found in a stratum about fifteen feet above the bed-rock, some of the blue gravel on and near the bed-rock being no richer than portions of the red dirt at the surface.

The inch of water is measured by the Milton Company in a different way from that adopted at Smartsville. The flow through an aperture twelve inches wide and twelve and three quarters inches high, when the water stands six inches above the top of the opening, is taken as two hundred inches.

The American Hill tunnel is 4,000 feet in length, and has a grade of ten to eleven inches in twelve feet. Outside the tunnel there are several sets of tail sluices, connecting the undercurrents, which are, according to my note-book, twenty in number, though in Raymond's report for the year 1875, page 95, the number is said to be "over forty." These undercurrents are generally from twenty to twenty-four feet wide; one of them is forty feet wide. The depth of the tunnel below bed-rock at shaft No. 1 is 190 feet, but the distance between tunnel and bed-rock will diminish rapidly as the banks are washed away gradually down the channel.

There is a great deal of low-grade gravel at this point, and quite large masses are still standing on the rims waiting for the time when water for washing can be obtained at a low rate.

The gold is generally fine in the gravel of this district. It is seldom that a nugget worth as much as five dollars is found. The finer gold at French Corral and at Manzanita Hill ranges between 935 and 950 thousandths in fineness, the coarser between 925 and 930. Further data upon this point are lacking.

The statistics as to yield of gold which I was able to collect are not very complete. At Manzanita Hill a "clean up" of from fourteen to twenty-two thousand dollars is usually made at the expiration of each run of thirty days of ten hours each, the final "clean up" in the fall of the year yielding a much larger sum. Statistics in regard to the gold saved by the undercurrents are given by Mr. Bowie in the paper to which reference has already been made.* I could find no data in reference to the cubic yards of gravel which have been removed from the different mines.

An approximate estimate of the yield of the gravel per cubic yard is from twenty-five to thirty cents. In Raymond's report for the year 1873, page 19, it is stated that the average yield of the American Hill gravel up to the close of the year 1871, at which time there had been six million cubic yards washed by the hydraulic process, had been thirty cents. From Mr. Miller I learned that in the years 1858 and 1859 the sum of $157,000 was taken from seven "claims" at the lower end of San Juan Hill, each claim being of rectangular shape, 180 feet by 80, and with an average depth of 120 feet. This would be a yield of thirty-five cents per cubic yard. The yield of the lower stratum of cemented blue gravel, wherever it has been worked by itself, has been many times the above amount. Some statistics are given in Raymond's report for 1873, on the page referred to above.

The principal bodies of gravel left untouched are those between French Corral and Birchville, and between Manzanita and American Hills, the former being about a mile and an eighth in length, the latter nearly a mile. The average thickness of the gravel over the deep channel between French Corral and Birchville is probably between 125 and 150 feet, and between Manzanita and American Hills as much as 200 feet, or something more.

The sulphurets found in the gravels are a source of gold which has been hitherto neglected. They are found in considerable quantity both on the bed-rock and in the gravel itself, and are caught in part by the riffles in the sluices. At Manzanita Hill the experiment of collecting the sulphurets at the "clean ups," stamping, roasting, and amalgamating them, has been attended with a good measure of success. At French Corral the sulphurets are more abundant than they are higher up on the ridge.

In the neighborhood of French Corral there are quartz veins in the bed-rock, cutting across the old channel, but none that have been worked to my knowledge. One outcrop, known as the Red Ledge, seemed to be attracting considerable interest, and I took the time to visit it. It lies upon the South Yuba side of the ridge, about a half-hour's walk from French Corral, and 150 feet lower than the bed-rock in the gravel mines. The outcrop has the appearance of a gossan. The ledge has not been worked enough to show precisely what it is. I took a few specimens of the ore for future examination. One peculiarity of this ledge is that the gold obtained from it is only from .640 to .650 fine, the remainder being silver.†

B. Lone Ridge and Montezuma Hill.

The Lone Ridge deposit of gravel lies about a mile and a half to the southeast of the town of San Juan, and a half-mile westerly from the Oak Tree Ranch, the position of which has been marked for correction on the Gravel Map. The gravel lies in the eastern half of Section 9, as the section lines are drawn upon the General Gravel Map. It is an interesting deposit, principally on account of its altitude, the bed-rock being fully 500 feet above that at the eastern end of San Juan Hill, or nearly as high as the site of the hotel at Cherokee. The gravel lies on the summits of two or three hills or short ridges at the head of a ravine leading down to the Oak Tree Ranch. Several shafts have been sunk upon the ridge, and have proved the existence of a bed of gravel with a thickness of at least seventy feet. The gravel is fine, loose, and shingly, with no cement, and carries fine gold all the way down to the bed-rock, which is granite. The crest of the main ridge, which is not very wide, is nearly level for a quarter of a mile below the tunnel mentioned further on, and follows a general southwesterly direction (S. 20 – 33° W., magnetic). From the lower extremity of the ridge a good view is obtained of Montezuma Hill, and down the valley of Shady Creek, but San Juan and its vicinity are shut off by the high ridge of country rock which lies to

* See note, p. 196.

† For information relating to this region I am indebted to Mr. N. C. Miller of French Corral, to Captain Thomas of the Manzanita Hill mines, to Superintendent McBride, Captain Bank, and Mr. John McCoy of American Hill, as well as to several other gentlemen whose names I have not preserved in my notes.

the north.   I took observations for altitude at the point of the ridge and at the mouth of a tunnel which a company of miners are now driving towards the centre of the deep gravel.   The course of the tunnel is S. 35° W. (magnetic), and it has a grade of nine inches to twelve feet.   Its length to the gravel will be about 550 feet, 460 of which are already completed.   The difficulty of getting a supply of water at that altitude will be a great hindrance in the way of any successful working of the deposit.   The gold in the ravines near the head of Sweetland Creek probably came from this Lone Ridge deposit.

The gravel at Montezuma Hill differs quite materially from any other that I know of on the ridge between the forks of the Yuba River.   The hill lies opposite Lone Ridge, on the divide between Shady Creek and the South Yuba River.   The hydraulic bank on the northwestern slope of the hill was plainly visible from Lone Ridge.   The most of the work done at Montezuma Hill has been drift-mining, the conditions not being favorable for the application of the hydraulic process. The drifts have been carried quite through the hill, from side to side, but they have been allowed to cave in and are not now accessible.   I was consequently unable to make that detailed examination of the place that I wished to.   The bed-rock, or body of the hill, is a granite, containing porphyritically disseminated hornblende.   Its altitude at the hydraulic bank on the northwest slope I made to be 2,356 feet, which is about 200 feet lower than the bed-rock at Lone Ridge, but at a point in the road near Malone's house, perhaps half a mile easterly from the point last mentioned, granite appeared again at an altitude of 2,529 feet, or 173 feet higher than at the hydraulic bank.   The height of bank exposed I found by the aneroid barometer to be 130 feet, and near that point was also the base of the lava capping.   The thickness of the lava capping is probably about 350 feet directly over the gravel, as I made the altitude of the top of the hill near one of the trees marked by the Keystone Company to be 2,853 feet.   The character of the lava capping is similar to that seen above Columbia Hill.

The bed-rock at the hydraulic bank was rapidly disintegrating into thin scales and sand.   The bank itself was made up of a fine sandy material containing more or less clay.   The clay in places is as much as seventy-five or a hundred feet in thickness.   There is a remarkable absence of boulders in this deposit, "nothing as big as one's fist being seen in the tailings."   Some fine quartz gravel was observed in the road at a point nearly level with the top of the exposed bank.

The following items of information about Montezuma Hill I obtained from Mr. Glassett, a miner who formerly worked in the mines at this place.   The bed-rock was nearly flat as a whole, with a slight fall from the northeast to the southwest.   Work was begun as early as 1853, and was discontinued in 1874 or 1875.   The Keystone Company attempted to employ the hydraulic process, but were stopped by the expense of moving the heavy bodies of pipe-clay.   The "pay streak" was only about one foot thick, next the bed-rock.   The drifts and breasts were run four feet high and four feet wide, and timbers were set every four feet in length.   A "set of timbers" would correspond, therefore, to sixty-four cubic feet, or about two and one third cubic yards.   The mine paid on the average " $20 to the set of timbers," which would make the yield per cubic yard between eight and nine dollars.   The gold was fine.

From the above description, it seems impossible to trace any simple or direct connection between the gravel of Montezuma Hill and that of Cherokee and points above ; the altitude of the bed-rock being only thirty-five feet below that at Badger Hill, to be described further on.

C. SAN JUAN TO NORTH BLOOMFIELD.

The gravel deposits of this portion of the ridge have been very well described, as a whole, by Mr. Hague, in the report from which quotations have been made in a previous chapter.*   Since the date of his report mining has been regularly carried on, the extent of bed-rock uncovered has been increased, and unfinished tunnels have been pushed forward to completion, but there has not been much, if anything, done, so far as I could learn, in the way of new explorations, nor have new

* See ante, pp. 206–207, passim.

discoveries been made, which would tend to throw light on any doubtful point. It will not be necessary for me to go afresh over the ground covered by Mr. Hague, Mr. Bowie, and others, whose reports have appeared in printed form, and my notes will relate principally to points of detail and to some questions in regard to which there is room for a difference of opinion.

On page 200 of this volume it is stated that the bed-rock between San Juan and Cherokee is slate. This is partly true, without doubt, for the bed-rock at Badger Hill is distinctly a slate; and it is probable that the high hill or ridge, which lies to the west, and which caused the old channel to turn towards the north near Cherokee, is also of slate; but it is certain that the granite which forms the bed-rock at San Juan and at Montezuma Hill extends, along the line of road from the Oak Tree Ranch to Cherokee, to within a few rods of the town. I did not attempt to trace the line of junction between granite and slate, but as there is a granite country-rock on the opposite side of the river, a little below Camptonville, I am inclined to think that the most of the rock for three or four miles above San Juan is also of that character. Above Cherokee the rock exposed is all slate as far as Bloomfield, with the exception of some of the rock at Malakoff.

The bed-rock exposed at the Malakoff mine is mostly a metamorphic slate, varying considerably in character within short distances. It is sometimes quite hard and again it is comparatively soft. According to Mr. Hamilton Smith, who had an excellent opportunity to become acquainted with this rock while the long tunnel was in process of construction, it frequently contains seams or belts of bluish quartz, which may be regarded as due to some local silicification or to some intrusion of foreign matter. There are no regular slips nor cleavage seams. A so-called granite belt, forty or fifty feet in width, cuts across the slate bed-rock in a zig-zag course. It is exposed to view in the mine, near shaft No. 8, and has been struck in the deep tunnel and at points on the rim. There is a sharp line of demarcation between the slate and the granite.*

That the old channel between San Juan and Badger Hill has been washed away admits of no doubt; there is no other path for it save that now occupied by the cañon of the Middle Yuba. The difference of level between the bed-rock at San Juan and that at Badger Hill I make to be 358 feet. The distance between the points in a straight line is four miles, which, allowing for windings, corresponds to a grade of about eighty feet to the mile. The position of the deep bed-rock is nowhere known between Badger Hill and the Malakoff mine, near Bloomfield. The bed-rock exposed at Grizzly Hill lies off the line of the main channel. The altitude of the bed-rock at Malakoff, shaft No. 8, I make to be 2,929 feet, or 538 feet above that at Badger Hill. The distance along the supposed course of the old channel is seven and a half miles. The grade, accordingly, was seventy-two feet to the mile.

As just stated above, the precise course of the old stream between Badger Hill and Malakoff is not known, but we may be sure that it followed approximately the line of gravel as laid down on the map. The only question about which there can be any serious doubt is as to the relation of the deposit at Grizzly Hill to the rest of the mass of gravel. On page 201 of this volume, in the quotation from Mr. Hague's report, a branch or tributary is spoken of as " coming from the southeast (Grizzly Hill)," and on page 202 there is an allusion to the slight inclination of the bed-rock between Badger Hill and Grizzly Hill. The difference of level between these points I cannot give as accurately as I would like to, but all the measurements agree in making this difference less than one hundred feet. The distance between the points is six miles. According to my barometric measurements the bed-rock at Grizzly Hill is ninety-three feet above that at Badger Hill, but the data obtained from Mr. McMurray, to which reference will again be made further on, make the difference only fifty-five feet. Again, my barometric measurements show that the Grizzly Hill bed-rock is only 445 feet below a given point at Malakoff, while the surveys of Mr.

---

* The specimen which was taken from a point scarcely five feet distant from that from which the slate specimen was taken, was examined microscopically by Mr. Wadsworth, who designated it as an "altered andesite," perhaps equivalent to "propylite (?)." It was difficult to get satisfactory specimens of this rock. There had evidently been pretty extensive surface changes through the action of water, even during the short time that the rock has been uncovered.

Perkins, or Mr. Hamilton Smith, make it 525 feet below the same point. These discrepancies could be in part reconciled on the assumption that my measurements make Grizzly Hill higher than it is in reality, but any change in the value I obtained for the altitude of that point would lead to other difficulties, for my determinations of the altitudes of Grizzly Hill and of Gopher Hill, on the opposite side of the South Yuba, agree almost exactly with what is known, by means of observations with the hand-level, of their relative positions. It is true there may be some error in regard to Gopher Hill, for I had only a very short series of observations with the mercurial barometer at Sailor Flat to depend upon, but it is impossible to tell where the error lies. Be these things as they may, however, the essential fact of the slight difference of level between the points under consideration remains unaffected, and we can only say that similar discrepancies may be expected when the barometer is employed to determine slight differences of level, unless an extended series of synchronous observations can be obtained. Badger Hill, as may be seen on the map, overlooks the cañon of the Middle Yuba, and Grizzly Hill is in a corresponding position on the opposite side of the ridge, but higher up, overlooking the South Yuba. Between them, and superficially connected with both, lies the immense deposit of gravel at Columbia Hill (or North Columbia, if the official name of the post-office is preferred). The only possible course for a tributary to the old stream from the neighborhood of Grizzly Hill is along the line now followed by Spring Creek (or Knapp Creek, both names being in use), for there are hills of bed-rock to the east and the west sufficiently high to prevent its going in any different direction. The junction of the main stream and the tributary, then, must have been at some point near the present site of Columbia Hill; certainly it could not have been much farther down the stream. The average grade of the old channel between Badger Hill and Malakoff has already been stated to be seventy-two feet to the mile, a grade which is in good accord with those ascertained for portions of the channel lower down the stream. From this it is clear that at the point of supposed junction, allowing the grade to be uniform between the extreme points, the bed-rock in the main stream would be considerably higher than that in the tributary at a point a mile and a half away. To make such a junction possible, the bed of the old stream must have been very steep for the first two or three miles below Malakoff, and then very nearly flat for the remainder of the distance to Badger Hill. In the absence of any facts that can be cited in support of this view, I am inclined to the belief that no *deep channel* will ever be found with a grade descending to the north or north-west from Grizzly Hill.

It is fair to say here that a generally accepted view among the mining men of this region is that such a deep channel does exist, and there may be reasons for the correctness of this view with which I am not acquainted. The balance of testimony, however, seems to me to incline in the other direction. A vast accumulation of tailings in Spring Creek has covered the original bed of the creek to a great depth, and there is now no easy method of telling just where the old rim-rock appeared on the surface of the ground. Slate bed-rock is now to be seen in the bed of the creek at the bridge on the road from Columbia Hill to Grizzly Hill, at an altitude certainly higher, though, to be sure, not much higher, than at the outlet of Grizzly Hill. The bed-rock continues in sight for a considerable distance up the creek on the right, or west, bank. On the left, or east, bank, gravel is seen immediately after crossing the bridge, and it forms a border to the tailings in the creek on that side, which grows gradually narrower as the high slate hills to the north approach nearer to the creek. At one point on the east bank of the creek, about half a mile above the bridge, I saw bed-rock at the present level of the tailings, and, of course, considerably higher than the bed-rock at Grizzly Hill. These projecting masses of slate narrow very much the limits between which any deep channel from the direction of Grizzly Hill can have come, and I was told that bed-rock was struck within thirty feet of the surface at all points where prospecting was carried on in the bed of Spring Creek in the earlier days of mining, before the tailings from the hydraulic banks had been deposited. Doubtless the gravel was continuous across some of the upper portions of Spring Creek, near the line between the Columbia Hill property and that next above, between Columbia Hill and Lake City, and in all probability the deep channel will there be found and some day worked.

I have dwelt thus in detail upon a question which may seem to be unimportant, on account of the bearing which it has upon any estimate of the probable depth of the gravel at Columbia Hill, and upon the view which is to be taken as to the relations of the gravels on the north to those on the south side of the South Yuba River, at Gopher Hill and Blue Tent.

The position of the deep channel between Columbia Hill and Lake City is not known; but the superficial area and the outline of the gravel deposit are given on the General Gravel Map with a pretty close approximation to accuracy, and I do not think it will be necessary to make any changes upon the lithographic stones until more detailed surveys are made. Local maps giving accurate information are few in number.*

For the ground between Columbia Hill and Lake City I have no fresh information to give. So far as I could judge from a couple of trips over the ground, the outlines of the gravel are given correctly on the Gravel Map. Above Lake City the gravel is not indicated quite as it should be, though the errors are not serious. The principal ones are in regard to Lake City, which stands upon lava and not upon gravel, and to the boundary lines between gravel and lava, or gravel and country-rock, respectively, in the neighborhood of Bloomfield. The gravel exposed to the south of Lake City is extremely narrow. Lake City itself stands upon a high spur of the lava ridge, being over four hundred feet higher than Columbia Hill, and nearly five hundred feet above the bed-rock at Malakoff. Standing at the hotel at Lake City I saw no higher point on the horizon through an arc of 160°, between the directions N. 80° W. and S. 60° E. (magnetic). In the latter direction the view was obstructed by a high hill of bed-rock, about half a mile distant, while in the former the level line strikes a point on the northern branch of the lava flow. To the north of the hotel the lava rises rather abruptly for about two hundred feet. Gravel begins to be seen at the surface in a southwesterly direction from the town at an altitude about two hundred feet lower.

Slate bed-rock is seen on both sides of Virgin Creek and in the stage road near the hotel at Malakoff, although there are extensive bodies of gravel for a considerable distance to the south of the creek and of the stage road. This high southern gravel has been worked in several places, but has no great depth. It is probably an overflow, or the result of the lateral expansion of the stream as the deep bed filled up.

Attention should be called to the occurrence of high bed-rock to the north and northwest of Malakoff between the gravel and the lava capping. I did not go to the spots myself where it is said to be seen in the beds of brooks or in ravines, but have given their positions on the Bloomfield map on the authority of Mr. Perkins. These spots are of importance in determining the probable position of the northwestern rim-rock.

The principal points at which regular mining has been carried on are at Badger Hill, Columbia Hill, and Malakoff, or Bloomfield.

The lower end, or outlet, of Badger Hill, as has already been stated, overlooks the cañon of the Middle Yuba. The descent from the mine to the river is exceedingly steep, almost precipitous. The bed-rock at the mine has been uncovered for a distance of nearly half a mile towards the south, and shows a very slight grade indeed. With the aneroid and hand-level alone it was not possible to make accurate measurements. This flatness of the channel at Badger Hill and its known steepness in the vicinity of Bloomfield lend some support to the view that there is an unusual change of grade at some intermediate point, and that the bed-rock at Columbia Hill may be low enough to admit a tributary from Grizzly Hill.

The deepest part of the channel at Badger Hill is quite narrow, being not over 200 feet in width, and the bed-rock rises rapidly both to the east and the west. The total width of the deposit at the outlet is scarcely more than 500 feet. The high bank at the south end of the mine I made to be 165 feet in height. The top gravel to a depth of eighty or a hundred feet is composed mainly of a fine quartz, well-rounded, white or reddish in color. The lower gravel for a thickness of from sixty to seventy-five feet is of a bluish cast of color, and contains, relatively to the upper gravel, a

* A comparison of the Eureka Lake Company's plats of their ground, as laid down by the United States surveyors, shows the general accuracy of the Gravel Map for the region near Columbia Hill.

much larger percentage of slate and other metamorphic rock. This lower gravel is for the most part a hard cement, excepting for a width of about 200 feet in the deepest part of the channel, where there is a stratum on the bed-rock which is soft and easily worked. On the rim the hard cement extends to the bed-rock. The best paying gravel is found within ten or fifteen feet of the bottom. Large boulders are not common. There was no one at work at the time of my visit.

For a distance of nearly half a mile farther to the south from the high bank just referred to, the top gravel, particularly on the west side of the deposit, has been removed pretty extensively. The general surface of the gravel also rises in this direction towards the Flat to the north of the town of Cherokee. Considerable prospecting has been done on the Flat, but no great depth has been reached at any one place.

To the south of Cherokee there has been no mining to speak of. Some shallow gravel is found along the crest of Pleasant Ridge, which is very likely an overflow from Columbia Hill.

Between Cherokee and Columbia Hill, south of the stage road, and near the junction of the two forks of Shady Creek, a great deal of work has been done in the Chimney Hill mines, but the creek is now choked with débris, and there was nothing doing when I was there. The Chimney Hill prospect shaft I did not visit. It is said to have been sunk to a depth of between 150 and 160 feet, and to have reached blue gravel, though work was suspended before reaching bed-rock, on account of the influx of water.

At Columbia Hill extensive mining has been carried on at the Laird claims, at the Farrell claims on the southern side of the deposit, and at the Consolidated mines on the right bank of Spring Creek. At the time of my visit the principal work was doing at the Farrell claim. The banks at this claim extend to the southern rim-rock, and vary very much in height. The gravel exposed to view was of a decidedly reddish color for a thickness of ten or twelve feet from the surface, but lower down it was much lighter, almost white, in color. The mass of the material was fine and sandy or clayey, with no regular and persistent strata of pipe-clay. The pebbles were small in size and mainly quartz, or of some variety of quartzose rock. Some of the clayey layers presented peculiarities of structure or color, which may be noticed. One layer, whitish in color, or yellowish white, was consolidated almost to the consistency of rock, and showed a well-marked slaty cleavage. Other layers were of a dark blackish-blue color, resembling some soft bituminous shales, and showing a cleavage structure. At one part of the claim, covering an area of thirty or forty acres, a dark-colored pipe-clay, six or eight feet in thickness, rests directly upon the bed-rock, with no gold in it or under it. I did not examine this locality with any especial care; it may be that the clay has resulted from a local disintegration of the bed-rock, instances of which are very common in other parts of the State. The gravel here is very easy to move, and very little use is made of powder or pick, except to break up the larger blocks of clay which the water, unaided, would not move through the sluices.

I did not make any close examination of the gravel at the Consolidated mines. It is very fine, sandy, and easily worked. The outlet is through Spring Creek. A pressure of 270 feet can be obtained in these mines, while in the Farrell and the Laird the pressures are 100 and 120 feet respectively. On the Consolidated ground two eight-inch nozzles will deliver 2,500 inches of water, while, on account of the difference of pressure, the same sized nozzles in use at the other claims use much less water. Nozzles as large as ten inches in diameter have been used, but only exceptionally. Eight inches is the ordinary size.

At Grizzly Hill the gravel has been removed for a few hundred feet back from the edge of the cañon, and the estimated height of the bank at the upper end of the workings is 125 feet. The lower fifty feet is blue in color, and the top, as usual, is red. No large boulders were seen here. The gold is said to be scaly. The property now belongs to the North Bloomfield Company.

Between Columbia Hill and Lake City there are some extensive surface openings, where the top gravel has been removed. But the tunnels were all too high, and the mines are for the present abandoned.

At Malakoff about fifteen acres of deep bed-rock have been exposed, along a line of 1,700 feet in length, extending from shaft No. 9 to the present high bank at the west end of the mine, besides a considerable extent at places on the southern rim. The rock is nearly level for three or four hundred feet, measuring across the channel from rim to rim. It is worn smooth, but does not show such striking forms as are to be seen at Smartsville, and there are frequent irregularities on the surface and knobs rising to the height of fifteen or twenty feet.

The high banks are from about 220 feet to 340 feet in height. The blue gravel, not much cemented, is about 130 feet in thickness. A striking feature of the Malakoff banks is the distinct stratification of the material,* which is made more noticeable by the occurrence of large bodies of pipe-clay, with intercalated red gravel. The clay is in some places as much as 150 feet thick. It falls in large blocks, and has to be broken up, like the rock boulders, with powder and sledges. The clay strata seem to have the same grade as the underlying gravel and bed-rock, but it cannot be said positively that they would show this character for any great distance. A great difficulty in the way of settling such a question is the constant removal of the exposed faces, and the absence of any records to show just what position the clay formerly had. At some distance above the heavy stratum of clay just referred to, another thick mass of the same material appears coming to view as the high banks are worked away to the north and northwest towards the lava ridge. The top gravel at Malakoff is mostly quartz, while the lower strata contain many varieties of rock and boulders. These latter are not so much worn and rounded as similar boulders are lower down the ridge.

In some parts of the mine the banks show a capping of volcanic material to a thickness of from two to twenty feet, but, in general, the gravel capped with tufa has not yet been reached.

The material is very easily moved by the hydraulic stream: indeed, sometimes it moves too easily. There is a constant dropping of fragments from the bank, and sometimes a slide which seriously interferes with the regularity of the working.

In addition to the main excavation, to which the preceding notes refer, there has been a large extent of top gravel removed to the south and west, both on the ground immediately adjacent to the Malakoff mine and farther down towards Lake City. Where a top bank has been long exposed to the weather the surface is grayish white in color, and very brilliant when illuminated by the bright sunlight.

I could not find any map of these outlying pieces of ground, nor learn of any measurements which could be used as a basis for calculating the amount of material already removed or yet to be removed from them. The surface of the property owned by the North Bloomfield Gravel Mining Company has been surveyed with great care, and laid off in small sections in such a way that it will be a comparatively easy matter to determine the number of cubic yards removed and the average yield per yard from year to year. Partial data of this kind are given on page 70 of Mr. Bowie's paper on " Hydraulic Mining in California." It is there stated, on the authority of Mr. Perkins, that for the years 1874 – 75 and 1875 – 76 the yield at the " No. 8 Claim " was respectively 3.9 and 6.6 cents per cubic yard.

The top dirt or volcanic tufa and the pipe-clay are barren of gold, though the latter is inter-stratified with thin layers of washed gravel which contain more or less of the precious metal. These thin auriferous layers amount in the aggregate, in the thickest stratum of pipe-clay, to as much as fifteen or eighteen feet, or from ten to twelve per cent of the whole thickness. There are not many large boulders at Malakoff, excepting on the surface or near the bed-rock. The boulders seem to be comparatively few in number; though, as it is the custom at this mine to blast the boulders and blocks of pipe-clay and run them through the sluice and tunnel, it is not easy to form any good estimate as to their relative quantity.

The underground explorations carried on from the Malakoff Company's Prospect Shaft No. 1, which is distant about half a mile from No. 8, make the grade for that distance about 125 feet to the mile, somewhat greater than the grade from Malakoff to Badger Hill, but not so great as that from the same point to Woolsey's Flat, as will be seen in what follows.

---

* See ante, Plates A and I.

The gold at Malakoff is usually in small grains, which exhibit a coarse or "nuggetty" exterior. There is no fine, scaly gold. Occasionally a heavy nugget worth as much as twenty dollars is found.

For the greater part of the distance in the long tunnel the current of water and gravel runs directly upon bed-rock, the appliances for saving gold being chiefly the sluice of 1600 feet in length at the upper end of the tunnel, and the undercurrents in the cañon below the tunnel's mouth. A "clean up" of the sluice takes place about twice a month; the undercurrents (some of them) are cleaned every six or eight weeks, but there is no general cleaning of the whole length of the tunnel oftener than once a year.

The following facts are compiled from the "Annual Report to the Stockholders of the North Bloomfield Gravel Mining Company, with Statement of Accounts for the year ending December 31, 1879."

Work was commenced at the North Bloomfield mine in 1866, but there was little profit until after 1874. The yield and profit for the first eight years and for each fiscal year since have been as follows:—

	Yield.	Profit.
1866 – 1874	$218,073.42	$2,232.84
1874 – 1875	83,078.63	22,072.45
1875 – 1876	200,366.54	98,476.28
1876 – 1877	291,125.42	148,172.09
1877 – 1878	311,276.70	140,635.61
1879 –	331,759.76	183,855.09
	$1,435,680.47	$595,444.36

There have been fifteen dividends paid since the start, aggregating $438,750, equal to $9.75 per share on 45,000 shares. There have been forty-three assessments collected, aggregating $1,545,000. The company has valuable property and property rights, but has bonds out amounting to $450,000, and a floating debt of $9,600.

The other mines and diggings in the vicinity of North Bloomfield are of small importance compared with the mine at Malakoff. It is not yet made out with certainty what their true relations are. Further explorations will have to be made under the lava or tufa which caps the ridge above this point for several miles before the question can be settled. In the Black — or Cadwallader — diggings, which lie on Humbug Cañon, near the centre of the town of North Bloomfield, there is said to be no blue gravel, but only the yellowish-red material which might be looked for on a rim-rock. The gravel is not much washed, and it may all belong to some small branch or tributary of the main stream. The gravel at the Marlowe ground also appeared to me to be lying upon a steep rim-rock or upon the steep slope of some tributary ravine. The bed-rock flume at the Cook and Porter ground, according to the measurements of Mr. Perkins, is 477 feet above the bed-rock at the Malakoff Shaft No. 8.

It has been already stated that the top of the ridge is covered with volcanic tufa in the neighborhood of Lake City. A reference to the General Gravel Map will show that the lava extends down as far as a point nearly north of Columbia Hill. The lowest lava that I saw in this vicinity was at a point on the Bloody Run ditch, from which the bearing of Columbia Hill was approximately S. 5° W. (magnetic). This observation indicates that the coloring for lava is carried about half a mile too far down on the map. This lava is supposed by some to cover a channel of auriferous gravel, different from and partly parallel with the main channel, on which the Columbia Hill and Malakoff mines are. Some support for this supposition is found in the occurrence of detached beds of gravel at altitudes considerably above that of the present upper surface of the main deposit. For example, on the road from Columbia Hill to Lake City, which lies near the line of junction of the slate and the lava, I noticed, while riding in the stage, a small amount of rolled gravel at too high an altitude to be looked upon as belonging to the principal deep channel. At the lower extremity of the lava-flow, and at several points on the northern slope of the ridge near the head of Grizzly Cañon, there are small bodies of gravel, near the junction of lava and slate-rock, which are said to have yielded well in early days, and would still be good hydraulic ground if the quantity of gravel were sufficient to justify the necessary outlay for securing a supply of water. The quantity, however, is very small. The region is attracting some attention from prospecters, and a company of miners from Forest City, organized as the Forest City Consolidated Gold Mining Company, were making arrangements at the time of my visit to undertake the working of some of these banks.

The altitude of the lower end of the lava-flow I made to be 3,469 feet, nearly 500 feet above the present surface at Columbia Hill, and more than twice that amount above the bed-rock at Badger Hill. The bed-rock exposed to view between the post-office at Columbia Hill and the lava is all slate. A comparison of the altitudes of this point and of the bed-rock at Montezuma Hill shows the difference of level to be a little more than 1,100 feet. The distance in a direct line is about five and a half or six miles, which would allow a grade of nearly 200 feet to the mile. If any connection is to be traced between Montezuma Hill and the gravel higher up on the ridge, it seems most probable that it will be found to be along this line.

At least one effort has been made to reach, by means of a tunnel, the supposed channel underneath the lava. Eurisco Tunnel is between two and three miles from Columbia Hill in a direction not far from N. 20° E. (magnetic). The mouth of the tunnel is in bed-rock, on the northern side of the ridge, near the head of Grizzly Cañon, and about 225 feet below the level of Bloody Run ditch. The course of the tunnel is S. 40° E. (magnetic) for a distance of 400 feet; then a change of direction carries the tunnel to the southwest for a distance of three or four hundred feet farther, and a second change to a more nearly south course carries it six or seven hundred farther still. The "gravel" is struck shortly after the first change of direction near the upper end of an incline raised about twenty feet through bed-rock. The altitude of this point is, according to my measurements, 3,440 feet, nearly thirty feet *lower* than the bed-rock at the extremity of the lava-flow, — a circumstance which surely does not point to any regular channel under the lava. The "gravel" at Eurisco is also peculiar. Occasionally a quartz pebble, not much worn, was seen, but the greater part of it was composed of slate or other metamorphic rock. The mass, as a whole, was remarkable for its clayey character. A great many of the faces of the small pebbles and of the exposed clay were slickensides, or, as the miners call it, "soapy." The deposit looked more like clay with an occasional imbedded pebble, than like gravel with spaces filled with clay. Near the farther end of the tunnel some large boulders are to be seen, which have not yet been completely uncovered, some of them being as much as eight or ten feet across. There were also smaller boulders, of two or three feet in diameter, of various kinds of rock. On the whole, the pebbles and boulders presented angular rather than rounded surfaces. In the breasts only about three and a half feet in thickness next the bed-rock has been removed. The deposit carries gold in small quantities, and there is also carbonized wood. I do not see any reason to believe that there is any deep channel there, but the deposit is interesting on account of its altitude and of its relation to some of the projecting spurs of light-colored gravel, to which reference has been made above.

The thickness of the lava capping between Columbia Hill and North Bloomfield is, on the average, not far from 600 feet. The ridge to the northwest of Malakoff I found to be at its highest part over 1,300 feet above the bed-rock at the mine. The precise position of the line of junction between the lava and the underlying bed-rock was not easy to make out, but the above estimate of 600 feet for the thickness of the lava must be nearly correct. There did not appear to be any material differences of composition in different portions of the lava, though there are some signs of a succession of flows. One of these signs is the existence of the projecting and prominent outcrops of nearly horizontal layers, which appear to withstand the disintegrating action of atmospheric agencies better than the rest of the mass.

D.   Between North Bloomfield and Eureka.

Immediately above North Bloomfield the lava capping on the ridge between the South and the Middle Yuba spreads out to a width nearly double that which it has below. Its surface for several miles forms a nearly flat table, with a gradual and regular grade rising to the northeast. The grade amounts to about ninety feet to the mile between a point on the ridge near Bloomfield and the summit, westerly from Shand's Ranch. The sloping sides of the volcanic stratum towards the streams are steep, though not precipitous. The slope to the southwest near the head of Humbug Cañon is also quite steep, and to the traveller approaching from the direction of Lake City the town of North Bloomfield seems to be built in a broad amphitheatre, with high walls nearly enclosing it on all sides, excepting towards the southwest, where Humbug Cañon has its outlet.

That the mass of gravel which covers the surface near Malakoff and North Bloomfield extends under the volcanic capping, there can be no doubt; but precisely where the deep channel lies, and whether or not the gravel is confined to any one single channel, are questions to which as yet no satisfactory answers can be given. There are deposits of gravel which have been worked by the hydraulic process both on the northern slope of the ridge, as at Woolsey Flat, and on the southern slope, at Relief Hill, which can be traced to the edge of the lava or even underneath it; and there have been shafts sunk through the volcanic capping from which important information as to the position and character of the underlying gravel has been obtained. In my description of this portion of the ridge I will first state what is known about the connection between the Bloomfield gravel and that of Woolsey Flat, and then take up the consideration of the gravel deposits on the northern slope of the ridge, reserving to the last the deposits on the southern side.

The two points between North Bloomfield and Woolsey Flat at which the rock lying under the volcanic stratum has been reached are at the Derbec Shaft and the Watt Shaft.* The Derbec Shaft lies about a mile to the north of the town of North Bloomfield, near the stage-road leading to the Backbone House. Work was begun on this shaft on the 1st of September, 1877, and was prosecuted without material interruption until the 18th of July, 1878, on which date the deep bed-rock was reached. The following data in regard to the materials passed through in the sinking are taken from one of the reports of the Derbec Company. The first gravel was struck at a depth of 167 feet from the surface, beneath volcanic cement. The first pipe-clay was reached at a depth of 208 feet, and the first gold-bearing gravel at 271 feet. The blue gravel was met at the depth of 310 feet. This gravel has a thickness of 150 feet, deep bed-rock being struck at the depth of 460 feet below the surface.

The altitude of the mouth of the shaft I made to be 3,813 feet above the level of the sea, which will give 3,353 feet as the altitude of the bed-rock at the bottom of the shaft. Between this point and the Prospect Shaft at Malakoff, referred to on page 396, the position of the bed-rock is not known, but there is a difference of level between the places of about 360 feet. The grade must be as much as 200 feet to the mile, or possibly more.

From the bottom of the shaft prospect drifts were run in several directions with the object of finding and tracing the course of the channel, and of opening the mine for work. By the kindness of the superintendent at the shaft, Mr. C. M. Cox, I was allowed to take from the plans of the workings the data necessary to show the evidence upon which rests the prevailing belief as to the course of the channel. They show that its direction was almost exactly from north to south for a distance of 500 feet from the north end of a drift run nearly parallel with and a little to the west of the channel, and that from there it followed a course curving a little to the eastward for about 400 feet farther. From these data it will be seen that rising bed-rock is known to exist at three points on what may be taken as the eastern rim of the channel. I was admitted to the underground workings, and found the blue gravel to be well cemented at the points where

* See General Gravel Map for the position of these points.

drifting was going on. A considerable quantity of white quartz pebbles was to be seen, and there were frequent boulders as large as five or six feet in diameter of an easily cleavable slate. In general appearance the gravel resembled that seen in corresponding positions in the open mines, and it is reported to be very rich. I have not had at my command any data upon which to base an estimate of yield per cubic yard.

A small deposit of clay with a few rolled gravel pebbles on the line of the Irwin ditch, a few rods from the Derbec Shaft, was pointed out to me by Mr. Perkins. The deposit is insignificant in extent, and it is worthy of notice principally on account of its altitude, which I made to be only seventy-two feet less than that of the mouth of the Derbec Shaft, or nearly one hundred feet higher than the base of the volcanic stratum. The pebbles at this point were probably picked up by the flowing mud at some higher elevation, and remained permanently enclosed in the volcanic stratum until brought to view by the digging of the ditch.

The Watt Shaft is on Bloody Run, at a point about a mile from Woolsey Flat in a southerly direction, and between half a mile and a mile easterly from where bed-rock first appears in the bed of the creek. At the time of my visit there was no one to be seen at the shaft except the watchman in charge of the property. Active work has been suspended for a considerable time. I could not go underground at this shaft, and can only give such data as I was able to get by inquiry and conversation. The shaft is said to be 414 feet deep, the first 200 feet being in the volcanic stratum, and the remainder in alternating beds of sand and pipe-clay of varying thicknesses. From the bottom of the shaft a drift was run in a direction a little south of east for a distance of 1,300 feet. Two winzes were sunk from this drift to bed-rock, each one through fifty feet of gravel. The gravel, however, was not rich enough to pay for drifting. My observation with the small aneroid barometer makes the altitude of the mouth of the shaft to be 4,262 feet. Deducting from this the depth of the shaft and the winze, and allowing two feet for rise in the drift, we have 3,800 feet as the altitude of deep bed-rock at this point. From these figures it appears that the fall of bed-rock between the Watt Shaft and the Derbec Shaft is 447 feet. The distance between the two points is nearly three miles, and the grade, if uniform, would be not far from 150 feet to the mile.

The exposed gravel deposits nearest to the Watt Shaft are on the opposite side of the ridge which lies between Bloody Run and the Middle Yuba, and are found on four projecting spurs leading down from the main ridge, and separated from each other by deep, cañon-like ravines. These places are known as Woolsey Flat, Moore's Flat, Orleans Flat, and Snow Point. I could not learn that the ravines in this neighborhood have any well-recognized names, but the one between Woolsey and Moore's Flat is sometimes called Blue Bank Ravine, the one between Moore's and Orleans Flats is known as Orleans Ravine, that above Orleans Flat as New York Ravine, and that above Snow Point as Golconda Ravine. The gravel at all these points is entirely free from lava capping except at Woolsey Flat. At this last-named point the old channel must have been deflected to the south, for the gravel at the present day can be seen to pass directly under the lava, and there is no high gravel now to be seen on any of the spurs of the Middle Yuba slope, below Woolsey Flat, until Badger Hill is reached.

I took observations to determine the grade of the bed-rock at three points: one near the base of the bank of the Boston Mine at Woolsey Flat; a second at the most easterly end of the Moore's Flat diggings, opposite Orleans Flat; and a third near the bank of the Shanghai Mine, at Snow Point. The altitude of the bed-rock at Woolsey Flat I found to be 3,890 feet, or ninety feet higher than that of the deepest bed-rock reached at the Watt Shaft, about a mile distant. The altitudes of the bed-rock at Moore's Flat and at Snow Point were 4,019 and 4,211 feet respectively. The precise distances of these points from each other, or from Woolsey Flat, cannot be given, but it will not be far from correct to say that Moore's Flat is a mile and a quarter above the Boston mine, and that Snow Point is a mile and three quarters above Moore's. From these figures it will be seen that there is a very regular grade of about one hundred feet to the mile between Snow Point and the Watt Shaft. As Snow Point is the highest point on this ridge at which gravel is found which can be supposed to belong to the old channel, whose course I have been following in these notes, it will be convenient to put here in tabular form a summary of what has been already written in regard to the question of grade of bed-rock. In regard to a supposed continuation of the old channel towards the north more will be said when the description of the Forest City divide is reached. The tabular statement is given below.

	Distance from preceding Station.	Total Distance.	Altitude.	Differences of Altitude.	Grade per mile between each Station and the one preceding.
Timbuctoo			473 feet		
Mooney Flat	2¼ miles	2¼ miles	757 "	284 feet	114 feet.
French Corral	9 "	11¼ "	1,579 "	822 "	91 "
North San Juan	6¾ "	18 "	2,033 "	454 "	67 "
Badger Hill	4¼ "	22¼ "	2,391 "	356 "	80 "
Malakoff	7½ "	29¾ "	2,929 "	538 "	72 "
Derbec Shaft	2 "	31¾ "	3,353 "	424 "	212 "
Watt Shaft	3 "	34¾ "	3,800 "	447 "	149 "
Snow Point	4 "	38¾ "	4,211 "	411 "	103 "

The average grade for thirty-eight miles and three quarters is ninety-six feet to the mile. The most noticeable deviation from the average grade is between Bloomfield and the Watt Shaft. Until further explorations are made no final explanation of this deviation can be reached. For the present I am inclined to the opinion that there will be found somewhere between Bloomfield and the Derbec Shaft, if work is ever so far prosecuted as to make an underground connection between the two places, either a fault in the strata, formed subsequently to the deposition of the gravel, or else evidence of the former existence of cascades or rapids. This opinion is based entirely upon the steepness of the grade at this point, and cannot be, so far as I know, corroborated by anything now observable in the rocks. This point will be referred to again in connection with the altitude of Relief Hill.

The bed-rock seen at Woolsey Flat is an easily cleavable slate, whose dip is usually nearly vertical, though in some places at an angle as low as from sixty to seventy degrees. At Moore's Flat the bed-rock is also slate, as it probably is at Orleans Flat. The last-mentioned place I did not visit ; the ground was worked out many years ago. In the second report of J. Ross Browne, dated in 1868, the diggings at Orleans Flat are spoken of as having been abandoned for several years. The bed-rock at the Shanghai diggings, which are on the northwesterly extremity of Snow Point, is a very soft, easily worked slate, which weathers very rapidly upon exposure to the air to a bluish or reddish clay, or else has suffered decomposition to a considerable depth while still covered with gravel. About five acres of bed-rock are uncovered at this point, and nearly the same amount on the northern slope of the hill. I made no attempt to examine this second exposure of bed-rock, for the reason that the diggings have not been worked for several years, and the slides of clay and sand have hidden a great part of the rock from view. I was told, however, that a few hundred feet to the east of the Shanghai diggings the bed-rock was a very hard slate, — the "hardest kind."

The precise position of the deep channel between Snow Point and the Boston mine cannot be given. The Snow Point gravel appears to have its longest axis in an easterly direction, and there are some indications of a rim on the northern side of the Shanghai diggings ; but the pitch to the south is very gradual, and there is nothing to be seen above Snow Point to lead to the belief that any old stream ever came down from an easterly or northeasterly direction. The Moore's Flat bed-rock at my point of observation seemed to be nearly level ; there was no well-defined rim to be seen. As before stated, this point was on the northeast side of the gravel deposit, opposite Orleans Flat. The gravel had been removed for a distance of from 800 to 1,000 feet back from the ravine, where work had to be suspended on account of the tunnels being "out of grade." The general course of the old channel from Moore's Flat to Woolsey Flat is about S. 20° W. (magnetic), but at the Boston mine it is S. 25° E. (magnetic), the deflection towards the south amounting to

45°. These statements are only approximations, however, for there is considerable doubt as to the position of the deep channel, where the gravel disappears under the lava capping. There are two companies interested in the Woolsey Flat gravel, — the Eureka Lake Company, which owns the Boston mine, and the owners of the Blue Bank mine, which joins the Boston mine on the east. As there was no work doing at the Blue Bank mine at the time of my visit, my observations were made principally at the Boston mine. The channel at that point is six or seven hundred feet in width. There is also some reason to believe that there is a second, or parallel, channel in the Blue Bank property, but the condition of the mine prevented any satisfactory examination, and I did not regard the question of the existence of such a channel of sufficient importance to justify any further outlay of time.

The only places where regular work was going on at the time I was there were at the Boston mine and at the Shanghai diggings, excepting such work as Chinamen were doing at Moore's Flat.

At the Boston mine the gravel has been removed, in places, nearly to the limit of profitable working by the hydraulic process, the slides of top dirt reaching nearly to the base of the steeper portion of the volcanic capping, though the perpendicular bank of gravel is still, where the washing is going on, a considerable distance away. I did not take any measurements for horizontal distances. The diagram (Plate X, Fig. 5) shows the relative thicknesses of the different strata, but it is not drawn to any horizontal scale. At each fresh slide from the top of the bank large masses of clay and of volcanic tufa fall to the bed-rock, where they have to be broken up, at considerable expense, by drilling and blasting. It was from some of the freshly fallen masses that the specimens, Nos. 31 and 32, were taken. In one of these specimens uncharred vegetable matter is distinctly to be seen. After two or three more heavy slides from the top of the bank it is probable that drifting will have to be resorted to, if mining is prosecuted any farther in this direction; though there still are large masses of gravel to the east and west to be moved by the water alone. The thickness of the volcanic tufa at this point, measuring from the top of the ridge to the pipe-clay, is 300 feet or more. Directly underneath the tufa there is a very large accumulation of pipe-clay, or of a sandy clay, probably 175 feet in thickness on the average. The clay carries here and there a few thin streaks of gravel, but is practically barren of gold. Below the clay is a stratum of blue gravel, hard and compact, though not cemented, varying in thickness from thirty to seventy feet. This variation is due to inequalities in the surface of the bed-rock, the upper surface of the gravel being nearly level. The lower half of the blue gravel is considerably coarser than the upper half, some of the boulders being as much as six or eight feet in diameter, while in the upper half there are scarcely any pebbles measuring more than six or eight inches through. The pay streak at the bottom contains some rather coarse gold, though in the form of flakes or scales. I was allowed by the superintendent in charge of the mine, Mr. H. A. Brigham, to take a few pieces of gold from the crevices in the slate near the head of the ground sluice. These have been examined by Mr. Wadsworth, who says of them that "some of the larger pieces show the quartz vein-stone, and are but little worn. Others are small and flattened into thin scales. One shows several little ends bent over upon the main mass. This gold, it would seem, came from unequal distances, or else was freed from the matrix in the bed of the stream later, the associated quartz preventing the wearing of the gold." The last remark quoted from Mr. Wadsworth receives some support from the fact that occasionally quartz boulders carrying free gold are found in the gravel at this mine.*

At the Blue Bank mine the order of succession of strata is similar to that at the Boston mine. The gravel has been removed so far back towards the ridge that hard lava, apparently in place, is seen on the face of the bank for a thickness of seventy-five feet, and resting upon the pipe-clay. Since the suspension of work the clay slides have covered up the gravel stratum to such an extent that I made no special effort to get access to it.

---

* One boulder in particular was quite rich, and was broken up for specimens by Mr. Brigham, who generously allowed me to select from his supply such as I wished to bring away with me.

The Moore's Flat gravel has been worked in former days both upon the eastern and the western side of the deposit.* On the east the gravel bank at present is as much as 135 feet in height. The gravel is light in color, rather fine, and of a quartzose character, with occasional streaks of sand and clay, the whole being capped with a reddish loam. The bed-rock at the present time is thickly covered with small boulders, some of the largest being six or eight feet in diameter, but the majority are scarcely more than a quarter as large. These boulders are mainly of white quartz, and do not seem to have suffered much wear. They are smooth upon the surface. The sharp angles and edges are gone, but the shape is irregular. I saw no blue gravel at this bank, though I have been told that blue gravel was met with in spots at the time the deposit was worked. The absence of the blue color is doubtless due to the fact that the gravel has not been so well protected from the action of air and water as has that at Woolsey Flat.

The gravel in the westerly and southwesterly excavations is similar in character to that just described, excepting that the strata of clay grow thicker and more important the more closely the main ridge is approached, though nowhere at Moore's Flat did I see such thick and persistent clay strata as are seen at Woolsey Flat.

According to my estimate, about two thirds of the whole body of gravel at Moore's Flat has been washed away. Deeper bed-rock tunnels will have to be run before the remainder can be washed with profit.

The gravel banks at Orleans Flat are said to have been very easily worked. They were from forty to sixty feet in height, and had little or no pipe-clay in or upon them. The gravel was very rich, and the Flat paid better returns than any of the other deposits in this vicinity.

The Snow Point gravel at the Shanghai Diggings differs in some respects from that occurring at the Flats below. The bank exposed at the time of my visit was 135 feet in height, the top, for ten or fifteen feet, being loam. Below the loam came from twenty to twenty-five feet of pipe-clay, which was followed by a bed of gravel, from seventy-five to ninety feet in thickness, consisting of fine, almost sandy, quartzose material, which carries some fine gold, barely sufficient in amount to pay "water money." The gravel gets gradually coarser towards the bottom, and in the last fifteen feet of its thickness pebbles of from four to six inches in diameter are met with. Here and there this lowest gravel is cemented, but not generally so. The color is reddish even to the bed-rock, no blue gravel nor blue cement being seen. Towards the bottom the gold is also coarser; the pieces being sometimes as large as grains of wheat or of corn. Boulders of from three to four feet in diameter are found upon the bed-rock.

The impression that I got from an examination of the Snow Point gravel was, that it probably lay to one side of the main channel, or in some lake or bay-like expansion of the stream, at a point where, perhaps, there was a long sweeping curve. This impression was strengthened later in the season, when I had an opportunity to get an extensive view of the whole slope of the ridge from a point near Minnesota, on the opposite side of the cañon.

I made inquiries at several times in regard to the yield of the gravel in this vicinity, but could not get any detailed statements of a satisfactory character.

Above Snow Point, near the head of Golconda Ravine, considerable prospecting has been done in the expectation of finding a north and south channel, — the supposed continuation of the Bald Mountain channel at Forest City, which is to be described further on in this report. Tunnels have been driven in to open the gravel deposits which are supposed to underlie the stratum of volcanic tufa. One of the gentlemen interested in these explorations is Mr. S. L. Blackwell (Moore's Flat Post-Office). I made his acquaintance at Moore's Flat, but did not get an opportunity to visit his mine. From all that I could learn from conversation with Mr. Blackwell and others, I concluded that nothing has yet been found to confirm the belief in the existence of any channel crossing the ridge at this altitude. It would not be surprising if gold in greater or less quantity were found under the lava, as, for instance, at the Eurisco Tunnel, described on page 398, but I should be very

---

* The former excavation has already been referred to; see page 101.

much surprised at the discovery and development of a well-defined channel of auriferous gravel running at right angles to the present course of the ridge.

If I had had a few more days to spare I presume that I could have had access to all the exploratory works referred to above. There did not seem to be any disposition to exclude me from them. Special appointments, however, would have been necessary, for the reason that little or no work was then going on.

Passing from the northern to the southern side of the ridge, we find two principal points to attract our attention, the hydraulic banks at Relief Hill, and the drifting at Mount Zion.

Relief Hill, in a direct line, is only about two miles from North Bloomfield, but in order to reach the place I climbed the steep slope of the ridge until I struck the Malakoff ditch, which I then followed in all its windings. I chose this route in preference to the wagon-road, because the latter, being built at a lower level, would give me no information in regard to the relations of bed-rock and lava. The ditch, from the point at which I first struck it, I found to be entirely in slate-rock, until I reached a point on one of the spurs above Missouri Cañon, and above the ditch-tender's house. From this point it is entirely in volcanic material as far as Relief Hill, and even a little beyond. The first bed-rock that I saw in the ditch after leaving Relief Hill was at a point bearing N. 20° E. from the town. The bed-rock was from that time visible in the ditch as far as I followed it; that is, to the beginning of the Washington trail.

Relief Hill is the highest large body of gravel on the right bank of the South Yuba River. Shallow gravel and surface diggings are found occasionally at higher altitudes, as, for instance, at Roscoe's ranch, five miles above Relief Hill, following the line of the ditch, and there may be gravel underneath the volcanic flow. Similar small deposits of shallow gravel are said to exist on some of the spurs lower down the river, between Relief Hill and Grizzly Hill, but there is no reason to suppose that they have any connection with the gravel of the old channel.

The gravel exposed to view at Relief Hill lies on the high spur which leads down from the main ridge between Rocky Ravine on the west and Logan Cañon on the east. The gravel, exclusive of what is covered by the volcanic tufa, spreads over about two hundred and fifty acres of ground, according to my estimates and measurements made on the spot. Considerable work has been done at this place from time to time, but hydraulic operations have been suspended since the fall of 1875, as I was told by Mr. Penrose. The few houses which make up the town are built upon the gravel, and the most of those few were unoccupied at the time I was there. I found Mr. Penrose, who acted as my guide for two or three hours, to be apparently well-informed as to what had been done in previous years. The three principal claims at Relief Hill are the Union claim, comprising about 120 acres of ground on the western and southern sides of the hill, the Eureka claim, in the centre of the town, and the Eagle claim, now belonging to the Eureka Lake Company, and lying to the east and northeast of the other two. The Eureka ground has been drifted, but never worked by the hydraulic process; in the Union claim water has been used to some extent, but for the last year or two drifting has been resorted to; the Eagle ground has been worked by the hydraulic process exclusively.

The bed-rock, where exposed to view, is slate. The position of the old channel is unknown, but the bed-rock, if the concurrent testimony of the miners is relied upon, "pitches into the hill," that is, it falls towards the north and west. Further explorations, however, will be needed before much that is decisive upon this point can be written. I could not get any special information in regard to the Eureka drifts, nor the Union drifts, except that they were driven in an easterly direction and had reached a distance of 1,200 feet from the face of the bank. The Great Eastern tunnel, which has been begun in Rocky Ravine and driven towards the supposed deep channel, I hoped would throw some positive light upon the question of the position of the bed-rock, but my hopes have not been realized. I took an observation for altitude at the mouth of the tunnel, which I make to be 3,713 feet above sea level. The following data concerning the tunnel I got from Mr. Penrose. Its course for 1,200 feet is N. 27° E. (magnetic), and then for 600 feet farther N. 15° W. (magnetic); the grade for 1,000 feet is two inches to twelve feet, and one inch to twelve feet for the remaining 800 feet. The tunnel was driven for the sake of opening a drift mine, not for hydraulic sluices. At a distance of 523 feet from the mouth of the tunnel a rise was made, and gravel was struck seventy feet above the tunnel. At 500 feet farther in gravel was found only seven feet above the top of the tunnel, and this distance seemed to remain nearly constant to the end of the tunnel. Some drifting was done in blue gravel mixed with large smooth boulders, for a few months, but for some reason the work has been discontinued. The bed-rock was flat and level.

The only other places at which I took observations for altitude were at Mr. Penrose's house and at the mouth of an old shaft in the Eagle claim, where it was said to be sixty feet to bed-rock. If this last statement be true, the bed-rock in the Eagle ground is fifteen feet lower than the mouth of the Great Eastern tunnel, the altitude at the mouth of the shaft being made to be 3,758 feet, and about fifty feet lower than the bed-rock at the farther extremity of the tunnel. So far as these observations have any value, they point to a probable course of the stream at this point more northerly than I had supposed, if it has a fall in the general direction of Bloomfield, or the Derbec Shaft, as is more likely the case. Any alternative supposition, which would select a more southerly course for the channel and make it sweep around more nearly parallel to the course of the present South Yuba, will have to be excluded, both on account of the topographical features of the country and on account of the great difference of altitude between Relief Hill and Bloomfield. If the gravel at the Derbec Shaft be supposed to be on the continuation of the channel from Relief Hill, the grade between the two places will be but little more than the average grade of the old stream for its whole length from Snow Point to Timbuctoo. The precise distance from Relief Hill to the Derbec Shaft cannot be given, but it will not exceed three miles nor fall much below it. The difference of level of bed-rock between those points I make to be 345 feet, — say 115 feet to the mile. The principal objection to the hypothesis of a connection between Relief Hill and the Derbec Shaft is that it requires the stream to follow a more northerly course than the old streams usually followed, — an objection less serious, it seems to me, than those which can be urged against other hypotheses.

The highest banks exposed at present, those in the Eagle claim, are from 100 to 150 feet in height. From twenty to forty feet of the top of these banks are made up of red dirt and volcanic boulders, the excavations not yet extending back to the lava in place. Below this there comes a stratum of white gravel, composed of fine quartz with small streaks of sand and clay, not cemented, which, certainly on the northeastern rim, reaches quite down to the bed-rock. In some places the white gravel must be as much as 140 feet thick. Below the white gravel there is said to be a blue gravel on the deepest bed-rock, reached only by means of the shaft.

The gold found at Relief Hill is mostly fine. In the Union claim nuggets worth from five to ten dollars are said to have been found, though rarely.

In regard to the drift mine at Mount Zion there is not much to be said. Work has been carried on here at intervals for more than twenty years, but no one seems to have had the courage to thoroughly test the property, and to settle the question whether there is or is not a deep gold-bearing channel under the lava. The main tunnel runs nearly due west for a distance of 1,400 feet, rising in that distance twenty-four feet. The tunnel is entirely in bed-rock, a hard metamorphic slate. At the farther extremity of the tunnel there is an incline with a vertical rise of fifty-two feet, followed by a short, nearly level drift to the gravel. From the point where the gravel is struck there has been a drift run in a northerly direction for six hundred feet. This drift is supposed to follow the rim, for the bed-rock has a strong pitch to the west. The material of the gravel in this drift is almost exclusively quartz, with occasional bits of slate, well washed and rounded. The gravel is not very coarse, pebbles of three or four inches in diameter being among the largest. The gravel is not cemented.

To the south of the main tunnel prospecting has been carried into what may be regarded as a branch or overflow. It was at this point that breasting was going on when I was there. The pay-streak is about three or three and a half feet thick, and on the bed-rock the gravel is frequently bluish in shade, though not cemented. There are also some large smooth boulders, up to six or eight feet in diameter. These boulders are usually of slate rock, with occasionally one of granite.

The altitude of the mouth of Mount Zion tunnel I made to be 4,297 feet, which is 625 feet less than the altitude of the ridge at the junction of Mount Zion and Eureka roads. Adding seventy-six feet for the rise in the tunnel and in the incline, we get 4,373 feet for the altitude of the gravel on the rim of the supposed channel. This point is 162 feet higher than the bed-rock at Snow Point, distant not far from two miles. The depth of the central bed-rock, if any such exists, is not known. Prior to the running of the present tunnel there was an older one run, seventy feet higher, and on about the same grade, which reached the gravel, and from which unsuccessful efforts were made to reach bed-rock by sinking through the gravel. The reason of the failure was the accumulation of water.

The presence of this gravel deposit at Mount Zion lends some support to the belief of Mr. Blackwell and others, that a high channel crosses the ridge near or above Snow Point ; but I am more inclined to the opinion that such deposits as that at Mount Zion, if they represent old channels at all, are in the places of former tributary streams, — like the supposed tributary by way of Relief Hill and the Derbec Shaft, for example. Whether this supposition is true or not can be easily settled, for the Mount Zion gravel, by pushing the tunnel forward a few hundred feet farther, and tracing the course of the deep channel under the most favorable conditions. That such an extension of the tunnel has not been made before this time excites the suspicion that either capital or confidence has been lacking.

The volcanic capping of the ridge, which, as has been said, extends in an unbroken line from Columbia Hill, reaches a culminating point, at an altitude of about 5,100 feet, a short distance above Mount Zion and Snow Point. Between this point and Eureka there is a gap in the ridge where the old lava-flow has been broken across by the cañons at the headwaters of Poorman's Creek. The line of junction between lava and bed-rock is to be seen on the road about half a mile to the southwest of Shand's Ranch. The bed-rock here is slate. About a quarter of a mile west of the town of Eureka, or Graniteville (as the post-office is called), the bed-rock changes from slate to granite.

Shand's Ranch is on a narrow ridge or backbone, which is the divide between Poorman's Creek and a nameless ravine emptying into the Middle Yuba. The altitude of this point is 4,627 feet. The three main ditches which bring the water from the lakes and reservoirs above Eureka here approach each other very closely, and run within a few feet of each other for a considerable distance. There are some indications of the presence of quartz gravel on this narrow ridge. One tunnel has been driven in from the Yuba side, about seventy-five feet below the road, and some gravel was found, though not in paying quantities. There is no reason to believe that this gravel is anything more than a small local deposit. If it were a remnant of an old continuous channel, we should expect to find accumulations of quartz and gold in the bed of Poorman's Creek. Such accumulations, I was told, have not been found.

E. Above Eureka.

The portion of the ridge which lies above Eureka I did not have time to examine in detail. My observations were confined to parts of two days. By the kindness of Mr. Perkins I was allowed to accompany a load of supplies which was to be sent to the Bowman dam. Leaving Malakoff in the morning, we reached Eureka by the middle of the afternoon and the dam about six in the evening. The next day we left the dam about eleven o'clock, in time for me to reach Moore's Flat before night.

The bed-rock above Eureka is all granite. The higher ridges are capped with volcanic material, similar in appearance to that seen lower down on the divide between the Yubas, or on the Washington Ridge, to the south of the South Yuba. To the northeast of the town, and a quarter of a mile distant, the lava capping sets in again, as a so-called Bald Mountain ; and this must be regarded as the extension of the same stratum, which, as we have seen, was cut off by the waters of Poorman's Creek just below Shand's. Some of the ridges, where bare of trees, are covered with a thick growth of manzanita chaparral. Seen from a little distance, the crest of such a ridge looks like a gently sloping smooth mass of rock without any vegetation at all.

The road from Eureka to the Bowman dam has to cross the ridge between Poorman's Creek and Cañon Creek, at the head of which the dam and reservoir are situated. The altitude of the highest point on the road I made to be 6,098 feet, and the thickness of the lava cap I estimated at from 350 to 500 feet.

Considerable search has been made at different times for beds of gravel under these lava ridges, in the belief that just those portions of the great lava-flow which now remain upon the ridges correspond to the position of the original depressions of the surface or old channels. About a mile and a quarter above Eureka, and to the left of the road to the dam, there was a small dump, at the mouth of a tunnel running towards the lava ridge which rises to the southeast. I made the altitude of the mouth of this tunnel to be 5,458 feet. There were no signs of gravel on the dump. The tunnel had caved in and was inaccessible. I suppose it to be the one which had been described to me by the name of Griffith's claim.

I also heard of another tunnel which could be reached by following the road which leads down the south fork of Poorman's Creek. My informant was Mr. W. C. Chase, who used to work in the tunnel. Mr. Chase is now in the employ of the Bloomfield Company, in charge of the Bowman dam. The tunnel was begun as far back as 1867, and has been worked at intervals since that time. It has penetrated nearly a thousand feet under the lava capping. It was run in a crumbly granitic bed-rock. The gravel found was in small quantities, filling the interstices between large crumbly granitic boulders. The quartz pebbles were seldom larger than a teacup. There was gold with the gravel, but not enough to pay expenses of working. The bed-rock appeared to pitch towards the centre of the ridge. A small amount of drifting was done, which proved to be unprofitable on account of the number and size of the boulders.

The altitude of the porch of the house at the dam I made to be 5,393 feet, a number a little higher than that adopted by Mr. Hamilton Smith. According to the figures on Mr. Smith's map, which I saw at Bowman's, the altitudes of some of the more prominent points in this vicinity are as follows. I have given the position of some of the points with reference to United States section lines, or where these lines ought to be if the surveys were extended over the region. The townships I am not able to give.

House at Bowman's dam	5,360 feet.
Top of dam	5,450 "
Lakes at head of South Fork of Cañon Creek.	
Shot-Gun Lake	6,410 "
Middle Lake	6,460 "
Crooked Lake	6,510 "
Round Lake	6,590 "
Island Lake	6,690 "
Fall Creek Mountain, S. W. ¼ of S. E. ¼ of Section 21	7,290 "
Grouse Ridge Mountain, S. W. ¼ of N. E. ¼ of Section 34	7,430 "
Milton Dam, in Middle Yuba River	5,670 "
Jackson's Ranch	5,870 "
Faucherie Reservoir, principally in S. E. ¼ of Section 13	6,060 "
English " " N. ½ " 4	6,140 "
Finlow Peak	7,020 "
English Mountain, N. E. ¼ of N. E. ¼ of Section 7	7,980 "
Eureka Lake (dam), principally in Sections 17 and 20	6,480 "
Meadow Lake, principally in E. ½ of Section 22 and N. W. ¼ of Section 23	7,040 "
Faucherie Mountain, near Section Corner 14|13 / 23|24	7,170 "

The big dam at Bowman's is just north of the half-mile stake between Sections 5 and 8.

The granite rock near Bowman's is distinctly scored with glacial markings. This is the only point on this ridge where I saw unmistakable signs of ice action. A careful examination of these higher regions would doubtless develop much of interest in relation to glacial phenomena, but time did not allow of my pushing inquiries any farther in that direction.

From several sources I heard of the existence of so-called "high gravel" on Fall Creek and Grouse Ridge. This is a district which is most easily reached from the Emigrant Gap station on the Central Pacific Railroad. It is hardly settled at all, and would be a difficult country to explore without animals or a camp outfit. I had the good fortune to meet, at Washington, a miner from the ridge who gave me such information about the deposits of gravel that I was satisfied I should not neglect anything of importance if I failed to go there. He spoke of shallow gravel banks, perhaps from ten to twenty feet in thickness, associated with granite boulders, and so situated that they might very well be of comparatively recent origin.

SECTION IV. — *The Washington Ridge, south of the South Yuba River.*

The careful exploration of any part of the gravel region lying to the south of the South Yuba River was not contemplated in the plan of my summer's work. A hasty trip to Omega and its neighborhood was all that I expected to have time to accomplish. By way of Relief Hill I reached the town of Washington, which lies low in the cañon on the left bank of the South Yuba, and has no very close relations with the high gravel deposits, on the evening of the day I left Malakoff. During the next six days I visited Alpha, Omega, and Diamond Creek, and returned to Malakoff by way of Sailor Flat and Blue Tent. The gravel deposits known as Phelps's Hill, Phelps's Point, Jefferson Hill, Gold Hill, and Cotton (or Colton, as it is called in Raymond's Report) Hill, which follow in order downwards from Alpha on successive spurs of the ridge, and which can be seen to good advantage from almost any prominent point on the ridge above Relief Hill, and from some of the high points near Omega, I was obliged to leave unvisited. The ravines are deep and steep, opposing very effectual barriers to easy communication between places which are in full view of each other, as, for example, Alpha and Omega, on opposite sides of Scotchman's Creek.

There can be no doubt of the former connection of the gravel between Omega and Cotton (or Colton?) Hill, and probably Relief Hill, prior to the erosion of the lateral valleys. Seen from any high point commanding a view of all the gravel, the deposits appear to be arranged in a regular series, nearly parallel with the present stream and having an easy grade downward towards Relief Hill. The series bears a very striking resemblance to that on the divide next north between Snow Point and Woolsey Flat. The deposits between Omega and Relief Hill are not capped with lava, the lower line of the lava at the present day being at a considerably higher altitude than the top of the gravel banks, excepting at Omega, where the difference of altitude is not so great. Indeed, if we were to regard alone the present appearance of these gravel beds, as they lie upon the high benches, there would be no difficulty in supposing them to have been deposited since the deposition of the lava, at the time when the present South Yuba followed a channel twelve or fourteen hundred feet above its present bed. The objections to such an hypothesis, however, are too many and too important to be overlooked. The principal of these objections are, the total absence of tufaceous material in the gravel, the probable extension of the gravel under the lava in place at Relief Hill, and the certainty of the extension of gravel, similar in character and presumably formed at the same time and under the same conditions, under the lava of Malakoff and Woolsey Flat.

My observations for altitude at Washington, Alpha, and Omega have not given very satisfactory results. The difference of bed rock level at the two places last mentioned is certainly much less than my computations make it to be. Either Alpha is made too low or Omega too high; probably the former. The greatest element of uncertainty is in the determination of the altitude of Washington, to which station alone the altitude of Alpha was referred. The total fall from Omega to Relief Hill, according to my computations, is 330 feet, allowing a grade of only sixty feet to the mile for five and a half miles, somewhat smaller than I had expected to find it.

The condition of affairs at the time of my visit was one of extreme quiet. No work was going on, the water season being over for the summer. The gravel deposit at Alpha is nearly exhausted, and very little work has been done there for several years. I have not been able to find any allusion to the place in either of Browne's or Raymond's reports, which cover the years from 1866 to 1875. In early days, as I have been told, Alpha was one of the most "lively camps" in the mountains, with two six-horse stages daily from Nevada City. When I was there I found but two men, who were occupying one of the half-dozen houses still standing. Some work has been done at Alpha the past year, under the superintendence of Mr. J. F. Perry, of Washington. There are several small bodies of gravel left which will undoubtedly pay for washing when water can be obtained at a cheap rate. The original gravel at Alpha covered an area of, as nearly as I could estimate, seventy-five acres, about four fifths of which has been removed. That which is left is either upon the borders of the deposit, or on its southern side, towards the main ridge. The bed-rock is slate. There are two principal varieties, differing very much in color. One is nearly black, and is cleavable into extremely thin laminæ which stand nearly vertical. The other variety is very light-colored, and is also easily cleavable. The bed-rock has a gradual fall from east to west, but is nearly level in a north and south direction; that is to say, there is nothing like a trough or gutter to mark the site of an old channel. On the south side the bed-rock is covered with boulders, chiefly quartz, not much worn, and measuring as much as five or six feet in diameter. To the north the boulders are fewer in number. The gravel at Alpha, where seen in the banks now standing, is light and sandy, growing coarser as the bed-rock is approached. There is no blue gravel to be seen, the lowest layers having a decidedly reddish color. The bank is about ninety feet in height, including twenty feet or so of pipe-clay at the top. The Alpha gravel is said to have yielded better to the cubic yard than that at Omega. The top gold was very fine; but nuggets of considerable size have been found on the bed-rock.

Between Alpha and the crest of the ridge high bed-rock is to be seen; the positions of lava, bed-rock, and gravel being represented in the accompanying section (Plate X, Fig. 6), which is not drawn to any accurate scale.

The gravel at Omega lies on a comparatively flat bench of ground, with a rather uneven surface, near the heads of Scotchman's Creek, Missouri Ravine, Baltimore Ravine, and Iowa Ravine, and covered originally, as nearly as I could estimate, about three hundred acres. The bed-rock has been uncovered over about one third of that area. There is a second "high bench" or "back channel," known as Gold Flat, which I have not included in the above estimate. The ridges between the ravines mentioned above rise to an elevation decidedly higher than the top of the present gravel bank, and effectually shut out the possibility of any channel coming in from or having an outlet in these several directions. There is also a small patch of gravel between Iowa Ravine and Missouri Cañon, which I did not find time to visit. It lies on the Missouri Cañon side of the ridge, and may be a remnant of an old northeasterly extension of the Omega channel, or it may be one of the small and local deposits, some of which are not over fifty feet square and six or eight feet deep, which, I am told, are frequently seen on the spurs leading down to the river.

The probable position of the old channel above Omega is a matter of doubt. I could see no reason for believing that any of the gravel higher up on the ridge, as, for example, at Diamond Creek, ever had any connection with that at Omega. It is the opinion of some persons that the extension of the old channel is to be looked for under the volcanic tufa to the southeast, the stream having come in by way of Gold Flat. This upper gravel has not been systematically explored either by tunnels or by shafts, although it has been cut into in places. I made as careful an examination as I could of the surface of the ground in that direction, and failed to find any strong support for such an hypothesis. The existence of bed-rock to the southeast of the town of Omega reduces very much the limits within which such a channel must have come, if at all. I am more inclined to the opinion that the channel came from the northeast. The long axis of the elliptical portion of gravel already worked down to the bed-rock, comprising nearly one hundred acres, has the direction S. 30° W. (magnetic), with a gradual fall also in that direction. The bed-rock is nearly flat in the transverse direction, showing no signs of a deep central channel or trough. From the lower or southwestern extremity there is an uninterrupted outlet for the channel, across Scotchman's Creek, to Alpha.

The bed-rock exposed to view at Omega is a slate, or a series of slates, which presents some remarkable variations in character and color within short distances. If the boulders which now cover the bed-rock could be removed, there would be seen a brilliant succession of highly colored parallel bands, having a northwesterly strike and varying in thickness from twenty to one hundred feet or more. The boulders on the bed-rock prevented my making out the precise thickness of the different bands.

The principal varieties of bed-rock were these : —

(1.) A light-colored, very fragile, clayey, thinly cleavable slate, with a silvery lustrous surface ;

(2.) A dark, bluish-black slate, — the two having a strong resemblance to the slates at Alpha ;

(3.) A light-colored, silicious variety of schistose rock, in which coarse grains of quartz were to be seen ;

(4.) A finer-grained silicious rock, almost silvery or pearly in lustre ;

(5.) A reddish-brown, thinly cleavable, very fragile rock, irregularly mottled and shaded on surfaces of cleavage, — when pulverized on the trail in the mine, it looked very much like scales of mica ;

(6.) A very fine-grained, almost impalpable, perfectly cleavable clay slate, of a bright pink or pinkish white color.

The most of these varieties were so fragile that I made no attempt to get specimens, excepting from the last.

The gravel at Omega has been known and worked, according to a statement in Raymond's report for the year 1874, page 125, since 1853. About the year 1869 the Omega Water and Mining Company became the owners of the greater part of the deposit, which they still retain in their possession. The name of R. W. Tully, of Stockton, was given me as that of one of the principal members of the company. The tailings have been deposited for the most part either in Scotchman's Creek or Missouri Ravine. The main tunnel is 3,000 feet in length, with an irregular grade varying from seven to twelve inches to the rod. The gravel at the lower end of the deposit resembles the gravel at Alpha in appearance, being of a light color, and irregularly interstratified with sand and sandy gravel, and here and there considerable layers of pipe clay, seven or eight feet in thickness. At the upper or eastern end the gravel in the lower stratum is decidedly blue in color when freshly exposed, but it changes very rapidly to a yellowish or reddish tint under the action of the atmosphere. The gravel is hard, but not cemented. The blue variety is said to be not so rich in gold as some of the yellow. The pebbles and boulders at the western end of the diggings are mostly quartz ; at the upper end some granite boulders are seen. The banks grow higher towards the upper end of the mine, being at present fully 130 feet in height, whereas the average height could not have been more than from sixty to eighty feet in the western portion. A thickness of from 125 to 140 feet may be expected over a considerable part of the gravel still left standing.

The Omega Company's ditch is about twelve miles long, and brings water from the South Yuba River. The supply is not sufficient to last all summer. In 1879 washing was stopped by the 1st of August. The nozzles are worked at present under a pressure of a hundred feet, which is not sufficient to cut the gravel away without the aid of drifting and blasting.

There was no one at Omega from whom I could get any detailed information as to the recent yield of the gravel. In Raymond's report for the year 1874, page 125, it is stated that the Omega Company had taken out $500,000 in five years. From Mr. John Goyne, the local superintendent, I learned that it used to be the custom at Omega to lay off "claims" one hundred feet square, and that an average yield for a claim was three or four thousand dollars. Assuming seventy feet as an average height of bank at the time those "claims" were worked, the yield per cubic yard was about 13½ cents.

There being some uncertainty about the area covered by the gravel between Omega and Diamond Creek, I thought it necessary to extend my examination in that direction. In so doing I followed first the line of the Omega ditch to a point opposite the Diamond Creek diggings, and subsequently, for a part of the way, the Blue Tent ditch, which runs nearly parallel to the other but at a higher altitude. The difference of altitude between the two ditches at a point near Omega I made to be 170 feet. On the Omega ditch I first saw bed-rock near the head of Iowa Ravine. Its altitude was about forty feet higher than the highest surface of the gravel at Omega. For the rest of the distance as far as Diamond Creek I saw in the ditch nothing but slate, or some other variety of schistose bed-rock. Upon the top of the ridge between Iowa Ravine and Missouri Cañon, near where the ditch is carried through a short tunnel, there was a small deposit of rolled gravel which very likely was once connected with the high bench at Gold Flat. There had been an opening made of perhaps 300 feet in diameter and fifteen in depth. The Blue Tent ditch was in lava until after the crossing of the spur between Iowa Ravine and Missouri Cañon, and from that point on in slate as far as the second or third branch of Diamond Creek, which was the limit of my walk in that direction. I went far enough to satisfy myself that the quartz gravel, said to be seen in the Blue Tent ditch above Missouri Cañon, had no obvious connection with the deposit at Omega.

The gravel at Diamond Creek extends for half a mile or more, principally upon the right or eastern bank of the creek, and about one hundred and fifty feet above its present bed. I took one observation for altitude upon the bed-rock near the old blacksmith-shop, just above the mouth of a tunnel. I made the altitude at this point to be 4,206 feet. I saw no one at work. The bed-rock has a steep pitch down the creek, and also a more gradual inclination under the bank of gravel lying to the east. At one of the lower banks, farther down the stream, the top of the gravel was on a level with my point of observation. Back of the gravel, bed-rock hills rise rapidly to a considerable height. The gravel differs very much in appearance from that at Omega. It looks more like an irregularly stratified drift, with small boulders distributed from top to bottom through a reddish sandy gravel. Many of the boulders were evidently of volcanic origin, though not all. At one of the lower banks, as seen from across the creek, the gravel appeared to be finer and lighter-colored.

Taking all these facts into consideration, I was led to the conclusion that the gravel at this point represents some former channel of Diamond Creek alone, and that there never was any direct connection with the Omega deposit.

From Omega I went by stage to a point about a mile below the Central House, and thence on foot to Blue Tent and Sailor Flat; but before reporting upon this portion of the ridge I will add a few words about a small deposit of low gravel, which probably owes its origin to the washing away of the old channel between Alpha and Omega.

The Hathaway claim, on the south bank of the South Yuba, at the mouth of Scotchman's Creek, about a mile above Washington, presents one or two features of interest. There is no precisely similar deposit that I have heard of along the river, either above or below. The gravel is reputed to extend back from the river for at least a quarter of a mile. Near the Yuba it has a thickness of twenty-five feet, the lower half being blue in color while the top is red. Farther from the river the thickness is considerably greater, a bank of two hundred feet being looked for in some parts of the claim. This deposit has been prospected by means of tunnels from the Yuba and from Scotchman's Creek, each about 1,200 feet in length. The mouth of the latter tunnel is now buried seventy feet beneath tailings. It seems almost beyond question that this deposit came from the washing away by natural causes of a portion of the old channel between Alpha and Omega, and it is probable that it has been caught where it is in consequence of a change in the courses of the ravines, for there is some evidence that there was formerly a different outlet, a little lower down the Yuba, for Krumbacher Ravine, or Scotchman's Creek, or both. The high spur of gravel on the east of the creek indicates that there was once a large body of gravel at this point. The ground is worked by a company, whose claim also covers the deep mass of tailings in the creek, which have come from the Alpha and Omega mines. Hydraulic mining can never be carried on here on a large scale, on account of lack of space for a dump.

The Washington ridge, from Diamond Creek downwards as far, at least, as Nevada City, needs a re-examination. It will have to be carefully surveyed and mapped before anything decisive can be published in explanation of the geological problems which there present themselves in connection with the gravel and the capping of volcanic tufa. The tufa when first exposed to view in the digging of the ditches is usually hard and firm, but it changes rapidly under the action of the atmosphere. Slides of lava down the steep sides of the ridge are of frequent occurrence. At the head of Washington and Jefferson Creeks there are some unusually steep and precipitous walls of volcanic rock, — the result of the rapid erosion. Efforts have been made to find high gravel under the lava of this ridge similar to those which have been referred to already in connection with the lava capping above Columbia Hill and Snow Point. I heard frequent mention made of the Centennial claim, and of the tunnel which had there been driven in under the lava until gravel was struck. The claim is said to be near the head of Jefferson Creek, a quarter of a mile northerly from the ridge-road. I could not learn that the gravel found was in paying quantities, and I did not have the time to attempt any personal examination.

The only other points on this ridge that I was able to visit in person were at and near Blue Tent and Sailor Flat. At the former place I was entertained by Mr. D. T. Hughes, at that time superintendent of the Blue Tent mines, and at the latter by Mr. B. D. Chadwick and Mr. O. B. Campbell, two of the owners of the Sailor Flat mines.

The map and sections on Plate O were prepared from documents at Mr. Hughes's office. The Blue Tent property is owned by an English company, known as the Blue Tent Consolidated Hydraulic Gravel Mining Company. The Sailor Flat property lies to the east of the Blue Tent ground, near the forks of Sailor Flat and Last Chance ravines. A small claim intervening between these two is owned by other persons. The gravel owned by the Blue Tent Company covers nearly five hundred acres, as will be seen by an inspection of the map, where the heavy lines are the boundary of the property. The broken line shows the boundary of the gravel, which extends both to the east and the west of the company's ground. The Sailor Flat Company owns about 360 acres of mining ground. I did not see any map of the property, but Mr. Chadwick gave me the approximate bounding lines, as follows: an east and west line of 7,000 feet in length along the centre of the ridge; a front line, also running east and west, of about 4,500 feet in length; and two side lines of about 3,000 feet each.

The surface gravel is continuous over the Blue Tent and Sailor Flat grounds, and, for the greater portion of the area, it is of unknown depth. About twelve acres of bed-rock, according to my estimate, are exposed to view at Gopher Hill, at the northeasterly end of the property, overlooking the South Yuba River. A larger area than that is represented on the map as "exposed bed-rock," but some of the lower gravel and some of the gravel on the edges is still unwashed. The bed-rock at this point is slate, and it has a gradual pitch to the south under the gravel. I took an observation for altitude on the bed-rock, near the point marked "B" on the map, about a thousand feet back from the river bluff, and made it to be 2,483 feet, practically the same as at Grizzly Hill, on the opposite bank of the river, — a result which agrees very well with the hand-level observations. The two deposits were doubtless connected with each other before the erosion of the present cañon of the South Yuba. The only other observation for altitude that I took in this vicinity was near an exposure of rim-rock at the Sailor Flat mine, on the spur between Sailor Flat and Last Chance ravines, about three quarters of a mile to the southeast of Gopher Hill. Here the altitude was 2,759 feet, — 276 feet higher than at the former point. There was no other exposure of bed-rock at which it seemed worth while to determine the altitude. At Sailor Flat the gravel has been removed for several hundred feet back towards the south from the point of which I took the

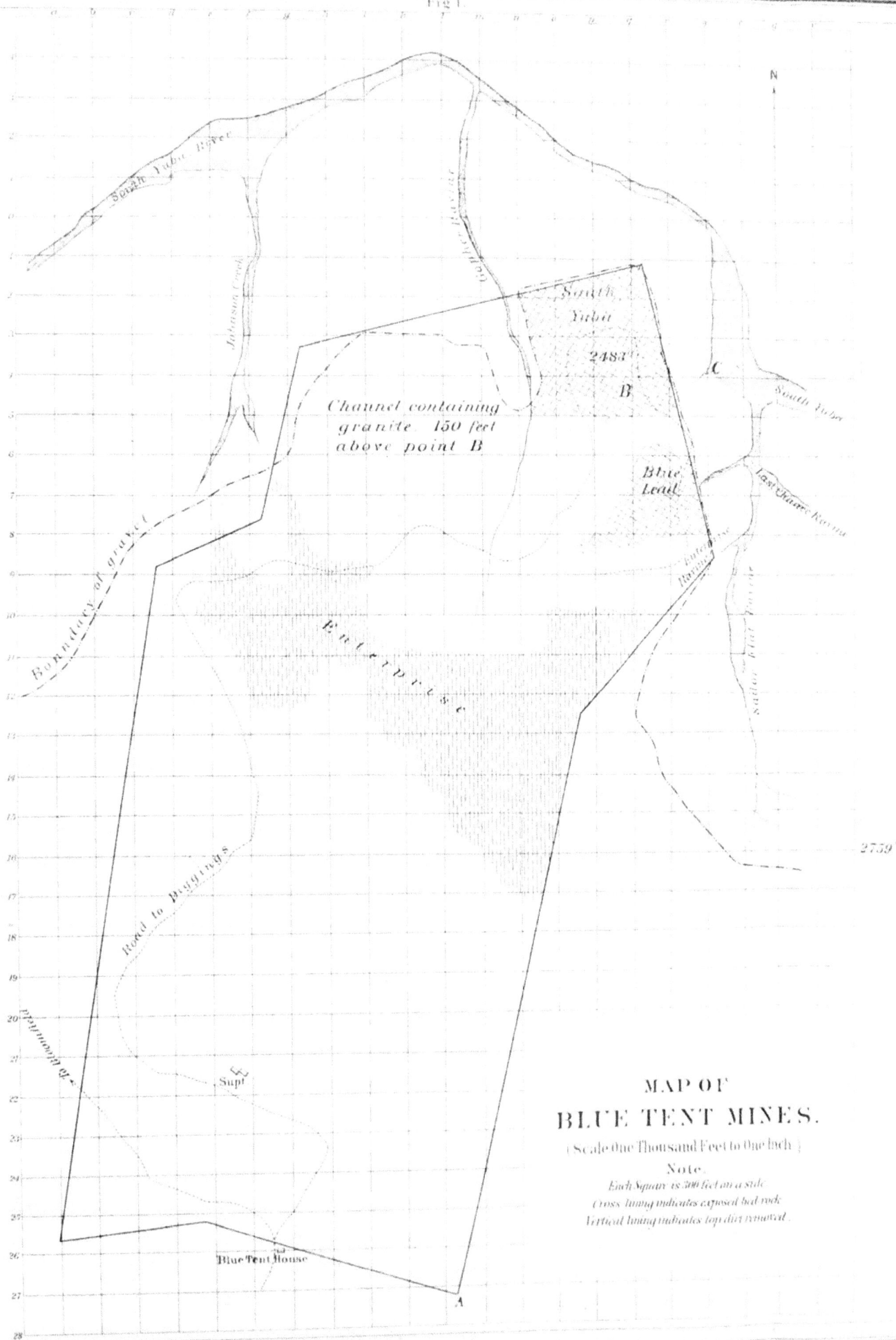

Fig 1.

N

South Yuba River

Johnson Creek

South Yuba

2483

B

A

South Yuba

Channel containing
granite 150 feet
above point B

Blue Lead

Boundary of gravel

Enterprise

Road to Diggings

2759

Plymouth Qr

Supt

MAP OF
BLUE TENT MINES.
(Scale One Thousand Feet to One Inch)

Note.
Each Square is 500 feet on a side
Cross lining indicates exposed bed rock
Vertical lining indicates top dirt removed.

Blue Tent House

A

PLATE 0.

Fig 3

SECTION TO ACCOMPANY MAP OF BLUE TENT MINES.
Approximately East and West. Scale same as that of Map.

Fig 2.

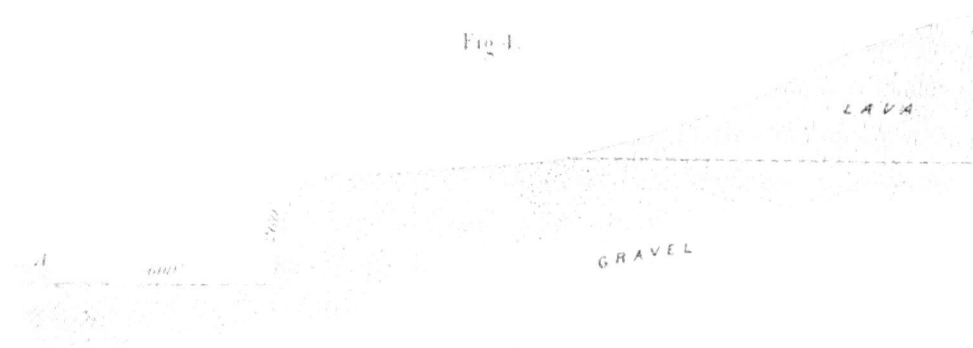

Fig 4.

LAVA

GRAVEL

SECTION NORTH AND SOUTH AT SAILOR FLAT MINE.

SECTION TO ACCOMPANY MAP OF BLUE TENT MINES.
Approximately North and South. Scale same as that of Map.

Fig 5.

BED ROCK AT CAMDEN CLAIM.

altitude, but no bed-rock has been reached. The present bank at one of the most advanced of the openings is 260 feet in height, and, with the exception of eight or ten feet of red dirt and volcanic debris at the top, is made up of alternating layers of sand, clay, and fine gravel, ranging from two to twenty feet in thickness. The material of the gravel and sand is almost exclusively quartz, and there is seldom a pebble more than three or four inches through. The water in the sluices, instead of being reddish in color, as is so common in gravel mines, was nearly as gray as if it carried nothing but tailings from a quartz mill. The lower hundred feet is of a bluish slaty color. There are no boulders excepting those of volcanic material which fall from the top of the bank. Some of the clay streaks are exceedingly rich in fossil leaves and impressions. There is also considerable pyritous fossil wood. The present bank at Gopher Hill is about 240 feet in height. As seen from a distance, the top gravel resembles that at Sailor Flat, just described, but much more free from sand and clay. Lower down the gravel is coarser and the boulders are larger. Towards the bottom the material of the gravel is for the most part metamorphic slate rock, without many quartz pebbles. It has a striking resemblance to the gravel at Smartsville, yet would be easily distinguished by any one familiar with the appearance of gravel banks. I cannot describe the difference in words, any more than I can the differences of the features of two persons who look alike. The lowest stratum is a hard blue cement.

At the Enterprise ground the top bench of light gravel has been removed over an area of about eighty acres. The high bank is there nearly three hundred feet high. At one point in the Enterprise ground there was a volcanic top-dirt as much as seventy feet thick in the thickest part. It thinned out rapidly on both sides, however, so that the cross-section resembled an old local ravine which had become filled by a slide from the central ridge.

At Sailor Flat there is no recognizable difference of value between the top and the bottom gravel. The gold is fine and scaly; but at Gopher Hill the bottom gravel is notably richer in gold, and the gold is of a coarser and more massive character.

There does not seem to be much use in trying to trace the course of any old channel at this point until more is known about the relations of the bed-rock. At Gopher Hill I paid some attention to the direction of the principal grooves and furrows in the bed-rock, and found it N. 70° W. (magnetic). This, however, is not an indication of much value. I will add here that, in the opinion of Mr. Hughes and of Mr. Chadwick, there are several nearly parallel gravel streams leading down obliquely from the direction of the main ridge, which did not all concentrate at one point. The data which I collected at New York Cañon, and which will be given on a subsequent page, have some bearing upon this view of the question.

The two sections on Plate O, Figs. 2, 3, are drawn on the same scale as the accompanying map, or nearly so. The north and south section (Fig. 2) extends from the South Yuba River to a point high up on the ridge above the Blue Tent House. It will be seen that if the bed-rock is level from the point " B," the gravel must be as much as 650 feet thick at the point where the house of the superintendent stands. I do not think that the gravel will prove to be of that thickness, for it seems most likely that the bed-rock will begin to rise in harmony with the general slope of the ridge before that point is reached.

The east and west section (Fig. 3) has one remarkable peculiarity. It shows a profile across the northern end of the deposit, where there is a " high channel " between Gopher Ravine and Johnson Creek, with its bed-rock 145 feet higher than the bed-rock at Gopher Hill. This high channel is said, furthermore, to have carried granite boulders, such as have not been met with yet at other places in this neighborhood.

In this connection another diagram (Plate O, Fig. 4) may be given to show a section, north and south, through the Sailor Flat mine to the centre of the ridge. I do not have the necessary data to make the section an accurate one. The heights and distances are estimated, but are fair approximations to the truth. The scale is 500 feet to the inch. The point A is the point on the rim at which the observation for altitude was taken. The surface on the flat bench above the gravel bank is covered with volcanic débris for a thickness of from five to twenty feet, and there is probably gravel underneath it. The main lava ridge rises to an altitude of about 400 feet higher than the upper bench.

The water for these mines is brought from the South Yuba River. The Blue Tent Company owns a ditch which, we have already seen, extends above Omega.* The Sailor Flat Company buys from the Blue Tent Company, or from the South Yuba Water Company, which has an office at Nevada City. At Sailor Flat 1,300 inches of water are used per day, under 300 feet pressure. The nozzles are five and six inches in diameter. At the Blue Tent mines the ordinary nozzles are from six and a half to seven or eight inches in diameter. Nozzles of nine inches in diameter have been sometimes employed.

Since the organization of the English company at Blue Tent, careful records have been kept of the amount of water used, the amount of gravel moved, and the yield of gold. Through the kindness of Mr. Hughes I was allowed to take the following data from the company's records : —

(1.) Between Sept. 1, 1876, and Aug. 15, 1877, the gravel removed at the South Yuba claim (Gopher Hill) amounted to 632,533 cubic yards, which was an average of $5\frac{83}{100}$ cubic yards per twenty-four-hour inch of water. The yield was $12\frac{6}{10}$ cents per cubic yard of gravel.

(2.) For the year 1878, at the same claim, the gravel removed amounted to 501,028 cubic yards, or an average of $6\frac{7}{100}$ yards per twenty-four-hour inch of water. The yield was 14 cents to the yard.

(3.) In 1878, at the "Blue Lead," there were removed 235,703 cubic yards of gravel, an average of $5\frac{32}{100}$ yards per inch of water, which yielded 7 cents per cubic yard.

(4.) From September, 1876, to August, 1877, there was a very large quantity of top gravel moved at the Enterprise ground. The gravel was fine, loose, and sandy, and easily moved. The amount removed was 1,398,963 cubic yards, at the rate of $10\frac{12}{100}$ cubic yards to the inch of water, but the yield was only $2\frac{6}{100}$ cents to the yard, barely sufficient to cover expenses.

(5.) In 1878, at Gopher Point, a mass of hard, indurated, clayey gravel, which could neither be washed nor blasted with ease, was removed at the rate of only $2\frac{38}{100}$ cubic yards per daily inch of water.

From the above figures it will be seen that the average yield has not been quite as much as was estimated at the time of the formation of the company. In Raymond's report for the year 1873, page 115, it is stated that up to that time there had been removed, according to the surveys of Mr. Bradley, 5,101,150 cubic yards of gravel. The yield up to that time was estimated at $770,000, equivalent to fully 15 cents per cubic yard, — and that to a large extent from the upper strata of the gravel, at some considerable distance from the bed-rock.

Mr. Hughes's estimate of the probable loss of gold in the hydraulic washings is fifteen per cent.

I have already † alluded to the difficulty of determining the precise position of the old channel at Blue Tent, and to the hypothesis adopted by Mr. Hughes, and I might have mentioned at the same time another view which is held by some well-informed persons, namely, that the channel extends under and across the ridge to a connection, by way of Scott's Flat, with the enormous deposits of gravel at Quaker Hill and You Bet. I was not able to make any personal examination of the intervening ground, but from what I could learn by inquiry of persons acquainted with the country, and from what I saw at Blue Tent and Quaker Hill, taken in connection with the relative altitudes of the bed-rock at the two places, I feel confident that no such channel exists. That there was, however, once a connection between the Blue Tent gravel and that at Columbia Hill, and, in general, that on the divide between the South and Middle Yubas, cannot be denied. The identity of bed-rock level at Gopher Hill and Grizzly Hill has already been alluded to. There is also a practical identity in the level of the upper surfaces of the gravel on the two sides of the South Yuba. Columbia Hill and Blue Tent are in plain sight of each other with no intervening

* According to Raymond's report for 1873, page 115, the Blue Tent Company brings its water through $27\frac{1}{2}$ miles of ditch from Culbertson's Bridge on the South Yuba River.

† See *note*, p. 411.

ridge or mass of bed-rock to obstruct the view. It seems most probable that this portion of the gravel field represents a broad estuary or lakelike expansion of water at the junction of two streams, or where two streams, by the filling up of their channels and the covering of the low intervening ridges, became practically one. If this latter view is correct, it is not impossible that there may once have been a current from Grizzly Hill towards Columbia Hill, even if the slope of the deep bed-rock is in just the opposite direction.

Above Sailor Flat there are other deposits of gravel whose relations I did not take the time to study in detail, though I was very glad to take advantage of Mr. Chadwick's kind offer to spend a part of a day with me in a visit to some claims in New York Cañon, about two miles east of Sailor Flat. Beyond this, and distant about half a mile, is Meeker's Cañon, which is followed, a mile farther east, by Lane's Cañon.

A complete profile section of the ridge from the river to the crest of the lava would show, first, a quite steep ascent, for two thirds or three quarters of the way, to a bench or flat, a half-mile or more in width, across which the ascent is much more gradual, and then a second steep pitch to the top. Along this bench or flat there are some surface indications of the existence of gravel below. Some of the indications are seen in the ditches, which are for the most part in the volcanic tufa, but with occasional exposures of clay and fine gravel, lying underneath the tufa. The marshy and swampy character of portions of this bench, especially in the winter, point to a substratum of clay, such as is usually found accompanying the gravel. The present surface dirt is to a great extent a volcanic wash which prevents any immediate examination of the supposed clay and gravel excepting at some expense. The wash is not a part of the main lava-flow, for there is a considerable belt of bed-rock here and there visible to the south of the bench or flat before the steep lava-cap is reached.

In the year 1875, or near that time, Mr. Chadwick, in company with others, began the sinking of a shaft in New York Cañon at a point below all the ditches and fully half a mile from the crest of the ridge. The mouth of the shaft is near to and on the left bank of the water-course in the cañon. I made the altitude of this point to be 3,149 feet. The first twenty feet of the shaft were in volcanic cement; below this there was as much as a hundred feet or more of clay and sand, with no gravel to amount to anything; and after this a stratum of clean, washed quartz gravel was penetrated for thirty feet without reaching bed-rock. The inflow of water was so great that further sinking was given up for the time. The position of this gravel is peculiar in this respect: to the west of the shaft there is high bed-rock, at least seventy-five feet higher than the mouth of the shaft, which shows that the gravel cannot be regarded as a portion of any deep-lying east and west channel. Possibly it has come into its present position through the action of the New York Cañon waters alone, though its depth from the surface affords a strong argument against that supposition.

Easterly from this shaft, at a distance of nearly a quarter of a mile, on the spur between New York Cañon and Meeker's Cañon, I saw rolled quartz gravel on the surface, some sixty feet higher than the mouth of the shaft, and Mr. Chadwick assured me that similar gravel could be seen at a point a quarter of a mile farther to the southeast, and at an altitude of from fifty to seventy-five feet above the point last mentioned, but I did not have the time to extend my observations any further in that direction. A mining company, of which the secretary is Mr. E. W. Bigelow of Nevada City, holds a claim of 500 acres in this vicinity, which will probably not be developed as long as the banks lower down need the amount of water they do at present.

I will close this section of my report with a few notes upon Round Mountain, which were given me by Mr. Hughes, who is personally interested in the gravel mines at that locality. Round Mountain is a lava-capped knob about three miles to the southwest of Blue Tent. The "channel" is on the northerly side of the ridge, and is 340 feet higher than the bed-rock at Gopher Hill. It has been explored by a shaft and by a tunnel. The shaft reached bed-rock at the distance of eighty-five feet from the surface. The first thirty feet were sunk through "soil"; the remainder through layers of coarser or finer pipe-clay and sandy gravel. The tunnel is at the lower end of the mountain. The pay stratum is about eight feet thick; it is made up of quartz which is very little worn, and it carries gold which is not much abraded. There is also some hard sand intermixed with quartz. The course of this channel is given as a little south of west.

SECTION V. — *The Gravel Deposits between Indiana Hill (near Gold Run) and Quaker Hill.*

Following the order which I at first proposed, I should have to defer to the last any consideration of the Gold Run, Dutch Flat, and You Bet gravels, but it seems best now to bring the matter in at this point, and thus complete all that I have to say about the gravels of Nevada County before passing to the more northern counties.

I postponed my trip to this region until after the other districts assigned to me had been visited, because I was already familiar, from my work in 1870 and 1871, with its main features. I reached Dutch Flat on the evening of the 15th of November, and left there for San Francisco on the 23d. During the week I reviewed the most of the ground from Indiana Hill to Quaker Hill that I had gone over in the former exploration, and spent as much time as possible at points where there was reason to expect new features of interest.*

The extracts from my former report, given on pages 143 to 181 of this volume, I am willing to let stand without essential alteration.† The statements there made are substantially correct, or were so at the time of writing, and no new light has been thrown upon the subject by subsequent mining operations which will require me to modify my former opinions, unless it be in regard to some points near Quaker Hill.

The principal changes which have been brought about in the last nine years in this vicinity are : —

(1.) The uncovering of deep bed-rock at the Indiana Hill and Cedar claims, near the southern end of the district ;

(2.) The extension of the washings between the railroad and Dutch Flat Cañon, so as to show more bed-rock than formerly, particularly in or near the Jehoshaphat claim ;

(3.) The removal of gravel to deep bed-rock at two claims on Dutch Flat (Gray's) Hill, the Polar Star, and the Southern Cross claims ;

(4.) The nearly complete removal of the Plug Ugly Hill gravel ;

(5.) The removal of great quantities of gravel at Little York, so as to make a connection through from the old Little York banks to those at Empire Hill ;

(6.) The extension of mining on Chicken Point toward the Sugar Loaf and Chalk Bluff ;

(7.) The washing away of the site of the old town of You Bet and the uncovering of extensive areas of bed-rock along the western rim ;

(8.) The enlargement of the opening at Waloupa ; and

(9.) The developments at Hunt's Hill and Quaker Hill.

These changes, together with a few points supplementary to what has already been published, form the main topics of the notes I collected during my week's stay. The changes and developments upon the ridge above Little York, in the neighborhood of Liberty, Lowell, and Remington Hills, I did not have time to pay any attention to.‡

---

* Several changes of names and of position have been made on the General Gravel Map, in accordance with the results of Professor Pettee's observations of the year 1879. The previously unpublished map of the Gold Run, Dutch Flat, Little York, and You Bet districts has also been corrected by him, and is now given with this portion of the Auriferous Gravels. To the map in question reference should be made, in order that the present section may be understood ; and it may also be used in connection with the previous account of this portion of the Gravel Region. (See *ante*, pp. 143 – 174.) — J. D. W.

† Except as to certain small errors of the press, the corrections for which I have marked upon the proof-sheets at Cambridge.

‡ For a map of the ridge above Little York I am indebted to Mr. D. W. C. Morgan, the superintendent of the Little York Company.

The bed-rock exposures are not yet sufficient in number or extent to put to rest all doubts in regard to the former position of the old channel or the mutual relations of the several channels, if more than one existed. Conflicting views are still held among the miners, which it would be useless for me to attempt to harmonize. I can only record the facts as I found them to exist on the ground, and leave to the future the settlement of contradictory theories.

The bed-rock exposed to view is, for the most part, an easily cleavable slate, with a southeasterly strike and a nearly vertical dip. The strike of the planes of cleavage at Indiana Hill is S. 45° E.; at the Polar Star claim, Dutch Flat, S. 35° E.; at Waloupa, S. 28° E.; and at Hunt's Hill, S. 20° E. (all magnetic). The bed-rock, however, is not entirely uniform in character. At Indiana Hill there is a portion which is less distinctly cleavable than the remainder, and considerably harder. At the Polar Star claim much of the rock has no very distinct cleavage, but is full of joints and seams. In some places masses of one variety of rock ten to fifteen feet in length by two or three feet in width appear to be enclosed, like "horses" in mineral veins, within other rock; some of the rock is quite hard, while other portions are so rotten that they can be broken up easily with the sledge. At Plug Ugly Hill the bed-rock is a slate, of a yellowish-brown, or, in places, greenish color, soft, and rapidly weathering to a clayey mass when exposed to the air. Some of the greenish rock, where weathered, shows a peculiar structure, — a structure, however, of which I saw a number of striking examples at the gravel mines of Plumas County. For a depth of from three to eight feet the slate has become converted into a succession of layers of reddish and brownish ferruginous clay, through which the vertical cleavage-planes of the still undecomposed slate are frequently distinctly traceable. The clayey mass itself looks as if it were regularly stratified, parallel with the inequalities of the original bed-rock surface. The rock is of such a fragile character, and the clay is so soft, that it was not possible to take specimens.

At the Camden claim, Hunt's Hill, there are two kinds of bed-rock, one considerably harder than the other, though both belong to the slate series. (See specimens Nos. 174 and 175.) The plane of junction between the two is nearly horizontal, the westerly dip amounting to only 10°. At one point in the claim the soft rock was seen to overlie the hard variety, as shown in the diagram (Plate O, Fig. 5), the seam between them being about an inch in thickness, and containing small fragments of quartz. The whole appearance is that of a fault; but I could not satisfy myself in regard to its age, whether it was older or younger than the gravel deposit. The hard rock has been worn into pot-holes and fantastic shapes since the removal of the gravel, and the rate of wear has been very rapid. At first sight it seemed to me impossible that so much erosion could have been produced since the rock was uncovered, and I thought I had been fortunate enough to find a fault of more recent age than the gravel, which might be of service in the discussion of some of the doubtful points connected with the position of the old channel. But the removal of a small quantity of the soft rock, so as to bring to view a surface that the sluice waters had not acted upon, showed that the original surface of the hard rock had not been water-worn, or that the evidences of wear had been obliterated at the time the rocks were faulted.

I took fresh barometric observations at or near seven of the points of which the altitudes were determined in 1870, though under conditions less favorable for accuracy than in that year. On the whole I was very well pleased with the closeness of the agreement of the different determinations, but I assign much the greater weight to those of 1870. The old and the new results for these seven cases were as follows : —

	1870.	1879.	Difference.
1. Point near southwesterly end of Plug Ugly Hill	3,072	3,090	18
2. Deep bed-rock in Little York mines	2,706	2,755	49
3. Point near site of store at Little York	2,839	2,862	23
4. Empire Hill bed-rock	2,666	2,693	27
5. Waloupa bed-rock	2,594	2,595	1
6. Bed-rock at Niece and West's old claim	2,625	2,613	12
7. Point near high flume, Sardine Flat	2,911	2,903	8

In regard to these stations a few words of explanation are needed : (1) the point visited on Plug Ugly Hill in 1879 was not the same as the 1870 station, — in both instances I looked for the average westerly rim-rock only ; (2) a similar remark may be made about the Little York bed-rock ; (3) the Little York store of 1870 has been destroyed by fire, and my point of observation in 1879 was a few rods above the old site, perhaps six or eight feet higher in altitude ; (4) the two points at Empire Hill were certainly not the same, though not far apart ; (5), (6), and (7) in the last three cases there was a very close approach to identity of position.

I tried also to get a fresh value for the altitude of the old cement mill at the Indiana Hill outlet, but could not fix upon the exact spot at which the old observation was taken. The altitude in 1870 was made to be 2,792 feet. In 1879 I made the altitude of the bed-rock at the Indiana Hill claim, about a quarter of a mile northerly from the old mill, to be 2,779 feet. If this determination could be trusted to within a few feet, it would indicate a downward grade of the bed-rock towards the north, in opposition to the general pitch of the rock between Indiana Hill and Dutch Flat, which is certainly towards the south, as will be seen later. There is also some evidence of a northerly direction for the channel at this point in the way that the pebbles of the gravel lie against each other ; and, indeed, within a distance of 150 feet, the rock exposed to view does fall to the north as much as six or eight feet. On the other hand, as I was informed by Mr. J. L. Gould, the superintendent of the Gold Run Ditch and Mining Company, there is a rise of bed-rock again, amounting to seven feet, before the Cedar claim, eight hundred feet farther north, is reached. My own observations made the Cedar bed-rock 2,784 feet above sea-level, or five feet higher than that at Indiana Hill. Mr. Gould told me further that the bed-rock pitches again to the south between Indiana Hill and the old mill. Mr. Uren's line of levels (see further on) still leaves some doubt as to the slope of the bed-rock between the Cedar claim and the old cement mill, the uncertainty perhaps arising from bad choice of terminal points. There is no great difference of level at best.

The next point, in a northerly direction, at which I took an observation for altitude was on the bed-rock exposed in or near the Jehoshaphat claim. I made its altitude to be 2,950 feet, but, according to Mr. E. C. Uren's statement, I was from seventy-five to one hundred feet above the lowest portions of the Dutch Flat gravel, the altitude of which, therefore, ought to be not far from 2,862 feet, or eighty-three feet above the bed-rock at Indiana Hill, — a value which is in very close accord with that obtained by Mr. Uren, who, by a recent spirit-level survey, made the deep bed-rock at the point where the old channel crosses Dutch Flat Cañon, in front of the Waukegan claim, to be seventy-seven feet higher than the mouth of the tunnel at the old mill. These points are nearly three and a half miles apart, and the average grade of the channel is but little more than twenty feet to the mile ; less than a quarter of that of the old stream on the San Juan divide.

The average rim-rock at the outlet of Thompson Hill, which Mr. Uren says is probably a little lower than the point he chose for a starting-point, I made in 1870 to be 2,848 feet above sea-level, or only fifty-six feet above the site of the old mill, according to the measurements of that year. From all these determinations it is clear that the Indiana Hill outlet is certainly lower than the lowest bed-rock at Dutch Flat, but not so much lower as we should expect when we take into account the distance between the places.

In striking contrast to this low grade is the steepness of the pitch along the long axis of Dutch Flat, or Gray's, Hill. Since 1870 deep bed-rock has been laid bare at two points on this hill, one nearly opposite the centre of the town, — the Southern Cross claim ; and the other a few hundred feet farther to the northeast, — the Polar Star claim. The altitudes of the bed-rock at these two claims I made to be, respectively, 3,054 and 3,075 feet ; the latter, therefore, being over two hundred feet higher than the low spot on Thompson Hill, scarcely a mile distant. There is no known evidence of a fault in the strata, and nothing in the appearance of the gravel indicates rapids or cascades. Further mining operations may be expected to bring to light some interesting features in regard to the position of the bed-rock.

Fig. 4

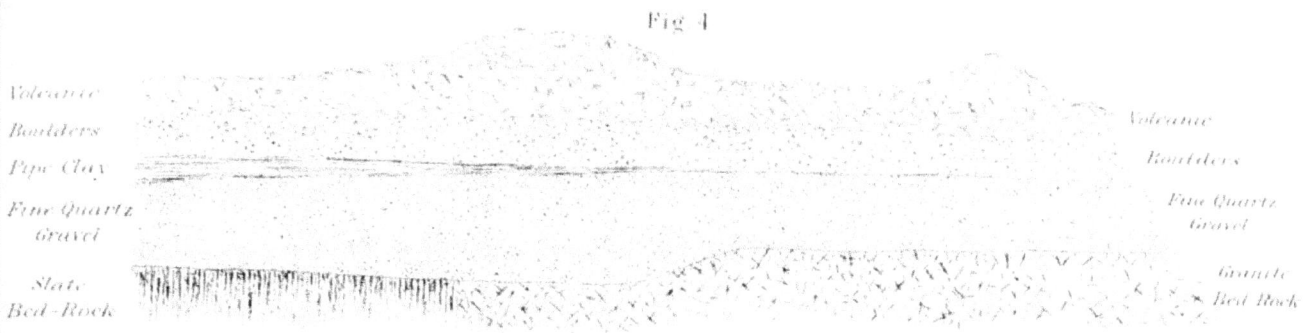

Volcanic
Boulders
Pipe Clay
Fine Quartz
Gravel
Slate
Bed-Rock

Volcanic
Boulders
Fine Quartz
Gravel
Granite
Bed Rock

LONGITUDINAL SECTION AT INDIAN HILL, SIERRA CO.

(Scale: Two Hundred Feet to one Inch)

Fig. 3.

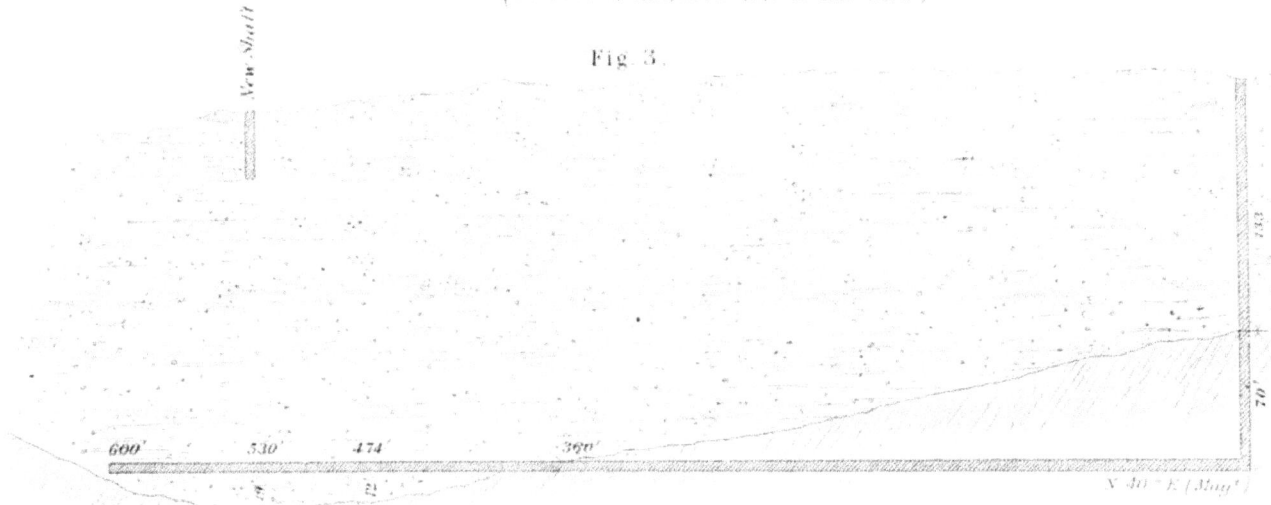

New Shaft

733'

70'

600'   530'   474'   360'

S. 40° W. (Mag')

N. 40° E. (Mag')

SECTION ACROSS QUAKER HILL CHANNEL.

(Scale: One Hundred Feet to one Inch)

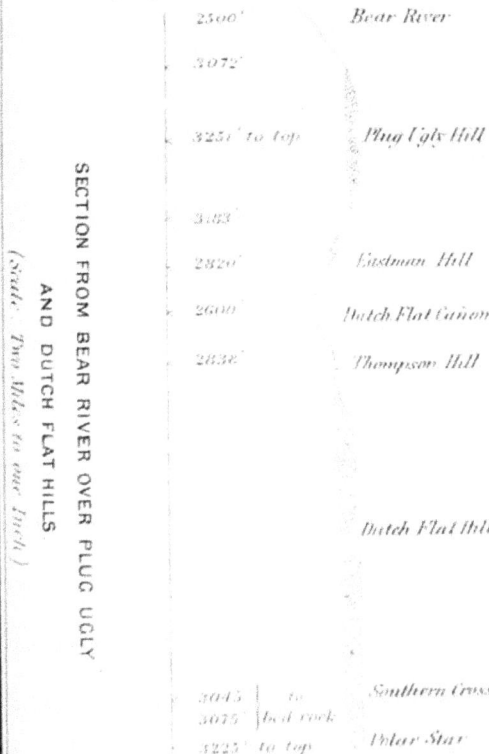

2500'	Bear River
3072'	
3251' to top	Plug Ugly Hill
3483	
2820'	Eastman Hill
2600	Dutch Flat Cañon
2838	Thompson Hill
	Dutch Flat Hill
3045 to	Southern Cross
3075 bed rock	
3225 to top	Polar Star

Fig. 1

SUGAR LOAF

Volcanic
(Tufa)

Fig. 2

Chalk
Yellow Sand
Green Gravel
Sand
Sand
and Gravel
Bed Rock

SECTION OF GRAVEL AT CHICKEN POINT

(Vertical Scale: One Hundred Feet to one Inch)

J. Bien lith N.Y.

The bed-rock at Plug Ugly Hill stands at a considerably higher level than that at Thompson Hill, on the opposite side of Dutch Flat Cañon, or that at Eastman Hill, on the same side of the cañon. The general surface of the Plug Ugly Hill bed-rock slopes off rapidly to the south or south-west, but at its northerly end it is over 300 feet above the level of Eastman Hill. The relations of these hills to each other are shown in the accompanying diagram-section (Plate P, Fig. 1). Plug Ugly Hill, therefore, cannot be considered as lying in the course of the old deep channel. Its bed-rock, indeed, presents neither a trough-like channel nor any very even surface. At the northern end of the hill the highest bed-rock is on the western side of the centre of the gravel deposit, with at first a gradual and then a steep pitch toward Dutch Flat and Squire's Cañon. At the southern end the transverse slope is in the opposite direction, the point of highest altitude being on the eastern side overlooking Squire's Cañon. The fall to the west within the limits of the gravel is as much as fifty feet. It will be seen from this description that the crest of the bed-rock ridge, under-lying the gravel, runs obliquely to the direction which the gravel seemed to follow before it was washed away.

The mining operations at You Bet since the year 1870 have thrown but little new light upon the difficulties which beset the problem of determining the position of any ancient channel between the outlets on the Steep Hollow side of the ridge and the mines at Red Dog. The principal diffi-culties have been already stated,* the chief among them being the lack of grade. A recent spirit-level survey by Mr. Uren corroborates entirely the conclusions drawn from our former series of barometric measurements. Mr. Uren selected for his starting-point the exposed bed-rock at Brown's claim, near the drain tunnel on Wilcox's Ravine, about 150 feet from the old incline, which has been allowed to cave in, and for his terminal point the bed-rock at the Red Dog mines. These two points have precisely the same level. The bed-rock at Brown's claim, according to Mr. Uren's surveys, was only four feet above that at Niece and West's ground, the level of which the baro-metric measurements made to be the same as that at Red Dog.†

On the west rim of the deposit, near the former site of You Bet, considerable bed-rock has been recently uncovered, which is seen to have a remarkably steep pitch to the east, disappearing under the gravel which still remains. I examined the locality with considerable care, but could find no evidence of a fault. Both the rock and the adjacent gravel are badly disturbed and broken up, though not more so than might be expected as the result of surface-slides at the time of or since the washing away of the main mass of gravel.

The course of the deep channel between Red Dog and Quaker Hill must be looked for in the direction of Hunt's Hill. Unfortunately, at my second visit as well as at my first in 1870, the conditions were unfavorable for getting precise and accurate information. While my barometric observations I am confident can be trusted within narrow limits of error, yet they could not be checked in such a way as to determine satisfactorily such slight differences of altitude as prevail in this neighborhood. At Hunt's Hill I made the altitude of the bed-rock in the Camden claim to be 2,620 feet, or five feet lower than at Red Dog. Above Hunt's Hill the bed-rock has either never been uncovered or it is now so deeply covered with tailings as to be inaccessible. At Quaker Hill a deep shaft has recently been sunk through the gravel to bed-rock. My observations were taken at the mouth of a new shaft, near the deep one just mentioned, and consequently at a known dis-tance above bed-rock, and at a distance of about a mile and a half above the Camden claim. The closest approximation that I can give for the altitude of bed-rock at this point is 2,650 feet. To settle the question of bed-rock grade in this vicinity careful spirit-level surveys ought to be made. Certain it is that the low bed-rock under the gravel between Quaker Hill and Indiana Hill, a dis-tance of about eight miles in a straight line, shows remarkable anomalies of position. These appear in a striking manner when presented in tabular form. The altitudes of deep bed-rock are as follows : —

* See *ante*, pp. 164–174.
† There is probably good reason for rejecting the value, enclosed in brackets, given on page 171, as the altitude of Brown's bed-rock.

Quaker Hill	2,650 feet.
Hunt's Hill	2,620 "
Red Dog	2,625 "
Niece and West's, You Bet	2,625 "
Waloupa	2,594 "
Little York	2,706 "
Thompson Hill	2,848 "
Indiana Hill	2,792 "

It does not seem possible that there was ever a deep channel flowing in either direction between the two extreme points. Dutch Flat, or Thompson Hill, must have stood at a parting of the ways; and it is very probable that there was another such parting between You Bet and Red Dog.

I formerly expressed the opinion, based upon the hasty and incomplete examination of 1870, that the Quaker Hill gravel would be confined to the south side of the ridge, and that there was no connection through, under the lava, with the gravel deposit at Scott's Flat, on Deer Creek.* I am now sorry to be obliged to say that at my second visit also I was so much restricted in time that I was not able to extend my observations to the north side of the ridge, nor to make any such survey as would be decisive. But since it has been proved by shaft-sinking and drifting that the deep bed-rock at Quaker Hill is even deeper than any of our former estimates made it to be, it seems to me more probable that the Scott's Flat and Quaker Hill gravels are really parts of the same deposit, and that ultimately a connection between the two places will be established. In harmony with this view I have represented the gravel as continuous through the west gap at Quaker Hill on the manuscript map of this region; for, even if the actual surface dirt is of a tufaceous character, the gravel or the superincumbent clay may be expected at a depth of a few feet only. This opinion, however, is not one in support of which I can give many reasons based upon my own observations in the field, and it does not carry with it any belief in the existence of a channel between Scott's Flat and Blue Tent, for which the advocates of the "blue lead" theory are apt to contend. A further survey of this part of the ridge is very much needed.

I have on a previous page alluded to the "deep channel" between Red Dog and Quaker Hill. There is also a second line of gravel deposits lying to the east, skirting the base of Chalk Bluff, and from three to four hundred feet higher than the deep channel, which demands attention. At You Bet the gravel is continuous across the channel from the site of the town to the high bed-rock at Chicken Point, but, to the north of Missouri Cañon, the high gravel is separated entirely from the low, the gravel of Darling's, Boston, and Buckeye Hills being quite distinct from that of Red Dog and Hunt's Hill. I made a careful examination of this portion of the gravel from Chicken Point to Buckeye Hill. At both these localities there are some evidences of the existence of a "high channel." The bed-rock along a line which can be traced for the greater part of the way with greater or less distinctness seems to have a slight pitch to the east, and thus to form a west rim. Along this rim at Chicken Point there are large quantities of blue quartz boulders, some of them from six to ten feet in diameter, not much washed; they differ very markedly in character from the other boulders of the neighborhood. The eastern rim of the high channel must be looked for near or under the bluff, where there is at the present time a continuous bank from the farther end of Chicken Point nearly to Hussey's mine, with the exception of one small projecting point of unwashed gravel near Timmons's house. On Buckeye Hill the gravel has been nearly all removed back to the face of the lava bluff. As previously stated,† the evidence in favor of any original high channel of well-defined character is not very strong, though it is not at all unlikely that at some time during the filling of the basin with gravel one of the shifting currents may have occupied this position.

Since the year 1870 there have been several changes in the ownership of gravel property in this district, and similar changes may be expected from year to year. It did not seem worth while

---

* See *ante*, pp. 179, 180.   † See *ante*, p. 170.

to attempt to collect detailed information upon this point, and I cannot give any list of incorporated companies or other mining organizations. The tendency is towards the consolidation of individual claims in the hands of large companies, with sufficient capital to control both the gravel and the water supply.

The published portion of my former report contains but little in regard to the lithological character and the stratigraphical features of the gravel of this region and its included boulders. The removal of the gravel to bed-rock at so many new places enables me now to present additional information upon these points, and, incidentally, upon the results of some of the more recent mining work.

The deep gravel at Indiana Hill, which is a part of the property belonging to the Gold Run Ditch and Mining Company, has been worked through a deep bed-rock tunnel from Cañon Creek. The main tunnel was driven 650 feet in the direction of the old 1849–54 shaft in Potato Ravine, which it will take 1,600 feet more to reach; and then a branch tunnel, 1,600 feet in length, was driven to Indiana Hill, which at its extremity was ninety feet below the deep bed-rock. A section of bed-rock, about 100 by 150 feet in dimensions, has been exposed and roughly cleaned up; and the gravel has been removed nearly to bed-rock over an area of about 500 feet square, or nearly six acres. For from four to six feet above bed-rock the gravel is a very hard blue cement, so hard that it does not get broken up by the water in the sluices, and consequently carries gold away to the cañon. On exposure to the weather, however, it gradually disintegrates and can be more easily moved. This cement Mr. Gould, the superintendent, proposes to treat in special mills to be built in the mine and be run with hurdy-gurdy wheels. Above this hard cement come from forty to sixty feet of a cement which can be easily worked with the aid of powder. Including these two strata, the blue gravel is about 200 feet in thickness. Like the other blue gravel of the district, this becomes rapidly red in color when exposed to the air. In this gravel there is an exceptionally small number of large boulders. Those which are met with represent several varieties of metamorphic rock, and the spaces between them contain a considerable quantity of quartz pebbles. Above the blue gravel comes the white or red variety, extending either to the original surface, or to the bottoms of the old workings, where the upper bench has been already removed. The top of the present highest bank is fully 400 feet above the deep bed-rock. The width of the channel a short distance from this point was estimated at 1,200 feet. The gravel of Indiana Hill is also remarkably free from strata, or even streaks, of clay. Here and there are local spots of a bluish sand, from a few inches to three feet in thickness, but not traceable in a continuous layer. Taken all in all, the Indiana Hill bank must be selected as the best specimen of a clean, uniform gravel deposit that I saw anywhere in the State.

At the Cedar claim an area of bed-rock of about 150 by 200 feet has been uncovered, which shows a decided pitch towards Indiana Hill. The gravel is similar to that at Indiana Hill, though there are more large boulders, particularly on the western rim.

The two claims on Dutch Flat Hill in which bed-rock has been reached — the Southern Cross and the Polar Star — have been opened by deep tunnels from Bear River, and at both places there are now deep, funnel-shaped excavations, in which can be seen about 150 feet in thickness of blue gravel at the bottom, and fifty feet or more of white quartz gravel above, reaching to the present surface of the ground. The original top gravel was washed away many years ago. At the Polar Star mine I saw several patches of red gravel, within the limits of the blue stratum, which had very peculiar shapes. They were not simply filling old depressions in the blue gravel, but were entirely surrounded by the blue, which by itself would not stand in place without supports if the red portions were removed. The blue gravel turns red very rapidly when water runs over it, as, for instance, where the waste water from the surface is allowed to run down the bank. The boulders in the blue gravel, like those at Indiana Hill, are almost exclusively of metamorphic rock; a piece of quartz as large as one's head is rarely seen. The interstices between the boulders, however, are almost always filled with white quartz. These blue boulders are large in size, up to ten or twelve feet in diameter, and pervade the whole of the blue gravel. They show plentiful signs of wear, but are not very much rounded. In some cases I saw broken and angular fragments of slate rock lying close by the side of smooth quartz. Some of the boulders at this mine are peculiar in composition. Some are conglomerates or breccias, and one, as Mr. Colgrove told me, which was found a hundred feet above bed-rock, was a mass of cemented gravel, with a smooth and rounded surface. As would be expected, where boulders are so common, clay streaks are very rare.

The white gravel above the blue stratum is made up almost exclusively of quartz. It would hardly be an exaggeration to say that nothing but white quartz is to be seen upon the present irregular surface of Dutch Flat Hill. The masses vary in size from an inch or less in diameter up to eight or ten feet. They are usually smooth, and in some places very much worn, though generally not enough to obliterate the original irregularities of surface. This quartz shows very frequently hollows, or vugs, containing good crystals and crystal facets. As a rule, it is barren of gold.

The gravel on Plug Ugly Hill was a white quartz, like the upper stratum at Dutch Flat, containing in addition a small quantity of easily decomposed pebbles, probably of slate.

At Little York the deep gravel carries some extremely large boulders, partly of white quartz, partly of blue quartz, and partly of a highly silicious rock which shows indistinct slaty characters. This rock is quite unlike any of the bed-rock of the vicinity, which is an easily cleavable slate. On the Empire Hill bed-rock there are also large piles of boulders of white or blue quartz together with two or three varieties of bluish slate, and in the gravel, at a considerable height above bed-rock, above much of the fine gravel, there are large angular boulders, which show but little trace of wear. It is difficult to account for all these phenomena by the action of running water alone, and it is also difficult to account for the local accumulations of special varieties of boulders. The presence of such large boulders in unusual quantity must have interfered very seriously with the regularity of the hydraulic washing.

The opening at the Waloupa bank is very much larger than it was in 1870, extending over an area of at least 800 feet in length, in a direction S. 40° E. (magnetic), by 300 in width. There is no bed-rock exposed excepting at the lower end of the mine. At the upper or northwestern end the top gravel alone has been worked, and the bank is about one hundred feet in height. At the lower end a little blue gravel appears near the bed-rock. The upper gravel is reddish in color and it carries considerable washed quartz, though composed principally of an easily decomposed metamorphic slate and other allied rocks. There is some sand, but no pipe clay of any consequence.

The principal mining operations at You Bet and Chicken Point are now carried on by the Nevada Hydraulic Mining Company, which owns in this neighborhood nearly one thousand acres of mining ground. The upper gravel — that at Chicken Point — is composed of white and blue quartz with some admixture of slate and other easily decomposed rock. There are a few irregular sand streaks, but there is no pipe clay, no cement, no blue gravel. Black sand occurs abundantly. The gravel has been washed away back to the base of the Sugar Loaf, under which it appears to take the shape of a wedge, which will soon come to an end if work is pressed in that direction. The face of the bank shows several distinct layers, the thickness of which I estimated by the eye alone, without stopping to make accurate measurements. Beginning at the bottom there are in succession strata of

	feet.
Gravel and sand	20
Sand	25
Green gravel	15
Yellow sand, or clay	12
Volcanic ash (the so-called "chalk")	40

The top of the Sugar Loaf is a hundred feet or more above the "chalk." The green gravel stratum is distinctly traceable along the face of the bank under Chalk Bluff. The diagram (Plate I', Fig. 2) shows the probable position of the gravel under the lava.

The Buckeye Hill gravel is practically exhausted, only a few Chinamen being at work there at present. The bed-rock is nearly level, but rises gradually towards the northeast. There is, however, a central depression, amounting to perhaps thirty feet, which may be looked upon as the representative of the old high channel. It is a little curious that there should be found on the western rim of this depression peculiar blue quartz boulders, like those on the supposed western rim at Chicken Point, which are not seen in the centre of the depression or on its eastern rim. The Buckeye Hill gravel appears to have been a rolled, white, opaque quartz, like that at Chicken Point, or the top gravel at Dutch Flat. There are also some sand streaks visible in the blocks of gravel left standing. At the upper end of the hill, under the bluff, there is a bank of about sixty feet in height, but the gravel has thinned out, and the pipe-clay has come in above. There are also layers of vegetable matter and many signs of former shifting currents in the arrangement of the layers and streaks of clay, sand, and gravel, — such as might be expected along the margin of an extensive lake.

Between the high gravel of Buckeye Hill and Greenhorn Creek there is nothing but a bed-rock slope, but on the opposite side of Greenhorn the low gravel of Hunt's Hill and Quaker Hill approaches very nearly to the present bank of the creek. Before the accumulations of tailings in the creek, there was a more marked eastern rim to the Hunt's Hill gravel and a steep slope down to the creek's bed. On the west, the bed-rock rises very rapidly, and forms an unmistakable rim to act as a barrier for the gravel in that direction. The Camden claim was the only one I visited at Hunt's Hill, where I had the advantage of the company of Mr. W. H. Wiseman, one of the owners of the claims. I have already referred (page 417) to the peculiarities of the bed-rock at this mine. A peculiar feature of the gravel here is, that it is cemented on the hard bed-rock, but not on the soft. On the western side of the claim, between two spurs of bed-rock, there is also a large body of pipe-clay, which reaches nearly or quite down to the soft bed-rock. A few large quartz boulders are seen in the gravel at this claim. On the Greenhorn side of the gravel there is one anomalous but interesting feature, which I will briefly describe. Lying near, but not on the bed-rock, there is a peculiar stratum about twelve feet in thickness, seventy-five feet in width (as has been proved by cutting across it), and over one hundred feet in length; one end is exposed to view, but the other is hidden under the gravel bank. The material of this stratum is broken and angular bed-rock, — "float bed-rock" as it is called, — similar to the ordinary cleavable slate of the country, with the cleavage planes arranged in no definite order in the different pieces, mixed with well-washed gravel. Above and apparently coterminous with this strange nest of material there comes a layer of fine sand from four to six feet thick. No ready explanation for the occurrence presented itself.

The gravel at Hunt's Hill used to be worked as drift diggings, and, according to Mr. Wiseman's statement, the deep gravel was extraordinarily rich, in places yielding as much as ten dollars for each square yard of bed-rock uncovered. The water for the hydraulic mining is brought from the Yuba River.

The extensive gravel mines at Quaker Hill are still owned, as they have been for many years, by Messrs. Jacobs and Sargent. The property extends from Hunt's Hill on the south to Scott's Flat on the north. The principal working place at the present time is near the head of Green Mountain Cañon. The cañons and ravines through which the Quaker Hill mines should find their natural outlet are now filled to a great depth with tailings. In some places, as, for example, near the old Empire Mill, on Gas Cañon, the tailings reach nearly to the top of the old hydraulic banks, where mining has been given up for several years. This filling of the cañons practically puts an end to hydraulic operations, excepting for the top gravel, of which, however, there is still a very large body left, if, as has been supposed, the deposit is continuous underneath the lava to Scott's Flat. The lower gravel will have to be worked by drifts, and extensive preparations have already been made with this end in view.

The top gravel at Quaker Hill appeared to me, at first sight, to be composed exclusively of very fine quartz, white or bluish in color, but a closer inspection showed the presence of a very large amount of slate or other metamorphic rock in small fragments, which, being easily decomposed, are washed away in the form of clay, leaving scarcely anything else to be seen on the bottom of the mine besides quartz. Masses of quartz as large as a man's head are said to be frequently met with in the fine gravel. The quartz is quite smooth and well rounded. As the original surface of the ground was very irregular and much cut up by ravines, the gravel has not been of uniform thickness. The highest banks at present exposed are between two hundred and two hundred and fifty feet high, and are nearly uniform in character for the whole thickness. On the northwestern side of the mine there are frequent streaks of sand, which do not appear to belong to any well-defined, continuous stratum. Where clay streaks occur, they are, as a rule, in the upper portions of the gravel, near the original surface of the ground. The gravel in the western bank, near the shaft-house, is noticeably lighter in color than that farther to the east. The lava gravel previously alluded to * is seldom seen excepting near the Sugar Loaf, to the west of which the channel is supposed to cross the ridge.

In regard to the lithological character of the lower gravel I have no direct information. It has been reached by means of a shaft and drift, as is shown in the diagram (Plate P, Fig. 3). For the details in respect to these underground workings I am indebted to Mr. Jacobs, with whom I spent an hour or two at the mine. The first shaft was sunk over the supposed centre of the channel, but reached bed-rock at a depth of only 133 feet. Sinking was continued for seventy feet further, and at that depth a drift was run in a westerly direction with a slight rising grade. For 360 feet this drift was entirely in bed-rock. At the distance of 474 feet from the shaft the first "pay gravel" was struck, and the bed-rock was twelve feet below the drift. The deepest gravel was found fifty-six feet farther west, bed-rock there being nineteen feet below the drift. The drift was continued seventy feet beyond the deepest channel, when it became evident that the western rim was near. The gravel was then only four feet in thickness. The width of the "pay channel" is 126 feet, and its working thickness is from six feet or less up to sixteen. The position of this shaft is nearly in the prolongation of the old Hotellen Incline. The true course of the deep channel having been found, it was decided to sink a new shaft, which will soon be completed. Through this shaft the bottom dirt will be hoisted to the surface, where it will be treated in cement mills to be built at the mine.

The new shaft is about a third of a mile above the old shaft at the Green Mountain mine,† and at an altitude about eighty-five feet higher, according to Mr. Jacobs's estimate. The difference of level between them by our barometric measurements, one made in 1870 and the other in 1879, is only fifty-eight feet, but these results cannot be regarded as anything more than an approximation, and the agreement with Mr. Jacobs's estimate is as close as could be expected.

In regard to the sources and the amount of water used, I did not have time to collect many data. The Gold Run Ditch and Mining Company, which formerly drew eight hundred inches of water from Bear River, now takes water also from the Yuba, and its ditches have a capacity of 2,150 inches, but the supply is not constant through the year. Last season, work did not begin in the mines until February, and it continued for only five months. In some years the mining season lasts for eight months. The company has $450,000 represented in ditches, flumes, reservoirs, and tunnels.

The water for the Quaker Hill mines is taken from Greenhorn Creek. Washing usually begins in the month of December; it has begun in November only four times in twenty years. When begun, it continues day and night until midsummer or into the month of September. The amount used is 800 inches per day through two $4\frac{1}{2}$ inch nozzles.

The character of the gold varies to some extent in the different mines and in the different strata of gravel. At Indiana Hill the upper gold is fine and floury or scaly, while that near bed-rock is smooth and rounded. Large nuggets are seldom found, but pieces worth one or two dollars are not uncommon. The mint value of this gold, as Mr. Gould tells me, is $18.50 per ounce.

The Polar Star gold is also coarser on bed-rock than it is in the upper gravel. A handy way of estimating the coarseness of the gold is to determine the value of a pound of amalgam. The Polar

* See ante, p. 150.   † See ante, p. 180.

Star amalgam from the washing of the deep gravel is said to be worth $190 per pound, whereas with fine gold a pound would not be worth much more than one half that sum.

On Chicken Point the gold is also fine and floury. It is said to have a high value, its fineness reaching .975.

The gold in the upper gravel at Quaker Hill has been compared to flattened pin-heads. On the bed-rock the gold is coarser; nuggets weighing as much as half an ounce were found, I was told, in the deep drift. The western side of the channel is richer than the eastern, the same as it is at Indiana Hill.

Statistics of yield are very hard to collect with precision and accuracy. I made frequent inquiries upon this point, but have but little that is satisfactory to present, in addition to what has been already published.

The drift diggings at Indiana Hill — at the lower end of the hill — have always yielded a large amount of gold, the amount varying from one dollar up to eight or nine dollars per car-load. The average yield of this mine for the mining seasons from 1872 to 1874 is given in Raymond's report for the year 1874, page 100, as $5.28 per cubic yard. (My own computations from the data contained in the table make the yield $5.22 instead of $5.28.) From 19,997 car-loads, each car holding $19\frac{1}{2}$ cubic feet, the yield was $75,422.47. No account appears to have been taken of the difference in the volume occupied by the gravel in place and the gravel in the car.

In order to get an approximate value for the Polar Star gravel, I estimated the dimensions of the funnel-shaped opening from which the gravel has been removed since the completion of the new tunnel. About an acre of bed-rock has been cleaned at the bottom of the mine, and over a much larger area the gravel has been nearly all removed. My estimate was that a block of gravel sufficient to fill a space 400 feet square by 150 feet deep, or 24,000,000 cubic feet, has been removed. The yield from this gravel, as I was told by Mr. Colgrove, has been not far from $100,000; which makes the yield per cubic yard about eleven cents. The best pay was found, not on the bed-rock, but from six or eight up to forty feet above it.

At Quaker Hill, in conversation with Mr. Jacobs, I endeavored to get some rough idea, at least, of the yield of the top gravel. Estimating the average expenditure of water at 800 inches per day, and that each inch would move five cubic yards of gravel, the amount of gravel moved per day would be 4,000 cubic yards. The average yield per day has been $200, or at the rate of five cents per cubic yard, — a value quite in harmony with the results of the calculations made in 1870 for the top gravel at Gold Run. If we assume that the average amount of gravel moved per day has been only 3,000 cubic yards, it will bring the yield per cubic yard up to $6\frac{3}{4}$ cents. As a partial check upon this calculation, I made a computation of a different kind. Mr. Jacobs pointed out the boundaries of the opening from which the gravel had been washed during the past six years, and estimated the volume of gravel removed to be equivalent to a block 1,000 feet long, 700 feet wide, and 150 feet thick, or 105,000,000 cubic feet. Assuming 225 as the average number of working days in a season, the daily average of gravel removed for the six years would be 2,881 cubic yards. The estimate of $6\frac{3}{4}$ cents as the yield per cubic yard cannot be far out of the way.

The bottom gravel is said to promise a much more abundant yield. A production of from ten to twelve dollars per ton is looked for.

In the cañons of Bear River, Steep Hollow, and Greenhorn Creek the tailings from the several mines are constantly accumulating to greater and greater depths. According to my barometric measurements, the tailings in Bear River, at the road-crossing between Dutch Flat and Little York, were 97 feet, and in Steep Hollow, at the crossing between Little York and You Bet, 156 feet deeper in 1879 than they were in 1870. I have no data for similar comparisons in Greenhorn. Where large quantities of tailings are brought down from lateral ravines, the effect is to build a dam across, or nearly across, the cañon, in such a way as to hold back considerable quantities of water and to form temporary lakelets. At the outlet of Dutch Flat Cañon there is a dam nearly completed across Bear River. At Wilcox Ravine, through which the tailings from Chicken Point are discharged, a dam has formed, which holds the water back for nearly a mile in the cañon of Steep Hollow. At the time I was there, the 21st of November, a part of the water from the mines was flowing "up stream" into this lake. A similar lake has been formed in Greenhorn Creek above the mouth of Gas Cañon; and if the mines at Hunt's Hill are worked, a similar dam will form near the junction of Greenhorn and Little Greenhorn creeks, which will work to the injury of Gas Cañon as a natural outlet for the Quaker Hill mines.

The rapid accumulation of tailings in the present cañons, in spite of their steep grade, shows conclusively that a comparatively short time will be sufficient for the formation of extensive deposits of gravel, provided only the conditions are favorable.

SECTION VI. — *The Forest City Divide, between the Middle and North Yuba Rivers, and the Vicinity of Sierra City.*

On the divide between the Middle and the North forks of the Yuba River the relations of the old gravel channels to the present topographical features of the country are radically different from those which prevail on the ridge between the Middle and the South forks. On the latter ridge, as has been seen from the preceding pages of this volume,* an old channel can be traced almost without interruption from Snow Point to French Corral, sometimes on the northern and again on the southern slope of the divide, but following in the main the same direction as the ridge. On the former the old channels are transverse to the ridge and cross it nearly at right angles, as if they were occupying the places of northern tributaries to the main northeasterly and southwesterly stream (on the ridge lying to the south), very nearly parallel to the course followed by the present North Fork, from a point southerly from Strawberry Valley to its junction with the Middle Fork, a couple of miles below North San Juan. This is true of the principal and best-defined channels, which are two in number. In addition to the gravel of these two channels there are other deposits, some large and some small, whose relations both to each other and to the old channels are a matter of considerable doubt. In the opinion of some there is a third channel, nearly parallel to the other two, the existence of which, however, I cannot regard as proven.

Any attempt to portray the topographical and geological features of this section will be to some extent unsatisfactory on account of the lack of good and trustworthy maps, to which the reader can be referred. A portion of the ridge lies in Yuba County, of which the most recent map bears date of 1861, and the remainder in Sierra County. Mr. Hendel's map of the latter county, though "compiled from official surveys" and dated 1874, still leaves much to be desired. Its errors, especially in those portions which have not been surveyed by Mr. Hendel in person, make the map practically unavailable as a background upon which to represent the geology of the county. Had Mr. Hendel been able to extend his instrumental surveys more generally over the county, there is no doubt that the most serious of the errors would have been detected and eliminated. My acquaintance with Mr. Hendel's work in Plumas County warrants me in making this assertion.

The detailed description of this ridge falls naturally into three parts. I will take up first the vicinity of Camptonville; second, the region near Forest City; and third, the higher portions of the ridge in the neighborhood of Sierra City.

A. CAMPTONVILLE AND VICINITY.

This region, lying partly in Yuba County and partly in Sierra County, does not fall within the limits of any published map that is constructed on a scale sufficiently large to be of practical value as a map of reference. For the western portion, the old Yuba County map may be used, though it is in some particulars inaccurate. I cannot state precisely what corrections should be made upon this map, for I had neither the time nor the means for making the necessary instrumental surveys. According to the information I received from several sources, and to the hasty observa-

* See *ante*, pp. 196–208.

tions that I was able to make upon the ground, the small water-course in Oak Valley empties into Willow Creek instead of Beaver Creek (or Brandy Creek, as it is also called), and the headwaters of Willow Creek are farther to the northeast than would appear from the map. If these changes could be made, the map would give a much more truthful representation of the facts than it now does.*

Leaving out of consideration for the present a few outlying deposits of gravel whose relations are questionable, I will begin with a description of the main north and south channel, which can be traced across the ridge between Camptonville and Depot Hill, or Indian Hill. The prolongation of the channel to the north of the North Yuba will be described in a subsequent section, in connection with the gravel of Brandy City.

The places at which hydraulic mining has been or is now carried on are, beginning with the most southerly, Camptonville, Galena Hill, Young's Hill, Weed's Point, Railroad Hill, Depot Hill, and Indian Hill. The first four places mentioned are in Yuba County; the last two are in Sierra County. The precise locality of Railroad Hill I cannot give. It is probably in Sierra County, the position assigned it on the map being erroneous.

The town of Camptonville stands on the ridge between Willow Creek and Oregon Creek, at the head of a ravine leading down to the latter. The gravel deposit lies to the west of the town, and at a lower altitude, the town being built upon the rapidly rising slope of the eastern rim of bedrock. The busiest days of mining at this place were between 1853 and 1860. It was one of the first places at which the hydraulic process was introduced and proved to be successful. Since 1874 no mining of any consequence has been done near the town, the deposit of gravel being practically exhausted; at some of the other places above-mentioned the supply is still sufficient to last many years.

The bed-rock at Camptonville is a slate which strikes a little to the west of north, and has a nearly vertical dip. About four miles to the southwest of the town, on the San Juan road, I observed the transition from slate to granite. There is also a similar change of bed-rock from slate to granite, about a mile to the west of the town, just below the junction of Horse Valley and Willow creeks. Some of the granite at this point has a very coarse-grained texture. At Galena Hill the bed-rock is an easily decomposable slate, while the slate at Depot Hill is of a harder variety and is considerably worn. The bed-rock relations at Indian Hill are in some respects peculiar. At the eastern or upper end there is a coarse, rotten, easily decomposed granite, containing mica crystals as much as a quarter or a half of an inch in diameter. At the lower end the bed-rock is again slate, like that at Depot Hill or beyond. The line of junction between the two kinds of rock has a northwesterly direction for the short distance along which I was able to trace it. It would be interesting to determine the boundary of this upper mass of granite, which appears to be separated from the granite already referred to by a broad belt of slate rock.

Between the gravel of Camptonville and that of North San Juan there is nothing to obstruct the view. The line of sight follows very nearly the course of Oregon Creek and then across the Middle Yuba. The river-bed at Freeman's Crossing is more than 1,100 feet below the Camptonville bed-rock. The character of the country is such that a very extensive view is obtainable of the northern slope of the San Juan ridge, and there is nothing in sight to suggest any opposition to the theory that the old channel from Camptonville followed approximately the line of the present Oregon Creek, and effected a junction with the more southern channel at some point near San Juan. The difference of altitude between the bed-rock at the southern end of the Camptonville deposit—2,657 feet—and that at the eastern end of San Juan Hill—2,033 feet—is 624 feet. The distance in a direct line is not far from seven miles, allowing a grade of eighty-nine feet to the mile, a grade which is in excellent agreement with that found for the average grade of the old

* A copy of the Yuba County map, which is referred to above, was given me by Mr. Joseph E. Young, an attorney-at-law residing at Camptonville. I am also indebted for information to Mr. Jason Meek, the present county surveyor. The alterations made upon the map in red ink, and the coloring for gravel, are made from my notes.

channel on the divide between the South and Middle Yubas. This is strong evidence in favor of the hypothesis that the old channels were once connected, and the hypothesis receives further support from the existence of small bodies of gravel high up on the spurs of the western slope of the Oregon Creek cañon, below Camptonville. These smaller deposits I did not have time to visit; I was first told about them by Mr. Bray.

For determining the grade of the old channel across the ridge I took barometric observations at Camptonville, at Galena Hill, and at Depot Hill. Observations were also taken at Indian Hill, to which reference will be made further on. The fall between Depot Hill and Camptonville, a distance of about four miles and a half, I make to be 463 feet, or 103 feet to the mile. The altitude of bed rock at Galena Hill is confirmatory of this result.

The old gravel mines at Camptonville covered an area of about 2,000 feet in length by 1,000 feet in width. Bed-rock ridges rise both to the east and west, which effectually preclude any other supposition than that the old channel had a southerly direction. The bed-rock is nearly level for the greater part of the width of the channel, rising rapidly as the rim is approached. The gravel appears to have been mainly quartz, of a reddish or a yellowish color. The banks were about ninety feet in height. The gold was fine and scaly. The richest portions are said to have been on the western rim.

The cañon of Willow Creek separates Camptonville from Galena Hill. By the route I followed between these places I was not able to trace in detail the probable course or connections of the old channel. The Galena Hill gravel, like that at Camptonville, is hemmed in on the east and west by high bed-rock, that on the west being on the opposite side of Horse Valley or Brandy Creek. The tailings from the Galena Hill mines are discharged into Willow Creek. The longest axis of the diggings has a course of N. 65° W. (magnetic), and is from 1,000 to 1,200 feet in length. The width of the gravel I estimated at from 600 to 800 feet, and the highest banks are from fifty to seventy feet in height. The gravel is of a red or yellow color, like that at Camptonville. Upon the bed-rock I saw scarcely anything excepting pebbles of white quartz. There was no one at work at Galena Hill at the time I was there, and the gravel appears to be nearly all gone.

Horse Valley Creek makes a shallow separation between Galena Hill and Young's Hill, the limits of the gravel on the two sides of the creek being barely a quarter of a mile apart. I did not make any personal inspection of the gravel at Young's Hill. I was told that the richest portions lay on the eastern rim, instead of on the western rim, as at Camptonville. If this statement is to be relied upon, its explanation is to be looked for in the curve that the old channel must have made near this point.

At Wood's Point I estimated the thickness of the workable gravel to be 125 feet, the lower twenty-five feet being blue in color, and the remainder red; but I did not stop to make any detailed examination of the bank.

Railroad Hill lies to some distance to the east of the road from Camptonville to Oak Valley, and was never a deposit of much importance. I was told that it is entirely worked out, and accordingly I did not think it worth while to make a special excursion to it.

At Depot Hill there is one of the largest bodies of gravel which remain to be worked in this vicinity. It occupies the crest of the ridge, or the divide, between Willow Creek and Indian Creek. It is worked both on the northern and the southern end, at the former discharging tailings into Indian Creek, and at the latter into Oak Valley Creek near its junction with Willow Creek. The deposit is about half a mile in length, along a course S. 17° W. (magnetic). Its extreme width at the northern end, where all my observations were made, is between five and six hundred feet, four hundred feet being in deep gravel. It is evident that the old channel must have crossed directly from Grizzly Hill (near Brandy City) or by way of Indian Hill. The Indian Hill, Grizzly Hill, and a portion of the Brandy City gravels are all in sight from the north end of Depot Hill, and are all enough higher than the latter to assure at least the average grade for the old channel between them. Grizzly Hill bears N. 15° W. (magnetic) from Depot Hill, while

Indian Hill lies more nearly northeast, and may very likely belong to some tributary stream, which came in from that direction. Indeed, there is some slight evidence at Depot Hill of the junction of two channels, there being a spur or ridge of bed-rock in the centre of the channel some twenty feet higher than the lower bed-rock to the east and west, though there is said to be no difference in the quality of the gold found in the two channels. The bank of gravel at present is one hundred feet in height. Upon the bed-rock the gravel is blue, and there is also a layer of boulders a few feet in thickness, which are seldom too large to be handled by a single man without the aid of powder. Many of these boulders are peculiar in appearance. Specimen 35 came from this locality. It appears to be serpentinous in character. I selected it on account of its resemblance to boulders that I had previously seen on Grizzly Hill. The gold in the top dirt at Depot Hill is said to have been very coarse and nuggetty, but in the gravel below it was fine and scaly, increasing in coarseness as the bed-rock is approached. The claim at the northern end is held by Mr. John Rule and one partner, who, neither owning water-rights nor being willing to pay the price charged by the water company, have adopted the plan of drifting in the richer blue stratum and panning the gravel by hand. At my suggestion Mr. Rule washed one pan of gravel, from which I was allowed to take a couple of small pieces of gold. Mr. Wadsworth says of them, that they are "flat, with rounded and rubbed edges. One grain is much lighter yellow in color [than the other], contains considerable of its quartz veinstone, and has been rounded on one side." None of the gravel between Camptonville and Depot Hill has any capping of volcanic material.

Indian Creek separates Depot Hill from Indian Hill. The latter occupies the lower extremity of the narrow ridge between the creek and the river, near their junction. The gravel can be reached by trail from Depot Hill or by wagon-road from points higher up on the ridge. The Indian Hill deposit is anomalous in many particulars. I have already spoken of the line of junction between slate and granite, which can be distinctly seen where the gravel has been removed. About midway of the length of the gravel now exposed there is a remarkable pitch in the granite bed-rock, where there might once have existed a cascade or rapids. The pitch is not vertical, and there are no reasons to suspect the presence of a fault in the rocks. The general grade of the rock is unusually high, there being as much as seventy feet fall within a quarter of a mile. These facts are illustrated in the diagram (Plate P, Fig. 4). The general course of the Indian Hill gravel may be taken as S. 60° W. (magnetic). Its width is from a thousand to fifteen hundred feet. The average thickness from bed-rock to surface cannot be given, partly on account of the great irregularity of the surface of the upper stratum, which is a volcanic cement as much as a hundred feet thick in some places, and falling off to almost nothing at others. The gravel has been removed to bed-rock on both sides of the hill, and one transverse excavation has been made near the upper end of the hill, so that the opportunities for examining the gravel at several points are unusually good. A very striking feature of this deposit is the stratum of coarse boulders, from thirty-five to forty-five feet in thickness, resting at the upper end of the hill, directly upon the lower stratum of fine white quartz gravel, which is about eighty feet thick. At the lower end of the hill a stratum of pipe-clay, thirty feet thick where it is prolonged toward the west, but thinning out to nothing at its eastern end, intervenes between the boulder stratum and the fine gravel. (See diagram.) In the boulder stratum there are also occasional streaks of fine sand, one or two feet in thickness, and ranging from twenty to forty feet in length. The banded appearance of the gravel caused by this unusual arrangement of the strata is distinguishable from a great distance. Seen from Depot Hill, the boulder stratum looks like a thick deposit of bluish pipe-clay. There is no sign of a fault in the gravel at the point where the clay streak disappears. The most natural explanation of the facts is, that after the deposit of the white gravel there came a period of quiet, favorable to the accumulation of fine mud or clay, which collected in the lower portions of the stream to a greater thickness than in the higher portions. Still later there was a second period of the accumulation of gravel, when the old channel received a large accession of material different in kind from that formerly deposited. This view is supported by the statement, which was made to me by Mr. Bliss, that gold in more than the average quantity is found along the surface of junction between the boulders and the fine gravel. This statement I had no means of verifying. The boulders which make up the great body of the boulder stratum are largely though not exclusively of volcanic origin. At the lower end of the hill, where the side gravel has not been removed to so great an extent as elsewhere, I had a good opportunity to see these boulders in place. A large number of them differed decidedly from the common type of tufaceous rock, which prevails at higher altitudes on the ridge. They presented rather the appearance of metamorphic crystalline rock, such as might be expected to compose the bed-rock somewhere in the mountains.

The principal pay-streak at Indian Hill is near the bed-rock. The gold is fine, the coarsest pieces seldom being worth more than two or three dollars. There is an abundance of large blue boulders in the pay-streak. At the upper end of the hill I noticed, resting upon the bed-rock, a stratum of about four feet in thickness of a black clay, containing abundant impressions of leaves and wood, but it was too fragile to furnish good specimens that would bear transportation.

The water used at Indian Hill is drawn from the heads of Indian Creek and from one of the branches of Humbug Creek, also a tributary of the North Yuba.

A little more than a mile to the east of Indian Hill, upon the spur which lies between Humbug Creek and the North Yuba, there is a mining claim known as Snowdon. I did not visit the locality, but was told that there is a boulder stratum there, like that at Indian Hill, which rests directly upon the bed-rock, without any fine gravel. It is suggested that the clean quartz gravel of the lower stratum of Indian Hill may belong to the old channel which flowed between Brandy City and Camptonville, and that the boulder stratum and volcanic capping belong to a later stream, which came from a different direction, and which may possibly be recognized again at Pittsburgh Hill, one of the outlying deposits to which I will next call attention.

The exact position of Pittsburgh Hill I am not able to give. It lies upon a long spur between Lost Creek and the North Yuba, about a mile below the mouth of Slate Creek, and seven or eight miles in a northwesterly direction from Camptonville. It is near the point where the course of the North Yuba changes from westerly to southerly. The course of Lost Creek from Pittsburgh Hill is first southerly and then westerly to the Yuba. There are four mining claims at this place. The most westerly is known as the Pittsburgh Hill claim; and this is followed by the Oshawa, the California, and still farther east by another, the name of which I do not know. There have been in the past some attempts to employ the hydraulic process, but at present, attention is directed solely to tunnelling and drifting. The mouth of the Oshawa tunnel is about a thousand feet in a direction a little north of west from the mouth of the California tunnel. The true course of the Oshawa tunnel is N. 49° E.; and that of the California tunnel is N. 2° W. for the first 468 feet, after which for nearly 500 feet farther the course is N. 22½° W. The altitude of the mouth of the California tunnel I made to be 2,860 feet; the Oshawa tunnel is eleven feet higher. The difference of level between Pittsburgh Hill and the boulder stratum at Indian Hill will allow a grade of nearly eighty feet to the mile between the two places. There is no direct evidence, however, of the existence of any channel between them: if such a channel did exist, the present North Yuba has cut it entirely away. To the east of Pittsburgh Hill bed-rock rises to a height of six or seven hundred feet above the level of the gravel, and thus shuts out the possibility of any channel coming in from that direction. I did not examine the Yuba slope of the hill, but I was told after my return to Camptonville that a channel of 150 feet in width can be seen entering the hill from that side.

The bed-rock at Pittsburgh Hill is a soft, easily worked granite. The gravel contains considerable rotten granite, and in the pay-streak, which is confined within two and a half feet of the bed-rock, the boulders are heavy. The body of the gravel, which is 137 feet in thickness, and barren of gold, is composed of washed and rounded volcanic pebbles mixed with granite. Some of the lava boulders are decomposed and soft, others are still hard.

The gold is black and rusty. I was allowed to take a few small pieces from the Oshawa dirt. They have been examined by Mr. Wadsworth. He says the gold is "covered partly by a dark clay resembling the clay derived from basaltic or andesitic material, is in somewhat rounded, worn, comparatively thick grains, but has (probably) not been carried far."

At the Oshawa tunnel the gravel is paying well, but at the California tunnel no rich spots have been struck. The bed-rock appears to be irregular and wavy, with no distinct channel. It seems as if the gold must be distributed in spots over an irregularly flat surface, which may or may not have been in the path of any stream, but which has been covered with volcanic material during the period of volcanic activity. The area of this surface cannot be stated. The California tunnel, over nine hundred feet in length, is not yet half-way through the hill. The gravel of Big Oak Flat, on the opposite side of the North Yuba, is said to be similar in character to that of Pittsburgh Hill. Further reference will be made to that locality on a subsequent page.

The present surface of Pittsburgh Hill, and to a depth of three or four feet, is said to have been rich in coarse gold. The old miners sank pits until they struck the lava clay, which they supposed to be bed-rock. It is difficult to account for the gold found on the top of the lava; perhaps it had the same origin as that reported from the top dirt at Depot Hill.*

The gravel deposits near Oregon Creek, at Tyler's Diggings and Tippecanoe, a few miles above Camptonville, are not of much importance. They are not much higher than the present bed of the creek, and they probably owe their origin to the action of the creek itself in connection with the cutting away of the Bald Mountain channel at Forest City. Tyler's Diggings are on the right bank of the creek, and lie on both sides of a small ravine. They seemed to be entirely deserted at the time of my visit, and I confined my observations to the western bank. The bed-rock at this point is a dark-blue clay slate with a northwesterly strike, and a nearly vertical northeasterly dip. At the southern end of the diggings the bed-rock is very irregularly intersected with clay seams, and with masses of a whitish or yellowish clayey material. The gravel has extended over an area of about 450 by 150 feet, the longer axis being nearly north and south. The present northern bank is about sixty feet in height, the lower half being a fine quartz-gravel, containing some fossil vegetation, and the upper a sandy clay with red dirt and volcanic boulders, such as might have come from the crest of the ridge above the diggings. A few quartz boulders, ranging from two to four feet in diameter, lay upon the bed-rock.

The altitude of the bed of Oregon Creek, where the trail crosses between Tyler's Diggings and Tippecanoe, I made to be 3,385 feet, 125 feet below the bed-rock at Tyler's, and 170 feet below that at Tippecanoe. The bed-rock at Tippecanoe is very variable in character within short distances. Some of it is soft and rotten, like a decomposed granite; some is slaty; and some is soft and fragile, breaking with an irregular fracture, like the serpentine of the higher portions of the ridge. There are also many seams with a clayey filling, called "talc" by the miners, similar to those seen at Tyler's. The gravel is said to extend for a length of 1,200 feet parallel with the course of the creek, and to be 600 feet in width in the centre, growing narrower at both ends. At the centre of the deposit the banks are about seventy feet in height. The bottom-gravel for a thickness of twenty feet is rather coarse, but without large boulders. It contains some rounded and some angular pieces of float bed-rock mingled with the blue and white quartz. Considerable petrified wood is also found in this lower stratum. Higher up, the bank is composed of finer and lighter gravel with a few thin sand-streaks.

The gravel at the head of Grizzly Cañon, two miles and a half above Pike City, is different in character from the gravel seen in the old channels. Grizzly Cañon is one of the tributaries of Oregon Creek. At its head is a broad flat at the base of the tufa ridge, which rises rapidly about a half-mile to the northeast of the mouth of the Grizzly Cañon tunnel. The tunnel is 1,300 feet in length, and the drifting extends over an area of about 250 feet in width. The gravel brought out to wash contains a great variety of boulders. There is also a peculiar cemented breccia made up of angular quartz; and, in addition, there are blocks of cemented bed-rock. The gold is sharp and angular, in spangles "like tea," or in thin flattened scales, not rolled. In short, there are none

---

* For a part of the foregoing information about Pittsburgh Hill I am indebted to Mr. L. A. Pelton, of Camptonville.

of the signs which ordinarily point to a river-wash. The fineness of the gold is also very low, at least much lower than that of any of the true river channels. There is no reason to believe that this deposit has any connection with the gravel higher up on the ridge between Oregon Creek and Kanaka Creek.

The following statistics in regard to the yield of the Camptonville district, and to the quality of the gold, I obtained from a trustworthy source at Camptonville.

From 1860 to 1875 the gross yield of the Camptonville gravel district, exclusive of the gold obtained by quartz mining, amounted to $8,000,000, an average of half a million dollars a year. Since 1875 the average yield has been about $150,000 a year. Since 1869 the quartz-mines have yielded in addition over $400,000. The average yield per cubic yard of gravel cannot be ascertained.

The fineness of the gold and value per ounce at the different mines have been on the average as follows :—

Locality	Fineness	Value.
Camptonville	.930 – .935	$18.35
Galena Hill	.940	18.40 – 18.50
Young's Hill	.940	18.40
Weed's Point	.925	18.25
Railroad Hill	.925	18.25
Depot Hill	.940	18.00
Indian Hill	.925	18.25 – 18.35

In the ravine diggings at Oak Valley and Dad's Gulch the quality of the gold has not been so good, ranging from .890 to .880, or from $17.50 to $17.00 per ounce.

The gold at High Point, on the southern side of Oregon Creek, has a fineness of .940, or from $18.50 to $18.75 per ounce. That of Oregon Creek will have an average fineness of .880 ; and that of Pike City of only .750, or $16.00 per ounce.

In this connection I will give, on the same authority, the fineness of the gold from two quartz-mines in the vicinity of Camptonville. The gold of the Alaska Company, Pike City, is only .740 fine, and that of the Brush Creek Company, near the new Mountain House, is .820.

I made frequent inquiries for animal or human fossils, but could not learn of their existence near Camptonville. I was shown a part of a mastodon tooth, which was said to have been found under thirty feet of gravel, near bed-rock, at Indiana Ranch in Keystone Valley, nine miles in a direct line to the southwest of Camptonville and on the opposite side of the North Yuba. It did not seem worth while to take the time for a visit to the locality. The tooth is in the possession of Mr. J. R. Young of Camptonville.

On the ridge road from Camptonville towards Forest City and Downieville the lower limit of the lava capping is seen at an altitude of 3,900 feet, over eleven hundred feet above the hotel at Camptonville. The road follows upon the lava capping with a steady ascending grade to the summit, near the old Mountain House, of which the altitude is 4,765 feet. The altitude of Nigger Tent I made to be 4,465 feet. Between the summit and Forest City there is a depression in the crest of the ridge, along the "backbone" which forms the water-shed between the tributaries of Oregon Creek and the water-courses which discharge into the North Yuba, near Goodyear's Bar. The lava capping disappears entirely before the new Mountain House is reached, the altitude of which is 4,140 feet.

It is the theory of many persons that the lava cap between the old Mountain House and Camptonville covers a gravel channel. I see no reason for adopting that theory, the preponderance of geological evidence going to show that the former drainage was not longitudinal, but transverse to the ridge. Considerable work has been done on the southern slope of the ridge in a search for such a channel. The only place that I found time to visit was Mr. D. H. Dahneke's tunnel. The altitude of the tunnel's mouth I made to be 3,875 feet, very nearly the same as that of the lower end of the lava cap. The tunnel is in bed-rock at its mouth, but at the distance of 525 feet the bed-rock pitched off to the north at a high angle, and the tunnel has been driven sixty-five feet

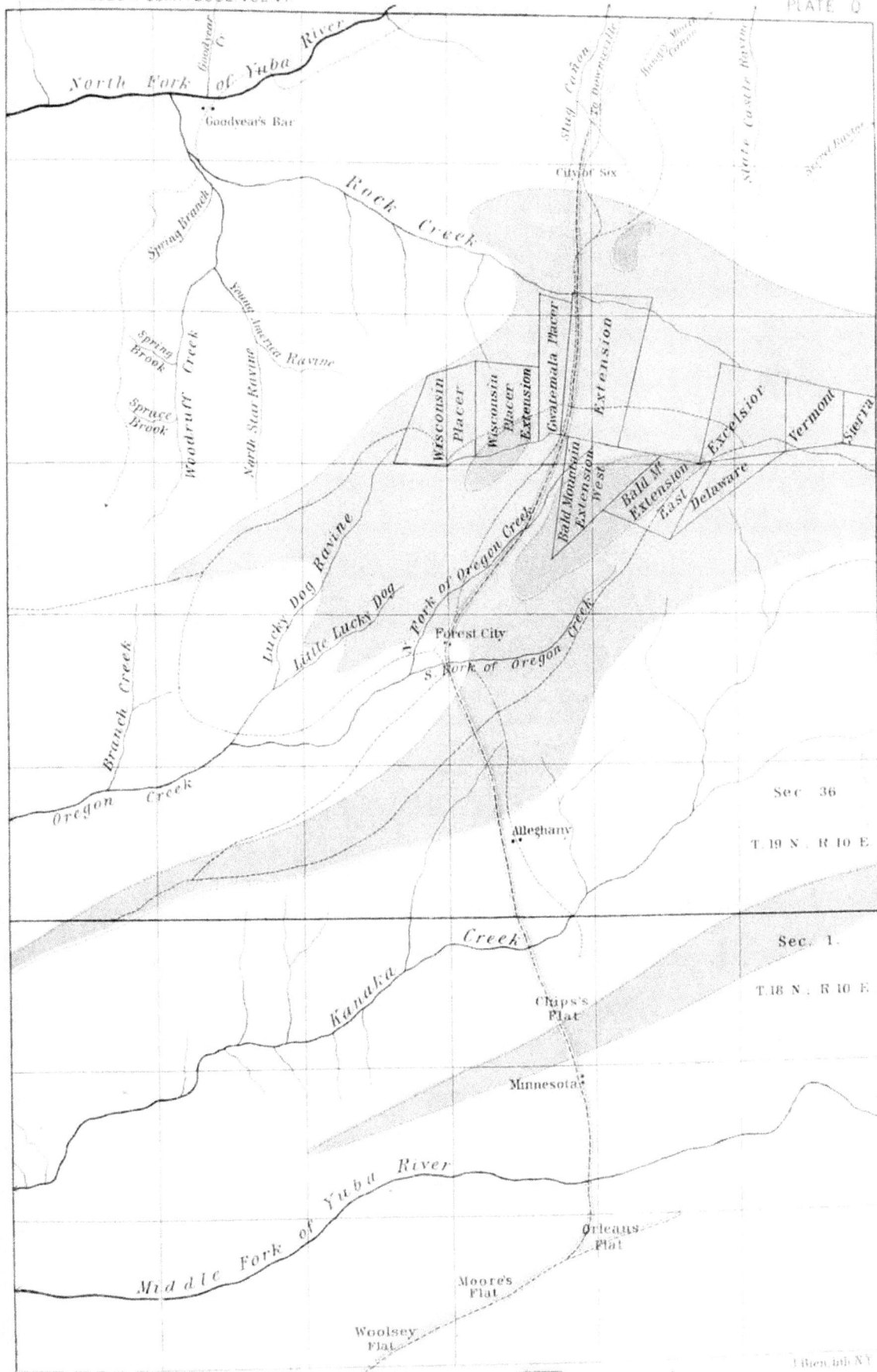

MAP OF THE MINING DISTRICT ADJACENT TO FOREST CITY

(Scale One Mile to the Inch)

---- Old Channel   Volcanic   Bald Mt Basalt

Julius Bien & Co. lith. N.Y.

further in lava cement. This tunnel is near Tyler's Diggings, described in a previous page, and not far from the Nigger Tent House, but I cannot give its precise position. The course of the tunnel is N. 10° E. (magnetic).

B. FOREST CITY AND VICINITY.

The portion of the ridge included under this head is comprised within the limits of the Mountain House on the west and the head of Wolf Creek, at American Hill, on the east. There is no map of the whole of this region to which I can refer, excepting that of Sierra County, but I have compiled one by transcribing the principal features of the public surveys of T. 19 N., R. 10 E., and a part of township 18 of the same range. (See Plate Q.) The ridge is cut up into several nearly parallel ridges by the deep cañons of Wolf, Kanaka, Oregon, and Rock creeks, the first three discharging their waters into the Middle, and the last into the North Fork of the Yuba. The steep northern slope is also deeply cut by Woodruff Creek, Slug Cañon, Slate, Castle, and Secret ravines, and Jim Crow Cañon, following the order from west to east.

At the Mountain House, as has already been stated, the lava capping of the ridge has been completely eroded. The separate, finger-like, longitudinal ridges have a capping of tufa or other volcanic material upon their higher portions, and above Forest City, near the head sources of Kanaka, Oregon, and Rock creeks there is a broad, nearly flat plateau of volcanic origin, which extends without interruption, though varying much in width, beyond American Hill, the easterly limit of my survey and observation in this vicinity. I have indicated upon the map the general position of the volcanic flow, but cannot vouch for the accuracy of its boundary in detail. There had been a heavy fall of snow just prior to the time I was in this region, and I was not able to reach all the points of interest.

The most of my observations upon this volcanic material were made in the immediate vicinity of Forest City and Bald Mountain, or along the trails between Forest City and Minnesota on the south, or City of Six on the north.

In its lower portion the volcanic material belongs, as a rule, to what is known as "cement"; that is, an aggregate of boulders or pebbles of greater or less size, though usually small, showing a rounded and smooth surface, and held together by an earthy cement as a rather loose and easily broken mass. The rounding of the pebbles may be due in some cases to the wear to which they were subjected in their journey from higher altitudes, or it may be explained by the existence of a concentric structure in the pebbles themselves. It is not uncommon to find lava pebbles with a marked tendency to break up, like an onion, into successive coats or shells. Together with the boulders, which are undoubtedly to be classed with volcanic tufa, and which resemble those met with on the San Juan ridge, there are also to be seen upon the surface, if not within the cement, boulders of granite and of several varieties of porphyritic and hornblendic rock, which do not so obviously belong to the volcanic series. They are sometimes very striking in appearance, and attract attention by their oddity, being quite different from the prevailing type. Mr. Wadsworth describes them as andesitic in character. I do not know of the existence of any similar rock in place.

Higher up in the volcanic mass there appears to be a persistent stratum, which may best be described as a mud conglomerate. I observed it near the base of the Bald Mountain bluff, and on the ridge between Forest City and Alleghany. A few rolled quartz pebbles are occasionally seen on the surface, and, as I was told by Mr. Wallis, the superintendent of the Bald Mountain gravel mine, similar pebbles are to be found in considerable quantity at a perennial spring of water, which breaks out under the lava of the bluff.

The highest volcanic stratum is a compact, fine-grained, rather light-colored basalt.* This

* This rock is thus described by Mr. Wadsworth: "A light gray, somewhat laminated rock, which shows fluidal structure, which has been the cause of the lamination. It is a surface specimen, evidently obtained from near the original surface of the lava. In the thin section the fluidal structure appears under the microscope to be strongly marked, and the rock is seen to be composed principally of basaltic plagioclase, innumerable granules, grains, and little crystals of augite, magnetite, olivine, and opacite. The base has been altered to a dirty brown fibrous mass, and considerable secondary feldspar formed."

stratum is about 250 feet in thickness at Bald Mountain. It occurs on both sides of the Downieville trail, but has been eroded entirely away along the line of the north fork of Oregon Creek. It is seen also as massive bluffs to the southeast of City of Six. Looking from the junction of the two forks of Oregon Creek, the general level of the ground is seen to rise quite rapidly to an elevated "flat," about 350 feet above the bed of the creek, beyond which the Bald Mountain bluff rises abruptly like a huge wall of rock. In climbing to the bluff, tufaceous and basaltic boulders are observed up to a point near the bluff's base. Above this altitude the tufaceous variety is entirely lacking. The rock of the bluff is jointed and fractured in such a way as to be easily mistaken for a metamorphic slate. The planes of jointing have a north and south direction, with a nearly vertical dip, sometimes to the east and sometimes to the west. The top of the mountain is barren of trees, but it is covered with a thick growth of manzanita chaparral, which makes progress in any direction, out of a beaten track, very slow and difficult. The precise boundaries of the basaltic cap cannot be given at present; it probably covers a greater area than I have marked upon the map.

The thickness of the volcanic capping is not constant. The altitude of the highest point of the Bald Mountain bluff I made to be 5,570 feet, or over 1,100 feet higher than Ellery's Forest City Hotel. The bed-rock rises very rapidly in a northeasterly direction, as is proven by the workings in the Bald Mountain mine, but there can be no doubt that, where thickest, there is as much as 900 feet of lava above the old river channel. To the west and southwest, and possibly also to the east, the thickness of the lava stratum diminishes. Between Forest City and Alleghany the thickness is nearly 600 feet; between Chips's Flat and Minnesota, about 250 feet; and near American Hill, between 600 and 700 feet.

The bed-rock of this portion of the ridge is in part slate, and in part serpentine or of a serpentinous character. I saw good exposures of slate at the Mountain House; at Minnesota; at Rock Creek, on the Downieville trail; on the northern slope of Kanaka Creek Cañon, above Alleghany; at the Crescent Tunnel; near Cornish Ranch; and at American Hill. The serpentine belt crosses the ridge in a northwesterly direction, and is exposed to view at a great many points between Minnesota and Goodyear's Bar. This same belt, or one nearly parallel, can also be traced to the north of the North Yuba, certainly as far as Plumas County. A westerly line of junction between slate and serpentine passes within a few rods of the Mountain House, and an easterly line can be traced near Minnesota and Chips's Flat. Whether the whole of the intervening belt is serpentine, or whether there are alternating strata of slate and serpentine, I cannot say with certainty. There are several reasons for supposing the latter to be the true arrangement. For example, at Minnesota there are several quartz veins, having nearly the same strike, which were described to me as contact deposits between slate and serpentine. I did not take the time to verify the statements. At Chips's Flat the line of junction between serpentine and slate has a northwesterly course, N. 50° W. (magnetic). A little to the west of the serpentine, the slate bed-rock changes abruptly from a nearly white to a black variety, the plane of junction having a dip of 80° to the southwest. Serpentinous and slaty bed-rock both occur under the gravel at the Bald Mountain and the North Fork drift mines at Forest City. The slate rock, in particular, is quite different in appearance from that exposed to view outside of the mines. For a depth of fifteen or twenty feet below the gravel the slate is almost always of a soft and clayey consistency, though still showing the planes of stratification or of cleavage. Below this depth the rock is again firm and hard. There are frequent changes in color and in texture. The frequency of the occurrence of "rotten bed-rock" under the gravel suggests the inquiry whether the waters which percolate the gravels may not have a greater disintegrating power than the surface waters do. Or, is the absence of clayey deposits, away from the gravel, to be explained solely by the fact that the products of disintegration are removed as rapidly as they are formed?

Hydraulic mining has not been carried on very extensively upon this portion of the ridge. Small banks have been washed away, as at Minnesota, Chips's Flat, Forest City, and City of Six, where the erosion of the modern creeks has exposed the gravel to view upon the sides of the cañons, and where the volcanic capping has also been worn away far enough to make the application of the hydraulic process profitable. The greater part of the yield of gold has been obtained by drifting. In my description I will begin at the southern end and advance towards the north.

Minnesota lies upon the northern slope of the cañon of the Middle Yuba, nearly opposite Snow Point or Orleans Flat. The town was built originally upon the exposed gravel. Hydraulic operations were confined to the eastern side of the spur, where the bed-rock has been uncovered over an area of about seven hundred by one hundred and fifty feet. The altitude of this bed-rock I made to be 4,220 feet. From my point of observation the whole of the southern slope of the cañon, from far above Snow Point to Woolsey Flat, was distinctly in view. According to the hand-level, I was nearly as high as the top of the pipe-clay at Snow Point, and considerably nigher than the bed-rock at Orleans Flat. The altitude of the Snow Point bed-rock, as determined earlier in the season, is 4,211 feet. The difference of level, therefore, between Minnesota and Snow Point, as determined by the barometer, is not quite as much as it should be; but as all my determinations of the altitudes of points to the north of the Middle Yuba are affected by the possible error in the estimated altitude of Oroville, it will be seen that the agreement is as close as any one could have a right to expect. It can safely be asserted that the old channel, while keeping its average grade, could easily have made connection with the channel to the south, at some point between Snow Point and Orleans Flat. Seen from Minnesota, the bank at Snow Point looks like a mass of clay, such as might have been deposited in some broad bay. The gravel banks at Minnesota are scarcely more than twenty feet in height. The material is, for the most part, a fine quartz, mixed with large white quartz boulders of from eight to ten feet in diameter. Silicified wood is very abundant, and there is a thick iron-cement on the bed-rock. Hydraulic mining was never carried on very systematically at this point. Each small claim of sixty feet front, extending back to the centre of the ridge, was worked by itself as far as it could be in this way, and then drifting had to be resorted to. The town of Centreville, of which not even the ruins are now left, lay a quarter of a mile to the west of Minnesota. Between the two places there was high bed-rock, which, however, did not rise to the upper surface of the gravel. A tunnel in gravel was formerly driven through the main ridge, connecting Centreville with Chips's Flat, and it is now utilized as an aid to ventilation by the Mammoth Company at Chips's.

Above Minnesota, on the ridge between the Yuba and Wolf Creek, there are said to be a few unimportant banks of gravel, which probably have some connection with the deposits of American and Bunker hills, which will be described on a subsequent page.

Upon the southern slope of the cañon of Kanaka Creek the principal mining town has been Chips's Flat. Balsam Flat is between a quarter and a half mile to the northeast of Chips's, and at a little higher level. At Oak Flat, farther down the creek, there is also said to be a deposit of gravel. I did not go to either of the two places last mentioned. At McNulty's mine, at Chips's Flat, the gravel has been washed away over an area of three or four acres. The face of the present high bank exposes next the bed-rock a stratum of six or eight feet of coarse white and bluish quartz, with quartz boulders of from three to five feet in diameter; this stratum is followed by about forty feet of fine gravel, upon which lies a thin stratum of a ferruginous cement. Above this there is a stratum of sandy and gravelly clay, thirty or forty feet in thickness, which reaches to the red dirt or volcanic capping. Silicified wood is common. There are also in the gravel of this mine some extraordinary masses of rock, known by the name of "grizzlies." They are like rough, unwashed boulders in appearance, and are sometimes as much as sixty feet in length by thirty in width or thickness. They are found at varying heights above the bed-rock. One of the largest of them has its opposite surfaces nearly parallel, and looks like a block from a mineral vein. The structure and mineral character of the grizzlies strengthen the belief that they are really fragments of vein-stone. Where they came from it is difficult to say, but quartz lodes are quite frequent in the vicinity, one in particular, the Rainbow, lying just to the west of the mine. The proximity of this lode may also account for the auriferous quartz boulders which are met with near the bed-rock.

The town of Alleghany lies on the opposite side of Kanaka Creek, and at a little higher altitude than Chips's Flat. The bed of the creek at the trail-crossing is over six hundred feet below the level of the town. The outlet of the channel from Forest City is at Smith Flat. There was a little hydraulic mining done at first, which was soon followed by drifting. Drifts or tunnels were driven many years ago through the ridge between Alleghany and Forest City, and the mines are now practically abandoned. I made no attempts to enter the old mines either at Alleghany or at Chips's. The pay-gravel is reported to have seldom exceeded three feet in thickness.

Forest City is built near the junction of the north and south forks of Oregon Creek, at the base of Bald Mountain. It owed its earlier prosperity to the rich drift mines under the Alleghany ridge; its present importance is a natural result of the successful mining operations, which have been conducted under the lava to the north and northeast of the town. There has never been an opportunity for much hydraulic mining. There used to be a large number of mining claims worked in this vicinity, to which it will not be necessary to refer in detail. At present the most extensive and profitable mine is that of the Bald Mountain Company. The North Fork Company, owning the adjoining ground to the west, has driven a long prospecting tunnel, and is still engaged in the prosecution of the work. By the kindness of the superintendents, Mr. H. Wallis of the Bald Mountain and Mr. Platt Ketchum of the North Fork Company, I was allowed to spend half a day in each of these two mines. The mining operations at Forest City have been described in considerable detail in Raymond's Report for the year 1874, pp. 151–155, and in that for 1875, p. 103. Extracts from these descriptions have been already given,* and my present report is in part supplementary to what is there stated.

The Bald Mountain mine is worked through a tunnel, which, though not straight, has a general northeasterly direction, and is now about a mile in length.† The mouth of the tunnel is in gravel, about twenty feet above the bed-rock. At a distance of 300 feet from the mouth a spur of serpentine bed-rock was struck, through which the tunnel had to be driven for 400 feet. Beyond the serpentine the tunnel for the rest of its length is entirely in gravel, or partly in gravel and partly in bed-rock, with the exception of 225 feet, where the old channel has been cut away by a peculiar flow of lava, or volcanic mud. The grade of the tunnel near its mouth is four feet to the hundred, or 211 feet to the mile; but farther in the grade is reduced, and at the extreme end it is only two feet to the hundred. Eighteen hundred feet from the mouth of the tunnel there is a shaft which was sunk from the surface. The first forty feet of the sinking was in the mountain-cement; then came two hundred and fifteen feet of irregularly stratified pipe-clay and sand, beneath which there were fourteen feet of gravel. The drifts and breasts are laid out both to the east and the west of the main tunnel, and are carried until rising bed-rock is met or the pay-gravel of the bottom is exhausted. The average width of the deep channel is about 500 feet. I tried to find out the width of the old channel at Alleghany and Chips's Flat, but with no good success. The estimates given me varied between 400 and 2,000 feet. The gravel is exclusively white quartz. Upon the bed-rock there are some small quartz boulders, but seldom more than two feet in diameter. There are occasional streaks of sand in the gravel, but no pipe-clay so far as I saw, either in the tunnel or in the drifts. Petrified wood is found, but no animal fossils. The drifts and breasts are about three and a half feet high on the average, including from a foot to a foot and a half of bed-rock. The softness and clayey character of the rock causes the gold to settle below the gravel, and the upper bed-rock is perhaps as rich as the lower gravel. The character of the bed-rock, it is asserted, has a remarkable effect upon the richness of the gravel. Where the rock is serpentine, the gravel is poor or barren; where it is a black slate, the yield is a maximum. The white slate is also a favorable rock, but not quite so favorable as the black. At Chips's Flat I was told that on the blue lead the white slate was the best rock for catching gold. There is such an unanimity of opinion in regard to the favorable or unfavorable character of certain varieties of

* See ante, pp. 213, 214.

† The plan (Plate 8, Fig. 1), copied from a map made by Mr. Wallis, and kept at the office of the company, shows only a part of the mine. The results of the more recent surveys have not yet been put upon the map.

bed-rock, that it seems clear there must be some foundation for the opinion in fact, though I confess I am at a loss for any rational explanation of the phenomenon. The lava mud-flow which cuts out the gravel for a distance of 225 feet is evidently of more recent origin than the gravel is. When the lava was struck, a prospect tunnel was run to the east, and drifts were carried to the west, until bed-rock was reached rising very rapidly. Upon this bed-rock there was pipe-clay, and above the pipe-clay lava. The main tunnel was finally carried across the lava with the same grade that it had been having below, and the gravel was found on the opposite side in just the position, with respect to grade, that the continuation of the lower gravel ought to have. Upon both sides the gravel dipped underneath the lava.

The gold found in the Bald Mountain gravel is very coarse. The same is true of the gold of the lower stratum at Chips's Flat, while that in the top gravel is fine. The secretary of the Bald Mountain Company allowed me to select a few specimens, which show well the general character of the gold. They were examined by Mr. Wadsworth, who says of them : "These grains are of very unequal sizes. One contains portions of the quartz vein-stone and has its edges rubbed down. Another is thick and well rounded. One is flat, and either composed of two pieces welded together, or else of one part bent over and upon a portion of the remainder. Whatever may have been the original form, if the gold was thin, it seems that it would easily be beaten into flat pieces with rounded edges. One queer form resembles a dress-hook. It is composed of quite a long, narrow strip of gold, that is bent partly upon itself twice. It has welded to it another smaller piece of gold, and it is easy to see how under the grinding, pounding action of pebbles it would form a rounded, thin gold grain." The fineness of this gold averages from .926 to .936. From the books of the company I obtained a few statistics in regard to the yield of the gravel between April, 1872, and July, 1879. The area worked amounts to about 1,500,000 square feet, and the total yield of gold is a little over $1,500,000, of which $664,000 have been distributed as dividends. The average yield per square foot has been $1.01½. The averages per square foot in the several years were as follows, $1.09, $1.01, $1.01½, $0.95¼, $0.99½, and $1.00. A yield of $1.01½ per square foot, with drifts three and a half feet high, corresponds to $7.83 per cubic yard. But according to the company's books the average yield per car-load has been $2.92. Each car is estimated to hold about one cubic yard of loosened gravel and rock, or one-half a cubic yard of rock in place. This corresponds, therefore, to a yield of $5.84 per cubic yard. These two results do not agree as closely as could be wished, but they are sufficient to confirm the statement that the bottom gravel is very rich. Probably the cars do not contain quite a cubic yard of loose rock ; their dimensions are given in Raymond's Report for the year 1874, p. 155, as "4½ feet long, 2 feet wide, and 2½ feet high." Assuming these to be the true dimensions, a yield of $2.92 per car-load corresponds to a yield of $7.01 per cubic yard.

The hauling of the gravel in the Bald Mountain tunnel is done by means of a locomotive similar to those in use in the coal-mines of Pennsylvania. The anthracite coal used costs, delivered at the mine, about $42 per ton.

The grade of the Bald Mountain channel is so much higher than is usual in the old gravel streams, that many persons have believed it to belong to some tributary, rather than to the main channel or blue lead, and that its continuation is to be looked for, not in the direction of Rock Creek and City of Six, but under the lava ridge to the east and northeast. In this belief mining claims have been laid out and explorations have been undertaken at several points higher up on the ridge, and also, in the hope of striking the main channel, at points to the west of the Bald Mountain mine.

In the latter direction the most important and extensive explorations have been conducted by the North Fork Company. A tunnel, of which the mouth is near that of the Bald Mountain tunnel, but on the opposite side of the north fork of Oregon Creek, has been driven in a general northwesterly direction for nearly a mile. A branch from the tunnel has at first a more northerly course, and then bears around to the west, in which direction it has been continued to and beyond the line of the first tunnel, and at a different level. There has been but little drifting done in this mine. I have not the full data for a detailed history of this exploration, and in the following description the distances given are only approximate. For the first 500 feet the tunnel was in gravel. Then there came 150 feet of clay resting upon gravel, followed by 150 feet of clay resting upon bed-rock. Beyond the clay the tunnel was in a soft slate for 300 feet, and then in pay-gravel, 325 feet farther, to the air-shaft. The shaft is 1,425 feet from the mouth of the tunnel. The strata met with in raising the shaft were, — twenty feet of gravel, forty feet of pipe-clay, and two hundred feet of volcanic cement. Beyond the air-shaft the tunnel was driven for 2,000 feet in bed-rock, with gravel from fifteen to twenty feet over-head. A coarse pay-gravel was then followed for 300 feet, to a so-called "lava cement," containing some coarse and some fine boulders, with about two feet of quartz gravel next the bed-rock. At the present face of the tunnel there is a thin gravelly cement on the bed-rock, covered with a sandy clay containing good impressions of leaves. The gold found is rather fine, and black sand is said to be plentiful. These explorations have so far failed to confirm the belief that the main channel lies to the west of that in the Bald Mountain mine. At present I am inclined to the belief that the gravel found in the North Fork mine, as well as that reported from other places, lying to the west of Alleghany and Chips's Flat, which I was not able to examine in person, does not belong to any continuous and independent channel, but was in some way once connected with the more easterly deposits.

Between Forest City and City of Six the gravel is nowhere exposed to view at the surface of the ground. At Rock Creek there used to be a mine worked by means of an incline, which is no longer accessible. About a quarter of a mile below the old incline Mr. William Irelan is driving a new tunnel, known as the Ruby Tunnel. The tunnel starts in a soft and clayey slate bed-rock, which requires but little blasting, and follows a southeasterly course for about 450 feet, when the course is changed to the east, and a branch is started to the northeast. Gravel has been found about twenty feet above the main tunnel, and near the roof of the branch. Between these points the gravel is continuous, and the bed-rock has a pitch to the east or northeast. My barometric measurements cannot, of course, be relied upon to determine with accuracy small differences of level, or to settle the question whether or not it is possible for a gravel connection to exist between Rock Creek and the Bald Mountain mine. The altitude of the mouth of the Ruby Tunnel I made to be 4,800 feet, and Mr. Irelan's estimate is, that the gravel in the Ruby Tunnel lies upon a side bench from seventy-five to one hundred feet above the deep channel, the level of which is known to be below that of the bed of the present Rock Creek. Taking into account the grade of the tunnel, and adopting the lower of Mr. Irelan's estimates, it may be said that the deep channel has an altitude of 4,750 feet. The mouth of the Bald Mountain tunnel I made to be 4,489 feet above sea level. If the rise in the tunnel is uniformly four feet to the hundred as far as the shaft, 1,800 feet from the mouth of the tunnel, the altitude at the bottom of the shaft is 4,561 feet. Between this point and the bottom of the Rock Creek incline there is a difference of level of 226 feet, according to the survey made by Mr. I. G. Jones, county surveyor of Sierra County. If these data are correct, the altitude of the Rock Creek channel is 4,787 feet, thirty-seven feet higher than I made it to be with the barometer. The distance between the shaft and the incline is probably between a mile and a half and two miles, which, supposing the grade of the Bald Mountain channel to continue to be as much as four feet to the hundred, would require from one to two hundred feet more difference of altitude between the places than really exists. The grade of the Bald Mountain channel, however, grows less as the tunnel increases in length, and I see no reason why there may not be an uninterrupted stratum of gravel between the present face of the Bald Mountain tunnel and Rock Creek. Against this hypothesis it is urged that the Rock Creek gravel lies in a narrow channel, only sixty feet in width, and that the gold of Rock Creek is different in character from that of Bald Mountain. I brought with me from Ruby Tunnel a few pieces of gold, given me by Mr. Irelan, which have been examined by Mr. Wadsworth, who says of them : "This gold shows the impress of the vein-stone, and while the thinner parts have been bent over and in upon the main mass, the grains as a whole seem to have suffered more from wear than from the direct pounding of the pebbles upon them. This came from a white clayey or fine sandy bed. The grains are thick, and owe their present forms largely to their original vein shape, the modifications occurring principally about the edges." He also says that the gold from Rock Creek and that from Bald Mountain "seem to have come from the same kind of gravel."

City of Six is at the head of Slug Cañon, near the Downieville trail, and about three quarters of a mile to the north of Rock Creek, from which it is separated by a lava ridge. The summit of the trail across the ridge is 230 feet above the mouth of the Ruby Tunnel. Tunnels or drifts used to connect City of Six with Rock Creek, and gold to the value of several hundred thousand dollars was taken from the gravel. The grade of the bed-rock is said to have been about two and a half feet to the hundred, sloping downwards towards Rock Creek. At City of Six there are open gravel banks, but they have been so long deserted that it did not seem worth while to stop to make any examination in detail. High bed-rock rises both to the east and the west of the cañon above the level of the gravel, indicating a northern source for the old stream. The view from the top of the gravel bank at City of Six is very extensive toward the north and northeast. In front there is the cañon of the North Fork of the North Yuba. To the right the Sierra Buttes are in full view. Monte Cristo to the left is just hidden from sight by a projecting spur. But over all the area which the eye can sweep there is nothing to guide one in a search for the continuation of the old river. There are no gravel banks, no towns, no mining camps in sight; nothing to give a hint as to where the stream came from. The gravels of Monte Cristo and Craycroft's, as will be seen later, are indeed at a higher altitude than that of City of Six, but the difference is not so great as to force one to a belief that these places were ever connected.

Above Forest City there has been nothing accomplished which can be called decisive as to the existence of a northeastern main channel or a northeastern tributary. The Bald Mountain Extension Company, which must not be confounded with the Bald Mountain Company, has acquired a large extent of territory immediately to the east of the Bald Mountain ground, and has begun a tunnel with the expectation of being obliged to go nearly four thousand feet before striking the rich channel. About one third of this distance is now completed. The mouth of the tunnel is at an altitude of 4,600 feet.

The Pliocene shaft, near the crest of the ridge, at the sag where the heads of Rock Creek and Kanaka Creek most nearly approach each other, was started with the expectation of reaching an underlying gravel channel beneath the lava capping. The present depth of the shaft, I was told, is about 180 feet. Further sinking has been suspended on account of the great influx of water, and no time can be set for the resumption of the work. The altitude of the mouth of the shaft is about 5,400 feet.

At several places on the southern slope of the lava ridge and about the heads of Kanaka Creek exploratory tunnels have been begun, and gold-bearing gravel has been found in small quantity; but, so far, there is nothing either in the character of the gravel, or in the relative positions of the tunnels, to prove the existence of any extensive continuous channel under the lava. That there is gravel in the Kanaka Creek cañon, at a higher altitude than the old channel at Alleghany and Chips's Flat, admits of no question; the deposits, however, so far as developed, have no necessary or even probable connection with each other further than this, that they may all have been on streams tributary to some main river.

The Crescent tunnel, at Buzan and Gauch's claim, was the only one I had an opportunity to visit. This tunnel is at the head of Barrett Creek, one of the forks of Kanaka Creek, between two and a half and three miles from Forest City. Its altitude is 4,855 feet, and the crest of the ridge above is about 600 feet higher. It lies in a depression between two spurs of bed-rock, which are about a quarter of a mile apart, and rise at least 150 feet above the mouth of the tunnel. The general course of the tunnel, which is not straight, is N. 30° W. (magnetic). Its length is 650 feet, and its grade, following the line between bed-rock and gravel, is about $2\frac{3}{4}$ feet to the hundred. At the shaft, which was sunk fifty-one feet to bed-rock, there were six feet of gravel covered first with a bluish clay, showing impressions of leaves, and then to the surface with mountain cement. The gravel which I saw contained considerable float bed-rock, together with quartz which shows comparatively little wear; there were no pebbles of volcanic origin. The yield of gold is not great, perhaps from twenty-five cents to one dollar per car-load. The indications are that this gravel lies at the head of a ravine which formerly led down from the ridge, and over which the volcanic flow extended.

The '76 tunnel is about a mile and a half farther east, at the head of Wild Joe Cañon. Seen from a point on the opposite side of Kanaka Creek, it appeared to be about 400 feet higher than the Crescent tunnel.

The lava capping grows very narrow at the sag, or divide, between Wolf Creek on the south and Jim Crow Cañon on the north. This latter cañon is reported to have been extremely rich in early days. At its head, though several hundred feet below the crest of the ridge, are the old Nebraska workings. There is no work now going on at this point, and partly on account of the lack of time, and partly on account of the snow, I made no attempt to study the region in detail. The descriptions given me by some of the older miners, however, lead me to think that a careful examination would bring to light several features of interest. It is still a question whether the gravel is continuous under the lava between Nebraska and American Hill.

The American Hill mining claim lies upon both sides of Wolf Creek, the claim next easterly on the southern side of the creek being known as Bunker Hill. The bed-rock is slate, but the line between the slate and the granite to the east is not more than a quarter of a mile away. The bed-rock surface where exposed to view is extremely uneven. I took an observation at the base of the most northerly bank at American Hill, and made its altitude to be 4,880 feet, or 700 feet less than that of the ridge road at the point where the road to the mine leaves it. The altitude of the bed-rock in the cut at Bunker Hill is only 4,700 feet. On the northern side of the creek there is apparently a rim of bed-rock nearly parallel with the present course of the creek; though farther to the west, below where a bad slide of lava has covered the rim, the channel is thought to recede farther back into the hill, and the spur between Big and Little Wolf Creeks is all gravel. The long tunnels, such as the Whiskey tunnel and the Dutch tunnel, which have been driven in the exploration of the American Hill gravel, are not now accessible. It is said that two channels, a front and a back one, have been recognized, the latter being considerably higher than the former, carrying a different kind of gravel, and yielding a finer quality of gold. There may be some connection between these two channels and the two which are reported as being distinguishable at Nebraska; but their mutual relations have not yet been made out. The front gravel at American Hill is a white quartz, mixed with some blue, and is not much worn; some of the pieces are still very angular. At Bunker Hill there is considerable float bed-rock in the gravel, and a slide of lava and clay had covered the eastern bank. The lower limit of the lava flow, on the ridge between Wolf Creek and the Middle Yuba, is just above the Bunker Hill gravel. Until mining operations are resumed and the old tunnels are reopened, or new tunnels are driven, it will be difficult to make any satisfactory report, or one that will clear up the doubts which exist in regard to the source of these gravels, or their connection with other deposits either above or below.

C. SIERRA CITY AND VICINITY.

Sierra City lies upon the South Fork of the North Yuba River, a short distance below the junction of a northerly and a southerly branch, thirteen miles above Downieville, and at the southern base of the Sierra Buttes. The town is built in the cañon, near the level of the stream. Its altitude, at the Yuba Gap House, is 4,175 feet. To the south of the town rises the high northeasterly extension of the Forest City ridge. I did not determine the altitudes of any of the highest points in this vicinity on account of the snow.

The bed-rock between Sierra City and Downieville, so far as exposed to view in the cañon and along the stage road, is an easily cleavable slate. About four miles above Sierra City, on the Sierraville road, there is a change to granite; and the easterly portion of the slate is, in places at least, much more compact and much less easily cleaved than are the portions farther west. In crossing the ridge which lies between Sierra City and Sierraville I saw no bed-rock at all on the eastern slope, unless possibly some of the granite was rock in place. Boulders of granite and of volcanic rock were abundant. Granite bed-rock is also to be seen near the Blue Gravel claim on the eastern side of Milton Creek, a tributary which joins the southern branch already mentioned about four miles above Sierra City. Upon Chips's Hill, which is the name given to the high point a couple of miles to the east of Sierra City, between the northern and the southern branches of the main stream, the relations of the bed-rock are not easy to make out. I saw schistose or slaty rock on the northwestern slope of the hill, and at the mouth of the 1001 tunnel, which is at the head of a ravine on the southeastern slope. The rock at an intermediate point is called porodite by Mr. Wadsworth, who has examined it microscopically.

The crests of the principal ridges and high points above and about Sierra City (I cannot speak positively about the Buttes) are covered with rock of volcanic origin, but it is not always clear that it was brought into its present position by volcanic agencies alone. Upon the granite near the Blue Gravel claim I saw distinct signs of glacial action. No report of much value can be made upon this portion of the volcanic capping without a prolonged study in the field. There is no good map to refer to, and the country is not an easy one to explore. The cañons are deep; the ridges are difficult of access; trails are very infrequent; there are no permanent settlements at the higher altitudes, and but few even temporary mining camps. One stage-road leads across the ridge from Sierra City to Sierraville, passing through the Yuba Gap at an altitude of about 6,625 feet, or 2,450 feet above Sierra City. To make a satisfactory geological survey of this region an explorer ought to go provided with camp outfit and all the instruments ordinarily used in a topographical reconnaissance.

The volcanic capping of the ridge between the Middle and the North Yuba rivers is in all probability continuous from Forest City upwards as far as Milton Creek; and it is possible that it is also unbroken along the narrow ridge in which this creek takes its rise. To the east and northeast of the creek, the ridge which forms the divide between the creek and the main stream (the two streams here following nearly parallel directions) is capped with lava. Farther north, upon the northern slope of the principal cañon above Sierra City, there appears to be a succession of bed-rock spurs, rising higher and higher towards the head of the stream, the whole capped with a volcanic deposit, of which one of the lower extremities is to be seen at Chips's Hill. Still farther north, beyond the line of the stage-road, the Mount Haskell ridge is said to be covered with the same kind of material, but I am not able to speak from personal observation. My own examinations were confined to Chips's Hill, and the ridge northeast of Milton Creek, and the trails leading to them from Sierra City.

Upon the trail to Chips's Hill I saw a great number of small and large boulders, porphyritic with feldspar or hornblende, or with both. Their weathered surfaces in particular attracted my attention. The highest portions of the hill, however, are free from boulders of this character. I saw none at any higher altitude than 5,800 feet, nearly a thousand feet below the crest of the hill as it rises above the 1001 tunnel. The top of the hill and ridge is said to be capped with hard and compact lava, but I saw nothing of the kind, though I climbed to within two hundred feet of the top, where the snow lay so deep as to make any farther ascent useless. The whole surface was covered with large masses of rock, apparently of volcanic origin. Some of these masses seemed to be quite different from the other rock in the vicinity, and it is not unlikely that they were brought into their present position through the agency of ice.

Upon my trip to the Blue Gravel claim I was accompanied and guided by Dr. J. J. Sawyer, of Sierra City. Our trail led us up the southern branch of the river to the mouth of Milton Creek, thence along the creek in a southeasterly direction, and finally up the steep slope on the northeast or right bank of the creek. Along the river trail there was no bed-rock to be seen, but there was an abundance of granitic and porphyritic boulders, similar to those seen on Chips's Hill. After entering the valley of Milton Creek we saw no more boulders of this character, which shows that their origin must be looked for more to the east or northeast. The bed-rock, which rises to great heights both to the east and the west, is very hard, compact, and dark-blue in color. About a mile from the mouth of the creek granite boulders were again met, which evidently came from the granite country at the head of the creek.

The altitude of the cabin at the Blue Gravel claim is 5,938 feet. The cabin stands upon the volcanic capping, near its lower limit. The capping at this point is composed mostly of boulders, some of which are of undoubted volcanic origin. Many of them are light-colored crystalline rocks, which at first sight have a striking resemblance to quartz. They generally show a partially rounded exterior. From the cabin I climbed about 500 feet up the slope of the ridge, and estimated that the crest was still between six and seven hundred feet higher. The surface was covered with boulders, similar to those seen lower down, in connection with boulders of granite, which were abundant. One persistent stratum, locally known as "pipe-clay," though having only a distant resemblance to the pipe-clay of the gravel mines, is seen cropping out in all the ravines. It is really a sandy streak, carrying much crystalline matter, brilliantly white when seen from a little distance, and reminding one in its general appearance of the Chalk Bluff at You Bet. Had I had more time, I would have climbed to the crest of the ridge. I wanted to inspect more closely what appeared to be castle-like masses of black and solid lava, extending as a broken and rugged ridge along the highest crest. Seen from below, this mass, which, judging from the appearance of a few scattered boulders, I suppose to be very different in character from the main body of the capping, appeared to be at least fifty feet in thickness.

There is a very general belief prevalent in the mining community of Sierra County that under these accumulations of boulders, sandy strata, and other volcanic material (if, indeed, it be strictly volcanic, and not in part glacial), there are old auriferous gravel channels much broader and more extensive than those existing at lower altitudes. Mr. Hendel appears to have adopted this view in the report prepared for Mr. Raymond, from which extracts have been given in a previous chapter.* It is thought by some that a channel will be traced from Haskell's Peak by way of Chips's Hill to a junction with another channel coming from a more easterly direction, and that the two united follow a course under the lava, by way of American Hill and Nebraska, to some point near Forest City, where a union is effected with the Bald Mountain channel. Others think that the high channel followed an independent course towards the south, and crossed the line of the present Middle Yuba River near the site of Milton, without making any connection at all with the lower channel, which passes by Forest City.

It was the reports of recent discoveries of gold and of extensive prospecting in this vicinity that made me decide to visit the region, and to get what information I could in regard to this question of the existence of high channels and in regard to actual mining operations. The partial examination that I was able to make in the short time at my disposal, at an unfavorable season of the year, is hardly sufficient to justify the expression of any decided opinion; but I may be allowed to say that I failed to see any very strong evidence upon which to rest a belief in the former existence of large rivers. I am not ready to assert that there are no old gravel channels in this part of the Sierra Nevada; I do not think it proved that there are any. The mining operations are still on too small a scale to settle the question.

I have already alluded to the Pliocene shaft and other exploratory works on the ridge near Forest City. In their unfinished condition these works show that persons can be found with confidence enough in the existence of a channel under the lava to spend money in hunting for it, but do not prove anything further. Undertakings similar in character have also been begun at higher points. The Savage Company, of which J. T. Mooney of Sierra City is the superintendent, has been engaged since the summer of 1877 in sinking a shaft and driving a tunnel at a claim on the southern slope of the ridge, about half-way between Forest City and Sierra City. The shaft has reached a depth of 139 feet. The present length of the tunnel is something over 325 feet. Owing to the lateness of the season, I was obliged to abandon my proposed visit to this claim.

At the Blue Gravel claim, east of Milton Creek, a small amount of work has been done, but all

* See ante, pp. 210, 211.

in the overlying material. From the bottom of the shaft, forty-two feet deep, some sandy matter carrying woody fibre and a few small rolled pebbles have been taken. Between the shaft and the creek, and along the sides of the cañon, there are frequent exposures of granitic rock, which, I was told, dip underneath the volcanic capping, and are regarded as indications of a rim-rock. I could not examine these points in detail.

The mouth of the 1001 tunnel is at the head of a small ravine upon the southeastern slope of Chips's Hill. Its altitude I made to be 5,938 feet, exactly the same as that of the Blue Gravel cabin. Its course is N. 42° W. (magnetic). The tunnel was started at the base of the rolled gravel, the bottom of the tunnel being in a schistose rock. The rock was so much weathered and broken that I could get no satisfactory observations for dip and strike. The tunnel is not driven upon or through a sloping rim-rock; if there is a true channel here, the tunnel follows its bed, or nearly so. At the point where the two branch tunnels were started and the raising of the air-shaft was begun, 335 feet from the mouth of the tunnel, the bed-rock pitches down a little, but soon rises again. By lowering the tunnel four feet, the lowest point could be drained. The northeasterly branch has a length of one hundred feet; the westerly, of fifty. The main tunnel has been extended to a total length of 500 feet. The air-shaft will be 179 feet deep. The gravel is made up of volcanic and granitic pebbles and boulders, with occasional bunches of micaceous sand and some carbonized wood. In the latter, the original woody structure is unusually well preserved. The boulders are of all sizes, up to six or eight feet through. I saw no quartz pebbles at all, with the possible exception of a few very small ones. The boulders are usually well rounded. The air-shaft has been raised forty feet, and without any essential change in the character of the gravel. It is evident that the channel — if one exists — was formed and filled under quite different conditions from those which prevailed at the time of the formation of the lower channels, which have been previously described. Some coarse gold has been found in this gravel, but I have no information as to its amount.

The prospecting tunnel at Haskell's Peak, or Mount Haskell, I should have tried to visit, if the weather had been more favorable; but, as the tunnel was still in bed-rock, not much information of a decisive character could have been expected. In regard to this tunnel I have been informed in a private letter from Dr. Sawyer, under date of March 30, 1880, that, after running through about 450 feet of granite bed-rock, the prospecters have struck the channel. "They found very nice washed gravel, more than half quartz. They are too high, however, and will have to sink down after running in some distance." This points very clearly to the existence of a rim-rock, but our knowledge of the district is so far from complete that it will not be worth while to speculate upon the probabilities of there being a connection between Mount Haskell and Chips's Hill, or the Blue Gravel claim.

The Gold Lake region, a few miles to the west of Mount Haskell, I had to leave entirely unexplored.

Gravel deposits of more recent origin may be seen at many points along the present cañon of the Yuba below Sierra City, certainly as far as Downieville. At Loganville, about three miles below Sierra City, there is a bench of gravel extending for more than a quarter of a mile, nearly parallel with the present stream, and about fifty feet higher. The bed-rock rises rapidly to the south, and is again visible beyond the gravel, upon the steep slope of the cañon, three or four hundred feet farther back from the river. At the time of my visit the face of the bank was about seventy feet in height. The gravel is well-rounded, but there is a great quantity of large boulders, some of them as much as twelve or fifteen feet in diameter. The boulders are largely hornblendic, granitic, or porphyritic in character, with but few of quartz. They must have been brought from the higher Sierra, the bed-rock in the vicinity being all slate.

The grade of the former bed of the Yuba, to which this and similar deposits evidently belong, was considerably less than that of the present stream, for the benches are at higher and higher levels above the river, the farther they are removed from its source. I noticed this fact particularly at Downieville, but did not take any measurements. The cañon of the North Yuba River, between

Downieville and Indian Hill, I did not explore at all; but it seems highly probable that the remarkable boulder stratum at the latter place was once in easy communication with such deposits as that at Loganville. The boulders, at least, may well have had the same origin. Over some portions of this district of country it is hard to tell where volcanic or plutonic agencies ceased to act, and glacial or fluviatile began.

SECTION VII. — *The Region North of the North Yuba River, and South of the Divide between the Yuba and the Feather Rivers.*

This extensive and important district lies partly in Sierra and partly in Plumas County. The boundary of the portion to be referred to in this report follows the ridge to the west of Slate Creek and La Porte as far as Pilot Peak, thence along the high crest, to the headwaters of the North Fork of the North Yuba River, and finally down the North Fork and the North Yuba, by way of Downieville, to the point of beginning.

There is no published map of the whole of this region to which the reader can be referred; none, that is, on a sufficiently large scale. At the court-house in Quincy, the county seat of Plumas County, there is a map in manuscript entitled "Official Map of Plumas County. By Arthur W. Keddie, C. E. Based on the work of the State Geological Survey by V. Wackenreuder, in 1866, under the direction of J. D. Whitney, State Geologist. Approved by Board of Supervisors, May 6, 1874. T. B. Whiting, Clerk of Board." From a tracing of this map in Mr. Hendel's office at La Porte, to which he had made additions from his own surveys, I made a second tracing of such portions as were necessary for the work in hand. The scale of this tracing is a mile and a half to one inch. Mr. Hendel also kindly allowed me to make use of data, which he had collected in his surveys, in the preparation of maps of parts of this district, to which reference will be made on subsequent pages.

The principal topographical features of this district are very striking and well-marked. The lowest point is at its southwestern corner, in the cañon of the North Yuba, where the altitude is not far from 2,000 feet. The highest peak is Alturas Mountain, near Howland Flat, of which the altitude — not precisely determined — is probably over 7,500 feet. The cañons, which have a general southwesterly direction, are deep and narrow; and the intervening ridges are also narrow and sharp. A line drawn in a southeasterly direction from a point near La Porte crosses, within a distance of less than twelve miles, four deep cañons, — those of Slate, Cañon, Little Cañon, and Goodyear creeks. The trail from La Porte to Downieville crosses the ridge between Slate Creek and Cañon Creek at an altitude of more than 1,000 feet above the stream on either side. Between Cañon Creek and Little Cañon Creek the trail by way of Craig's Flat reaches a height of 800 feet above the level of the water; and a ridge nearly a thousand feet high has to be crossed in getting from Little Cañon Creek to Goodyear Creek. Beyond Goodyear Creek the sag in the ridge at Monte Cristo is still higher. The relative positions and altitudes of several points on the Downieville trail are given on the diagram section. (Plate G, Fig. 3.) There are a few good wagon-roads in the western part of the district, connecting places on the opposite sides of Slate Creek; but, for the rest, the chief dependence has to be placed on trails. The extreme irregularity of the surface and the lack of easy means of communication make it difficult to get anything more than a general outline of the geological structure without the expenditure of a great deal of time.

The prevailing bed-rock is a slate, similar in character to that which makes up the greater part of the western slope of the Sierra Nevada. Its local variations and modifications will be referred to in the descriptions, which are to follow, of special localities. The slate is crossed by a belt of serpentine, which follows a course a little to the west of north, and which has in places a width of at least four or five hundred feet. I cannot say whether this belt is a continuation of that which has already been referred to as crossing the Forest City divide, or is a separate belt, nearly parallel with the former. At present I am inclined to the latter opinion, for the Forest City serpentine, unless

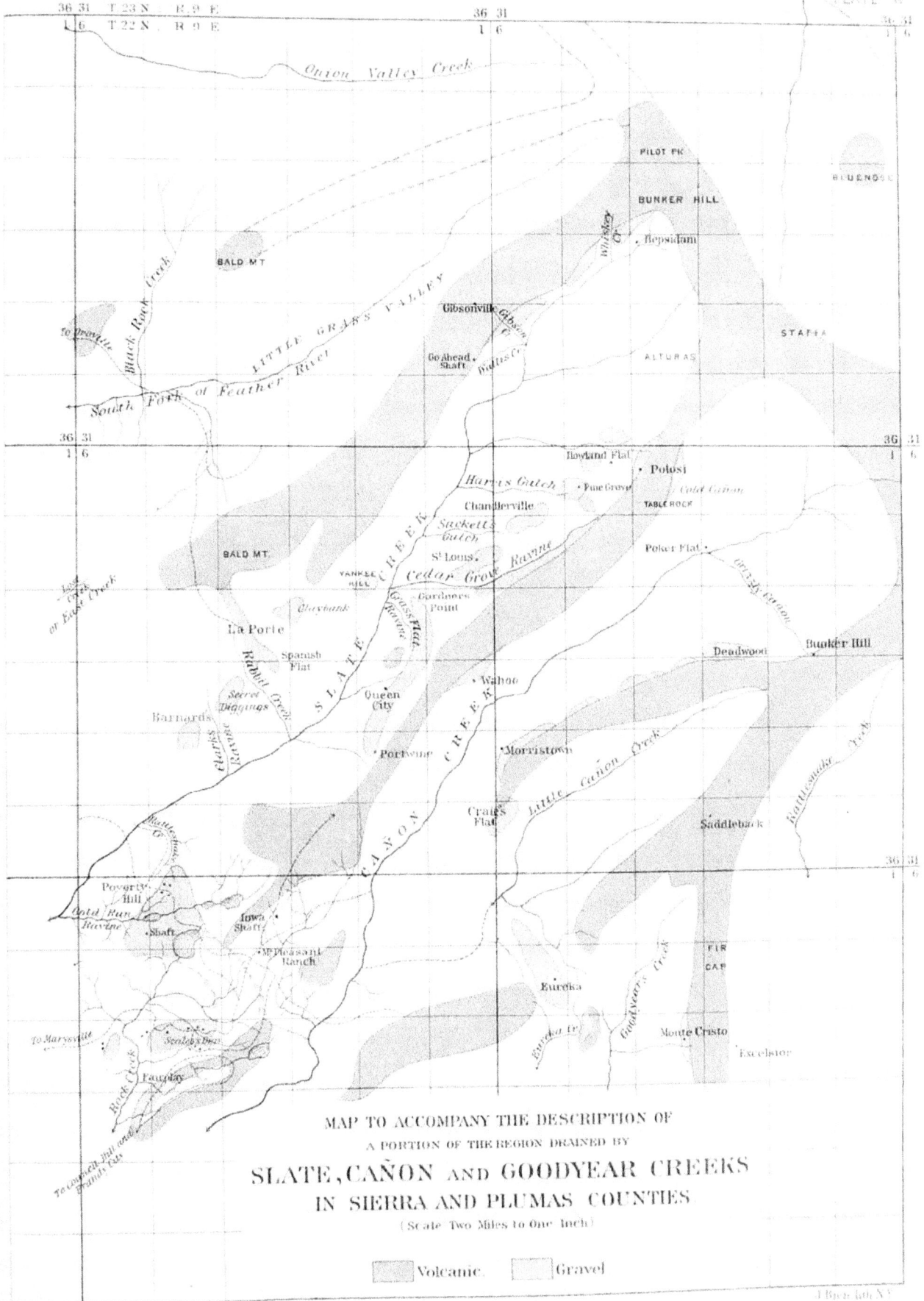

36 31 T 23 N R 9 E
1 6 T 22 N R 9 E

36 31
1 6

36 31
1 6

Onion Valley Creek

PILOT PK

BLUENOSE

BUNKER HILL

Hepsidam

BALD MT

Whiskey Cr

LITTLE GRASS VALLEY

Black Rock Creek

Gibsonville

Gibson

STAFFA

to Oroville

Go Ahead Shaft

Willis Cr

ALTURAS

South Fork of Feather River

36 31
1 6

Howland Flat

Polosi

Harris Gulch

Pine Grove

Cold Cañon

Chandlerville

TABLE ROCK

Sackett's Gulch

St Louis

Cedar Grove Ravine

Poker Flat

BALD MT

YANKEE HILL

Grass Flat

Gardners Point

Grizzly Cañon

Lost Creek or East Creek

Claybank

La Porte

SLATE

Deadwood

Bunker Hill

Rabbit Creek

Spanish Flat

CREEK

Wahoo

Secret Diggings

Queen City

Clarks Ravine

Barnards

CAÑON

Little Cañon Creek

Morristown

Rattlesnake Creek

Portwine

CREEK

Saddleback

Craigs Flat

36 31
1 6

Rattlesnake Cr

Poverty Hill

Cold Run Ravine

Iowa Shaft

Shaft

FIR CAP

M⁹ Pleasant Ranch

Eureka

Goodyear's Creek

To Marysville

Monte Cristo

Scales Dict

Eureka Cr

Excelsior

Rock Creek

Fauday

To Connell Hill and Brandy Cit

MAP TO ACCOMPANY THE DESCRIPTION OF

A PORTION OF THE REGION DRAINED BY

SLATE, CAÑON AND GOODYEAR CREEKS

IN SIERRA AND PLUMAS COUNTIES

(Scale Two Miles to One Inch)

Volcanic.   Gravel

it changes its direction would lie farther to the west than the serpentine that I saw on the trail from Downieville to Excelsior, on the trail from Deadwood to Poker Flat, and near Whiskey Diggings or Newark.

The highest points are capped with a basaltic or an andesitic lava, and upon the crests of the ridges between the cañons the lava extends in almost unbroken streams nearly to the southern and southwestern limits of the district. The mountain ridges at the head of the cañons are much more irregular in outline than are the ridges between the forks of the Yuba River in Nevada and Sierra Counties, which have been described in the previous pages. There is very little of that smoothness and evenness of grade which is so marked a feature of the more southern country. Sharp peaks alternate with deep sags, and their sides are deeply furrowed with secondary cañons and gulches. There are also many evidences of the frequency of land slides. I could not find time to climb any of the higher peaks, nor to make a much needed examination of the country at the head of Cañon Creek. The principal end in view in such an examination would be to ascertain the relations of the volcanic material to the bed-rock. I was told by persons well acquainted in the region that the bed of Cañon Creek is crossed by volcanic rock near Poker Flat. In the short time at my disposal I was unable to satisfy myself in regard to this matter. The serpentine belt appears, it is true, to abut directly against volcanic rock; at least I saw none of the characteristic slate at the point where I crossed the creek, but it is possible that the appearance is due to the effects of land slides, and that the volcanic rock seen was not in place. I must leave this question unsettled.

In the descriptions which follow it will be convenient to take up first the more northwesterly portions of the district, beginning with the vicinity of La Porte and Gibsonville, and then to take in order the successive ridges which lie to the southeast.

A. The Vicinity of La Porte and Gibsonville, West of Slate Creek.

I reached La Porte on the evening of the 8th of September by stage from Oroville, by way of Forbestown and Brownsville. The bed-rock observed by the roadside after leaving Brownsville was nearly all slate. No volcanic rock was seen. The highest point on the road is about a mile to the west or southwest of La Porte, on the crest of the ridge which separates the waters of Slate Creek from those of East Creek and other affluents of the South Fork of the Feather River. The town of La Porte lies on the southeastern slope of this ridge, upon Rabbit Creek, by which name the town also was formerly known. The altitude of the second story of the Union Hotel at La Porte I made to be 4,993 feet.

The map (Plate R), which is one of those for which I am largely indebted to the kindness of Mr. C. W. Hendel, shows the principal geographical features of the region, and the relative positions of bed-rock, volcanic tufa, and exposed gravel. My explorations did not extend farther south than the trail from Barnard's to Poverty Hill. Near this point the old channel crossed to the opposite side of Slate Creek, and its continuation in that direction will be one of the subjects taken up in the next section of this report. It will furthermore be convenient to consider the vicinity of La Porte as a sub-district by itself, not much being known about the half-dozen miles which intervene between La Porte and Gibsonville.

(I.) A little more than a mile to the northwest of La Porte may be seen upon the ridge the lower extremity of the volcanic capping, which extends from that point, without interruption, in a northeasterly direction to Pilot Peak. The La Porte Bald Mountain is a prominent object on the ridge, and lies a little to the east of north from the town. I determined the altitude of the sag to the west of Bald Mountain, at a point near the line of the proposed ditch tunnel, to be 5,500 feet, and estimated the top of the mountain to be two hundred and fifty feet higher. The altitude of the base of the lava, as nearly as could be ascertained, is 5,175 feet; the thickness of the volcanic capping is, therefore, nearly 600 feet. The lava is an andesite, according to Mr. Wadsworth's determination. It is frequently spoken of as "gray lava," to distinguish it from the black basalt which is found upon some of the neighboring ridges.

At the southern base of the lava ridge is the commencement of the exposed gravel deposit, which extends by way of Secret Diggings to Barnard's, being cut off only at Rabbit Creek and Clark's Ravine. The gravel apparently extends backwards underneath the lava, and a bank similar in character would naturally be looked for on the northern slope of the ridge, near Little Grass Valley. I saw no signs of any bank, however, upon that side, and I was told that the drifts, which have been run under the ridge from the south, were cut off abruptly by a wall of gray lava, and that all efforts to find the continuation of the channel in that direction had been unsuccessful. It was not possible for me to examine the old workings, and, consequently, I cannot verify these statements from personal observation. There is a theory held by some, that here was one of the original fissures from which the volcanic material was ejected. I am not ready to accept this theory, though I see no reason to doubt the statements made relative to the cutting off of the gravel. There is abundant additional evidence that this section of country has been the scene of unusual disturbance since the deposition of the gravel.

It was not easy to make out the character of the gravel at the northern extremity of the deposit —at the base of the lava—on account of the heavy slides from the top which have taken place since active mining operations were suspended. As nearly as I could estimate, there was a stratum of fine white quartz gravel at the bottom, about fifty feet in thickness, upon which, and reaching up to the base of the lava, was a stratum of equal or greater thickness of a yellowish or grayish pipe-clay. From this bank downwards through the La Porte and Secret Diggings mines, for a distance of over two miles, the greater part of the deep gravel has been removed. There are still considerable masses of gravel in places along the rims, which will ultimately be washed away; and the amount still remaining at Barnard's is also probably large, but I am not able to make any specific report. The gravel in the banks which remain unwashed, and the boulders — large and small — which have been left upon the bed-rock, are both almost exclusively either white or bluish quartz. The gravel is fine, the pebbles seldom exceeding four or five inches in diameter, though some of the larger boulders are as much as six or eight feet through. At Secret Diggings there is some sand, and occasional bunches of pipe-clay are mixed with the gravel near the top. The boulders seen upon the bed-rock other than quartz doubtless came from the surface, and are mostly of a volcanic character. Some of them contain porphyritically disseminated hornblende. The quartz is well-washed and rounded, and it generally has a brilliant lustrous surface. This latter feature is so characteristic of the gravel of the old channels in this vicinity that all gravel, however well smoothed and rounded, which is dull in lustre or lacks clearness of color, is spoken of with contempt as "bogus," "river wash," or "bastard gravel," and is said to be barren of gold. I saw considerable "bogus gravel" upon the surface of the ground to the east of the main La Porte channel. In the upper portions of the La Porte banks the gravel exhibits a well-marked, small-lenticular stratification, or has, using the terms employed in Dana's Manual of Geology, a "sand-drift" or a "flow-and-plunge" structure, in striking contrast with the more uniform stratification of the banks at Malakoff and other points in Nevada County.

The general course of the channel at La Porte is southeasterly, but at Secret Diggings there is a long, sweeping curve to the southwest. The rim on the west is usually at a lower level than that on the east, which is due in part to the erosive action of the western creek and ravine, and in part to the disturbances whose effects are more noticeable on the east. For a part of the distance there seem to be two parallel channels, the more easterly one running upon a higher bench of bed-rock. To the north of the town of La Porte the bed-rock is quite uneven, the channel being divided by island-like masses, which rise to heights of thirty feet or more above the deeper portions. Below the town the channel is more regular. At Secret Diggings, in the widest portion, there is a well-defined, gutter-like channel along the centre. The width of the channel or of the gravel deposit is not uniform.

The observations which I took to determine the grade of the channel are not in all respects satisfactory. I made the altitude of the deep northern bed-rock, at the base of the lava ridge, to be 5,077 feet. At the southeastern outlet of the La Porte mines, where the channel is supposed to have crossed to Secret Diggings, the bed-rock is covered to a great extent by slides of earth, and it is not possible to tell the precise position of the deepest portions. As nearly as I could judge, the deep rock has an altitude of only a little more than 4,900 feet, hardly as high as the rock at the upper end of Secret Diggings, where I made the altitude to be 4,913 feet. These measurements were made with the aneroid barometer, but in both cases there was a direct comparison of the aneroid with the mercurial barometer at the Union Hotel, within less than two hours of the time of observation. There is thus a little uncertainty as to the precise mode of connection between La Porte and Secret Diggings. There is apparently a low ridge of bed-rock between the places, which, however, does not rise high enough to shut off all possibility of connection, and there can be no doubt that a connection formerly existed. The altitude of the bed-rock at the southwestern end of Secret Diggings I made to be 4,828 feet, which gives a total fall of two hundred and forty-nine feet in a little more than two miles.

The thickness of the gravel has been on the average between forty and sixty feet. I could not learn of the existence of any animal fossils in the gravel. Silicified and carbonized wood have both been found in great abundance. The latter has been used as fuel in the blacksmith-shops.

The bed-rock under the gravel is in some places a hard bluish slate, but in others it is soft and clay-like, showing for a variable depth alternations of color and arrangement similar to those which have been referred to already in connection with Plug Ugly Hill, near Dutch Flat. Below the soft stratum the rock is moderately hard and bluish in color, resembling in composition and structure a chloritic slate. The depth to which the alteration extends is from nothing to eight or ten feet. The altered rock is grayish, bluish, reddish, yellowish, or nearly white in color. The line of demarcation between the soft and the hard rock is always sharp and distinct, even when wavy and uneven, and the stripes of differently colored clays run parallel with this line. The planes of cleavage or lamination are independent of the color, and the joints and fractures pass without interruption indiscriminately through all the varieties. In some places this altered rock shows a fine horizontal lamination ; and, again, when seen in mass, it looks as if it were made up of brightly colored concentric shells or layers, resembling the structure of agate. This appearance is probably the result of the gradual decomposition of bed-rock boulders.

The evidences of disturbance subsequent to the deposition of the gravel are of various kinds. Allusion has already been made to the cutting off of the La Porte gravel under the lava ridge to the north of the town. I will now call attention to one or two other cases. The eastern portion of what is included on the map (Plate R) under the name of Secret Diggings is known as Illinois Hill. The bed-rock at this hill is from fifty to one hundred feet higher than that in the adjacent Secret Diggings mines. The diagram (Plate S, Fig. 2) shows an east and west section across Illinois Hill and Secret Diggings, as sketched on the spot. No accurate measurements were taken either for horizontal distances or for altitudes. The top of the gravel which remains unwashed at the northern end of Illinois Hill has an altitude of 4,993 feet, and that of the slate bed-rock at the base of the lava knob is 4,916 feet. The lava knob here referred to covers an area of two or three hundred feet square, and rises to a height of about sixty feet above the adjacent slate. It stands nearly opposite the middle of the Secret Diggings mines. Upon its northern, eastern, and southern sides there used to be a deposit of clean quartz gravel, the greater part of which has been washed away. Its western side forms a part of the steep, precipitous eastern rim of Secret Diggings. The bed-rock in Secret Diggings is said to have been much disturbed and broken up where it abutted against the lava ; but I saw nothing unusual either in the position or in the character of the slates where they joined the lava on Illinois Hill. Enclosed within the lava at intervals, even to its top, there were small streaks and bunches of quartz gravel, ranging from one or two up to eight or ten feet in the longest dimension. The quartz pebbles were, many of them, partially or entirely coated with a red or a black scale, and were brittle, as if they had been exposed to a high heat. In a conversation about this (or some other) body of intrusive lava at Secret Diggings, Colonel B. F. Baker of Gardner's Point told me that in early days, when the lava was first struck, the miners ran a tunnel into it at bed-rock level, and then sank a shaft sixty feet in depth without again reaching bed-rock. He also told me that in the neighborhood of the main mass sheets of lava, from one foot to four or five feet in thickness, and lying in nearly horizontal positions, were found penetrating the gravel beds for as much as fifty feet, and that in many cases blocks of detached gravel were completely encased in lava. The only explanation consistent with the observed facts is that after the deposition of the gravel, it was broken through forcibly by an eruption from below. Other interesting facts of a similar nature will be given on a subsequent page.

The eastern rim-rock at La Porte also shows signs of having been disturbed, so much so that in some places it actually overhangs the gravel. During the period of the removal of the gravel this abnormal position of bed-rock was very noticeable. I have the authority of Mr. Hendel for saying that the layers of gravel and sand could easily be seen to conform to the displaced position of the rock. The section (Plate 8, Fig. 3) was taken from one of Mr. Hendel's old note-books, having been made by him when the curvatures could be distinctly seen in the mines.

The disturbance of the eastern rim is further illustrated in the prospecting tunnel upon which Mr. Ellis was at work at the time of my visit. Mr. Ellis holds some peculiar theoretical views as to the origin of the gravel deposits, and has commenced a tunnel in the eastern rim-rock in the hope of proving their correctness. The tunnel is 115 feet long. It starts in the broken rock which has fallen from above, but the rock soon gets harder, and for fifty feet is of a dark color, with the exception of one light-colored clayey seam. Fifty feet from the mouth of the tunnel there is to be seen a narrow strip of rolled gravel in the roof; and twenty-five feet farther in there is again gravel in the roof, and bed-rock also above the gravel. The slates above the tunnel are nearly horizontal, instead of having a dip of from 65° to 70° to the east as they do a short distance away. Everything points to the conclusion that the overhanging bed-rock in its fall has caught and enclosed occasional bunches of gravel.

It seems probable also that the position of the Spanish Flat gravel has some relation to these disturbances. Spanish Flat lies just to the southeast of the main La Porte deposit, but its bed-rock is about one hundred and fifty feet lower than that at the outlet of the La Porte mines. Upon the sloping surface of slate rock between the two places I saw quartz gravel at almost every step. The dip of the bed-rock, particularly on the northwestern side of the Flat, is very irregular, changing within a few feet from the vertical to forty-five degrees on either side, and being in places nearly horizontal. The rock in the central portions of the Flat is more nearly normal in appearance. The gravel, particularly in the upper twenty feet, shows the same "sand-drift" arrangement which is so noticeable in the upper layers of the La Porte gravel. This deposit suggests many interesting inquiries, for which no easy answers are ready. May it not be possible, for instance, that the La Porte channel, either before or after the deposition of the main mass of gravel, occupied this lower level? Or, may it not be possible that a connection once existed by way of the Claybank gravel with some channel now concealed under the lava to the northeast? The Claybank deposit lies about a quarter of a mile to the east of the central portions of the La Porte mines, and its bed-rock is said to be 160 feet lower than that at La Porte. See Mr. Hendel's section (Plate 8, Fig. 3). I could not make any satisfactory examination of this deposit. There has been no bank exposed by hydraulic washing, and it did not seem worth while to try to penetrate the old tunnels, in which no work has recently been done. It is probable that they are inaccessible. The ground is said to have been prospected by a tunnel 2,400 feet in length "nearly all through gravel, with 1,250 feet of cross or side drifts to the right and left, all in gravel." The current reports are that good rolled gravel was found, and that the pitch of the bed-rock was such as to lead to a belief in the existence of a deep channel under Bald Mountain. It is indeed supposed by some persons that the Claybank deposit is the natural southwestern outlet of the Gibsonville channel. If this supposition be true, some very difficult questions will arise in regard to the mutual relations of the two channels, which here so nearly approached each other, and yet flowed at such different levels. The question, From what direction did the old channel come? is one of practical importance to the owners of mining property in the vicinity of La Porte. To this question I will not attempt to give a final answer. In the time at my disposal I could not become

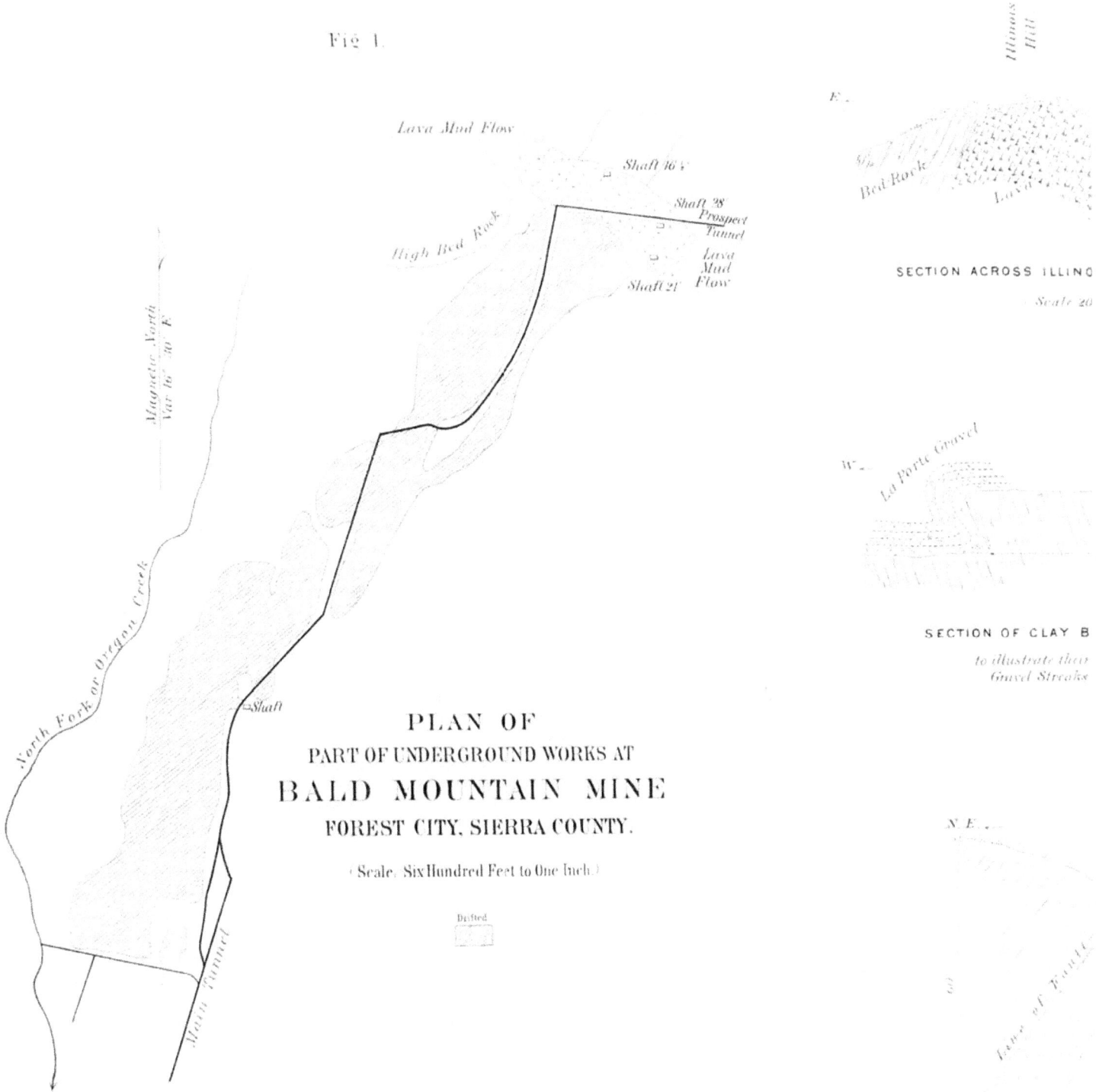

Fig 1.

Lava Mud Flow

Shaft 16 ?

Shaft 28
Prospect
Tunnel

High Bed Rock

Lava
Mud
Flow

Shaft 21

Shaft

PLAN OF
PART OF UNDERGROUND WORKS AT
BALD MOUNTAIN MINE
FOREST CITY, SIERRA COUNTY.

(Scale, Six Hundred Feet to One Inch.)

Drifted

Shaft

Main Tunnel

North Fork of Oregon Creek

Magnetic North
Var 16° 30' E.

E.

Illinois
Hill

Bed Rock

Lava

SECTION ACROSS ILLINO

Scale 20

W.

La Porte Gravel

SECTION OF CLAY B

to illustrate their
Gravel Streaks

N. E.

SECTIO
A

Fig 6

SECTION ACROSS PINE GROVE CHANNEL.

Fig 5.

FAULT IN GRAVEL; MONTE CRISTO SLOPE.
ABOVE PORTWINE, SIERRA CO.

ND SECRET DIGGINGS

Fig 3

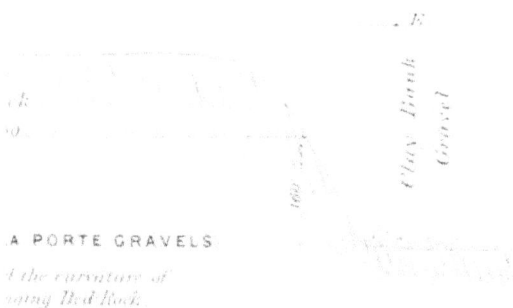

A PORTE GRAVELS

Fig 4.

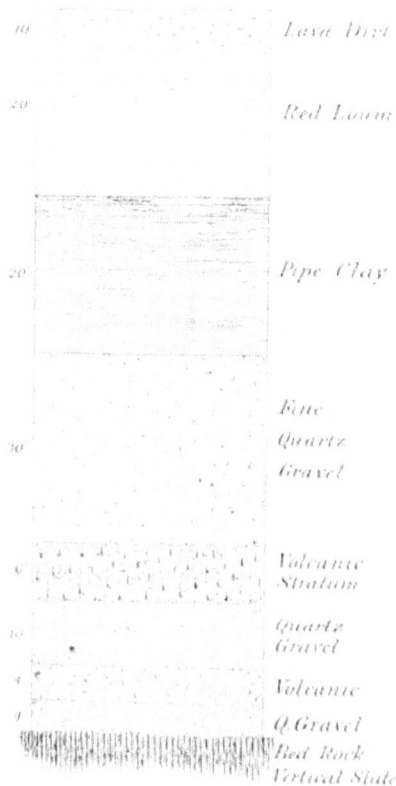

SECTION OF GRAVEL BANK
AT PORTWINE

ANDY CITY FAULT

thoroughly acquainted with all the geological features of the district, nor could I make the careful and detailed survey upon which the decision of such an important question ought to be founded. I must be content with describing as clearly as I can those things which I actually observed, with cautiously expressing the opinions to which my observations led, and with suggesting some of the more important points as to which further information must come from further explorations. My present opinion is that the ridge of lava between La Porte and Gibsonville covers an old channel, and that none of the gravel deposits in the valley of the Feather River can be regarded as belonging to the system of drainage which had its outlet by way of La Porte.

Trustworthy statistics in regard to the amount of gravel that has been removed from the mines of this vicinity, and in regard to the yield of gold in gross or per cubic yard, are hard to get. All accounts agree in estimating the yield of the district by millions of dollars, though it is not possible to say in just what way the aggregate yield ought to be credited to the different mining camps. La Porte has been one of the principal shipping centres for the gold obtained, not only from its own immediate vicinity, but also from the extensive deposits of gravel near Gibsonville and on the opposite side of Slate Creek between Potosi and Portwine. To give some idea of the past value of these mines, I will quote a few sentences from a sworn statement of Dr. S. T. Brewster, the proprietor of the stage-line from La Porte to Marysville from 1855 to 1871. Dr. Brewster's statement as to the business of those years is : " . . . . from books and accounts in my possession I find that there has been transported for the banking-house of John Conly & Co. [of La Porte] over $40,000,000 worth of gold-dust and bullion. I get this account valuing gold-dust at $20,000 for the 100 pounds, making 100 tons of treasure for that house alone." Other banking-houses shipped about half as much more, and a large amount was taken by private hands ; so that the total shipments must have amounted to at least $60,000,000, or an average of nearly $4,000,000 a year. In more recent years the yield has not exceeded $1,000,000 per year.

The following data in respect to the yield of gold per cubic yard of gravel I obtained from Mr. Hendel, who personally assured me of their accuracy. In one of the La Porte claims a piece of ground 250 by 100 feet square and with banks of about thirty feet in height yielded $87,000, an average of $3.13 per cubic yard of gravel washed. In another claim a piece of ground 125 by 100 feet square was drifted to a height of six feet, and yielded $57,500 ; this corresponds to the enormous value of $20.87 per cubic yard. At another drift mine, not in the immediate vicinity of La Porte, however, the yield amounted to $4.20 per cubic yard. When these values are compared with those obtained at Forest City, they do not seem to be overestimated.

From another source I learned that in the year 1864 or 1865 the sum of $300,000 was obtained by the washing of a portion of the Secret Diggings gravel, 300 by 100 feet square. The bank was 140 feet high. The average yield per cubic yard, therefore, was nearly two dollars for the whole bank.

(II.) I have already expressed the opinion that the lava ridge above La Porte covers a channel of gravel. The grounds for this opinion are mainly hypothetical, for, so far as I could learn, no one has any positive knowledge of what underlies the lava for a distance of at least five miles, that being about the distance from La Porte to the "Go-ahead shaft." I went over the road between La Porte and Gibsonville twice in each direction. On one occasion I was the guest of Mr. S. Wheeler, one of the proprietors of the Bank of La Porte, and a gentleman interested in many of the mining operations in the vicinity. I am also indebted to him for many other favors.

The stage-road from La Porte, after crossing the gravel mines, follows up one of the forks of Rabbit Creek, and soon gets above the level of the exposed bed-rock. For the rest of the way, until a point near Gibsonville is reached, nothing but volcanic rock is seen along the road. The summit of the road has an altitude of 5,583 feet, but this is several hundred feet below the crest of the ridge. From several points striking and impressive views of the old channel on the opposite side of Slate Creek can be obtained.

About a mile from La Porte there is a remarkable erratic boulder of large size by the side of the road. At first sight it appears to be a mass of slate rock; but upon inspection it proves to possess a highly crystalline structure.

The Go-ahead shaft was sunk several years ago. It is said to have reached gravel and bed-rock at a depth of 352 feet, when further operations were suspended on account of the great influx of water. The surface works can still be seen a little to the northwest of the road. I did not stop to go to the mouth of the shaft, but made its altitude to be approximately 5,450 feet. If the above-mentioned depth of the shaft is correctly given, the bed-rock has an altitude of only 5,100 feet, hardly twenty-five feet higher than the bed-rock at the upper end of the La Porte mines, though a little more than two hundred feet higher than that at Claybank. Compared with the bed-rock at Gibsonville, the altitude of which I made to be 5,420 feet (estimating at eighty feet the difference between it and the hotel), there is a fall of 320 feet within a distance of but little, if any, over a mile. Either the statements made about the Go-ahead shaft are incorrect, or there is here a fresh puzzle to be solved by further surveys.

The Gibsonville sub-district embraces a region about three miles in length on the right bank of the northerly fork of Slate Creek, between the creek and the high lava ridge already mentioned as extending from La Porte to Pilot Peak. The lower portion of this district is shown on the map (Plate T), which is a reduced copy of a manuscript map made by Mr. Hendel in 1876. The crest of the lava ridge is from six to nine hundred feet above the hotel at Gibsonville, the altitude of which I made to be 5,500 feet. A part of the gravel is covered by the volcanic material, and has been worked by drifting only. Where the gravel is not so covered, the hydraulic process has been used to some extent; but lack of water has prevented its general adoption, and drifting has been a necessity. With increased facilities for obtaining a supply of water, some of the banks from which the lower stratum has already been removed will ultimately be washed away.

The old custom used to be to lay out claims four hundred feet wide in front, and extend them back to the centre of the ridge. The tunnels were usually driven from 2,000 to 2,500 feet before the back rim was struck. The following list comprises the most important of the claims in this vicinity, beginning with those near Gibsonville: Chalcedonia, Boot-jack, Wild Boar, Blue, Sierra, Enterprise, Mount Pleasant, Ditch Company, Union, Nip and Tuck, Michigan, Nevada, Goodshaw, Swiftsure (at Whiskey Diggings), Kepner and Johnson, Washington, Redding, Gem, Phœnix, New York, Niagara Consolidated (including several different claims near Hepsidam, the point farthest to the northeast from Gibsonville and near the base of Pilot Peak). I have omitted the most of these names from the map (Plate T), for lack of time to ascertain the precise positions of the several claims.

There was no work doing in the hydraulic mines at the time of my visit. Even those which can command a supply of water are able to run only for from four to six months in the year, beginning usually in the month of February. During the fall and early winter work has to be suspended. There was also little or nothing doing in any of the drift mines, with the exception of the Niagara. Many of these mines, indeed, were worked out several years ago. It did not seem worth while to make the attempt to gather information — uncertain at best — about these old workings, nor to try to get access to tunnels in which work had been stopped. The most important of my observations were made at the Mount Pleasant hydraulic mine, where I was hospitably treated by Mr. F. A. Gourley, and at the Niagara drift mine, through which I was taken by the superintendent, Mr. Miles Schofield.

At Mount Pleasant a tunnel was started about twenty years ago, and drift mining was carried on for eight or ten years. Since the introduction of the hydraulic process about ten acres of bed-rock have been uncovered. The bank where highest shows at bottom a stratum of blue cement, four feet in thickness, above which come six or eight feet of red cement, and about eighty feet of clean quartz gravel.

The property of the Niagara Consolidated Company is at the head of the north fork of Slate Creek, and extends back to the crest of the ridge which connects the high point called Bunker Hill with Pilot Peak. It comprises over four hundred acres of mining ground, about one tenth of which

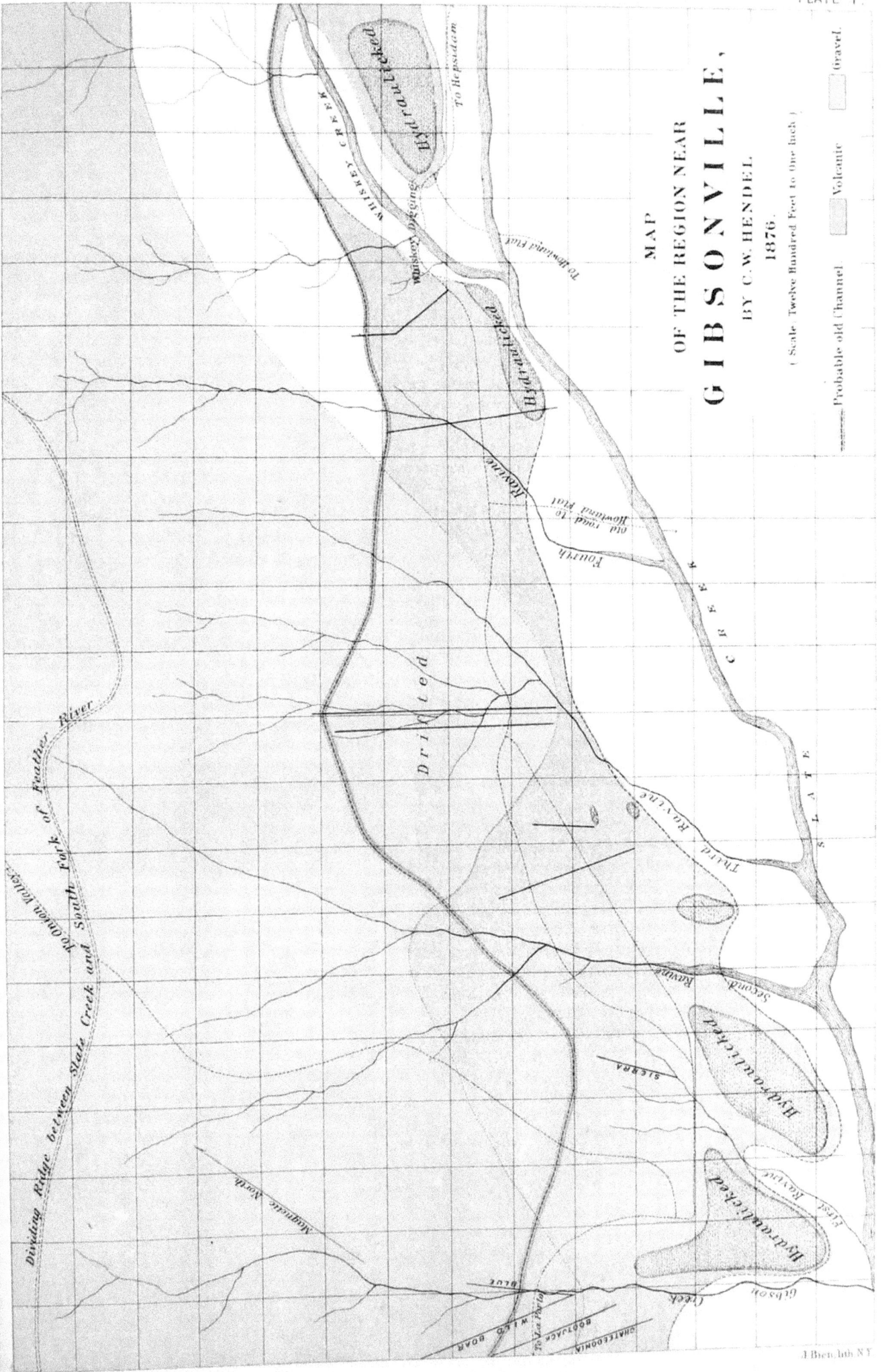

MAP
OF THE REGION NEAR
GIBSONVILLE,
BY C. W. HENDEL.
1876.

(Scale Twelve Hundred Feet to One Inch)

Probable old Channel   Volcanic   Gravel.

J. Bien, lith. N.Y.

has been explored and worked by drifting. The works extend nearly to the northeastern boundary of the property. There are two principal tunnels, the upper one being nearly a mile in length, though not straight. The main thoroughfare in the upper tunnel follows for 1,700 feet a course N. 60° E. then N. 73° E., for 800 feet, N. 43° E., 900 feet, and, finally, N. 17° W., 1,300 feet. (The bearings are all magnetic.) The grade of the tunnel in the four sections is, respectively, $1\frac{3}{4}$, $7\frac{1}{2}$, $5\frac{1}{2}$, and 6 inches for every twelve feet. The total rise in the tunnel, according to these data, would be 152 feet. At a point 1,500 feet from the mouth of the tunnel a shaft was raised 337 feet to the surface. The strata in this shaft were reported to me to be as follows: Sixty to seventy feet of white quartz gravel, forty feet of clay with gravel streaks, seven to eight feet of gravel and sand, one hundred and eighty to one hundred and ninety feet of pipe-clay, and forty feet of surface dirt, volcanic tufa, and the like. The drifts take about three and a half feet of the gravel, and from six to fourteen inches of bed-rock. The gravel is a white quartz, rather smooth, but not so well rounded and polished as that of La Porte. There is also a great deal of clayey material mixed with the gravel, amounting in places to as much as a quarter or a half of the whole deposit. The pay-streak contains frequent fragments of float bed-rock and clay boulders, which appear to be the result of the decomposition of slate boulders. Specimens 93 and 94 are from this place. These clay boulders vary in size from a few inches up to several feet in diameter. Large quartz boulders, from twenty to forty feet in diameter, are also met with. The width of the channel, where drifted, is five or six hundred feet, but is from seven hundred to one thousand feet between rims. The gold is coarse, pieces weighing as much as ten ounces being sometimes found.* The average fineness of the gold is said to be from .916 to .918. The yield of this gravel was reported to me as $1.50 per car-load, reckoning everything brought from the mine. The cars hold about a cubic yard of loose dirt, representing perhaps half a cubic yard of gravel in place. I was also told that a shaft had been raised in the gravel of the upper workings, by Mr. Morgan, through seventy feet of gravel, the average yield of which was $62\frac{1}{2}$ cents per cubic yard.

The bed-rock of this district is slate, with the exception of the serpentine belt, which has already been referred to as crossing in a northerly direction by Whiskey Diggings. Under the gravel the slate is usually very soft and full of crevices, or is changed to a depth of several feet to a clay similar to that described as seen at La Porte and Secret Diggings.

The fall in the bed-rock is quite rapid from Hepsidam to Gibsonville. The altitude of the upper tunnel, at its mouth, at the Niagara mine, I made to be 6,000 feet; the mouth of the lower tunnel is 106 feet lower. Changes in the level of bed-rock are frequent and sudden. The upper tunnel for 1,500 feet is all in slate bed-rock; though a point a hundred feet to the northwest of the mouth, and on the same level, would be forty or fifty feet above bed-rock. In the mine there are sudden changes of level, or "jumps," amounting to three or four feet at a time, and indicating the former existence of cascades or rapids. At Mount Pleasant I found the altitude of the bed-rock to be 5,560 feet, and at Gibsonville, 5,420. There is thus a fall of over 600 feet in about three miles. The gravel is nearly continuous for this distance, excepting where it has been cut across by Whiskey Creek. The opinion is stoutly held by some among the miners that there is also a back channel farther to the northwest, and hidden under the lava. It is said, for instance, that in the Union claim the northwestern rim is crest-like, and on the farther side has a pitch to the northwest, and that upon this sloping surface coarse gold was found in connection with boulders, sand, vegetable matter, and some quartz gravel, the whole being quite different in appearance from the deposit in the main channel. On the whole, I do not regard the evidence in favor of the existence of a back channel as very strong.

The two difficult questions which arise in regard to this piece of channel are, — where did it find an outlet? and from what direction did it come? There is high slate rock to the southeast

---

* A specimen selected from the stock on hand, as a "fair average nugget," weighs a quarter of an ounce. It is three quarters of an inch long by half an inch wide, and is of irregular thickness. It has numerous small angular fragments of quartz imbedded in it, and its angles and edges are all more or less rounded by abrasion.

between the forks of Slate Creek, and also between Slate Creek and Wallis Creek. There is, therefore, no natural outlet now open, except through the narrow cañon of the North Fork of Slate Creek. Possibly some light might be thrown on this question by a careful examination of the Chalcedonia and Boot-jack tunnels, which lie to the southwest of the town of Gibsonville, on the opposite side of Gibson Creek. Work is now suspended in these tunnels, and I did not make any attempt to enter them. From the best information I could get I judge that the bed-rock maintains its downward grade to the southwest beyond the natural outlet above mentioned; and if so, the channel must extend under the volcanic capping, to reappear at some point lower down, possibly as far down as La Porte. No more positive information can be given at the present time, nor is anything definite known in regard to the extension of the channel above Hepsidam. The crest of the ridge, a half mile to the northwest of Bunker Hill, has an altitude of 6,750 feet, the top of Bunker Hill being about 250 feet higher. The bed-rock directly underneath is probably 150 feet above the mouth of the upper tunnel. This leaves 600 feet for the thickness of gravel, clay, and volcanic material at this point. The crest between Bunker Hill and Pilot Peak is a part of the water-shed between the Yuba and the Feather rivers. To the east and northeast the surface of the ground falls off rapidly into the valleys and the cañons of the head-waters of the Feather. Blue-nose Mountain is the only prominent high point to the east, of an altitude sufficient to justify any expectation that the extension of the gravel channel might be found under it. Standing upon this crest I could see nothing to serve as a guide in any further explorations; I seemed to have reached the end. If the sources of the old channel were still farther to the east or northeast, the evidences of their existence are all gone. The course of the channel, even with only a slight upward grade, would pass quite above the tops of almost all the peaks in sight. I did not make any attempt to study the northeastern slope of the ridge, and do not know of any outcropping gravel beds on that side. Having traced this channel to its highest known point, I must leave it. If we suppose the channel came in under Pilot Peak, we shall be equally at a loss to tell where it came from.

The water-supply of this region, including both the La Porte and the Gibsonville sub-districts, is small. In winter there are heavy falls of snow, which frequently accumulates to a depth of twenty feet or more. With the melting of the snow active mining operations begin. Some efforts have been made to bring water from the Feather River by piercing the ridge with tunnels. One tunnel, 500 feet in length, cuts through the ridge, near the Onion Valley road above Gibsonville, and another has been begun, to take water from Little Grass Valley through the ridge to La Porte. When completed this latter tunnel will be over 4,000 feet in length. At Mount Pleasant, in the height of the season, a supply of between 1,500 and 2,000 miner's inches of water from all sources may be counted upon. The smallness of the supply of water affects unfavorably the mode of working the gravel. The sluices are short, the sluice-boxes are narrow, and the grade is high. There are no such extensive arrangements for saving gold, no such long sluices nor elaborate systems of undercurrents, as are to be seen in Nevada County.

Another drawback to success in mining seems to be the uncertainty of mining titles, to which may be added the lack of concentrated effort on the part of the miners. If all litigation were brought to an end, and the smaller claims were consolidated under one management, the ground could be worked with much greater profit than it now is.

B. From Council Hill to Howland Flat; the Divide between Slate Creek and Cañon Creek.

Slate Creek and Cañon Creek are nearly parallel to each other from their sources, on opposite sides of Alturas Mountain, to their junctions with the North Yuba. The ridge between them is narrow and steep, the streams being sometimes less than two miles apart, while the ridge is more than a thousand feet in height. The gravel deposits on this ridge, together with those near La Porte and Gibsonville, make up what is frequently called the Slate Creek Basin, though below Poverty Hill the natural outlet of the mines is into Cañon Creek. The map (Plate R) shows the principal features of the ridge. The vicinity of Poverty Hill and Scales's Diggings is shown on a

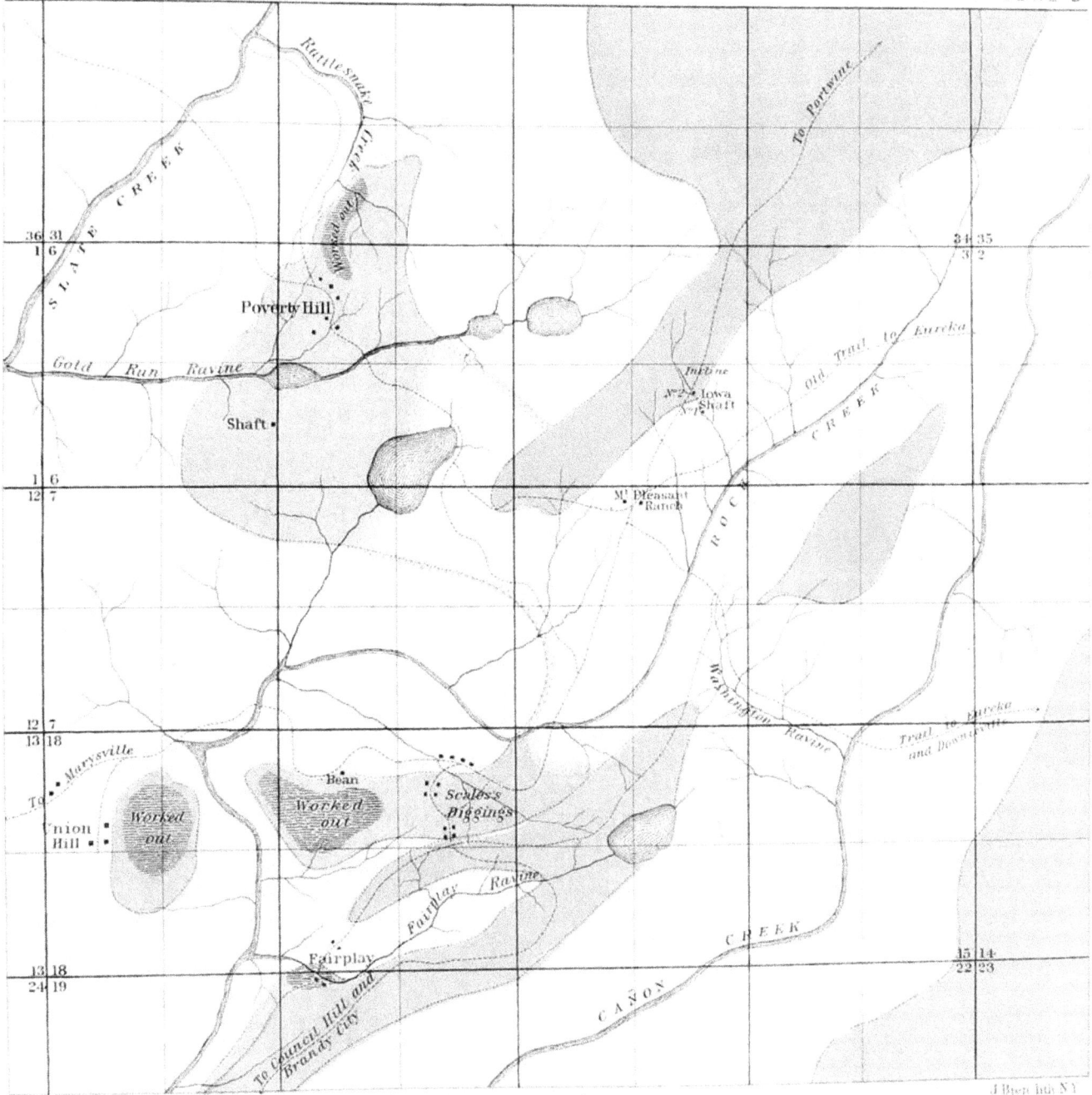

Volcanic.   Gravel.

J. Bien lith N.Y

MAP OF POVERTY HILL, SCALES'S DIGGINGS AND VICINITY,

FROM SURVEYS BY C.W. HENDEL.

(Scale: One and a Half Inches to the Mile.)

larger scale on Plate U. These maps, based in part upon actual surveys, show the main features of the region with a fair approach to accuracy. Some of the details have had to be sketched by the eye alone, especially such as relate to the boundaries of the gravel deposits and those of the volcanic capping, and I cannot vouch for the correctness of all the minor features. I did the best I could with the time and means at my disposal to get trustworthy information.

The principal points at which mining operations have been carried on, and to which reference will be made in the next few pages, are, beginning at the southwest, Council Hill, Fairplay, Union Hill, Scales's Diggings, Iowa Shaft (all lying upon the Cañon Creek slope of the ridge); Poverty Hill, Portwine, Grass Flat, Gardner's Point, Cedar Grove Ravine, St. Louis, Chandlerville, Pine Grove, Howland Flat, and Potosi, upon the Slate Creek slope. Big Oak Flat, to which reference has already been made,* is also on this ridge, about five miles below Scales's Diggings and on the opposite side of Rock Creek. I did not take time to explore that portion of the ridge, having been told that no one was working there, and that the old shafts were filled with water. Reports were contradictory as to the character of the Big Oak Flat gravel. It was described to me at one time as resembling the gravel of Pittsburgh Hill, and at another time as being " clean quartz gravel."

No satisfactory survey of such a ridge as this can be made in a short trip of a few days' duration. The partial list of places just given, the extremes being about thirteen miles apart, is large enough to show that an explorer who wishes to get more than a superficial glimpse of the country must rather extend his examination over weeks than compress it into days. Where so much mining and prospecting has been carried on in tunnels and drifts, a surveyor's progress is necessarily slow if he undertakes to see all that can be seen in regard to the gravel, its position, its course, and its character. I was obliged to pass by many places with only a hasty glance, or no glance at all, and my report, in consequence, will lack in fulness and completeness.

The country rock is mainly slate, though not uniform in character. Where it has been covered with gravel, the upper portions of the slate have sometimes been altered in appearance, to a depth of several feet. The colors of the upper surface frequently change abruptly, and range from dark-blue, through reddish or yellowish shades, to a silver gray, even when the same rock, where exposed in the deeper cuts and in the tunnels, has a uniformly deep-blue color. The strike of the slates where observed was northwest and southeast. The dip was nearly vertical at Portwine, but made an angle of about 40° to the east at Scales's Diggings.

The volcanic capping upon the crest of the ridge is continuous to within a mile of Scales's Diggings. Below this point it is seen only on the higher hills. At Portwine the cap is very narrow, and, where the Morristown trail crosses the ridge, is less than 200 feet in thickness. Between Portwine and Poverty Hill it becomes much broader and thicker. The most prominent points on the ridge are Alturas Mountain and Table Rock, both in the vicinity of Howland Flat. The altitude of the gap between them, upon the road from Howland Flat to Poker Flat, I made to be 6,370 feet. I did not go to either summit. There are two kinds of lava upon this ridge. At Council Hill, at Fairplay, and at several places near Portwine I observed the lava to be tufaceous in character, and to resemble in mineral composition the andesitic lava already described as occurring in the vicinity of La Porte. The lava of Table Rock is darker in color, solid, and compact. The specimens which I brought from the Sugar Loaf, near Pine Grove, and which are doubtless the same in character as Table Rock, are called basalt by Mr. Wadsworth. The relations of the different lavas to each other I did not have time to study. Some interesting questions have arisen in this connection, which cannot be definitely settled without further work in the field. I have already alluded to the absence of bed-rock exposures at Poker Flat; other difficulties will appear when I come to the description of the country northwest of La Porte, in the valley of the Feather.

The Council Hill deposit is near the trail from Scales's Diggings to Brandy City, and overlooks the deep and almost precipitous cañon of Cañon Creek. I estimated the depth of the cañon to be 1,500 feet, but did not cross it at this place. The southward continuation of the old gravel channel

* See *ante*, p. 431.

from Council Hill is seen at Brandy City, and will be described in a subsequent section of this report. Only a small area of bed-rock has been uncovered at Council Hill. Its altitude is 3,963 feet. The face of the highest bank shows at bottom a stratum of gravel, bluish in color and containing some sandy streaks. The large boulders in this stratum are granite and slate, the latter being in relatively greater quantity than at Fairplay or at Scales's Diggings. Above the blue gravel there comes first a yellowish streak, and then fine quartz gravel extends up to the volcanic capping. Under this capping the gravel is in all probability continuous to Fairplay, a deposit on the left bank of Rock Creek, about a mile south of Scales's Diggings. The Fairplay bed-rock has an altitude of 4,123 feet, or 170 feet above that at Council Hill. The bottom stratum of gravel is from five to twenty feet in thickness, blue in color, not cemented, though hard enough to require the use of powder drifts. It contains many boulders of granite and some of slate or other kinds of crystalline rock. The interstices between the boulders are filled with a fine quartz gravel. Above the blue stratum there is reddish gravel for about ten feet, and then nearly 120 feet of fine white quartz. The volcanic capping at the highest bank was fully a hundred feet thick. It is tufaceous in character, resembling mud or ashes, and is very easily moved. The line between the quartz and tufa is wavy and irregular. The gravel carries considerable charred and petrified wood, and some impressions of leaves. No bones or animal remains have been found. The season of active mining is confined to four or five months in winter and spring.

Scales's Diggings is a place of more importance than the two just mentioned. The gravel deposit is nearly a third of a mile in width and about a mile in length, not including the gravel of Union Hill, which lies on the opposite side of Rock Creek and is probably on the line of the old channel between Scales's and Fairplay. The altitude of the low bed-rock at the Rock Creek outlet of the Cleveland mine I made to be 4,253 feet. A large part of the top gravel has been removed, especially at the southwestern end of the deposit, though there are still some high banks yet to be washed away on the eastern side of the present main opening. The blue gravel on the bed-rock is of varying thickness, being as much as thirty feet in some places. The overlying white quartz gravel is said to have had a maximum thickness of 150 feet. The lower strata of the gravel abound in large granitic or syenitic boulders, which frequently measure six or eight feet in diameter. They are tightly cemented together, and powder-blasts are necessary to loosen the mass. Many of the boulders are covered with dendritic markings. I saw no boulders like these at any higher altitudes in this vicinity, nor did I see any granitic rock in place. There is said to be a granite country to the north and west of Scales's Diggings, from which the boulders might have come, but I have no more definite information about it. The average fineness of the gold is .935.

To the north of Scales's Diggings, on the divide between Rock Creek and Slate Creek, and near the heads of Rattlesnake Creek and Gold Run Ravine, there is a large body of gravel, the precise extent of which is not known. The mines at its northern end are known as Poverty Hill. This point has been mentioned on a previous page as the probable continuation of the La Porte channel by way of Secret Diggings and Barnard's. The bed-rock at the northern end of the mines has an altitude of 4,563 feet. The southern bank is nearly a hundred feet in height; the gravel is a fine and sandy white quartz, with a few small streaks of clay. On the bed-rock there is a little iron cement, quite different in character, however, from the lower strata at Scales's and at Fairplay. A half-mile or more to the south of the mine there has been a shaft sunk in gravel. The depth of this shaft I could not ascertain. The altitude of the mouth of the shaft is 4,553 feet, nearly the same as that of the bed-rock just referred to. Upon the surface there was a great deal of "bogus" gravel. This body of gravel is supposed to extend as far back as the base of the lower end of the lava flow, where it is probably concealed from view at the surface by a comparatively thin covering of volcanic material. I have represented upon the maps (Plates R and U) by dotted lines a possible connection of Poverty Hill with Scales's Diggings. In so doing I have adopted in part the views of Mr. Hendel. It seems clear that some such connection must have existed, and high knolls of slate rock appear to stand in the way of any more direct connection. The crookedness of the path suggested cannot be urged as any objection to its probable accuracy, for there is abundant evidence that the old channel made frequent curvatures between La Porte and Brandy City. I regret that I was unable to revisit this district and give it a more thorough examination after my first hasty reconnaissance.

Between Poverty Hill and Portwine, a distance of about three miles, there are no indications of gravel on the Slate Creek slope of the ridge, with the possible exception of a small deposit at Poverty Point (not shown on the map). This absence of gravel may be accounted for either by supposing the old channel to have had nearly the same position as the present Slate Creek, or on the hypothesis that the channel crossed from Portwine to the Cañon Creek slope, or possibly lies concealed under the present lava crest. Against this latter view may be urged the narrowness of the ridge for a mile or so below Portwine. It does not seem possible that a channel could be so effectually concealed under this lava as to give no signs of its existence, either at its point of disappearance or at its emergence from the lava on the opposite side. Whichever view be chosen, it must be admitted that a junction of the two channels, now traceable on opposite sides of Slate Creek, was effected at some point near Poverty Hill or Scales's Diggings. The Iowa shaft and incline, two miles northeasterly from Scales's, and near the southeastern base of the lava-flow, were sunk with the hope of finding the Portwine channel on the more easterly of the two positions which it has been supposed it might have taken. It was lack of time only that prevented my visiting these works. From all that I could learn, however, about the character of the gravel and of the strata next the bed-rock, I judge that the extension of the channel has not been found.

At Portwine there has been some hydraulic mining done at the southerly end of the gravel deposit. The extent of bed-rock uncovered is not large, but the highest bank shows a vertical face of about one hundred feet. In one important respect this bank is remarkable, if not unique. It furnishes evidence of the intrusion of lava into the gravel after its deposition. (See section, Plate S, Fig. 6.) The bed-rock at my point of observation pitched to the northeast at an angle of nearly twenty degrees, showing that the centre of the channel was not exposed to view. The lowest bed-rock accessible at the base of the bank has an altitude of 4,853 feet. Resting upon the bed-rock there is a stratum about four feet in thickness of a rather coarse quartz gravel, which has been drifted extensively and proved to be rich in gold. Above this there is a compact rock of volcanic origin, three and a half or four feet thick, which the miners call "cap" rock. When freshly broken the surface of this cap is bluish in color. Then there comes a stratum, ten or twelve feet in thickness, of quartz gravel, neither so coarse nor so rich in gold as the one below. Above this gravel there is a second cap rock, about six feet in thickness, which, upon a fresh surface, is rather light colored. Above this second cap there come in succession twenty-five to thirty feet of very fine quartz gravel, twenty feet of variegated pipe-clay, twenty feet of red loam, and, at top, a few feet of the common volcanic surface dirt. I obtained on the spot specimens from both cap-rocks. They have been examined by Mr. Wadsworth.* Both layers of cap-rock appear to come to an end towards the southeastern edge of the bank, the upper one extending a little farther in that direction than the lower. They are sensibly parallel to each other and conform to the pitch of the bed-rock. Were it not for the fact that they come to an end within the limits of the exposed gravel, it would be easy to believe that they were parts of ancient lava-flows — more ancient than the gravel.

For a distance of two miles above Portwine, as far as Gardner's Point, there has been little or no hydraulic mining done. From the prospecting tunnels and the drift diggings the general course of the old channel is known to be approximately parallel to the present Slate Creek, though there may be some doubt as to the precise position of the deep channel for a part of the way. This is the portion of the old channel to which I paid the least attention. I got little more than a bird's-eye view of the surface from high points on the Morristown trail and on the La Porte side of the creek. The altitude of the bed-rock at Gardner's Point I made to be 4,845 feet. This, it will be

---

* The specimen from the so-called upper cap-rock was found on examination to be a decomposed olivine-bearing basalt. That from the lower cap-rock proved to be an undecomposed basalt, and no olivine was observed in it.

noticed, is eight feet *lower* than the bed-rock at Portwine. The measurement at Portwine was made, it is true, upon the western rim, but it would be reasonable to expect the altitude, even high up on the rim, to be less than that of the deep bed-rock at Gardner's Point. Without any knowledge of the intermediate channel, the actual condition of things would be quite difficult of explanation. A partial if not a complete key to the difficulty is said to be found in the old "Monte Cristo slope," a drift mine about three quarters of a mile above Portwine. I could not examine the mine in person, and shall have to record hearsay testimony as to what was observed there when the mine was worked, several years ago. The coarse pay-streak, which the miners were following on bed-rock, with a steady rise from the direction of Portwine, suddenly changed to a fine gravel, poor in gold. The bed-rock as suddenly dropped over fifty feet. The wall of rock, overhanging to the north, was smooth and polished, showing a slickensides surface. At the lower level the pay-streak was again struck, and followed upon a rising grade for 500 feet, the gravel and water being all hoisted over the ledge of rock. This description, illustrated in the diagram (Plate 8, Fig. 5), can refer to nothing else than a fault in the strata, of a date more recent than the deposition of the gravel.

It will not be worth while to attempt to enumerate the different mining claims which have been worked or laid out along this part of the ridge. The positions of nearly twenty such claims, within a distance of a little more than a mile, are given on Mr. Hendel's map of the Pioneer Company's ground. As a result of this extensive drifting, the richer parts of the gravel have been, to a large extent, removed. What remains would probably yield well, if water for hydraulic mining could be obtained at a low rate and in sufficient quantity. The most of the water at present obtainable from Slate Creek is used at Pine Grove, but there seems to be no good reason why some of this water, if not all, may not be used over again at the lower mines. To bring water from Cañon Creek or from the sources of Feather River will require the construction of more expensive ditches and tunnels. Considerable work has been done with this end in view, for the sake of washing a large body of untouched gravel at Grass Flat. At the last-mentioned place I visited, in company with Colonel B. F. Baker and Mr. G. W. Cox of Pine Grove, the large, unfinished sluice-tunnel in bed-rock, and the upper prospecting tunnel in gravel. Since the death of Mr. Ralston, who was one of the principal owners of the property, all work has been suspended. There is some prospect, however, of the resumption of operations at an early day. My examination of this deposit was not sufficiently minute to justify my making any estimate of the amount of gravel left to be washed, or of its probable yield of gold. I was told by Colonel Baker that the gravel in the prospecting tunnel, though from twenty-two to thirty feet above bed-rock, yielded at the rate of one dollar per cubic yard, and that, by the sinking of a shaft, the gravel has been proved to be one hundred feet in thickness. It is also claimed that the channel, where it crosses Grass Flat, is 1,400 feet wide.

Hydraulic mining at Gardner's Point, where the average thickness of gravel worked was between twenty and thirty feet, has exposed to view an overflow of volcanic material, a thousand feet or more in width, which in some places has cut away the gravel to within four feet of the bed-rock. The gravel is a clean white quartz, with the exception of two or three feet in thickness of an iron cement upon the rim. There are no large boulders, and there is but little pipe-clay, though more may be expected as the bank is washed farther back. The gold is said to resemble flax-seed, never being found in large nuggets. It was here that Colonel Baker, in 1872, found a diamond, which, when cut, weighed a carat. Perhaps the small rough diamond in Mr. Hendel's possession, found in the Slate Creek tailings, also came from this point.

Between Gardner's Point and St. Louis there is a high bed-rock spur, which must have thrown the old channel to the west. Yankee Hill, a small deposit of gravel on the right bank of Slate Creek, has an altitude, as estimated by sighting across the cañon with the hand-level, a little higher than that of Gardner's Point. I did not go to Yankee Hill, but it is evident that its gravel belongs to the St. Louis channel, and has no connection with the channel from Gibsonville.

To reach St. Louis from Gardner's Point, Cedar Grove Ravine has to be crossed. Near this crossing there must have been at one time a junction of a tributary from the northeast with the

THE DIVIDE BETWEEN SLATE AND CAÑON CREEKS.

main stream from Howland Flat, for, between the ravine and the crest of the ridge, gravel has been found, and drift mining has been carried on at several places. I cannot speak of this branch channel from personal observation, but it seems to have taken its rise somewhere near Table Rock. The St. Louis gravel, between Cedar Grove Ravine and Sackett's Gulch, has been worked to a considerable extent by the hydraulic process, particularly at the northern and the southern ends of the field. The central portion, where the town stands, has hardly been touched. The gravel is favorably situated for mining, but the process of washing will go on at a slow rate, unless the available quantity of water be in some way increased. The altitude of the bed-rock at the lower end of the mines, near Cedar Grove Ravine, is 4,993 feet.

Between Sackett's Gulch and Harris Gulch the town of Chandlerville used to stand, but there are no houses there now, and the gravel deposit is exhausted. Pine Grove also was once one of the principal mining towns of this vicinity; but since its destruction by fire it has never been rebuilt. It is at this place that the most of the Slate Creek water is at present used. Mr. G. W. Cox, of the Sears Union Company, says that in the busy season the supply of water amounts to three or four thousand inches. The altitude of Mr. Cox's business office I made to be 5,486 feet, and that of the bed-rock in the Pine Groves mines, about midway of the present opening, 5,396 feet. The gravel in these mines has been disturbed in several ways since it was first deposited. Just back of Pine Grove there is a small sugar-loaf knoll of volcanic material, standing by itself, quite detached from the main volcanic capping at Table Rock. Whether this represents an ancient slide from the higher ridge or not cannot be told with certainty, but there is evidence that there has been a flow of lava or mud to cut away a part of the original gravel. There is also a so-called "horse" of similar material on the old rim of the channel, on the side opposite to the Sugar Loaf. The section (Plate S, Fig. 4) is not drawn to scale, but shows the relations of the gravel and lava as I sketched them roughly on the spot. I have Mr. Cox's authority for saying that the presence of gravel under the eastern body of lava has been proved by drifting. In this connection I will add that I was told by Mr. Wallis, the superintendent of the Bald Mountain Company at Forest City, of a mass of lava at Pine Grove, "like a bell upside down," which covered gravel under its rim on all sides, without there being any bed-rock discoverable under the apex of the bell, even though tunnels were driven in and shafts sunk. I did not make the acquaintance of Mr. Wallis until after I had left this part of the country, and could not return for any revision of my work. Another interesting occurrence at Pine Grove is that of gravel in thin seams, two or three inches wide, in bed-rock. The quartz fragments which this gravel contains are angular.

Above Pine Grove the old channel is traceable through Howland Flat to Potosi, though its precise winding course cannot be shown upon any published map. A detailed topographical map on a large scale is needed. Howland Flat is not quite a mile from Pine Grove, and Potosi is about half a mile to the east of Howland Flat. The altitude of the lower floor of Becker's Hotel, at the latter place, I made to be 5,600 feet, and that of the mouth of the Bonanza tunnel, at Potosi, 5,655 feet. There are everywhere signs of former mining activity. At Howland Flat there have been both hydraulic and drift mines. The banks now exposed to view to the north and west of the town are small and low, and do not appear to have been worked very recently. The drift mines at Potosi have been developed pretty extensively. Tunnels have been driven in a northeasterly direction, in the hope of finding the continuation of the old channel under Alturas Mountain, and other tunnels, like the new Bonanza, have been made to follow a more easterly or southeasterly course, on the hypothesis that the channel came across the ridge near where the low gap now is between Table Rock and Alturas. To go through all the mines and collect data for a satisfactory independent judgment would require several days' time. The Bonanza tunnel was the only one I was able to visit, though I could easily have obtained access to others. This tunnel follows a course N. 83° 30' E. (magnetic) for 1,547 feet, and then S. 65° 51' E. (magnetic) for 2,289 feet farther, rising, in the whole distance, twenty feet. The tunnel was started in volcanic material, the probable position of the gravel being pretty well known from the developments in the adjoining claim, the Empire. The successive rock formations met with in the tunnel are as follows :—

lava, 680 feet ; hard rim-rock, 300 feet ; soft rim-rock, 200 feet ; clay, 700 feet. For the last 1,790 feet the tunnel has been entirely in gravel. In the uncompleted air-shaft there were thirty-three feet of gravel. Above the gravel the shaft has been in clay for 142 feet. When finished, the shaft will be 352 feet deep. The gravel seen in the tunnel was a good clean quartz, carrying some charred wood and some poor impressions of leaves. I tried to secure some good specimens to show the character of the vegetation, but was not successful.

I have now traced the old channel from Council Hill to Potosi. In what direction shall I look for its higher sources? I am not at all satisfied with the evidence advanced in favor of the theory that this channel came from the direction of Pilot Peak and Alturas, — the theory evidently entertained by Mr. Hendel at the time of the preparation of the report from which an extract is given on a previous page.* It is true I did not examine in detail the whole of the ridge between Potosi and Hepsidam ; I only followed the wagon-road between the two places. But I made frequent inquiry of such miners as seemed to have the most thorough acquaintance with the whole district, and, as a result of these inquiries, combined with my own observations, making proper allowance for any warping of the judgment due to considerations of personal interest on the part of those from whom I sought information, I am inclined to the opinion that the gravel channel will be traced from Potosi through the ridge to its southeastern slope above Poker Flat. The only piece of direct evidence that I have, which bears upon this question, is the altitude of the mouth of an old tunnel, which is to be seen a little to the south of the Poker Flat road. I made the altitude to be only 5,526 feet, or nearly thirty feet below the mouth of Bonanza tunnel and fifty feet below the gravel at the present farther end of that tunnel. When making this measurement I had not heard of the Cold Cañon mines, and had nothing to guide me to the proper spot at which to take the observation. But whether I stumbled by accident upon the old tunnel through which the Cold Cañon gravel was worked or not, the parole testimony goes to show that in the Cold Cañon mine the bed-rock had a dip towards Potosi, and that the mine was abandoned only when the accumulation of water in the drifts made further work unprofitable.

But if the channel is traced to Cold Cañon, what then? I cannot answer the question. Standing at the mouth of the old tunnel, one has before him the rough, jagged, open amphitheatre of the head of Cañon Creek, the opposite side of which he scans in vain for any indication which shall guide his future search. As upon the ridge above Hepsidam, so here, I reached a limit beyond which I could not go. There is gravel on the opposite side of Cañon Creek, which will be described in the next subsection, but I cannot see any connection between it and the gravel of Cold Cañon.

The total fall between Potosi and Council Hill is nearly 1,600 feet ; the distance is not far from thirteen miles. The average grade, therefore, is about 122 feet to the mile. This grade is a few feet less than the average grade of Slate Creek between the bridge on the road from Potosi to Gibsonville, where the altitude is 5,480 feet, and the Poverty Hill trail at the mouth of Rattlesnake Creek, where the altitude is only 4,000 feet. Taking the distance as eleven miles, the grade is nearly 135 feet to the mile. The altitude of the bed of Slate Creek at the bridge, on the road from La Porte to Pottwine, I made to be 4,318 feet.

The tailings in Slate Creek, which have been accumulating for so many years, must be very rich in gold. The appliances for saving gold in this vicinity have been by no means so efficient as those in use in the more southern counties, and the average yield of the gravels has been much more to the cubic yard washed. At some time in the future a large profit may be expected from them. A specimen taken from the Slate Creek tailings showed partially amalgamated gold.

Upon the southeastern slope of the ridge, below Poker Flat, there are deposits of gravel, as at Wahoo, for instance, which would have been examined had time allowed. Their relations to the other gravels upon this and other ridges have not yet been well made out, and I must dismiss them with this simple allusion to their existence.

* See *ante*, p. 211.

C. The Ridge between Cañon and Little Cañon Creeks.

Little Cañon Creek takes its rise in the high mass or knot of mountains in which are found also the sources of Goodyear Creek and of many of the feeders of the North Fork of the North Yuba River, and it joins Cañon Creek at a point about midway between Eureka and Scales's Diggings. The ridge between these two streams is of comparatively small extent. In former years there has been considerable mining carried on in this section, but at present there is very little doing. Craig's Flat, Morristown, Deadwood, and Bunker Hill are the only points which I was able to visit. I crossed Cañon Creek at two places, Poker Flat, where the altitude is 4,854 feet, and on the trail from Craig's Flat to Portwine. At the latter place the bed of the stream has an altitude of 4,252 feet. I also crossed Little Cañon Creek on the trail from Craig's Flat to Eureka, and made the altitude of the bed of the creek to be 4,345 feet. The crest of the ridge is lava-capped as far down as Craig's Flat. I did not make any examination of the lava in place, but it seems probable, judging from the boulders seen at Morristown, that both the basaltic and the andesitic types occur here. Two specimens from this locality have been examined microscopically by Mr. Wadsworth, who calls one a basalt and the other an andesite.

The bed-rock, where exposed to view after the removal of the gravel, is a very soft slate. At Morristown a shaft was sunk to a depth of eighty-six feet in a rock so soft as not to require blasting. The color of the slate is not uniform, it being in some places white or silvery, and in others reddish or bluish. The strike is to the northwest, and the dip is nearly vertical. At Deadwood the serpentine belt, to which frequent allusions have been made already, crosses the ridge. The spur leading down from Deadwood to Poker Flat is entirely of this kind of rock. The serpentine belt is only a few hundred feet in width, and the slate sets in again about a half mile to the southwest of the Bunker Hill tunnel.

At Craig's Flat an area of perhaps ten acres of bed-rock has been uncovered. Its altitude I made to be 5,100 feet. No work has been done here for several years, and the claim appears to be abandoned. It is said to have been extremely rich in gold at the time it was worked. From this point the gravel banks of Eureka, on the opposite side of Little Cañon Creek, are in plain sight and at a little lower level. It is clear that the two deposits were at one time connected.

The Morristown gravel is about a mile above Craig's Flat and on the Cañon Creek slope. The ridge between the two places is capped with volcanic material, beneath which, in all probability, the old channel lies. The bed-rock at Morristown is uneven, though, on the whole, nearly level. I made its altitude to be 5,160 feet. The Wahoo deposit, on the opposite side of Cañon Creek, appeared to be about eighty feet lower, and probably belongs to a different channel. The area of bed-rock uncovered at Morristown I estimated at thirty-five acres. The gravel is all smooth, rolled quartz without any cement. Some heavy boulders of smooth quartz, five or six feet in diameter, are found, but they are confined to within ten feet of the bed-rock. The richest stratum, near the bed-rock, is about eight feet thick on the average. Above this come sixty feet of quartz gravel and eighteen or twenty feet of pipe-clay. The banks are nowhere yet worked back to the volcanic capping. Between the gravel and the pipe-clay there is a peculiar black stratum, which, though not constant in thickness nor universally met with, has some features of interest. It appears to be rich in sulphur and in iron, and looks as if it had been exposed to a roasting heat. The statements in regard to the yield of the Morristown gravel are conflicting. According to one statement, the product of the mine in fifteen years, in which time the gravel was removed from twenty acres of bed-rock, was $460,000; according to others, there have been more than two and a quarter million dollars taken from this deposit. The gold is fine in quality, the mint returns for the year 1875–76, which were shown to me by Mr. Joel Pike, the superintendent in charge of the property, ranging from .942 to .965½. The gold found in the present bed of Cañon Creek is said to be coarser, and to have a fineness of only .800.

Of the ridge between Morristown and Deadwood I know nothing from personal examination. I was told that gravel has been found in tunnels on both sides of the ridge, and also on the spurs between the lateral ravines leading down to Cañon Creek. This last statement is in accord with what I could see from my point of observation at Deadwood. At the latter place there was no one at work, and I made only a short stop. The bed-rock, where it was exposed over an area of three or four acres, was quite irregular. Its average altitude I determined to be about 5,700 feet. The gravel banks were thirty or forty feet in height. I could not feel sure that the deposit was in the path of any old channel; in some respects, it had the appearance of a slide from some higher point on the ridge.

At Bunker Hill there has been no hydraulic mining done. The mouth of the tunnel is at the head of Grizzly Cañon, at an altitude of 5,900 feet, and near the base of an amphitheatre of volcanic rocks. The two principal peaks are known as Bunker Hill and Democrat Mountain, rising respectively, as well as I could estimate by the eye, 500 and 800 feet above the mouth of the tunnel. The course of the tunnel, S. 15° E. (magnetic), is such as to carry it towards the sag between these two high points. The tunnel is nearly two thousand feet in length, the last third of the distance being in gravel, in which some drifting has been done, though the works are not extensive away from the line of the tunnel. The bed-rock, as has been stated already, is a soft slate. In the parts of the mine that I visited, there were no very large boulders. Occasionally pieces of slate or of other metamorphic rock were seen, but as a whole the gravel was composed of quartz. Like that at Hepsidam, it carries considerable clay. There was a little charred wood in the clay, and I was told that impressions of leaves were sometimes seen. I saw nothing resembling the rotten boulders of Hepsidam. The gold is coarse and in the form of nuggets, sometimes flattened.

From what I saw in this mine it seems clear that there is here a channel-like depression containing gravel. As to the extent and the connections of the channel, there is nothing definite known. There are no good maps of the region; none that show the topographical features with any pretensions to accuracy. It is possible that a connection once existed by way of Deadwood and Morristown with Craig's Flat and Eureka; and it is also possible that a connection exists through the ridge with the gravel found near Sebastopol.

The precise position of Sebastopol I cannot give upon the map. Possibly the peak which I have spoken of as Bunker Hill is the same as the one which in other parts of Sierra County goes by the name of Sebastopol Mountain. I expected to be able to visit the Sebastopol tunnels later in the season in connection with my work near Downieville, but was prevented by the lack of time and by an unusually early fall of snow. The site was pointed out to me from Craycroft's Diggings (to be described on a subsequent page), and I was told that a tunnel, some 2,000 feet in length, had been driven in the direction of Deadwood, on a course a little to the west of north, and that quartz gravel had been found. The gravel, however, though frequently rich, does not form a continuous deposit, the "mountain cement" or tufa frequently resting immediately upon the bed-rock.

D. THE BRANDY CITY AND EUREKA RIDGE.

The district here included lies between Cañon Creek on the west and Goodyear Creek on the east. The position of Eureka is given on the map (Plate R); Brandy City would lie a little below the southern edge of that map. The distance between the two places is about ten miles. Seen from points which command an extensive view of this ridge, its crest appears as a nearly uniform line, with the regular southwesterly downward grade characteristic of the lava-capped ridges of this portion of the Sierra Nevada. Towards the lower extremity of the ridge the volcanic capping grows comparatively thin, and near Brandy City very little of the gravel has any capping at all. At Eureka also the volcanic material has been eroded almost entirely at the point where the old gravel channel crosses the ridge.

Brandy City is approached most easily by the trail from Camptonville. The trail from Scales's Diggings has the reputation of being one of the steepest in the State. The former trail crosses the North Yuba at Cherokee Bridge, and then follows up Cherokee Ravine. The altitude of the floor of the bridge I made to be 2,225 feet. On the eastern spur, between the ravine and the river, lies the Grizzly Hill gravel deposit. A high knob of bed-rock, lying southwest of the gravel, marks the place of the former rim, and shows that the natural outlet of the old channel was towards Indian Hill or Depot Hill, places which have been described on former pages. The altitude of the top of the Grizzly Hill bank I made to be 3,560 feet. On the western side of the hill the bed-rock was covered with slides from the banks; on the side towards the river there are high hydraulic banks and probably ten acres or more of uncovered bed-rock. I did not examine the gravel in detail.

Cherokee Ravine cuts off the Grizzly Hill gravel from any immediate connection with that of Brandy City. The trail from Camptonville crosses the bed-rock of the lower end of the Brandy City deposit at an altitude of 3,435 feet. From where the trail crosses the gravel the old channel can be traced up stream along a winding course, first northwesterly, then northeasterly, and again more to the east, with an average width of five or six hundred feet, for something more than a third of a mile, to the crossing of the Eureka wagon-road, near the lower part of the town of Brandy City. I could not find any good map of this mining district, nor much that was serviceable in the way of maps of individual mining properties. The town of Brandy City is built a little off from and above the gravel channel, on the bed-rock which lies between the lower mines and those of Windyville. The altitude of the hotel I made to be 3,650 feet.

In the lower mines there has been a great deal of gravel removed, and there is still a large quantity left. The bed-rock seems not to have been entirely cleaned off, excepting over a small part of the workings. The highest banks are as much as 150 feet in height, including in some places forty or fifty feet of pipe-clay, and twenty or thirty feet of volcanic cement. The original surface of the country was quite irregular and much cut up by small ravines. This makes it impossible to tell what the average height of bank has been. A fault in the bed-rock is an interesting feature of these mines. The diagram (Plate 8, Fig. 7) is from a sketch taken on the spot. My first impression on approaching the place was that there had been a cascade or fall in the old channel, amounting to sixty or eighty feet. The bed-rock does not exhibit any marked differences of external appearance on the two sides of the fault, but the plane of contact can be easily followed by aid of the slickensides. The rock in place is quite dark in color; it has become much lighter since removal, and probably cannot be kept for any great length of time. The line of the fault can also be distinctly seen in a tunnel, which is intended to serve as an outlet for the claims above the fault. My sketch was taken in a narrow cut, a few feet only in width. The superficial covering of gravel prevented me from tracing the fault any farther in the direction of its strike. The hanging wall of the fault, as I was told by a miner at work in the tunnel above-mentioned, used to project twenty or thirty feet beyond its present position. The gravel enclosed between the two bed-rock walls looked exactly like that on the outside, and showed no appearance of having been disturbed. This position of the gravel is peculiar, and at first sight seems hard to explain, if it be supposed that the fault was formed subsequently to the deposition of the gravel. I feel confident, however, that the formation of the fault is the more recent of the two events. One ground for this confidence is that the evidence goes to show that the bed-rock falls off again above the fault in such a way as to leave a depression in the bed of the channel,—a depression that it would be equally difficult to account for, if the present high rock marks only the site of a former waterfall.

The bed-rock of the Brandy City mines and of those at Windyville is generally a slate of some kind, with frequent alternations of color and of texture. Some of the rock is very easily cleavable, while some is more compact and crystalline. It is often nearly black or dark-blue in color, though sometimes lighter and sandy. It is worthy of remark that such frequent changes of color are seldom seen away from the gravel banks, and it is probable that they are due in part to the action of the waters which percolate the gravel, and that they may not be permanent in depth.

At Windyville my observations were confined principally to the mine of Arnott & Co. The altitude of the average bed-rock, of which about one acre is exposed to view, I made to be 3,510 feet. The strike of the slates is to the northwest; the dip is to the northeast, at an angle between sixty and seventy degrees. There are some deep pot-holes in the bed-rock. The bank shows next the bed-rock a stratum of twenty or twenty-five feet of blue gravel, which is so compact that it has to be loosened by the aid of powder drifts. Above this there are about seventy-five feet of fine quartz gravel of a reddish or whitish color. The quartz pebbles, taken by themselves, are usually either blue or white. There is but little pipe-clay and no volcanic cement within the present limits of the mine. A quarter of a mile to the west the volcanic capping is nearly one hundred feet thick. Granitic and metamorphic boulders, like those seen at Scales's Diggings and Council Hill, though generally smaller in size and not requiring much blasting, are very common. There seems to be no doubt that the old channel crossed from Council Hill to Windyville, and thence by way of Brandy City to Depot Hill, Camptonville, and San Juan. The precise position of the deep channel at Windyville is not so clear. Mr. Arnott's view, in which I am inclined to coincide, is that the connection between Windyville and Brandy City was by way of Little Cherokee ravine. This would make the channel rather crooked, though scarcely more so than it is known to be in other places. The width of the channel from rim to rim at Windyville is from six to seven hundred feet. Impressions of leaves are occasionally seen, but no evidences of animal remains have been brought to light.

Another instance of a fault formed after or during the deposition of the gravel is to be seen at Arnott's mine. The diagram (Plate V, Fig. 1), is from a sketch taken on the spot. The fault crosses the mine in a northwesterly direction, nearly, though not quite, coinciding with the strike of the slates. To the northwest of the blacksmith-shop the hanging wall of the fault is a black slate, while the foot-wall is a sandy, schistose rock, light in color. The sketch shows a section at this point. To the southeast of the shop there is a different variety of rock upon the hanging wall. A little easterly from the fault the bed-rock is light-colored, crystalline, and granular, resembling some varieties of granite or diorite. The amount of the fault is about twelve feet and the dip is such that the gravel is found underlying the overhanging rock for eight or ten feet. The stratum containing the heavier boulders on the upper side of the fault was not continuous at the same level on the lower side. In connection with these obvious movements of the bed-rock, Mr. Arnott pointed out to me in the gravel of the northwestern bank a peculiar soft, clayey seam, which, where I saw it, was not more than two inches in width, and was not very distinct. The clay-like filling of the seam showed smoothed surfaces, such as would be expected if there had been a sliding of one part upon another. There were no marked differences in the character of the gravel on opposite sides of the seam. Farther to the southeast, where the gravel has been washed away, this seam, as Mr. Arnott says, was more nearly vertical than it is at the spot where I saw it, and as I have represented it in the diagram. It followed more nearly the general dip of the rocks and of the fault, and was eighteen inches in width. One peculiar feature of this seam, which has long attracted the attention of the miners employed at the mine, is that it has never been traced into any of the overlying, horizontally stratified beds of pipe-clay. If all these different phenomena are related, the fault must have been formed at some time during the deposition of the gravel.

I could not learn anything of much value in regard to the yield of the mines in this vicinity. The water used is brought in a ditch which follows the line of Cañon Creek, and has a carrying capacity of twelve hundred miner's inches. The washing season usually lasts about five months. If rains come before snow in the fall of the year, some mining can be done in the month of November; otherwise, the ditch cannot be used until the melting of the snow in the spring. The fineness of the Brandy City gold was given me as .940 for the bottom gravel and .950 for the top, or, in value, from $18.60 to $19.00 per ounce. The gold at Arnott's is fine and scaly, with the exception of an occasional coarse nugget. By Mr. Arnott's kindness I was allowed to take with me the gold and sands from a pan of the blue gravel, which he washed in my presence. The mixture has been examined microscopically by Mr. Wadsworth, who says that it is made up of "a few flattened and worn scales of gold associated with fragments and some crystals of zircon, quartz, feldspar, augite (?), magnetite, pyrite, gray grains (platinum?), etc. The forms are such that in a majority of cases it is difficult to ascertain their true nature."

Fig 1

SW →

NE →

Soft Streak
Slickensides

Boulders

Boulders

Worn Black Slate
with Pot-holes

Worn light colored
Slate with Pot-holes.

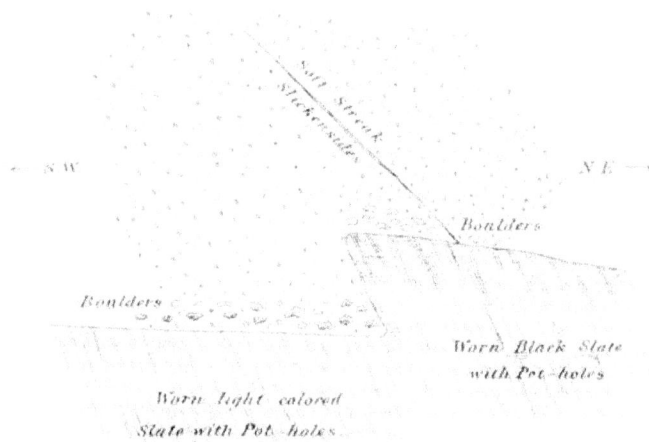

SECTION SHOWING FAULT IN THE BED-ROCK AT BRANDY CITY

Fig 2

← SW

NE →

Ravine

Exposed
Bed-rock

Gravel

Lava

Bed-rock

Bed-rock

SECTION ACROSS THE CHANNEL AT EUREKA

(App Scale 1320 Feet to One Inch)

Fig 6.

Spanish Creek

Original Surface

Bed-rock

Line of Fault

Worked out

Gravel

Tunnel

SKETCH OF FAULT AT BEAN HILL CLAIM

(After and while washing the Bed-rock caved to line _ _ _)

PLATE V

Fig 4

Fig 5

ON AT GOPHER HILL
(Looking North)

SECTION AT GOPHER HILL
(Looking South)

Fig 3

SECTION OF GRAVEL DEPOSITS NEAR CRAYCROFT'S
(Scale One Thousand Feet to One Inch)

Fig. 7

→ S.E.

I Basalt
II Pipe Clay
III Gravel
IV Rotten Boulders
V Blue Gravel
VI Gravel
VII Pipe Clay
VIII Gravel

Bed - rock

SECTION ACROSS UPPER MINE AT CHEROKEE FLAT,
FROM SUGAR-LOAF TO POINT NEAR SPRING VALLEY HOTEL.
(App. Scale Five Hundred Feet to One Inch)

J. Bien, lith. N.Y.

It is a favorite idea with many of the miners that there is an old gravel channel underlying the volcanic capping between Brandy City and Eureka, and the claim is made that the ridge has "never been half prospected." I was told, however, of several tunnels that had been driven at different points, both on the Cañon Creek and on the Fiddle Creek side of the ridge, in which gravel was found, but in which, for some good reason or other, further operations had to be suspended. I did not have time to examine the ridge in detail, but from all that I could see or learn, I am led to the conclusion that no continuous channel exists there.

The gravel deposits at Eureka and Mugginsville cross the ridge at a gap where the volcanic capping has been almost entirely removed; and, were it not for the cap at Chaparral Hill, it would be easy to believe that the gravel channel is of more recent origin than the lava-flow. The diagram (Plate V, Fig. 2), representing a section across the Eureka gravel from northeast to southwest, gives the general features of the topography as they were sketched by me at the time of my visit; there was no time for detailed measurements of altitudes and distances. My observations for altitude in this vicinity were made under unfavorable conditions; for I had to depend upon the small aneroid alone, without any opportunity for comparison with the mercurial barometer, for an interval of more than thirty-six hours. I cannot tell how great the probable error is, but there is good reason for believing that my results are all too high. I am led to this conclusion by a comparison of the values obtained for the altitudes of Craig's Flat and Monte Cristo, places whose relative heights I was able to ascertain by observations with the hand-level.

The town of Eureka, built originally directly upon the gravel, comprises now only a few houses, and these are doomed to speedy destruction if hydraulic mining is pushed any further. The altitude of Wolfe's Hotel I made to be 5,138 feet. The gravel extends in an unbroken mass, Mugginsville being separated from Eureka only by the narrow water-course of Eureka Creek, in a southeasterly direction for more than a mile, and with an average width of fully one thousand feet. The bed-rock has been uncovered over a large part of this area, but there is still a very large body of gravel left to move.

In the Eureka mines the clean quartz gravel, carrying a little clay or sand near the bed-rock, is nearly seventy feet in thickness. Above this there is a heavy deposit of pipe-clay, which in places is as much as fifty feet thick. At Mugginsville the gravel has been removed from a semi-circular opening, five or six acres in extent, on the northwestern slope of Chaparral Hill. The altitude of the bed-rock I made to be 5,090 feet, which is undoubtedly too high. The bed-rock is slate, and has the usual northwesterly strike and vertical dip. The most of it is yellow in color, and is quite soft for at least forty feet in depth. There is also some banded clay, a few feet in thickness, looking like decomposed bed-rock, and similar to that previously described as occurring at Plug Ugly Hill and at Secret Diggings. The highest bank is nearly two hundred feet in height, and consists of twenty-five feet of gravel at bottom, covered by pipe-clay, which in places is as much as thirty feet in thickness, fifteen feet of a finer sandy gravel, twenty feet of clay and loam, and a hundred feet of a very easily moved volcanic dirt. About one half of the quartz pebbles are bluish in color. They are also considerably coarser than the pebbles at La Porte or Morristown. Large, worn boulders are also found in abundance. There is some petrified wood in the gravel, but there are no impressions of leaves and no animal remains. The gold is coarse and like shot, and, according to the statement of Mr. W. A. Morse, has a fineness of .930–.934. A small fault of about eight inches was observed in the face of the bank, where the break in the strata of differently colored clays made it easy to detect the slight movement.

The altitude of the top of Chaparral Hill I made to be 5,275 feet. According to the hand-level observation, this should be the same as that of a point near the base of the bank at Monte Cristo, on the opposite side of Goodyear Creek. The hydraulic banks on the southeastern slope of the hill I did not have time to visit. I went as far as the mouth of Hardee's tunnel, where I determined the altitude to be 5,038 feet. The course of this tunnel is S. 45° W. (magnetic) for 400 feet, and then more westerly for 200 feet farther. The rise in the tunnel is two inches in twelve feet for 300 feet, and for the rest of the way ten inches to twelve feet. Gravel lies about twelve feet above the tunnel. The bed-rock brought out from the tunnel was very soft and of a green color, resembling a chloritic or talcose schist.

The portion of the ridge lying to the northeast of Eureka I made no attempt to examine.

E. MONTE CRISTO AND CRAYCROFT'S.

These two places can be easily reached by good trails from Downieville. The former is upon the ridge between Goodyear Creek and the North Fork of the North Yuba River, and the latter upon the spur between the middle and west branches of the North Fork of the North Yuba. The position of Monte Cristo is shown upon the map (Plate E), though perhaps not with accuracy. Craycroft's would also fall within the limits of that map, but I have no good data for fixing its position. On my trip to Monte Cristo I followed first the western trail from Downieville, then crossed the ridge to Excelsior, and returned by the Excelsior trail. A subsequent day was taken for the visit to Craycroft's.

The country rock near Downieville is an easily cleavable slate. It shows frequent alternations of color, and the fresh surfaces of cleavage are often fantastically variegated. Across the slate there is a belt of serpentine, which strikes a little to the west of north. It is probably a part of the belt which is seen at Deadwood and at Whiskey Diggings. The eastern border of the serpentine can be seen at several points on the Excelsior trail; its western border crosses the mines at Monte Cristo. At the Empire tunnel the rock at the mouth is slate; the serpentine appears about 400 feet from the mouth. At Excelsior the bed-rock is slate.

Twenty years ago Monte Cristo was a flourishing mining camp. At an election in 1859 nearly a thousand votes were cast; now the total population does not exceed thirty persons. The most of the gold obtained came from drift mines; hydraulic mining has amounted to but little. The claims used to be laid out parallel to each other, in a direction N. 20° E., fronting on the cañon of Goodyear Creek, and extending back to the middle of the ridge. Within a distance of a third of a mile there were a half-dozen or more long tunnels. The lengths of some of these were given me by Mr. Thatcher. Beginning at the southeast, they are as follows: Empire, 1,400 feet; Swallow, 900; Poodle, 700; Exchange, 1,300; Cold Spring and Bigelow, 700. Many of these tunnels have been allowed to cave in and are no longer accessible. The only one that I entered was the Empire. I determined the altitude of the mouth of this tunnel to be 5,010 feet. The gravel that I saw was a clean white quartz. It was from this mine that the "petrified knot-hole," specimen 122, came. I had the specimen taken from the charred trunk of a large tree, which was lying in a horizontal position in the gravel and could be traced for at least thirty feet. The old Empire incline is said to have passed through one hundred feet of fine quartz gravel, forty feet of pipe clay, then a little more gravel, and finally the volcanic cement to the surface, the total distance being 500 feet.

My determinations of altitude at Monte Cristo I regard as more trustworthy than those made at Eureka and Mugginsville. The observations with the hand-level at Monte Cristo agree with those made at Chaparral Hill so far as to place beyond doubt the fact that the Monte Cristo bed-rock is higher—probably more than a hundred feet higher—than that at Hardee's tunnel or Mugginsville. The altitude of Mr. Thatcher's house at Monte Cristo I made to be 5,056 feet. The southwestern rim of the channel is at a little lower altitude than this. Its course is about S. 62° E. (magnetic) along the front of the mines; it then turns to the east and crosses under the volcanic cement to Excelsior, which is on the opposite side of the ridge and nearly due east from Monte Cristo. That the gravel extends through the ridge, has been proved by at least one underground connection. The pay channel, between the rising rims on the north and south, is about 700 feet in width; beyond that limit the gravel was not rich enough to pay more than three or four dollars a day to the man. In regard to the direction of the flow of the old stream between Monte Cristo and Excelsior, the evidence is conflicting. Some persons told me that the fall of the bed-rock is from east to west; others, that it is in just the opposite direction. My determination of the altitude of the mouth of one of the tunnels at Excelsior, 5,020 feet, is not sufficient to settle this question.

The position of this channel, being, as it is, directly under a comparatively low sag in the volcanic capping, suggests some interesting inquiries. Just above and to the north or northeast of the mines there is a steep and abrupt bluff of solid and compact rock, which Mr. Wadsworth's microscopic examination has shown to be an andesite. The altitude of the crest of the sag, above the channel, I made to be 5,492 feet, not quite 500 feet above the bed-rock in the mines. The volcanic material at the sag is much broken up, and the common surface boulders have more the appearance of an altered rock or a tufa than of a compact andesite. At first sight it seems as if it might be possible that the excavation of the channel was posterior to the lava-flow; but the absence of pebbles of volcanic rock from the deep gravel furnishes a very strong argument against this supposition. Whatever may be the explanation, it is certainly a very striking fact that many of the gravel deposits, like those at Hepsidam, Potosi, Bunker Hill, Eureka, Monte Cristo, and the Bald Mountain channel between Forest City and City of Six, either lie uncovered between high points of rock of volcanic origin, or, covered only with a loose tufaceous material, can be traced beneath some low sag in the ridge, where frequently there are bluffs of hard and compact lava on either side.

The gold at Monte Cristo is usually coarse in the lower strata and fine in the upper. There are also two grades of fineness: the gold of the "old mine" used to bring from $17.80 to $18.00 per ounce, while that of the so-called "back channel" is worth only $16.75.

About a mile below Monte Cristo some prospecting has been done at Mount Holly, at the head of Sailor Cañon, the richness of the cañon having led to the belief that there was a lower channel to be found somewhere in the vicinity. The mines at Fir Cap, about two miles up the ridge from Monte Cristo, are said to have been extremely rich. The stories told of them sound like fable. One report put the yield at 120 ounces to the car-load. No work has been done in them since 1874. Between Monte Cristo and Fir Cap there has been no successful mining on the west slope of the ridge. Wood's mine is on the eastern slope, nearly north from Excelsior and below Fir Cap. I saw the bank only from Craycroft's, on the opposite side of the cañon; I did not have time to visit it.

Upon the hill known as Craycroft's there are two deposits of gravel which appear to be distinct from each other. Their relative positions are shown in the diagram (Plate V. Fig. 3), which represents a longitudinal, or north and south, section of the ridge, with a vertical projection of the lower gravel upon the plane of the section. The upper gravel lies directly upon the crest of the ridge, which is here between three and four hundred feet in width. The gravel covers in all an area of not over five acres. It seems to lie in a basin-like depression, towards whose centre the bed-rock pitches from all directions. The bank exposed to view was about eighty feet in height. With the exception of some pebbles of a bluish color, the gravel was composed of clean, well-washed, bright, white quartz, resembling very closely the gravel of La Porte. There was nothing, however, to resemble the "bogus" gravel of La Porte.

The lower deposit, owned by Messrs. Eggleston and Mowry, is ninety rods from the upper gravel, and is seventy-five feet lower in altitude. It lies upon the northwestern slope of the ridge, something more than 200 feet below the crest. The highest point of the Downieville trail, where it crosses the ridge from the southeastern side, has an altitude of 5,370 feet. The altitude of the mouth of the tunnel I made to be 5,137 feet. The bed-rock is an easily worked slate, whose strike is nearly north and south, and whose dip makes a high angle to the east. At the principal opening the deep channel in the bed-rock is irregular and narrow, not exceeding one hundred feet in width, and has a course of S. 25° W. (magnetic). The greatest thickness of gravel seen was fifty feet. There is said to be a thickness of nearly ninety feet in a shaft near the southwestern extremity of the deposit. The gravel is reddish in color, and contains, besides some large boulders, considerable angular quartz, some pieces of float bed-rock, some lava pebbles, and a very little clay, thus differing in almost every particular from the upper gravel. It contains neither fossilized wood nor impressions of leaves. The deposit can be traced along the ridge towards the southwest for more than half a mile. The surface gravel at the lower extremity of the deposit is a mixture of rolled and of angular quartz. Some of the gold is very coarse, the nuggets not unfrequently being worth from ten to fifty dollars. Good "pay" has been found high up on the southeastern rim rock, where the dirt was only two or three feet in depth. The supply of water is brought in two ditches, four miles and eight miles in length respectively. The working season extends from the last of April to the middle of July.

These deposits have no obvious or necessary connection with any of the others that have been described, though there is nothing to prevent there having been a former channel from Craycroft's to Excelsior. On the higher points to the north or east I know of no gravels, and have heard of none, that can be supposed to have any connection with those at Craycroft's, except possibly those in the vicinity of Gold Lake. At the Germania claim, as Mr. Eggleston told me, a mile above Craycroft's, on the southeastern slope of the ridge, a tunnel was run, but without finding gravel; the volcanic cement rested immediately upon the bed-rock. The gravel on Rattlesnake Creek, nearly due north from Craycroft's, is, as I also learned from Mr. Eggleston, quite different in character, and is, moreover, at a considerably lower altitude. I did not have time to make any personal examination of that district.

In regard to the possible connection of these gravels with those lying to the south of the North Yuba, nothing definite can be said. City of Six, it is true, lies at a lower level than either Eureka, Monte Cristo, or Craycroft's, but the intervening distances are so great that the grade of any connecting stream would have been much less than we know the average grade of the old channels to have been. I am more inclined to think that the outlet from these deposits was by way of Goodyear Creek and the North Yuba, and that a junction may have been effected with the more westerly streams near Indian Hill.

The following statements in regard to the fineness of the gold from the mines treated of in this section, and from a few other localities, were given me by Mr. Briggs of Downieville.

Eureka	.920–.930
Fir Cap	.836
Monte Cristo	.914
Craycroft's	.939
Gold Lake	.925
North Fork of North Yuba	.885
South Fork of North Yuba	.865
Hog Cañon	.864
Jim Crow Cañon	.926
American Hill	.934

From another source I learned that the gold of the North Yuba ranged in fineness from .835 to .890, that of Goodyear's Bar being .884.

Section VIII. — *A Portion of the Feather River Basin.*

The only portions of the basin of the Feather River that I had time to examine were a small district near the head-waters of the South Fork; the region lying between Spanish Peak and Quincy; and, very hastily, a few points near the stage-road in the valley of the Middle Fork between Quincy and Sierraville. It will be convenient to treat these different districts in separate sub-sections.

A.   Little Grass Valley and Vicinity.

Little Grass Valley is easily reached from La Porte by the trail or road which crosses the narrow divide that separates the streams belonging to the Yuba system of drainage from those that are tributary to the Feather River. Its position is shown on the map (Plate R). The position of the Davis Point gravel on the ridge between the South Fork and Fall River is given on the diagram-map (Plate X, Fig. 1). This map is on the same scale as that on Plate R, and includes a piece of territory lying to the west of the northern part of that map. The district is one of which very little is known with accuracy. The United States surveys have been extended over it only in part. There are but few inhabitants, and but few good roads or trails. The chief importance of the district, from a mining point of view, lies in the possibility that somewhere in this quarter the higher portions of the La Porte channel may ultimately be traced. Extensive mining claims have been laid out, and small amounts of work have been done in several places. In the short time that I had to spare, I could not make any satisfactory examination of the country; I saw only the outline of a geological problem, which promises interesting results to some future explorer. The routes I followed were these: from La Porte to Davis Point and return, from Gibsonville to the Monitor shaft and return, and along the stage-road which leads from Gibsonville to Quincy.

One very interesting feature of the geological problem is presented by the distribution of the volcanic rocks and their mutual relations. Above the Little Grass Valley Bald Mountain, which is evidently capped with lava, the whole of the valley of the South Fork, as seen from a point on the stage-road near the base of Pilot Peak, appears to be free from volcanic rock. On the road itself bed-rock is seen shortly after crossing the first ridge above Gibsonville, the volcanic rock at the same time rising with the ridge to the east. Frequent exposures of bed-rock at high altitudes are also to be seen upon the upper trail from the Monitor shaft to Gibsonville. The serpentine belt is crossed by the road several times between Gibsonville and Onion Valley, but no more lava is encountered until the crest of the ridge north of Onion Valley is reached. The altitude of the highest point on the road between Gibsonville and Onion Valley I made to be 6,390 feet; that of the hotel at Onion Valley, 6,160; and that of the summit between the last-named place and Nelson Point, 6,430. The dotted red lines on the map on Plate R, from Pilot Peak to Bald Mountain, are intended to indicate only a probable former connection.

The crest of the ridge from La Porte to Gibsonville has already been spoken of as covered with an andesitic lava, grayish in color. On the descent from the crest of this ridge to the Monitor shaft I saw boulders of black, basaltic lava near the bottom of the cañon, at an altitude some two or three hundred feet above the shaft. The shaft is about four miles above Little Grass Valley, at a point opposite La Porte; it was started a few rods from the bed of the stream and scarcely five feet above the level of the water. The bed of the stream at this point is in gravel. The altitude of the mouth of the shaft I made to be 5,282 feet. About an eighth of a mile down stream from the shaft a basaltic dyke of at least twenty feet in width crosses the stream. Similar basaltic dykes are frequently met with in running bed-rock tunnels in this vicinity, as I was told by Mr. Winchell, of Gibsonville.

Westerly from Bald Mountain, Blackrock Creek gets its name from the color of the rock through which its channel has been eroded. Seen from a little distance, the black rock has all the appearance of basalt. Upon the divide between the South Fork and Fall River, known as the Mooretown ridge, both andesitic and basaltic lavas are to be seen. The trail which I followed crossed this ridge from a point near the junction of Blackrock Creek and the South Fork to the head of Fall River and thence down that stream as far as Davis Point. The altitude of the bed of the South Fork at the trail crossing I made to be 4,930 feet. The ascent is quite steep from the base of the ridge. A little less than half-way up the slope, at an altitude of 5,240 feet, there is a narrow, flat bench, above which a second slope rises, steeper than the first. The summit of the trail has an altitude of 5,660 feet. Owing to the presence of soil and vegetation, it was not possible to tell just where the lower limit of the volcanic capping was, nor whether there was more than one kind of lava on the ridge at this point. The surface fragments seen at the summit resembled those on the ridge above La Porte. On the Fall River slope of the ridge there are good exposures of rock in the ditch which has been dug to bring water to the gravel claim at Davis Point. The ditch is two miles and a quarter long, and it follows very nearly the line of junction between the bed-rock and the volcanic capping. From a point a half-mile to the northeast of the claim I got good specimens of basalt. The microscopic examination by Mr. Wadsworth has confirmed the field observation. Near the point from which these specimens were taken, there were also good exposures of bed rock. On the crest of the ridge, just above the mining claim, on the Oroville road, at an altitude of 5,060 feet, there is again a gray, andesitic lava. If these observations are correct, this part of the Mooretown ridge shows basalt at the base and andesite above. This, I believe, is an order of arrangement just the reverse of what has been supposed to be the rule in the Sierra Nevada. In further support of the observations here described, I will give the substance of the information I obtained upon this point from Mr. D. Post, a miner and prospector whom I came across by accident upon the ridge. Mr. Post showed himself to be an unusually intelligent observer, and was evidently very familiar with the principal topographical and geological features of the vicinity. I took the time to go to one or two places in regard to which Mr. Post's descriptions had excited my curiosity, and found that his statements were fully in accord with the facts. According to Mr. Post's observations, the ridge between Rock Creek, a creek lying to the southwest of the limits of the map on Plate X (Fig. 1) and the South Fork, is capped with " black lava"; the Mooretown ridge is in the main capped with black lava, similar in character to the specimens taken from the ditch, but it is crossed by a flow of " gray lava " at an angle of between forty and sixty degrees. This flow is about a mile in width, and can be traced in a similar way across other ridges, in a direction from southeast to northwest.

There is also evidence in the bed of Fall River of a deeper flow, or of a broad dyke, of gray lava. Under the guidance of Mr. Post, I went to examine the locality where the deep channel was reported to cross the river, which is about a half-mile below the mouth of the east branch of the river, the branch on which the Davis Point gravel lies. Near the junction of the branch with the main stream a dam has been built to turn the water of the stream into the China Gulch Company's ditch. The altitude of the dam I made to be 4,615 feet. From the dam I followed the line of the ditch down the stream. The bed-rock exposed is granite. About a half-mile below the dam the granite suddenly gave place to what looked like a gray volcanic tufa. Between the granite and the tufa there was a layer of clayey material, six feet in thickness. This belt of volcanic rock is said to extend for nearly half a mile along the ditch, after which the clayey streak and the granite come again in the reverse order. I did not have time to verify this statement by personal observation. When first excavated, the volcanic rock was hard and required blasting. The specimen which I brought from this point has been examined microscopically by Mr. Wadsworth, who calls it " trachyte (?) or an allied somewhat altered andesite." In the bed of the stream, below the ditch, volcanic rock takes the place of granite for a distance estimated at from ten to twelve hundred feet. A specimen taken directly from a good exposure of this rock, where it was kept clean by running water, has been called andesite by Mr. Wadsworth after a microscopic examination. The altitude of the river-bed at this point I made to be 4,540 feet. Upon the map (Plate X, Fig. 1) I have marked the approximate position of my point of observation only, without connecting the lava with that of the ridge. Nothing is known as to the depth to which this lava extends. Mr. Post's examination of the country leads him to regard it as the filling of an old channel eroded at some time subsequent to the formation of the basalt, and to estimate its depth at the river crossing to be 300 feet. A few facts which support the hypothesis of the existence of such a channel will be given further on.

It has been stated already that granite bed-rock is found below the dam across Fall River. At the Davis Point mine the bed-rock is a very soft, green, talcose or chloritic slate. It strikes N. 40° W. (magnetic), and is nearly vertical in dip. This slate-belt, however, cannot be more than 200 feet in width. To the southeast of the mine the bed-rock seen in the ditch, near the point from which the specimens of basalt were taken, is of a granitic or dioritic character. A similar rock is seen near Post's cabin in Wilson's ravine, a ravine on the southeastern slope of the ridge leading to the South Fork of the Feather River.

The gravel of Davis Point lies on the northwestern slope of the ridge, a short distance above Fall River. The altitude of the bed rock at the mouth of the tunnel I made to be 4,800 feet.

There has not been work enough done to prove the existence of any large body of gravel. A tunnel seventy feet in length has been driven, showing that the gravel is continuous for that distance, and that the bed-rock pitches to the southeast; that is, towards the centre of the ridge. Messrs. Gard and Makins, to whom I am under obligations for a very hospitable reception, were about bringing water in for the commencement of hydraulic operations at the time I was there. The gravel, when exposed to view, is composed of small, rounded quartz pebbles, together with some thin sandy layers, and thin clay seams with impressions of leaves. These were too fragile for preservation. Charred wood is also found. The gold is said to be coarse, resembling flax-seed. There is no scaly gold. The altitude of the mouth of a shaft, said to have been sunk in 1857 to a depth of fifty feet through pipe-clay and sand to bed-rock, I made to be 4,900 feet.

The gravel at Post's claim in Wilson's ravine presents some features of extraordinary interest. The claim lies a few hundred feet back from the South Fork of the Feather River, and about 150 feet above the bed of the stream. The altitude of the bed-rock at the mouth of the tunnel I made to be 4,425 feet, — nearly 400 feet lower, it will be observed, than that of the bed-rock at Davis Point, on the opposite side of the ridge. The tunnel has been driven for 200 feet in a northwesterly direction; but, as the bed-rock pitches also in that direction, the accumulation of water prevents any further prosecution of the work without the use of pumping machinery. Lateral breasts have been driven from the tunnel sufficient to indicate a width of 200 feet of gravel. High bed-rock exposures are seen both to the north and the south of the gravel. The bed-rock in the tunnel shows many signs of wear. The thickness of the gravel has been proved to be as much as twenty feet in some places. Above the gravel there is pipe-clay, and the surface of the ridge above the deposit is covered with gray lava. The composition of the gravel is widely different from that of any other deposit that I saw anywhere else in this vicinity. In addition to the ordinary fine white and blue quartz with some large boulders, there are several varieties of bed-rock represented, such as metamorphic slate and granite. The pebbles are all well washed and rounded though those composed of bed-rock are much decomposed. Bunches of micaceous sand are quite frequent. But the most striking and unusual occurrence is that of rolled pebbles of basaltic rock. To avoid all error of field observation, I picked from the gravel near the farther end of the tunnel a few small pebbles of this character, — such as could be easily brought away in the pocket, — and put them into Mr. Wadsworth's hands for microscopic examination. He describes them as basaltic rock which has undergone some decomposition, — as might be expected when the circumstances under which the pebbles are found are taken into consideration. Among the pebbles there are also some which look as if they came from some previously existing gravel bed. The gold in this gravel is fine and scaly.

The facts above presented lead irresistibly to the conclusion that the gravel in this deposit is geologically younger than that of La Porte and other mining towns on the Yuba slope of the divide. My information is too meagre to justify my hazarding any theory as to the origin of this deposit or its relations to others. There is, however, something very plausible in Mr. Post's belief that it has some connection with a deep channel crossing the ridge in a northwesterly direction, and now filled with gray lava. The position of the lava, as above described, in the bed of Fall River, and its altitude, lend some support to this belief. In this connection I will add (on Mr. Post's authority, not from personal observation) that there are also other gravel deposits, not far distant from this one, but at a higher altitude, in which the gravelly material is quartz and bed-rock with no basalt, in which the gold is thick and heavy, and upon which a basaltic capping rests. I regret that I was unable to make any further study of so promising a region.

At the Monitor shaft, previously alluded to, bed-rock has been reached since the date of my visit, at a depth of fifty-two feet from the surface. For the first eight feet the shaft was in the common surface dirt; the rest of the distance was in gravel. Charred wood was found ten feet from the bottom of the shaft. From what I saw and could learn about the explorations at this point, I was led to conclude that the deposit is one of very limited extent, and of purely local interest, without any connection with any larger body of gravel.

The country to the north of Fall River, including the gravel of Franklin Hill and Cammel's Peak, I made no attempt to explore. From the descriptions I received of those places, I thought my time could be employed more profitably elsewhere. The only important deposits of gravel in this vicinity that I passed by without examination were those of Saw Pit Flat, Richmond Hill, and Washington Hill near Onion Valley. I got a glimpse of the diggings, and a general view of the course of Onion Valley Creek, from the stage-road on my way from La Porte to Quincy. Had I not been travelling by stage, I could have made a stop of a few hours, and thus have seen something of the details of the situation. It did not seem worth while to make a stop of three days at Onion Valley for the sake of so little. The chief interest of these deposits lies in the possibility of their representing an upper portion of the same old channel, whose lower course can be traced from Hepsidam or La Porte down the valley of Slate Creek. My own opinion, based, however, upon imperfect observations, is opposed to any hypothesis of that kind. It seems to me that the Onion Valley gravels stand by themselves, quite distinct from any of the more southern channels. Indeed, I may go further, and say, as the final generalization from all the observations I was able to make on both sides of the divide, that I can see no reason for regarding the gravels in the region drained by the Feather River as in any way related to those in the basin of the Yuba. The present dividing ridge between the rivers is a boundary which was equally well-marked at the time the gravels on either side were deposited.

B. QUINCY AND VICINITY.

The region to be described in this sub-section lies to the north of the Middle Fork, and south of the east branch of the North Fork of the Feather River. I approached it by stage from La Porte by way of Onion Valley. In the preceding sub-section I gave the altitude of the summit of the road near Onion Valley as 6,430 feet. From that point there is a very rapid descent down the cañon of Nelson Creek to the Middle Feather at Nelson Point. The rock exposed along this portion of the road is almost entirely a very thinly laminated slate. The altitude of the bridge across Nelson Creek I made to be 4,120 feet; that of the bridge at Nelson Point, 3,950. Along the road between the river and Quincy I saw, at frequent intervals, considerable detached volcanic rock; though the prevailing rock, so far as it could be recognized, was slate. I could not fix the relations of the two to each other, but I was satisfied that the so-called "Little Volcano," near Nelson Point, has nothing to do with igneous phenomena.

Near Nelson Point there is good evidence that the Feather River once flowed in a deeper channel than it does now. On the northern bank of the stream there is a small gravel deposit, in which, as I was told, a shaft has been sunk to a depth of fifty feet below the present river-bed. A quarter of a mile northerly from the river the stage-road crosses a belt of lava cement, in which a shaft has been sunk to a reported depth of 350 feet without reaching bed-rock, though a high rim of such rock is seen cropping out only a short distance farther along on the road. The general belief is that this deep channel is traceable from Bell's Bar to Nelson Point. I did not learn anything about the character of the gravel; its position shows that it can have had no connection with the so-called "high gravels."

The gravel deposits which lie within a few miles of the town of Quincy, to the north, west, and south, present a great many puzzling geological questions, the solutions to which cannot be reached in a hasty reconnaissance of a week. It is very evident, for instance, that the topographical features of the country have been much changed since the date of the deposition of the earlier gravels. There is not that harmony between the former and the present systems of drainage which is so prominent a feature in parts of Sierra and Nevada counties. In some cases, indeed, the direction of drainage is precisely reversed. It is also evident that the gravel deposits do not in all cases, if in any, antedate the flows of volcanic rock; and that some of them, at any rate, are of comparatively recent origin. Some of the older gravels may, very likely, have been subjected to a re-arrangement, or a re-deposition, in connection with the changes of surface topography. From these statements it will be seen at once that the problem here presented to the geologist is quite different from that to be studied in the basins of the Yuba and the American rivers. It may perhaps forever remain an unsettled question, whether or not the old streams of the southern counties once had their sources at points farther to the east and north than their channels have yet been traced; but I feel sure that there is nothing in that portion of Plumas County which came under my observation, north of the Middle Feather, to support the belief that the gravels of that county, in their present position and condition, and those of the Yuba basin, are representatives of the same old rivers.

The best description of this district which has yet been published is included in a paper, by Mr. J. A. Edman of Mumford Hill, in Raymond's Report for the year 1875, pages 109–128. The geographical information on the map accompanying Mr. Edman's paper was taken from the official map of Plumas County, previously alluded to.* My own observations do not cover quite the same ground, nor are they as extensive as those of Mr. Edman, but they agree in almost every particular with his so far as they relate to the same localities.

The most striking topographical feature of this vicinity is the high mass of syenite or granite, known as Spanish Peak, nine miles west of Quincy. The precise altitude of this peak is not known; it is probably between 6,500 and 7,000 feet high.† The western slope of the mountain is comparatively regular and gradual; the eastern face is more steep and abrupt. Seen from the east, this face looks like a nearly vertical wall of rock or escarpment, several hundred feet in height. Directly east of Spanish Peak lies the elevated valley of Spanish Creek, shut off from the two branches of the Feather River on the north and south by well-marked high ridges, which abut on the west against the granite of the peak. The course of the creek from its head in the small lakes near the summit of the mountain, for a distance of about twelve miles, is nearly due east,—a course just opposite to that of the principal drainage of the country. The town of Quincy stands near the point where the creek emerges into American Valley through a narrow gap between the northern and southern ridges. The altitude of the town I made to be 3,383 feet, about 3,500 feet lower than the top of Spanish Peak. My observing station was in the second story of the Plumas House. Just below Quincy the creek makes a sharp curve to the north and west, and joins the east branch of the North Fork of the Feather near Soda Bar. The waters of the creek make a circuit of nearly thirty miles in order to reach a point less than four miles distant from where they started. It is in this valley, and on its enclosing ridges, that the gravel deposits to be described on the following pages are found. In my descriptions I shall make no attempt, except in a few cases, to show how the different deposits are, or ever have been, connected together; though I shall follow, so far as possible, a systematic geographical order in my treatment of them.

The Elizabethtown Flats are between one and a half and two miles northwest of Quincy, and on the northern side of Spanish Creek. Elizabethtown was once the rival of Quincy in size and importance: now it is nearly deserted, excepting by a few persons who are engaged in mining. The gravel lies in the trough, along the sides, and at the head of a small stream or ravine, in which the water flows easterly and southeasterly towards Spanish Creek. High bed-rock rises on the north and south. At the west there is a low saddle, near the top of which beds of gravel, clay, and sand are exposed to view. The altitude of the crest of the saddle I made to be 3,750 feet: a quarter of a mile westerly, on the ridge, bed-rock is seen at an altitude fully 150 feet higher. I cannot say with certainty whether or not the gravel is continuous across the saddle towards Newtown. The principal body of gravel in the Flats is not much above the general level of Spanish Creek and American Valley. The long tunnel, nearly half a mile in length, which has been driven from the lowest available point to the southeast, winding in such a way as to follow the line of separation between gravel and bed-rock, is still not deep enough to drain the central portions of the deposit. Both gravel and water have to be raised to the surface through a shaft more than fifty feet in depth. The conditions are evidently not favorable for hydraulic mining. On the

* See ante, p. 444.
† See ante, p. 216. My measurement gave 7,058 feet as the altitude of Spanish Peak.— J. D. W.

southern slope of the ravine there is some evidence in the lay of the bed-rock that a narrow channel, thirty or forty feet wide, used to run at an altitude of from ten to sixty feet above the present bed of the stream. The gravel in this old channel is but slightly worn, and only a few of the many large boulders, which, as a rule, are metamorphic rock, are well rounded. To the north of the gravel in the ravine there is a large quartz ledge. Large, heavy nuggets of gold have been found at Elizabethtown, and the bed-rock, in spots, is said to have been wonderfully rich. Taken as a whole, this gravel deposit must be looked upon as of comparatively recent origin, and not as belonging to any of the old channels.

A mile and a half to the north of Elizabethtown, on the spur between Little Blackhawk Creek and Jackass Gulch, there is a peculiar deposit of gravel known as the Deadwood Channel. It has not been worked at all by the hydraulic process. In early days a large number of prospecting shafts were sunk on or near the course of this channel, but no one of them appears to have been sunk to a profit. More recently, tunnels have been driven as a preliminary to drifting. The new tunnel on Little Blackhawk Creek has an altitude, at its mouth, of 3,500 feet. There are three unusual and interesting features of this deposit which are deserving of mention. The first of these relates to the grade of the bed-rock. I had no opportunity to verify the statements made to me, and I give them on the authority of Mr. J. M. Keller, one of the owners of the Deadwood property. The gravel, which appears on both sides of Jackass Gulch, the bed of which has been eroded entirely through the gravel and into the bed-rock, sweeps around in a long curve, and can be traced, with a falling grade, up the western bank of Little Blackhawk Creek ; that is to say, in a direction contrary to that of the present drainage. Here, then, is a case in which the drainage of the country has been reversed, unless some local disturbance, not affecting the general drainage, has deranged the position of this gravel alone. The second point of interest is that the bed of the channel is much more irregular than is usual in such deposits.* There are rapid changes of level, suggestive of cascades, and changes in the cross-section which point to the existence of a higher bench, distinct from and parallel with the deep trough. The width of the channel is about 130 feet. The bed-rock is a talcose slate, shiny, soft, and fragile. Small hand specimens, which will not crumble to pieces, are not easily obtainable. The third point relates to the character of the gravel. There is here an incompatible mixture of large, well-worn, smooth quartz pebbles or boulders, and sharply angular blocks of slate bed-rock. These two varieties cannot owe their shapes to the same agencies. The most obvious explanation of the phenomena is that the Deadwood channel is of secondary origin, resulting from the breaking up of some previously existing gravel deposit.

Spanish Ranch, on Spanish Creek, a short distance below its junction with Silver Creek, is nearly due west from Quincy, and about six miles distant. The altitude of the express-office at this place I made to be 3,585 feet, almost precisely 200 feet above the hotel at Quincy. Between the two places, upon the ridge to the north of Spanish Creek, there are extensive deposits of gravel upon three at least of the spurs which make down from the main ridge. These are known as Shores Hill, Badger Hill, and Gopher Hill, naming them in order from east to west. For much detailed information in regard to these deposits, and to others farther west on the same ridge, I am under obligations to Mr. N. Cadwallader, of Spanish Ranch, the manager of the property of the Plumas Mining and Water Company. Shores Hill I did not go to ; but I determined an approximate altitude for its bed-rock by an observation on Hungarian Hill, on the opposite side of the creek. With one exception, my determinations of the altitude of the bed-rock at these three places agree very well with what Mr. Cadwallader told me of their known differences of altitude, calculated from the grade of the ditch, and the amount of pressure at the different mines. My measurements made the altitude of the Shores Hill bed-rock 3,988 feet ; that of the lowest bed-rock exposed at Badger Hill (about twenty feet below the point to which Mr. Cadwallader's data refer), 3,880 feet ; and that of the low bed-rock, near the face of the bank at Gopher Hill, 3,835 feet. The differences between these numbers ought to be, according to the miners' measurements, 105 and 135 feet respectively, instead of 108 and 45. The discrepancy can be easily explained by supposing an error on my part of one hundred feet in reading the barometer at Gopher Hill. The distance from Shores Hill to Gopher Hill is about a mile and a quarter. If the fall between the two places is nearly two hundred and forty feet, the grade of the old channel was unusually high, and, what is still more remarkable, in a direction just the reverse of that of Spanish Creek, though conforming to the main drainage of the country, as illustrated in the forks of Feather River.

Near the western end of Gopher Hill there are two strong pieces of evidence of extraordinary disturbances since the deposition of the gravel. On Plate V are given two sketches as they were made on the spot. At one place (Fig. 4) the bed-rock distinctly overhangs the gravel, the projection amounting to about six feet at the point where I took my sketch. The arrangement of the small boulders and flat pebbles in the gravel in the vicinity of this fault was peculiar, and led to the belief that the fault was formed after the gravel was deposited. A few rods to the south of the point just referred to the bed-rock is seen to fall off nearly vertically to the west (Fig. 5). What the extent of this fault is, no one knows. In a well which was sunk in the gravel to the west of the fault, to a depth of forty feet, no bed-rock was reached. From this point on in a westerly direction, across and on the opposite side of Wapousa (or Waponseh) Creek, gravel is seen at the surface, without much change of appearance, and but little above the level of Spanish Creek, for the distance of a mile. Near the mouth of Wapousa Creek the southern rim of the channel is seen between the gravel and Spanish Creek, pitching rapidly toward the north; the northern rim is seen in the form of high bed-rock, a quarter of a mile distant. Between Wapousa Creek and Spanish Ranch the surface-gravel spreads out over a much wider area, as a low gravel ridge, from a half-mile to a mile in width. The thickness of this gravel deposit has never been ascertained, nor is it known precisely where the bed-rock rises again to the west. At Hepsidam and Potosi I traced the channels until they disappeared in the air; here I have run one into the ground. An attempt to suggest an explanation for the facts above described is deferred until additional details in regard to these gravels, and others in the vicinity, have been given.

The bed-rock, where exposed to view upon these hills, is slate. Upon Badger Hill I made the strike of the slates to be N. 75° W. (magnetic): the dip is nearly vertical. About a quarter of a mile westerly from Wapousa Creek the easterly edge of the great serpentine belt is seen. This is probably a continuation of the same belt that has previously been traced from Downieville, by way of Deadwood and Whiskey Diggings, to Onion Valley.

The gravel of these deposits is not clean white quartz, like that of La Porte and Howland Flat. It is made up principally of pebbles of metamorphic or eruptive rock, flattened rather than spherical, and small in size. There are hardly any pieces larger than a man can lift. At Badger Hill I saw pebbles of volcanic rock, similar in character to that which caps some of the neighboring ridges, distributed through the gravel all the way down to bed-rock. The occurrence of these pebbles in this position is sufficient to prove the comparatively recent origin of the deposit. Owing to the great irregularity of the bed-rock, — a fact which seems to indicate that the old channel was not uniformly eroded over its whole width of 1,200 feet before the deposition of the gravel began, — it is hard to say what the average height of bank is. At different places on Badger Hill I estimated the height and character of the bank as follows: Fifty feet of clean gravel; twenty-five feet of gravel covered with fifty feet of clay and sand; one hundred feet of gravel and one hundred feet of clay, with thin gravel-streaks near the top. Some of this irregularity may be due to the fact that the southern parts of Badger Hill and Shores Hill have been cut away by Spanish Creek. Upon Shores Hill there is said to be no pipe-clay, excepting here and there in small patches. No petrified wood is found in these gravels. These deposits were worked as drift-mines as long ago as 1854, and were regarded as very rich. The best gravel is found on and near the bed-rock; but all the coarse gravel, up to forty or fifty feet in thickness, carries gold in paying quantities. The gold is, on the whole, coarser than that of Nevada County, though there is also considerable fine gold. Occasionally a nugget of good size is found; the largest one saved last year was worth twenty dollars. The average fineness of the gold is .940, ranging from .925 for the coarse bottom gold to .953 for the finest. The average yield per acre of hydraulic ground is stated to be $10,000.

The water for these mines is brought from two small lakes upon the northwestern slope of Spanish Peak, near its summit. The capacity of the ditch is 1,500 miners' inches, measured with a head of six inches.

Along the ridge to the north and northwest of Spanish Ranch, for a distance of five or six miles, there are said to be deposits of high gravel, upon which more or less mining has been done. In Quien Sabe Ravine the gravel is said to be capped with lava; and lava also appears in the ditch at intervals for a couple of miles farther eastward. Above Bean Hill, which is about two miles distant from Spanish Ranch, the gravel is free from any volcanic capping. These deposits, with the exception of Chaparral Hill and Fales's Hill, I did not find time to visit. As nearly as can be estimated from the known grade of the ditch between Badger Hill and Bean Hill, the bed-rock at the latter place must be about 800 feet higher than the store at Spanish Ranch. At Bean Hill there is further evidence that faults have been formed since the deposition of the gravel. See the sketch (Plate V, Fig. 6) which was drawn for me by Mr. Edman.

Between Spanish Ranch and Chaparral Hill I followed the so-called wagon-road, — a road, which, taken all in all, is one of the worst I have ever ridden over. The surface boulders seen on the route were in part serpentine, which is probably the country rock; in part metamorphic rock of several different kinds; in part a porphyritic rock of peculiar structure; and, to some extent, volcanic tufa. At several places, on the higher spurs between the small ravines, the surface was covered, over a limited area, with well-washed gravel. The gravel deposit known as Chaparral Hill lies at the head of one of the branches of Spanish Creek, near the old Mountain House. The bed-rock exposed to view is serpentinous in character. Its altitude at the eastern end of the hill I made to be 4,980 feet, which is about 1,400 feet higher than Spanish Ranch. The mining done here must have been almost entirely confined to shallow diggings, for there are no high banks to indicate extensive hydraulic operations, if the large, sloping bank on the southern side of Spanish Creek, light in color and looking like clay or sand, be excepted. Chaparral Hill has been practically deserted for several years, save by Chinamen, who work the ground over and over again every winter, with aid of the small supply of water obtained from the melting snows. Large amounts of gold are said to have been taken from the shallow diggings. Fales's Hill lies on the opposite side of the ridge, near Mill Creek, in a westerly or northwesterly direction from Chaparral Hill, and about one mile distant. The gravel is generally supposed to be continuous across the ridge; at least, I was told that no bed-rock is to be seen between the two places. The width of this supposed channel, which may be taken as 1,000 feet at Chaparral Hill, narrows to 500 feet or less at Fales's Hill. The bed-rock at the latter place is very uneven; the supposed southern rim stands at an unusually high angle. I made its average altitude, where exposed to view, to be 4,880 feet, indicating a grade of a hundred feet to the mile from east to west. The gravel both here and at Chaparral Hill is quite different in character and appearance from any that is seen in the Yuba basin. The general color of the exposed bank at Fales's Hill, for at least ten feet from the surface of the ground, is a bright crimson, something like the color of cinnabar. The boulders and pebbles represent many different kinds of rock. Some of the layers of fine crimson quartz sand and rounded pebbles of various kinds have a dip as high as 15°. Neither pipe-clay nor lava is to be seen here. Of the extent or probable value of these deposits I can say nothing. Mr. Keller, of Quincy, told me that in 1854 a shaft was sunk on Fales's Hill, to a depth of ninety feet, without reaching bed-rock. There is some prospect of an early resumption of mining operations at one or both these hills.

South of Spanish Creek the first locality to be mentioned is Hungarian Hill, a point about three miles west of Quincy, and nearly opposite Shores Hill. Within the area included under this name gravel is found in several places at widely different altitudes and under conditions so diverse that it is difficult, if not impossible, to see how the several deposits can ever have belonged together, or have been formed at the same time or by the same agencies. The map of Hungarian Hill (Plate W) I was allowed to copy from one belonging to Messrs. Peter and James A. Orr, the owners of a large extent of mining ground on the southern side of the hill. Hungarian Hill is

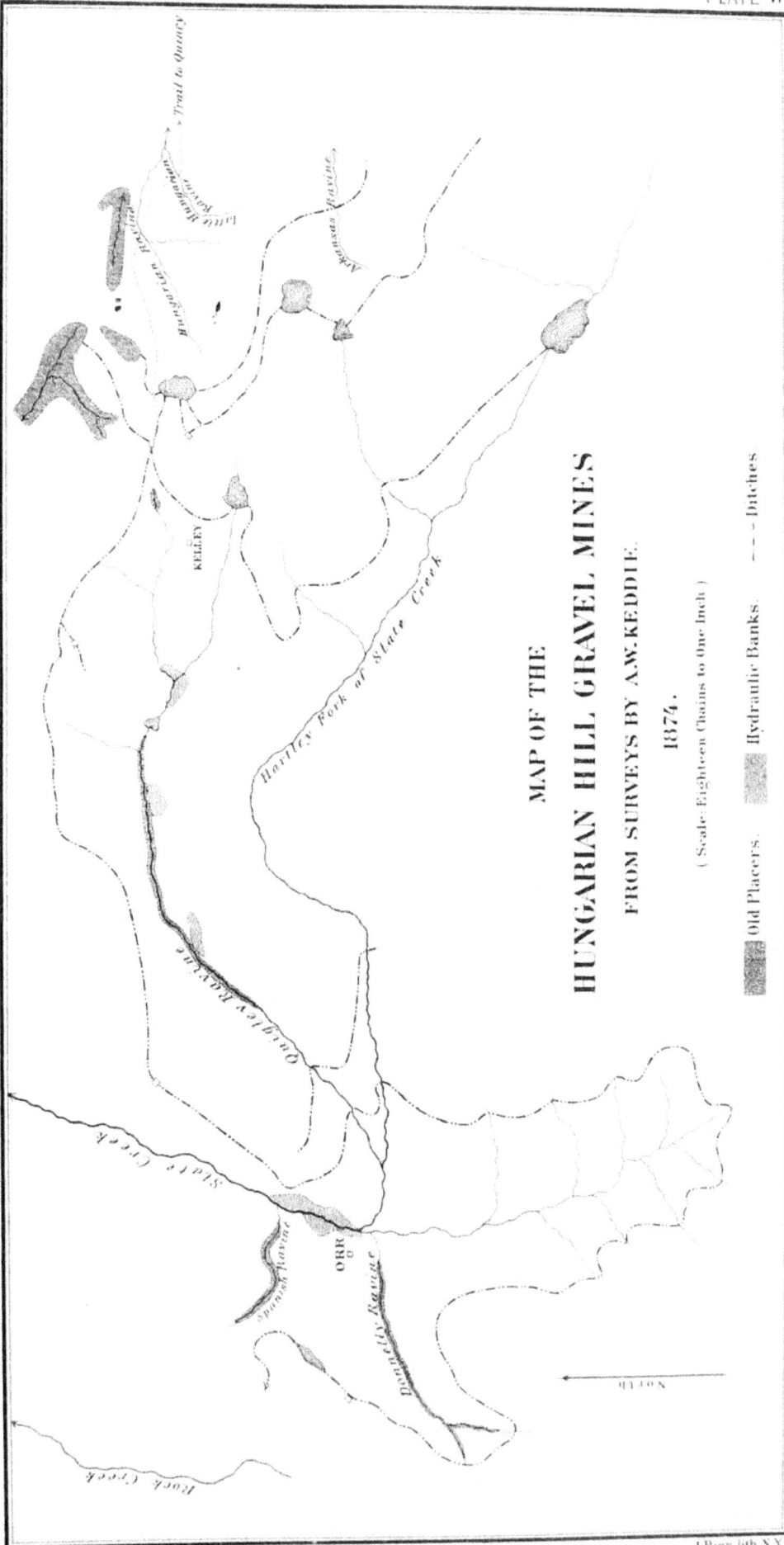

MAP OF THE

HUNGARIAN HILL GRAVEL MINES

FROM SURVEYS BY A.W. KEDDIE.

1874.

(Scale Eighteen Chains to One Inch)

Old Placers.

Hydraulic Banks.

— — Ditches.

Trail to Quincy

Little Hungarian Ravine

Hungarian Hill

Arkansas Ravine

KELLEY

Hartley Fork of Slate Creek

Onalley Ravine

Slate Creek

Spanish Ravine

Donnelly Ravine

ORR

Rock Creek

NORTH

reached from Quincy by a good trail, which, leaving the stage-road a short distance west of the town, follows up Hungarian Ravine. The highest point of the trail near the reservoir has an altitude of 4,280 feet. At this point bed-rock is known to exist within a few feet of the surface, being covered only with the red surface-dirt. To the north of the trail, and about one hundred feet lower, there has been considerable mining done on the divide between two ravines, which, starting quite near together, lead down, the one to the west and the other to the east, towards Spanish Creek. The material washed at this place looked like surface-dirt alone, without any admixture of rolled gravel. The altitude of this deposit is such that it may possibly have had some connection with Shores Hill and Badger Hill, but the character of the dirt is so much different from that seen at the place last named, that I am inclined to the belief that no such connection ever existed. On the southern side of the hill, Quigley Ravine, starting from near the reservoir, follows, for a little more than a mile, a westerly course to its junction with Slate Creek, east of Mr. Orr's house. The Flat at the head of the ravine is said to have been quite rich in gold; and the present bed of the ravine shows signs of having been worked to a depth of three or four feet. The material in the bed of the ravine at its upper end is peculiar in its arrangement and its structure, there being at the bottom thin layers of clay, fine gravel, and "brown sediment" carrying no gold, which are covered with the clean gravel above. From Mr. Kelley's Arkansas claim I obtained a few grains of gold recently panned from the gravel. Of these Mr. Wadsworth says : " They are much worn. On three of the grains some of the pits are not filled, and on the fourth we find these forms deep, numerous, and bright. They are part of the original impressions produced by the intergrowth of the matrix and the gold. They came from veins, probably quartz." In addition to the shallow diggings there are at different points along the course of the ravine, particularly on its left bank, beds of gravel sufficient in extent to justify the employment of the hydraulic process of excavation. The general appearance is as if the ravine had once followed a deeper track a few rods to the south and east of its present position, from which it was deflected after the accumulation of the gravel. I made the altitude of the bed-rock at the base of one of these banks to be 4,003 feet. The bank is of a brick-red color and about seventy-five feet in height. The bed-rock pitches rapidly toward the southwest, but the beds of gravel and clay are nearly horizontal. The material of the bank for the first twenty feet from bed-rock is lava and slate, with scarcely any quartz. The pebbles are small, and not much worn ; the lava pebbles have suffered much from weathering. Above this lower stratum the bank is lighter in color, and consists mostly of clay, sand, and fine gravel. The body of gravel near the outlet of Quigley Ravine is known as the Five Points. The altitude of bed-rock at the upper end of the present excavation is 3,740 feet. The width of the gravel deposit here I estimated at 400 feet. The thickness is, in places, as much as eighty feet. On the opposite bank of Slate Creek there is also a bed of gravel upon which a little hydraulic mining has been done on the northwestern side, where the tailings are discharged into Rock Creek. The altitude of the bed-rock at this opening I made to be 3,698 feet, about 500 feet below the old workings near the summit of the Quincy trail. The bank exposed to view is sixty feet in height, and it contains pebbles of the so-called "greenstone porphyry," a little quartz, and considerable gray, tufaceous lava, the last being found all the way down to the bed-rock. None of the pebbles are very large, and what boulders there are, are not much worn. There are a few streaks of sand and clay, but no cement in this gravel. The concentric structure of the disintegrating lava pebbles was quite noticeable. The bed-rock, which is a slate, with a northerly strike and a vertical dip, appears to rise on all sides excepting towards Rock Creek and Meadow Valley. Badger Hill is distant about a mile and a half in a northwesterly direction, and is about 200 feet higher in altitude. An intervening high ridge of bedrock also effectually shuts out any possibility of a connection between the two places. It seems to me most probable that this gravel is of comparatively modern origin ; that it was accumulated in Quigley Ravine when its outlet was dammed up by morainal matter, and then finally redistributed by the action of the waters of this ravine alone, without any connection with other channels.

The region to the southeast, between Hungarian Hill and Clermont, I did not visit. From Mr. Orr I learned that gravel, which has not yet been prospected to any extent, is to be seen on the surface for a considerable distance to the south and southeast of the Five Points claim; that, after crossing the low divide between Slate Creek and Rock Creek, gravel is found along the right bank of the latter stream; and that it can be traced beyond Deer Creek, a small stream which crosses the gravel without cutting down to bed-rock. The high points of the ridge are said to be capped with lava, as, for instance, at the head of Deer Creek, though no lava is to be seen along the line of Orr's ditch for the first five or six miles above Five Points. From Mr. D. L. Haun, whom I met at Quincy, I learned that, since the date of Professor Whitney's visit to Clermont,[*] a second tunnel, 800 feet in length, has been driven under the lava cap, on the western side. The gold-bearing gravel channel was found in this tunnel, but the gravel was too thin for profitable working.

The only other points that I visited in this vicinity lie nearer to the base of Spanish Peak, at the headwaters of some of the southern tributaries of Spanish Creek. Mumford's Hill, the most southerly point reached, occupies the spur between Big Creek and Eagle Gulch. The altitude of Mr. Edman's house I made to be 4,710 feet, which is about thirty-five feet higher than the bed-rock at the upper end of the gravel deposit. The bed-rock is, in general terms, a slate, with a strike of N. 55° W. (magnetic), though changing very frequently in color, in lustre, and in composition. One striking variety is made up of alternate silicious and clayey layers, the former being from one eighth to three eighths or a half an inch in thickness, while the latter are not more than one third as thick, which, though not distinguishable on a fresh surface, appear very prominently as easily separable layers, alternately light and dark in color, when the rock is exposed to the action of the weather. The effect is very pretty, but cannot easily be preserved in a hand specimen. Directly under the gravel, and forming a part of the bed-rock of the mine, there is a dolomitic belt, or fissure vein with a dolomitic gangue, carrying fine gold. The dip of this "ledge" is 60° to the northeast; in strike, it follows very nearly that of the enclosing slates; its average width may be taken as nearly fifty feet. The following description of this peculiar deposit is taken from Mr. Edman's paper in Raymond's Eighth Report, to which reference has already been made. "The vein-stone of this ledge is magnesian limestone, penetrated by a network of small quartz veins, and filled with lumps of talcose slate, rich in crystals of oxide of iron. Through this mass course larger veins of quartz, often swelling to large masses; it is frequently divided by bands or lenticular masses of an altered talcose slate, impregnated by decomposed sulphurets. Near the walls are often found alternating layers of clay and broken quartz, and a ferruginous quartz-conglomerate. The magnesian lime-stone is decomposed to a great depth, forming a soft, porous mass of brick-red earth, full of quartz and slate fragments, with rounded boulders of unaltered vein-stone reposing therein." The ledge has been worked by an incline to a depth of 100 feet, where the dolomitic gangue is solid and distinctly recognizable. The average yield of this ledge is from three to four dollars per ton; the richer portions run much higher. The length of the Mumford Hill gravel deposit, between Big Creek and Eagle Gulch, is about 800 feet, following the general direction of the strike of the bed-rock, until the curve west of Mr. Edman's house is reached. The grade of the bed-rock is towards the northwest. At its upper end the channel is 500 feet in width; at the lower end, about 250 feet. The eastern rim of bed-rock is very steep; the western rim has a more gradual slope. The banks of gravel average from forty to fifty feet in height. The gravel is made up of different varieties of country rock, intermixed with a little quartz. Neither pebbles nor boulders show many signs of wear. Among the boulders are some of quite peculiar character, which appear to be completely changed to clay; though the original rock, when sound, is fine-grained and crystalline, as is proved by the finding of specimens in a state of only partial disintegration. The rock in place from which these boulders come Mr. Edman thinks he has found, at a point about two miles to the south of Mumford Hill. Some of the gravel shows manganese stains.

* See ante, p. 215.

The gold occurs principally on the bed-rock, the top gravel being poor. That which comes directly from the gravel is coarse; the gold of the underlying ledge is finer. The average fineness of the Mumford Hill gold is .945. Since the deposition of this gravel there has been a fault or a slip in the bed-rock amounting to at least six feet. The fault is traced up through the gravel by means of a clayey slickensides, the face of which is smooth and striated with parallel lines. It was not easy to secure specimens which would show these markings, and at the same time be firm enough to bear transportation. That this really marks the place of a crevice in the gravel is further shown by the presence of rootlets of trees to a depth of twenty-five feet below the surface of the ground, — a depth several times greater than that to which these rootlets make their way in the undisturbed gravel. The strike of the fault or slip corresponds very closely with that of the bed-rock.

From Mr. Edman I obtained a small quantity of concentrated sands from Rock Island, a gravel deposit the precise position of which I cannot give. These sands have been examined microscopically by Mr. Wadsworth, who says of them: "The white crystals here are zircons, the majority having the form figured in Dana's System of Mineralogy, page 273, except that the crystals are longer and the prismatic planes correspondingly developed. Magnetic iron is abundant. Some pale greenish fragments occur, which may possibly be broken augites. A grain or two of rutile was seen."

Between Mumford's Hill and the toll-gate on the stage-road from Quincy to Oroville the trail crosses Eagle Gulch and Deadwood Gulch. Between these two gulches lies Scad Point, "scad" being in miner's parlance the equivalent of nugget. The Scad Point gravel is about a third of a mile north of Mumford's Hill, but is nearly 200 feet lower, the altitude of its bed-rock being only 4,510 feet. It is evidently more recent in origin than the gravel of Mumford's Hill, for it contains, together with a little quartz and some clay, a large proportion of volcanic material and representatives of the country rock which is found in places in the near vicinity. The bed-rock has been designated as a "greenstone porphyry," and forms part of a belt of similar rock, which can be traced for several miles in a northwesterly direction. The grade of the channel is from east to west, or directly towards the high escarpment of Spanish Peak. The thickness of the gravel is, on the average, nearly fifty feet. The gold is said to be distributed through the whole mass. In the beds of Eagle Gulch and Deadwood Gulch gold used to be found as far up as their intersection with the two gravel deposits last described; higher up, the gulches were poor or barren. On the toll-gate ridge, as I was informed by Mr. Edman, there is a small gravel deposit, supposed to be on the continuation of the Scad Point channel, in which further evidence of faultings of the country rock, since the formation of the gravel, may be found. I did not visit the locality.

The country, for five or six miles along the eastern base of Spanish Peak, between the stage-road and Chaparral Hill, is said to contain several gravel deposits, of greater or less extent and of doubtful date and uncertain origin. It is a district of which, on account of the abundance of vegetation and the infrequency of good exposures of rock, but little can be learned in a hasty trip of a few hours. I went as far as Tucker's Ranch, at the head of Scales's Gulch, where the altitude is 4,100 feet. Near this point, according to Mr. Tucker's statement, there are two old channels, running in different directions. A third channel, at an altitude several hundred feet higher, is supposed to exist near the base of Spanish Peak. Further evidence of faulted bed-rock is reported from a tunnel in Scales's Gulch. These interesting points had to be left without further examination on account of lack of time.

The lower portions of this district, about Meadow Valley and Spanish Ranch, are covered, over hundreds of acres, with gravel to an unknown depth. The low grade of Spanish Creek prevents the application of the hydraulic process in the mining of these gravels, and the accumulation of water in the workings would probably make shaft-sinking and drift-mining unprofitable.

From the previous descriptions it can be easily seen that the gravel problem in the vicinity of Quincy is not the same as it is in the country south of the Feather River. Irregularity takes the place of a comparative uniformity; a much-broken, faulted, and disturbed bed-rock, the place of one in which even small slips are of rare occurrence. No easy solution offers itself for the problem, unless it can be admitted that Spanish Peak itself is an eruptive mass, which has been upheaved since the time when the channel of Gopher Hill and Badger Hill was making its accumulations of gravel. Further explorations will have to be made before the question can be considered settled.

C. BETWEEN QUINCY AND SIERRAVILLE.

My examination of the country lying to the east of Quincy was far from satisfactory. The setting in of the winter's snow at an unusually early day in October prevented my doing anything more than spend a few hours at Jamison City, and a part of a day at Mr. Andrew Jackson's diggings on the ridge to the south of Mohawk Valley, near Wash P. O. The reported gravel deposits at Argentine, Alturas, and other places lying to the north of the stage-road I had to pass by entirely.

Along the stage-road from Quincy to the Twenty-mile House boulders and large masses of volcanic rock, apparently hard and solid, are of frequent occurrence. Farther east, between the Twenty-mile House and Knott's Ranch, or Mohawk Valley P. O., the whole surface of the country is covered with an immense, uninterrupted deposit of comparatively loose material, resembling volcanic ash. The flanks of the ridges, on either hand, frequently show nothing but steep or precipitous gray and barren slopes, laid bare by extensive land-slides, some of them evidently of quite recent origin. There are no exposures of bed-rock. The waters of the Middle Feather run upon a mass of river gravel or lava, the depth of which is not known. The altitude of Knott's Ranch I made to be 4,325 feet, which is probably not far from correct, though, on account of the approaching storm, my barometric determinations of altitude in this vicinity are less to be trusted than usual. The highest point on the road from Knott's Ranch to Jamison City, near Mr. S. Babb's house, has an approximate elevation of 4,960 feet. From this point the road descends into the cañon of Jamison Creek, along the banks of which Jamison City is built. The altitude of the town at the Miners' Home is about 4,800 feet.

The whole of the country which came within my field of observation between Jamison City and Johntown, the latter place being a couple of miles or more farther up the creek, and at an altitude 350 feet higher than the former, is covered with an accumulation of so-called "cement" and boulders. Neither in the bed of the creek nor upon the sloping sides of the cañon was there any rock in place to be seen. I was told that I could find such rock in the creek a little way above Johntown, which is probably true ; but the most contradictory statements were made in reference to outcrops of bed-rock below Jamison City. Some persons said that there was no rock exposed to view for two miles down the creek, while others asserted that I could see it within a quarter of a mile of the town. Eureka Peak is said be a mass of slate-rock, even to its summit. As to the extent of territory covered by the loose material above referred to, I can give no definite information. It is a common belief among the miners of the vicinity that the cement and boulders fill the deep and broad channel of an old stream, which once flowed in or from a northwesterly direction along what is now the slope of the ridge, and that they cover a stratum of rich auriferous gravel. Some even go so far as to say that the genuine "blue lead" passes directly beneath Jamison City. Supported by their belief, an association of about twenty miners, organized as the Enterprise Company, has begun the sinking of a shaft in the bed of the creek a few rods below the bridge at Jamison City. If current report be true, between twenty and twenty-five thousand dollars have already been expended in this undertaking. It was impossible to get any very satisfactory information about operations at this shaft. The altitude of the mouth of the shaft is not far from 4,750 feet. After reaching a depth of about 160 feet, the projectors found it necessary to enlarge the cross-section of the shaft in order to make room for heavier pumps. The enlargement was not completed at the time I was there, and I have no means of knowing whether bed-rock or pay-gravel has ever been reached. The proprietors keep their own secrets.

Among the pebbles and boulders which cover the surface between Jamison City and Johntown,

I saw no quartz at all, and almost no volcanic rock of the basaltic or tufaceous types, which are so common on the ridges farther west and south. The prevailing type of rock is coarsely crystalline, and frequently porphyritic, with feldspar or hornblende. The feldspathic crystals, sometimes an inch or more in length, show very plainly as large white spots upon a weathered surface. When freshly broken, the rock ingredients are not so strikingly different in color. Besides these porphyritic rocks there are others, which resemble syenite in composition; hornblende is more abundant than any other mineral. Very few of the boulders are fine-grained. But few of them show many signs of having been worn by water action, and some are decidedly angular. There are also numerous large and angular erratics — ten to fifteen feet or more in diameter — scattered over the surface. Along the creek the exposed faces of the banks sometimes show signs of stratification. There are no beds of clay or sand, but the gravel in these banks is finer than elsewhere. In other places the exposed banks more nearly resemble a moraine in structure. The whole deposit is more probably due to glacial than to river action. The upper cañon of Jamison Creek, to the southeast of Eureka Peak, is very deep and steep, and glacial scorings have been reported from the rocks of that and other peaks. The gold that has been found above the "cement" in Jamison Creek probably came from the erosion of the Plumas Eureka quartz ledge.

The gold-bearing gravels at Jackson's Diggings do not throw much light upon the question of the distribution of such deposits in the Feather River basin. The claims are in the N. W. ¼ of Sec. 1, T. 21 N.; R. 12 E., some seven or eight miles southeasterly from Knott's Ranch, and about one mile from the Wash P. O., or Mrs. King's ranch. The altitude of the last-named place I made to be 4,450 feet; that of Jackson's cabin at the mines, 5,025 feet. The claims consist principally of shallow gulch workings upon Wash and other creeks. No mining had been done here prior to the year 1868. The lower portion of this gulch, which has been worked to bed-rock, is now filled with the heavier boulders which could not be carried along by the water. The highest point visited was about 300 feet above the cabin. The most interesting feature of this deposit is the abundance of porphyritic boulders. Mr. Jackson informed me that his ditch crossed a belt, fifty or a hundred feet in width, of the peculiar rock from which the large supply of spotted porphyry boulders might have come. The gold from the gulches between Wash and Ohio creeks is very coarse.

Between Jackson's and Sierraville, to which place I had to go in order to reach Sierra City, I passed over the road in the stage, and did not see much that would add anything to the value of this report.

SECTION IX. — *The Vicinity of Oroville and Cherokee Flat.*

The district to be described in this section is in the southern part of Butte County, northwest of the junction of the Middle and the North Forks of Feather River. The Official Map of the County of Butte, published in 1877, on the scale of one mile to the inch, will be convenient for reference in the study of this region; a portion of it, showing the region about to be described, is given on Plate Y. The most striking feature in the topography of the region is Table Mountain, on or near which the principal deposits of gravel, to which my attention was directed, are found. Leaving out of mind for the present a few localities of less importance, to which reference will be made at the close of the section, I will first take up the description of this mountain and its immediate vicinity.

Table Mountain, so called on account of the flatness of the broad sheet of basalt of which its higher portions consist, may be said to occupy an area of more than six miles in length from north to south, by nearly three miles in breadth, though the actual area covered by volcanic rock would be considerably less. At several points, particularly on the western side of the mountain, deep ravines, extending nearly across to the opposite side, have been eroded quite through the volcanic rock, and at the southern extremity a small portion of the cap, known as South Table Mountain, has been separated entirely from the main body by Schermer's and Morris ravines. At Cherokee

Flat, also, the connection between the rest of the mountain and the Sugar Loaf has been broken, and it is possible that a detached mass of lava remains between the forks of Schermer's Ravine. With these exceptions, the basaltic layer appears to be continuous and unbroken. At its highest point, the Sugar Loaf just mentioned, the mountain reaches an altitude of 1,500 feet above the level of the Feather River at Oroville. The top of the mountain has a gentle, uniform grade to the southwest of about one hundred feet to the mile. The thickness of the basalt is from one hundred to one hundred and fifty feet, or possibly a little more in some places. Had it been practicable for me to make precise measurements of the height of the bluff at a large number of points on all sides of the mountain, it is possible that a greater variation from uniformity of thickness than is indicated above might have been detected; still it cannot be far from correct to say that the under surface of the basalt, or the surface of the material upon which the basalt now rests and over which it originally flowed, has nearly the same grade — one hundred feet to the mile — as the present upper surface of the mountain. This naturally leads to the inquiry, What is the character of the underlying material? The question cannot be answered completely, for no openings have been made, no shafts have been sunk down through the basalt to lay bare what is there concealed; the only way of approaching a solution to the question is to examine with care the material at the base of the bluff, or within a short distance of the base, at as many points as possible. Such an examination, however, as is here contemplated is not easy to make. At the base of the bluff the large accumulations of broken rock and débris pretty effectually conceal all underlying rock from view. At the best, it will be only here and there that really satisfactory information can be obtained. The number of my own observations which bear directly upon this question is small, — so small, indeed, that it will not be worth while to pursue this branch of the inquiry any further, except as it comes up incidentally in connection with the descriptions of the different places visited.

At the north end of Table Mountain there is an extensive deposit of gravel near Cherokee Flat, which in many particulars is one of the most interesting deposits that I am acquainted with in the State. Owing to its comparatively elevated position with respect to the surrounding country, unusual outlays have had to be made, and unusual engineering devices resorted to, in order to secure a permanent and ample supply of water for hydraulic purposes. In the earlier days of mining in California this ground could only be worked on a very small scale, and during the rainy season. In the year 1870 a pipe, in the form of an inverted siphon, was laid to bring water across the cañon of the west branch of Feather River; and in 1873 other pipes were laid, for a similar purpose, across the deep cañons of Feather River and Little Butte Creek. Since the latter date hydraulic mining has been carried on with great energy and success. According to measurements made in the summer of 1879, under the direction of the State Engineer, there have been 22,000,000 cubic yards of gravel removed since the commencement of mining at this place. The hydraulic operations, up to the spring of 1880, were carried on principally by the Spring Valley Mining and Irrigating Company, which, by consolidations and purchases, had acquired a large extent of mining ground, variously estimated as comprising from 320 to 440 acres. The differences of these estimates are to be accounted for by the differences of opinion in regard to the proper classification of some of the ground owned by the company. Since the date of my visit this property, together with nearly 900 acres of ground adjoining it on the southwest, belonging to the Cherokee Flat Blue Gravel Company, has been purchased by a new organization, the Spring Valley Hydraulic Gold Company, for the reported sum of $2,000,000. As no mining of consequence has been done on the ground formerly owned by the Blue Gravel Company, my field observations were made principally at the mines of the Spring Valley Company; and, in what follows, reference will be made to the property of that company alone, unless the contrary is stated.

The portion of this ground which has been worked lies near the base of the Sugar Loaf, partly to the north of it, in Campbell's Ravine, and more extensively in Saw Mill Ravine, between the Sugar Loaf and the main body of Table Mountain. In the former, or upper mine, the direction of the long axis of the excavation is S. 35° W. (magnetic); in the lower mine, S. 30° E. (magnetic). A portion of the northern and eastern boundary of the deposit, as indicated by the pres-

Clear Creek

St Clair Flat

Fence

Saw Mill Ravine

Cherokee Ravine

Cherokee Flat

Saw Mill

Dry Creek

Valley Gulch

Sydenham Gulch

36 | 31
1 | 6

Blue Gravel Co.

SYDENHAM CO.

Gold Run

Coal Cañon Cave

Coal Cañon

TABLE MOUNTAIN

Oregon
City

BEATSON'S HOLLOW

Schermer's Ravine

Mt DE
ORO

W. C. Hendricks & Co.

Morris Ravine

Oregon Gulch

SOUTH
TABLE MOUNTAIN

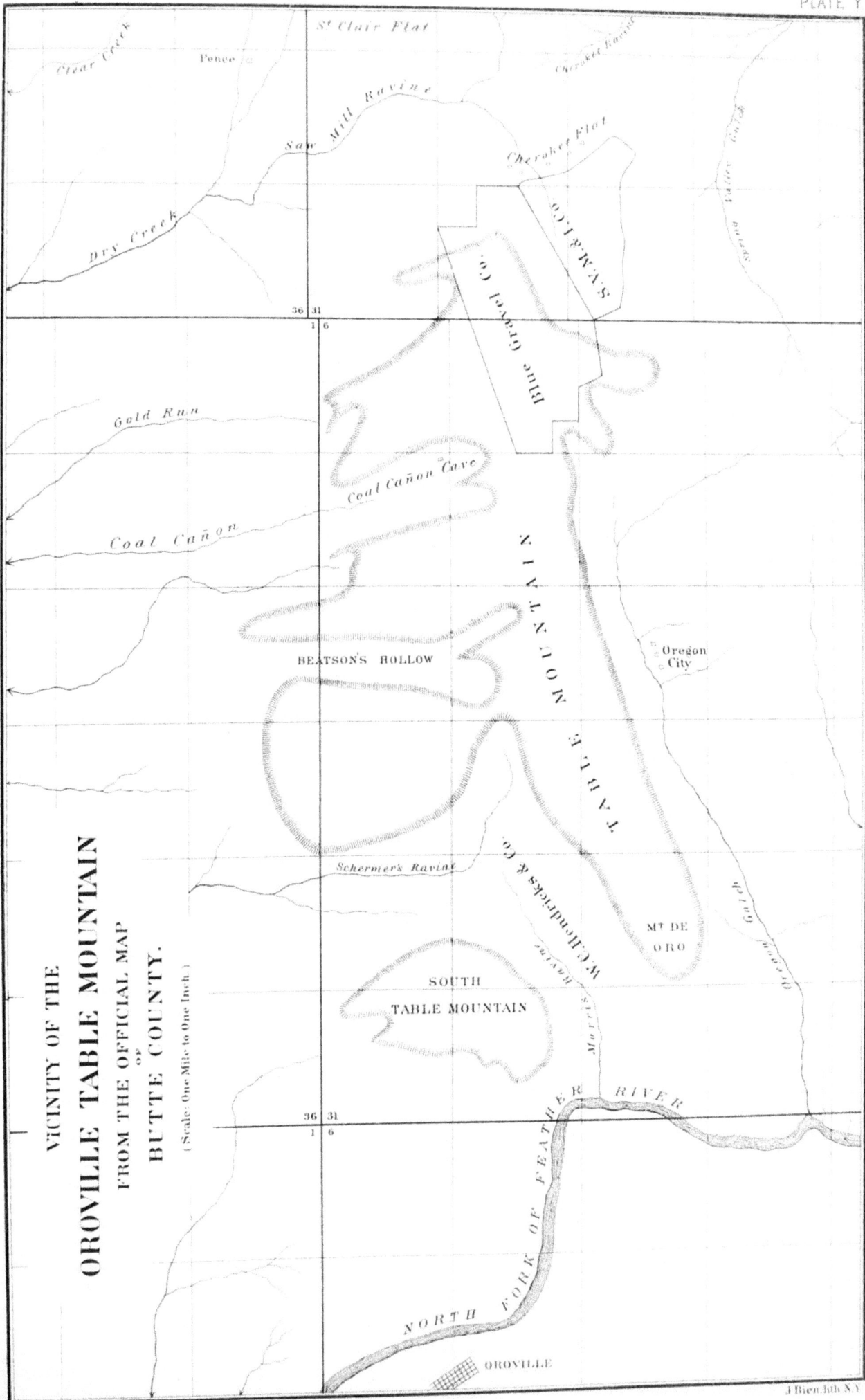

VICINITY OF THE
OROVILLE TABLE MOUNTAIN
FROM THE OFFICIAL MAP
OF
BUTTE COUNTY.
(Scale: One Mile to One Inch.)

36 | 31
1 | 6

NORTH FORK OF FEATHER RIVER

OROVILLE

J. Bien, lith. N.Y.

ence of rim-rock, may be described as follows: Beginning at a point in Saw Mill Ravine, near the Spring Valley House, the rim has a general northeasterly direction for about a mile; it then sweeps around and returns parallel to its former direction for nearly half that distance, enclosing a bay-like mass of gravel 1,500 feet in width, and finally follows a southerly direction, passing a quarter of a mile to the east of the Sugar Loaf. The grade of the bed-rock where exposed to the north of the Sugar Loaf, is very steep from northeast to southwest. At the road-crossing, near the high northeastern extremity of the workings, I made the altitude of the bed-rock to be 1,300 feet, while that of the bed-rock near the blacksmith-shop in the lower mines is only 1,087 feet, the distance between the two points of observation being less than a mile. To the south and west of the blacksmith-shop the position of the bed-rock is not known, but it is certainly fully as low as that whose altitude has just been given. The bed-rock northwest of Cherokee Flat is principally slate, but at the mine several distinct varieties of rock, not all of them slaty in structure, are represented. The best place to study the relations of these rocks would be in one of the tunnels, at a time when there was no work doing in the mine. The rocks met with in driving the Eureka tunnel were, as I was told by the superintendent, as follows: At the mouth of the tunnel in Saw Mill Ravine the rock was first a soft slate, which gradually grew harder for a distance of seven hundred feet; then came four or five hundred feet of an extremely tough rock, something like pudding-stone; after this for five or six hundred feet the rock was soft, easily picked, and greenish in color; the remainder of the two thousand feet of the tunnel was again in a very hard rock. Welch's tunnel, which is higher and shorter, is mostly in slate.

At the bottom of the gravel next the bed-rock there is a stratum, from ten to thirty feet in thickness, blue in color, resembling the "blue gravel" of the mines of Yuba and Nevada counties, more or less cemented together, especially near the bottom, and containing in its lower portions large boulders, which are not much rounded. Some of these boulders, indeed, are decidedly angular, and some of them are quite flat. The color of this stratum as a whole is due largely to the prevailing bluish color of the boulders; for between them, and frequently cemented to them, the gravel is composed of fine, white, rolled quartz. The upper third of the stratum is lighter in color and carries no large boulders. Above the blue gravel there is a remarkable stratum, varying from three to ten feet in thickness, of the so-called "rotten boulders." These have evidently been exposed to decomposing agencies, for, though still retaining their shape, they contain a great deal of yellowish-red iron oxide, and are very easily broken up. The line between the rotten boulders and the blue gravel frequently passes directly through a pebble or a boulder in such a way that the upper half has to be reckoned with one stratum and the lower half with the other. Above the rotten boulders there are generally, though not always, a few inches of a hard iron cement. Above this comes the main mass of fine gravel and sand, reaching to the surface or to pipe-clay, which in its turn is sometimes capped with basalt. The material of this stratum is almost exclusively quartz. The pebbles are seldom as large as a hen's egg, and they are all exquisitely rounded and smoothed, like the pebbles on a beach. The stratum varies in thickness in different parts of the mine. In some places it is fully 200 feet thick. Upon this point, however, it is not easy to get accurate data. I made the altitude of the top of the stratum in the high bank at the northern base of the Sugar Loaf to be 1,377 feet; its base was concealed from view. The color of this body of gravel is not constant. The banks nearest the Sugar Loaf have a peculiar pink or a delicate purple color, which is not so noticeable on the banks to the west and south. The cause of this color is to be looked for in the presence of an unusual amount of rose quartz in the gravel, or to a surface discoloration possibly originating from the decomposition of the overlying basalt. In structure, also, this stratum of gravel is quite different from the great majority of banks seen in other gravel-mining districts. It does not show, for instance, the regular stratification of the Malakoff gravel, nor any approximate parallelism of flat, lenticular masses. On the contrary, it is commonly and almost universally the case that limited portions of the gravel, homogeneous in themselves, are cut off abruptly by other portions, differing in color, in fineness, or in thickness, and in which the lines of lamination make large angles with those of the first.

The heaviest bed of pipe-clay rests upon the purple gravel. Where exposed to view on the face of the bank it is nearly one hundred feet thick, and is probably thicker than that a little way back from the face. The altitude of the highest point of the big bank I made to be 1,460 feet; that of the base of the Sugar Loaf, 1,550 feet; and that of its summit, 1,647. The gravel in Saw Mill Ravine, when mining was begun, had no pipe-clay upon it, and the beds now visible on the western bank do not appear to have any connection with those under the Sugar Loaf. The distribution of the pipe-clay on the northwestern side of the upper mine deserves a detailed notice on account of the bearing it has upon the question of the antiquity of the stone mortars and other implements which have been found in the gravel. In some cases this clay appears to take the place of the white gravel almost entirely; in others, it rests upon the rotten boulders; and, at one point, it has been seen to rest upon the blue gravel. It thins out rapidly towards the high bank of purple gravel on the opposite side of the mine. The lamination of this clay is not horizontal, but is highly inclined, in conformity with the general sloping surface of the gravel on which it rests. These circumstances, taken in connection with the great differences of altitude, show that these masses of clay have had no direct stratigraphical connection with those previously referred to. A cross section of the upper mine, from S. E. to N. W., through the Sugar Loaf, shows these differences of arrangement in a very striking manner. See section on Plate V, Fig. 7. Attention is called to the position of the clay (VII), between two layers of gravel, and their inclined position. I have little doubt that the northwestern portions of the deposit have been subjected to some rearrangement since the deposition of the main mass of the gravel.

The relations of these facts to the finding of the stone implements will appear from the following considerations.* In a conversation with Mr. Eaholtz in regard to the mortar found in 1858, I learned that he was then working where the bank was about forty feet in height, the lower ten or fifteen feet being gravel, the remainder pipe-clay and sand. In working the bank, the top stratum was allowed to fall as a "cave." Mr. Eaholtz was confident that the mortar came from the gravel, but he could not remember positively whether the gravel was blue or white. The place of working was about 300 yards above the junction of Saw Mill Ravine and Campbell's Ravine, that is to say, in the neighborhood where the strata of gravel and clay show signs of rearrangement. The result of my inquiries and my examination of the locality inclines me to the opinion that this particular discovery cannot be depended on as proving beyond question that the mortars were in use among men during the period of the deposition of the gravel, and prior to the time of the volcanic overflow. Of the other mortars which have been found here, but few are supposed to have come from any remarkable depth below the surface, and, as they were mostly found below the point between the two ravines, where an Indian "rancherie" used to stand, as I was told by Mr. Pulliam, they need not be of very great antiquity. Doubtless they are prehistoric; the Indians of the present day, at any rate, deny all knowledge of them, and refuse to use them. I made many inquiries in the hope of hearing of the discovery of the works of man in the gravels of the higher banks, but without success. The universal testimony is, that nothing of the kind has been found since the era of deep mining began.

The thickness of the basaltic cap at the Sugar Loaf has been already alluded to. There can be no doubt that the gravel beds extend underneath the basalt at this point. At no other points have the hydraulic banks been washed back so near to the base of the lava bluff; but there is every reason to believe that the beds are continuous to the west and southwest beyond the face of the bluff. The practical questions are, How far does the gravel extend? and How far does it carry gold in paying quantities? To these questions only probable answers can yet be given. It is evident that the difficulties of working by the hydraulic process will increase from year to year, and that finally drifting will have to be resorted to. I understand that it is the intention of the new company to commence drifting at once on the lower part of its property.

The gold at Cherokee Flat is very fine and scaly, excepting near the bed-rock, where grains worth from twenty-five cents to a dollar are not uncommon. It is remarkably fine in quality, usually ranging between .958 and .968. Gold from the upper gravel alone has been known to reach .980. An examination of the returns from the mint for five years shows that in only a very few cases has the fineness of this gold fallen below .950, and in a few cases it has exceeded .970. It is well known that diamonds have been found with the gold at this point; many of them are microscopic, but Mr. Abbey informed me that fifty-six stones had been saved, one of which had a value of two hundred and fifty dollars. A sample of the fine gold of this locality has been examined microscopically by Mr. Wadsworth. He reports upon it as follows: "The grains are all small and flaky, except one, which is of fair size, and shows the abrasion and polish which travel only gives. The remainder show but little wear, and were probably all of about the same size in the vein-stone. Associated with these are numerous glassy crystals, unworn and generally unbroken, which I regard as zircons. Forms identical with them occur in the Sierra granites, and, if my memory is not at fault, they are the same as the strange minerals which Professor Zirkel regarded as zircons (Micro-Petrography: 40th Parallel Survey). They are abundant in some of the gneisses, schists, etc., described by him (Neues Jahrbuch, 1880, 89–92). Magnetite is abundant; besides this, black worn octahedral crystals occur, non-magnetic, which I refer with doubt to spinel. Minute, rounded, gray, malleable, magnetic grains were frequently seen (platinum?). Other minerals of unknown nature exist, but not abundantly; they are generally in angular fragments, and probably belong to feldspar, etc. No transparent mineral was found that was isotropic."

The water-supply for these mines is drawn from Fall Creek, the North Fork of Feather River, Concow Creek, the West Branch of Feather River, and Butte Creek. The supply from these sources amounts to 2,210 miners' inches the year round, and cannot easily be increased without increasing the capacity of the iron pipes in which the water is taken across the deep cañons. During the rainy season a larger supply is obtained for a few months by storing in reservoirs the water collected from Table Mountain.

The yield of the Spring Valley mines for the first five years after the full supply of water was secured, that is, from 1873 to 1878, amounted to $2,073,628.69. The cost of operating the mines has been about $145,000 per year. In order to secure a proper dumping-ground for the mine tailings, the Spring Valley Company has purchased over 19,000 acres of ranch land in the Sacramento Valley. The lower end of the tailings canal, where it enters the sink of Butte Creek, is twenty-five miles from Cherokee Flat. The water then has to make its way for fifteen miles farther through the sloughs and the tule swamps, before reaching the Sacramento River. It enters the river free from sediment.

The Spring Valley gravel can be followed in a southeasterly direction as far as the flat at the head of Saw Mill Ravine. It then seems to disappear under the basalt, and no other deposits of gravel are known to exist from that point on along the northern and eastern side of Table Mountain, at least not until Monte de Oro, at the southern extremity of the mountain, is reached. Oregon Gulch, which lies to the east of Table Mountain, is said to have been extremely rich in gold, though it contains no deposits of gravel. It is said also that the ravines leading into this gulch from the side of Table Mountain have all yielded richly in gold, while those from the opposite direction have been nearly always poor or barren. If the enrichment of Oregon Gulch really came from Table Mountain, it is somewhat remarkable that the accompanying erosion has left no sign in the shape of basaltic boulders on the surface of the ground. To the east of Table Mountain such boulders are scarcely if ever seen, though they are extremely abundant in the valley to the west and southwest of the mountain.

The deposit at Gregory's claim, near the base of the basalt on the east side of Monte de Oro, is unique. The bed-rock, which is a very rotten slate, has an altitude of 1,033 feet. The so-called gravel is at first sight indistinguishable from a clay; but a little examination reveals fragments of rough quartz and of some ferruginous variety of rock disseminated through the mass. The gold, which is said to be present in paying quantities, is coarse and scraggy. The whole thickness of the deposit is not more than twenty feet. In its upper half a few streaks of gravel one or two inches thick are to be seen. Above the gravel there is a layer of undoubted clay, covered with blocks of loose and broken basalt. The mass resembles a slide from some higher level or a decomposing rock stratum. Possibly it may be connected with the outcrop of some quartz vein in the vicinity. Certainly it sheds no light upon any of the doubtful questions which have arisen in the study of Table Mountain.

On the west side of Monte de Oro there have been efforts made, both by surface works and by drifting, to find a gravel channel under the basalt, and gold to a considerable amount is said to have been taken from the locality. I could get no authoritative details in regard to the work here done. By some persons it is thought that gravel was found, though not in paying quantities; by others, that the basalt rests directly upon the bed-rock. The occurrence of bed-rock at a high altitude in Good's Ravine, the first ravine of prominence to the north of Monte de Oro, lends some support to the latter belief. I determined the altitude of a point on Hendrick's ditch, near the head of this ravine, to be 964 feet. Bed-rock was seen in the ravine at a little higher altitude than this, but still not within a hundred feet of the base of the basaltic bluff. Both the sloping surface of the ground and the bed of the ravine being covered to a large extent with loose blocks of basalt, it was impossible to tell exactly what the material immediately underlying the volcanic cap is. I saw no bed-rock, neither did I see any gravel. This high point of rock is important to notice, for it is said to be the only bed-rock visible at the western base of the main mass of Table Mountain between Monte de Oro and Cherokee Flat. The highest bed-rock seen on the opposite side of Morris Ravine, on the eastern slope of the South Table Mountain, is considerably lower in altitude, not much, if any, over 500 feet. The bed-rock evidently pitches rapidly to the west, and disappears under the detrital material of the Sacramento Valley, not to reappear again east of the Coast Range. See section (Plate X, Fig. 2). The bed-rock seen in Morris Ravine is not easy to name; partly because, when exposed to view, it is apt to be in an advanced state of clayey decomposition, and partly because of certain peculiarities of structure and bedding. The decomposed rock is sometimes yellowish-brown in color, and sometimes a bluish-green, resembling in appearance glauconite or green earth. In general terms, however, the most of the rock must be regarded as belonging to the slate family. A few rods below Mr. Hendrick's house there is some evidence of the existence of a fault in the bed-rock; but, on account of the accumulations of clay and other loose material along the supposed line of strike, it was not possible to tell with certainty.

Morris Ravine is one of those places from which extraordinarily large quantities of gold are said to have been taken in the early days of placer mining, and it still contains an enormous quantity of undoubtedly auriferous material. The deposit is not exactly like that at any of the other gold-bearing localities in the vicinity; at least, it is not in the same condition. At the "Point," as it is called, about a mile south of Mr. Hendrick's house, there are banks, sixty to eighty feet in height, composed of sand and clay of various colors, with horizontally interstratified fine quartz gravel, the pebbles being of the size of beans or peach-stones. In fineness and purity these pebbles remind one of the pebbles at Cherokee Flat, but there is a total lack of the "sand-drift structure," common at the latter place. Possibly there has been a modern rearrangement of the gravel in Morris Ravine since its original deposition. The bed-rock shows none of the ordinary signs of wear common in the beds of old channels. Near the head of the ravine there is a moving and sliding mass, covering from 120 to 150 acres (estimated) of mixed clay, sand, and gravel. It has a slow motion down the ravine in a southeasterly direction. The effects of the slide are seen in several ways. Wave-like ridges are heaped up in front of the moving mass, and upon its surface there are huge cracks, like glacial crevasses, signs of which can be seen even as far back as the base of the basaltic bluff. Mixed with the finer material of the slide there are many boulders, measuring from three to four feet in diameter, and representing many varieties of metamorphic rock, which I could not trace to their origin. Some of these boulders show a peculiar concentric spherical structure. Large sums of money have been spent in the effort to bring water into the ravine in sufficient quantity to work the moving material with profit. When small quantities of water are used, or when the pressure

PLATE 8

Fig 3.

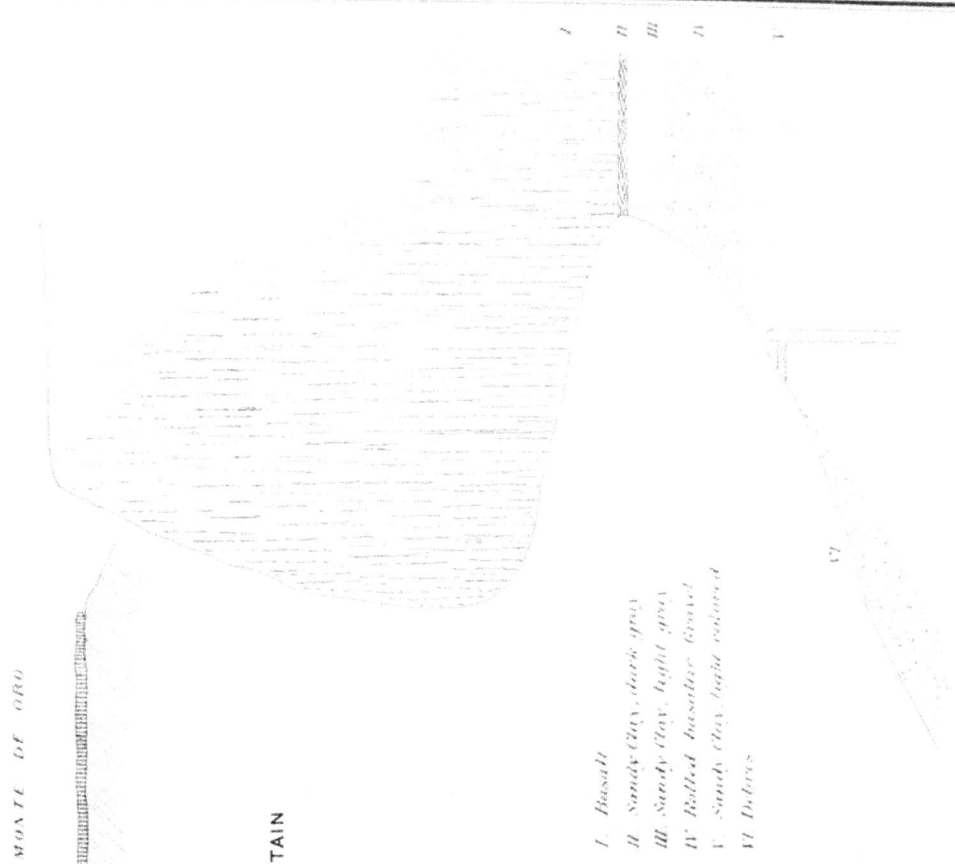

SECTION OF CAVE IN COAL CAÑON.
OROVILLE TABLE MOUNTAIN
(Scale Fifty Feet to one Inch)

I. Basalt
II. Sandy Clay, dark grey
III. Sandy Clay, light grey
IV. Rolled basaltic Gravel
V. Sandy Clay light coloured
VI. Debris

Fig 2.

SOUTH TABLE MOUNTAIN

MORRIS RAVINE

MONTE DE ORO

SECTION ACROSS MONTE DE ORO, MORRIS RAVINE AND SOUTH TABLE MOUNTAIN

To illustrate supposed relations of Bed rock, Gravel and Basalt.

(Scale Two Thousand Feet to one Inch)

Fig 1.

CAMELS PEAK

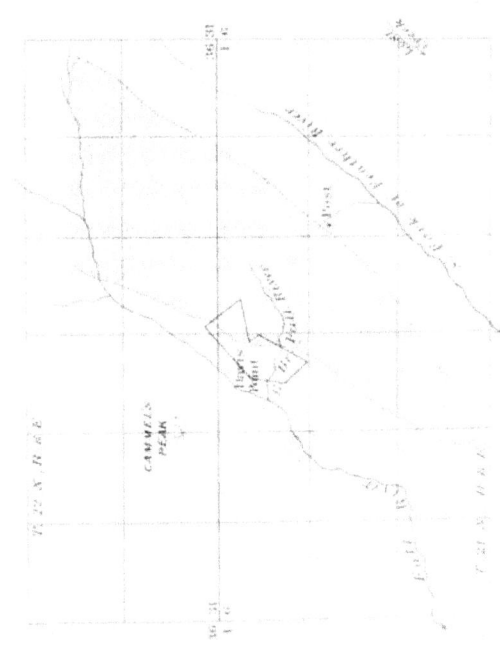

DIAGRAM TO ILLUSTRATE DESCRIPTION OF
DAVIS POINT AND VICINITY
(Scale Two Miles to one Inch)

is small, the slide advances about as rapidly as the face of the bank is removed, and no progress is made towards getting at the pay gravel, which is supposed to lie next the bed-rock. The water used is brought from the west branch of the Feather River. The ditches, flumes, pipes, and natural water-courses, through which the water has to pass before reaching the mine, amount to fifty miles in length. Two grades of gold are found in this ravine, — one fine, scaly, and floury; the other, coarse and nuggetty. I have no data in regard to yield or expenses.

The ridge dividing Morris Ravine from the upper portion of Schermer's Ravine is so low that at first sight the two ravines appear to belong together. In Schermer's Ravine no bed-rock is to be seen. The ravine is supposed to be filled with the same kind of material as that which gives Morris Ravine its value, though it has never been so rich, and no profitable mining has been carried on there.

On the crest of the saddle between Table Mountain and South Table Mountain, at the head of one of the branches of Schermer's Ravine, I made the altitude of the base of the basalt to be 900 feet. At this point there was a stratum of a peculiar sand-rock underlying the basalt and resting upon a mass of gravel. It seems to be an imperfectly compacted mass, made up of grains of volcanic rock. On the western slope of this saddle, and about 250 feet below the crest, tunnels have been driven into the gravel just mentioned. One of the tunnels has reached a length of 250 feet. The gravel is coarse, pebbles from six to eight inches in diameter being common. There is almost no quartz to be seen in the tunnel; the pebbles are of easily decomposable varieties of metamorphic rock. Gold was not found in paying quantity. In this gravel supposed remains of organic life have been found, but the specimens that I saw were not well enough preserved to be identified.

In Coal Cañon, on the western side of Table Mountain and between two and three miles south of Cherokee Flat, the basalt and the underlying strata have been eroded in such a way as to leave a good exposure, a few feet in thickness, of the upper portion of the material on which the basaltic cap rests. The altitude of the base of the lava at this point I made to be 1,049 feet. The thickness of the lava is about 150 feet. The width of the "cave," as it is called, I estimated at 300 feet, and its depth, under the overhanging cliff, at 100. The bluffs of basalt are nearly vertical, and show very well indeed the prismatic columnar structure so common in that rock. In the cañon there are huge piles of basaltic débris, much of which looks as fresh as if it had only recently fallen. See section (Plate X, Fig. 3). The strata, marked with Roman numerals on the section, are as follows: —

I. Basalt.

II. Dark gray, sandy clay, two feet in thickness.

III. Light gray, sandy clay, eight to ten feet in thickness, the lower half having a yellowish tint.

IV. Rolled volcanic (basaltic) gravel, with an occasional quartz pebble, fifteen to twenty feet in thickness. Quartz pebbles are very scarce; the basaltic pebbles are nearly uniform in size, ranging from three to six inches in diameter.

V. Light-colored, sandy clay, very compact, and very much resembling a sandstone. The thickness of this stratum down to the pile of débris is twelve feet. In it a shaft has been sunk to an estimated depth of thirty or forty feet. The material in the dump seems to be all of the same kind. There is said to be another shaft, lower down the cañon, which has been sunk to quartz gravel, but my information is not positive on this point.

At the base of the lava in Morris Ravine, where I made the altitude to be 1,084 feet, basaltic pebbles, similar in character to those from stratum IV above, were found. No specimens were brought away.

From the preceding descriptions it will be seen that the eastern edge of the Table Mountain basalt probably rests for the whole length of the mountain, with the exception of a short distance at Cherokee Flat, upon the bed-rock, or with only a thin layer of gravel intervening; that the bed-rock has a more rapid westerly pitch than the surface of the basalt; and that under the cap there is probably a wedge-shaped mass of detrital material, of unknown thickness on the west and gradually growing thinner towards the east. At Cherokee Flat the gravel is certainly very rich in gold, and the former richness of the Morris Ravine placers, the material of which probably came from under the basalt, indicates that the whole mass of detritus is auriferous. But the fact that, while the gravel of Cherokee Flat, with the exception of the large boulders, is almost exclusively quartz, there is scarcely any quartz to be seen in the gravel of the western slope of the South Table Mountain, is sufficient to raise a doubt as to the uniformity of the distribution of the gold. This doubt is still further increased by the differences in the surface-dirt of Schermer's and Morris ravines, which stand in almost identical relations to the cap of basalt. By some persons it is supposed that there is a true gold-bearing, old river channel, with a north and south course, connecting Cherokee Flat with Morris Ravine. I do not feel satisfied that this is the case. In the first place, the gravel of Cherokee Flat lacks the peculiar arrangement and structure of the river gravels; and, in the second place, there are no similar gravel deposits, so far as I could learn, within a radius of several miles at least, with which the Cherokee Flat gravels could have ever had any connection. From the top of Table Mountain an excellent and extensive view of the high mountains to the north-east can be had; but nowhere in that direction for many miles are there any gravel deposits at all. Walker's Plains, or "Walker's Lava-beds," distant about eighteen miles in an air-line, are said to be covered with basalt like that of Table Mountain, and it is very probable that they once formed parts of the same lava-flow. From Walker's Plains on a clear day the water in the Table Mountain reservoirs can be seen. At higher altitudes in the mountains, at the "gravel range," and at Lott's Diggings, near the line between Butte and Plumas counties, there are reported to be gravel deposits in some respects resembling that of Cherokee Flat, though not identical with it. The higher portions of Butte County are much cut up by deep cañons, and the region is one difficult of access; an entire season might well be taken for the study of its detrital formations.

It was a part of my plan to extend my trip as far north, at least, as Dogtown (Magalia P. O.), fifteen miles from Cherokee Flat, but the commencement of the rainy season prevented my putting the plan into execution. The gravels at St. Clair Flat, between Cherokee Flat and Pence's Ranch, were the only ones in that direction to which I paid any attention. The bed-rock along the stage-road from Cherokee to St. Clair Flat is an easily cleavable slate. To the west and south of the road the bed-rock soon disappears under the surface and is not seen again, either on the flank of Table Mountain or in the valley. The gravel deposits of the Flat, and along Dry Creek, where it emerges from the mountains, are of considerable extent, superficially, but they belong to a different class of deposits from that at Cherokee Flat. They do not seem to be confined to particular channels, but to form part of the general detrital material of the valley. The altitude of the bed-rock at one opening I made to be 765 feet, at another 735. The first large excavation to the north of the road, that I saw on my way down from Cherokee, was in a bank of gravel resting against the base of a bed-rock hill. The opening was nearly a quarter of a mile in length, but of no great width. The gravel was lightly cemented, and resembled a breccia rather than a conglomerate. In a few places it was blue in color, though generally reddish and brownish. It was not quartzose. The boulders, some of them as much as six or eight feet in diameter, were heterogeneous in character. Specimen 162 represents a common variety. Welch's Bank, on Dry Creek, lies at the foot of or under one of the beds of black volcanic conglomerates of which so many are seen, like beach-marks on the hills of this vicinity. The face of the bank showed at bottom fifteen feet of gravel, containing but little if any quartz; then came about thirty feet of a dark drab or slate-colored clay, with here and there thin streaks of gravel; above this there was a stratum, about thirty-five feet in thickness, of a very dark, nearly black, finely cleavable or horizontally laminated, clayey, carbonaceous material, which might be described as a sandy and sulphurous impure lignite. It effloresces on exposure to the air. The bank does not carry gold enough to pay for moving the whole mass by the hydraulic process. Drifting operations have been begun. On the bed-rock the gravel is reported to contain about fifty cents worth of gold to the car-load.

The low gravels near Oroville I did not examine with sufficient care to justify any extended report upon them. Along the banks of Feather River, for a distance of perhaps 400 feet below the bridge, bed-rock is now exposed to view. Below that point a few spurs of high bed-rock used to be seen in the bed of the river before the accumulations of the tailings. Bed-rock is also seen in the eastern part of the city. With these exceptions, it may be said that the whole river-bottom is filled with gravel. On the right bank of the river the gravel of Thompson's Flat, extending back to the base of South Table Mountain, has been worked near the river at a great many places. The faces of the banks are visible from the opposite side of the river for a long distance. At the claim of Mr. O. P. Powers the banks are from fifty to sixty feet high. The lack of a good outlet makes it difficult to work these banks to advantage. The gravel contains but little quartz; its mass is made up of crystalline rocks, similar to those seen in the bed of the present streams. The gold is said to be very fine, and to be pretty evenly distributed through the gravel from top to bottom. No bed-rock is to be seen in the mine at Powers's claim. The gravel also carries an unusual quantity of black sand. It seems probable that this came from the wear of the neighboring basalt. On the left bank of the river, in the city of Oroville, the gravel mine of Mr. J. B. Hewitt is the most extensive. Owing to the lack of grade, the banks farthest from the river can now be worked to a depth of only twenty or twenty-five feet. The gold is fine and scaly, and uniformly distributed. It has a fineness of .947 on the average. The gravels of the vicinity of Oroville have recently attracted considerable attention, for the reason that Mr. Edison has found that they contain a larger percentage of platinum than is usual in these deposits. At the present time Major McLaughlin, an associate of Mr. Edison, is engaged in erecting works at Oroville, with the expectation of saving the platinum, and at the same time recovering a large part of the gold which is carried off with the black sands.

B.

REVIEW OF FIELD-NOTES OF 1871, AND DISCUSSION OF GENERAL TOPICS CONNECTED WITH THE GRAVEL QUESTION.

By W. A. GOODYEAR.

SECTION I. — *Review of Field-Notes.*

In the following pages I propose to write first a sort of running review of the mass of my field-notes for 1871, herewith presented, making such additional remarks as may be applicable to special localities, with references, so far as needful, to the pages where these localities are more particularly described; and afterwards to discuss to some extent a variety of more general topics connected with the gravel question.

It is necessary to premise in the first place that, in the section of country to which these notes relate, although a considerable number of instances may be found of distinct and unmistakable *channels* worn in the solid bed-rock, and marking for certain distances the exact location and course of the ancient streams which accumulated the gravel that now fills their beds, and though in a few cases these channels have been actually traced and *proved* continuous for distances of from one to two miles; yet, when compared with the vast aggregate of the surface over which the ancient auriferous gravel was spread, the *channels* which can be thus definitely and certainly traced, even for the shortest distances, are few and far between. It will also appear more fully, I think, in the sequel, that this state of things is not only a natural, but a necessary result of the mode in which the gravel was accumulated, and the situation in which it has been left since the excavation of the modern cañons.

Furthermore, wherever mention is made hereafter of the *course* of any of these ancient channels, or of the direction of flow of the water in them, unless otherwise distinctly specified, it must be understood as referring to the direction of its flow at that time only, while the water was running upon the naked surface of the rock, or when the lowest stratum of gravel was beginning to gather in the channels; for there is plenty of evidence that the period during which the gravel accumulated was long, and that during this period the streams were constantly shifting their beds. Thus it is often the case that the water has flowed successively in very different directions over the same ground, and the direction of its flow at first upon the surface of the rock is frequently no indication whatever of the subsequent courses which it may have followed over the same spot while the upper portions of the banks were accumulating.

The estimate on page 119 would give in round numbers about 26,000,000 cubic yards of auriferous gravel in the Iowa Hill ridge between the first Sugar Loaf and Independence Hill. Of this quantity I should estimate that about 20,000,000 yards were concentrated in the deep channel which crosses the crest at Iowa Hill itself, and the balance scattered along the sides of the ridge at various points, but chiefly along the southeastern side near Indian Cañon, and in the shallow patch which caps the ridge near the Catholic church a little northeast of Iowa Hill. The total quantity already washed away is probably between two and three million yards, of which I estimate nine tenths to have come from the deep channel at Iowa Hill, and the balance from smaller pits at other points. This would leave between twenty-three and twenty-four million cubic yards yet capable of being hydraulicked. How much of this it will ever *pay* to wash, is a difficult question to answer. But probably the greater portion of it will do so whenever the supply of water shall be plentiful, reliable, and cheap.

Such estimates as these must be taken only for what they are worth. In this case I estimated the area by pacing, and had the barometric determinations of the maximum heights of the banks to guide me, together with such information respecting the lay of the bed-rock under the gravel as my other explorations, and the statements of men who have worked here in shafts and tunnels, gave.

The deep channel at Iowa Hill is perfectly well defined, crossing the crest of the ridge in a direction a little to the east of south. It is very rare in the country through which I have travelled to find an ancient channel cut so deeply into the bed-rock as this, — the rims on either side being fully 200 feet above the bed of the channel. The width of the latter across the top from rim to rim, at Iowa Hill, appears to be something over 2,000 feet, and may be half a mile, — the exact locality of the highest part of the southwestern rim being somewhat uncertain.

The barometric observations will show that the bottom of this channel at Iowa Hill is considerably lower than the bed-rock at Indiana Hill and Gold Run on the opposite side of the river.

I think there can be little doubt that Dutch Flat, Gold Run, Indiana Hill, Iowa Hill, Wisconsin Hill, and some points in the Forest Hill divide mark the ancient course of a river which corresponded in the gravel period to the present North Fork of the American.

Though the ancient stream appears to have crossed the present one between Indiana Hill and Iowa Hill, yet there is a striking approach to parallelism between the general course of the modern stream and that of the ancient one as thus indicated from Dutch Flat to the Forest Hill divide; and if, as is more than probable, the ancient stream, on reaching the latter point, curved rapidly again to the west of south, this parallelism becomes more striking still.

This is the only instance I have seen in which the course of so large a stream at the commencement of the gravel period can be traced with so much approximation to certainty for so many miles. It is probable also that throughout the greater portion, if not the whole, of the gravel period, this stream, between Dutch Flat and the point where its course is intersected by the present Shirt Tail Cañon, was subject to far less change and shifting of its bed than was generally the case with the streams elsewhere. The depth to which this channel was excavated in the solid rock, and the correspondingly great and uncommon depth of the auriferous *metamorphic* gravel which has been accumulated between such narrow limits along its course, all speak strongly in favor of comparative permanence in the course of this stream through a very long period of time.

But, prominent as the general outlines of this stream appear to be, and clearly as it seems to be defined at some localities, yet, even in this case, when we attempt to trace its course in detail from one locality to another, we encounter many difficulties. From Iowa Hill to Wisconsin Hill, for example, though these two places are so near together, the exact course which was followed by the stream is somewhat doubtful. According to the barometric observations, it would appear that while the lowest bed-rock in the claims at Wisconsin Hill is a little lower than it is at Iowa Hill, yet at Elizabeth Hill it is some forty or fifty feet higher than it is at Iowa Hill; while all the way along the crest of the ridge from Elizabeth Hill towards the east as far as the Wisconsin Hill school-house the bed-rock is higher still. It appears, therefore, that while the stream was running in its lowest bed, it could not have passed through Elizabeth Hill, nor anywhere through the present ridge between there and the Wisconsin Hill school-house; nor is it probable that it passed to the west of Elizabeth Hill. Yet the depth of the metamorphic gravel on the crest of the sag at Elizabeth Hill is very considerable, and it is perfectly similar in its general appearance and character to that at Iowa and Wisconsin hills. To the east of the Wisconsin Hill school-house the crest of the main ridge rises rapidly, and becomes at once covered with a heavy mass of volcanic débris; and the extent of the tunnelling has not been sufficient to either prove or disprove the existence of a deep channel through here. But on the northwest side of the ridge the Morning Star tunnel, driven a considerable distance into the hill, shows that as far as it has gone the surface of the bed-rock here is at very nearly the same level as in the deep channel at Iowa Hill. Moreover, the depth of the shaft once sunk in New York Cañon, and the barometric observation at its mouth, show that there also the bed-rock is very low. I think the probability strong, therefore, that the old deep channel, on leaving the Iowa Hill ridge, curved slightly to the east, entering the ridge on the southeast of Indian Cañon at or in the vicinity of the Morning Star mine, and passing to the east of the school-house, reached the head of New York Cañon, and so around to Wisconsin Hill. If this be correct it could only have been at a later period, when a considerable depth of gravel had already accumulated in the deeper channel, that the water could have flowed over the crest of the sag at Elizabeth Hill, and deposited the gravel there. When this happened, the bed-rock crest of the present ridge between Elizabeth Hill and the Wisconsin Hill school-house would still have remained an island in the broad gravel surface, till at last the accumulation became so great as to cover even its crest. This is the most plausible theory that I can now frame for this particular locality. One of the strongest objections which can be urged against this view is the fact that neither at the Morning Star mine, nor anywhere else along the *northwestern* side of the ridge between Indian and Shirt Tail cañons, is there any visible proof of the present existence of so great a *thickness* of *metamorphic* gravel as exists at Wisconsin Hill and around the vicinity of New York Cañon, as well as at Iowa Hill. The volcanic capping immediately southeast of the Morning Star mine is very heavy, and the thickness of the gravel beneath it uncertain. But it seems not at all unlikely that after the cessation of the flow of the main stream here, and before the coming of the volcanic epoch, there may have been considerable excavation and irregular re-distribution of the previously accumulated gravel by subsequent and smaller streams.

The fact is not without interest in this connection that the top of the metamorphic gravel near the Wisconsin Hill school-house, as well as at another point between there and the Morning Star mine, is, in round numbers, 350 feet higher than the lowest bed-rock in the hydraulic claims at Wisconsin Hill and Iowa Hill, — or 150 feet above the tops of the highest banks at Iowa Hill.

The quantity of gravel distributed along the sides of Indian Cañon for several miles above Iowa Hill, the mode of its distribution, and the developments made by the tunnels at Strawberry Flat and still farther northeast, point to a strong probability that Indian Cañon itself occupies pretty nearly the site of a branch of the ancient stream which joined the latter at a point not far from where the present cañon cuts it, or possibly a little farther southeast towards Wisconsin Hill.

There seems also to have been at one time a small but well-defined channel running southerly in the spur on the east of New York Cañon; — but whether this channel had any connection with Grizzly Flat is very doubtful.

Any further attempt to trace at present any distinct and well-marked *channels* in the bed-rock underlying the gravel in the ridges between Shirt Tail Cañon and the North Fork of the American River would probably be useless. Yet the area over which the gravel is spread is very large. The ridges above Iowa Hill and Wisconsin Hill, as well as between Elizabeth Hill and King's Hill, have generally a heavy capping of volcanic matter, beneath which, however, there is proved by the numerous tunnels to exist almost everywhere a thin layer of metamorphic auriferous gravel on the rock. As a general thing this layer is but a few feet in thickness, frequently not over a foot or eighteen inches, and sometimes it vanishes entirely, the volcanic matter closing down tightly upon the solid rock. But everywhere, without exception, wherever the surface of the rock has been exposed beneath the gravel or the volcanic matter, either by hydraulic washing or by drift-ing, and the rock has been hard enough to retain and distinctly exhibit the character of its original surface, it has been found smoothed and worn into rounded forms by running water.

This last remark is also true, almost without exception, wherever the ancient gravel has been mined in any way, through all the ridges in the country over which my work extended, between the North Fork of the American and the Mokelumne River.

There is a *possibility* that a channel may run through from Grizzly Flat to the Morning Star mine.

The main channel at Damascus, that is, the "Mountain Gate" channel, appears to be well defined, running into the ridge in a direction nearly due southeast (magnetic). How deep this channel may be cut in the bed-rock, and how wide it may be across its top from rim to rim, is very uncertain. But from such indications as can be seen along the sides of the hills, it seems likely that its breadth is considerable, and that the maximum height of its rims above its bed may reach, in places at least, from 200 to 300 feet. The barometric observations show that the bed of this channel passes through the present hill at a depth of not less than about 800 feet beneath its crest; of which depth the whole mass, with the exception of the comparatively thin layer of gravel which fills the bottom of the channel, appears to consist of volcanic conglomerates and breccias.

The accompanying notes will show how far this channel has already been followed in the tunnels. But how much farther it may continue to hold its southeasterly course, and where it ultimately goes to, are points which remain to be proved, and upon which there are many different opinions among the miners.

The main ridge beneath which it passes in the vicinity of the Forks House is two or three miles wide, and there are several possibilities in the case. But I shall add a few remarks on this point further on.

Where this channel came from towards the northwest has also been a much-vexed question. The character of its gravel is peculiar, it being by far the purest *quartz* gravel which I have seen in the country, and containing the most numerous and the largest *white quartz* boulders, among which masses of forty to fifty tons in weight are not uncommon. The grade of this channel is not uncommonly heavy, and I do not think such enormous boulders can have travelled very far along its bed.

I can only explain their presence here, in connection with a gravel of which the sand as well as the smaller pebbles consist almost exclusively of quartz, by supposing that somewhere along the course of the ancient stream, at no great distance from their present resting-place, there once existed an enormous mass of quartz in the region which has since been excavated to form the tremendous cañons of the branches of the North Fork of the American; and that, while this quartz was broken up to form the present gravel, the slates and other rocks which were associated with it being softer, and liable also to disintegration by chemical means, were entirely ground to sand and silt and completely washed away.

I am told no indications of the existence of such a channel as this have ever been found, either on the opposite side of the North Fork of the river or higher in the mountains among its branches to the northeast. But since, in the near vicinity of Damascus, the North Fork of the American River splits up into three or four large branches, and the cañons here are very deep and the ridges narrow for some distance above the forks, there is a considerable area here in which the ridges and spurs, though very precipitous, do not rise high enough to touch the plane of the ante-volcanic gravel, and any vestiges of ancient channels which may have once existed within this area have since been swept away. Under such circumstances, I think the most probable solution of the question, Whence came the Damascus channel? is the supposition that, if we could have followed up its ancient course for a short distance farther to the northwest over the region now occupied by the basin of the Humbug Cañon and the branches of the river, we should have found it then to curve gradually around to the north and northeast and probably to split up into various smaller channels, and thus to completely lose both its size and its distinctive character long before it would reach any of the present ridges in the higher northeastern country at points where any remnants of these branches might now be traced.

It will be noticed as an exceptional fact that at Canada Hill there seems to have existed a well-defined channel with a heavy fall running in an *easterly* direction. This channel evidently could not have held this course, however, for any great distance towards the summit of the range; but probably bent rapidly around to the north and west over the basin of Sailor Cañon to a larger stream running somewhere over the present cañon of the North Fork of the American.

It is very possible, though by no means certain, that the Basin channel at the Devil's Basin above Deadwood, and near the trail from there to Last Chance, may be a continuation of the same channel as that upon which is located the Slab claim, on the opposite side of the river below Last Chance. This channel appears to enter the Basin ridge at the point where the trail, descending from its crest towards the river, meets the bed-rock, and to run at first for several hundred feet southwesterly, and then more nearly south, for a distance of some 1,700 or 1,800 feet altogether to where it meets the Basin tunnel. At this point it seems to make a pretty sharp bend to the west, and continue thence in a general course about S. 70° W. (magnetic) some 2,500 feet, or as far as worked in the Basin tunnel. Beyond this point it is uncertain whether it continues the same course, passing under the head of Black Cañon, and so on through the ridge beyond towards Deadwood, or whether it bends sharply again to the south, and continues under the Basin ridge in a direction nearly south (magnetic) for half a mile or more to the upper end of the Reed mine. Mr. Reed thinks the latter will prove to be the fact. But I consider it in the absence of further proof nearly an even chance one way or the other. But whether the Reed mine be upon the same channel as the Basin claim or not, there can be little doubt that the Rattlesnake claim is upon a continuation of the Reed mine channel, and is either the main channel itself or a branch of the same. If the Reed mine connects with the Basin channel, then the Rattlesnake will be only a branch; but if not, it is probably the main channel. From the point at which the Rattlesnake and the Basin channels will unite if they prove to be branches of the same, the channel in the Reed mine runs first some 2,000 feet, about N. 85° W. (magnetic), then some 1,400 or 1,500 feet about S. 35° W. (magnetic), to the lower end of the workings in Reed's mine. Within the four or five hundred feet yet left unworked between here and the upper end of Hornby's mine it seems to bend rapidly again to the west, and to hold a course of about S. 80° W. (magnetic) till nearly under the town of Deadwood, where it again curves to the south, its course in the lower part of Hornby's mine being about S. 35° W. The length of Hornby's workings on this channel is some 1,500 feet. Immediately below the mouth of Hornby's mine this channel appears to have crossed the heads of some little gulches, which now run southerly to the North Fork of the Middle Fork near Bogus Thunder,—and a portion of it seems to have been carried away. But it is believed that the same channel re-enters the main spur a little farther southwest and passes through it, coming out on the northwest side towards El Dorado Cañon at the Cobb claim, a little to the southwest of the trail from Deadwood to Michigan Bluff.

There are evidently fragments of two or three other channels at and about Deadwood, which were probably well defined at one time in the surface of the bed-rock. But this is the only one which can now be traced with any certainty for any considerable distance. And yet, wherever tunnels have been driven, the surface of the bed-rock beneath the volcanic capping has always been found worn smooth by water, and almost always there is more or less metamorphic gravel upon it.

It is believed by some that Weske's mine in the Michigan Bluff divide is upon the continuation of the Reed mine channel. But, though this is certainly among the possibilities, yet I think the probabilities are against it.

It is difficult to say whence came the great mass of auriferous gravel at Michigan Bluff. The heavy banks of deep gravel, however, which still exist in the spurs flanking the main ridge on the west side of El Dorado Cañon for several miles above Michigan Bluff, would seem to indicate the probable existence here in the gravel period of a considerable stream whose general course — for these few miles at least — was nearly parallel with the present cañon. But this series of heavy banks appears to terminate at Gas Hill; the claims above that point being only tunnel-claims.

When the termination here of this series of heavy banks is considered in connection with the additional significant fact that the west branch of El Dorado Cañon is stated to have been rich in early days up to a point nearly opposite to the Ayers claim, but not above, it becomes eminently probable, I think, that the stream which accumulated these heavy banks, from which the greater portion of the gold originally found in El Dorado Cañon seems to have been derived, issued somewhere in this immediate vicinity from the ridge on the west, and here first entered the area now occupied by the modern cañon. There is little chance, I think, of its having come from the east or northeast across the present cañon instead of from the western side; because no correspondingly heavy banks of gravel have ever been found either in the spurs on the opposite side of this branch of the cañon, or anywhere along the eastern branch above the near vicinity of Deadwood.

The southeasterly direction in which the Mountain Gate channel at Damascus is running, and the northwesterly direction from which the channel at the Dam claim appears to be coming, so far as it has yet been followed, together with the short rectilinear distance between the two localities, might lead one who should judge from the map alone to infer that the Dam claim was on the continuation of the Mountain Gate channel. But I do not think that such can be the fact.

The relative heights of the two places indeed are such as might permit of it; though in that case, the distance being some two miles (and probably more than this, as the stream would run), the average grade for the whole distance would be considerably less than the grade of the Mountain Gate channel has proven to be for nearly three quarters of a mile from the mouth of the tunnel at Damascus. But the strongest argument against any connection between these two channels lies in the totally different character of the gravels which fill them. What is the real size of the channel at the Dam claim northwest of the "party-tunnel" is not known, as its eastern rim has not been seen. But it is not probably so large as the Mountain Gate channel.

In any case it is almost impossible to conceive how a channel like that at Damascus could be filled for a distance of at least three quarters of a mile with a gravel so purely quartz, and full of such enormous white boulders of it, and then within the next mile and a half have the character of its gravel so completely changed that quartz should become almost a rarity, more than nine tenths of the boulders being dark-colored metamorphic rocks.

Nor do I think it probable, either, that the Dam claim has any connection with the "black channels" at Damascus. This, indeed, is possible. But the peculiar "gray cement" of Deadwood is plenty at the Dam claim, while I saw none of it at Damascus; and furthermore I have an impression (although I cannot now find any note of it) that the grade of the black channels at Damascus descends the other way, that is, towards the northwest.

But if the Mountain Gate channel of Damascus, though running now straight towards the Dam claim, does not go there, where then does it go? I answer that it seems to me the probabilities are strong that after holding on for a certain distance farther its present southeasterly course, it will be found to curve to the south and southwest, and continue along under the ridge between the heads of Shirt Tail and the western branch of El Dorado cañons, finally breaking out into the latter in the vicinity of Gas Hill, thus proving itself identical with the stream which furnished the gravel along the spurs from there to Michigan Bluff.

One other item in addition to those already mentioned, which goes to strengthen the probability of this, is the fact that in all the heavy banks along these spurs, as well as at Michigan Bluff itself, the gravel is pretty highly quartziferous, though mingled to a greater or less extent with other matters; and it is only in the *tunnel claims* along this part of the ridge that the gravel is generally dark in color and contains but little quartz.

If this conclusion be correct with reference to the course of the Mountain Gate channel, we thus have, during a large portion at least of the gravel period, a stream of considerable size corresponding moderately well with the present El Dorado Cañon. Indeed, it is not impossible that this ancient stream may also have forked at a point not far from where the modern cañon does, an easterly branch coming in from the vicinity of Last Chance by way of the northern edge of the Devil's Basin and the western side of the ridge at Deadwood. And when we remember that at all three of these points there are more or less heavy masses of highly quartziferous gravel which appear to be entirely distinct in origin, and probably also in time, from the darker metamorphic gravels in adjacent channels in the same ridges; and when we note the additional fact that the thickness of these *quartziferous* banks wherever they do occur at these localities is generally far greater than that of the dark-colored gravel which fills the other channels, and seems to underlie so extensively in thin and irregular sheets the volcanic capping,—we may be tempted to infer that such was actually the fact. But this would perhaps be going somewhat too far. For I have not been able to trace any well-defined channels here of this quartziferous gravel for any distance; and furthermore, I do not know its relative age, compared with that of the darker gravels, nor is it certain, indeed, that this relative age is the same at all the different localities. I suspect the quartziferous gravels to be generally older than the others, but I do not know.

Where the large stream went to next from Michigan Bluff is also a question of much uncertainty, and it seems very probable that it may have followed, at least to some extent, different courses at different times. But there are many facts which lead me to think it probable that there was at least a period of time during which, after continuing its course a little to the west of south for a short distance farther, through Sage Hill, and passing out over the site of the modern cañon in the vicinity of the present mouth of the North Fork of the Middle Fork, and meeting near this locality with one or more other streams, which probably then as now drained the central and southeastern portions of the present upper basin of the Middle Fork of the American, it curved to the west, and, rounding the spurs whence now come Mad and Ladies' cañons, turned somewhat then to the north of west, and, crossing Volcano Cañon, entered the Forest Hill divide at Bath, and then, after following a little way farther on this course, turned sharply again southwest and followed that course along the great channel which seems to exist in the Forest Hill divide, probably meeting also somewhere in this divide with the northern stream from Iowa and Wisconsin hills.

One of the little facts which, so far as my knowledge of this country yet extends, seems to lend strong probability to the assumption that the drainage of the central portion of the upper basin of the Middle Fork once found its way to Bath, is the occurrence in the banks at that locality of numerous boulders of a peculiar variety of rock of the same class as the "white lava" of Placerville, of which there are great quantities in place in the country about Long Cañon, but of which I saw none at all in any form in the country to the north of the Middle Fork of the Middle Fork.

Furthermore, no stream could ever have passed from the hydraulic claims at Michigan Bluff directly through the ridge to Volcano Cañon and so across to Bath; for the bed-rock in this ridge is everywhere far too high to admit of that. It must therefore have rounded the spurs below.

But again, in the Forest Hill divide itself, it seems impossible (at all events, with the aid of only such facts as can now be gathered) to trace out any exact system for the streams which seem to have once united in this vicinity, and say, "This is the exact course they followed," at any particular time. If the North and Middle Forks of the American River ever really united in this divide, as the probability seems very strong that once they did, there is plenty of further evidence to lead to the inference that both their point of union and the detailed courses of the streams themselves have shifted about extensively during the long period of the slow accumulation of the auriferous gravel, aside from all the wanderings to which they may have been subject afterwards, during and after the volcanic period.

The distribution of even the deepest accumulations of metamorphic gravel here among the branches of Brushy Cañon, at Yankee Jim's, Todd's Valley, the Dardanelles, and other localities, is such as no single system of streams, which other facts can render either plausible or admissible, is capable of accounting for. The evidences of change are everywhere. It is probable, indeed, that a large stream once flowed for a long period of time southwesterly along the great deep channel which seems to underlie the central portion of the ridge at Forest Hill, and it *may* be that this channel is continuous to-day beneath the ridge as far southwest as Peckham Hill. But again, there is evidence in the furrowing of the bed-rock at Yankee Jim's, that two large streams once met here, one flowing westerly and the other one northwesterly. Moreover, the arrangement of the metamorphic gravel in the great channel itself is such as I can explain only on the theory of extensive accumulation followed by extensive excavation before the volcanic epoch. The crest of the southeastern rim of this channel at Forest Hill is some 700 or 800 feet back in the hill from the mouths of the tunnels; and all over this rim the rock is smoothly water-worn, and there is more or less auriferous gravel. But back, so far as the works have yet extended towards the central and lowest portions of the channel, the stratum of metamorphic gravel is not very thick, and the volcanic roof here descends far below the level of the southeastern rim. There are also evidences of similar excavation of the previously accumulated gravel, and subsequent refilling with volcanic matter at Yankee Jim's, the Dardanelles, and other points. But the most interesting locality in this vicinity for the exhibition of a series of changes is perhaps the Paragon mine at Bath.

It appears that at this locality * there was first accumulated a heavy mass of crushed and angular fragments of the bed-rock which, by whatever means it was produced, cannot have travelled any great distance, or have been exposed for any great length of time to the wearing action of running water, though it does contain *occasional* water-worn and smoothly rounded pebbles. After this came the accumulation of the great mass of smoothly water-worn gravel which underlies the "pay-streak" or "lead" of the Paragon mine. Judging from the indications now visible at this locality alone, it would be difficult, if not impossible, to say whether the stream which brought this mass of gravel here came from the northwest or the southeast. But taking into consideration also the facts at other localities, I am strongly inclined to think that it came from the southeast, in accordance with the view above presented of the course of a river from Michigan Bluff to Bath and Forest Hill.

When the great mass of this gravel had already accumulated, there came another change by which the comparatively thin sheet of far richer gravel was spread a few feet thick with curious uniformity over the broad surface of the older and poorer gravel. The question *whence* this pay-streak came, presents some curious difficulties, as will be seen from a study of the notes relating to it. But, all things considered, I think it probably came from the northeast, over the high bed-rock so near at hand in this direction.

After this came the stream from up towards Weske's claim in the Michigan Bluff divide, crossing Volcano Cañon in the deep channel where the shaft is sunk near the upper road, following on for a certain distance nearly parallel with the cañon along its right bank, then passing more westerly into the ridge, cutting across the back end of the Paragon mine, excavating its channel to an unknown breadth and depth through the pay-streak and the underlying gravel, and afterwards refilling that channel with volcanic matter in the form of that called "gray cement" at Deadwood, containing much semi-carbonized wood, etc. Thus at this locality there were no less than four distinct and well-marked changes during the accumulation of the gravel itself. Next in order, and overlying all previous accumulations, comes the gray, micaceous, volcanic sand (forming the roof of the mine), whose thickness is so great and whose distribution so extensive in this ridge. Then came another comparatively thin stratum of metamorphic gravel, and finally, over all, a great thickness of the ordinary dark-colored volcanic 'cement,' filled with boulders of all sizes, which forms the usual capping of the ridges here.

From what has been already stated,† it appears that at the New Jersey mine the southeastern rim of the great channel is about 900 feet back in the hill, and that beyond this point the distance already prospected on the slope of the rock going down into that channel is about 1,900 feet, and that the lowest point here attained is about 157 feet below the level of the crest of the southeast rim at that point. Yet the bottom of the channel has not yet been reached, and no one knows its depth. It also appears that a little to the northeast of the New Jersey mine this rim rock rises much higher, in the form of a ridge, which extends some distance in that direction towards Bath. Moreover, a large portion of the breasting and drifting in the tunnel mines at Forest Hill has been

---

* See *ante*, p. 94. † See *ante*, pp. 94, 106.

done on the southeastern or outward slope of this southeast rim. And I think it not at all improbable that this outer slope (which in the New Jersey mine is 900 feet long) may once have formed a portion of another great channel, parallel, or nearly so, with that which now seems to exist farther back in the hill. In that case, however, this front channel must have extended far out to the southeast over the present cañon of the Middle Fork of the American, and a great portion of its breadth must have been carried away in the excavation of that cañon. But this is speculation.

In order to appreciate the full force of the evidence furnished by the boulders of white lava in the banks at Bath, of their having come from the Long Cañon country, and the adjacent region in the upper basin of the Middle Fork of the American, the attention should be more particularly directed to one point which I omitted to mention, that is, to the *texture* of the rock which forms these boulders. A few words here with reference to the character of the white lava in general will set this point in clearer light. The statement previously made* must not be understood as meaning that I saw nothing in the country north of the Middle Fork of the Middle Fork belonging to the same great class of material as the white lava, but only that I saw nothing at all nearly resembling that particular variety of it which forms so many of the boulders at Bath, and which is common in the country farther south. The gray, or grayish white, granitic-looking sand cement, which forms the roofs of the mines so extensively at Bath and elsewhere in the Forest Hill ridge, belongs undoubtedly, I think, to the same general class of material as the white lava farther south. The same may be said of the material of an outcrop on the road near Independence Hill, in the Iowa Hill ridge, of which, I believe, I took specimens.

Moreover, the roof of the tunnel I entered at the Strawberry mine near Monona Flat, though a very dark-colored substance, and pretty hard, is probably a similar material. I take it all to be volcanic ash. But wherever seen to the north of the Middle Fork of the Middle Fork of the American (and it was not seen in any *great quantity* in any part of this region, excepting from Bath southwesterly in the Forest Hill ridge), it was always comparatively *coarse-grained*, rather soft, of rather loose texture, and sharp, harsh, gritty feel, and generally contained a notable quantity of *black mica* in scales, often the twentieth, and sometimes the tenth of an inch in diameter, giving the whiter varieties of it that peculiar pepper-and-salt appearance so strikingly suggestive of granitic sand. But the white lava varies from this coarser-grained and loose material all the way down through finer and finer grained varieties to a rock of perfectly compact and almost flinty or semi-opaline appearance, with small transparent crystals — probably of glassy feldspar — scattered through it here and there. This most compact variety is not always white. It sometimes assumes a brownish, and not unfrequently a pinkish tinge. And it is precisely this *compact* variety which forms many of the boulders at Bath, and of which there are large quantities in place in the southwestern end of the ridge between Long Cañon and the Middle Fork of the Middle Fork, at a rectilinear distance of only some six or seven miles from Bath, and of which I did not see a particle elsewhere north of the Middle Fork of the Middle Fork.

It is curious that to the north of this stream the white volcanic ashes, so far as seen, should be almost universally coarser-grained, — this most compact variety being nowhere seen, — while in the country south of it, where also the mass of the white lava occurring is vastly greater, the compact variety is not uncommon, and the rather finer-grained varieties with little mica seem the rule, and the coarser-grained micaceous ones are far less common.

The proportion of mica in the white lava varies as greatly as it does in volcanic rocks in general. Some varieties are almost absolutely free from it, while in some the proportion is very large; and there is every grade between.

It is worthy of note that the white lava, wherever it occurs, so far as I have seen (and in many places its quantity is simply enormous), is always *segregated* by itself, that is, it never contains large boulders or much foreign matter of any kind. I do not remember having ever seen imbedded in it a stone so large as the two fists. Yet it almost everywhere contains *occasional*

* See *ante*, p. 494.

(though rare) *small* pebbles. It never forms the matrix of volcanic breccia, conglomerate, or gravel; that matrix being always a darker-colored material of altogether different texture and appearance. Where the white lava is in heavy masses it often shows a tendency to columnar forms, and at a few localities, mentioned in the preceding notes, well developed prismatic columns may be seen. There is a striking analogy in physical character and appearance between some varieties of the white lava and the material of the volcanic table-land north of the head of Owen's Valley, though it would probably not be difficult in any case to distinguish the one from the other.

The presence of the *granite* boulders in the banks at Bath would seem to furnish additional evidence also, pointing to the same direction for the course of the stream which brought them. For everywhere excepting at Bath, in the country where I travelled to the north of the Middle Fork of the Middle Fork, granite is *extremely* rare among the boulders either of the metamorphic gravels or of the volcanic accumulations; and by far the nearest point to Bath where granite bed-rock exists higher up in the mountains is in the Long Cañon country, or between Long Cañon and the South Fork of the Middle Fork, and granite boulders are common throughout the Long Cañon country.

Moreover, if it be true, as Mr. Kates believes, that a ridge of very high bed-rock stretches from Volcano Cañon near Baker's ranch entirely across the ridge to Shirt Tail, this would seem to imply of necessity that the flow of the great channel at Forest Hill could not have proceeded for any great distance above that town from the northeast, in spite of the fact that *at* Forest Hill its course appears to have been southwest.

I think, therefore, the evidence is very strong that at *some* time during the gravel period a large stream once flowed in the general course that I have indicated above. And yet, the shifting character, even of the largest streams engaged in piling up the gravel, must not be forgotten.

Mr. Pond's estimate of the ground that has been hydraulicked off at Todd's Valley, and the amount of gold obtained from it,* would give, in round numbers, about 8,000,000 cubic yards washed away, with an average yield of about fifty cents per cubic yard for the whole. This yield per cubic yard would correspond tolerably well with the yield from the ground washed by Mr. Pond within the last few years, as already given.

I think it not at all improbable that the great channel at Forest Hill may be continuous under the central portion of the ridge, passing a little northwest of Todd's Valley, to a point as far southwest as Peckham Hill. But at or near Peckham Hill, if not before, it must pass out of the ridge; and below (that is, southwest of) there the country is so low that neither gravel nor volcanic matter remains upon the present hills.

There is little doubt, I think, that a great field yet remains for mining enterprise in this Forest Hill ridge. But one great drawback there now is lack of water for hydraulic and other purposes. Furthermore, the "back channel" there can never be worked to advantage without the previous expenditure of large sums of money in the driving of new tunnels, etc.

The general character of the upper basin of the Middle Fork of the American, immediately before the modern cañons began to be excavated, appears to have been that of a broad smooth plain, with a gentle southwesterly slope, just about equal in amount to the average southwestern slope of the Sierra Nevada, while at the same time the surface of the basin probably sloped very gently inwards from the edges in all directions towards its central portions. The surface of the country here at that time seems also to have consisted chiefly of volcanic matter in the form of ashes, breccias, and conglomerates or gravels which had slowly accumulated to depths varying from one or two hundred to 600 or 800 feet over the buried auriferous gravel. Here and there isolated patches of bed-rock might have been seen at the surface, the accumulation of volcanic matter, great as it was, being insufficient probably to entirely bury all the older hills, which, however, seem to have very rarely projected much, if any, above the general level of the volcanic surface.

Through this plain the cañons of the modern branches of the Middle Fork have been slowly

* See *ante*, p. 118.

excavated to depths ranging from 1,500 to 2,500 feet below the ancient surface, leaving a series of ridges whose crests (wherever the space between adjacent cañons is wide enough to allow of it) almost invariably rise to nearly a uniform plane, and over a considerable area of country never rise much above it. These ridges are now almost everywhere heavily timbered, and so, when this region is seen from a favorable point of view to-day, as, for example, from the Canada Hill Bald Mountain, which overlooks it and is near enough to give a good view of it, and yet too far removed to enable us to see down into the cañons, it presents far more the appearance of the ancient basin-plain than it does of the series of narrow ridges and tremendous cañons which it really is to-day. There is perhaps no point from which so good an idea can be obtained of the general appearance of this portion of the country at the close of the volcanic period as from the summit of this mountain.

From here the Georgetown ridge — that is, the ridge which forms the divide between the main South Fork and the South Fork of the Middle Fork of the American River — is very conspicuous, bounding the basin of the Middle Fork, and limiting the view in that direction in the form of a great spur of the Sierra Nevada, running far southwesterly from high up towards the summit of the range down almost to Georgetown itself, with a crest which is not smooth like those of the ridges *in* the basin of the Middle Fork, but rough and peaky with bed-rock hills, and whose general height all the way to Georgetown is considerably greater than that of the adjacent country for a long distance either to the northwest or southeast of it on a line parallel with the axis of the range.

It was when I caught this view from the Canada Hill Bald Mountain that my previous suspicion first deepened into conviction that, in this portion of the country at least, the outlines of the larger drainage basins were *essentially the same* in the gravel period as now. I felt confident, even then, that the higher parts of that Georgetown ridge were bed-rock hills, and so it proved.

As *in* the basin of the Middle Fork the volcanic-capped ridges do not rise above a certain level, so in the ridge between the South and Middle Forks the volcanic matter does not rise above a certain level on the sides of the higher hills. Wherever I went for a distance of thirty miles east from Georgetown in this ridge I found everything above a certain horizon to consist of bed-rock hills. I do not think these hills have ever been covered either with gravel or volcanic matter, unless indeed by possible showers of ashes dropped in the air, which were readily washed away by rains to lower ground.

This distribution of the volcanic matter here, together with the fact that the total quantity both of gravel and of volcanic matter scattered over this ridge, (though the aggregate, especially of the volcanic matter, is large) is yet far less than it is in the country to the north — both seem to be necessary consequences of these features of the ancient topography of the region.

Strong additional evidence in favor of the view above presented of the general topography of the country here in the gravel period is furnished by the courses of the ancient channels in the region immediately north and northeast of Georgetown, and between it and the river, wherever such channels were excavated to a sufficient depth in the bed-rock to make them unmistakable, and have been since preserved, and at last definitely traced and exposed by mining operations.

The Castle Hill channel, near the Clipper Mill, runs about N. 20° W. (magnetic). The Roanoke Channel, near Bottle Hill, runs about N. 50° W. (magnetic) for more than a mile through the ridge. It then zigzags for some little distance in a general westerly direction, and from the last point seen of it here to Jones's Hill (if in reality the latter be, as I think it probably is, a continuation of the same channel) the course is about S. 80° W. (magnetic), — still a little north of west (*true* course), — while *in* Jones's Hill it seems to make, for a short distance at least, another bend to the north, and pass out northwesterly from that hill over the present cañon of the American. Moreover, the little branch of this channel, which has been drifted entirely through the ridge beneath or just west of Bottle Hill, comes into the main channel from the *south*.

Moreover, the channel upon which Flora's mine is situated, though a considerable portion of it has indeed gone down the river cañon, seems to have been pretty well defined, and to have had, from a point a little northeast of Volcanoville, a course somewhat to the south of west, for some distance just cutting across the tops of the modern spurs, but afterwards entering the ridge which it then seems to follow as far as Flora's mine, whence it evidently passed out over the modern river cañon and has been washed away. All these channels have not simply a perceptible, but a considerable grade, falling in the directions of their flow as indicated above, and all point to the same general conclusion respecting the ancient topography of the country, namely, that the great spur of the Sierra, which for so many miles above Georgetown now forms the divide between the South and Middle Forks of the American River, existed in the gravel period as well as now, with its relative prominence above the adjacent country, perhaps, even greater then than now, owing to the subsequent filling of the basin of the Middle Fork with volcanic matter, etc.; and that then, as now, it formed the divide between the *South and Middle Forks of the American River*. It may indeed be *possible*, for aught I know yet to the contrary, that during the whole or a part of the gravel period the South Fork of the American may have been an independent stream, not joining the North Fork, but reaching the valley separately somewhere southeast of Folsom. But I know of no proof of this, and the topography of the country below Shingle Springs leads me to think it improbable. There is much more plausibility in the idea that at one time before the modern cañon of the South Fork was excavated to any considerable depth that stream may have flowed across the present ridge at Centreville (that is, Pilot Hill), and joined the North Fork somewhere in the vicinity of Lacey's or Rattlesnake Bar. There is a depression here across the ridge which might have permitted of such a course, and the presence of the *granite* pebbles in the gravel at Pilot Hill may seem to speak in favor of it; but the evidence that I saw is not very strong, after all, though some of the miners are confident that it was so.

To return to Georgetown. A line drawn from the junction of Long Cañon and the South Fork of the Middle Fork of the American, southeasterly along the crest of the Tunnel Hill ridge, and curving westerly a mile or two south of Work's ranch by Tipton Hill, and thence to Georgetown, then following the Spanish Dry Diggings road and the crest of the ridge to the Middle Fork of the American at a point between Spanish Dry Diggings and the mouth of Cañon Creek, would include between it and the river all the volcanic matter and all the unquestionably ante-volcanic gravel that exists, so far as I know, between the South and Middle Forks to the west of Tunnel Hill. To the south from Georgetown the country slopes gradually towards the cañon of the South Fork of the American, and here, as well as to the west and southwest of the ridge between Spanish Dry Diggings and the branches of Cañon Creek, it consists entirely of the bed-rock, slates, etc., covered only by the varying quantities of soil and comparatively recent gravel which have accumulated over it. The ridge last referred to stretches southeasterly from the river west of Cañon Creek for several miles, to within a mile or two of Georgetown, where it connects with the main watershed between the South and Middle Forks of the American. This ridge, though not high enough to be conspicuous, is yet an easily distinguishable feature in the topography, and overlooks the country for several miles to the eastward from it. As already stated, it is bed-rock to the crest, neither gravel nor volcanic matter being found upon it; and it seems to have formed a barrier to the southwestern course of some of the smaller ancient as well as modern streams, and to have turned them northerly towards the Middle Fork. The entire absence of volcanic matter to the *westward* from it is certainly an argument in favor of this idea. It thus appears that not only the great general skeleton, but even some of the minor features of the ancient topography may be found, which remain to-day almost as they were then. In connection with this idea of similarity between the bed-rock topography of then and now, another locality, which may be worthy of a passing notice, is the vicinity of Wilcox's claim (otherwise known recently as the China claim), on the North Fork of Long Cañon.* Here a channel of considerable magnitude appears to come through the ridge in a south-east direction and cross the present North Fork of Long Cañon; after which, in all probability, it curves southwesterly. The gravel in this channel is mainly of metamorphic rocks, and yet all the bed-rock *here* is granite; and it is several miles from here, either northwesterly or northerly, to the nearest point at which I have any information of any considerable quantity of metamorphic rocks

* See *ante*, p. 97.

in place. So far as these facts go, therefore, they point to at least a probability that a stream of some magnitude once flowed here for several miles in a general southeasterly or southerly direction before bending southwesterly to the normal course of the Sierra streams; and, whenever the region about and above here comes to be accurately mapped, it will be seen that this is precisely what is done by several of the longest modern branches of the Middle Fork of the American. Duncan Cañon, the Middle Fork of the Middle Fork, and the South Fork of the Middle Fork, all do the same thing. And, in fact, the upper drainage basin of the Middle Fork of the American expands northeasterly, around and to the east of the Canada Hill Bald Mountain, in such a way that the streams which drain its farthest limits in this direction *must* flow southeasterly and southerly here for a number of miles before they can bend southwesterly to their normal course. Probably a similar state of things in this respect existed here also in the gravel period.

From all that I was able to learn, it appears that to the south of the line of water-shed between the South and Middle Forks of the American, and between it and the South Fork, there is nowhere any considerable quantity either of ancient gravel or of volcanic matter. In other words, almost all the gravel and volcanic matter between the South and Middle Forks is on the northern drainage slope, where the water runs to the Middle Fork.

But, on crossing the South Fork of the river, we find again, in the region about Placerville, extensive and very complex accumulations both of gravel and volcanic matter. I spent considerable time in this vicinity, and the preceding detailed notes relating to it are very full.*

In passing from this mass of detail to general considerations, the most salient point, and the first conclusion which presents itself, will probably be the complexity and apparent confusion of structure in the gravel and volcanic banks throughout this region, and the extreme difficulty, if not indeed the impossibility, of tracing out to-day any definite system of *contemporaneous* streams which may have existed at any one particular time during the long period of the slow accumulation of these banks.

Indeed, I may say that with all the time and study that I gave to this amphitheatre of hills surrounding Placerville, and all the detail I was able to gather respecting them, I do not consider it practicable yet (with a single exception only) to trace continuously with any definite *certainty* for any considerable distance the course of even a single stream at any particular time.

There are, indeed, at various localities where the bed-rock is hard, as at Webber Hill, Little Spanish Hill, etc., systems of parallel furrows worn in the surface of the rock, which show, as I conceive, with tolerable certainty, the directions of latest flow upon the surface of the rock at these localities. And if the bed-rock were generally hard enough beneath the gravel so that the miners would not disturb it in their operations, we should probably see these furrowings wherever it is uncovered, and their study might become an extremely interesting as well as complex problem. But, unfortunately for this, the bed-rock beneath the gravel is far oftener than otherwise thoroughly decomposed, and so soft that in drifting, as well as in hydraulic work, from six inches to a foot of its surface is generally taken up on account of the gold which has worked its way into its crevices, etc.

Moreover, it is by no means certain, and, indeed, I consider it altogether improbable, that all these furrowings, even in the surface of the bed-rock itself, are contemporaneous in their origin. I think it more than probable that, if the bed-rock were everywhere hard, and if the gravel and volcanic matter overlying it were all removed, so that its ancient surface could be minutely studied, we should find in these furrowings an intricate system of directions, — perhaps with a *general* westerly and southwesterly trend, — but still intricate enough to be capable of explanation only by the theory of *shifting* streams, the gradual filling of the deeper channels, and the frequent local turning of the water in new directions over higher and still higher portions of the rocky surface till finally all was buried.

In all the complex structure of the higher portions of the banks themselves above the surface of the rock, I find, almost everywhere, what seems to me indubitable proof, not only that these banks

* See *note*, pp. 98, 101, 108, 111, 112, 119, 120.

are the work of streams of fresh and running water, but also that their slow accumulation progressed through a still more frequent series of local shiftings and changes in currents and channels than was possible in the nature of the case while the water flowed immediately upon the naked rock. This process during the accumulation of the upper portions of the banks would necessarily (and especially if, as is more than probable, the streams were subject from any cause to alternate freshets and droughts) be accompanied by the frequent local excavation of new channels to greater or less depths through the previously accumulated matter, and the subsequent refilling of these, either with similar or with different materials. The frequent occurrence in these banks of nearly horizontal lenticular masses — sometimes of great volume and entirely isolated — of white lava and of other materials, entirely different from anything immediately adjacent to them around, above, or beneath, is a fact which I can explain upon no other theory than this. So also of the peculiar style of local bedding exhibited in the matter overlying the gravel in the Rocky Mountain claim at Negro Hill.

The bed-rock hills between the two upper forks of Hangtown Creek, which head respectively at Smith's Flat and Oak Grove Flat, are generally so high that I think the probabilities are there has never been any gravel or any considerable quantity of volcanic matter on them. But the whole of the immediate vicinity of Placerville and the valley below the union of the two forks of the creek above referred to, as well as a large portion of the smaller ravines, are below the level of the gravel-plane, and it is more than probable that a considerable portion of this area was once covered both with gravel and volcanic matter. Indeed, it is probable that the present gravel hills represent but a very small proportion of the total area over which the ancient streams wandered, and which once was covered with similar material; by far the greater portion of it having been carried away in the excavation of the modern cañons of the South Fork of the American, and of Webber Creek and Hangtown Creek and their branches. But there can be little doubt, I think, that this gravel was accumulated by a system, or series of systems, of streams which drained essentially the same country that is now drained by the South Fork of the American and its branches, and that the general course of the drainage was nearly the same as now.

A single one among the ancient channels in the vicinity of Placerville appears to be defined with considerable certainty for a distance of several miles. I allude to the so-called "blue channel," which seems to pass from White Rock in a general direction a little to the east of south (magnetic) beneath the hills to Smith's Flat and beyond.* There are uncertainties, indeed, even about this. But, if asked to trace its course as well as I can, I should say that, coming from the northeast over the region now occupied by the cañon of the South Fork of the American, it entered the present ridge at or a little to the east of Georgia Hill (just east of White Rock Flat), and passing southwesterly in a somewhat zigzag course through White Rock Flat and around White Rock Point, it then turned southeasterly across White Rock Cañon, and, entering the present ridge again on the southwest side of that cañon, passed on beneath the volcanic-capped ridge to Dirty Flat, thence through beneath another spur (called Granite Hill) to Smith's Flat, whose whole length it traversed, passing on into the ridge beyond. Whether it then continued its southeasterly course to the Try Again tunnel, or, curving southwest again, passed under Prospect Flat and so on beneath the ridge toward the head of Cedar Ravine, it is impossible now to tell; but I am rather inclined to think that it followed the latter course. In either case, I consider it extremely probable that it was but a branch of a larger stream flowing southwesterly somewhere not far from the present Webber Creek. It does not seem to have been a very broad channel, nor very deeply excavated in the bed-rock, though sufficiently so to define it well. Both at White Rock and at Smith's Flat it appears to have two well-marked benches or terraces along its sides, as if there had been three distinct periods in the excavation of its rocky bed. But all is now filled with gravel, and all except a little in the vicinity of White Rock is covered with volcanic matter. The gravel shown on the map at Smith's Flat is only a layer of surface gravel overlying a portion of the white lava that here covers the deeper channel, as well as a portion of the bed-rock rims,

* See Plate C, opposite page 98.

though there seems also to be one or more layers of gravel at about this height or above it, extending into the hill eastward from Smith's Flat. The streaks of metamorphic gravel running through the white lava high above the bed-rock at Prospect Flat are worthy of notice as bearing upon a general point of which I shall speak again hereafter. It may be noted, too, that most of the diamonds that have been found in the vicinity of Placerville have come from the gravel of this blue lead channel, — those at Webber Hill being the only exceptions that I learned of, — while, indeed, it is not impossible that the continuation of this same channel may once have flowed over the bed-rock along the southeastern foot of Webber Hill itself.

In turning from this blue lead channel, the attempt to trace through the intricate structural mazes of the gravel and volcanic hills of Placerville any other definite channels becomes far more difficult, and full of uncertainty.

I think the probability very strong, however, that for a long time during the gravel period a stream of considerable magnitude followed a nearly true west course, not far from the line of the present Webber Creek, for seven or eight miles at least, from the vicinity of Newtown to a point as low down as Placerville. The lay of the bed-rock in Hangtown Hill and Coon Hill, sloping southerly from the northern edge of Hangtown Hill to Coon Hollow, and seeming to form here, for nearly half a mile in width, but a portion of a very broad channel, the southern half of which has been carried away by the excavation of Webber Creek, the lowness of the bed-rock in Webber Hill, and the quantity and depth of the auriferous gravel scattered along the hills on the southern side of Webber Creek from the vicinity of Fort Jim to Newtown and above, all speak strongly in favor of this idea. Moreover, the occurrence of such heavy masses of "mountain gravel" (that is, smoothly rounded volcanic gravel) in Hangtown Hill and Coon Hill, and around the head of Chili Ravine and in Webber Hill, and scattered in smaller patches along the southern edge of the ridge for at least a mile farther east, when taken in connection with the comparative rarity — in fact, almost the absence — of this material in the hills farther north between here and Hangtown Creek, and to the *east* of Oregon Point, seems to point in the same direction. The origin and course of the stream, or streams, which formed the isolated deposit of the Diamond Springs Sugar Loaf, and the adjacent one of Bean Hill, at a much lower level than the former, are very problematical. It is possible that this Webber Creek stream may once have curved far enough southerly here to do it, and then bent northwesterly again. The bed-rock in the Sugar Loaf itself is low enough to have permitted this stream to come there, even after leaving Coon Hill. I do not think it likely, however, that this stream ever crossed the present divide in this vicinity between Webber Creek and the Cosumnes River to discharge itself in the latter direction towards the valley, though this, too, may be among the possibilities. It is certain that the deposits of Bean Hill and of the Diamond Springs Sugar Loaf, though so close together yet having so great a difference of level between them, could not have been accumulated simultaneously. Yet which is the older of the two I do not know. There is no difficulty in seeing how either one may have been the first deposited.

If we grant that during a portion of the gravel period a large stream flowed westerly, as above supposed, near the line of the present Webber Creek, the question then at once suggests itself whether, at that time, this was the main stream, draining all the adjacent region as well as the country far above in the mountains, and thus corresponding to the present South Fork of the American, or whether it was only a branch of the larger stream, and thus corresponded almost exactly with Webber Creek. But I know of no definite means of answering this question. The tops of the present cañons are so broad, the portions of the ancient channels still left in the hills are relatively so very small, the structure of the banks that are left is so very complex, and the possibilities as to what may once have existed over the areas now occupied by the cañons, etc., are so varied and so great, that any opinion now on such a point could be no better than an idle guess.

In an article published in the Placerville Mountain Democrat, I expressed a leaning to the opinion that Big Spanish Hill, the eastern portion of Little Spanish Hill, and Coon Hill were all on the same channel, which probably flowed from Little Spanish Hill across Cedar Ravine to Cedar Hill, and then traversed the latter obliquely in a southwesterly direction to Coon Hill.

There are some difficulties, or at least some points which are very unsatisfactorily explained, in the way of such a theory. Yet the heavy banks at Spanish Hill and Coon Hill resemble each other so closely in their general character, and in so many of the details of their structure too, while the furrowings of the bed-rock on the south side of Little Spanish Hill are in the right direction for it, and the bed-rock in Cedar Hill at the proper point is known to be low enough for it, and the banks of metamorphic gravel which are visible at the latter point, though bearing no comparison in point of magnitude with the masses of Coon and Spanish hills, are yet the heaviest masses of that material to be seen anywhere along the northern side of Cedar Hill, that, so long as I know of nothing in Cedar Hill that is positively prohibitory of it, I am still inclined to hold that theory. But if it be true, then it is possible, either that this stream may have been a branch of the one flowing westerly near Webber Creek, or that it may have been not at all contemporary with it, but entirely an independent stream at some other time. It is, in any case, impossible now to say where this stream came from to the east or northeast of Spanish Hill. There are no more banks resembling these now visible in any of the hills around the head or on the other side of Hangtown Creek.

There are many facts which seem to point to a probability that a considerable stream once flowed southwesterly through Negro Hill, and the little outliers beyond, known as Little Indian Hill, Indian Hill, and Clay Hill. The predominant part which a certain dark-colored volcanic sand, free from pebbles and boulders, plays in the formation of most of the banks at these localities speaks strongly in favor of their close connection in origin, and the lay of the bed-rock is such as to accord with this idea. But the exact localities whence this stream came, and where it went to, are alike uncertain. It was in all probability a considerably more recent stream than the one which flowed in the blue lead channel above described, from White Rock to Smith's Flat, etc.; and it is possible, though by no means certain, that it may have flowed transversely over and across that channel, in the ridge just south of White Rock.

Any further attempts to trace out definite channels in the vicinity of Placerville would probably be useless now. The banks are everywhere full of unmistakable evidence of the multiplicity of changes undergone by the streams which gathered them, and the process must have been a slow and long one.

The frequent white lava boulders which occur in the gravel in the deepest portions of the channel beneath Smith's Flat, as well as the occasional ones found in the lower gravels at Webber Hill and elsewhere, are interesting as proofs of the fact that the volcanoes were already active in the High Sierra long before the great masses of volcanic matter now capping these hills and ridges came here, and that, even while some of the earlier gravels about Placerville were gathering, there existed somewhere in the higher northeastern country masses of these white volcanic ashes, already consolidated into a rock from which boulders could be made. The white lava also, wherever found, appears to be the earliest kind of volcanic matter that was afterwards spread over this region.

And the occurrence at Prospect Flat (a fact, by the way, which has its parallel perhaps at Smith's Flat also, and certainly at many other localities in the country I studied last summer) of thin strata of metamorphic, quartzose, and auriferous gravel, intercalated between bodies of white lava, high above the bed-rock, proves that, after the first heavy bodies of volcanic matter reached here, the streams still flowed, passing some of the way in their new courses over the naked surface of the higher and still uncovered bed-rock, and again over the surface of the volcanic matter already accumulated, and so, here and there, spreading thin local sheets of metamorphic gravel far and wide over the volcanic floor already laid above the older gravel.

Then came other floods of volcanic matter, and sometimes other sheets of gravel, and so on through the slow-moving changes. At other localities, instead of spreading gravel over the earlier beds of volcanic ash, the streams seem to have excavated new channels through them, sometimes cutting deep enough to take them all away, and isolate greater or smaller patches of them here and there, thus giving origin to those curious lenticular masses, like the heavy one of white lava, which has already been entirely washed away in the hydraulic work at Coon Hill.

The channels so excavated were subsequently filled, either with similar material, or with gravel, or with a different kind of volcanic matter, and so the long work went on. After the white lava had ceased to gather, then came to Placerville the mountain gravel, the streams which brought it excavating also to a greater or less extent their own channels through the previously accumulated matter, gathering their own materials from the dark-colored compact volcanic rocks, which had made their appearance higher in the mountains, grinding a portion of these materials to powder, rounding and smoothing the rest to a perfect gravel, and picking up occasionally a pebble of quartz or granite or slate by the way, and finally piling up all this stuff, to depths sometimes of a hundred feet. Lastly came the successive deluges of black lava, or volcanic breccias, which overspread the whole, and crown the highest crests of the modern hills and ridges.

Such seems to be the outline of the most prominent features in the history of the gravel hills of Placerville. The vast complexity of their detail forms a labyrinth whose windings it is hopeless to expect will ever be completely traced, and whose history, if fully known, it would probably require many a ponderous volume to relate.

Yet the general *modus operandi* of their formation seems evident; the order of succession of the leading kinds of material which form the great mass of the hills is also seen, and the general fact seems clear that these materials came from the northeastern country, which formed then as now the higher portion of the mountains. And, though the order of succession of the different kinds of material is sometimes different at different localities, and there is no end of variety in the details, yet in all the facts last stated, that is, in almost all the *leading, general* features of the subject, these hills of Placerville present an epitome of the whole gravel region so far as I studied it in 1871.

There is a range of country running from near the South Fork of the American at the mouth of Webber Creek, and passing just west of Shingle Springs in a direction a little to the east of south, and passing just east of Latrobe to the edge of the cañon of the Cosumnes just below the mouth of Big Cañon Creek, which is so high that I do not believe it has ever been crossed by any of the large streams coming from the mountains above. And I think there is far more probability that the South Fork of the American may once have crossed the ridge at Centreville and joined the North Fork higher up than now, than that it ever found its way to the valley as a separate stream. Indeed, it seems but natural that, when the locality of the junction of two mountain streams is shifted, it should be changed to points lower down the mountain slope rather than to those which are higher up. Such seems to have been the case at Forest Hill, and such may very possibly have been the case here.

The immense quantities of gravel and other river débris accumulated in the vicinity of Folsom speak volumes in favor of the idea that the *débouchement* of the American River from the mountains has never since the earliest gravel period been far from where it is to-day; while the absence of any corresponding quantity of similar materials between these and the vicinity of Michigan Bar is equally strong evidence that since that time no very large stream has reached the valley at any point between the *débouchement* of the Cosumnes and that of the American.

I can probably add little of any moment to what is given in the detailed notes respecting the region I traversed in going from Placerville, *via* Fairplay, Grizzly Flat, Brownsville, Indian Diggings, Volcano, Fiddletown, Mud Springs, and Latrobe, to Michigan Bar. It will be seen that in the country passed over on this trip, from Fairplay through the localities named to Fiddletown, there are great quantities of volcanic matter, and in places considerable gravel beneath it in the ridges. But I have been able to trace no definite channels here. I saw nothing, however, to conflict with the theory that the general course of the streams was southwesterly here as well as elsewhere during the gravel period, that from time to time they shifted their beds, and therefore, that if the actual system of channels could be fully traced, it would prove a complex network here as elsewhere, and finally, that all the volcanic matter which afterwards buried them so deeply (every particle of which is fragmentary) was brought here *in one way or another* by the *aid* of *water*, from the high Sierra far up towards the summit of the range. On the contrary, I can see no other theory that is capable of accounting for the facts throughout this region, any more than elsewhere, where I went last summer.

It is worthy of special note that wherever the ancient gravel rests on limestone bed-rock, there it is arranged in most peculiar and utter confusion. No such scenes of disorder can be found in the structure of gravel banks anywhere else, as at these localities. This fact may perhaps be partially explained by the extremely irregular erosion of the limestone previous to the deposition of the gravel upon it. But I think also that in many instances it is largely due to subsequent erosion and land-slips since the volcanic period, and to the slow excavation by the percolating waters of subterranean caverns in the limestone, whose roofs have occasionally broken in, and allowed the overlying mass to settle down. Another interesting phenomenon is the occasional finding of such subterranean caverns of considerable size, perfectly filled with the finest of silt, which has found its way in through cracks and crannies too small to admit the passage of pebbles or even of the coarser sand.

My estimates of the gravel in the vicinity of Michigan Bar would give for the total amount already washed away in the first hill visited, 775,000 cubic yards, with some three or four million yards remaining yet to wash ; in the second hill 200,000 washed away, with a couple of millions yet remaining ; and in the third hill between two and three hundred thousand washed away ; while the total original aggregate amount of gravel in this vicinity, in the hills to the south of the Cosumnes River, I fix at from twelve to fifteen million cubic yards. These figures are simply round numbers, of course, and being only rapid estimates made chiefly by the eye, they must be taken only for what they are worth. On the north side of the river there is comparatively little gravel in the hills.

My examination of the gravel in the vicinity of Folsom was only cursory, and I made no attempt to estimate its amount. But its quantity is immense (including the volcanic gravels), and a more careful investigation of it could hardly fail to develop many facts of interest which I failed to learn.

With reference to the region between Michigan Bar, Fiddletown, Volcano, the Mokelumne River and the Valley, I can add little on any special points to what is contained in the preceding notes.

I may notice, however, the existence at Butte City, at the southwestern foot of the Jackson Butte, of what seems to be a curious gravel-basin, whose outlet is not known, and which is probably worthy of further study if time and opportunity should offer. And I am tempted to refer once more to the intense resemblance which the Jackson Butte bears to a peak of metamorphic rock when seen from a little distance northwest of it.

I now proceed to the consideration of some more general questions.

Section II. — *General Observations on the Occurrence of the Gravel.*

The history of the Sierra Nevada, even if fully known, would be neither a simple nor an easy thing to write. With reference to its earlier history I shall venture to say little here, my notes relating chiefly to the later period, when the mountains were already formed, and after the auriferous gravel had begun to accumulate upon their slopes. A few words, however, may not be out of place respecting some of those structural features of the range which date back to the earliest period of its history.

Though in a certain general sense it may be said, perhaps, that the Sierra consists essentially of an axis of granite flanked by sedimentary rocks, yet there is another and a stricter sense in which this is not true. By far the greatest bodies of granitic rock which occur in the range are found, it is true, in its higher and central portions. But the granite is by no means confined to this portion of the range, nor does it seem to be confined to any two definite belts, one occupying the central portion, and the other the western foot-hills of the range. On the contrary, it occurs in broad, irregular patches, scattered over the western slope of the range, here and there, from base to summit.

The stratigraphy of the metamorphic rocks of the Sierra is a puzzle to me. On the western slope their strike is almost universally northwest and southeast, and their dip almost as universally at very high angles towards the northeast, that is, towards the crest of the range. These rocks are locally disturbed, of course. Their strike sometimes varies considerably within short distances, and sometimes their dip is to the southwest of the vertical; but these local disturbances are not so great but that the whole series of auriferous slates may be said to be conformable from the base of the range to the highest points at which they occur toward the summit. Yet the aggregate thickness of these strata is, in many parts of the range, so enormous as to make it hardly credible that they can constitute a single consecutive series, deposited between the earliest epoch to which any of the stratified rocks of the Sierra have been proven to belong, and the upheaval of the range in the Jurassic.

But if these rocks be not a consecutive series, then how and by what means have they been folded in such perfectly parallel masses, presenting only their broken and eroded edges at the surface, and dipping at such universally high angles to the northeast? There is, indeed, no lack of *possible* causes and conceivable forces which *might* have produced such a result; but I know of no positive proof that any particular force or set of forces is the one that *has* produced it, even if such be the fact at all. In fact, so far as my own knowledge is concerned, the whole question of the stratigraphy of the metamorphic rocks of the Sierra is yet a mystery. One fact which adds yet more difficulty to the problem is the frequent absence of any noticeable amount of local disturbance in the immediate vicinity of heavy masses of granitic rock. This fact is so frequent that I know not how to account for it on the supposition that the granitic rocks are all eruptive; and, taken in connection with the occasional distinctly bedded structure of the granite, and the occurrence in it — in some localities at least — of hornblendic nodules of lenticular form, with a general parallelism in the directions of their axial planes, it seems to me at least very suggestive of the question whether a portion at least of the Sierra granite may not be metamorphic in origin. In this connection I may allude once more to that characteristic Mount Whitney form which is repeated in so many of the culminating peaks of the range on the west of Owen's Valley. There has surely been some special cause for the repetition of this same form so frequently there, nor can I readily believe that it has been exclusively the action of external forces. Whatever may be due to the sculpturing power of ice and snow, or to the character and direction of other erosive forces which have aided in carving the peaks, I cannot help thinking there is probably something in the structure or texture of the rock itself which has contributed in no small degree to the production of this peculiar form.

With reference to the greater morphological features in the structure of the Sierra at large, I believe that these, except in so far as they have been subsequently modified by volcanic and erosive action, existed from the beginning, or at least from a very early period in the history of the range; that is to say, I believe that the greater and loftier spurs which stretch here and there so far down the western slope of the range, and mark the dividing lines between the greater of the modern drainage basins, as, for example, the ridge between the South and Middle Forks of the American, and also the greater of the isolated ridges which, in places far down on the western slope, rise high above everything else around them, as, for example, the Calaveras County Bear Mountain ridge, are probably as old in their elevation as the crest itself, and were uplifted with it to heights above the adjacent regions.

The relative height of these grander spurs above the adjacent regions may, indeed, have been far less originally than it now is, for the tremendous denudation which has since taken place would naturally be felt in its greatest power in the lower regions, while the more elevated ridges, being free from the devastating floods and currents, would wear away more slowly, and thus the difference of altitude between adjacent higher and lower regions would be, in many localities at least, increased instead of diminished by time. But I think that the relief of the range in its earliest history was sufficient and of such a character as to mark out, even then, a drainage system for its western slope, which, so far as the outlines of its greater basins at least are concerned, has remained essentially unchanged from that day to the present time.

At all events, the evidence that such was the case with the basin of the American River, not simply as a whole, but even with the separate basins of its North, its Middle, and its Southern Forks, is very strong.

I know of no proof that any portion of the western slope of the Sierra *which I have seen* has been beneath the sea at any time subsequent to the Jurassic Period. Nor do I know of any evidence that the range has been greatly disturbed in altitude since the Cretaceous, while the evidence seems to me conclusive that it has *not* been subject to any considerable change in this respect since the earliest of the ancient auriferous gravels began to accumulate on its western slope.

What vast periods of time may have elapsed between the original upheaval of the range in the Jurassic (if then it was) and the time when the auriferous gravels now underlying the volcanic matter began to accumulate, as also what changes may have taken place within those periods in the surface sculpturing of the range, I do not know. But I think it was within this period that the auriferous quartz veins themselves were formed. They were certainly formed before the gravel period began; while it also seems eminently probable, if not altogether certain, that the date of their formation was entirely subsequent to that of the great upheaval of the rocks.

But, if asked my opinion as to whence came the gold contained in the quartz veins, I should say from the enclosing and surrounding rocks themselves. There are abundant facts to prove beyond all doubt that gold is distributed in minute quantities far and wide throughout the great mass of the metamorphic rocks themselves, entirely independent of all the quartz veins. Indeed, it may well be questioned whether there is so much as a single ton of metamorphic rock of the auriferous series above the level of the sea in the Sierra Nevada in which a sufficiently delicate chemical analysis might not detect a trace of gold; and it can hardly admit of a doubt that the aggregate quantity of the gold which is thus distributed through the mass of the rock is incomparably greater than that which exists in the quartz itself. There is thus no lack of resource from which the gold in the quartz may have easily been derived; while, on the other hand, the general character of the quartz veins themselves appears to me to furnish strong evidence in favor of the view that the gold in them has been thus derived. They are not often what are technically called fissure veins. They are generally intercalated between the planes of stratification of the slates, and irregular bunches and lenticular masses of limited extent are common among them. Furthermore, the rocks in many localities are penetrated in every direction by little irregular quartz veinules and stringers, which often carry gold, and are sometimes in spots extremely rich, even when less than an inch in thickness.

As examples of many of the facts just stated above, we need only glance at such localities as Spanish Dry Diggings, Georgia Slide, Greenwood, and the country between Johntown and Georgetown in El Dorado County, Quail Hill in Calaveras County, Whiskey Hill (the locality of the Harpending mine) in Placer County, etc.

Since the whole theory of the ancient gravel streams of the Sierra is so intimately connected with the topography of the range during and immediately preceding the gravel period, it may be well before going further to notice some of the theories previously held upon this subject, together with some of the more prominent objections against them.

It has been believed by some that the ancient gravel banks are the work of an ocean or inland sea, which was supposed to cover the country during the time of their accumulation. Against this idea may be urged the conclusive objections: First, that the gravel is not of salt-water origin, no marine fossils having ever been found in it, while the bones of land-animals are frequently found, and the remains of trees and land-plants are plenty in it everywhere; second, the internal structure of the banks themselves is in general not such as is produced by the waves of a sea, but is precisely like what running streams produce; third, the manner of distribution of the gravel over the surface of the country is not only entirely unlike anything that a sea could produce, but is so far removed from it that it is hardly possible to imagine any natural cause or combination of causes which, even by redistribution, could have brought about the present state of things, if the gravel had been originally spread beneath a sea; fourth, the idea that this work was accomplished by either a salt or a fresh water sea involves of necessity the additional idea of the subsequent upheaval of the mountains, — an idea in favor of which there is not a particle of evidence, so far as I know, while there is the strongest proof against it. No sea, however deep, could spread such gravel in such a way through a range of not less than six thousand feet of altitude, over the flanks of even a submerged mountain-range, resembling in form and outline the Sierra Nevada of to-day.

Another theory which has been broached, I believe, with reference to the ancient gravel is that it was the direct work of glaciers. Aside from the fact that there is not a particle of evidence to support this idea, nor any positive evidence, so far as I know, that glaciers ever existed in any part of the Sierra at a period so ancient as this, nor any proof that even in their greatest development in later times they ever extended, as the ancient gravel does, to the foot of the range or anywhere near it, — aside from all these facts, it is probably a sufficient answer to such a theory to mention the fact, so palpable everywhere in the gravel region, that the whole character and structure and distribution of these gravel banks is utterly unlike anything that glaciers ever produce, but is exactly like what we may see produced to-day, wherever circumstances favor it, by streams of fresh and running water.

Against any idea that the gravel can have been distributed by floating ice, dropping its débris as it melted, may be urged almost all the objections which hold against seas and against glaciers combined, together with other and no less potent ones peculiar to itself.

In fact, wherever we go throughout the gravel region, the evidence constantly presented is cumulative and overwhelming, that these gravel banks are really the work of streams of fresh and running water. And so much as this has, indeed, been from the earliest times the accepted belief of all well-informed observers who have had any opportunity of studying for themselves the ancient gravel.[*] But yet the great questions remained unsolved, — Whence came these streams, and where did they go?

In answer to these questions, the theory which first gained extensive credence was that of the former existence of a single great river, flowing southeasterly for hundreds of miles along what is now the southwestern slope of the Sierra, and nearly parallel with the axis of the range, before the mountains were uplifted. And the supposed channel of this great stream, now filled with gravel, was called the "blue lead."

This theory was founded, of course, upon the supposed existence of a narrow belt of deep gravel banks, stretching across the country in a southeasterly direction. But later, when it came to be more generally understood that the deep gravel banks were not all confined within so narrow a belt as this theory would require, the theory was modified far enough to admit the existence in the gravel period of two or three large rivers with their branches, the courses of the main streams, however,

---

[*] Mr. Goodyear had not taken pains to make himself acquainted with what had been published in regard to the Gravel Region, or he would not have written the above paragraph. If the reader will compare what is stated in a previous chapter in regard to this point (see pages 66–72), he will see that the theory of the *marine* origin of the gravel deposits was the one adopted by most, if not all, of the professional geologists who had examined the Sierra Nevada previous to the beginning of the work of the Geological Survey, especially by Messrs. W. P. Blake, Laur, and Hector. — J. D. W.

being still supposed to be nearly parallel, and the general direction of their flow to be south-easterly.

Such, if I understand it correctly, is the general outline of the blue lead theory. It was necessary to state it, in order to present more clearly and definitely the objections against it. So great was the confidence in this theory that Richthofen has drawn important inferences from it [*] in relation to great disturbances of the Sierra Nevada, and more than one prominent gentleman has believed that he had actually traced the blue lead from Sierra County to Placerville, if not beyond.

The first objection to this blue lead theory lies in the fact that from the Middle Yuba to the Mokelumne River, including the best known and most important gravel mining region in the State, there is no well-defined belt or belts of deep gravel banks stretching northwesterly and southeasterly, such as must necessarily underlie the very foundation of this theory.

In one case, indeed, there is for a few miles — that is, from Gold Run to the Forest Hill ridge — a well-defined belt or line of deep banks, extending, not southeast, but nearly south. But beyond the points named in this direction it cannot be traced. Going south from the Forest Hill ridge, there are no more heavy banks of ancient auriferous gravel until we reach Placerville ; and at Placerville the character of the banks is in some important respects different from anything we find between Gold Run and the Forest Hill ridge.

Moreover, in going southeast from Placerville we find no more heavy banks until we get at least beyond the Mokelumne River. It must be noted that in this connection I am speaking not of thin sheets of gravel, or of smaller channels in which the gravel may be from one to six or eight feet in thickness, or even sometimes eighteen or twenty feet, but of great accumulations, from seventy-five to one hundred and fifty or two hundred feet in thickness, of metamorphic gravel, such as may be supposed to mark the courses of the greater of the ancient streams.

Again, in going north or northwest from Gold Run, the " belt " in question can be no longer traced, for another reason, namely, it becomes lost in the broad extent of country over which the heavy banks are spread from Smartsville, almost at the foot of the mountains, to Dutch Flat, 3,400 feet above the sea, and occurring at too many points within this range of altitude, as at San Juan, French Corral, etc., to justify any inference of a general southeasterly flow of the streams which gathered it. I speak still of heavy banks and not of lighter deposits and smaller channels. But the latter are scattered far and wide over the western slope of the Sierra beneath the volcanic matter, covering a vast aggregate area, through a range of more than 6,000 feet of altitude and over a breadth of from thirty to fifty miles of country, and distributed in such a way as can hardly with possibility, and certainly not with any approach to probability, be accounted for on the blue lead theory.

The single fact of the non-existence of any such definite northwest and southeast belt or belts of deep gravel banks as have been supposed, though perhaps insufficient of itself alone to positively disprove the blue lead theory, does nevertheless completely remove the very foundation on which it was built. But there are other and more positive objections to it. Before speaking of them, however, I will state incidentally another important fact connected with this theory. The term " blue lead " derived its origin, in part at least, from the blue gravel, and there has always been an intimate association in ideas between the two, many even going so far as to believe that the blue gravel was characteristic everywhere of the blue lead, and was not found outside of it. Now the fact is that the blue gravel is scattered here and there all over the gravel region, though its quantity is ordinarily the greatest where the banks are heaviest. It cannot be said to be characteristic of any particular channel or set of channels. And I am thoroughly satisfied that the peculiar dark-bluish color which gave rise to the name " blue gravel " is due, in the vast majority of cases at least, not to any original peculiarity in the character of the rocks from which the gravel was derived, nor to any peculiarity in special streams which once flowed in particular channels, but simply to local chemical action in the mass of the gravel itself subsequent to its deposition.

[*] See The Natural System of Volcanic Rocks, pp. 86 and 87.

Exceptions to this statement may be found, of course, and bodies of gravel are by no means rare whose general color to-day is largely due to the original character of the rocks from which they were derived. But the general rule is as stated above.

Another objection to the whole blue lead theory, and one which to my mind is insuperable, lies in the simple fact that this theory presupposes or of necessity involves a great subsequent upheaval of the whole Sierra Nevada, as it is evident that at no time since the upheaval of the range could any river have flowed far in a southeasterly direction along its southwestern slope. Moreover, the subsequent upheaval which this theory requires is not merely a moderate lifting, by which the altitude of a previously existing range might have been somewhat increased, but a grand and general uplift, essentially creating and calling into new existence the Sierra as a mountain-range; for, otherwise, the same objection would still hold good against the direction of its rivers. However low may be the crest of a range of mountains, if it exist as such at all, or if the country has even a very light general slope, the rivers do not run far in the directions of the contour lines.

Now I do not know of any evidence whatever to prove that any such upheaval as this has occurred since the gravel period. On the contrary, there is abundance of evidence which appears to me very strong as tending to show not only that no such upheaval has occurred, but also that no considerable disturbance of any kind whatever has occurred in the auriferous slates or in the mass of the range itself since the earliest gravel period. Indeed, I do not believe that since that time there has been sufficient disturbance of the bed-rock in any part of the Sierra where I have been to influence to any perceptible extent the general flow of even the smaller streams. The streams, both large and small, have changed their beds indeed, but it has been in another way and from another cause.

A fact which, in connection with other things, appears to me conclusive as evidence that no considerable disturbance of the bed-rock has occurred since the gravel period, is the almost total absence of local disturbances since that date. It is difficult to believe that any considerable general disturbance of such a range of mountains could occur at any time without producing at the same time numerous smaller and local disturbances in the way of faults, dislocations, bendings of the strata, etc.

If such disturbances as these had occurred since the deposition of the gravel banks, then these banks themselves would have been correspondingly disturbed and faulted, and their sections would exhibit it when they came to be worked. Such things do indeed exist, but they are extremely rare; and I do not know of a single instance where such a disturbance exists of so great a magnitude that it may not be reasonably attributed to earthquake shocks not greatly exceeding, perhaps, in violence some which are already on record as having occurred within the State. I may say that, in all my travels in the gravel region, I met with but a single case of such a disturbance (the one at Yankee Jim's) in which I considered the proof absolute and unquestionable that the dislocation was more recent in date than the gravel period, and here the amount of displacement was only some twelve or fifteen feet. Other possible localities are King's Hill, between Indian and Shirt Tail cañons, the Dardanelles, Flora's claim below Volcanoville, the Castle Hill ridge a little way above Georgetown, and Spanish Hill near Placerville, though I consider some of these as very doubtful, to say the least, and there has certainly been no great displacement in any case. Moreover, the bedding of the gravel is almost everywhere horizontal, or very nearly so; and, excepting at Yankee Jim's and perhaps two or three of the other localities named, I never saw this bedding distinctly broken or otherwise disturbed or bent to such an extent as could not be readily accounted for by the action of the shifting streams which deposited it, or else by special and wholly local causes, as in the cases where the gravel rests upon the limestone.

When these things are considered in connection with the facts respecting the present directions and grades of the ancient channels in the bed-rock, wherever these have been distinctly and definitely traced, and with a multitude of other facts in the distribution and structural arrangement of the gravel itself, all pointing in the same direction, I consider it proved beyond all reasonable doubt or question, first, that the blue lead theory is not only utterly inadequate to account for the facts, but is directly in conflict with them; second, that there has neither been any great upheaval nor any considerable disturbance of any kind in the bed-rocks of the Sierra since the gravel period; third, that the general course and direction of the drainage of the range was the same in the gravel period that it is to-day; and fourth, as shown by evidence already adduced, that the situation and general outlines of the larger drainage basins themselves were also then essentially the same as now, while the instances are not rare in which the resemblance may be traced even somewhat further yet into detail.

But though the general drainage system was thus the same, yet many of the most important circumstances attending its action were vastly different.

Before going further, I may simply mention one other theory, occasionally ventilated in the past, respecting the ancient gravel, and remarkable only for the peculiar facility with which its inventor compels grand rivers to follow the dictates of his fancy. This theory supposes that during the gravel period the Columbia River, by some mysterious and occult means, found its way to the western slope of the central California Sierra, and then flowed southeast for an indefinite distance, thus furnishing the requisite water, and constituting the great blue lead. I merely notice two objections to this theory: the first one is that the blue lead never existed; and the second is, that, even if it had existed, I know of no definite correlative facts which could furnish any better reason for identifying it particularly with the Columbia River than with the Colorado, the Mississippi, or the Ganges.

I propose now to consider in a few words the general surface and aspect of the western slope at the commencement of the gravel period, and then to sketch as briefly and definitely as possible the general outlines of what I conceive to be the history both of the gravel and the volcanic periods in the region of country where I travelled in 1871. After this, there are certain special questions which I will consider somewhat more in detail.

At the commencement, then, of the gravel period there were no great cañons in the gentle southwestern slope of the Sierra, nor in all probability any volcanic matter spread over the extensive region through which the ancient gravel is now distributed. If any volcanic matter then existed here at all, its quantity must have been extremely small, and all the evidence known to me points to the conclusion that there was none. Whatever may have been going on at that time along the summit or in the country to the east of the Sierra, or whatever may have been the exact date of the *first* volcanic outbreaks there, it was certainly not until long ages afterward, and near the close of the gravel period, that the volcanic energy of the Sierra reached its grander development, or that the materials ejected began to find their way to any considerable distance down the western slope. Indeed, I am much inclined to think that the earliest volcanic phenomena of the Sierra did not occur until very late in the gravel period; for there are many facts which appear to me indicative of a strong probability that, after the first outbreaks, the development of the volcanic energy was rapid as well as great towards the maximum of its grandeur, though we know that its after subsidence was slow and long. But more of this hereafter.

The southwestern slope of the Sierra Nevada, then, at the commencement of the gravel period was a broad, gently undulating, and moderately hilly country, having in all probability exactly the same gentle average slope which exists to-day southwesterly towards the valley,—a slope which for great distances ranges from one hundred to one hundred and thirty feet to the mile,—and having at the same time its greater drainage basins, already marked out where they exist to-day, by the occasional ridges and spurs which traversed, sometimes for considerable distances, the rolling hilly country, though rarely rising so much as a thousand feet above the general plane. High up towards the summit of the range, indeed, the country grew gradually rougher; and here there was probably more or less of cañon and of precipice, though it is not at all likely that even here there was anything comparable in magnitude with many of the modern cañons.

There may possibly, indeed, have been among the peaks occasional gorges two or three thousand feet in depth ; but it seems hardly possible that any such gorges then could have been continuous for many miles, and they were the results in all probability, not so much of denudation or erosion, as of the original upheaving forces which built the range and raised the lofty peaks.

Among these loftier summits the streams which slowly gathered the ancient auriferous gravel took their rise. From thence they flowed southwesterly across the smoother sloping country, the larger streams draining respectively nearly the same regions as now and in the same directions, and finding their way by the shortest practicable routes to the valley, which, however, may then have been a lake or sea.

These streams must have been rapid, too, as the grade was heavy for running water ; but the velocity of their flow seems rarely if ever to have approached that of the modern rivers in a flood, nor did they possess any such excavating power, because of the simple fact that they were not confined in the bottoms of deep and narrow gorges, but, wandering through the rolling country, were free, whenever their volume was large, to spread out over broader areas, their waters still remaining shallow, or even to take new courses through adjacent hills. They were also, without doubt, subject to freshets and droughts, as well as the modern streams, swelling at one time to floods which might suddenly transport for certain distances large quantities of gravel of varying coarseness, spreading it far and wide over their broad, pebbly beds, and again dwindling to far smaller dimensions, or even to little rivulets, which might wander isolated here and there over the expanse of gravel already accumulated, depositing only here and there little patches and bars of sand. When a certain depth of gravel (a depth varying of course with every little detail of the local topography) had thus been slowly accumulated in the original channel, a point would be reached when a flood would cause the stream at some point in its course to overflow its banks, and seek a new course through the low rolling hills. This new course the stream might retain after the flood had passed away, and then, in the new channel, the same process would be repeated ; and this would occur again and again, while yet the general direction of the stream would remain the same. It is easy to see how, in the constant repetition of this process, a stream in seeking a new channel would often strike its old course again farther down, or even intersect and cross it, running in some new direction. And thus it often happened that streams flowed in new directions over bodies of gravel previously accumulated, either by themselves or by some other streams. The result of such an incident might be either a simple continued accumulation of gravel over the same spot, but coming from a new direction, or it might be first the excavation of a new channel to a greater or less depth through the previously gathered mass, and then, when this new channel came to be filled again, it might be filled either with material perfectly similar to that through which it had been excavated, or with something of entirely different character.

This, it appears to me, is the general outline of the manner and means by which the ancient auriferous gravel was slowly accumulated, its materials — that is, its gold as well as its boulders and pebbles and sand — being derived from the bed-rock over which the streams ran, and from the quartz, etc., which it contained. The longer such a process continued, the more the streams would wander, the greater would be the quantity of gravel accumulated, and the greater the actual area over which it would be spread. I think this process did continue until the advent of the volcanic era, when the enormous quantities of volcanic matter, which were spread far and wide over the western slope of the Sierra, buried beneath their mass the previously accumulated gravel, as well as the rocks from which it was derived, and were, I think, the chief and direct cause of the final ending of that gravel period.

Yet the process of the accumulation and distribution of this volcanic matter, too, was slow, complex, and long. The boulders of white lava occurring in the gravel near Placerville, beneath the horizon of all the heavy accumulations of volcanic matter in that vicinity, and also the not infrequent intercalation of sheets of auriferous gravel between heavy masses of volcanic matter high above the bed-rock, prove that in some sense the gravel period passed gradually into the volcanic, and that there were long interruptions between successive deluges of volcanic matter, during which there was time for gravel to gather and spread over the earlier volcanic beds. The two periods are nevertheless perfectly well marked, and it is evident almost everywhere that the auriferous gravel had been slowly gathering for a very long period of time before the first volcanic matter came. The occurrence of the white lava boulders in the lowest gravel, as at some localities in the vicinity of Placerville, is a rare and abnormal thing; and therefore, though this gravel be the lowest at these places, and rests immediately upon the bed-rock, I am nevertheless disposed to consider it of somewhat more recent origin than the great mass of the gravel which underlies the volcanic matter. The general fact is that, although metamorphic gravel may here and there be found interstratified with the volcanic matter at all heights from the bottom of the latter up to the present surface, yet there is a horizon below which nothing volcanic can be found in the banks, and everything between this horizon and the surface of the bed-rock is metamorphic gravel.*

As already repeatedly stated, every particle of volcanic matter in the region through which I travelled in 1871 is fragmentary. There are no lava-streams, and not a pound of all the volcanic matter that I saw between the Mokelumne River and the North Fork of the American (unless it be the Jackson Butte) appears to rest to-day where it was when it last cooled from igneous fusion. I think the whole of it has been transported and distributed in one way or another through the agency of water; and I consider it certain that, at least during the earlier portion of the volcanic era, a very large proportion of it was brought down the mountains and spread over the country by means precisely similar to those which had previously distributed the gravel, that is, by shallow and shifting streams of running water; and the immense quantities of smoothly washed and perfectly rounded volcanic gravel which exist near Placerville, and also at sundry localities farther south and southeast, prove also that the same agency was still active at a far later period in the history. But there are also vast accumulations of volcanic breccias whose material has evidently not been subjected for any considerable length of time to the action of running water. And there are also enormous quantities of volcanic matter which is neither a breccia nor properly speaking a gravel, but which I have often called conglomerate, and which is known all through the southern part of Placer County as simply "cement," or sometimes "mountain cement." It is filled with hard boulders of volcanic rock of all sizes up to many tons in weight, these pebbles and boulders being generally but partially rounded; and the finer material which constitutes their matrix is dark colored, and has apparently been produced by the trituration and pulverization and decomposition of rocks of similar kinds to the boulders which it encloses. This material can hardly have been distributed, I think, by the ordinary action of ordinary streams, but there is no lack of agencies by which both its distribution and that of the breccias may have been brought about. The volcanic energy which once displayed itself on so grand a scale in the High Sierra can hardly have been unattended by a class of phenomena so common in active volcanic regions now. The sudden melting of great masses of snow; the occasional ejection from beneath the surface of bodies of water, either hot or cold; the sudden deluges of rain which are among the meteorological effects of eruptions; the quick, overwhelming flow or the slow, majestic creep of great volumes of volcanic mud, etc., — all these, and more, are probably among the agencies which have contributed from time to time to the distribution of these materials. The scattering of ashes in showers through the air may also of course have aided in the work, although the probability is that little if any of the material so distributed remains at present where it fell, having been quickly washed off the hillsides by the rains and redistributed by the streams.

Thus it is probable that the slow and ordinary working of the streams was from time to time interrupted by grander paroxysmal action, during which great quantities of material were rapidly spread over extensive areas, while during the intervals between these paroxysms the streams again resumed their action and their general course, though rarely perhaps following exactly the same

---

* Wherever I use the words "metamorphic gravel" without further explanation, I mean by them not gravel which has been metamorphosed since its deposition, but simply gravel which has been derived from the metamorphic rocks.

channels as before. The working of the streams themselves was doubtless as varied and complex as the work of such shifting streams, on such a slope and in such circumstances, could be, and excavation along some portions of their channels was probably almost as constant a fact as accumulation was at others, though the localities of both were constantly shifting. Yet the prevailing tendency was still to accumulation, until the era of great eruptions had passed away, and the volcanic period was drawing toward its close. Then, and not till then, did the excavation of the great modern cañons in reality begin.

It is a common idea that, because the volcanic matter upon the western slope of the Sierra is now found almost exclusively upon the crests of the present ridges, the capacity of this volcanic capping to resist erosive and denuding action is greater than that of the hard metamorphic slates themselves, and therefore that this capping, wherever it occurs, has acted as an effectual protection to the bed-rock beneath it against the erosive action of the modern streams. Almost a necessary corollary to this idea was the inference that the areas now occupied by the modern cañons had never been covered by any considerable quantity of volcanic matter and some have gone so far as to believe that the volcanic matter only flowed along the beds of ancient cañons, between which, and exactly over the modern cañons, then existed great and high-projecting bed-rock ridges, so that since the volcanic era the whole topographical relief of ridge and cañon on the western slope has been completely reversed, — exactly inverted.

Now I believe that nothing could be further from representing the real facts and the actual history of the case, as a general rule, than the above suppositions are. There are instances, indeed, — like the Table Mountain of Tuolumne County, — where actual lava-streams have solidified into hard and compact basalt, and wherever this has occurred the basalt is of course far harder than the slates, and has afforded an effective protection to them. But even in this case, though it is self-evident that the molten lava must have followed in its flow the lowest channel it could find, it does not follow, from this alone, that the adjacent country need have been much, if any, higher than what was required to simply confine the stream within its banks. At all events, there is no necessity for assuming on this account, without other facts to prove it, that above the plane of the surface of the lava-stream the adjacent country towered in high mountain masses where now the whole surface of the region is hundreds of feet below the level of the flow.

But again, I am satisfied that such lava-streams, which have flowed for any considerable distance down the southwestern slope of the central and southeastern Sierra, are few and far between. Indeed, from Plumas County southeast to Mariposa, or even to Walker's Pass, throughout the whole region over which the ancient auriferous gravel is distributed on the southwestern slope between these limits, a region in places from thirty to fifty miles in width, I do not believe that the total volume of the lava-streams constitutes even so large a proportion as one per cent of the vast aggregate of volcanic matter which is spread over the country. It is not likely that there is a lava-flow anywhere below the altitude of 6,000 feet on the southwestern slope, between the Central Pacific Railroad and the Mokelumne River. At least, I did not see one that I could recognize as such. And of all the varieties which help to make up the enormous aggregate of fragmentary volcanic matter spread over that region, there is not one, so far as I have ever seen, that would offer a greater resistance to erosive action than the bed-rock itself. Of all these varieties, so far as my observation extends, the hardest are some of the breccias. Yet even the hardest of these would be eroded by running water with greater ease and greater rapidity than even the upturned edges of the greater portion of the hard, tough bed-rock would; while vastly the greater portion of all the volcanic matter here is far softer, and would be eroded with much greater ease. I believe, therefore, that, instead of the volcanic matter's protecting the bed-rock beneath it, the very ease and rapidity with which the volcanic matter was first eroded has, in many parts of the country at least, contributed in no small degree to increase the subsequent rapidity of erosion of the bed-rock itself in the excavation of the tremendous modern cañons. The basin of the Middle Fork of the American River, east and southeast from Michigan Bluff, is a good illustration of this.

Few streams in the Sierra have deeper cañons than the various forks of the American River.

In this portion of the basin of the Middle Fork, also, there can be little doubt that the volcanic matter once covered the areas of the present cañons as well as the tops of the modern ridges. All this region at the close of the volcanic period appears to have been a smooth and gently sloping plain, consisting mainly of volcanic débris to depths of from 300 or 400 to 600 or 800 feet, underlaid by the auriferous gravel resting on the rock. When now the streams came, after this, to follow once more their quiet courses across the surface of this loose material, they would naturally excavate little channels in it ; and the supplies of new material from higher in the mountains being partially at least shut off by the cessation of the eruptions, the tendency of every freshet afterwards would be to deepen more and more the channels already begun. From this time forward the excavation would proceed more and more rapidly till a depth of channel was reached such that the greatest floods could no longer raise the streams sufficiently to make them overflow their banks ; and, finally, when the streams in their gradually sinking beds had reached the surface of the solid rock, they would be already confined in the bottom of narrow cañons from 300 to 800 feet deep, with sides as steep as the material would lie, and would have reached their maximum of excavating power.

When the volcanic period passed away, and the streams were first left free to choose anew their courses over the newest surface of the ground, the slightest inequalities in the form of that surface would of course suffice to influence or determine the local courses of the streams. These courses would therefore be chosen and determined without any reference whatever to the network of older channels, now so deeply buried under them, beyond the single fact that they would drain nearly the same regions of country, and in the same general direction. The newer channels would therefore cross the courses of the older ones at innumerable points and in every possible relative direction, cutting them up into detached pieces and patches of network, whose sections, when afterwards exposed, would appear as complex and intricate as the work and the history of the earlier streams had been.

The courses of the new streams over the smooth surface might have been at first easily changed, and it seems not at all unlikely that, if man had been here then, it might have been even within his power to have guided their courses, and thus remodelled the whole present system of cañons in this portion of the basin of the Middle Fork. But when the streams, in a later period of their history, reached the surface of the solid rock beneath, then they had already gathered a power which in flood-time nothing could resist, and which even projecting knobs and spurs of most compact and solid rock could hardly turn aside from its direct and onward course. From that time forward the excavation of the solid rock itself went on, as it does now, varied only, so far as we know, by variations in the quantity of water, due to climatic changes attendant upon the glacial epoch, etc. When the streams were low, they possessed, like the present ones, little or no excavating power ; but whenever the floods came, and they were swollen to great and roaring torrents, carrying not simply sand and pebbles, but even great boulders of tons in weight, thundering along their beds, then the wear was great and the excavation rapid.

The actual direct work of the streams themselves has always been confined, of course, to the bottoms of their cañons. This is the work which has deepened the cañons, while their corresponding widening has been chiefly due to another cause, namely, the land-slips and slides, which from time to time rushed down from the too steep sides, often filling the bottoms of the cañons with enormous masses of débris, which, however, were quickly swept away.

Such is a sketch of what I conceive to be the general outline of the history of the gravel and volcanic periods in this portion of the country, together with the subsequent excavation of the modern cañons. I now propose to consider further some special points of interest in connection with the general subject.

The axis of the Calaveras County Bear Mountain range lies northwesterly and southeasterly, and I know of no volcanic matter or ancient gravel among the hills to the southwest of this ridge. The length of this ridge is some twelve or fifteen miles ; and if it be true that there is neither gravel nor volcanic matter in the hills southwest of it, while both are found to the northeast, this would be an evidence of the existence of that ridge in the gravel period, and that it acted then as a barrier which the streams could not cross, but were obliged to pass around.

It is the general fact throughout the gravel region,—to which there are indeed occasional though rare exceptions,—that in any given bank the richest portion of the gravel, as well as that containing the coarsest gold, is that which lies at the bottom, immediately upon the surface of the bed-rock. In order to account for this fact, the idea has been advanced that it was not originally so, but that the gold was distributed through the upper as well as the lower portions of the bank with much greater approach to uniformity, so that there was at first but little difference in richness between the gravel immediately upon the surface of the bed-rock and that higher up towards the top of the bank. Afterwards, however, as is supposed, the coarser particles and lumps of gold, by virtue of their high specific gravity (and apparently also by virtue of some further and mysterious faculty which they must also in that case have possessed, of pushing aside the sand and pebbles in a firm and compact bank), gradually settled down through the underlying gravel until at last they reached the surface of the bed-rock, and so became concentrated there. Now this explanation appears very unsatisfactory to me, because, in the first place, I do not believe that, high as is the specific gravity of gold, it is sufficient to cause particles and pellets of it, no larger than those which are commonly found in tunnel and hydraulic mining, to displace the firm gravel beneath them, and actually thus sink through it, however slowly, to the rock; and in the second place, because I further notice that it is also a general rule, wherever very large boulders occur in a bank of gravel, that these, too, are at the bottom, either immediately upon or very near the surface of the rock. It is very rare indeed to find very large boulders at any great height above the bed-rock in banks of metamorphic gravel; and for these boulders, at least, the sinkage theory will not do. I do not deny, of course, that earthquake shocks, or anything else which severely jarred the ground, might cause large lumps of gold, if such were imbedded in the gravel, to settle a little. Nevertheless, I do not believe that any great portion of the gold has sunk very far through the mass of gravel in this way or in any other. I can see but two plausible methods of accounting for the almost universal fact. The first of these is by supposing that the earliest gravel streams flowed at first for a long period of time over the naked bed-rock, deriving large quantities of auriferous débris from somewhere, though without excavating their own channels to any considerable depth in the rock, and at the same time without accumulating any considerable quantity of gravel in their channels, the rocky débris, wherever it came from, being thoroughly disintegrated or ground up, and carried on by the streams, while the gold which it contained was gradually concentrated in their beds,—the era of accumulation only commencing later in the history. This supposition, if admissible, might account perhaps for the bottom gravel being the richest in gold; but there are serious difficulties in the hypothesis itself, aside from the fact that it does not account in the least for the general absence of all large boulders from the upper portions of the deeper banks of metamorphic gravel, while they are plenty enough in the same banks immediately upon and near the bed-rock.

A much more plausible, and I am strongly inclined to think the true, explanation of both these facts lies in the hypothesis (if, indeed, it be not something better already than a hypothesis) that, as a general rule, neither the coarsest gold nor the largest boulders were ever transported very far from the spots whence they were originally derived. It is easy to see how, if this were true, the two facts named would be necessary consequences of it. For as the gravel at any particular locality, or in any particular channel increased in depth, so also would it increase in lateral extent, and more especially would the length of channel-bed already covered rapidly increase, and thus would increase with exactly equal rapidity the distance through which any additional material must roll over the surface of the gravel already accumulated before it could reach a position over the already deeply buried spots.

But there is another point in connection with the gold of the ancient gravel, which is far more difficult to explain. It is the uniform testimony of mining men throughout the gravel region that, as a general rule, the coarser the gold from the gravel banks, the less is its value per ounce. Now, if this statement merely referred to the value of the crude gold before melting, it might easily be due to the fact that large pieces of gold are liable to contain a higher percentage of bits of quartz or other rocky matter than are the finer particles. But it refers to a very different thing. The gold is not assayed at all until after melting. And the fact as stated is, that the coarser gold actually contains on the average a higher percentage by weight of silver than the finer gold does. I know of no explanation of this, and simply mention it as an interesting and curious fact, whose significance remains to be studied hereafter.

I will also note here, though it be a repetition, the fact that among all the hydraulic mines I visited, and all the miles of tunnels and drifts, etc., that I traversed in 1871, I never yet saw a spot where the bed-rock was exposed in these mines, and was hard enough to show anything of its original surface, where it did not exhibit unmistakably the wearing and smoothing effects of water. Whether the rock were high or low, it was always water-worn. In fact, the water seems to have run at one period or another over almost if not quite every square foot of the bed-rock that now lies buried under either the gravel or the volcanic matter.

I may as well say here a few words upon the forms of the depressions and furrowings in the water-worn bed-rock, and the extent to which they may be relied upon as indications of the direction of flow of the water over it. Both Mr. Pettee and Mr. Bowman called my attention to these things before I started out in the spring of 1871, and if I understood them aright, they were then inclined to place considerable reliance upon a peculiar and not uncommon form of depression, which Mr. Pettee once not inaptly characterized as slipper-shaped, that is, an elongated depression or trough, with a steep descent at one end and a gradual rise towards the other. I then understood that wherever these slipper-shaped depressions were numerous, their steeper ends were almost all turned in the same direction; and the question was: In such cases did the water flow in the direction of the less steep slope, or *vice versa?* I do not know whether they were satisfied upon this point or not; but I afterwards paid particular attention to these depressions at Iowa Hill, and numerous other localities, and though I could find plenty of the slipper-shaped depressions, yet I could never find any uniformity in the directions towards which their steeper ends were turned. There was often, indeed, a general approach to parallelism in the directions of the axes of the elongated depressions, whatever might be the forms of their ends; but if, for example, the direction of the axes was northeast and southwest (which is the commonest direction), it generally happened that some of the slipper-shaped depressions had their northeast ends the steepest, and some the southwest ends, and I could never tell, at any one locality, which of these two were the most numerous. My conclusion therefore was that the shape of the ends of the depressions was of no special value as an indication of the direction of the flow.

Moreover, I did not find well-defined slipper-shaped depressions by any means the most common form. So far as my own observations went, the most common form of depression, and the only one which seemed to me anything like a regular and uniformly intelligible result of the direct action of the water, was a simple furrow. In various localities these little furrows were very numerous. They vary in length from a foot or less to eight or ten feet, and in width and depth from two or three inches to a foot, the width, however, being generally considerably greater than the depth. They are not often either exactly straight or regular. But their little irregularities are always smoothed and rounded, and there is generally a tolerable approach to parallelism in the average directions of their axes. There is no regularity to be traced in the forms of their ends so far as I could see. Where I first noticed these little furrows they were mostly very nearly parallel with the strike of the slates, and I had of course a suspicion that the bedding of the rock might have had something to do with their origin. But I afterwards found them almost at right angles with the stratification, and at other localities crossing it obliquely, and still preserving their general parallelism and characteristic appearance, and I therefore concluded that they were really due to the action of the water in such a way as to be indicative of the direction of its flow. I afterwards noticed the same thing in the bed of the present Dry Creek, below the village of Amador City. I think that where these little furrowings are numerous, it is safe to infer that the water flowed in the general direction of their axes; but in which of the two opposite directions thus indicated, I do not know that there is any means of determining from the forms of the furrows alone. For this, recourse must be had to the general grade of the rock-surface.

With reference to the value of barometric determinations of the grades of the ancient channels, as also with reference to the extent to which these grades themselves, even if accurately determined with the spirit-level, can be relied upon as pointing out the exact direction of flow upon the bed-rock, I will make a few remarks.

The grades of the earliest gravel channels, though they undoubtedly varied in certain localities between the widest limits, and there were probably sometimes falls, that is, moderate cascades or cataracts in their courses, yet did not *generally* exceed three feet in a hundred, and from two to two and a half or even less was probably a commoner grade.

Now, in the beds of the ancient streams there are many irregularities, and it is not at all uncommon, in the bottoms of the greater channels, to find irregular depressions and projecting knobs and ridges of bed-rock, with differences of twenty or twenty-five feet in height, and sometimes more, between the highest and the lowest points, though these are close together and in the very bottom, or even in the central portion of the deep, broad channel. Moreover, these irregularities are often extremely numerous. It is evident, therefore, that even the most accurate determinations of relative heights could generally be relied upon as definitely determining the direction of the flow only when the extreme points were taken at a considerable distance apart. And with the errors to which the barometer at best is liable, and the rarity with which any single ancient channel can now be certainly traced for any considerable distance in the bed-rock, it is evident that its results can in ordinary cases be only of general value, and that the instances in which it alone can determine with certainty the original direction of flow in particular channels are comparatively few and far between. But believing, as I do, that there has been no considerable and generally no appreciable disturbance of the bed-rock since the gravel period, I also believe, of course, that the present grades of the ancient channels, wherever the latter can be distinctly traced, — and a sufficient distance is taken into account to determine with certainty the direction of their fall, — are generally safe and certain evidence as to the direction of the ancient flow.

It should have been mentioned in another connection that thin streaks of rich gravel, with coarse gold, do sometimes occur in the banks at considerable heights above the bed-rock. But such cases are somewhat rare, and I believe that, where they do occur, they were spread over the surface of the older gravel by new streams coming from new directions, in which the bed-rock generally was not far off.

Another thing which, though by no means universal, is nevertheless an extremely common fact throughout the ancient gravel hills, is that the gravel is richest upon the rim-rock. By the "rim-rock," however, in this connection, is not meant the actual rim of any particular ancient channel or channels, but simply the rock which underlies the gravel along its edges in the spurs and hill-sides fronting the present gulches and cañons. So common is this fact that it is often stated by the miners as a general if not almost the universal rule.

There can be little doubt that this impression among the miners is *partially* due, at least, to the fact that on the rim-rock the gravel is necessarily shallow, however steeply the banks may rise behind it, and it therefore can be more cheaply worked, and at far less cost per square yard of bed-rock surface, than farther back in the hill. On this account alone the profits of working the rim-rock would, of course, even on the supposition that there was no perceptible difference in the actual richness of the gravel, be much greater than those of working farther back beneath the hill. Yet, considering all the testimony, it can hardly be doubted that the rim-rock gravel has, oftener perhaps than otherwise, proved actually somewhat richer in gold than that farther back in the hills; and if this be indeed the fact, it is a difficult thing to account for. We can easily understand, indeed, that, even on these steep hillsides, the rains, etc., in washing away the upper portions of the sloping banks, would have some tendency to concentrate and leave behind, upon or near the rim-rock, a portion at least of the gold which they contained. But this does not seem adequate to account for half the statements made, and I know not what other cause to assign.

It is not uncommon to see the bedding of the gravel in the lower portions of the banks conform here and there, within certain limits, to the contour of the rock on which it rests, the strata curving gently upwards as they approach the rising rock. But I do not consider this as indicative of any disturbance since its deposition, except, perhaps, in certain cases on or near the rim-rock where land-slides may have affected it. In a little miner's reservoir, a perfect section of which was afterwards exposed by the breaking of the bank through its centre, I have seen the laminæ of mud and silt deposited in it, which were perfectly horizontal across its central portions, bend sharply upward — at angles as high, I think, as 15° or 20° — on approaching the banks at the edges of the reservoir.

It is worthy, perhaps, of special notice, that what has been called, rather inaptly, I think, the "typical form" of the ancient gravel ridges, that is, the form which appears to be so well exemplified in the Tuolumne County Table Mountain, and is illustrated in cross-section in Geology I. (page 248), is a thing of extremely rare occurrence. In fact, this Table Mountain is the only case I know of in which a lava-stream has followed for any considerable distance the course of one of these ancient channels, and covered it with a continuous protective capping of solid, compact rock, so that the axis of the ancient channel corresponds, after all the subsequent denudation, with the axis of the present ridge. And it is evident that, except in the case in which these two axes do essentially coincide, the typical, or at least the most frequent and characteristic forms of the ridges as exhibited in cross-section, cannot be that which is shown in the section above referred to. This form of a single great trough in the bed-rock, with rim-rocks rising high on either side, and running continuously parallel with the crest of the present ridge, is, indeed, I suppose, the typical or characteristic form for the cross-section of that Table Mountain; but elsewhere it is rarely seen, and only for short distances, and probably as the result of accident in the relative courses of the ancient and modern streams. The width of the ridges, so generally and heavily capped with volcanic matter in the gravel region, varies from a hundred yards or less to several miles, and the surface of the bed-rock, as shown in the cross-section of one of these ridges, between the edges of the great cañons, would generally exhibit simply an irregularly undulating wavy line. This line would occasionally present tolerably distinct cross-sections of channels (generally of no great width or depth), and occasionally long, smooth stretches (where the line of cross-section might happen to strike and run for a little distance parallel with the bed of one of the earlier channels), while elsewhere it would be varied by knobs and irregularities of every kind, only with nothing sharply angular, but all smoothed and rounded by the action of running water. In fact, the ordinary form exhibited would be precisely that of the profile of a gently undulating, rocky country, over every foot of which the water has at one time or another held its course. And in making this statement I am not theorizing, but simply giving the fact, as shown by the vast extent of drift and tunnel mining wherever I have been throughout the gravel region. Yet it must not be by any means supposed that all the bed-rock buried beneath the volcanic matter is so smooth as the preceding statement alone might seem to imply. Even in the broad basin of the Middle Fork of the American, where the above statement is eminently applicable, there were occasional bed-rock hills, rising several hundred feet above the beds of the earlier streams; and sometimes these ancient hills rise now to the surface of the present volcanic capping, and so present themselves to-day as patches of bed-rock covered with only a partial coating of soil, and destitute alike of volcanic capping and of any considerable amount of gravel.

A common and striking feature almost everywhere of the ancient gravel, and also to a great extent of the volcanic matter which overlies it, is the vast amount of chemical action which has taken place in the banks since they were deposited. The more or less complete decomposition and thorough softening of the hardest metamorphic and volcanic boulders, the formation at many localities of great quantities of semi-opal, which encrusts and fills the interstices between the pebbles, the frequent silicification of fossil-wood, its transformation at other localities into sulphuret of iron, the irregular cementing together of the originally loose materials, and the formation in many places of the large quantities of sulphuret of iron which often contributes so largely both to the hardness and to the peculiar color of the "blue gravel," are all examples of its effects. So also is the commonest condition of the bed-rock immediately underlying the gravel, which, far oftener than otherwise, is more or less thoroughly softened and decomposed.

I do not think this chemical action has been in general either the effect of thermal springs, or of anything else which has issued from the bed-rock beneath. Nor do I see the necessity of supplying any other agency to accomplish it than filtering atmospheric waters. The atmospheric waters, in filtering through such masses of volcanic debris as cap these ridges, would undoubtedly slowly attack them and derive from them certain substances which they would carry forward in solution, and which might be peculiarly adapted to act upon the metamorphic material beneath. When we consider also the presence of the organic matter in the banks, and the very complex chemical constitution of the banks themselves as a whole, I think that, with the atmospheric waters and the lapse of ages, we have all that is necessary to account for a vast amount of chemical action. Even the metamorphism (for it is really such, I think) of some of the fine and white volcanic ashes, included in the popular name "white lava," into a compact, almost clinkstone-like material, enclosing small crystals of glassy feldspar, may probably have been effected by similar means. And, in connection with this point, it is interesting to note the fact that at certain localities, as, for instance, in the bluff near the toll-house above Smith's Flat on the Placerville and Carson road, at the Jackson Valley Buttes, etc., the sharply granular white lava has become incrusted on its weathered surface by a crust from an eighth of an inch to half an inch thick, of perfectly similar compact material, which sometimes has almost a semi-vitreous lustre and appearance, and which can have been produced *in loco* only by a change in the texture of the material.

In this connection it may be well to notice an impression which is not uncommon among the miners, that "the gravel is always of the same color as the bed-rock on which it rests," that is, that if the bed-rock immediately underlying the gravel be hard, undecomposed, and dark-bluish in color, then the gravel immediately upon it will be "blue gravel"; but if the bed-rock be decomposed and white, or of any light color, stained reddish or yellowish, then the gravel upon it will be red or yellow. Now it is true, indeed, that this is very often the case, and I have no doubt that in many instances both the color of the bed-rock and that of the gravel upon it have been due to the same cause, namely, the character and extent of the local chemical action in the banks. But it is by no means always so, and there is certainly no lack of instances, as at Yankee Jim's, where red or yellow gravel rests immediately on dark-blue bed-rock.

A curious question relating to the western slope of the Sierra is the origin of its diamonds. They have been found in circumstances which prove that they came from the ancient gravel, and therefore originally, in all probability, from the bed-rock itself. But from what particular variety of rock, and under what circumstances, were they formed? Also, why is it that none have yet been found (so far as I know, at least) farther northwest than White Rock, near Placerville, while by far the greatest masses of the ancient auriferous gravel are northwest of the North Fork of the American?

In may be noted that not infrequently, at different localities not far removed from each other, the gravel is so entirely different in both its lithological and physical characteristics that this fact would be sufficient of itself alone to prove a difference of local origin. This is eminently the case at Muletown and Irishtown, northwest of Ione Valley.

I do not know how to account for such large accumulations as exist at some localities of nearly pure quartz gravel, except upon the supposition that this gravel has not travelled very far, and that the enclosing rocks from which it came were soft enough to disintegrate and be carried off by the water as sand and mud.

It is an interesting fact that, so far as I was able to learn, there are along the edge of the Sacramento Valley, skirting the foot-hills between the American and the Mokelumne rivers, no large accumulations of gravel at all, except in the immediate vicinity of the points where the present streams leave the mountains. Yet at every such point there is more or less of it, and the quantity is greatest near the largest streams. This appears, therefore, to point very strongly towards the conclusion that the main streams issued from the mountains in the gravel period at very nearly the same points where they do now.

The bed of fine volcanic breccia underlying a heavy bank of metamorphic gravel in the hills near Folsom is a fact of much interest, and possibly of much significance in some directions, which will better appear when it shall be further investigated.

It appears to have sometimes happened during the excavation of the modern cañons that a gulch, — like the little Golden Gate Cañon at Damascus, — after being excavated to a certain depth, has had its bed filled by a heavy slide, and has then excavated a new channel alongside of it, leaving its old bed still buried in the hillside.

It is certain that, unless it be in very rare instances, as, for example, that of the Jackson Butte, the belt of volcanic activity in this part of the country never extended to any considerable distance down the western slope of the Sierra. I saw no signs of volcanic action in any part of the country which I traversed in 1871.

With reference to any arguments for a change of grade or other disturbance of the bed-rock on the western slope of the Sierra since the gravel period, based upon the assumption that any particular amount of grade is necessary in order that gravel may accumulate, and the conclusion which might be drawn from them that the alternate prevalence at different times of denudation and accumulation has been determined by changes of grade in the mountains, I will only remark, in addition to what I have already said upon the subject of such supposed recent disturbances in the Sierra, that any such argument as this appears to me extremely ill-founded, because of the fact that the degree of slope upon which gravel will accumulate and rest varies under different circumstances between very wide limits indeed, and is dependent between these limits almost entirely upon the character and force of the streams which carry it. We all know that a torrent like the American in flood would move boulders of considerable size over grades of less than fifty feet to the mile, while a stream of 300 or 400 miner's inches of water, when allowed to spread, as it will on leaving a flume, to a stream four or five feet in width, can only just roll the ordinary débris from a hydraulic claim down a slope as high as twelve degrees, which was the actual slope of one pile of tailings, which I noted under such circumstances.

It has also been often supposed that the quantity of water in the Sierra during the gravel period must have been vastly greater than it is to-day in order to enable it to accumulate such immense quantities of gravel. I do not see any necessity for this supposition either. It may have been so, or it may not. But other things being equal, an increased quantity of water would produce an increased tendency to excavation. Whereas, given a moderate quantity of water, and given the alternations of drought and freshet, it is easy enough to see how accumulation might have gone on, even with no more water than is furnished by the present climate, and how, if it did go on under such circumstances, it would be productive of exactly that kind of structure which we see in the banks, and, with time enough, might accumulate even greater quantities of gravel than were accumulated. But there was certainly a far later period, namely, that of the glaciers, during which the quantity of water in the Sierra was greater.

The low, terraced hills, which form a belt of greater or less width, skirting the foot-hills of the Sierra proper for a long distance southeasterly from Folsom along the eastern margin of the Sacramento and San Joaquin Valley, present some interesting facts and problems. The whole mass of these hills has undoubtedly been derived at some time from the slopes of the Sierra. Moreover, a very large portion of their volume has been gathered since the commencement of the volcanic era, as is proven by the very large extent to which volcanic materials enter into their structure. Furthermore, there are in the history of these hills two well-marked periods, — the period of accumulation, and the period of erosion. But whether the material of these hills continued to accumulate until the close of the volcanic era, and then their erosion commenced simultaneously with the beginning of the excavation of the modern cañons in the mountains, or whether the two localities were entirely independent of each other as to the times when accumulation ceased and erosion began, I do not know.

The character as well as the extent of the denudation which has taken place on the western slope of the Sierra, since the close of the volcanic era, has evidently varied considerably at different localities, and probably owing often to special causes which we do not yet understand. For example, in the basin of the Middle Fork of the American the denudation seems to have been chiefly limited to the excavation of the modern cañons. At least, I do not know at present of any evidence to prove that the crests of the broader ridges and volcanic-capped areas here have been perceptibly lowered since that time. But in the country adjacent to the Tuolumne County Table Mountain it appears that broad areas have been lowered for hundreds of feet at least, or sufficient to leave that Table Mountain prominent above the surrounding region.

In the eastern portion of the valley the volcanic materials, in the shape of pebbles, sand, etc., are scattered many miles beyond the foot-hills of the mountains proper, and indeed it seems probable that they may be found almost anywhere, to a greater or less extent, until we reach the eastern margin of the true adobe soil which covers all the central portion of the valley; and one point of interest in connection with the borings from that artesian well at Stockton would be to discover whether there is any volcanic matter here beneath the adobe soil, and if so, to how great a depth it may be found, as this would give interesting information as to how great a depth of filling has accumulated in the central portion of the valley since the commencement of the volcanic era.

It may be noticed as an interesting fact that in the vicinity of Placerville wherever volcanic gravel and breccia occur together, the breccia always overlies the gravel, while in the region about Sutter Creek and below there the reverse is generally the fact.

One point of no little general interest and importance in connection with the distribution of the volcanic matter over the western slope of the Sierra is the fact that the earliest volcanic beds consist in general of the finest materials. Indeed, at many localities the first warning of the approach of the volcanic era seems to have been a sudden thickening of the waters of the streams with volumes of finest mud, of which the so-called "chocolate" at Deadwood and elsewhere is an example. Moreover, the white lava, enormous as its quantity is at many localities, whenever it occurs in close association with other and coarser volcanic materials (which is oftener than otherwise the case), always underlies them, forming the lowest volcanic stratum. The same rule holds with other varieties of fine volcanic sediment and ash. Wherever they occur, their position is low down, at or near the bottom of the volcanic matter.

This fact is especially conspicuous with the white lava, as its quantity is so great. These finer sediments are generally entirely free from boulders, and almost entirely free from pebbles of any kind. They do indeed contain occasionally small pebbles, which, however, are just as often of quartz or metamorphic rock as they are of anything volcanic. It is by no means easy to understand exactly how such enormous masses of this white lava have been so perfectly segregated as they are from all other kinds of material. It has certainly, I think, been done by water. But why should not the water which transported such vast quantities of this material, fine and light as it doubtless was, have brought more frequently little pebbles or pellets of some kind with it? Was it indeed brought here as sediment, freely suspended in and deposited by running water at all? Or did it come in the form of semi-fluid mud-flows? (It could hardly have been viscid at all. It is too sharply granular for that.) Or was it first distributed over the country in the form of showers of ashes through the air, and afterwards collected in the valleys by the rains? The last, indeed, seems hardly likely; for in that case it is difficult to see how so few pebbles could have been mixed with it in the process. But there are some difficulties, also, in either of the first two hypotheses, and nothing seems altogether satisfactory in explanation of it. I feel sure, however, that it came here in some way through the agency of water, and in the condition of volcanic ash, and not in a state of igneous fusion; for not only does the whole texture and appearance of the material speak strongly against the last idea, but I consider the presence of even the few little pebbles and bits of quartz and metamorphic slates which it contains enclosed, as proof positive against it. Some of them are smoothly water-worn, and some of them almost angular, but all clean and showing no effects of any high degrees of heat.

Whether the fact that these fine materials form so largely the lowest beds of volcanic matter here to-day is any proof that they were the earliest ejected among the products of the volcanic action of the Sierra, I do not know. I can easily understand how the lightest materials may have been most quickly transported to considerable distances. And yet they must have been at least among the earlier products; for it is extremely rare (if, indeed, it ever occurs) to find a bed of white lava overlying any of the coarser varieties of volcanic débris whose mass is so enormous in this country.

There may possibly be another point of interest in this connection too. If the white lava and its allied materials are essentially rhyolitic in character, and the gray and darker-colored compact rocks which constitute the boulders and solid fragments in the conglomerates, the breccias and the volcanic gravels, are generally trachytic* in their nature, then this fact would seem inconsonant with Richthofen's general theory; for it would then appear, not as an isolated instance, but as a general fact, all over this section of the country, that trachyte had succeeded rhyolite in point of time of its ejection. I believe he applies his theory in its strictness, however, only to "massive eruptions." And if all the volcanic matter spread over this region be indeed the result of what he calls "volcanic action proper," it might then be held, perhaps, that no anomalies which it may present can affect the general theory.

The transition from these lower beds of ash and volcanic mud to the coarser overlying materials is not gradual but sudden. There are in general no intermediate steps or varieties. Immediately upon the white lava rests the volcanic gravel, the breccia, or the coarser conglomerate, as the case may be. Among the latter, indeed, there are beds of coarser and beds of comparatively fine material, and sometimes in different portions of the same bed variations may be noticed in the coarseness of the material. But these variations are rarely great, and I have never noticed them except in a few cases of comparatively thin beds.

When I was in San Francisco in July and August, 1871, my attention was called to the description of some curious facts in the structure of the volcanic ridge near Mokelumne Hill in Geology I. (p. 267), and after leaving town again in August, I watched constantly to see if I could find a parallel to it. I found plenty of instances which corresponded in part with that description, but not one which did in all. I found numerous cases where "the upper part of the ridge" was "a mass of boulders or fragments of trachytic lava, not polished nor smooth, but roughly rounded," and this is the general character of the larger boulders in the coarser volcanic conglomerate throughout the country where I have been. But I never yet saw a case of a heavy mass of this material (and there are plenty of heavy masses of it almost everywhere) in which I could satisfy myself at all that the boulders were in reality "largest at the top" and grew "quite small towards the bottom of the bed." It is often true, indeed, especially where the crest of a ridge of this material is narrow, that the number of boulders visible, both large and small, is far greater on the crest and brows of the ridge than lower down its sides; but it seemed to me that this was only a natural result of the slow action of the rains and melting snows, which might slowly remove the finer soil from even the crests of the ridges where it could not move the boulders, while lower down on the sloping sides of the ridge, the finer material could not be removed without at the same time undermining the boulders and leaving them free to roll down into the cañons. If this be correct, its direct result in time would be a great increase in the number of visible boulders on the crest and brows of the ridge, while the original proportion of boulders which the mass contained might be better represented lower down its sides. I also noticed the fact that when the ridges are

---

* Microscopic examinations have shown that the dark-colored volcanic rocks to which reference is here made are chiefly andesitic in character. Trachytes are very uncommon in the Sierra. So far as relates to the order of succession enunciated by Richthofen, the bearing of the facts mentioned by Mr. Goodyear is not changed by the substitution of andesitic for trachytic. — J. D. W.

broad (and they are sometimes several miles in width) though there are occasional patches where the boulders are very thick, especially about the slopes of the little gulches, etc., yet the general rule is that the boulders visible on the flat surface of the ridge are not so very numerous ; but almost without exception, on approaching the cañons, I found them very numerous about the brow of the hill, while again they were generally not so numerous lower down its side. Moreover, wherever the boulders at the top were large, I hardly ever failed on hunting to find occasional large ones also sticking low down in the sides of the ridge. Furthermore, I have seen a good many sections of this kind of material from thirty to fifty feet in height, and sometimes considerably more, exposed by hydraulic mining, and I never yet saw such a section in which the boulders appeared, as a rule, any larger near the top than they did lower down. I think, therefore, that the structure of this ridge at Mokelumne Hill (which I did not visit) must be very peculiar in this respect.

But there is another statement respecting that locality to which I have not been able to find a parallel anywhere else. It is this : " There are no other kinds of rock than volcanic represented in this bed, and no stray pebbles even of quartz or slate ; this fact is not peculiar to the lava ridges of Mokelumne Hill, but has been observed in many other places in this region." I cannot help thinking the fact stated in the last clause quoted above a very extraordinary one, for I have never yet found a locality on any of the volcanic ridges over which I travelled last year where there were " no stray pebbles, even of quartz or slate." I have, indeed, seen places enough where they were scarce ; in fact, they are generally so, and oftentimes extremely rare ; but often when, in riding along for some little distance, I could see none of them, I have dismounted and looked for them, and in such cases, rare as they might be, I have never failed to find them after a little search. There are plenty of them, that is, comparatively speaking, on the ridge between Sutter Creek and Jackson. And if there is any resemblance in the general physical character of the volcanic matter in the ridges on the opposite sides of the Mokelumne River, I can hardly think that their entire absence is a common thing in the region southeast of that stream. In fact, I should really be more disposed to question the thoroughness of any ordinary search which might have been made for them, than I should be to believe in their entire absence.

Nor have I seen any place in this part of the country where it appeared to me necessary, or even admissible, to suppose that a " mass of lava had been broken up while flowing and before it was entirely consolidated," the fragments being " pushed over each other by the pressure of the mass from behind," and thus becoming rounded by their own friction alone. I acknowledge that there are difficulties, which I cannot remove, in the way of fully and satisfactorily explaining just how so many volcanic boulders, of such huge dimensions as some of them are, have been transported so far, and distributed without sorting at every height above the bed-rock from top to bottom, through such masses of finer fragmentary material ; but considering the stray pebbles of quartz and metamorphic rocks and granite, which, so far as my own observation has extended, the mass of this material always does contain, I know of no means which seems to me capable of having accomplished the work without the aid and instrumentality of water in some way. And when we consider the vast period of time which has elapsed since this material came here (the modern cañons have all been excavated since then), in connection with the slow effects of weathering, etc., on even the hardest rocks, I think it will hardly do to argue against the agency of water in some way in the matter, simply because the boulders are " not polished or smooth " to-day, and appear but roughly rounded.

From the description given * of the material underlying this mass of boulders at Mokelumne Hill, it evidently belongs to the same class of material that I have constantly called " white lava," after the local name by which it is known at Placerville.

With reference to the manner in which the breccias were distributed over the country, I am much inclined to think that they travelled in the form of mud-flows. I know not how else to account for the sharp angularity of their rocky fragments, together with the presence in them of

* See Geology, Vol. I. p. 268.

fossil leaves, so delicately preserved as those at Negro Hill near Placerville. There are, indeed, some difficulties connected with this supposition, one of which is the difficulty of understanding how mud of such a character, that is, so full of angular, rocky fragments, could have flowed so far over such gentle grades, and in such thin sheets as we sometimes find among these breccias. Yet this is decidedly the most plausible idea that occurs to me, all others being fraught with greater difficulties still.

The volcanic gravel (known at Placerville under the name of "mountain gravel," and at other localities as "black gravel," etc.) is of course the work of running streams, precisely like the metamorphic auriferous gravel, only consisting of volcanic instead of metamorphic materials.

The coarser volcanic conglomerates (I call them such for want of a better name, though oftentimes they are not consolidated into a rocky whole, but only form rather compact banks), with their roughly rounded and sometimes enormous boulders, still remain. Their materials are not sorted, so far as I have seen, in any way, with respect to coarseness or fineness, and I am by no means certain as to the means by which they were transported. They can hardly be the work, I think, of constantly running streams, of any magnitude whatever; for it seems hardly possible that any such streams, of sufficient force to have brought these enormous boulders over such gentle grades, should not have carried all the fine material farther on, instead of leaving it here to form the matrix in which the boulders are imbedded. Nor does it seem likely that they were simple mud-flows; for if the mud were fluid enough to flow so far over such gentle slopes, it appears probable that such heavy boulders would have forced their way downward through the mass, and sunk to the bottom.

May they not, perhaps, have been brought here by a series of occasional sudden and heavy floods, bringing down with them vast quantities of débris, and spreading it quickly over considerable areas at a time, each flood of this kind being succeeded by a period of quiet, no matter how long, if only long enough to allow the surplus water to ooze and drain out from the newest arrived material, and the mass to become comparatively dry and somewhat compacted together? In this way the largest boulders of the latest flood might be left on top of even the finest material of the preceding one, and thus, as the mass gradually accumulated, these boulders might become imbedded in it at every height above the bed-rock from the bottom to the top. It is true that, if this be the case, I do not know what caused the floods. But it is not difficult to point out agencies which might produce them in a high volcanic mountain country. The only positive evidence that I know of against this idea (if indeed it be such) is the rarity of any distinct traces of definite bedding or stratification in heavy masses of this material. But of all the ideas which have occurred to me in connection with this difficult problem, this supposition appears to me to be liable to the least objections, and in reality the most plausible; especially when we consider the fact that in this peculiarity of distribution of the heaviest boulders throughout the banks, from the bottom to the top of the mass, we may perhaps find an exact parallel to this material in the "sage-brush slopes" of granitic débris in Owen's Valley, where, at all events in many places, huge boulders rest upon the very surface of the slopes at points far distant from the foot of the mountains, and where I feel confident that some such agency as this has aided in their distribution.

The question, how far up the mountains towards the summit the ancient gravel extends, is one to which a definite answer can hardly perhaps be given. Mr. Sterrett, near Canada Hill, told me of one acquaintance of his who had found well-washed metamorphic gravel beneath the volcanic matter at the very summit; and I consider this by no means an impossibility, though perhaps hardly probable; but it certainly extends very far up, and may be found here and there beneath the volcanic matter to within short distances of the summit, if not actually to it. It is the general fact, however, that after we rise above a certain level, the aggregate quantity of the ancient gravel diminishes, and then grows less the higher we go; and, though there is hardly a rule of any kind relating to the ancient gravel which has not here and there its apparent exceptions, and though we may occasionally find very smoothly washed gravel very far up toward the summit, yet it is also a general rule that above a certain altitude the ancient gravel is less smoothly washed, and grows more and more angular the higher we go. And both the facts last stated appear to me to furnish strong additional evidence as to the source of the ancient streams and the general direction of their flow. It is also worthy of note that universally, so far as I have yet seen, wherever there is much volcanic matter on the western slope of the Sierra Nevada, its quantity increases as we ascend, and the farther and the higher we go toward the summit of the range, the heavier does its mass become.

I have seen but little of the actual summit of the range at any portion of its course northwest of the Sonora Pass, under such circumstances as enabled me to recognize the materials of which it is formed. But from the little that I have seen, I should be inclined to infer that the comparative scarcity even here of compact solid lavas, and the great prevalence of enormous quantities of breccia and other fragmentary materials, often stratified in nearly horizontal beds, was a curious fact, and not altogether an easy one to satisfactorily explain. I do not know, indeed, that it is the general fact. But, if it be so, it is one from which much might possibly be learned by careful study of the summit.

The greatest hindrance at present to mining operations, wherever I have travelled in the gravel region, is the scarcity and uncertainty of water for hydraulic purposes, and the high prices which the miners are obliged to pay for what they do get of it. If water were plenty, reliable, and somewhat cheaper, the extent of mining operations here would be rapidly and greatly increased, and in many localities with profit to all.

I do not know that I have used in these notes any popular terms or expressions which require any further explanation than has already been given, unless it be the single word "cement." Under this designation is included, in different parts of the country, almost everything which occurs in the gravel region except the solid bed-rock ; and its meaning varies in different localities. But throughout the southern part of Placer and the northern part of El Dorado County it is used to designate anything and everything which overlies the auriferous gravel, and which *does not contain any appreciable amount of gold.* Here therefore it corresponds, of course, in general, to volcanic matter of every kind. But by far the commonest form of volcanic matter in the region examined by me is the variety which I have elsewhere called a "coarse conglomerate" ; this consists essentially of a mass of fine earthy material, more or less thickly filled from bottom to top with volcanic pebbles and roughly rounded boulders of every size, from the smallest to the greatest, which exhibit much variety of texture, and also, among darker colors, a large variety of shades.

I am thoroughly satisfied that a very large proportion of the shallow-placer gold, that is, the gold which made the beds of the modern streams so rich in the early days of California history, was derived from the ancient gravel which those streams have washed away; though another portion, and undoubtedly a large one too, came directly from the quartz and the bed-rock which have been cut away in the excavation of the modern cañons. Furthermore, I greatly suspect that the aggregate quantity of gold which, in the process of excavating these cañons, has been ground to almost impalpable dust, and scattered by the water, as finest flour-gold, all over and throughout the mass of the whole Sacramento Valley, may be incomparably greater than all that the early miners found remaining in the gulches and the cañons.

C.

ALTITUDES OF POINTS IN THE AURIFEROUS GRAVEL REGION OF THE SIERRA NEVADA, IN FEET, ABOVE THE SEA-LEVEL.

TUOLUMNE COUNTY.

Street level, front of Hathaway's drug-store, Columbia	2137
Street level at Post-office, Shaw's Flat	2111
Top of Table Mountain, above Valentine's shaft	2214
Top of Table Mountain, about a mile and a half, S. 45° W. (magnetic), from Columbia	2247
Top of Table Mountain, about a mile and a half, S. 65° W. (magnetic), from Columbia, left bank of Stanislaus River	2369

CALAVERAS COUNTY.

Lowest water in Stanislaus River, at Abby's Ferry	879
W. H. Robinson's house	1761
Summit of Table Mountain on stage road from Vallecito to Abby's Ferry	2175
Summit of trail over Table Mountain from Murphy's to Pine Log	2515
Lowest wheel at shaft, Douglass Flat	1986
Summit of Bald Mountain	1801
Floor of hotel at Altaville	1577
Lower floor of hotel at Angel's	1394
Summit of hill of volcanic breccia, about two miles north (magnetic) of Vallecito	2569
Street level front of Sperry's hotel, Murphy's	2191

AMADOR COUNTY.

J. D. Mason's house, Buena Vista, Jackson Valley	295
Summit of Jackson Valley Buttes	829
Arcade House, Ione City	310
Bed-rock, south end of Tunnel Hill, near Jackson	1418
Bridge over Mokelumne River, between Jackson and Mokelumne Hill	676

Globe Hotel, Jackson 1243

American Exchange Hotel, Sutter Creek 1197

Crest of McIntyre's Hill, near Sutter Creek 1759

Bed-rock in Humbug Hill, near Sutter Creek 1533

Highest crest of ridge south of Sutter Creek, half a mile southwest of Humbug Hill . 1754

Hotel, Amador City 862

Express office, Drytown 612

Summit of gravel hill, near Forest Home 675

Puckerville 1037

Brow of hill just southeast of Enterprise, overlooking Cosumnes River 1428

Toll-house at suspension bridge, just above Enterprise, Cosumnes River . . . 799

United States Hotel, Fiddletown 1693

Mouth of slope, Jameson mine, Loafer Hill, near Fiddletown 1880

Summit of Loafer Hill, near Fiddletown 1964

St. George Hotel, Volcano 2100

Aqueduct City 2435

Bottom of "white lava," between Indian and Jackass gulches, in ridge just north of Volcano   2628

Store at Indian Diggings 3162

SACRAMENTO COUNTY.

Toll-bridge, Michigan Bar 227

Boarding-house, summit of gravel hill, Michigan Bar 380

Central House, Folsom 200

Bed of American River, one mile below Folsom 142

Top of bluff, north side of American River, opposite the preceding station . . . 228

Broad gravel bench, opposite Folsom 199

Crest of bluff, north of American River, just opposite the preceding station, and east of

     Sacramento Valley R. R. cut 295

Surface of ground at Bunker Hill, near Folsom 267

Volcanic breccia floor, underlying gravel bank at Willow Springs Hill, near Folsom . . 267

EL DORADO COUNTY.

POINTS IN THE SOUTHERN PART OF THE COUNTY.

Buck's Bar, Cosumnes River 1628

Crossing of Middle Fork of Cosumnes River 1671

Remick's Hotel, Fairplay 2385

Edner's Saloon, Brownsville 3499

Bridge over North Fork of Middle Fork of Cosumnes River, on trail from Brownsville *via*

    Henry's diggings to Grizzly Flat 2845

Under bridge at Steely's Fork, on road from Henry's diggings to Grizzly Flat . . 3367

Schneider's Hotel, Grizzly Flat 3949

Bed-rock, Young and McClellan's hydraulic claim, three miles east of Pleasant Valley . 3029

Burns's ranch, between Pleasant Valley and Newtown 2518

O'Brien's tunnel, half a mile above Newtown 2380

Newtown 2482

Bed-rock, Deadhead claim, below Newtown 2416

Logtown 1939

Mud Springs 1658

Miller's Hotel, Latrobe 790

Planter's House, Shingle Springs 1459

Marble Valley 925

Bed-rock, Diamond Springs, Sugar Loaf 1870

Bed-rock, Bean Hill, near preceding station 1777

PLACERVILLE AND NEIGHBORHOOD, SOUTH OF THE CARSON ROAD.

Cary House, Placerville 1873

Bed-rock, southwest extremity of Webber Hill spur 2027

Bed-rock, hydraulic pit, south side of Webber Hill 1920

Mouth of tunnel in "white lava," south side of Webber Hill 2026

Crest of Webber Hill 2184

Bottom of "black lava" in spur next east of one at Webber claim, north of Chili Ravine 2135

Bottom of solid "black lava" at head of little gulch, just east of preceding station . . 2169

Top of "black lava" cap of hill, at mouth of old shaft, 300 feet south of Mr. Hardy's house 2282

Mouth of Twin tunnel, near head of Chili Ravine 2037

Bed-rock, Texas Hill, south of flume, south of head of Cedar Ravine . . . 2049

Sag under flume south of head of Cedar Ravine 2245

Bed-rock, Excelsior hydraulic mine, Coon Hill 2036

Top of present bank (Oct. 1871), Excelsior hydraulic mine, Coon Hill . . . 2212

Bed-rock, Webber claim, Coon Hill 2080

Crest of Coon Hill ridge, northeast of the Webber claim 2232

Crest of Hangtown Hill at city reservoir, near Placerville 2155

Rim-rock, north side of Hangtown Hill, 400 feet west of Oregon Point . . . 2115

Rim-rock, north side of Hangtown Hill, a quarter of a mile west of Oregon Point, and near

    west end of banks 2108

Crest of Hangtown Hill, opposite preceding station 2198

Rim-rock, north side of Cedar Hill, just east of Oregon Point 2170

Rim-rock, north side of Cedar Hill, four or five hundred feet east of preceding station . 2112

Rim-rock, north side of Cedar Hill, an eighth of a mile west of Dickerhoff's mill . . 2049

Small ditch opposite preceding station 2244

Level of South Fork Canal, opposite last station but one 2286

Crest of main ridge, opposite last station but two 2327

Track in tunnel at Dickerhoff's mine 2033

Rim-rock, north side of Cedar Hill, first hydraulic pit east of Dickerhoff's mine . . 2132

Lowest point of sag, under the flume, south of head of Cedar Ravine . . . 2246

Highest crest of Cedar Hill, west of the flume 2449

Bed-rock, Franklin claim, Little Spanish Hill 2170

Bed-rock, east of Franklin claim, Little Spanish Hill 2148

Crest of Little Spanish Hill 2321

Crest of ridge, Placerville race-track 2576

Mouth of shaft, Prospect Flat 2214

NEIGHBORHOOD OF PLACERVILLE, NORTH OF THE CARSON ROAD.

Bed-rock, Clay Hill 2177

Bed-rock, east end of Indian Hill 2202

Bed-rock, Pennsylvania claim 2254

Average bed-rock, Henderson claim 2257

Top of bank, Henderson claim 2408

Bed-rock, Wick's claim, Negro Hill 2146

Bed-rock, Stevens's claim, near Negro Hill reservoir 2312

Top of highest bank, Stevens's claim 2383

Ward Brothers' mill, White Rock, on about the same level as the deep channel here . 2243

On level of South Fork Canal, at a point bearing N. 65° E. about a quarter of a mile from

    White Rock store 2377

Near mouth of old shaft on crest of White Rock Point ridge 2564

Level of bottom of channel at Georgia Hill, just east of White Rock Flat . . . 2340

Mr. Peer's house, White Rock Point 2457

POINTS ON THE RIDGE ABOVE PLACERVILLE.

Cruson's house, Dirty Flat 2355

Crest of hill between Dirty Flat and Smith's Flat 2499

Saul's Hotel, Smith's Flat 2239

Toll-house, Carson road, near Smith's Flat 2400

Mouth of Try Again tunnel 2186

Summit of Mt. Pleasant, northeast of Smith's Flat 2811

Sportsman's Hall, Carson road, twelve miles above Placerville 3706

Crest of main ridge, Carson road, a mile and a quarter above Sportsman's Hall, nearly on

    the same level with the 13-mile house 3995

Menafee's house, on South Fork Canal 2504

Bottom of cañon of South Fork of American River, opposite Menafee's house . . 1859

Chili Bar, South Fork of American River 931

POINTS IN THE NORTHERN PART OF THE COUNTY, BETWEEN THE SOUTH AND THE MIDDLE

FORKS OF THE AMERICAN RIVER.

Suspension Bridge (65 feet above bed of river) just below junction of North and Middle

    Forks of American River 609

Pilot Hill post-office, Centreville 1191

Summit of Pilot Hill 1857

Forks of road to Johntown, between Michigan Flat and Coloma 958

Johntown or Garden Valley 1951

Corner toll-house, Greenwood 1610

Berry's Hotel, Georgetown 2674

Creque's Hotel, Georgetown 2673

Summit of Mt. Oliver, near Georgetown 3221

Highest crest south of Cañon Creek 3249

Robbins's tunnel, Castle Hill, above Georgetown 2813

Store at Georgia Slide 2330

Hunter's store, Spanish Dry Diggings 2158

Mouth of Columbia tunnel, Jones's Hill 2114

Summit of Jones's Hill 2343

Ford's Bar, Middle Fork of American River 866

Crest of saddle, Bottle Hill 2570

Mouth of Roanoke tunnel, Bottle Hill 2337

Summit of Bald Mountain, between Otter and Cañon Creeks 3097

Crossing of Cañon Creek, on trail from Georgetown to Volcanoville . . . 2481

Darling's house, between Otter and Cañon Creeks 3011

Forks of Otter Creek, on trail from Georgetown to Volcanoville 2254

Crossing of Cañon Creek on road from Georgetown to Kentucky Flat . . . 2910

Crest of ridge between Otter and Cañon Creeks, on trail from Georgetown to Kentucky Flat	3327
Crossing of Otter Creek, on trail from Georgetown to Kentucky Flat	2781
Bed-rock, hydraulic ground, Kentucky Flat	3233
Lowest bed-rock in quartz boulder pit, Kentucky Flat	3220
Bed-rock, Tipton Hill	3186
Works's ranch	3386
Crest of sag, where road crosses Tunnel Hill, over line of proposed ditch	4043
Crest of Tunnel Hill, highest point reached by the road	4205
Dark Cañon Hotel	4229
Forney's, Pilot Creek	4173
Bed-rock, Flora's mine	2530
Volcanoville	3081
Mouth of Oak Grove tunnel, above Volcanoville	2867
Crest of ridge, back of Oak Grove tunnel	3385
"Shoe Fly" claim, Missouri Cañon	3096
Buckeye Sucker claim, near Mt. Gregory	3206
Mt. Gregory	3525
Crest of ridge between Kentucky Flat and the South Fork of the Middle Fork	3513
Bottom of cañon of the South Fork of the Middle Fork of the American River, just below mouth of Long Cañon, on trail from Kentucky Flat to Blacksmith Flat	1242
Crest of main spur between mouth of Long Cañon and the Middle Fork of the Middle Fork of the American River	3480
Charles Jarrad's house, Blacksmith Flat	3831
Crest of main ridge, at point bearing N. 5° W. three quarters of a mile from Blacksmith Flat	4840
Wilcox's, Forks of Long Cañon	4342
Bed-rock, Wilcox's claim, Long Cañon	4466
Wilcox's Meadows, Long Cañon	5344
Crest of high peak, south of Wilcox's Meadows	6057

PLACER COUNTY.

POINTS ON THE SOUTH SIDE OF THE FOREST HILL DIVIDE, AS FAR AS MICHIGAN BLUFF.

United States Hotel, eight miles below Forest Hill	2156
Mouth of "Big Channel" shaft, Peckham Hill	2349
Todd's Valley, centre of town	2750
Crest of ridge, near North Star House, one mile below Todd's Valley	2851
Bed-rock, bottom of Pond's claim, Todd's Valley	2770

Top of bank, Pond's claim, Todd's Valley 2855

Penstock, head of pipe, Pond's claim 2999

Bed-rock, Dardanelles claim 2677

Highest point of top of bank, Dardanelles claim 2911

Oro Flat; near line between Dardanelles and Oro claims, 25 or 30 feet above the gravel 2842

Oro Flat; top of gravel near line between Hope and Green Spring claims . . . 2829

Mouth of lower tunnel at San Francisco claim 2703

Forest House, Forest Hill 3237

Forks of road on crest of ridge near Forest Shades 3308

Mouth of New Jersey tunnel, Forest Hill 2850

Road on crest of ridge over New Jersey tunnel 3207

Mouth of Gore tunnel, Forest Hill 2948

Mouth of Paragon tunnel, Bath 2990

Mouth of air-shaft, crest of ridge, back of Paragon mine 3483

Mouth of Allen's shaft (153 feet deep), Volcano Cañon 3269

Bed-rock, crest of ridge, forks of road, near Baker's house 3720

Volcano Cañon crossing on lower road from Bath to Michigan Bluff . . . 2870

Express office, Michigan Bluff 3491

Skating Rink, Michigan Bluff 3482

Summit of Michigan Bluff Sugar Loaf 3790

Bed-rock, Van Emmons claim, Michigan Bluff 3320

Mouth of Weske's tunnel 3335

Crest of ridge over Weske's tunnel 3974

Mouth of Bowen's tunnel 3407

Mouth of Landcraft tunnel, Franklin claim 3422

Bed-rock, hydraulic bank, Franklin claim 3349

Bridge at crossing of Volcano Cañon, on McBride road from Michigan Bluff to saw-mill 3638

Crest of ridge back of Ayer's mine 4077

Mouth of Ayer's tunnel 3488

Bed-rock, lower edge of workings, El Dorado Hill 3212

Bottom of El Dorado Cañon, on trail from Michigan Bluff to Deadwood . . . 1835

Bed-rock, Sage Hill 3125

POINTS NEAR AND ABOVE DEADWOOD AND LAST CHANCE.

Ferguson's Hotel, Deadwood 3951

Mouth of Reed's tunnel, Deadwood 3642

Crest of ridge, over Reed's tunnel, Deadwood 4261

Mouth of Hornby's tunnel, Deadwood 3611

Mouth of Robertson's tunnel, Deadwood 3555

Mouth of Kaylor's tunnel, Deadwood 3564

Bed-rock, quartz channel, Deadwood 3717

E. T. Darling's house, Deadwood 3702

Crest of ridge, near Devil's Basin, above Deadwood 4345

Bed-rock, on trail near Devil's Basin 3916

Mouth of tunnel, Rattlesnake claim, below Devil's Basin 3790

Brow of hill overlooking Devil's Basin 4380

Mouth of Basin tunnel, Devil's Basin 3797

Bed-rock, little hydraulic pit, near Last Chance trail 3878

Bed-rock under quartz gravel, northeast of Devil's Basin 3908

Bottom of cañon of the North Fork of the Middle Fork of the American River, between

       Deadwood and Last Chance 2718

Jansen's Hotel, Last Chance 4582

Mouth of tunnel, Nick Anderson's claim, Last Chance 4241

Home Ticket mine, near Last Chance 4556

Mouth of Keystone Flat tunnel, near Last Chance 4310

Level of track in lower Slab tunnel 3999

Lower Tunnel, Fiddler's Green 4123

San Francisco mine, between Last Chance and Startown 4418

Upper tunnel, San Francisco mine 4427

Level of ditch immediately back of San Francisco mine 4708

Mouth of tunnel, John Yule's claim, Startown 4542

Crest of ridge, back of Startown 4872

Dinsmore's tunnel, Grouse Cañon, about a mile and a half south of Startown . . . 4524

Big Trees, above Last Chance 5207

Barney Kavanagh's cabin, Miller's Defeat 5740

Yank's cabin, Canada Hill 6217

Bed-rock, Sterrett's claim, Sailor Cañon 5251

Northwest summit Canada Hill, Bald Mountain 7091

POINTS ON NORTH SIDE OF FOREST HILL DIVIDE, NEAR YANKEE JIM'S.

M. B. Tubbs's saloon, Yankee Jim's 2578

Top of highest bank, Yankee Jim's 2761

Bed-rock, Smith's Point, between First and Second Brushy Cañons 2673

Crest of hill, saddle at head of Young America Cañon 3126

Highest crest between Second Brushy and Black Hawk Cañons, below Young America

Cañon 3176

Crest of southwest spur between Second Brushy and Black Hawk Cañons, just back of

hydraulic bank 3082

Bed-rock, hydraulic bank, southwest extremity of spur between Second Brushy and Black

Hawk Cañons 2761

POINTS BETWEEN SHIRT-TAIL CAÑON AND THE NORTH FORK OF THE AMERICAN RIVER, AS FAR AS GREEN VALLEY GORGE.

Toll-house, Rice's Bar, North Fork of American River 1184

Bed-rock, King's Hill 2538

Crest of ridge between King's Hill and Elizabeth Hill 3054

Bed-rock, southwest hydraulic pit, Elizabeth Hill 2750

Bed-rock, middle pit, Elizabeth Hill 2744

Bed-rock, eastern pit, Elizabeth Hill 2749

Top of gravel immediately southwest of Elizabeth Hill 2847

Bed-rock, crest of ridge, half-way between Elizabeth Hill and school-house . . 2946

Junction of Shirt-tail and Brushy Cañons 1499

Bed-rock in hydraulic pit, head of Refuge Cañon 2917

Mouth of old shaft in New York Cañon, below McCullough's . . . 2806

Mouth of tunnel, north side of New York Cañon 2757

Teasland's house, Wisconsin Hill . . . .(?) 2936

Bed-rock, Ingersoll and Vaughn's claim, Wisconsin Hill . . . . (?) 2659

Mouth of Lebanon tunnel, near Wisconsin Hill 2867

Top of gravel near school-house, Wisconsin Hill 3027

Crest of high spur, back of Morning Star mine 3295

Top of gravel spur on road between Wisconsin Hill and Iowa Hill . . 3023

Mouth of tunnel, Grizzly Flat 2982

Sucker Flat 3410

Parker House, Iowa Hill 2873

Top of bank, Weissler's hydraulic claim, Iowa Hill 2853

Bed-rock, Weissler's claim, Iowa Hill 2647

Summit of first Sugar Loaf, Iowa Hill 3087

Top of bank, Sailor Union claim, Iowa Hill 2849

Bed-rock, Sailor Union claim, Iowa Hill 2657

Mouth of Morning Star tunnel, Iowa Hill 2640

Bed-rock, Reno claim, Independence Hill 2972

Mouth of Wolverine tunnel, above Independence Hill 3025

Bed-rock, Metcalf's claim, above Independence Hill 2943

Crest of ridge about a quarter of a mile northeast of Independence Hill 3469

Rim-rock, Roach Hill 3037

Strawberry tunnel, just above Monona Flat, Indian Cañon 3160

Crest of main ridge, just south of Green Valley gorge 4125

Mouth of Nabor's tunnel, Green Valley gorge 3727

Bed-rock, spur next northeast of Green Valley gorge 3881

POINTS NEAR AND ABOVE DAMASCUS, SOUTH OF THE NORTH FORK OF THE AMERICAN RIVER.

Mouth of Humbug Cañon 2080

Hotel at Damascus 4006

Mouth of Mountain Gate upper tunnel, Damascus 3867

Mouth of tunnel, Cement Mill claim, Damascus 3827

Bed-rock, hydraulic pit, Damascus 3922

Crest of main ridge, over Mountain Gate tunnel, just back of Damascus . . . 4641

Highest bed-rock in road, half a mile northwest from Damascus 4073

Bed-rock, on ditch a mile and a half above Damascus 4177

Forks House, above Damascus 4764

Forks of road on crest of ridge immediately back of the Dam claim, a mile and a half from

   Forks House 4560

Mouth of tunnel, Dam claim 3671

Crest of ridge between Forks House and Hog's Back 5436

Secret House 5423

Summit of Secret Hill 6536

GOLD RUN AND NEIGHBORHOOD.

Level of tailings on North Fork of American River, near outlet of Indiana Hill mines,

   October, 1870 1418

Summit of hill between cement mill and Cañon Creek 3099

Bed-rock at entrance of tunnel at cement mill, near Indiana Hill 2792

Highest pressure-box, Indiana Hill mines, northwest side 3179

Point on bed-rock spur, one eighth of a mile S. S. E. from Fiddler's ranch ; base of Cold

   Spring Mountain, lava and cemented sands 3284

Top of Cold Spring Mountain 3679

Top of old '49 shaft, Potato Ravine 2956

Edge of rim-rock in Rattlesnake Ravine, near Gold Run 3200

DUTCH FLAT AND NEIGHBORHOOD.

*W. & P. Nicholls's banking office, Dutch Flat 3159

*Junction of white and blue gravel at Polar Star mine 3225

*Bed-rock at Polar Star mine 3075

*Bed-rock at Southern Cross mine 3054

Mouth of Teaff's shaft 3137

Mouth of Dutch Flat Tunnel, on Bear River side of diggings 3099

Average rim-rock at Thompson Hill 2848

Mouth of deep shaft in Dutch Flat Cañon 2931

Edge of west rim-rock, Squire's Cañon 3047

Point on ridge between Squire's Cañon and Bear River, near the end of the gravel . 2963

Rim-rock at most westerly gravel opening on "Plug Ugly" Hill 3072

Top of "Plug Ugly" Hill 3251

*Top of gravel at north end of "Plug Ugly" Hill 3183

Level of tailings in Bear River (November, 1879) at Dutch Flat and Little York road

    crossing 2549

Level of tailings in Bear River (November, 1870) at Dutch Flat and Little York road crossing 2452

POINTS NEAR SUMMIT STATION AND LAKE TAHOE.

Top of Devil's Peak 7758

Level of spring at divide between Summit Valley and ice-houses . . . 6941

Point above and east of Summit Station, level with base of lava in Summit Gap . 7341

"Glacial Point," near Summit Station, level with base of lava on mountain south of railroad 7722

Top of lava cliff bearing N. 75° E. (magnetic) from Summit Station . . . 8154

Base of lava on Snowy Mountain, southerly from Summit Station . . . 7537

Top of Snowy Mountain 8425

Smith and Redding's soda springs 6175

Base of main lava cliff, "Three Prong" 7755

Top of "Three Prong" 9000

Top of "Granite Chief" 9144

Squaw Valley Pass 8771

Ranch house in Squaw Valley 6304

Highest summit of Knoxville cliffs, opposite Squaw Valley 7900

Volcanic bench between Truckee River and "Pluto" 7178

Top of Pluto 8633

Point on spur from cliffs on ridge next south of Blackwood's Creek . . . 7489

Blackwood's Cliffs, four miles southwest of McConnell's 8278

Top of "Lava Castle," near mouth of Blackwood's Creek	6524
Mt. Anderson	9000
Highest silicious schists, head of Ward's Valley	8459
McConnell's sheep station, Ward's Valley	6664
Top of "Bonaparte's Hat, No. 2"	8661
Base of summit cliffs, Bonaparte's old hat, "Twin Peaks"	8724
Sugar Pine Pass	7130
Top of mountain, three-fourths mile east of, and overlooking, Sugar Pine Pass	7868
Point of rock, estimated to be 275 feet above the beginning (western end) of moraine at Sugar Pine Point	7745

PLACER AND NEVADA COUNTIES.

POINTS NEAR LINE OF C. P. R. R., FROM SUMMIT TO BLUE CAÑON.

Level of small lakes in valley on trail from Summit to Mt. Stanford	7219
Spring on south slope of Mt. Stanford	7966
Top of Mt. Stanford	9102
Crystal Lake House, second floor	5955
Point one-fourth mile west of South Yuba tunnel (leading from Bear River to south fork of Steep Hollow)	4472
Water level, Yuba River, Cisco Flats, June, 1870	4696
Tompkins's Hotel, Bear Valley	4588
Bear Valley, foot of trail from Emigrant Gap station	4665
Point on road from Emigrant Gap to Bear Valley, level with low moraine on north side of Bear Valley	4913
Summit of moraine junction, above Wilson's Valley	6049
Top of highest moraine, below Emigrant Gap station	5490
Lower level of lava near saw-mill in Wilson's Ravine, about a mile westerly from Emigrant Gap station	4900
Bed-rock in Boston claim	4214
Bed-rock in Slumgullion claim	4386
Lower limit of lava, as seen in C. P. R. R. cuts, above Blue Cañon station	4870
Top of lava ridge east of Blue Cañon station	5071
Highest slate-rock seen up Blue Cañon creek, three eighths of a mile from the station	4799
Brow of Grizzly Hill, one mile southwest of Blue Cañon station	4911

NEVADA COUNTY.

LITTLE YORK AND NEIGHBORHOOD.

Bed-rock at Missouri Hill	2756
Deserted house on eastern slope of Christmas Hill, near road	3160
Christmas Hill bed-rock, at west end of mines above Missouri Ravine	3065
Top of Christmas Hill	3225
Rim-rock on Christmas Hill	3073
Street level in front of Scott's store, Little York	2839
Bed-rock, at junction of sluices in bottom of mine, Little York	2706
Top of Manzanita Hill	3054
Top of bank, at stake near flume	2899
Deepest bed-rock, near old mill, Empire Hill	2666
Mouth of Cariboo tunnel, on Steep Hollow side	2413
Top of Cariboo shaft	2487
Top of saddle between Cariboo tunnel and Nigger Ravine	2709

LOWELL HILL AND NEIGHBORHOOD.

Average bed-rock at outlet of Liberty Hill mines	3349
Top of lava mountain above Liberty Hill	4461
Level of road front of Lowell Hill hotel	3951
Bed-rock under eastern bank of Lowell Hill diggings	3824
Bed of Steep Hollow, at trail from Lowell Hill to Remington Hill	3342
Bed-rock at Remington Hill	3869
Southeast corner of Klipstein's house	4117
Bed-rock outside of Klipstein's tunnel	4023
Bed-rock at Excelsior mine	4080
Top of ridge above Klipstein's house, where trail meets road	4723

YOU BET, QUAKER HILL, AND NEIGHBORHOOD.

Level of tailings in Steep Hollow, at Little York and You Bet road crossing (Nov., 1879)	2419
Level of tailings in Steep Hollow, at Little York and You Bet road crossing (Nov., 1870)	2283
Level of tailings at junction of Bear River and Steep Hollow (Nov., 1870)	2213
McManus & Co.'s ditch on " '49 road" from Little York to You Bet, near Steep Hollow crossing	2407
Point on bed-rock range where a little lava gravel was found ; a little below junction of old Little York and You Bet road with '49 road	2863

Squirrel Point, bed-rock 2651

Average bed-rock in lowest part of channel at Niece and West's diggings 2625

Track level at mouth of tunnel of Washington Mining Company 2648

Bed-rock near road (opposite Savage's) where it disappears under the gravel . . . 2833

Bed-rock at bottom of Mallory's incline 2646

Street level, front of Grand Hotel, You Bet 2985

End of dump at Hubbard's tunnel, Pine Hill 2651

Bed-rock level, end of Pine Hill, near Hubbard's 2678

*Junction of roads from You Bet to Little York and to Niece and West's old mine . . 2613

*Top of northwest bank at "Waloupa" 2832

Average bed-rock at "Waloupa" 2594

West end of range near "Copper District" 2672

Bed-rock in mines on Snake Ravine, west of White House 2723

Mule Cañon, on rim of channel, near Major Park's 2818

*Nevada Company's boarding-house 2931

Rim-rock near high flume, Sardine Flat 2911

Point near old shaft on Chicken Point 3075

High bed-rock on Chicken Point 3070

Top timber of Cozzen's and Garber's old shaft 2742

Bed-rock at outlet of mines, opposite Cozzen's and Garber's old mill 2762

Blue line of sandy deposit in main channel, Wilcox Ravine 2799

Sluice near old blacksmith shop, at mouth of Brockmann's old mines, Missouri Cañon . 2838

Mouth of Brockmann's bed-rock ravine 2805

High bed-rock in cut at head of north fork of Missouri Cañon, near Timmens's . . . 3056

Bed-rock near Hussey's 2917

Point on Bunker Hill ditch, where slate was first seen in the bottom of the ditch, going

    from Hussey's 2913

Top of gravel knolls, near high flume, opposite Bunker Hill 2826

Bed-rock at William's diggings on Bunker Hill, near Red Dog 2641

Average deep bed-rock at Red Dog 2625

Rim-rock at small opening on Independence Hill 2728

Average bed-rock, Independence Hill, opposite Red Dog mines 2626

Level of tailings in Greenhorn Creek, near Gas Cañon (Nov., 1879) 2696

Level of tailings in Greenhorn Creek (Dec., 1870) at outlet of small opening on Inde-

    pendence Hill 2496

Level of tailings in Greenhorn Creek, half a mile lower down (Dec., 1870) . . . 2446

Level of tailings in Greenhorn Creek (Nov., 1870) at crossing of road from Red Dog

    to Nevada City 2428

Foot of lowest fall in Arkansas Cañon	2498
Level of tailings at junction of Greenhorn Creek and Missouri Cañon (Nov., 1870) . . .	2356
Level of tailings in Greenhorn Creek (Nov., 1870) at bridge on You Bet and Grass Valley	
road	2230
*Bed-rock in Camden claim, Hunt's Hill	2620
*Mr. W. H. Wiseman's house, Hunt's Hill	2965
Gouge Eye mine, Hunt's Hill, bed-rock	2679
Hunt's Hill, opposite blacksmith-shop	2898
Bed-rock on Darling's Hill	3071
Head of tunnel in lowest part of Duryea's mines, Buckeye Hill	3054
Duryea's bank, at pipe head, Buckeye Hill	3182
*Bed-rock at northwest end of Buckeye Hill	3107
Highest point on ridge, near road from Nevada City to Quaker Hill	3749
Gap at Quaker Hill, east end of flume	3300
Bed-rock on west side of Quaker Hill deep channel, near road on Gas Cañon . .	3006
*Mouth of shaft at Quaker Hill, 215 feet above bed-rock	? 2865
Top of highest bank of gravel worked in 1879 at Quaker Hill mines . . .	? 3130
*Boarding-house, Quaker Hill	? 3265
Head of pipe in ditch above Green Mountain mine, Quaker Hill	2992
Mouth of shaft Green Mountain mine	2807
Blacksmith's shop in Railroad mine, Quaker Hill	2963
Bed-rock, Empire mine near Quaker Hill	2695
End of Chalk Bluffs, near You Bet	3695
Top of " Sand Line " on Sugar Loaf, Chalk Bluffs	3355
Northwest corner of tool-house in Sugar Loaf mines	3072
Bed-rock, near small swampy place north of Sugar Loaf mines	2985
Stranahan's house, near Chalk Bluffs	3375
Sugar Loaf, near Chalk Bluffs	3537
Point on Chalk Bluffs, ridge road, three fourths of a mile above Stranahan's . .	3712
Point on the ridge road, in the woods, about six miles above Stranahan's . . .	4420
Point on ridge road, about ten miles above Stranahan's, and about two miles east of Voss's	
old saw-mill	4697
Junction of ridge road with road from Washington ridge	5025
Point on ridge opposite Omega	5187

NEIGHBORHOOD OF NEVADA CITY AND GRASS VALLEY.

South Yuba Water Office, Nevada City	2480
Bed-rock in Manzanita diggings, Nevada City	2645

Top-sill of Kansas shaft 2890
Top of Live Oak shaft 2840
Mouth of Dean's tunnel 2667
Mouth of Ragon's old "Empire shaft" 2814
Top of Nevada City Sugar Loaf 3111
Lower level of lava on Cement Hill, opposite Sugar Loaf 2824
Point on Cement Hill 3011
Summit west of Cement Hill 2946
Average bed-rock at Peck's diggings, Native American Ravine . . . 2631
Mouth of tunnel at Rolfe and Stranahan's diggings, Long Tom Ravine . . 2633
Lower level of lava between Long Tom and Native American ravines, just above Peck's house 2704
Lower level of lava on ridge between Native American and Know Nothing ravines . . 2614
Point on ridge west of Nevada and Grass Valley toll-house, where the Marysville road
     reaches top of ridge 2925
Toll-house, top of ridge between Nevada and Grass Valley 2841
Junction of lava and slate-rock 450 feet south of toll-house 2808
Mouth of tunnel in slate-rock at head of ravine, below the road from toll-house to Bannerville 2986
Top of hill above the tunnel 3231
Point on continuation of ridge towards Bannerville, where lava was again seen . . 3321
Bed of Deer Creek under Suspension Bridge, Nevada City 2441
Level of tailings (October, 1870) in Deer Creek, at junction with Slate Creek . . 2094
Knoll in bare field near McKune's house 2936
Hughes's house 2790
Mouth of "El Dorado Boys" tunnel, one third of a mile west of Hughes's . . 2564
Top of hill one eighth to one quarter of a mile S. S. E. from Crocker's . . . 2837
Bed-rock in Grass Valley Ravine at "Grass Valley Slide" 2490
Mouth of tunnel at upper end of Grass Valley Ravine 2546
Sill of shaft-house, Alta No. 2, Alta Hill, near Grass Valley 2758

ABOVE NEVADA CITY, ON WASHINGTON RIDGE.

Determined in 1879.

Blue Tent, Superintendent's house 3108
Sailor Flat, B. D. Chadwick's house 3050
Bed-rock at Gopher Point 2483
Rim-rock on spur between Sailor Flat and Last Chance Cañons . . . 2759
Mouth of old shaft in New York Cañon 3149
Washington, Griswold's Hotel, second story 2633
Bed-rock at Alpha 3852

Bed-rock at Omega 4028

Hotel at Omega, second story 4201

Bed-rock in Omega ditch, near head of Iowa Ravine 4238

Bed-rock at Diamond Creek mines, near tunnel 4206

BETWEEN NEVADA CITY AND SMARTSVILLE.

Lola Montez diggings 2489

Mouth of Rock tunnel 2482

Mouth of North Star tunnel 2414

Mouth of Virginia tunnel 2444

Summit on Rough and Ready road above Virginia tunnel 2615

Lower edge of lava on hill in northeast quarter of section six, T. 16 N.; R. 7 E. . . 2133

Summit between Owl and Kentucky creeks, on line due south from Keystone saw-mill . 2323

Summit of road between Beasly's and Keystone saw-mill 1870

Beckman Hill; mouth of upper tunnel near road from Novey's to Beasly's . . . 1754

Beckman Hill; summit on road between Novey's and Beasly's 1950

Lower level of lava on northerly end of Beckman Hill 1770

Mouth of tunnel on south side of road one quarter to one half of a mile above Walling's Hotel 2032

Walling's Hotel, Rough and Ready, second story 1901

Mouth of shaft on Goshen Hill 1838

Summit on road between Whitesel's ranch and Rough and Ready 1957

Whitesel's ranch 1686

Water-level in Deer Creek (July, 1870) at Texas Flat 1303

Bed of Deer Creek (July, 1870) at the Anthony House 1146

*Summit on road from Anthony House to Bridgeport 1340

*Bridgeport, thirty feet above level of tailings in South Yuba River 700

Mouth of Stark's tunnel 1570

Lower edge of gravel on hill northeasterly from Pearl's and northwesterly from the An-

    thony House 1549

Summit of preceding hill 1641

North side of Pearl's Hill, on old '49 road 1536

Top of ridge above Pearl's diggings 1612

Mouth of Pearl's new tunnel 1470

Lower edge of lava capping on Pearl's Hill (first lava seen on road from Fiene's to the

    Anthony House) 1528

Lower edge of gravel on northerly side of same hill near where the road crosses the sag . 1225

Lower edge of gravel slide in ravine on east side of preceding hill 1168

Mouth of old shaft to the north of the preceding station 1244

Summit of bed-rock hill, northeasterly from Fiene's toll-house 1276

Bed of Deer Creek under bridge at Mooney Flat 643

Porch of Schmidt's, Mooney Flat 732

Summit of road on Pet Hill 1631

Hyatt's toll-house 1259

POINTS ON THE DIVIDE BETWEEN THE SOUTH AND MIDDLE YUBA RIVERS.

Determined in 1879.

Mouth of old shaft at "coal pits" northwesterly from French Corral 2120

N. C. Miller's house, French Corral 1575

Lowest bed-rock at outlet of French Corral mines 1579

Mrs. Farrelly's house, French Corral 1704

Top of hill on San Juan road, where ditch crosses, southwest of Birchville . . . 1812

Bed-rock at upper end of Birchville mines, near Woodpecker Ravine 1734

Top of high hill north of Woodpecker Ravine 2137

Bed-rock near shaft No. 1, Manzanita Hill 1871

Bed-rock at north end of American Hill 1981

Bed-rock at east end of San Juan Hill 2033

National Hotel, San Juan, second story 2143

Mouth of tunnel at Lone Ridge 2510

Top of Lone Ridge, opposite Montezuma Hill 2619

Bed-rock at hydraulic bank, north side of Montezuma Hill 2356

Top of Montezuma Hill 2853

Glassett's house, Nevada City and San Juan road 2460

Bed-rock at Badger Hill outlet 2391

Top of gravel bank at Badger Hill 2556

Top of gravel between Badger Hill and Cherokee 2700

Hotel at Cherokee, second story 2575

Columbia Hill, R. C. McMurray's house 2986

Lowest lava seen in ditch at lower extremity of ridge near Columbia Hill . . . 3469

Mouth of Eurisco tunnel 3411

Bridge over Spring Creek on road from Columbia Hill to Grizzly Hill 2536

Bed-rock at Grizzly Hill outlet 2484

Tailings in South Yuba River, at mouth of Kennebec Ravine 1988

Kennebec House 2783

Hotel at Lake City 3418

Bed-rock at shaft No. 8, Malakoff mine	2929
Malakoff, H. C. Perkins's house	3173
North Bloomfield Mining Company's datum plane	2099
North Bloomfield, stage office	3278
Bed-rock flume, Cook and Porter claim, North Bloomfield	3406
Mouth of Great Eastern tunnel, Relief Hill	3713
Relief Hill, Mr. Penrose's house	3911
Mouth of old shaft at Relief Hill, sixty feet above bed-rock	3758
Malakoff ditch at trail from Relief Hill to Washington	4165
Derbec Shaft, office	3813
Back-bone House	4093
Mouth of Watt shaft, Bloody Run	4262
Top of highest bank, Boston mine, Woolsey Flat	4326
Lowest bed-rock seen at Boston mine	3890
Junction of roads, at ditches, above Moore's Flat	4563
Moore's Flat Hotel, second story	4231
Bed-rock at Moore's Flat, opposite Orleans Flat	4019
Bed-rock at Snow Point, near Shanghai bank	4211
Junction of Bloomfield and Eureka road with Henness Pass road on ridge . .	4461
Junction, on ridge, of roads to Eureka and to Mt. Zion	4922
Mouth of Mt. Zion tunnel	4297
Summit on Eureka road, west of Shand's	5054
Shand's	4627
Mouth of Griffith's tunnel, above Eureka	5458
Junction of roads, above Eureka, leading to Bowman's and down the south fork of Poorman's Creek	5583
Summit on road from Eureka to Bowman's	6098
Porch of house at Bowman Dam	5393

YUBA COUNTY.

POINTS NEAR AND BELOW SMARTSVILLE.

*Mouth of Mooney Flat tunnel, Deer Creek	642
Empire Ranch	840
Blacksmith-shop, Mooney Flat mine, estimated fifty feet above bed-rock . .	807
Telegraph pole 525 feet northwest of Empire Ranch	883
Base of lava cement on Campbell's Hill	1081
Top of Mooney Flat Hill	1170

Gravel hill southwest of the preceding, upper end of series trending from Park's Bar . . . 398

Gravel hill below Brady's hill 346

Top of Brady's south hill 544

Brady's northern hill 684

Brady's 211

Top of hill near Spinney's diggings 266

Reed's, where Bear River leaves the foot-hills 433

Top of knob, where the Yuba enters the valley of the Sacramento 389

Water-level in Yuba River, at mouth of big ravine on old road below Pritchard's . . 242

Left bank of river, at Pritchard's Ferry 251

Water-level in river, at Pritchard's Ferry (May, 1870) 217

Floor at Captain Pickens's house 325

Water-level in Yuba River, 1000 paces above Captain Pickens's ravine (June, 1870) . . 213

Water-level in Yuba River, at Sand Flat, east of Chinatown (June, 1870) . . . 172

Water-level in Yuba River, north side of knob in Schroeder's garden (June, 1870) . . 161

Water-level in Yuba River, at Smith's (June, 1870) 133

Floor at Gurley's 172

Rolling plains two thirds of a mile south of Fossil Hill 159

Bahrenberg's Turkey Ranch 154

Lava flow in valley near Winkelsett's 148

Lowest bottom in Salisbury's fields 147

Point near Riley Lane's field barn 119

Floor at Riley Lane's 137

The altitudes for the remainder of the list were all determined in 1879.

YUBA AND SIERRA COUNTIES.

POINTS ON THE DIVIDE BETWEEN THE MIDDLE AND NORTH YUBA RIVERS, AND NEAR SIERRA CITY.

Bridge over Middle Yuba River, on road from San Juan to Camptonville . . . 1515

Mouth of California tunnel, Pittsburgh Hill 2860

Camptonville Hotel, second story 2770

Bed-rock at outlet of Camptonville mines 2657

Bed-rock at Galena Hill 2755

Top of bank at Weed's Point 2910

Oak Valley school-house 3040

Summit of Brandy City trail, above Oak Valley 3335

Cherokee Bridge, over North Yuba River, on trail to Brandy City	2225
Mr. Bishop's house, Oak Valley	3100
Bed-rock at north end of Depot Hill	3120
Indian Creek, at crossing of trail from Depot Hill to Indian Hill	2810
Indian Hill, store	3355
Bed-rock at Indian Hill, lower end, near junction of granite and slate	3217
Ranch above Camptonville, near lower limit of lava	3900
Mouth of Dahneke's tunnel	3875
Oregon Creek, at trail from Tyler's diggings to Tippecanoe	3385
Bed-rock at Tyler's diggings	3510
Bed-rock at Tippecanoe	3555
Summit of trail from Tippecanoe to Grizzly Cañon, above Pike City	3890
Mouth of tunnel, Grizzly Cañon	3680
Nigger Tent House	4465
Summit on road from Nigger Tent to Mountain House	4765
Mountain House	4440
Point on ridge above Lucky Dog Ravine, near Henness Pass road	5153
Point on ridge above North Fork mine, Forest City	5020
Top of bank at City of Six	4893
Mouth of Ruby Tunnel, Rock Creek	4800
Summit of Downieville trail, between Rock Creek and Forest City	5404
Mouth of Bald Mountain tunnel	4489
Forest City, Ellery's Hotel, second story	4465
Top of Bald Mountain Bluff, above Forest City	5570
Mouth of Bald Mountain Extension tunnel	4605
Summit on road from Forest City to Alleghany	5020
Golden Anchor House, Alleghany	4375
Bridge over Kanaka Creek, on trail from Alleghany to Chips's Flat	3750
Bed-rock at McNulty's hydraulic mine, Chips's Flat	4235
Mouth of Rainbow shaft	4360
Summit of trail from Chips's Flat to Minnesota	4570
Bed-rock at Minnesota	4220
Mouth of Crescent tunnel, near head of Kanaka Creek	4855
Point on road near Pliocene shaft, estimated to be fifty feet above mouth of shaft	5424
Bed-rock on road above Nelson's mill, one mile east of preceding station	5480
Junction of ridge-road with road to American Hill, on sag between Jim Crow Cañon and Wolf Creek	5575
Low bed-rock near bank at American Hill	4880

Bed-rock at Bunker Hill	4700
Bed-rock at Loganville	3886
Sierra City, Yuba Gap Hotel, second story	4188
Mouth of 1001 tunnel	5938
Cabin at Blue Gravel claim, above Sierra City	5938

YUBA, SIERRA, AND PLUMAS COUNTIES.

POINTS NORTH OF THE NORTH YUBA RIVER AND SOUTH OF THE DIVIDE BETWEEN THE YUBA
AND THE FEATHER RIVERS.

(a.) *Near La Porte and Gibsonville.*

Strawberry Valley Hotel	3640
Low bed-rock at southwest end of Secret Diggings, near La Porte . . .	4828
Bed-rock at Illinois Hill, at base of lava	4916
Top of gravel at Illinois Hill	4993
Point near lowest bed-rock at south end of La Porte mines	4928
Bed-rock near lower end of Spanish Flat	4756
La Porte, Union Hotel, second story	4993
Bed-rock near bank at upper end of La Porte mines	5077
Summit of ridge between La Porte and Little Grass Valley, on line of projected tunnel .	5500
Big boulder on road from La Porte to Gibsonville	5168
Summit on road from La Porte to Gibsonville	5583
Point on road estimated to be fifteen feet below mouth of Go-Ahead shaft . .	5435
Gibsonville Hotel, first floor	5500
Average bed-rock at Mt. Pleasant mines	5560
Mouth of upper Hepsidam tunnel	6000
Ridge above Hepsidam, between Pilot Peak and Bunker Hill	6750
Top of ridge between Gibsonville and Little Grass Valley, on trail to Monitor shaft .	6281

(b.) *On St. Louis Ridge.*

Bed-rock at Council Hill	3963
Bed-rock at Fairplay	4133
Lowest bed-rock at Rock Creek outlet of mines at Scales's Diggings . . .	4253
Ditch crossing near head of Rock Creek, on road from Scales's Diggings to Poverty Hill .	4718
Mouth of shaft below Poverty Hill	4553
Top of bank at south end of Poverty Hill	4603
Bed-rock at north end of Poverty Hill	4563
Bed of Slate Creek at Poverty Hill trail	4000

Bed of Slate Creek at bridge on road from La Porte to Portwine 4318

Lowest bed-rock near bank at Portwine 4853

Summit on trail from Portwine to Morristown 5345

High bed-rock visible on Morristown trail, on Cañon Creek side of ridge . . . 5165

Bed-rock at Gardner's Point, opposite Yankee Hill 4845

Bed-rock at outlet of St. Louis mines, near Cedar Grove Ravine 4993

Pine Grove, G. W. Cox's office 5486

Bed-rock at Pine Grove mines, above Mr. Cox's house 5396

Howland Flat, Becker's Hotel, second story 5610

Mouth of Bonanza Tunnel, Potosi 5655

Bridge over Slate Creek on road from Potosi to Gibsonville, estimated twenty-five feet
    above bed of creek 5505

Summit of trail from Potosi to Poker Flat 6370

Mouth of tunnel in Cold Cañon 5626

(c.) *On Morristown and Eureka Ridges.*

Bridge over Cañon Creek on trail from Portwine to Craig's Flat, estimated ten feet above
    bed of creek 4262

Bed-rock at Craig's Flat 5100

Bed-rock at Morristown 5160

Bed-rock at Deadwood 5707

Mouth of Bunker Hill tunnel, head of Grizzly Cañon 5900

Bed of Cañon Creek at Poker Flat 4854

Top of bank at Grizzly Hill, near Brandy City 3560

Bed-rock in Brandy City mines, on Camptonville trail 3435

Brandy City Hotel, first floor 3650

Bed-rock at Arnott's mine, near Windyville 3540

Bridge over Little Cañon Creek on trail from Craig's Flat to Eureka, estimated fifteen feet
    above bed of creek 4360

Eureka, Wolfe's Hotel, first floor 5138

Bed-rock at Mugginsville 5090

Mouth of Hardy's tunnel 5038

(d.) *Near Downieville.*

Downieville, St. Charles Hotel, first floor 2806

Monte Cristo, Mr. Thatcher's house 5056

Mouth of Empire tunnel, Monte Cristo 5010

Summit of sag above Monte Cristo, on trail to Excelsior 5492

Mouth of Excelsior tunnel	5020
Top of ridge, near Craycroft's, on trail from Downieville	5370
Mouth of Craycroft tunnel	5137

PLUMAS COUNTY.

POINTS NORTH OF THE DIVIDE BETWEEN THE YUBA AND THE FEATHER RIVERS, AND SOUTH OF THE MIDDLE FORK OF THE FEATHER RIVER.

Lava in bed of Fall River below Davis Point	4540
Dam across Fall River near Davis Point	4615
Bed-rock at mouth of cut in gravel at Gard's claim, Davis Point	4800
Top of ridge above Davis Point	5060
Mouth of old shaft between Gard's claim and the "milk ranch"	4900
Bed-rock in the ravine west of the "milk ranch"	4835
Mouth of Post's tunnel in Wilson's Ravine	4425
Top of ridge on trail from Davis Point to Little Grass Valley	5660
Bench of basaltic lava	5240
South Fork of Feather River, at crossing of trail from Davis Point to Little Grass Valley	4930
Webber's house, Little Grass Valley	5025
Mouth of Monitor shaft	5282
Summit on road from Gibsonville to Onion Valley	6390
Onion Valley Hotel	6160
Summit on road from Onion Valley to Nelson Point	6430
Bridge over Nelson Creek, on road from Onion Valley, estimated twenty-five feet above bed of creek	4120
Bridge over Middle Fork of Feather River, at Nelson Point, estimated fifteen feet above bed of river	3950

POINTS NEAR QUINCY AND IN SIERRA VALLEY.

Tucker's ranch, near Spanish Peak	4100
Bed-rock at Scad Point	4510
J. A. Edman's house, Mumford Hill	4710
Bed-rock at Fales's Hill	4880
Bed-rock at Chaparral Hill, near old Mountain House	4980
Spanish Ranch, store	3585
Low bed-rock near bank at Gopher Hill	?3835
Top of ridge, above Gopher Hill, overlooking Badger Hill	?4050
Bed-rock at Badger Hill, lowest seen	?3880
Point level with bed-rock at Shore's Hill	3988

Bed-rock at northwest end of Orr's mine, Hungarian Hill   3698

Bed-rock at foot of high gravel bank in upper part of ravine, near trail to Quincy .   .   4003

Summit on trail from Hungarian Hill to Quincy, near reservoir   4280

Quincy, Plumas House, second story   3383

Crest of ridge above mines in Elizabethtown Flat   3750

Mouth of Keller's Deadwood tunnel, on Little Blackhawk Creek   3500

Mohawk Valley; Knott's ranch   4325

Top of hill, near Babb's house, on road from Knott's ranch to Jamison City .   .   .   4960

Jamison City, Porch of Miners' Home   4800

Mohawk Valley, Mrs. King's, second story   4450

Cabin at Andrew Jackson's mine   5025

Sierraville, Globe Hotel   4950

BUTTE AND YUBA COUNTIES.

POINTS IN THE NEIGHBORHOOD OF OROVILLE.

Oroville, McDermott's drug-store, — assumed altitude   150

Boston Ranch   1500

Summit on stage road between Boston Ranch and Forbestown   3060

Forbestown Hotel   2625

Brownville, Post-office   2125

Morris Ravine, W. C. Hendricks's house   524

Point on Hendricks's ditch, near head of Good's Ravine, northerly from Mt. de Oro .   .   964

Base of basaltic cap near divide between Schermer's and Morris Ravines .   .   .   1084

Bed-rock at Gregory's claim, Oregon gulch   1033

Base of basaltic cap at cave, head of Coal Cañon   1049

Point on stage road near Hutchinson's house where road crosses upper end of Cherokee

      Flat mines   1300

Top of Sugar Loaf, Cherokee Flat   1647

Base of basaltic cap at Sugar Loaf   1550

Top of highest bank at Spring Valley mines   1460

Top of pink gravel, below Sugar Loaf   1377

Low bed-rock near blacksmith shop in Spring Valley mines   1087

Cherokee Flat, Spring Valley House, second story   1187

Bed-rock at St. Clair Flat   765

Top of first bench above Welch's diggings   935

INDEX.

University Press : John Wilson & Son, Cambridge.

www.ingramcontent.com/pod-product-compliance
Lightning Source LLC
Chambersburg PA
CBHW062020210326

41458CB00075B/6232